Signals and Communication Technology

This series is devoted to fundamentals and applications of modern methods of signal processing and cutting-edge communication technologies. The main topics are information and signal theory, acoustical signal processing, image processing and multimedia systems, mobile and wireless communications, and computer and communication networks. Volumes in the series address researchers in academia and industrial R&D departments. The series is application-oriented. The level of presentation of each individual volume, however, depends on the subject and can range from practical to scientific.

Indexing: All books in "Signals and Communication Technology" are indexed by Scopus and zbMATH

For general information about this book series, comments or suggestions, please contact Mary James at mary.james@springer.com or Ramesh Nath Premnath at ramesh.premnath@springer.com.

More information about this series at http://www.springer.com/series/4748

Keonwook Kim

Conceptual Digital Signal Processing with MATLAB

Keonwook Kim
Division of Electronics
and Electrical Engineering
Dongguk University
Seoul, Korea (Republic of)

ISSN 1860-4862 ISSN 1860-4870 (electronic)
Signals and Communication Technology
ISBN 978-981-15-2583-4 ISBN 978-981-15-2584-1 (eBook)
https://doi.org/10.1007/978-981-15-2584-1

This Springer imprint is published by the registered company Springer Nature Singapore Pte Ltd.
The registered company address is: 152 Beach Road, #21-01/04 Gateway East, Singapore 189721, Singapore

To my parents

To Saea, Saeyun, and my lovely wife Gumhong Kim

To all of my family

Preface

Digital signal processing (DSP) is an extensive study area to understand the signal and noise in the digital domain. The advanced digital technology in the 60s and 70s started to accelerate the theoretical signal processing for the feasible engineering over the academia. In the 80s, the research in articles were organized and consolidated as practical textbooks for the graduate program. There were great textbooks such as "Discrete-Time Signal Processing" by Oppenheim and Schafer and "Digital Filter Design" by Parks and Burrus. The decade of the 90s further enjoyed the processor technologies along with programming environments for DSP. MATLAB (founded in 1984) has provided a reliable and convenient computer language (also tools) to realize and verify the mathematical algorithms. Many textbooks appreciate the benefits of MATLAB, for instance, "Signals and Systems Using MATLAB" by Chaparro. For DSP, books and tools are existing.

This book addresses DSP in novel approach based on reconfiguration. The intuitions and theories along with applications are the basic philosophy of the book. The author tries to avoid the dictionary style (bottom-up structure) and to place theories within applications. Students often experience a loss of motivation in the middle of a semester due to the fragmented knowledge of digital filter design. The author believes that the great researchers in DSP did not establish solid theories from pure mathematics. The imagination initiates the research, and the equation finalizes the theory. This book tries to provide the theories with derivations, illustrations, and/or applications. Once the student has a certain picture of the procedure and intelligence, the knowledge can be extended and maintained for a long time. MATLAB also helps to design the top-down structure for the self-motivated learning process. The book is organized as follows.

Chapter 1

- Basics of digital filters based on the intuition,
- Simple weight and sum of recent inputs for desired output,
- Design the low, high, and band pass filter.

Chapter 2

- Definition of frequency in continuous and discrete time domain,
- Sampling theory,
- Discrete time signal representations.

Chapter 3

- Fundamentals to find the frequency magnitude from time domain,
- Discrete-time Fourier transform (DTFT) for non-periodic signal,
- Discrete Fourier transform (DTF) for periodic signal.

Chapter 4

- Reason to have the linear time invariance property,
- Finite impulse response (FIR) filter,
- Simple FIR filter design from the specification.

Chapter 5

- Extension of DTFT and DFT to Z-transform,
- Z-transform for infinite impulse response (IIR) filter,
- Intuitive IIR filter designs.

Chapter 6

- Understanding the filter specification,
- Advanced FIR filter designs,
- Advanced IIR filter designs.

Chapter 7

- Quantization effect for fixed-point number system,
- Implementation matters for FIR and IIR filters,
- Frequency domain filter realization.

Chapter 8

- Various filter realizations in MATLAB,
- Fixed-point number for MATLAB,
- C code generation for MATLAB.

In addition to above, there are two more sections for MATALB fundamentals and Symbolic Math Toolbox as an appendix. This book includes comprehensive information from basic DSP theories to the real-time filter realization of digital computers and processors. The DSP class from this book is intended to provide one- or two-semester program at the undergraduate level. The instructor can design the DSP class for theory or practical intensive program as follows. Also, a two-semester class can be handled using the complete contents of this book.

One-semester DSP program (theory intensive)

- Chapters 1+2+3+4+5
- Chapter 6
  - Specification + FIR window method + FIR types,
  - IIR Butterworth method with bilinear transform.

One-semester DSP program (practical intensive)

- Chapters 1+2+3+4+5
- Chapter 6 (Specification + IIR Butterworth method with bilinear transform),
- Chapter 7 (numerical representation + filter implementation components),
- Chapter 8 (whole).

Two-semester DSP program

- First semester: Theory-intensive DSP program as shown above.
- Second semester: Chapter 5 to end.

DSP is realized by programming languages which can be chosen from the rudimentary level such as assembly language to the advanced level, for example, MATLAB. Unlike the low-level language, MATLAB provides a high degree of freedom for syntax, variable, function, execution, etc. Also, the robust support for symbolic mathematics can solve the calculus problems without using the hand derivations. The conversion between the symbolic and numerical mathematics is seamless in MATLAB; hence, the equation establishment is the only requirement for DSP realization, in most of the situations. Examples in this book demonstrate the visualization, verification, and realization of DSP algorithms based on the MATLAB programming. The evolution of DSP is continuously happening to the next-generation applications such as data science, deep learning, etc. The author of this book hopes that the readers can grasp the fundamentals of DSP and employ the understandings of DSP to real-world applications.

Seoul, Korea (Republic of) Keonwook Kim

Contents

About the Author

Dr. Keonwook Kim received the B.S. degree in Electronics Engineering from Dongguk University, Seoul, Korea in 1995 and the M.S. and Ph.D. degrees in Electrical and Computer Engineering from the University of Florida, Gainesville, United States in 1997 and 2001, respectively. He is presently Professor in the Division of Electronics and Electrical Engineering at the Dongguk University. Prior to joining Dongguk University, he worked as Assistant Professor in the Department of Electrical and Computer Engineering at the Florida State University from 2001 to 2003. His primary research interest is acoustic localization via using the multi-aural architecture in order to mimic the aural system of animals which include human.

Chapter 1
Preliminary Digital Filter Design

The general filter passes various matters to separate out unwanted things. In electrical engineering, the filter is known as the device for minimizing or suppressing the noise frequencies to obtain the better signals. The digital filter handles the input and output in discrete time domain as well as the quantized magnitude form. This chapter introduces the digital filter from scratch. The filter equation is derived from the general engineering concept. We assume that the input signal to the digital filter presents the real-time property which represents the past, present, and future tense in the signal. The conceptual derivation performed in this chapter actually operates filter function in the real field based on the trial and error approach. In the following chapters, the abstract design is followed by the further delicate analyses to meet the accurate filter performance.

1.1 Discrete Time Signal

The continuous time is t as real number and discrete time is n that is the integer number. The () handles the continuous arguments and [] deals with discrete arguments. Any functions can be continuous and discrete form with arguments. The given signal is $x[n]$ as shown in Fig. 1.1.

The discrete signal with integer argument has no tense as past, present, and future unless we specify the current time. That is the sequence of signal which shows the relative position of numbers. The higher index number shows later, and the lower index appears earlier. The real-time signal with n index provides the present time position n which is the variable integer number. Therefore, the $n + 1$, $n + 2$, and etc. are the future of the signal and $n-1$, $n-2$ … are the past of them. Based on the time schedule, we only have the data up to the n index and $n + 1$ data will be gathered next. The magnitude of the data is real number and any numbers are possible for received value. The frequency components of the signal are not numerically analyzed yet, but we presume that the signal with rapid fluctuation

K. Kim, *Conceptual Digital Signal Processing with MATLAB*,
Signals and Communication Technology,
https://doi.org/10.1007/978-981-15-2584-1_1

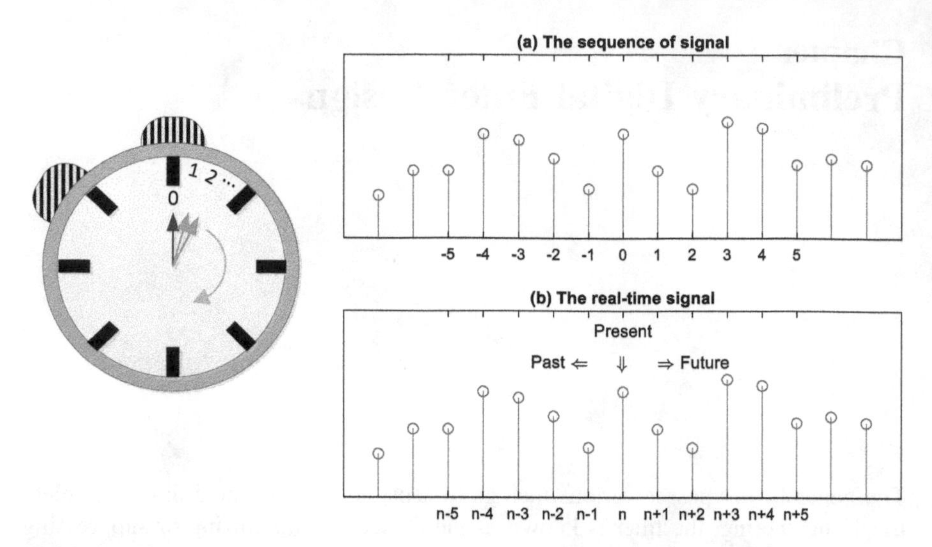

Fig. 1.1 **a** Example of discrete time signal with absolute time index number. **b** The corresponding discrete time signal with real-time representation with *n*. The clock shows the current time *n*

contains the additional high frequency components. In contrast, the signal with gentle movement includes the dominant low frequency components. The source of the signal can be anything such as dice toss and heart rate. Conventionally the discrete time signal is obtained from the continuous information by regular sampling. We assume that the sampling distance for time is even; hence, the actual time difference between adjacent indexes are always identical. After every fixed time, the number for the discrete index n will be updated to the next integer. In the digital signal processing, the $x[n]$ has two meanings those are whole signal sequence (variable n) and current signal value (given n). The discrete signal can be defined as the signal sequence as shown below.

$$x[n] = \cos\left(\frac{\pi}{2}n\right) \tag{1.1}$$

If the sequence is used to build another discrete time signal, the input sequence is employed as real-time signal as shown below.

$$y[n] = \frac{1}{3}x[n] + \frac{1}{3}x[n-1] + \frac{1}{3}x[n-2] \tag{1.2}$$

The $y[n]$ function uses the current x value as $x[n]$ and two previous values as $x[n-1]$ and $x[n-2]$.

1.2 Design the Digital Filters

The real-time discrete time signal is given, and you are asked to design the filter to reduce the noise frequencies and emphasize signal information. I believe that this is the initial point of digital signal processing. Let's design the filter.

The filter output $y[n]$ is the operated results from the filter input $x[n]$s, and the signal is obtained up to the n index value, as shown above Fig. 1.2. Any idea? Note that the signal is continuously received by the system in every constant interval; therefore, the system is willing to have the significant amount of data in short time. It is impossible to process the whole obtained data for filtering; hence, the limited length of data is considered to provide the filter output. The simple idea is to take average the recent inputs for low pass filtering (smoothing). The recent N samples are denoted as below.

$$x[n], x[n-1], x[n-2], \ldots, x[n-N+1] \tag{1.3}$$

The current filter out $y[n]$ is computed as below.

$$y[n] = \frac{1}{N}\{x[n] + x[n-1] + x[n-2] + \cdots + x[n-N+1]\} \tag{1.4}$$

The filter output with input data is illustrated in Fig. 1.3. The recent three data samples are considered to create average value for output $y[n]$. In Fig. 1.3, the n values (n is the variable and not the current index here) are demonstrated from

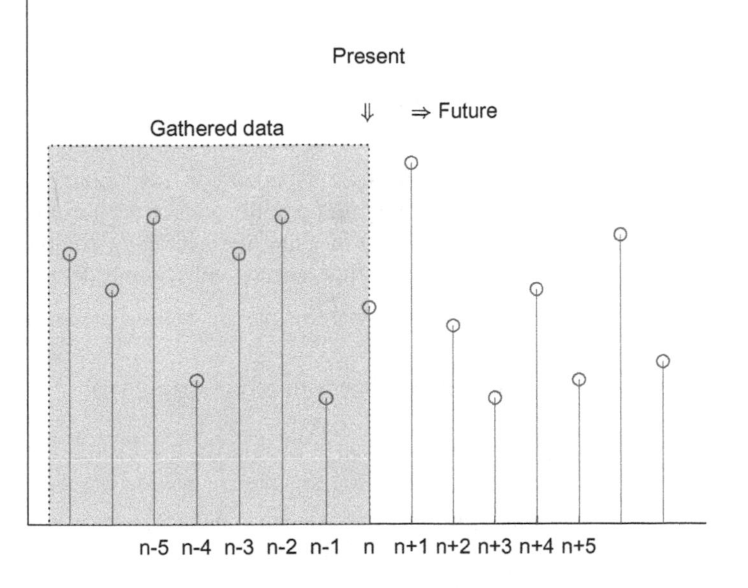

Fig. 1.2 One example of discrete time signal with real-time representation

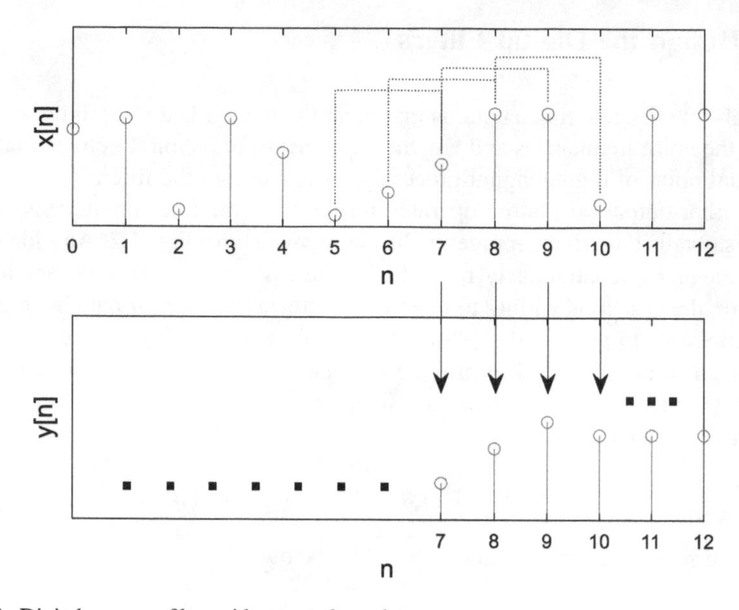

Fig. 1.3 Digital average filter with recent three data

seven to ten and the output values are noticeably smoothed comparing to the income signal. This is the primitive low pass filter (LPF).

Example 1.1
Write the equation for digital average filter with recent three data.

Solution

$$y[n] = \frac{1}{3}\{x[n] + x[n-1] + x[n-2]\}$$

■

The independent parameter for the LPF is operation (averaging) and length (three). As shown in Eq. (1.4), the averaging operation can be seen as the weighted sum for the recent incoming data. Instead of using the averaging computation, the different weight values can be applied on the data set. First of all, let's change the length of the averaging in the LPF as Fig. 1.4.

Example 1.2
Write the equation for digital average filter with recent seven data.

Solution

$$y[n] = \frac{1}{7}\{x[n] + x[n-1] + \cdots + x[n-5] + x[n-6]\}$$

■

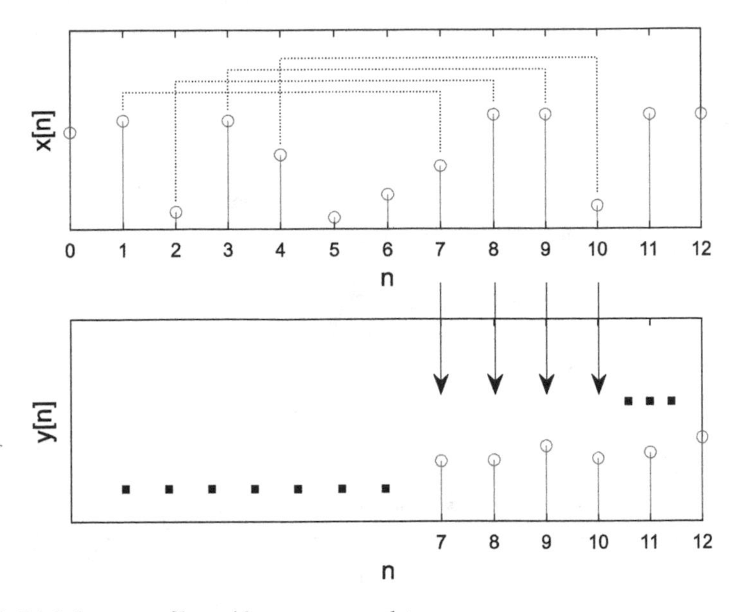

Fig. 1.4 Digital average filter with recent seven data

For the increased length (seven) of averaging operation, the filter output $y[n]$ delivers further smoothed outcome than the shorter version (length three). Therefore, depending on the filtering length, smoothness of the output can be decided proportionally. In other word, the longer length averaging passes the narrower range frequencies from zero frequency, and the shorter length averaging passes the wider range frequencies that demonstrates the high fluctuation in filter output. The length of the filter is important parameter to determine the frequency range of the output.

Example 1.3

Write the equation for $y[7]$ digital average filter in Fig. 1.5.

Solution

$$y[7] = \frac{1}{7}\{x[7] + x[6] + \cdots + x[2] + x[1]\}$$

∎

The other parameter to be considered is weight values in the filter. The averaging operation multiplies the constant $1/N$ to each data values for limited N length and accumulates for filter output. The averaging range, latest N data, can be seen as the window with weight for the given time sequence, and the window slides to the next for new output. Figure 1.5 illustrates the explained procedures in terms of window with seven window length. We understand that the filter length controls the output

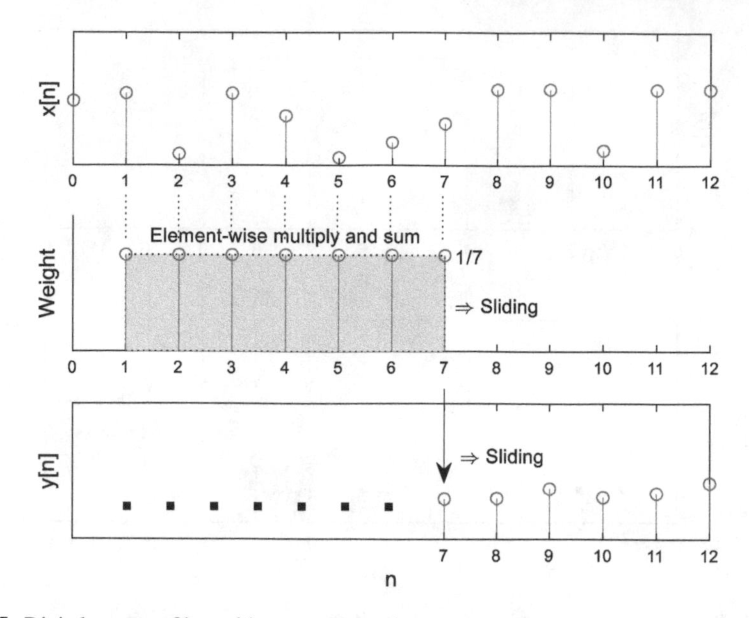

Fig. 1.5 Digital average filter with recent seven data

frequency range. The shape or weight of the window significantly contributes to manipulate the frequency component of the output as well. Let's do another experiment.

Example 1.4
Write the equation for high pass filter in Fig. 1.6.

Solution

$$y[n] = \frac{1}{3}\{-x[n] + x[n-1] - x[n-2]\}$$

■

Unlike averaging filter, the filter output shows the higher variation in magnitude than the original signal; therefore, the high frequency components are exaggerated via the filtering process. The designed filter is high pass filter (HPF) with three window length. The values in the window decide the characteristics of the filter which prefers to pass the high frequency components; hence, we can think that the window shape plays an important role. Comparing to the LPF, the HPF also uses the same window length but the filter output is completely reverse in action. Applying the longer HPF window derives same variation as LPF? The longer HPF length generates concentrated filter output to the high frequencies than the three HPF length situation as Fig. 1.7.

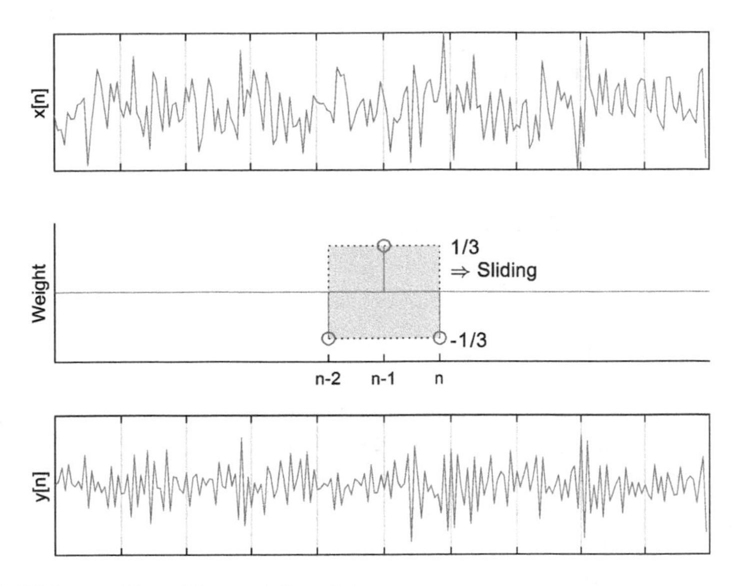

Fig. 1.6 High pass filter with recent three data

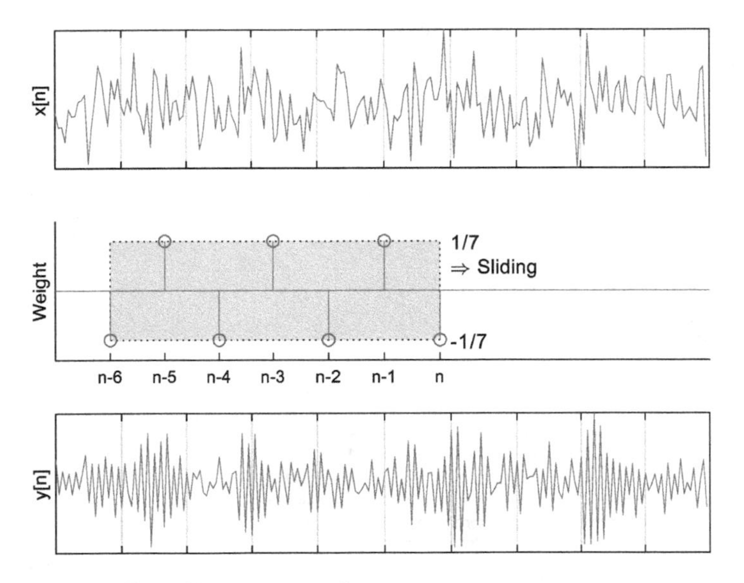

Fig. 1.7 High pass filter with recent seven data

Example 1.5
Write the equation for high pass filter in Fig. 1.7.

Solution

$$y[n] = \frac{1}{7}\{-x[n] + x[n-1] + \cdots + x[n-5] - x[n-6]\}$$

■

The window length controls the frequency focus of the filter output in inversely proportional manner. The longer window creates the further concentrated output for the specific filter type which is determined by the window shape. The relationship between the output frequencies and window length is derived from the above simple experiments. How we can decide the filter type from window shape? Let's draw the two previously used filters in Fig. 1.8.

Can you figure out the filter type by observing the filter shape? Yes, the filter output follows the window shape; therefore, smoothness and roughness of the window provide the filter type that specify the designated frequency you want to pass. Other than LPF and HPF, let's perform another example. The window shape of this filter is certain periodic signal which shows the 6-sample period and 19-sample length. The filter output is similar to the window shape as illustrated in Fig. 1.9. The longer length expects to emphasize the window shape on the output based on the intuition of previous lessons.

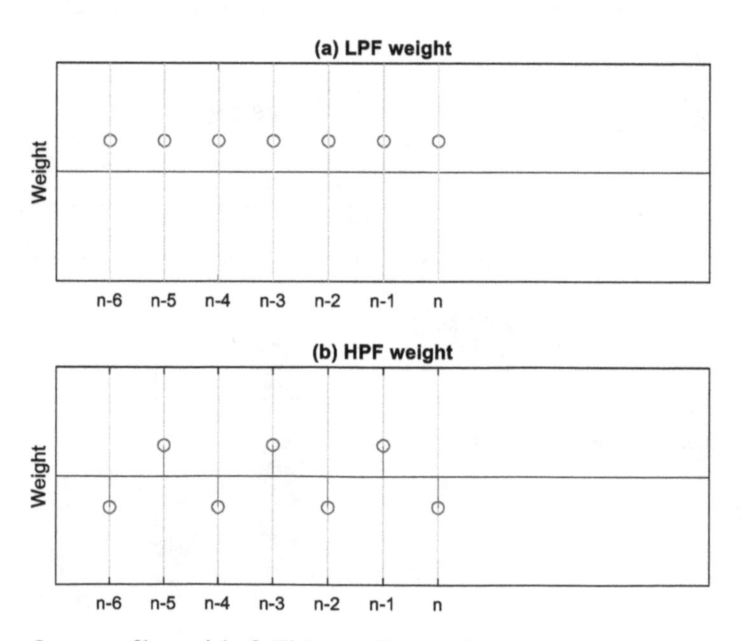

Fig. 1.8 **a** Low pass filter weight. **b** High pass filter weight

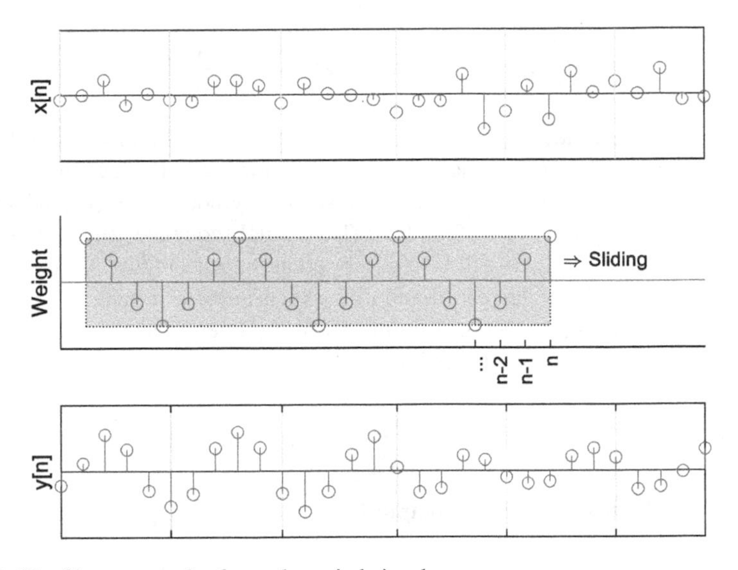

Fig. 1.9 The filter to pass the 6-sample period signal

Example 1.6
Write the equation for digital filter in Fig. 1.9.

Solution

$$y[n] = \sum_{k=0}^{18} x[n-k]\cos\left(\frac{2\pi k}{6}\right)$$

■

Up to now, we have designed primitive filters for LPF, HPF, etc. This is the fundamental of the digital filter theory. The idea and computation initiated the filtering system and signal processing in digital domain. The design method based on the trial and error does not lend the solid foundation for building the intended filter. The shown intuitive methods should be formularized in engineering area for further analysis and application. This book not only provides the mathematical representations of the algorithms but also illustrates the physical meanings of the system extensively in the coming sections and chapters.

1.3 Filter Architecture

The illustrated filter system provides approximate methodological principle of the filtering. This section converts the pictured procedures into the mathematical formula. Let's consider the LPF as below.

$$y[n] = \frac{1}{N}x[n] + \frac{1}{N}x[n-1] + \frac{1}{N}x[n-2] + \cdots + \frac{1}{N}x[n-N+1] \tag{1.5}$$

The recent N samples of data sequence is averaged for the low pass filtering in terms of weighting the $1/N$ on each sample. Since the weight values in the window can be various for intended filtering purpose, the window is separated and slid sequentially in every interval as Fig. 1.10. The signal sequence $x[n]$ starts from the zero n value and the window will be overlapped at the zero index. As the time goes by, overlap range will be increased and completed after $N-1$ sample shift; hence, we expect to have proper averaged output from the operations.

The window function is specified as below.

$$q[0] = q[-1] = q[-2] = \cdots = q[-(N-1)] = \frac{1}{N}$$

Therefore, the first appropriate output by complete overlap is defined as;

$$\begin{aligned} y[N-1] = {} & q[-(N-1)]x[0] + q[-(N-2)]x[1] + q[-(N-3)]x[2] + \cdots \\ & + q[-1]x[N-2] + q[0]x[N-1] \end{aligned}$$

The second and third proper outputs are shown below with sliding input sequence.

$$\begin{aligned} y[N] = {} & q[-(N-1)]x[1] + q[-(N-2)]x[2] + q[-(N-3)]x[3] + \cdots \\ & + q[-1]x[N-1] + q[0]x[N] \end{aligned}$$

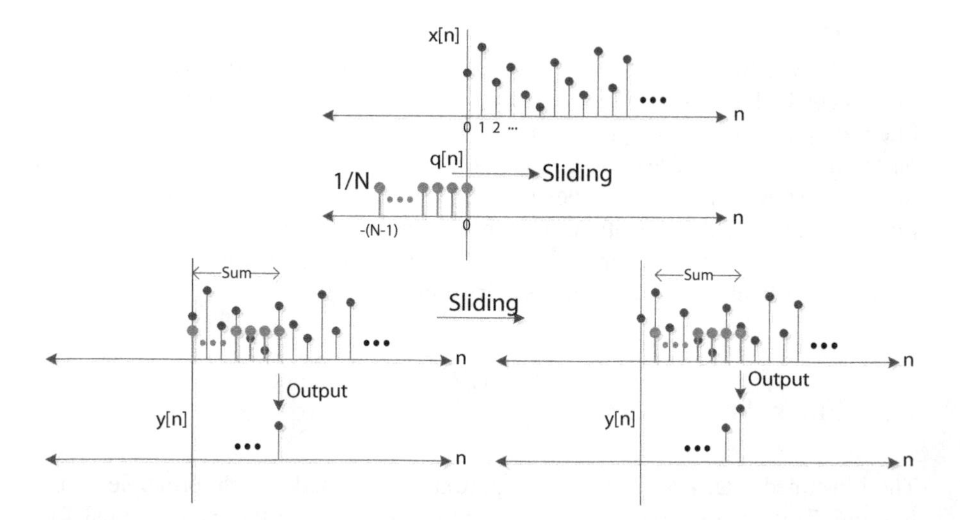

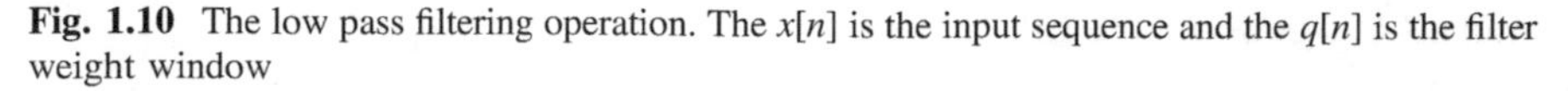

Fig. 1.10 The low pass filtering operation. The $x[n]$ is the input sequence and the $q[n]$ is the filter weight window

$$y[N+1] = q[-(N-1)]x[2] + q[-(N-2)]x[3] + q[-(N-3)]x[4] + \cdots \\ + q[-1]x[N] + q[0]x[N+1]$$

Generalizing the above equation by summation as below.

$$y[n] = \sum_{k=0}^{N-1} x[n-k]q[-k] \tag{1.6}$$

The n is the current time in the equation and the integer value n starts from zero to infinite. Based on the recent N input sequence, the filter output is computed by weight $q[\cdot]$ function. This is the nice equation to formulate the intuitive concept of the primitive filtering. Since the input data sequence initiates from the zero-time index to positive number, the weight window is also flipped over the vertical axis in order to start from the zero as below.

$$q[0] \to h[0], q[-1] \to h[1], q[-2] \to h[2], \ldots, q[-(N-1)] \to h[N-1]$$

Due to the relocation, the tense of the weight window is changed completely as shown in Fig. 1.11. The original window function $q[\cdot]$ represents the $q[0]$ as present moment and the its tails on the left as past time. Note that the $q[0]$ is always multiplied with the present $x[\cdot]$ value in the filtering computations shown at Eq. (1.6). The further to the left indicates the earlier times in weight window. In the flipped window function $h[\cdot]$, the further to the right is past and $h[0]$ is present time. Therefore, the equation with $h[\cdot]$ is below.

$$y[n] = h[(N-1)]x[n-(N-1)] + h[(N-2)]x[n-(N-2)] + \cdots \\ + h[3]x[n-2] + h[1]x[n-1] + h[0]x[n]$$

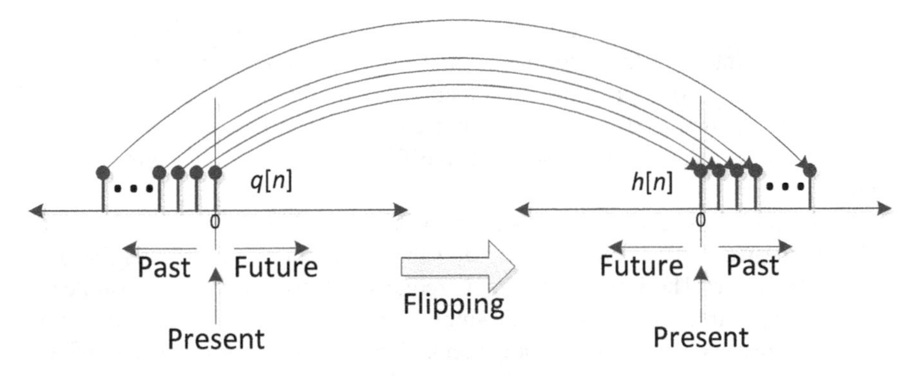

Fig. 1.11 Flipping the filter weight window for $h[n]$

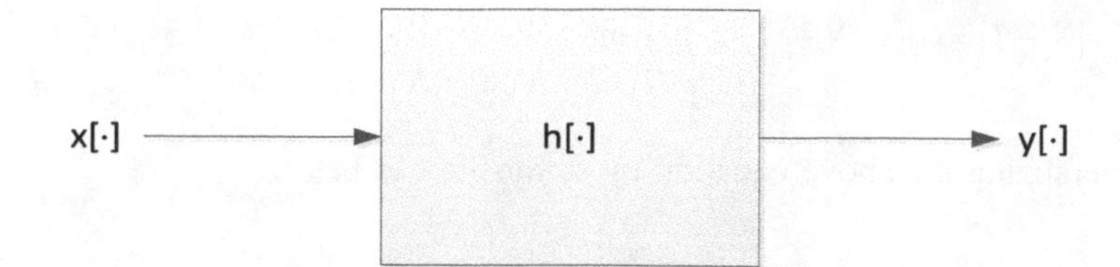

Fig. 1.12 Digital filter block diagram

Generalizing the above equation by summation as below.

$$y[n] = \sum_{k=0}^{N-1} x[n-k]h[k] \tag{1.7}$$

The only difference with non-flipped window filter is that there is no minus sign in the window function. This is the digital filter equation also known as convolution sum. The equation is the time domain filter with given input signal $x[\cdot]$ and produces the filtered output $y[\cdot]$. For the specific purpose, the shape and length of $h[\cdot]$ have to be determined in analytical manner to improve the signal and reduce the noise component from incoming discrete data sequence on Fig. 1.12.

The diagram for the filter is given as above. The real-time input sequence $x[\cdot]$ is provided to the filter system represented by $h[\cdot]$. The corresponding filter output is $y[\cdot]$.

1.4 Digital Filter Definitions and Requirements

By definition, the filter is a process to eliminate the unwanted frequencies, which are classified as noise, from an input signal. The designed filters were explained in terms of emphasizing the certain frequencies; however, a range of frequencies are suppressed to intensify the information in fact. The important thing in the filter realization is that 'do not add any information by filtering.' You can increase and decrease certain frequency magnitudes but cannot create new frequencies in the filtering process. This is the reason that the filter is explained by elimination of frequencies.

One example of digital signal processing is shown in Fig. 1.13. The input signal is the cosine wave with certain period and the processing output limits the input magnitude to one. The input and output frequency distributions are illustrated as well. The input frequency consists of single dominant component and the output frequency presents at least three components. The additional prominent frequency components are pointed by the circles in the Fig. 1.13. The given digital signal processing is disqualified from the filter requirements according to the frequency component non-creation rule. However, the reverse situation is valid as the digital

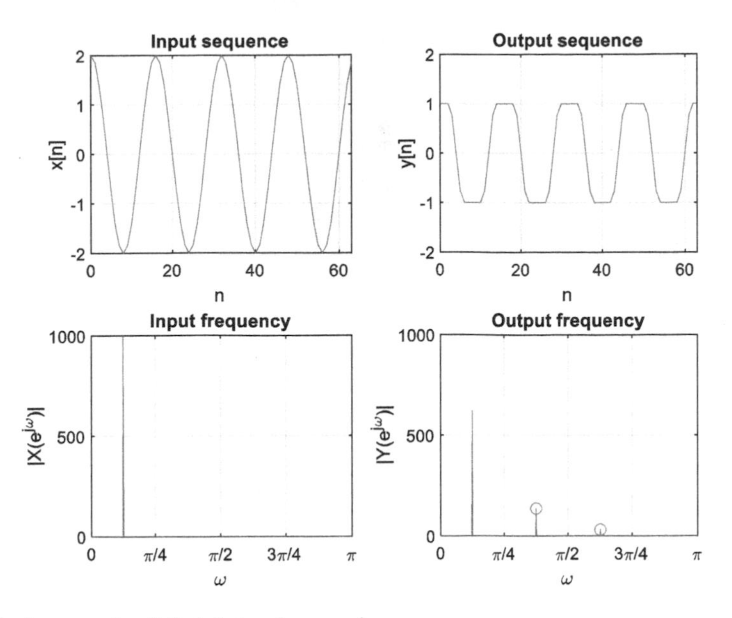

Fig. 1.13 An example of digital signal processing

filter. The output sequence placed at the filter input provides the output which only includes the single frequency component. The multiple frequency components are removed to obtain the signal frequency; therefore, the filter dose not insert any frequencies to the output.

Example 1.7
Write the equation to add the zero frequency component over the signa $x[n]$.

Solution

$$y[n] = x[n] + \text{constant}$$

Since the constant values indicate the zero frequency, the output $y[n]$ contains the additional zero frequency component.

■

Now we know that the designed filter by $h[\cdot]$ can compute the output by Eq. (1.7). When you have the unknown filter, how we can find the $h[\cdot]$ and what is the relationship with $y[\cdot]$ from $h[\cdot]$? What signal figures out the $h[\cdot]$? Do not look at the equation. Observe that we are dealing with discrete time signal. The primitive and fundamental element of the discrete signal is the function that has single value in the certain time index. Let's see the example at Fig. 1.14.

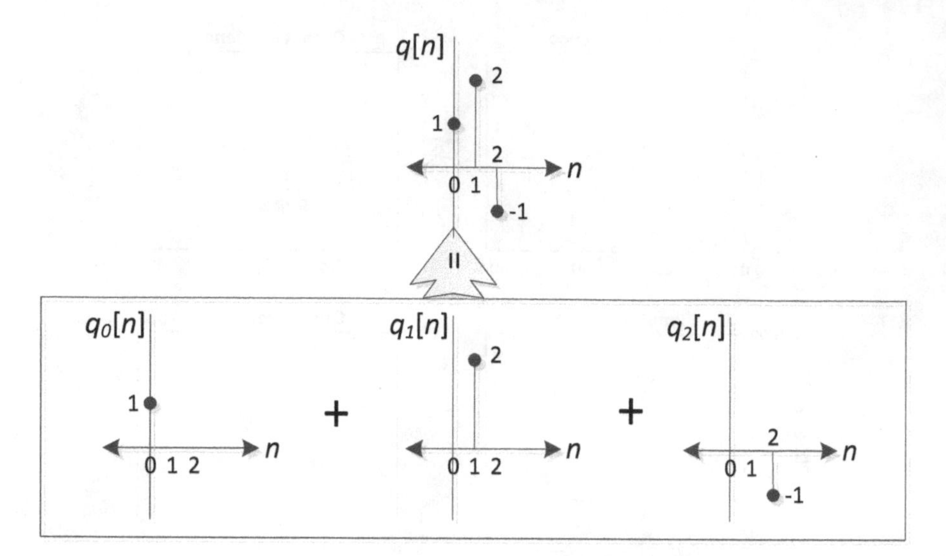

Fig. 1.14 An example of function decomposition

The function $q[\cdot]$ is the linear combination (simple addition) of $q_0[\cdot]$, $q_1[\cdot]$ and $q_2[\cdot]$. The individual $q_1[\cdot]$ and $q_2[\cdot]$ can be represented by the shift location and magnitude multiplication of $q_0[\cdot]$ as shown below.

$$q_0[n] \xrightarrow{\text{Shift right by 1}} \xrightarrow{\text{Multiply by 2}} q_1[n]$$

$$q_0[n] \xrightarrow{\text{Shift right by 2}} \xrightarrow{\text{Multiply by } -1} q_2[n]$$

Therefore, the manipulations and combinations of $q_0[\cdot]$ can provide the $q[\cdot]$ and any arbitrary functions. Based on this idea, we can think about the filter output by analytical combinations. If the filter output of $q[\cdot]$ is the linear combination of the filter outputs from $q_0[\cdot]$, $q_1[\cdot]$ and $q_2[\cdot]$, then the $q[\cdot]$ output can be decomposed into the simpler form. Furthermore, the individual filter output of $q_1[\cdot]$ and $q_2[\cdot]$ can be described by the shift location and magnitude multiplication of $q_0[\cdot]$ output just as relationships of $q_0[\cdot]$, $q_1[\cdot]$ and $q_2[\cdot]$, then the filter output will be expressed by linear combination of representative output like $q_0[\cdot]$ in this example. The generating output from individual inputs is illustrated in Fig. 1.15.

Figure 1.15 shows that the filter output can be derived from the linear combination and time shift of the primitive output. This kind of the system is called as linear and time invariant system. The linear system denotes the filter outcome by addition over scaled version of individual element outputs. The time invariant system preserves the filter output shape for the time shifted input with identical time relocation. With linear and time invariance condition, the system can be characterized by very simple format that is the output of the single value in single time.

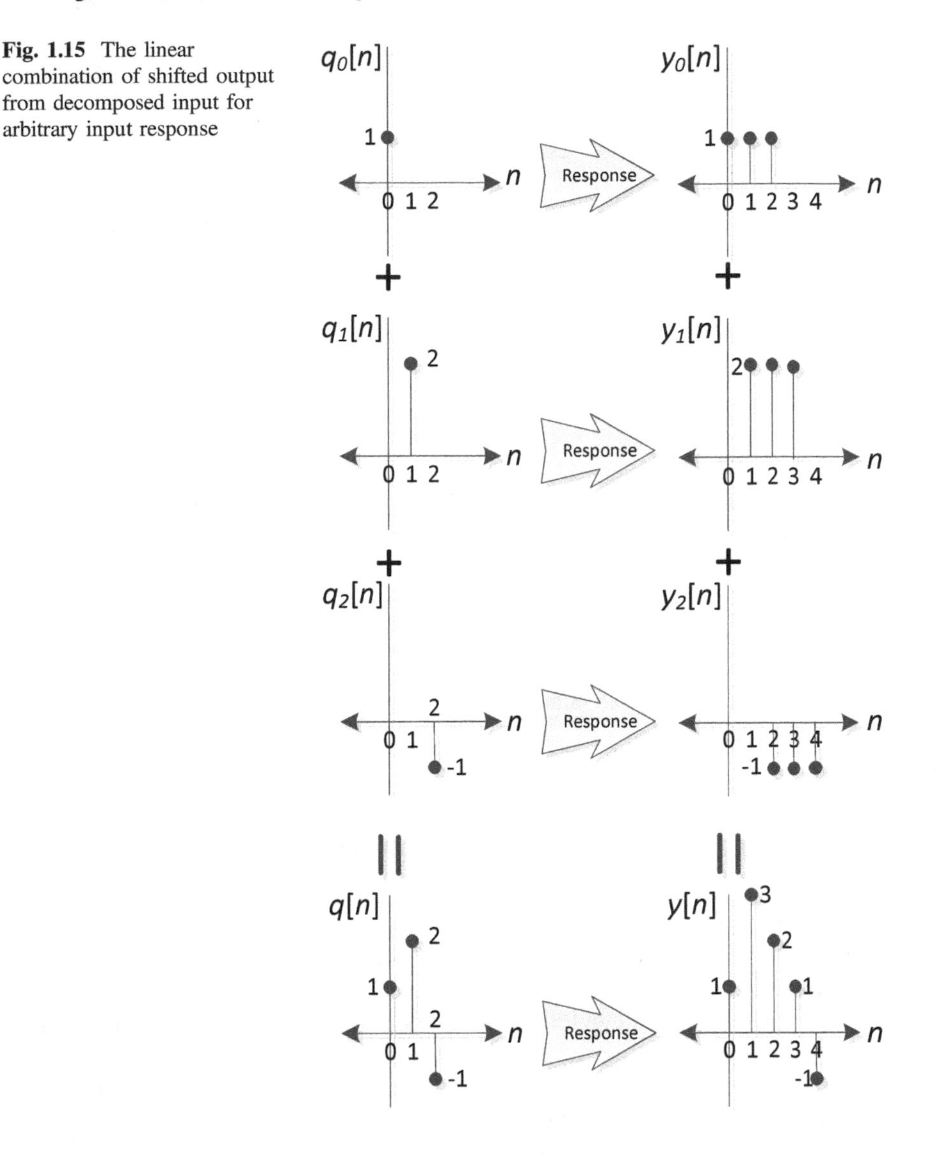

Fig. 1.15 The linear combination of shifted output from decomposed input for arbitrary input response

Any other outputs can be produced by the scaled and shifted combinations of the characterizing output. This is very important property since the characterizing output represents the complete feature of the system described as $h[\cdot]$ in the equation. How we can find the $h[\cdot]$ from the unknown linear and time invariant system? Let's perform Fig. 1.10 example in reverse manner as below.

As shown in Fig. 1.16, the impulse signal (one magnitude at time zero), which is the characterizing input, sequentially scans the $h[\cdot]$ of the system and provides the $h[\cdot]$ of the filter. The discovered $h[\cdot]$ tells us the property and performance of the filter in detail. Therefore, the user can find the system feature of the unknown and linear

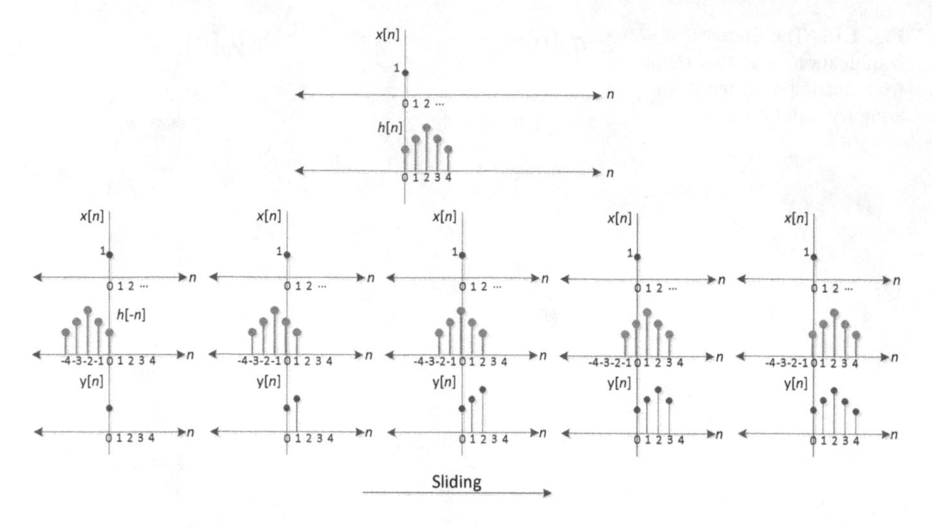

Fig. 1.16 Place the impulse signal to derive the $h[n]$

& time invariant filter by impulse signal input. Now, we understand that the $h[\cdot]$ can be designed and found for the specific filtering purpose. However, we have to explore further to study the relationship between the $h[\cdot]$ and frequency components in coming chapters.

1.5 What We Need for Further (Optional)

If you are going to build the digital filter by using the trial and error method, the chapter one is enough information for you. The information above shows about window shape and filter computation intuitively. The low pass filter is realized by the constant weight window for filter equation and the length is determined by your trials until you obtain the satisfactory outcome. Other types of the filters can be implemented by the derived window and numerous executions. The fundamental concept is delivered but certain analytical approaches are missed in above.

For the engineering aspects, the systematical methodology is required for approximate closed form solutions in filter design. The frequency is fundamental elements of the system input/output and the signal is the limited or unlimited combination of frequency components in general. The definition of the frequency is introduced in the first part of this book. The digital signal processing manages the signal in discrete time domain; however, the frequency is well understood in the continuous time domain. The frequency for the digital signal is explained with the sampling theory that describes the signal transformation between the continuous and discrete time domain. The digital signal can be decomposed into the simple elements with linear and time shifting combination; therefore, the fundamental and essential signal elements are described and defined as well.

Typically, the given signals are delivered in time domain and do not represent the frequency information directly. To understand the signal and analyze the performance, the designer needs to see the spectral distribution in the signal. The Fourier analysis provides the mathematical tool to transform the information between the time and frequency. The Fourier analysis computes the magnitude and phase (or delay) of the frequencies from the periodic and non-periodic signal. However, the Fourier analysis does not generate all-round solutions to all transform matters. The users need to note that there are numerous limitations and conditions to apply the analysis over the signals. The physical meaning and overall constraints of Fourier analysis are described in this book.

The time domain filter is already introduced in this chapter. The filter equation is primitive but essential fundamental of the signal frequency managing. Upon the understanding of the frequency definition and Fourier analysis, the filter equation is revisited to comprehend further based on the mathematical study. The frequency response of the filter from the window shape and length is derived to meet the intended filter requirements. Also, the finite window is extended to the infinite length by recursive filter architecture and its frequency response is explored as well. By using the frequency modulation, the low pass filter can be located in any frequency position for designated filter specification. The simple filters based on the frequency modulation are provided for practical filter design.

The Fourier analysis can be extended to the Z-transform to understand the signal and system with additional perspective. The Z-transform converts the subject between the time domain and the Z domain known as complex number plane. The rational polynomial of complex number Z represents limited and unlimited length signal and system with constant coefficients. The solutions to the polynomial provide the various information such as filter type, stability, and etc. The beauty of the Z-transform is that the transform presents powerful mathematical tool to handle and visualize the system in simple manner. Also, the Z-transform can be adopted to design the any type filters.

The digital filter is implemented by the digital processor (or logic) with analog to digital converter (ADC) and digital to analog converter (DAC). Since the capability of the processor and ADC/DAC are limited, the filter algorithm and signal representation should be optimized in terms of computation and dimension. Numerous concerns should be exercised to realize the digital filter in real-time processing. In the software perspective, the algorithm can be executed quickly via using the corresponding fast algorithms. In the other hardware viewpoint, the numbers can be handled in the fixed- or floating-point representation that can change the execution accuracy and speed. Therefore, the algorithm structure and format significantly affect the filter performance in various ways.

This book extensively uses the MATLAB to understand the digital signal processing theory throughout the chapters. In the last chapter, the comprehensive digital filter design and implementation are presented via using the MATLAB from theoretical fundamentals described by the previous chapters. The MATLAB toolboxes provide the effective methods to apply the complicated theory to the actual filter system such as personal computer and embedded system. Choice for the hardware is extensive and the MATLAB supports the full- and semi- automated procedure in order to realize the digital signal processing system. For your information, the basic tutorials of the MATLAB and symbolic math toolbox are organized in the appendices.

The first chapter of this book tries to inspire the readers toward the DSP world with simple explanation. As stated above, the subsequent chapters develop the advanced and detailed description for individual part of the theories with supportive illustrations. From the time domain signal to frequency analysis, the major principles are demonstrated to design the practical digital filter system. The author challenges to visualize the underlying fundamentals of the DSP mathematics based on the figures and illustrations for further understanding. Once the reader grasps the idea, the theory can be realized by the MATLAB and its extensive toolboxes. The author expects that the readers obtain the implementation-oriented DSP knowledge from this book, hopefully.

1.6 Problems

1. Indicate the tense of signal $x[.]$ below.
   $x[n]$:
   $x[n-2]$:
   $x[n+3]$:
2. Organize the x[.] sequence in time order from older appeared on left and newer shown on right.
   $x[5]$, $x[6]$, $x[3]$, $x[7]$, $x[-2]$, $x[9]$
3. Write the simple low pass filter with recent 5 samples.
4. Write the simple high pass filter with recent 9 samples.
5. Write the simple digital filter to pass the 4-sample period signal.
6. The generalized filter equation is given below.

$$y[n] = \sum_{k=0}^{N-1} x[n-k]h[k]$$

   Is it possible to realize the filter with infinite length ($N = \infty$)?
7. Limiting the magnitude of the signal provides the additional frequency components as shown below. To reverse back to input sequence, you suggest the filter method.

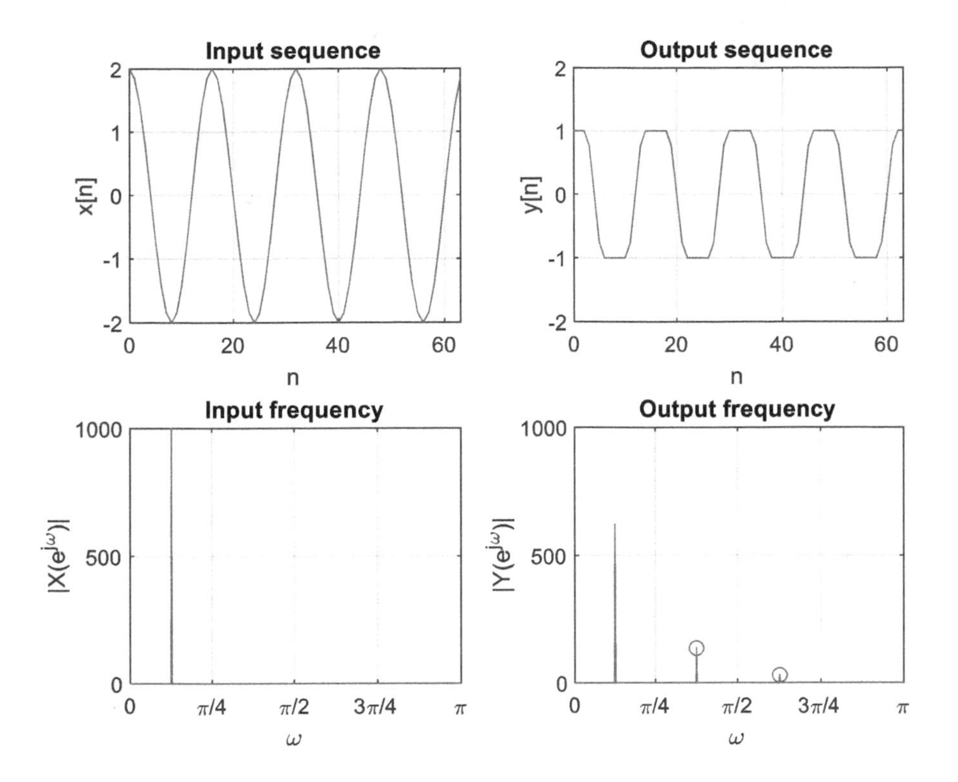

8. According to the decomposition, what is the primitive signal? In other word, what basic signal shape describes any other arbitrary signal by shifting and scaling?

Chapter 2
Frequency and Signals in Discrete Domain

Our digital system receives and produces signal which contains the information in predictable manner. The additive noise to the signal contaminates the information in random way; however, we assume that the DSP system in this book only considers the discrete signal which is the sampled version of the continuous counterpart. The signal can be represented by the combination of periodic sinusoid know as frequency component. The frequency of the discrete time signal is somehow tricky to understand directly; therefore, we borrow the frequency concept from the continuous domain then apply to the discrete study. The transition between both domains is performed by the sampling operation. The discrete time signal is further decomposed into the primitive function that can be used for the filter characterization in Chap. 4.

2.1 Continuous Sinusoid Signals

The continuous signal is denoted by the parentheses of the function for independent variables as $x(\cdot)$. The unit of frequency f is the cycles per second (or equivalently Hz); therefore, the frequency represents the number of rotations within the second. The rotation is efficiently illustrated in the complex domain with exponential power as $re^{j\theta}$. The r is the radius of the circle and j is the complex number. The vector shown below Fig. 2.1 is the $re^{j\theta}$ and the function follows the circle as the radian angle θ changing.

Since the signal and frequency are the function of time, the t is placed at the input variable as below.

$$x(t) = re^{j\Omega t} \tag{2.1}$$

K. Kim, *Conceptual Digital Signal Processing with MATLAB*,
Signals and Communication Technology,
https://doi.org/10.1007/978-981-15-2584-1_2

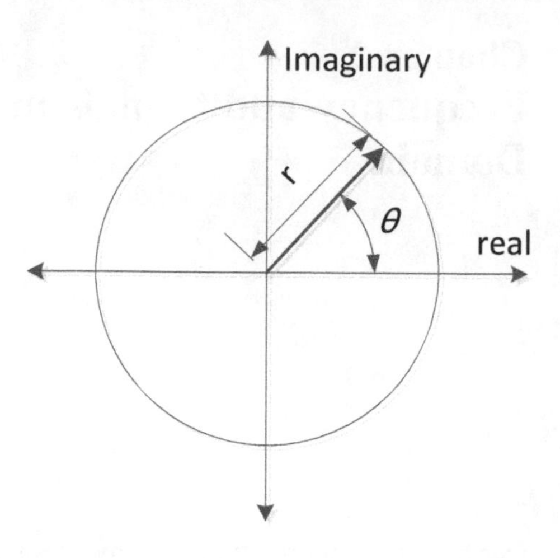

Fig. 2.1 The $re^{j\theta}$ circular rotation at complex number domain

The exponential power Ωt is the radian angle and, in every second, the vector rotates for Ω radian. The Ω is known as the radian frequency that indicates the radian per second. The conventional frequency f is related to the Ω as below.

$$\Omega = 2\pi f \qquad f = \frac{\Omega}{2\pi} \tag{2.2}$$

Unless using the animation, the 2D polar coordinate (complex domain) barely expresses the location variation over the time. Figure 2.2 represents the circular rotation in 3D plot over time according to the Eq. (2.1). The figure shows all information about the circular motion; however, it is difficult to identify the time and radian value due to the limited viewpoint.

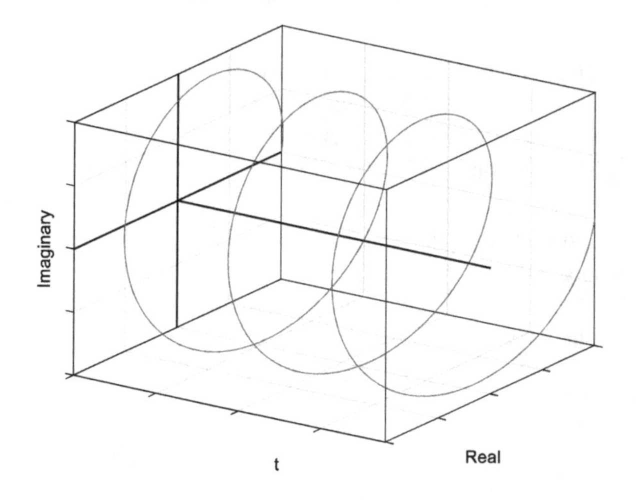

Fig. 2.2 The $re^{j\Omega t}$ circular rotation over time t

Prog. 2.1 MATLAB program for Fig. 2.2.

```
xt = @(t) t;
yt = @(t) cos(t);
zt = @(t) sin(t);
fplot3(xt,yt,zt,[0 6*pi]);          %3-D plot
xL = xlim;
yL = ylim;
zL = zlim;
line([0 0], yL, [0 0],'LineWidth',1,'Color','black');    %y-axis
line(xL, [0 0], [0 0],'LineWidth',1,'Color','black');    %x-axis
line([0 0], [0 0], zL,'LineWidth',1,'Color','black');    %z-axis
...
```

The Cartesian coordinate solves this problem via projecting the vector onto the real and imaginary axis. The values on each axis are located with time information as Fig. 2.3. The projection on the real and imaginary axis are derived from the cosine and sine function respectively.

$$x(t) = re^{j\Omega t} = re^{j2\pi ft} = r\cos(2\pi ft) + jr\sin(2\pi ft) \tag{2.3}$$

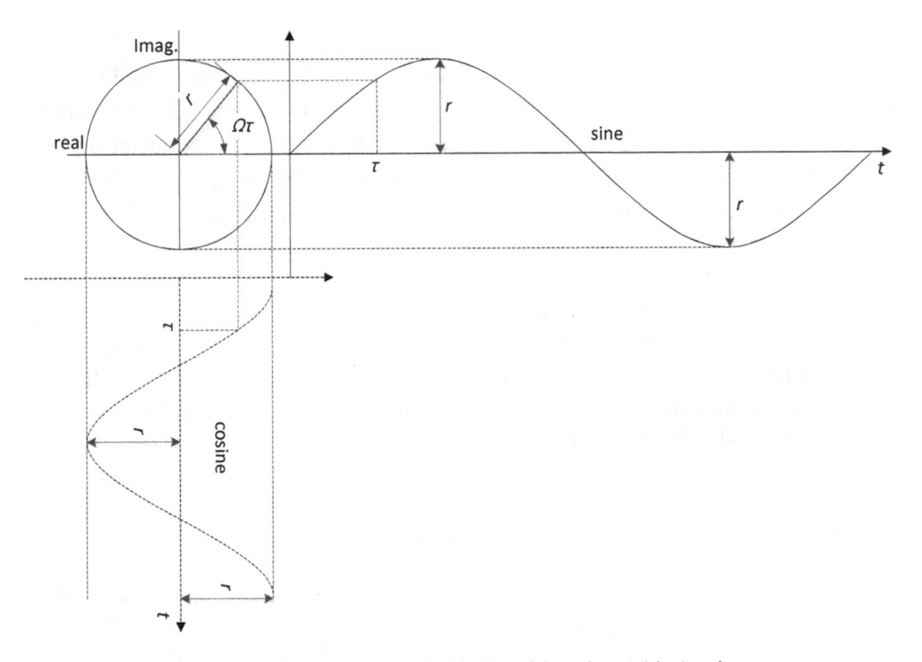

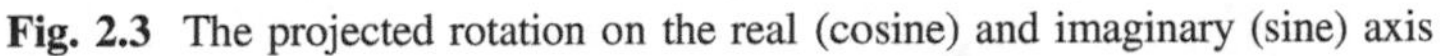

Fig. 2.3 The projected rotation on the real (cosine) and imaginary (sine) axis

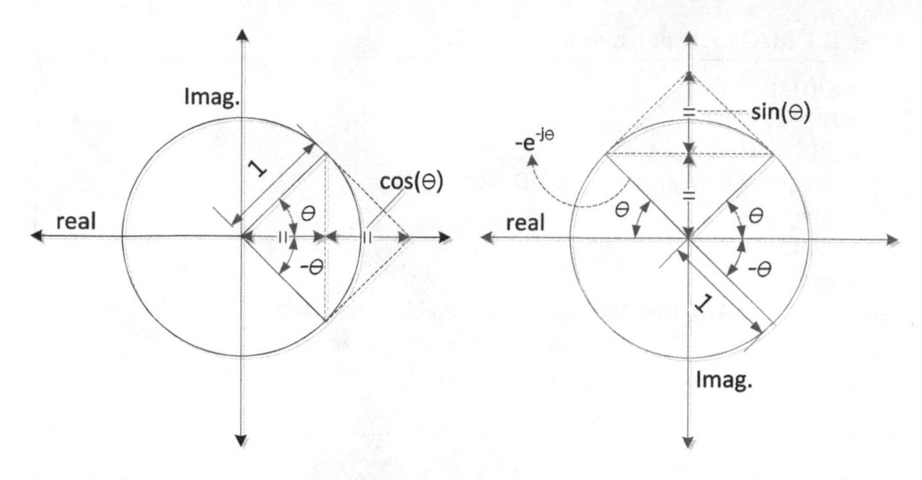

Fig. 2.4 Graphical representation of Euler formula for cosine and sine function

The projection can figure out the cosine (real part) and sine (imaginary part) component of the complex number. In similar manner, the conjugated complex number (switch the imaginary part sign or change the phase sign) provides the straightforward equations known as Euler formula as below.

$$\cos(\theta) = \frac{e^{j\theta} + e^{-j\theta}}{2} \quad \text{and} \quad \sin(\theta) = \frac{e^{j\theta} - e^{-j\theta}}{2j} \tag{2.4}$$

The graphical explanation is demonstrated at Fig. 2.4. The conjugated complex number $e^{-j\theta}$ is the reflection over the real axis; hence, the vector addition with the $e^{j\theta}$ leaves the twice of the real part as $2\cos(\theta)$. Divide by two on vector addition complies with the Euler formula. The negative conjugated complex number $-e^{-j\theta}$ is the reflection over the imaginary axis of the coordinates. The vector addition with the $e^{j\theta}$ provides the twice of the imaginary part as $2j\sin(\theta)$. Divide by $2j$ on the addition results in the above Euler formula as well.

It is important to note that the complex rotation can be transformed to the corresponding cosine (real part) and sine (imaginary part) component by simple linear operation with negative angle counterpart. The pair of the complex numbers is changed to the real number. The negative frequency is used for the real valued signal in the following chapters.

Example 2.1
Show the frequency components of $r\cos(\Omega t)$ and $r\sin(\Omega t)$ by Euler formula.

Solution

$$r\cos(\Omega t) = \frac{re^{j\Omega t} + re^{-j\Omega t}}{2} = \frac{r}{2}e^{j\Omega t} + \frac{r}{2}e^{-j\Omega t}$$

$$r\sin(\Omega t) = \frac{re^{j\Omega t} - re^{-j\Omega t}}{2j} = \frac{r}{2j}e^{j\Omega t} - \frac{r}{2j}e^{-j\Omega t}$$

Note that each sinusoid signal contains the positive frequency Ω as well as negative frequency Ω component.

∎

The conventional sinusoidal signal is represented by the cosine and sine function. The phased sinusoid can be decomposed into the non-phased cosine and sine function as Eq. (2.5). The $Re(\cdot)$ operator compute the real part value from complex number.

$$\begin{aligned} A\cos(\Omega t + \theta) &= Re\left(Ae^{j(\Omega t + \theta)}\right) = Re\left(Ae^{j\theta}e^{j\Omega t}\right) \\ &= Re\left((A\cos(\theta) + Aj\sin(\theta))e^{j\Omega t}\right) \\ &= Re\left(A\cos(\theta)e^{j\Omega t} + Ae^{j\frac{\pi}{2}}\sin(\theta)e^{j\Omega t}\right) \\ &= Re\left(A\cos(\theta)e^{j\Omega t} + A\sin(\theta)e^{j\left(\Omega t + \frac{\pi}{2}\right)}\right) \\ &= A\cos(\theta)\cos(\Omega t) + A\sin(\theta)\cos\left(\Omega t + \frac{\pi}{2}\right) \\ &= A\cos(\theta)\cos(\Omega t) - A\sin(\theta)\sin(\Omega t) \end{aligned} \tag{2.5}$$

Observe the signal shift in Fig. 2.5. By adding or subtracting, the $x(t)$ moves to left or right respectively with assumption of positive τ.

Note the following equivalences in trigonometry for reference.

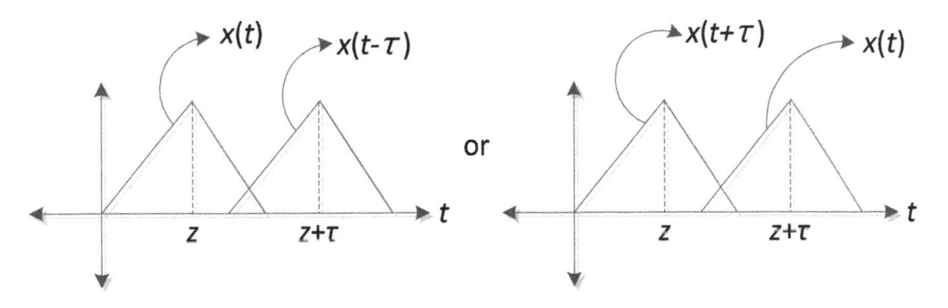

Fig. 2.5 The time shift of signal $x(t)$

Prog. 2.2 MATLAB program for Fig. 2.6.

```
ome = -2*pi:pi/100:2*pi;
x1 = cos(ome);
x2 = sin(ome);
str = {'-2\pi','-3\pi/2','-\pi','-\pi/2','0','\pi/2','\pi','3\pi/2','2\pi'};

figure,
subplot(221), plot(ome,x1,ome,x2), grid
xticks([-2*pi -3*pi/2 -pi -pi/2 0 pi/2 pi 3*pi/2 2*pi])
xticklabels(str);
xlim([-2*pi 2*pi])
ylim([-1.5 1.5])

subplot(222), plot(ome,x1,ome,x2,ome,-x2), grid
xticks([-2*pi -3*pi/2 -pi -pi/2 0 pi/2 pi 3*pi/2 2*pi])
xticklabels(str);
xlim([-2*pi 2*pi])
ylim([-1.5 1.5])

subplot(223), plot(ome,x1,ome,-x1), grid
xticks([-2*pi -3*pi/2 -pi -pi/2 0 pi/2 pi 3*pi/2 2*pi])
xticklabels(str);
xlim([-2*pi 2*pi])
ylim([-1.5 1.5])
xlabel('\theta')

subplot(224), plot(ome,x2,ome,-x2), grid
xticks([-2*pi -3*pi/2 -pi -pi/2 0 pi/2 pi 3*pi/2 2*pi])
xticklabels(str);
xlim([-2*pi 2*pi])
ylim([-1.5 1.5])
xlabel('\theta')
```

Example 2.2

What is the meaning of below equation?

$$A\cos(\Omega t + \theta) = A\cos(\theta)\cos(\Omega t) - A\sin(\theta)\sin(\Omega t)$$

Solution

The linear combination of weighted $\cos(\Omega t)$ and $\sin(\Omega t)$ provides the any arbitrary magnitude and phase cosine function. The weights are derived from the constant magnitude and phase. The time dependent terms present the frequency of the signal. ■

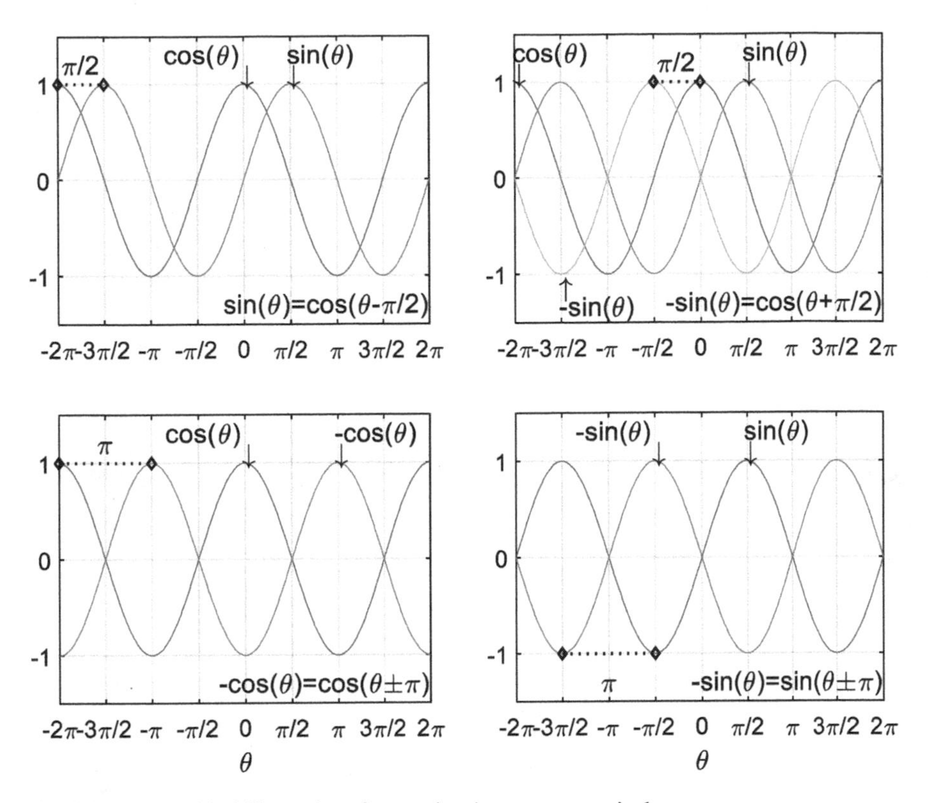

Fig. 2.6 The graphical illustrations for certain trigonometry equivalences

As shown in Eq. (2.5), the phased cosine function is divided into the non-phased cosine and sine function with modified magnitude for $A\cos(\theta)$ and $-A\sin(\theta)$ respectively. Without the phase, the cosine and sine function are inherently even and odd function that are described as below.

$$x_e(t) = x_e(-t); \;\; x_o(t) = -x_o(-t) \tag{2.6}$$

The even function is symmetric with respect to the y axis and the odd function is symmetric with respect to the origin of the coordinates as shown in Fig. 2.7. Every function can be decomposed into the even and odd function that can be found by Eq. (2.7). The conventional phased sine and cosine function can be written as the sum of the non-phased cosine (even function) and non-phased sine (odd function) as well.

$$x(t) = x_e(t) + x_o(t); \;\; x_e(t) = \frac{x(t) + x(-t)}{2}; \;\; x_o(t) = \frac{x(t) - x(-t)}{2} \tag{2.7}$$

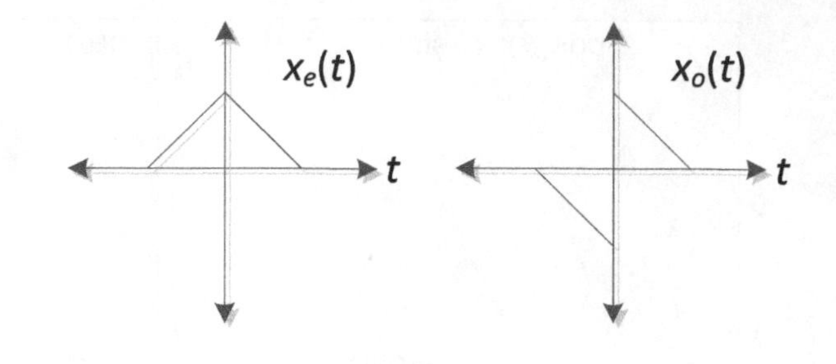

Fig. 2.7 Even and odd function example

Example 2.3
Find the even and odd function of $x(t) = A\cos(\Omega t + \theta)$.

Solution

$$
\begin{aligned}
x_e(t) =& \frac{x(t) + x(-t)}{2} = A\frac{\cos(\Omega t + \theta) + \cos(-\Omega t + \theta)}{2} \\
& A\frac{e^{j\Omega t}e^{j\theta} + e^{-j\Omega t}e^{-j\theta} + e^{j-\Omega t}e^{j\theta} + e^{j\Omega t}e^{-j\theta}}{4} \\
=& A\frac{e^{j\Omega t}\left(e^{j\theta} + e^{-j\theta}\right) + e^{-j\Omega t}\left(e^{j\theta} + e^{-j\theta}\right)}{4} = A\cos(\theta)\frac{e^{j\Omega t} + e^{-j\Omega t}}{2} \\
=& A\cos(\theta)\cos(\Omega t)
\end{aligned}
$$

$$
\begin{aligned}
x_o(t) =& \frac{x(t) - x(-t)}{2} = A\frac{\cos(\Omega t + \theta) - \cos(-\Omega t + \theta)}{2} \\
& A\frac{e^{j\Omega t}e^{j\theta} + e^{-j\Omega t}e^{-j\theta} - e^{j-\Omega t}e^{j\theta} - e^{j\Omega t}e^{-j\theta}}{4} \\
=& A\frac{e^{j\Omega t}\left(e^{j\theta} - e^{-j\theta}\right) - e^{-j\Omega t}\left(e^{j\theta} - e^{-j\theta}\right)}{4} = Aj\sin(\theta)\frac{e^{j\Omega t} - e^{-j\Omega t}}{2} \\
=& -A\sin(\theta)\sin(\Omega t)
\end{aligned}
$$

■

On the contrary, the multiple sinusoids can be converted to the single sinusoid term by using the phasor. The combining the sinusoids is only applied to the identical frequency functions; in other words, the different frequency sinusoids cannot be merged together to the single term. The phasor indicates the static vector of the sinusoid as below.

$$
Ae^{j\theta} = \text{Phasor}\{A\cos(\Omega t + \theta)\} \tag{2.8}
$$

For the identical frequency, the sum of multiple sinusoids uses the phasor as shown below.

$$\begin{aligned} A_1\cos(\Omega t+\theta_1)+A_2\cos(\Omega t+\theta_2) &= \mathrm{Re}\{A_1e^{j(\Omega t+\theta_1)}\}+\mathrm{Re}\{A_2e^{j(\Omega t+\theta_2)}\} \\ &= \mathrm{Re}\{A_1e^{j(\Omega t+\theta_1)}+A_2e^{j(\Omega t+\theta_2)}\} \\ &= \mathrm{Re}\{e^{j\Omega t}(A_1e^{j\theta_1}+A_2e^{j\theta_2})\}=\mathrm{Re}\{e^{j\Omega t}A_3e^{j\theta_3}\} \\ &= \mathrm{Re}\left\{A_3e^{j(\Omega t+\theta_3)}\right\}=A_3\cos(\Omega t+\theta_3) \end{aligned}$$

Using the phasor, we can directly compute the magnitude and phase of the sinusoid sum as below.

$$\begin{aligned} &\mathrm{Phasor}\{A_1\cos(\Omega t+\theta_1)\}+\mathrm{Phasor}\{A_2\cos(\Omega t+\theta_2)\} \\ &=A_1e^{j\theta_1}+A_2e^{j\theta_2}=A_3e^{j\theta_3}\rightarrow A_3\cos(\Omega t+\theta_3) \end{aligned} \tag{2.9}$$

The addition of multiple complex numbers can be illustrated by the vector addition as shown in Fig. 2.8. The phasor in polar coordinates should be transformed to the cartesian coordinates for addition and transformed back to the polar coordinates to apply on the sinusoid form.

The complex addition of $A_1e^{j\theta_1}$ and $A_2e^{j\theta_2}$ means the new magnitude and phase for the combined sinusoid as $A_3e^{j\theta_3}$. Then, simply apply on the derived sinusoid like $A_3\cos(\Omega t+\theta_3)$ to produce the single term. Once the sinusoid frequencies are equal, the phasor addition simplifies the linearly combined sinusoids in uncomplicated

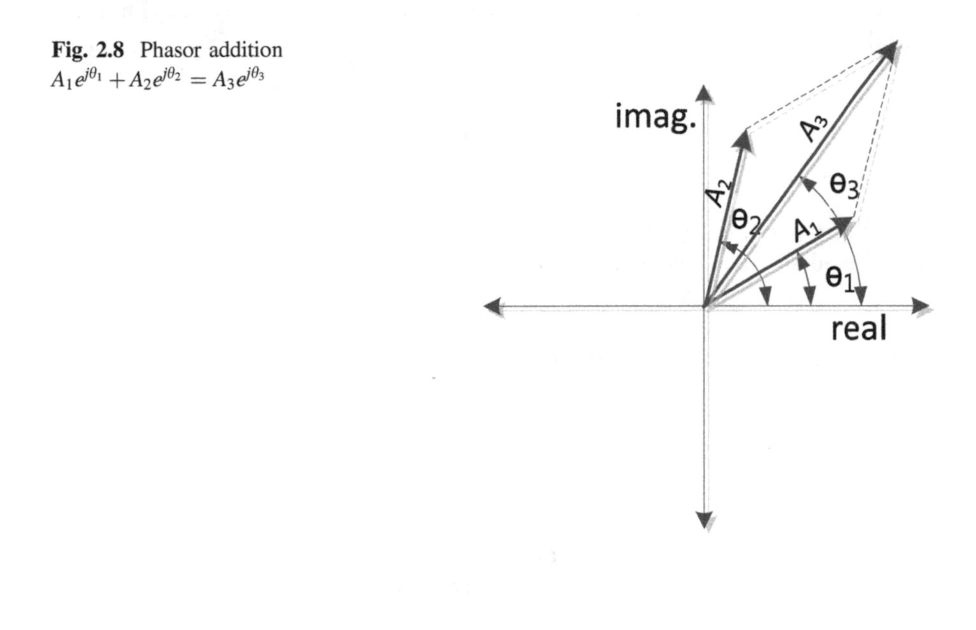

Fig. 2.8 Phasor addition $A_1e^{j\theta_1}+A_2e^{j\theta_2}=A_3e^{j\theta_3}$

form. Provided that the frequencies are not identical, the further simplification is not applicable; therefore, the present equation is the last step for mathematical maneuver as below.

$$\begin{gathered}\text{If } f_1 \neq f_2, \text{ then}\\ A_1\cos(2\pi f_1 t+\theta_1)+A_2\cos(2\pi f_2 t+\theta_2)\\ \neq A_3\cos(2\pi f_3 t+\theta_3)\\ \forall A_{1,2,3},\theta_{1,2,3},f_{1,2,3}\in\mathbb{R}\end{gathered}$$

Example 2.4
Add the sinusoid $\cos(\Omega t)+\sin(\Omega t+\pi)$.

Solution

$$\begin{aligned}\cos(\Omega t)+\sin(\Omega t+\pi) &= \cos(\Omega t)+\sin\left(\Omega t+\tfrac{\pi}{2}+\tfrac{\pi}{2}\right)\\ &= \cos(\Omega \mathrm{t})+\cos\left(\Omega \mathrm{t}+\tfrac{\pi}{2}\right)\\ &\rightarrow \text{Phasor}\{\cos(\Omega \mathrm{t})\}+\text{Phasor}\left\{\cos\left(\Omega \mathrm{t}+\tfrac{\pi}{2}\right)\right\}\\ &= e^{j0}+e^{j\pi/2}=1+j=\sqrt{2}e^{j\pi/4}\\ &\rightarrow \sqrt{2}\cos\left(\Omega t+\tfrac{\pi}{4}\right)\end{aligned}$$

■

The continuous linear combination of the sinusoids can present any arbitrary signal as shown below.

$$x(t)=\int_0^\infty m(\Omega)\cos(\Omega t+\theta(\Omega))\,d\Omega\in\mathbb{R} \tag{2.10}$$

Using the Eq. (2.5), the phased cosine function can be represented by the non-phased sinusoid functions as below.

$$\begin{aligned}x(t) &= \int_0^\infty m(\Omega)\cos(\theta(\Omega))\cos(\Omega t)\,d\Omega-\int_0^\infty m(\Omega)\sin(\theta(\Omega))\sin(\Omega t)\,d\Omega\\ &= \int_0^\infty a(\Omega)\cos(\Omega t)\,d\Omega-\int_0^\infty b(\Omega)\sin(\Omega t)\,d\Omega\end{aligned} \tag{2.11}$$

$$\text{where}\begin{cases}a(\Omega)=m(\Omega)\cos(\theta(\Omega))\\ b(\Omega)=m(\Omega)\sin(\theta(\Omega))\end{cases}$$

The Euler formula further decomposes the equation into the analytic representation as below.

$$x(t) = \int_0^\infty a(\Omega)\frac{e^{j\Omega t} + e^{-j\Omega t}}{2}d\Omega - \int_0^\infty b(\Omega)\frac{e^{j\Omega t} - e^{-j\Omega t}}{2j}d\Omega$$

$$= \int_0^\infty \left(\frac{a(\Omega)}{2}e^{j\Omega t} - \frac{b(\Omega)}{2j}e^{j\Omega t}\right)d\Omega + \int_0^\infty \left(\frac{a(\Omega)}{2}e^{-j\Omega t} + \frac{b(\Omega)}{2j}e^{-j\Omega t}\right)d\Omega$$

$$= \int_0^\infty \frac{a(\Omega) + jb(\Omega)}{2}e^{j\Omega t}d\Omega + \int_0^\infty \frac{a(\Omega) - jb(\Omega)}{2}e^{-j\Omega t}d\Omega$$

For negative radian frequency Ω, the integrand of the integral should maintain the even function for symmetrical purpose as below.

$$m(\Omega)\cos(\Omega t + \theta(\Omega)) = m(-\Omega)\cos(-\Omega t + \theta(-\Omega))$$

The individual functions in the integrand must demonstrate the following symmetric properties to meet the above equation condition.

$$m(\Omega) = m(-\Omega)$$

$$\theta(\Omega) = -\theta(-\Omega)$$

The magnitude and phase function in the sinusoid present the even and odd symmetric property, respectively. The corresponding $a(\Omega)$ and $b(\Omega)$ show the below symmetric properties as well.

$$a(\Omega) = m(\Omega)\cos(\theta(\Omega)) = a(-\Omega) = m(-\Omega)\cos(-\theta(\Omega))$$

$$b(\Omega) = m(\Omega)\sin(\theta(\Omega)) = -b(-\Omega) = -m(-\Omega)\sin(-\theta(\Omega))$$

By applying the symmetric properties, the integrals are combined together as below.

$$x(t) = \int_0^\infty \frac{a(\Omega) + jb(\Omega)}{2}e^{j\Omega t}d\Omega + \int_{-\infty}^0 \frac{a(-\Omega) - jb(-\Omega)}{2}e^{j\Omega t}d\Omega$$

$$= \int_0^\infty \frac{a(\Omega) + jb(\Omega)}{2}e^{j\Omega t}d\Omega + \int_{-\infty}^0 \frac{a(\Omega) + jb(\Omega)}{2}e^{j\Omega t}d\Omega$$

$$= \int_{-\infty}^\infty \frac{a(\Omega) + jb(\Omega)}{2}e^{j\Omega t}d\Omega = \int_{-\infty}^\infty c(\Omega)e^{j\Omega t}d\Omega$$

In short, the derived equation is organized as below.

$$x(t) = \int_{-\infty}^{\infty} c(\Omega)e^{j\Omega t}d\Omega \text{ where} \begin{cases} a(\Omega) = m(\Omega)\cos(\theta(\Omega)) \\ b(\Omega) = m(\Omega)\sin(\theta(\Omega)) \\ c(\Omega) = \frac{a(\Omega)+jb(\Omega)}{2} \end{cases} \quad (2.12)$$

The arbitrary signal $x(t)$ can be divided into the continuous linear combination of $c(\Omega)e^{j\Omega t}$. Once we can find the complex magnitude $c(\Omega)$ of the analytic signal $e^{j\Omega t}$ for radian frequency Ω, we can represent the time domain signal $x(t)$ by using the integral as below.

$$x(t) = \int_{0}^{\infty} a(\Omega)\cos(\Omega t)d\Omega - \int_{0}^{\infty} b(\Omega)\sin(\Omega t)d\Omega \quad (2.13)$$

Equivalently, the integral of phased cosine function can be used as below.

$$x(t) = \int_{0}^{\infty} m(\Omega)\cos(\Omega t + \theta(\Omega))d\Omega$$

Equations (2.10), (2.11), and (2.12) are synthesis equations of the Fourier transform which tells us that infinite combination of the frequencies provides the arbitrary time function of interest. The $c(\Omega)$ is derived by the following equation by using the orthogonality of the sinusoid.

$$\begin{aligned} \int_{-\infty}^{\infty} x(t)e^{-j\Omega t}dt &= \int_{-\infty}^{\infty}\left\{\int_{-\infty}^{\infty} c(\Psi)e^{j\Psi t}d\Psi\right\}e^{-j\Omega t}dt \\ &= \int_{-\infty}^{\infty}\int_{-\infty}^{\infty} c(\Psi)e^{j(\Psi-\Omega)t}dt\, d\Psi \\ &= \int_{-\infty}^{\infty} c(\Psi)\left\{\int_{-\infty}^{\infty} e^{j(\Psi-\Omega)t}dt\right\}d\Psi \\ &\quad \text{where } x(t) = \int_{-\infty}^{\infty} c(\Psi)e^{j\Psi t}d\Psi \end{aligned}$$

The inner term of the above equation is the integral of the periodic signal $e^{j(\Psi-\Omega)t}$ over the infinite range. The signal frequency is the $(\Psi - \Omega)$ radian per second and the corresponding period T is defined as $2\pi/(\Psi - \Omega)$. The integral is divided into the two cases as $\Psi \neq \Omega$ and $\Psi = \Omega$. For the non-equal case, the periodic signal

$e^{j(\Psi-\Omega)t}$ has finite period T and the integral is segmented & repeated into the individual periods with infinite number k multiplication. The other case, which is the equal case, the infinite period T provides the single long infinite period for the integral range. The complex exponential function can be written as sine and cosine function in segmented form as below.

$$\int_{-\infty}^{\infty} e^{j(\Psi-\Omega)t}dt = \int_{-\infty}^{\infty} \cos((\Psi-\Omega)t)dt + j\int_{-\infty}^{\infty} \sin((\Psi-\Omega)t)dt$$

$$= \lim_{\substack{k\to\infty \\ k\in\mathbb{Z}}} k\int_{-\frac{T}{2}}^{\frac{T}{2}} \cos((\Psi-\Omega)t)dt + j\lim_{\substack{k\to\infty \\ k\in\mathbb{Z}}} k\int_{-\frac{T}{2}}^{\frac{T}{2}} \sin((\Psi-\Omega)t)dt$$

The period T depends on the sinusoid radian frequency $(\Psi-\Omega)$ as below.

$$T = \frac{2\pi}{(\Psi-\Omega)}$$

Figure 2.9 demonstrates the approximated equivalence between the entire range and one period sinusoid integral. The one period integral on sinusoid denotes the zero; therefore, the entire range integral should be zero for conventional case.

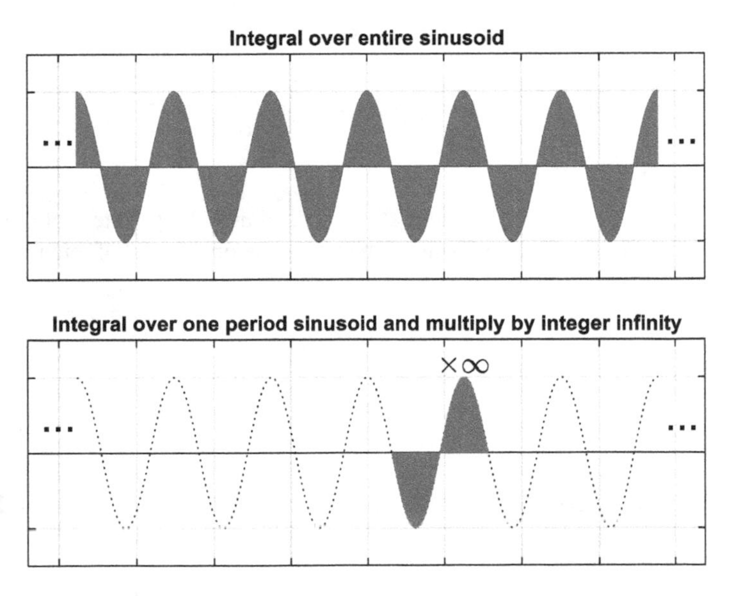

Fig. 2.9 Approximation of entire sinusoid integral with one period integral

Prog. 2.3 MATLAB program for Fig. 2.9.

```
t = -6*pi:pi/100:6*pi;
t1 = pi/2:pi/100:5*pi/2;
x = cos(t);
x1 = cos(t1);

figure,
subplot(211), plot(t,x,'Color','k'), grid
hold on
stem(t,x,'Marker','none')
hold off

subplot(212), plot(t,x,'LineStyle',':','Color','k'), grid
hold on
stem(t1,x1,'Marker','none')
hold off
```

The one period sinusoid provides the following integral values on the frequency conditions.

$$\int_{-\frac{T}{2}}^{\frac{T}{2}} \sin((\Psi-\Omega)\mathrm{t})dt = \begin{cases} 0 & \text{for } \Psi \neq \Omega \\ 0 & \text{for } \Psi = \Omega \end{cases}$$

$$\int_{-\frac{T}{2}}^{\frac{T}{2}} \cos((\Psi-\Omega)\mathrm{t})dt = \begin{cases} 0 & \text{for } \Psi \neq \Omega \\ T & \text{for } \Psi = \Omega \end{cases}$$

The integral on complex exponential is arranged as below. Note that the $\Psi = \Omega$ condition produces the infinite T value. Therefore, the one period of integral is valid.

$$\int_{-\infty}^{\infty} e^{j(\Psi-\Omega)\mathrm{t}}dt$$

$$= \begin{cases} \lim\limits_{\substack{k\to\infty \\ k\in\mathbb{N}}} k\left[\int_{-\frac{T}{2}}^{\frac{T}{2}} \cos((\Psi-\Omega)\mathrm{t})dt + j\int_{-\frac{T}{2}}^{\frac{T}{2}} \sin((\Psi-\Omega)\mathrm{t})dt\right] = 0 & \text{for } \Psi \neq \Omega \\ \int_{-\frac{T}{2}}^{\frac{T}{2}} \cos((\Psi-\Omega)\mathrm{t})dt + j\int_{-\frac{T}{2}}^{\frac{T}{2}} \sin((\Psi-\Omega)\mathrm{t})dt = T \text{ for } \begin{cases} \Psi = \Omega \\ T = \frac{2\pi}{(\Psi-\Omega)} = \infty \end{cases} \end{cases}$$

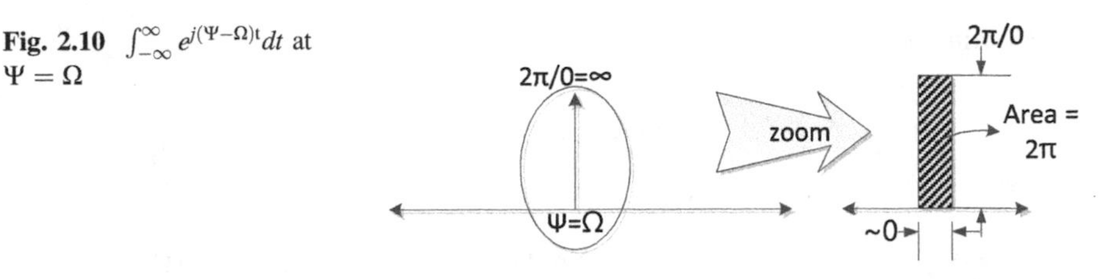

Fig. 2.10 $\int_{-\infty}^{\infty} e^{j(\Psi-\Omega)t}dt$ at $\Psi = \Omega$

Both case integral results are shown below.

$$\int_{-\infty}^{\infty} e^{j(\Psi-\Omega)t}dt = \begin{cases} 0 & \text{for } \Psi \neq \Omega \\ T = \frac{2\pi}{(\Psi-\Omega)} = \infty & \text{for } \Psi = \Omega \end{cases} \tag{2.14}$$

The non-zero outcome T is appeared at the equal moment between Ψ and Ω. The instantaneous value at the equal situation is the infinite as shown below.

The area of the given equation over the $\Psi - \Omega$ axis is 2π when the Ψ and Ω are equal. The rectangular base shown in Fig. 2.10 is approaching to the zero because of the instant moment ($\Psi = \Omega$) and the height is advanced to the $2\pi/0$ from the integral computation. We cannot directly compute the rectangular area by multiplication due to the 0/0 condition; however, the result is 2π in mathematics (L'Hospital's rule) [1]. Note that this intuitive and graphical illustration is not valid for the mathematical proof in a strict sense; therefore, the reader should refer the Dirac delta function [2] for further information. In simple form, the result is presented below.

$$\int_{-\infty}^{\infty} e^{j(\Psi-\Omega)t}dt = 2\pi\delta(\Psi - \Omega) \tag{2.15}$$

The $\delta(\Psi - \Omega)$ is Dirac delta function (in short delta function) that only provides the infinite output at the zero-input condition. Also, the total area of the delta function is one. The characteristic is shown in below.

$$\delta(t) = \begin{cases} \infty & \text{for } t = 0 \\ 0 & \text{for } t \neq 0 \end{cases} \quad \text{and} \quad \int_{-\infty}^{\infty} \delta(t)dt = 1 \tag{2.16}$$

With the continuous domain ($t \in \mathbb{R}$), the delta function provides the method to specify the value on the particular position. The inevitable interference with adjacent positions is avoided by the zero width on continuous domain. The value cannot be built on the empty space; hence, the area of zero width is designed to be one as shown in Eq. (2.16) and Fig. 2.10. The application of delta function is shown at Eq. (2.17).

$$x(t)\delta(t - t_0) = \begin{cases} \infty & \text{for } t = t_0 \\ 0 & \text{for } t \neq t_0 \end{cases} \text{and} \int_{-\infty}^{\infty} x(t)\delta(t - t_0)dt = x(t_0) \tag{2.17}$$

Continuously solve the double integral problem.

$$\begin{aligned}\int_{-\infty}^{\infty} c(\Psi)\left\{\int_{-\infty}^{\infty} e^{j(\Psi-\Omega)t}dt\right\}d\Psi &= \int_{-\infty}^{\infty} c(\Psi)2\pi\delta(\Psi-\Omega)d\Psi \\ &= 2\pi c(\Omega)\int_{-\infty}^{\infty}\delta(\Psi-\Omega)d\Psi = 2\pi c(\Omega)\end{aligned} \tag{2.18}$$

Note that the $\delta(\Psi - \Omega)$ shows the value only at the $\Psi = \Omega$ situation; therefore, Ψ and Ω are interchangeable. The $c(\Psi)$ is escaped from the integral by replacing the input argument. Hence, this procedure presents the analysis equation of the Fourier transform which presents the magnitude of the individual frequency component as $c(\Omega)$. The derived equation is shown below.

$$c(\Omega) = \frac{1}{2\pi}\int_{-\infty}^{\infty} x(t)e^{-j\Omega t}dt \tag{2.19}$$

Once we know the frequency magnitude $c(\Omega)$, the time domain function can be synthesized by below.

$$x(t) = \int_{-\infty}^{\infty} c(\Omega)e^{j\Omega t}d\Omega \tag{2.20}$$

The general form of Fourier transform [3] is given as below.

$$\text{Let the } X(\Omega) = 2\pi c(\Omega)\begin{cases} X(\Omega) = \int_{-\infty}^{\infty} x(t)e^{-j\Omega t}dt \\ x(t) = \frac{1}{2\pi}\int_{-\infty}^{\infty} X(\Omega)e^{j\Omega t}d\Omega \end{cases} \tag{2.21}$$

The pair of the Fourier transform equations provides the frequency distribution of the continuous time signal. The analysis equation computes complex magnitude $X(\Omega)$ which represents the height and position of the cosine function at the radian frequency Ω. From the given $X(\Omega)$, the synthesis equation derives the corresponding time signal. Observe that the $x(t)$ is generally real value continuous time signal; hence, $X(\Omega)$ should be distributed in certain rule as real part in even symmetric and imaginary part in odd symmetric.

Example 2.5

Compute and draw $X(\Omega)$ of below signal.

$$x(t) = \begin{cases} 1 & \text{for } -1 \le t \le 1 \\ 0 & \text{Otherwise} \end{cases}$$

Solution

$$X(\Omega) = \int_{-1}^{1} e^{-j\Omega t} dt = \frac{e^{-j\Omega t}}{-j\Omega}\bigg|_{-1}^{1} = \frac{e^{j\Omega} - e^{-j\Omega}}{j\Omega} = \frac{2\sin(\Omega)}{\Omega}$$

Prog. 2.4 MATLAB program for Fig. 2.11.

```
syms t;
b = [-100:1/100:100];
x1 = heaviside(t+1) - heaviside(t-1);
y2 = 2*(sin(b+eps))./(b+eps);                    %Add eps to avoid NaN
figure,
subplot(211), ezplot(x1, [-2 2]), grid
axis([-2 2 -0.2 1.2]);
subplot(212), plot(b, y2), grid
axis([-100.0000   100.0000   -0.5   2.2])
```

■

The frequency component of the given time window is $2\sin(\Omega)/\Omega$ that is known as the (unnormalized) sinc function [4]. The synthesis equation from Fourier transform provides the original time window as well. Due to the requirement of

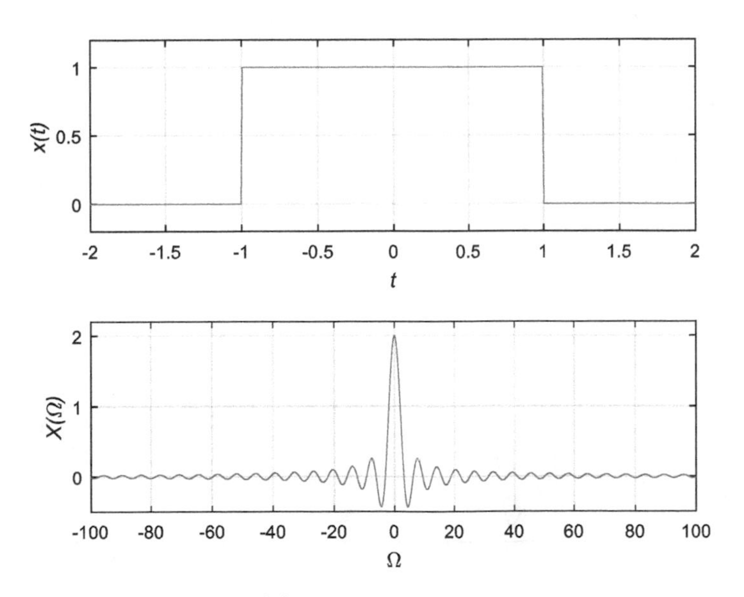

Fig. 2.11 $x(t)$ and corresponding $X(\Omega)$ by Fourier transform

further understanding on Fourier transform, the inverse transform is performed with numerical method as below.

$$
\begin{aligned}
x(t) &= \frac{1}{2\pi}\int_{-\infty}^{\infty} X(\Omega)e^{j\Omega t}d\Omega \\
&= \frac{1}{2\pi}X(0) + \frac{1}{2\pi}\int_{0^+}^{\infty} X(\Omega)e^{j\Omega t}d\Omega + \frac{1}{2\pi}\int_{-\infty}^{0^-} X(\Omega)e^{j\Omega t}d\Omega; X(\Omega) = \frac{2\sin(\Omega)}{\Omega} \\
&= \frac{1}{2\pi}X(0) + \frac{1}{2\pi}\int_{0^+}^{\infty} X(\Omega)e^{j\Omega t}d\Omega + \frac{1}{2\pi}\int_{0^+}^{\infty} X(-\Omega)e^{-j\Omega t}d\Omega \\
&= \frac{1}{2\pi}X(0) + \frac{1}{2\pi}\int_{0^+}^{\infty} X(\Omega)(e^{j\Omega t} + e^{-j\Omega t})d\Omega \because X(\Omega) = X(-\Omega) \\
&= \frac{1}{2\pi}X(0) + \frac{1}{2\pi}\int_{0^+}^{\infty} 2X(\Omega)\cos(\Omega t)d\Omega \\
&\approx \left\{\frac{1}{2\pi}X(0) + \frac{1}{\pi}\sum_{k=1}^{N} X(\Omega_k)\cos(\Omega_k t)\right\}2\pi\Delta_f = \tilde{x}(t)
\end{aligned} \tag{2.22}
$$

In the last step of the above derivation, the numerical integration is adopted to compute the integral part where Ω_k is the discrete radian frequency as $2\pi\Delta_f k$ with Δ_f frequency distance. Figure 2.12 illustrates the $X(\Omega) = 2sin(\Omega)/\Omega$ and inverse transformed $\tilde{x}(t)$ by numerical method for $-10\pi \leq \Omega_k \leq 10\pi$.

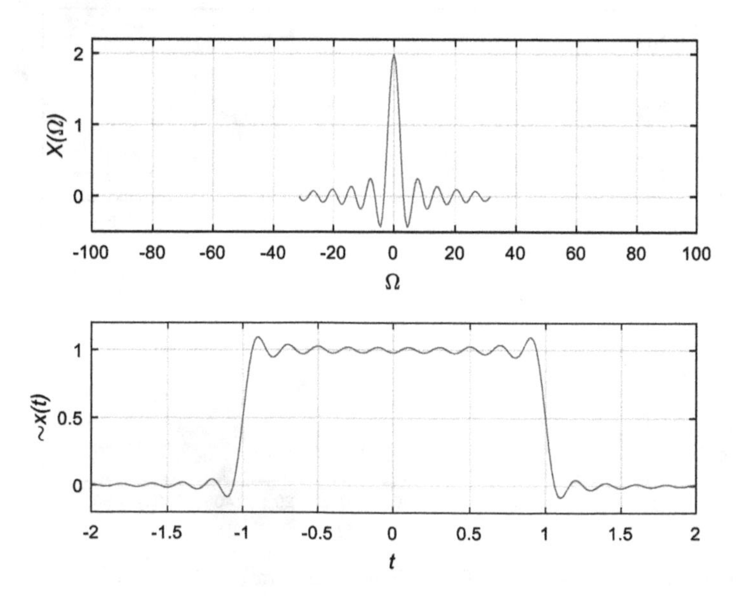

Fig. 2.12 $X(\Omega)$ and corresponding $\tilde{x}(t)$ by numerical Fourier transform for $-31.4 \leq \Omega_k \leq 31.4$

Prog. 2.5 MATLAB program for Fig. 2.12.

```
delf = 0.1;                    % Frequency interval
delt = 0.001;                  % Time interval
fmax = 30/(2*pi);
tmax = 2;
ff = 0.1:delf:fmax;
tt = -tmax:delt:tmax;
out = [];
f1 = 2*pi*(-fmax:delf:fmax);
xc = 2*sin(f1+eps)./(f1+eps);

for k1 = tt          % time
     temp = 0;
     for k2 = ff        % frequency
          temp = temp + sin(2*pi*k2)*cos(2*pi*k2*k1)/(2*pi*k2);
     end
     out = [out ((temp*2/pi)+(1/pi))*2*pi*delf];
end

figure,
subplot(211), plot(f1,xc), grid
axis([-100.0000   100.0000    -0.5      2.2])
subplot(212), plot(tt,out), grid
axis([-2 2 -0.2 1.2]);
```

The reconstructed $x(t)$ presents the approximated window shape in the time domain. The wide fluctuations and transitions are observed due to the limited frequency range in the inverse Fourier transform. Figure 2.13 illustrates the $X(\Omega)$ and $\tilde{x}(t)$ by numerical method for extended frequency range. The further accurate window shape is realized by the inverse transform with wider spectrum range. The fluctuations and transitions are both narrow to follow the precise window shape in time domain.

Prog. 2.6 MATLAB program for Fig. 2.13. Exactly identical to **Prog. 2.5** except following.

```
...
fmax = 100/(2*pi);
...
```

Figures 2.12 and 2.13 illustrate the simulated inverse Fourier transform to obtain the relationship between the frequency range and reconstructed shape. The wider spectrum of the given equation provides the more accurate representation of the original time signal. The inverse transform expects to deliver the exact time signal

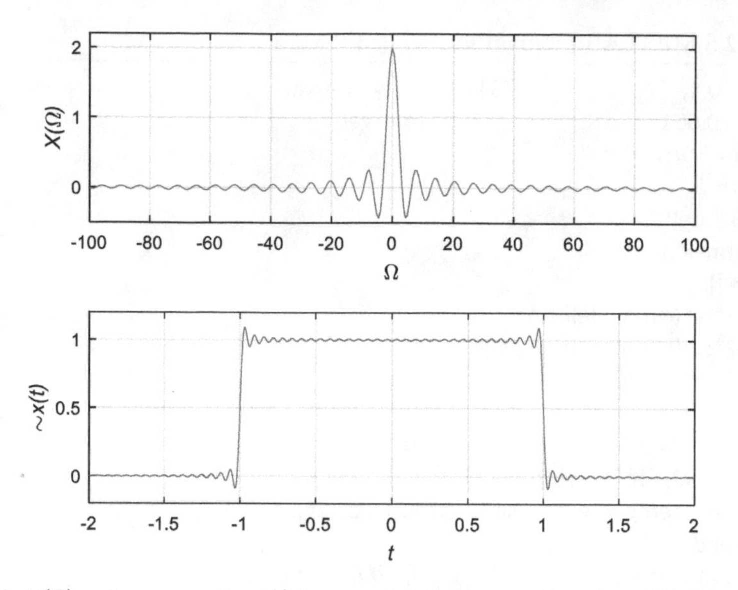

Fig. 2.13 $X(\Omega)$ and corresponding $\tilde{x}(t)$ by numerical Fourier transform for $-100 \leq \Omega_k \leq 100$

once the infinite range and continuous frequency is applied to the integral. The infinite distribution of the frequencies generates the designed arbitrary signal in time domain and vice versa. Therefore, the frequency is the primitive elements of the signal.

2.2 Sampling

The sampling is the procedure to convert the continuous signal into the discrete version as data sequence. The discrete time signal samples the continuous signal as real number at the specific time locations which are the integer multiple of specific period. Therefore, the converter reads the instantaneous magnitude of the continuous time signal in every constant interval and generates the data sequence for further processing. The mathematical and graphical representation are below in Fig. 2.14.

$$x[n] = x(nT_s) \quad n \in \mathbb{Z}; x[\cdot], x(\cdot), T_s \in \mathbb{R} \tag{2.23}$$

The T_s is the sampling period and $f_s = 1/T_s$ is the sampling frequency which denotes the number of reading or sampling per second. Usually, the sampling is specified by the sampling frequency and the higher f_s describes the signal in further detail. What is the minimum sampling frequency? Let's consider the sinusoid signal in Fig. 2.15.

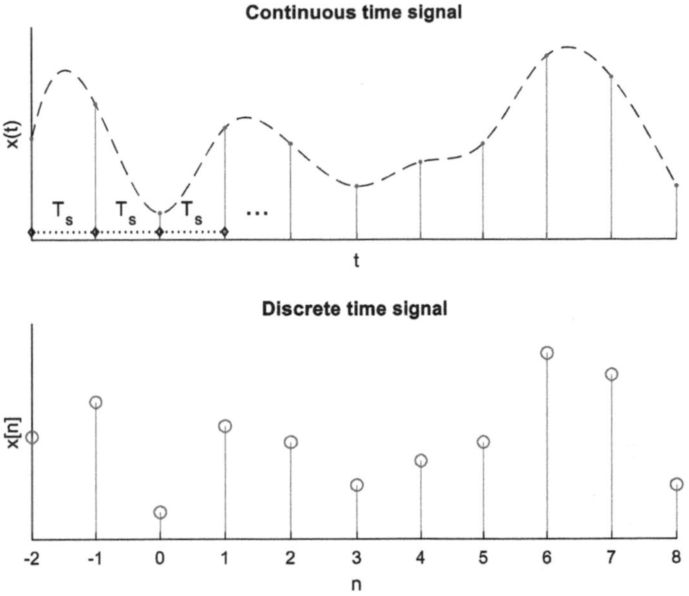

Continuous time signal

x(t)

T_s T_s T_s ...

t

Discrete time signal

x[n]

-2 -1 0 1 2 3 4 5 6 7 8

n

Fig. 2.14 Continuous and discrete time signal example with T_s sampling period

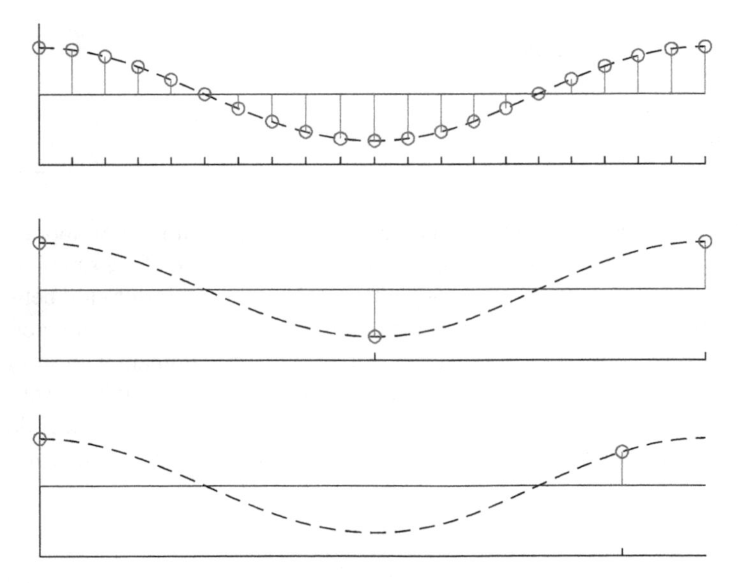

Fig. 2.15 Sampling the sinusoid signal with various sampling frequencies f_s (or sampling period T_s)

Prog. 2.7 MATLAB program for Fig. 2.15.

```
nn0 = 0:pi/100:2*pi;
nn1 = 0:pi/10:2*pi;
nn2 = 0:pi:2*pi;
nn3 = [0 (3*pi/2+2*pi)/2];

xn0 = cos(nn0);
xn1 = cos(nn1);
xn2 = cos(nn2);
xn3 = cos(nn3);

figure,
subplot(311)
plot(nn0,xn0,'k','LineStyle','--');
hold on
stem(nn1,xn1);
hold off
subplot(312)
plot(nn0,xn0,'k','LineStyle','--');
hold on
stem(nn2,xn2);
hold off
subplot(313)
plot(nn0,xn0,'k','LineStyle','--');
hold on
stem(nn3,xn3);
hold off
```

The given sinusoid $\cos(\Omega t)$ is sampled by the three f_s in high on top, minimum on middle, and low on bottom in Fig. 2.15. The performance of the sampling can be defined by the representation accuracy after conversion and restoration between the domains. For the time being, we assume that the restoration to the continuous signal is performed by connecting the adjacent values in discrete domain with straight line. The high sampling frequency depicts the sinusoid with fine detail; however, the low sampling frequency cannot recover the original signal by connecting dots. The improper sampling changes the signal frequency to the lower value; hence, the

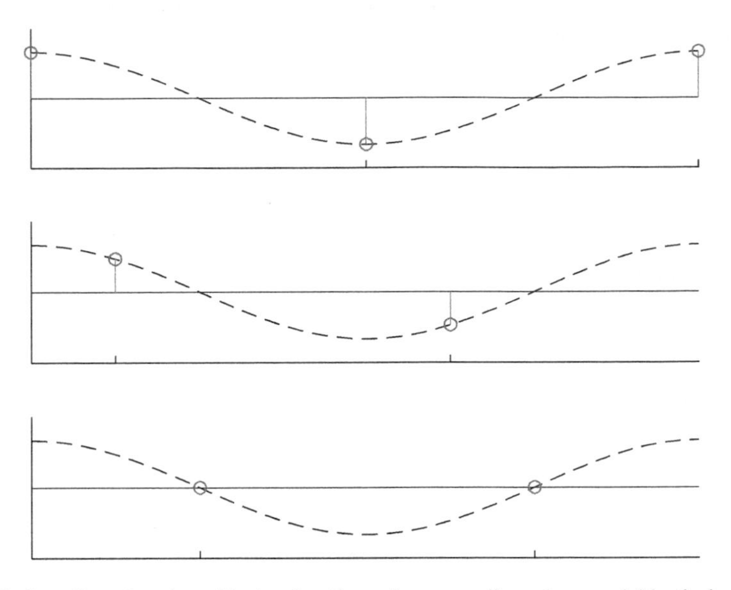

Fig. 2.16 Sampling the sinusoid signal with various sampling phase and identical sampling frequency f_s

original signal is contaminated by that frequency shift. The sampling rate in the middle figure provides the minimum proper sampling frequency for the given sinusoid as twice sampling reading in one signal period. The minimum rate may cause reduced signal magnitude due to the out of phase sampling as Fig. 2.16 but the signal frequency is preserved. The decision between the proper and improper sampling is performed based on the frequency shift from the conversion process. In general, the sampling is executed in marginally higher rate than the minimum requirement.

Prog. 2.8 MATLAB program for Fig. 2.16.

```
nn0 = 0:pi/100:2*pi;
nn2 = 0:pi:2*pi;
nn4 = nn2+pi/4;
nn5 = nn2+pi/2;
xn0 = cos(nn0);
xn2 = cos(nn2);
xn4 = cos(nn4);
xn5 = cos(nn5);

figure,
subplot(311)
plot(nn0,xn0,'k','LineStyle','--');
hold on
stem(nn2,xn2);
hold off
ylim([-1.5 1.5])
xlim([0 2*pi])
subplot(312)
plot(nn0,xn0,'k','LineStyle','--');
hold on
stem(nn4,xn4);
hold off
ylim([-1.5 1.5])
xlim([0 2*pi])
subplot(313)
plot(nn0,xn0,'k','LineStyle','--');
hold on
stem(nn5,xn5);
hold off
ylim([-1.5 1.5])
xlim([0 2*pi])
```

Let's apply the above theory to the more typical signal, the bandlimited continuous signal, as Eq. (2.24). The signal $x(t)$ contains frequency distribution up to the Ω_{max} the highest frequency. As shown Fig. 2.15, the higher signal frequency requires the higher sampling rate correspondingly. Once we select the proper sampling parameter for Ω_{max}, the signals in lower frequencies fit in well with the sampling condition.

$$x(t) = \int_0^{\Omega_{max}} m(\Omega)\cos(\Omega t + \theta(\Omega))d\Omega \tag{2.24}$$

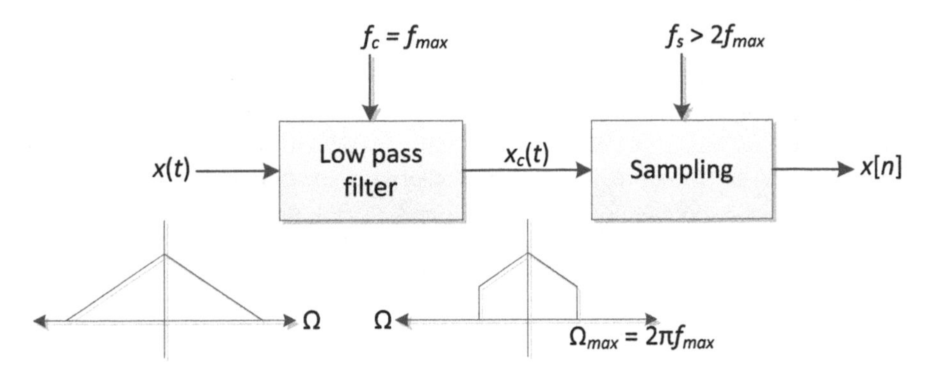

Fig. 2.17 Sampling the band non-limited continuous signal with preprocessing

The general theory of sampling frequency known as Nyquist rate [5] is specified below.

$$f_s > 2f_{max} \qquad \Omega_{max} = 2\pi f_{max} \tag{2.25}$$

In order to satisfy the sampling condition for the highest frequency, the sampling reading should be performed at least twice within the highest frequency signal period. In other words, the sampling frequency must be higher than the twice of the frequency maximum used in the bandlimited signal. This is the Nyquist sampling rate. The lower than the Nyquist rate produces the frequency shift and causes spectral contamination in high signal frequencies. Upon the cases such as non-limited frequency band signal and limited sampling capability system, the preprocessing is required to removes frequencies above Ω_{max}. The low pass filter tailors the frequency distribution and the sampling process does not initiate the signal representation problems. The preprocessing is illustrated below in Fig. 2.17.

Example 2.5
Find the minimum sampling rate for following signal.

$$x(t) = \cos(100\pi t) + \sin(200\pi t) + \cos\left(500\pi t + \frac{\pi}{4}\right) + 7$$

Solution
The highest frequency of the $x(t)$ is $\Omega_{max} = 500\pi$. Therefore, f_{max} is 250 Hz. According to the Nyquist rate, the minimum sampling rate is below.

$$f_s > 2f_{max} = 500\,\text{Hz}$$

■

2.3 Signal Representation

The fundamental of the discrete time signal is delta function (Kronecker delta function [6]) as Fig. 2.18. Whenever the input argument in the square bracket is zero, the delta function provides the one; otherwise, the output is always zero.

$$\delta[n] = \begin{cases} 1 & n = 0 \\ 0 & \text{Otherwise} \end{cases} \tag{2.26}$$

The delta function is very useful signal in order to specify the sequence location. Let's consider the arbitrary discrete time signal as Fig. 2.19.

The discrete time signal $x[n]$ is the finite or infinite combination of individual values such as $x[0], x[1], x[2], x[3], x[4]$, and etc. However, the simple linear combination of the individual values cannot represent the $x[n]$ due to the missing sequence location for each value as below.

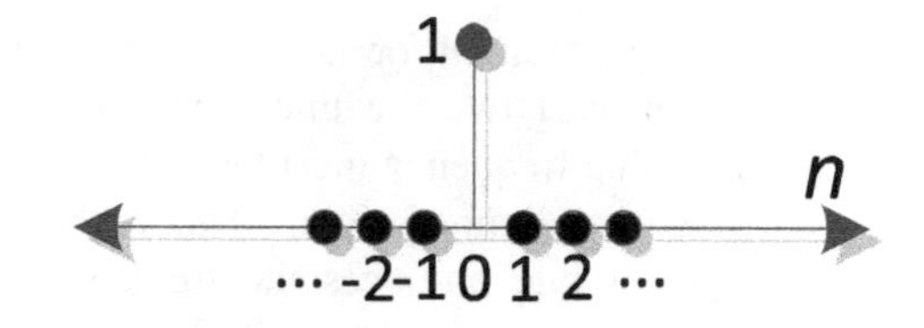

Fig. 2.18 Delta function (Kronecker delta function)

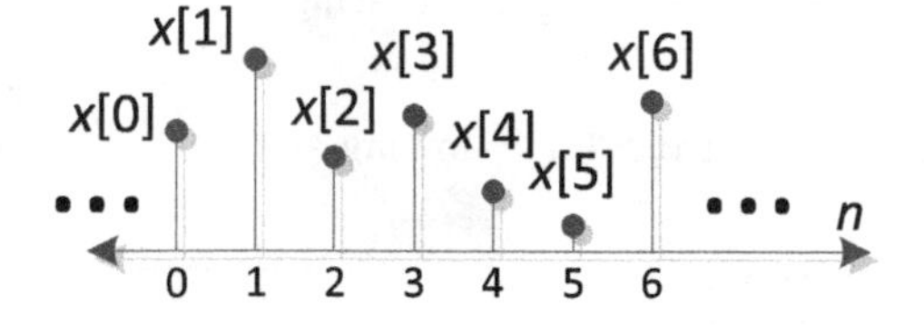

Fig. 2.19 Discrete time signal example $x[n]$

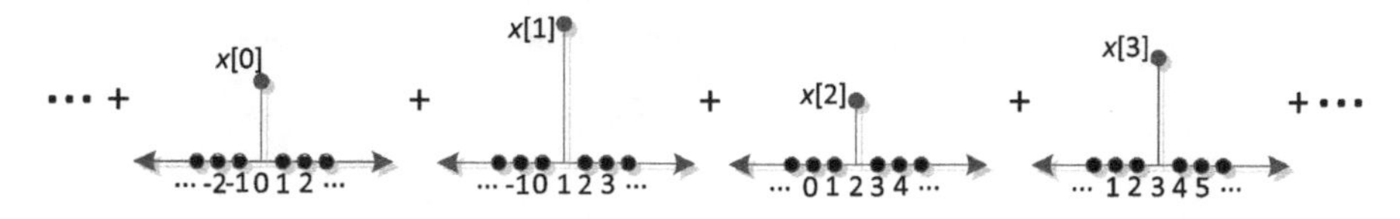

Fig. 2.20 Combination of weighted and shifted delta functions for arbitrary $x[n]$

$$x[n] \neq \cdots + x[0] + x[1] + x[2] + x[3] + x[4] + \cdots = \sum_{k=-\infty}^{\infty} x[k] = \text{Constant}$$

The simple sum of the discrete values is only constant value that cannot stand for the time varying sequence $x[n]$. The delta function supports to indicate the sequence location as Eq. (2.27).

$$\begin{aligned} x[n] = \cdots &+ x[0]\delta[n] + x[1]\delta[n-1] + x[2]\delta[n-2] + x[3]\delta[n-3] \\ &+ x[4]\delta[n-4] + \cdots = \sum_{k=-\infty}^{\infty} x[k]\delta[n-k] \end{aligned} \tag{2.27}$$

The graphical illustration of Eq. (2.27) is shown in Fig. 2.20.

The multiplication with delta function presents the sequence position in n domain. For example, $x[3]\delta[n-3]$ shows the value $x[3]$ only at the $n = 3$ because of the delta function. Therefore, each value has the corresponding position specifier as delta function to illustrate part or whole of the given function. The delta function is very fundamental for the discrete time signal to demonstrate and decompose into the individual values.

Example 2.6
Describe the following signal with delta function.

$$x[n] = \begin{cases} 2 \text{ for } -2 \leq n \leq 1 \\ 0 \quad \text{Otherwise} \end{cases}$$

Solution

$$x[n] = 2\delta[n+2] + 2\delta[n+1] + 2\delta[n] + 2\delta[n-1]$$

∎

The derived signals from the delta function are unit-step function and ramp function as below.

$$u[n] = \begin{cases} 1 & n \geq 0 \\ 0 & n < 0 \end{cases} \tag{2.28}$$

$$r[n] = \begin{cases} n & n \geq 0 \\ 0 & n < 0 \end{cases} \tag{2.29}$$

The graphical illustration of Eqs. (2.28) and (2.29) is shown in Fig. 2.21.

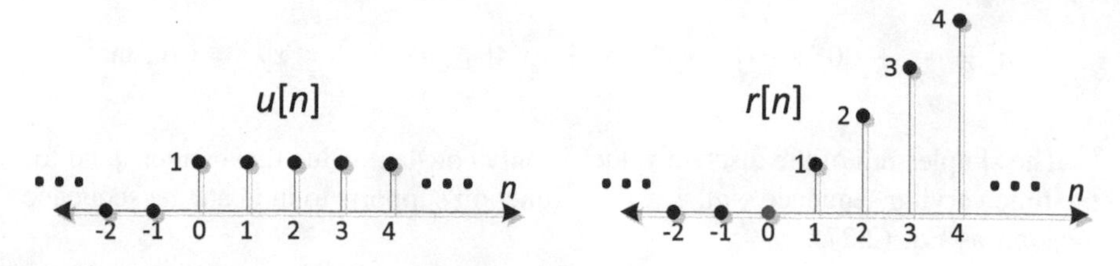

Fig. 2.21 Unit-step function $u[n]$ and ramp function $r[n]$

The unit-step function has the value one after the n is greater than and equal to zero. The ramp function shows the linearly increasing output when the n is positive and zero. The relation between those functions are organized as Eqs. (2.30) and (2.31).

$$\delta[n] \xrightarrow{\sum_{k=-\infty}^{n} \delta[k]} u[n] \xrightarrow{\sum_{k=-\infty}^{n} u[k-1]} r[n] \tag{2.30}$$

$$\delta[n] \xleftarrow{u[n]-u[n-1]} u[n] \xleftarrow{r[n+1]-r[n]} r[n] \tag{2.31}$$

The summation from the minus infinite to the present time n transforms the delta function to the unit-step and ramp function in each time. The difference between adjacent values convert the ramp function to the unit-step and delta function in consecutive time.

From the given signal $x[n]$, we can change the signal position by shifting, reflecting, and scaling in sequence. The signal shift can be performed by adding or subtracting the constant value in the sequence variable as $x[n+d]$. The shift operation depicts the given function by changing the sequence variable to the $n+d$; however, the actual horizontal axis is indexed by n variable. Overall, the given function moves to the right for the negative d and to the left for the positive d value as shown in Fig. 2.22.

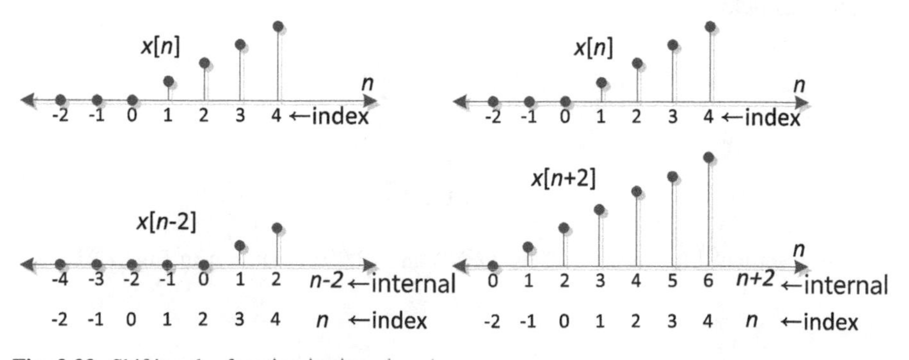

Fig. 2.22 Shifting the function in time domain

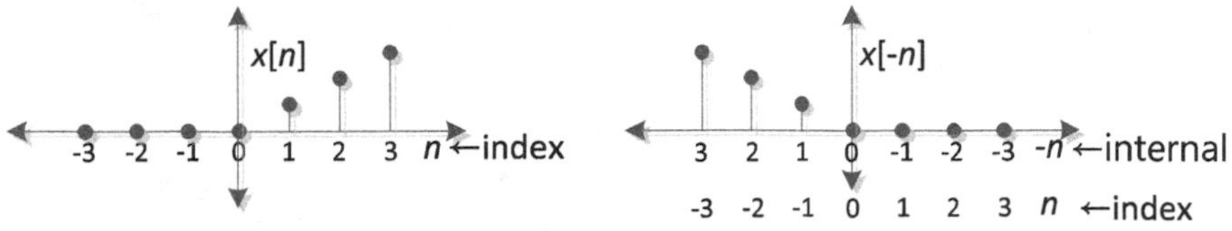

Fig. 2.23 Signal flipping over the vertical axis

The signal can be flipped over the vertical axis by reflection which can be performed by changing the sign of the sequence variable. For example, the signal $x[-n]$ exchanges the position between the positive and negative index values from the $x[n]$ as Fig. 2.23.

The discrete time signal can be expanded or compressed over the horizontal axis by multiplying the constant value on the sequence variable. As stated below, the q value decides the shape of the signal after the operation.

$$x[qn] = \begin{Bmatrix} q\,\text{times compressed}, & \text{for}\, q > 1, q \in \mathbb{N} \\ 1/q\,\text{time expaned}, & \text{for}\, 0<q<1, \;\; q \in \mathbb{Q} \end{Bmatrix} \tag{2.32}$$

$$qn \in \mathbb{Z};\, \text{otherwise}\, x[qn] = 0$$

The sequence index is scaled by q and the corresponding qn position values are placed at the n index. For the q as natural number greater than 1, the $x[qn]$ is the compressed version of $x[n]$. For example, the q is equal to 2 then the $\ldots x[-2], x[0], x[2], x[4], \ldots$ moves to the $\ldots x[-1], x[0], x[1], x[2], \ldots$, respectively. For the q as positive rational number less than 1, the $x[qn]$ is the expanded version of $x[n]$. For example, the q is equal to 1/2 then the $\ldots x[-1], x[0], x[1], x[2], \ldots$ moves to the $\ldots x[-2], x[0], x[2], x[4], \ldots$, correspondingly. In the expansion, the unassigned values between the relocated signal are zero and further processing computes the median values from the adjacent magnitudes upon the request. The illustration of compression and expansion is given in Fig. 2.24.

The combination of shifting, reflecting, and scaling in sequence should be executed with operation priority. The general order is the reflection, (time) scaling, and shifting for the $x[-q(n-d)]$ form; therefore, the sequence variable is collected by the scaling constant and reflection sign first.

$$x[n] \xrightarrow[\text{Reflection}]{} x[-n]] \xrightarrow[\text{Scaling}]{} x[-qn] \xrightarrow[\text{Shifting}]{} x[-q(n-d)] \tag{2.33}$$

Example 2.7
Draw the $x[-2(n-1)]$ of given $x[n]$.

$$x[n] = 5\delta[n] + 4\delta[n-1] + 3\delta[n-2] + 2\delta[n-3] + \delta[n-4]$$

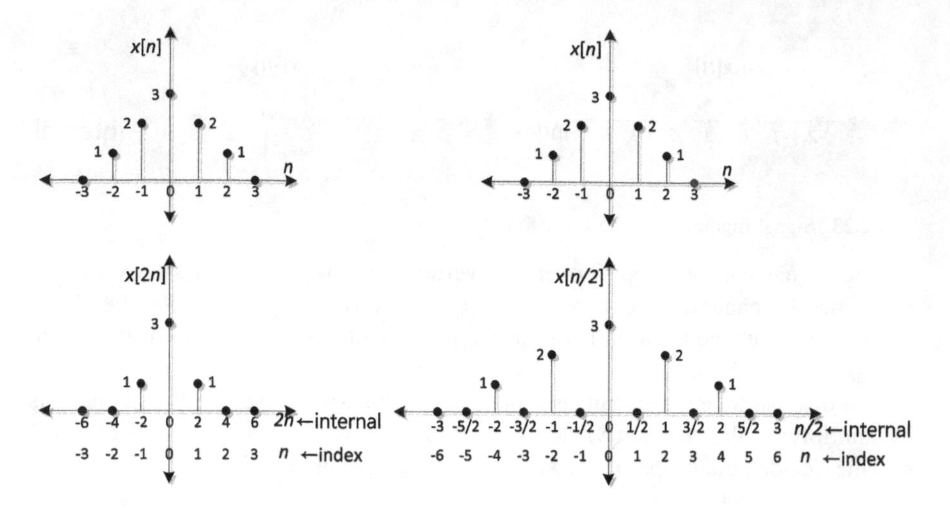

Fig. 2.24 Signal compression and expansion by time scaling

Solution

The combination is below.

$$x[n] \xrightarrow[\text{Reflection}]{} x[-n]] \xrightarrow[\text{Scaling}]{} x[-2n] \xrightarrow[\text{Shifting}]{} x[-2(n-1)]$$

Below figure shows the above procedure.

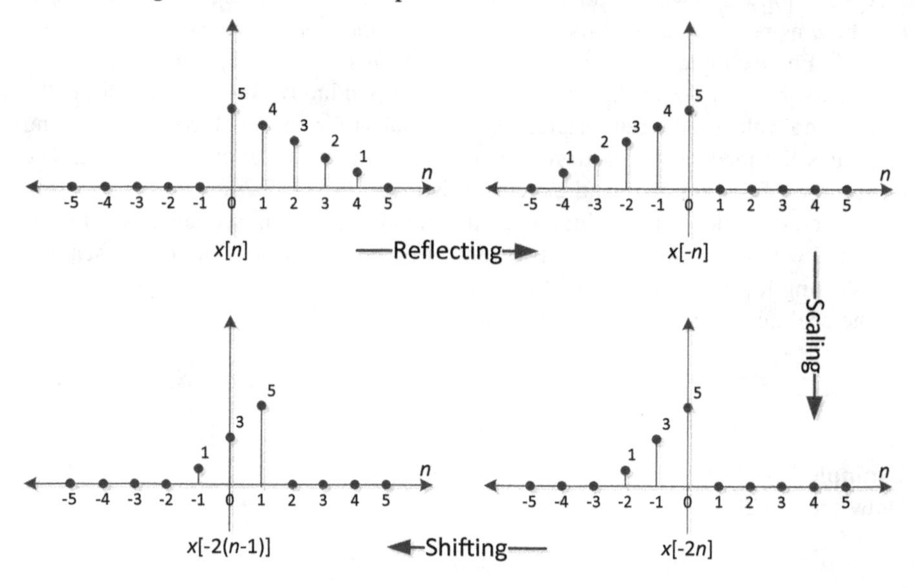

■

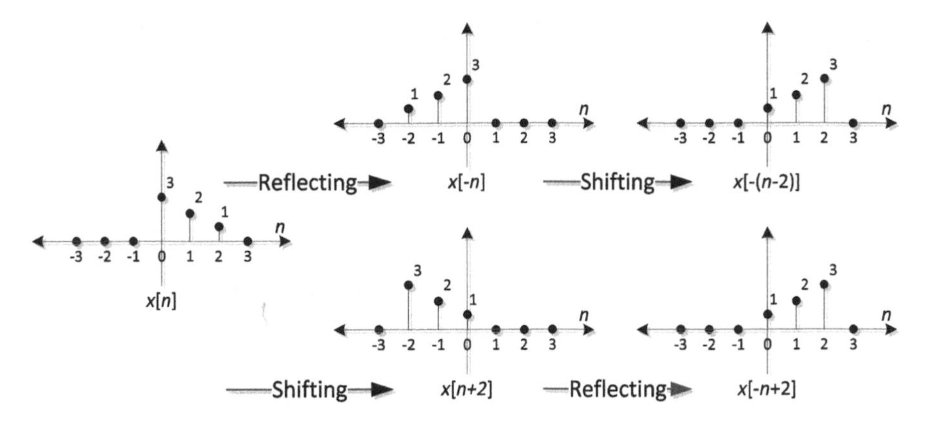

Fig. 2.25 Combination of reflecting and shifting in sequence

If you do not have scaling in the sequence, you may shift the function directly without collecting the variable. The shifting direction depends on the processing sequence as shown in Fig. 2.25. Overall, the given function moves to the left for the positive d and to the right for the negative d in expanded form as $\pm n + d$. Observe that the shifting, reflecting, and scaling operation replace the n with $(n - d)$, $(-n)$, and (αn), respectively.

Previously, we decompose the arbitrary discrete signal into the delta functions. The sequence position is defined by the shifted delta function and magnitude is specified by the function scale factor as below.

$$\begin{aligned} x[n] = \cdots &+ x[0]\delta[n] + x[1]\delta[n-1] + x[2]\delta[n-2] + x[3]\delta[n-3] \\ &+ x[4]\delta[n-4] + \cdots = \sum_{k=-\infty}^{\infty} x[k]\delta[n-k] \end{aligned} \tag{2.34}$$

The most elemental part of the discrete time signal is the delta function which can describe the arbitrary function by linear combination of shifted versions. Provided that the system satisfies the shifting, scaling, and linearity for the output, we can represent the system by simple function. The required properties for the system are time invariance and linearity as Fig. 2.26.

The time invariance is the property that the time shift in input signal provides the identical shift in output signal without modifying any output distribution. The linearity consists of two properties which are scalability and additivity. The scalability is provided by the proportional output magnitude from the input scale in exact manner. For instance, if you increase the input signal by two, the output is amplified twice scale for all values. The additivity is related to the signal decomposition. The system output from the whole input signal is equal to the sum of the individual outputs from decomposed inputs. Usually, the additivity allows you to separate the input signal in time domain; therefore, the individual sequence output is accumulated for the final output. The overall time invariance and linearity property are organized in Eqs. (2.35) and (2.36).

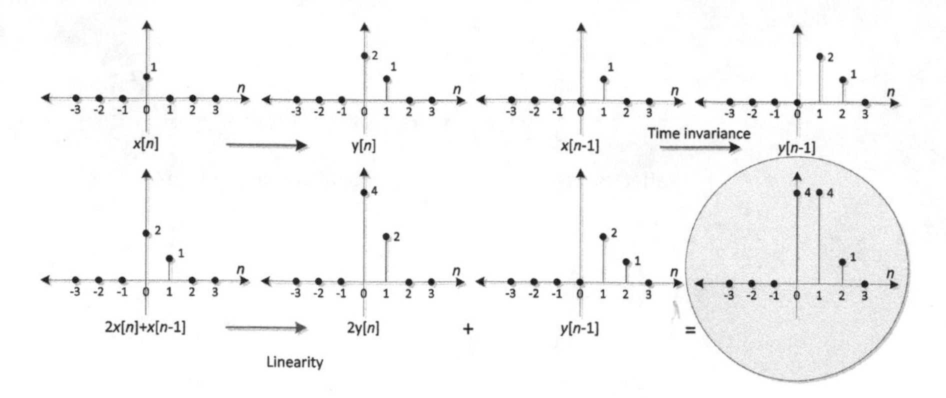

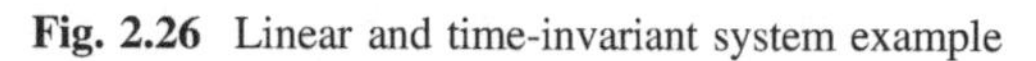
Fig. 2.26 Linear and time-invariant system example

$$y[n] = S\{x[n]\} \text{ then } y[n-d] = S\{x[n-d]\},\ \forall d \in \mathbb{Z} \tag{2.35}$$

$$\begin{aligned} x[n] = \alpha x_1[n] + \beta x_2[n] \text{ then } y[n] = S\{x[n]\} \\ = S\{\alpha x_1[n] + \beta x_2[n]\} = \alpha S\{x_1[n]\} + \beta S\{x_2[n]\} \end{aligned} \tag{2.36}$$

The linear and time invariant (LTI) system simplifies the output computation due to the input signal property. The input is linear combination of shifted delta functions; therefore, the output is also linear combination of shifted output from delta function. Once we obtain the response of the delta function, the output of any input can be computed by simple operation. Since we call the delta function as impulse, the output of the delta function is known as impulse response. The output computation from the impulse response is illustrated in Eq. (2.37).

$$\begin{aligned} h[n] &= S\{\delta[n]\} \\ y[n] &= S\{x[n]\} = S\{\ldots + x[0]\delta[n] + x[1]\delta[n-1] + x[2]\delta[n-2] + \cdots\} \\ &= \cdots + x[0]S\{\delta[n]\} + x[1]S\{\delta[n-1]\} + x[2]S\{\delta[n-2]\} + \cdots \\ &= \cdots + x[0]h[n] + x[1]h[n-1] + x[2]h[n-2] + \cdots \\ &= \sum_{k=-\infty}^{\infty} x[k]h[n-k] = y[n] \end{aligned} \tag{2.37}$$

Example 2.8
From the given impulse response $h[n]$ as below, determine the LTI system output of shown input $x[n]$.

$$h[n] = \delta[n] + \delta[n-1]$$

$$x[n] = \delta[n] - \delta[n-2]$$

Solution
Let's decompose the input $x[n]$ as below.

$$x[n] = x_1[n] + x_2[n] \text{ where } \begin{cases} x_1[n] = \delta[n] \\ x_2[n] = -\delta[n-2] \end{cases}$$

Each $x_1[n]$ and $x_2[n]$ generates $y_1[n]$ and $y_2[n]$, respectively.

$$y_1[n] = \delta[n] + \delta[n-1]$$

$$y_2[n] = -\delta[n-2] - \delta[n-3]$$

The final output is the combination of individual outputs as below.

$$y[n] = y_1[n] + y_2[n] = \delta[n] + \delta[n-1] - \delta[n-2] - \delta[n-3]$$

■

The derived equation is the convolution sum to compute filter output for the given signal $x[k]$. The impulse response tells us everything about the filter including the filter frequency characteristics. The convolution sum follows the commutative law between the impulse response and input signal; hence, the input to the impulse response provides the identical output for the impulse to the input system as Fig. 2.27.

$$\begin{aligned} y[n] &= \sum_{k=-\infty}^{\infty} x[k]h[n-k] = \sum_{k=-\infty}^{\infty} h[k]x[n-k] \\ &= x[n] * h[n] = h[n] * x[n] \end{aligned} \tag{2.38}$$

The system output by convolution sum is equal to the intuitive filter equation from Chap. 1. As described in the previous chapter, the convolution sum computes

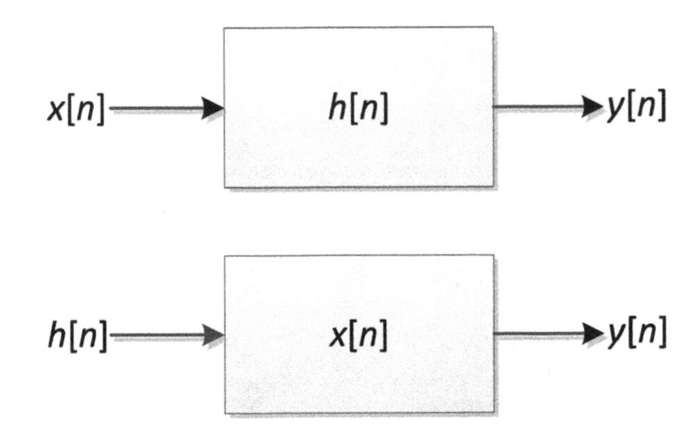

Fig. 2.27 Commutative law for convolution

the present output from the recent weighted input samples likewise the input averaging for low pass filter. Also, the filter delivers the output similar to the impulse response profile with high likelihood. Now, we understand the convolution sum equation and basic properties of the filter. Once the filter system guarantees the linear and time invariant property, the user fully describes the system by reading the impulse response. Furthermore, we can design the designated filter by carefully deriving the impulse response for reducing or amplifying the range of the frequencies.

Sometimes, we need to compute the power and energy of the signal to analyze the system. The energy is the sum of all squared discrete input signal as Eq. (2.39).

$$E_x = \sum_{n=-\infty}^{\infty} |x[n]|^2 \tag{2.39}$$

The power is the average of squared discrete input signal in range as Eq. (2.40).

$$P_x = \lim_{N\to\infty} \frac{1}{2N+1} \sum_{n=-N}^{N} |x[n]|^2 \tag{2.40}$$

If the signal is periodic and limited magnitude with infinite sequence length, then the energy is infinite, and power is finite since the power is the average of signal period. If the signal is aperiodic and limited magnitude with finite sequence length (or infinite converge sequence), then the energy and power are finite values. Unless one of the signal values is not infinite over entire infinite sequence, the power exists with finite value. Figure 2.28 shows the conditions to exist the energy and power.

Example 2.9
Decide the energy and power existence of below signals.

$$x_1[n] = \cos\left(\frac{2\pi}{4}n\right)$$

$$x_2[n] = \tan\left(\frac{2\pi}{8}n\right)$$

$$x_3[n] = 0.5^n$$

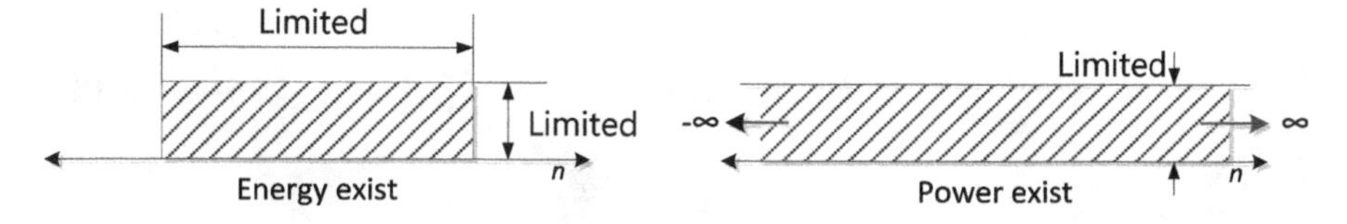

Fig. 2.28 Signal energy and power exist condition

$$x_4[n] = 2^n u[-n]$$

$$x_4[n] = 1$$

Solution

$$E_{x_1} = \sum_{n=-\infty}^{\infty} \left| \cos\left(\frac{2\pi}{4} n\right) \right|^2 = \infty$$

$$P_{x_1} = \lim_{N\to\infty} \frac{1}{2N+1} \sum_{n=-N}^{N} \left| \cos\left(\frac{2\pi}{4} n\right) \right|^2 = \frac{1}{4} \sum_{n=0}^{3} \left| \cos\left(\frac{2\pi}{4} n\right) \right|^2 = \frac{1}{2}$$

For $x_1[n]$, the energy is not existed, and the power is existed.

$$E_{x_2} = \sum_{n=-\infty}^{\infty} \left| \tan\left(\frac{2\pi}{8} n\right) \right|^2 = \infty$$

$$P_{x_2} = \lim_{N\to\infty} \frac{1}{2N+1} \sum_{n=-N}^{N} \left| \tan\left(\frac{2\pi}{8} n\right) \right|^2 = \frac{1}{8} \sum_{n=0}^{7} \left| \tan\left(\frac{2\pi}{8} n\right) \right|^2 = \infty$$

For $x_2[n]$, the energy is not existed, and the power is not existed.

$$E_{x_3} = \sum_{n=-\infty}^{\infty} |0.5^n|^2 = \infty$$

$$P_{x_3} = \lim_{N\to\infty} \frac{1}{2N+1} \sum_{n=-N}^{N} |0.5^n|^2 = \lim_{N\to\infty} \frac{1}{N+1} \sum_{n=0}^{N} 0.25^n + \lim_{N\to\infty} \frac{1}{N} \sum_{n=-N}^{-1} 0.25^n$$

$$= \lim_{N\to\infty} \frac{1}{N+1} \sum_{n=0}^{N} 0.25^n + \lim_{N\to\infty} \frac{1}{N} \sum_{n=1}^{N} 4^n$$

$$= \lim_{N\to\infty} \frac{1}{N+1} \frac{1-0.25^{N+1}}{1-0.25} + \lim_{N\to\infty} \frac{1}{N} \frac{4(1-4^N)}{1-4} = \infty$$

For $x_3[n]$, the energy is not existed, and the power is not existed.

$$E_{x_4} = \sum_{n=-\infty}^{\infty} |2^n u[-n]|^2 = \sum_{n=-\infty}^{0} |2^n|^2 = \sum_{n=0}^{\infty} 0.25^n = \frac{1}{1-0.25} = \frac{3}{4}$$

$$P_{x_4} = \lim_{N\to\infty} \frac{1}{2N+1} \sum_{n=-N}^{N} |2^n u[-n]|^2 = \lim_{N\to\infty} \frac{1}{2N+1} \frac{3}{4} = 0$$

For $x_4[n]$, the energy is existed, and the power is existed.

$$E_{x_5} = \sum_{n=-\infty}^{\infty} |1|^2 = \infty$$

$$P_{x_5} = \lim_{N\to\infty} \frac{1}{2N+1} \sum_{n=-N}^{N} |1|^2 = \lim_{N\to\infty} \frac{2N+1}{2N+1} = 1$$

For $x_5[n]$, the energy is not existed, and the power is existed.

■

The energy of the signal delivers finite value when the signal shows the limited sequence length (or infinite converge sequence) and bounded magnitude value. The power derives the finite value at the limited magnitude only over entire infinite sequence. The LTI system cannot handle the instantaneous infinite value; therefore, the input and output signal power should be existed for realization. The energy is usually used for system analysis that will be shown in Chap. 4. Next step is to discuss the frequency for the continuous and discrete time domain.

2.4 Frequency in Discrete Time Signal

The frequency is the number of rotations in second. The cyclic motion is defined in the complex number domain with polar or Cartesian coordinates as Eq. (2.41).

$$\begin{aligned} x(t) &= A_1 e^{j\Omega_1 t} = a_1 e^{j(\Omega_1 t + \theta_1)} \\ &= a_1 \cos(\Omega_1 t + \theta_1) + j a_1 \sin(\Omega_1 t + \theta_1) \text{where } A_1 = a_1 e^{j\theta_1} \end{aligned} \tag{2.41}$$

In every second, the $x(t)$ rotates the Ω_1 radian or $\Omega_1/2\pi$ cycles in count clockwise direction for the positive Ω_1. Strictly speaking, the Ω_1 radian frequency signal is $A_1 e^{j\Omega_1 t}$ with the A_1 complex magnitude that specify the amplitude and phase (delay) of the signal. If the Ω_1 is negative number, then the cycle is rotated in opposite direction as clockwise. As mentioned earlier, the distinct frequencies cannot be combined together for the single term; therefore, the multiple frequency signals are described by the linear combination of the exponentials as Eq. (2.42).

$$\begin{aligned} x(t) &= \cdots + A_1 e^{j\Omega_1 t} + A_2 e^{j\Omega_2 t} + A_3 e^{j\Omega_3 t} + A_4 e^{j\Omega_4 t} + \cdots \\ &= \begin{cases} \int_{-\infty}^{\infty} A(\Omega) e^{j\Omega t} d\Omega, & \text{for continuous } \Omega \\ \sum_{-\infty}^{\infty} A_k e^{j\Omega_k t}, & \text{for discrete } \Omega_k \end{cases} \end{aligned} \tag{2.42}$$

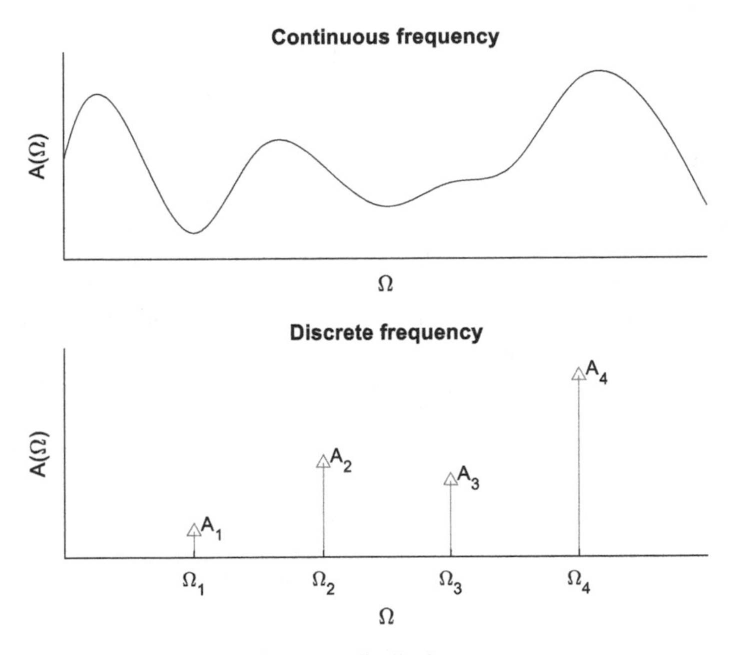

Fig. 2.29 Continuous and discrete frequency distribution

From the Eq. (2.42) and Fig. 2.29, the frequency $\Omega_1, \Omega_2, \Omega_3, \Omega_4, \ldots$ has the magnitude $A_1, A_2, A_3, A_4, \ldots$ respectively. The signal in discrete time domain is expressed by the linear combination of the delta function as Eq. (2.43). The delta function specifies the location of the sequence. In similar manner, the exponential in the frequency equation defines the position of the frequency Ω; therefore, the $e^{j\Omega t}$ is the fundamental component in frequency domain. The sum or integral of all weighted frequency components is our arbitrary signal.

$$\begin{aligned} x[n] = \cdots &+ x[0]\delta[n] + x[1]\delta[n-1] + x[2]\delta[n-2] + x[3]\delta[n-3] \\ &+ x[4]\delta[n-4] + \cdots = \sum_{k=-\infty}^{\infty} x[k]\delta[n-k] \end{aligned} \tag{2.43}$$

The rotation from the exponential produces the complex numbers that cannot be realized in the actual system. In general, the complex number is projected onto the real or imaginary axis by using the cosine or sine operation, respectively. However, the cosine and sine are not the linear operators that represent the frequency components in simple manner. The Euler formula is employed to project the complex number on the axis as Eqs. (2.44) and (2.45).

$$\begin{aligned} x_1(t) &= \frac{A_1 e^{j\Omega_1 t} + A_1^* e^{-j\Omega_1 t}}{2} = \frac{a_1 e^{j(\Omega_1 t + \theta_1)} + a_1 e^{-j(\Omega_1 t + \theta_1)}}{2} \\ &= a_1 \cos(\Omega_1 t + \theta_1)\,, \quad \text{where } A_1 = a_1 e^{j\theta_1} \text{ and } A_1^* = a_1 e^{-j\theta_1} \end{aligned} \tag{2.44}$$

$$x_2(t) = \frac{B_1 e^{j\Omega_1 t} - B_1^* e^{-j\Omega_1 t}}{2j} = \frac{b_1 e^{j(\Omega_1 t+\theta_1)} - b_1 e^{-j(\Omega_1 t+\theta_1)}}{2j}$$
$$= b_1 \sin(\Omega_1 t + \theta_1), \quad \text{where } B_1 = b_1 e^{j\theta_1} \text{ and } B_1^* = b_1 e^{-j\theta_1} \tag{2.45}$$

A pair of the exponential is required to make the complex rotation into the real number. The pair should have the complex magnitudes with conjugated form (denoted with superscript *) which exhibits the complementary phase on the exponential. Depending on the even (cosine) or odd (sine) time function, proper conjugated complex magnitudes are necessary on the frequency component to obtain the real numbers in time. In other words, the given time function must contain the positive frequency Ω as well as negative frequency $-\Omega$ in pair with conjugated complex magnitude. Therefore, the negative frequency is introduced in order to neutralize the complex number as shown in Fig. 2.4.

Note that period of arbitrary function does not stand for the pure frequency. The function contains the frequency component from the period inverse in significant amount but also the numerous side frequencies exist in neighbor. For example, the square wave with 0.01 s period includes the 100 Hz component as well as the integer multiple of the 100 Hz components as shown Fig. 2.30. The cosine function with 0.01 s period only provides the 100 Hz signal. Observe also the negative frequencies.

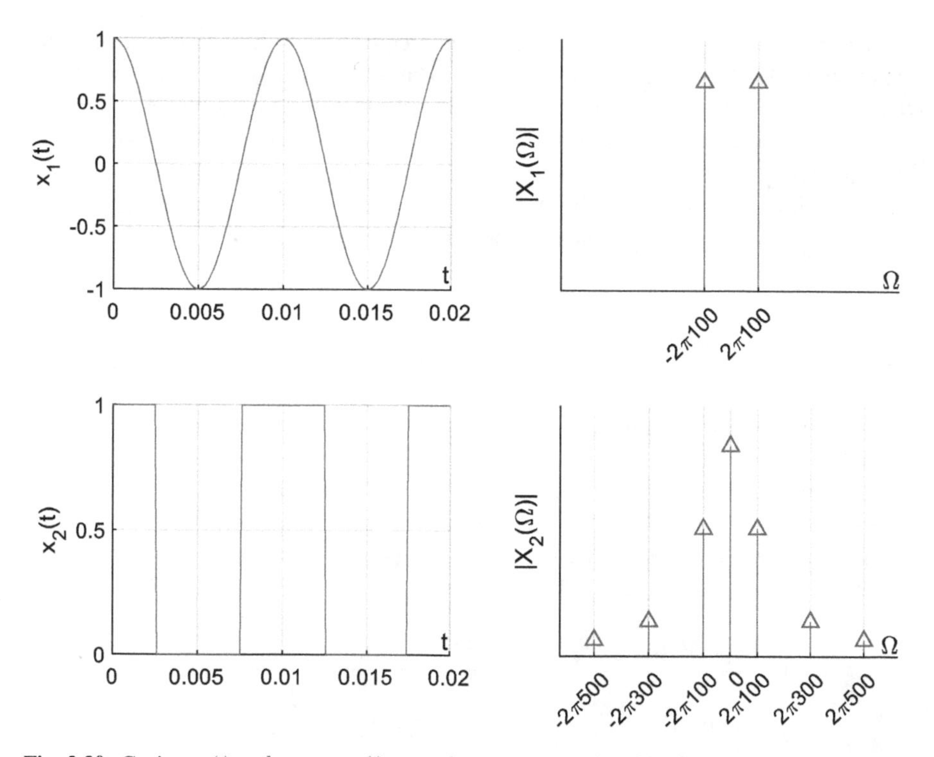

Fig. 2.30 Cosine $x_1(t)$ and square $x_2(t)$ wave for 0.01 s period and its frequency distribution

The frequency from the discrete time signal presents the completely different notation due to the sequence n instead of the time t. Let's see the example below.

$$x(t) = e^{j(2\pi 10)t} \text{ and } x[n] = e^{j(2\pi 10)n} \quad t \in \mathbb{R}, n \in \mathbb{Z} \tag{2.46}$$

The $x(t)$ shows the 10 rotations per second. Every moment of the time can be illustrated since the t is the real number. However, the $x[n]$ rotates the 10 cycles in every sequence because of the integer sequence. There is no way to obtain the values in between the adjacent sequences. It does not matter how many rotations the function does. In fact, the minimum angular displacement between the adjacent sequences is significant. In this case, the $2\pi 10$ is equal to the 0π, 2π, 4π, 6π, and etc.; hence, no angular movement is observed with given parameter as shown in Fig. 2.31.

The sampling procedure reads the continuous time data in every T_s interval to obtain the discrete time sequence. The replacement of t with nT_s derives the discrete frequency ω that is the radian frequency based on the ratio between the given frequency f and the sampling frequency f_s as shown in Eq. (2.47).

$$\begin{aligned} x(t) = e^{j\Omega t} &\xrightarrow{\text{Sampling};t=nT_s} x[n] = e^{j\Omega nT_s} \\ = e^{j2\pi fnT_s} &= e^{j2\pi \frac{f}{f_s}n} = e^{j\omega n} \text{where } \omega = 2\pi\frac{f}{f_s}. \end{aligned} \tag{2.47}$$

According to the Nyquist rate, the sampling frequency should be at least twice greater than the given frequency. Hence, the discrete frequency ω is ranged up to π as shown in Eq. (2.48). Also, note that negative frequencies are included for real value sequence as well.

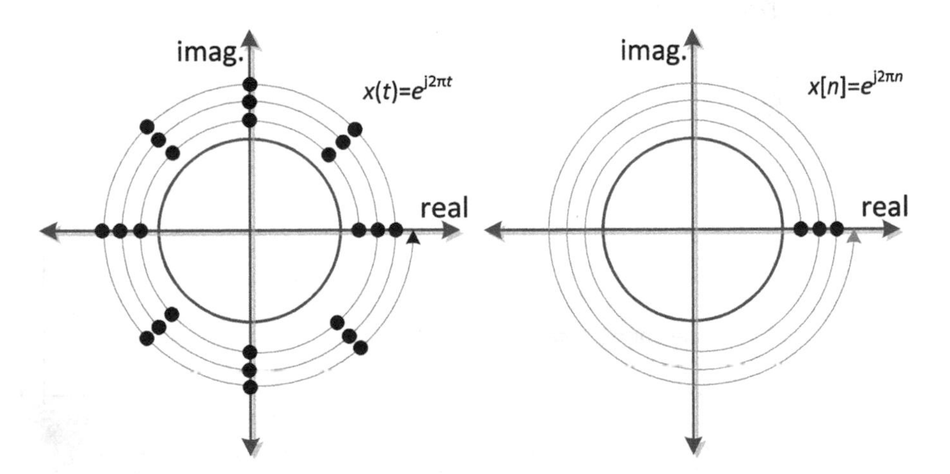

Fig. 2.31 $x(t) = e^{j2\pi t}$ and $x[n] = e^{j2\pi n}$ representation for real t and integer n. Assume that t is the integer multiple of 1/8 in this example

$$f_s > 2f \text{ then } \frac{f}{f_s} < \frac{1}{2}; \text{therefore, } \omega = 2\pi\frac{f}{f_s} < \pi \tag{2.48}$$

Figure 2.32 presents the three sampling cases with ω as $\pi/4$, π, and $3\pi/2$. The ω with $\pi/4$ represents that the sampling frequency is eight times greater than the given frequency *f*. Therefore, the one period signal is sampled eight times in cosine function as well as the complex circle. The sampling is dense enough to recover the given continuous signal by connecting the neighbor dots, so it is proper sampling. We can say the ω is valid. The ω with π provides that the sampling frequency is twice greater than the given frequency *f*. Therefore, the one period signal is sampled twice in cosine function as well as the complex circle. The sampling is on the edge to recover and the ω is narrowly valid.

The last case for $\omega = 3\pi/2$ shows the ratio as $f/f_s = 3/4$ and three periods of the signal is sampled four times. In the complex circle, after the four sampling by $3\pi/2$ angular distance, the sampling position is placed back to the initial location as $(3\pi/2)4 = 6\pi$. Hence, the three period 6π is read by four sampling process. The $3\pi/2$ frequency misses some important signal information such as peaks and valleys, so the connecting procedure does not recover the given continuous signal as shown in Fig. 2.33. The feasible representative frequency by the given ω is the $\pi/2$ denoted by the dotted line in Fig. 2.33. Likewise, the $\pi/2$ angular space is depicted on the complex circle sampling in Fig. 2.32. The sampling procedure changes the frequency of sampled signal; therefore, the ω is not valid. Above the π radian frequencies cannot stand for discrete frequency ω in proper manner.

Prog. 2.9 MATLAB program for Fig. 2.33.

```
T = 0.01;
T1 = 0.03;
fs = 10000;
fss = 1/(3*T/4);
t1 = 0:1/fs:T*4.5;
t2 = 0:1/fss:T*4.5;
x1 = cos(2*pi*t1/T);
x2 = cos(2*pi*t2/T);
x3 = cos(2*pi*t1/T1);

figure,
plot(t1,x1,'k',t1,x3,'k--'), grid
hold on
stem(t2,x2,'LineStyle','none','MarkerFaceColor','k')
hold off
```

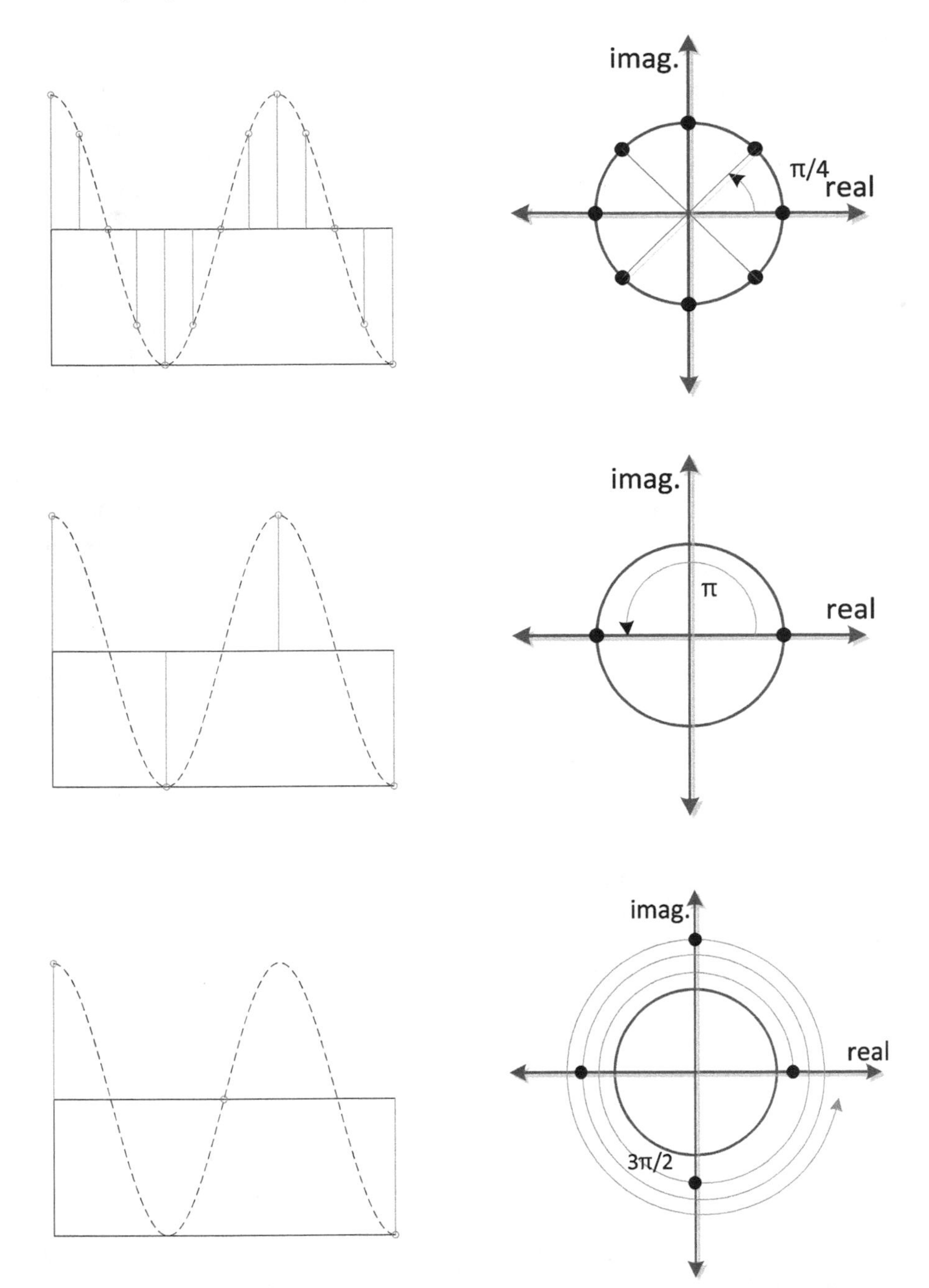

Fig. 2.32 Sampling the sinusoid with different sampling frequencies to present the radian frequencies ω as $\pi/4, \pi$, and $3\pi/2$

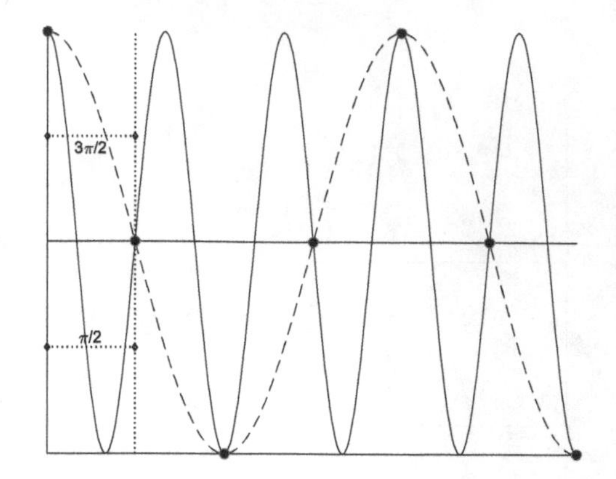

Fig. 2.33 Improper sampling example with 3π/2 radian frequency (equivalently $\pi/2$)

The effective ω derived from the invalid radian frequency can be computed by adding the integer multiple of 2π in order to obtain the less than or equal to π radian frequency as Eq. (2.49).

$$\text{For } k \in \mathbb{Z} \begin{cases} \omega = 2k\pi \overset{\text{effective } \omega}{\longrightarrow} 0 \\ \omega \neq 2k\pi \overset{\text{effective } \omega}{\longrightarrow} 0 < \min|\omega + 2k\pi| \leq \pi \end{cases} \tag{2.49}$$

For example, the $3\pi/2$ is added by the -2π to deliver the $\pi/2$ effective radian frequency by absolute value. The $5\pi/3$ is added by -2π to present absolute $\pi/3$. The $4\pi/3$ provides the $2\pi/3$ effective frequency. Note that the 2π radian frequency changes to the zero radian that illustrates the non-moving constant signal.

Figure 2.34 shows the two discrete frequencies as $5\pi/3$ and $4\pi/3$. The appearance sequence is out of order in terms of circular motion but the recovery by connecting the adjacent values generates the count clockwise cycle increased by $\pi/3$ and $2\pi/3$, respectively. The effective frequency is the only representative frequency used by the discrete processing since the other frequencies are ambiguous to illustrate from sampling procedure. Figure 2.35 shows the projected real values of the circular motion from the two discrete frequencies as $5\pi/3$ and $11\pi/3$. The angular differences between the marked points are distinct for each frequency described by the caption at Fig. 2.35; however, the location is identical to the sampling point circles in Fig. 2.35. The direct connection between the points represent the effective frequency $\pi/3$ derived from Eq. (2.49). Therefore, the system designer should choose the proper sampling frequency based on the Nyquist rate to avoid the unwanted spectral shift by effective frequency.

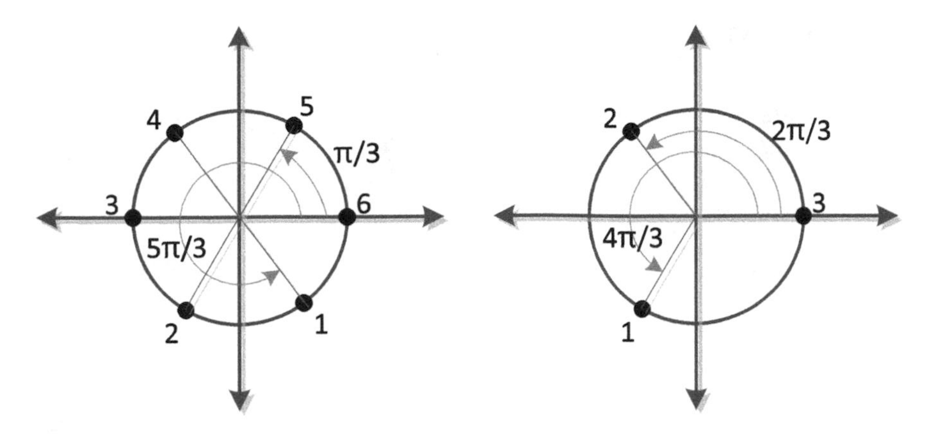

Fig. 2.34 Two discrete frequencies $5\pi/3$ and $4\pi/3$ (effective radian frequencies $\pi/3$ and $2\pi/3$, respectively)

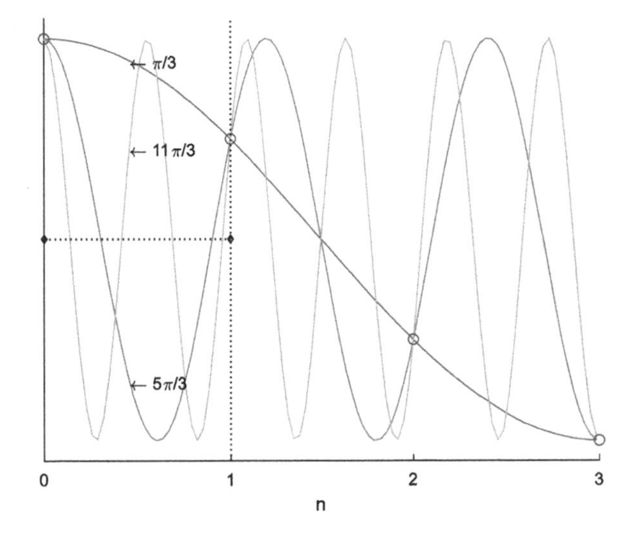

Fig. 2.35 Projected real value of circular motion from the three discrete frequencies as $\pi/3, 5\pi/3,$ and $11\pi/3$.The sampling points are illustrated by circles

Prog. 2.10 MATLAB program for Fig. 2.35.

```
w1 = pi/3;
w2 = 5*pi/3;
w3 = 11*pi/3;

n1 = 0:pi/100:3;

x1 = cos(w1*n1);
x2 = cos(w2*n1);
x3 = cos(w3*n1);

figure,
plot(n1,x1,n1,x2,n1,x3)
hold on
scatter ([0 1 2 3], cos(w1*[0 1 2 3]));
hold off
```

Figure 2.36 shows the identical situation as Fig. 2.35 for two discrete frequencies as $4\pi/3$ and $10\pi/3$. The direct connection between the points represent the effective frequency $2\pi/3$.

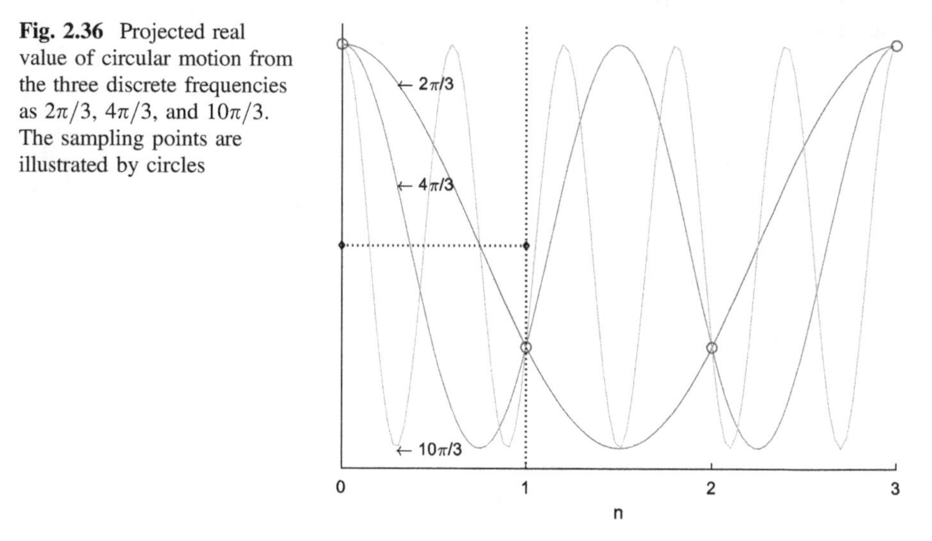

Fig. 2.36 Projected real value of circular motion from the three discrete frequencies as $2\pi/3$, $4\pi/3$, and $10\pi/3$. The sampling points are illustrated by circles

Prog. 2.11 MATLAB program for Fig. 2.36.

```
w1 = 2*pi/3;
w2 = 4*pi/3;
w3 = 10*pi/3;

n1 = 0:pi/100:3;

x1 = cos(w1*n1);
x2 = cos(w2*n1);
x3 = cos(w3*n1);

figure,
plot(n1,x1,n1,x2,n1,x3)
hold on
scatter ([0 1 2 3], cos(w1*[0 1 2 3]));
hold off
```

The relationship between the continuous frequency Ω and discrete frequency ω is given by Eq. (2.50). Simply the effective ω is multiplied with f_s to compute the processing frequency Ω from the sampling procedure.

$$2\pi\frac{f}{f_s} = \frac{\Omega}{f_s} = \omega \rightarrow \omega_{\mathrm{eff}} = 2\pi\frac{f_{\mathrm{eff}}}{f_s} = \frac{\Omega_{\mathrm{eff}}}{f_s} \tag{2.50}$$

The effective frequency from the sampling procedure is produced by the aliasing due to the improper sampling. Developed from Eq. (2.50), the linearly increasing processing frequency f and fixed sampling frequency f_s provides the non-linear effective frequency f_{eff} because of the sampling procedure. If the f is less than or equal to the f_s half then the effective frequency f_{eff} is equal to the f, otherwise not equal. The conversion plot is shown in Fig. 2.37.

Above the $f_s/2$, the effective frequency starts to decrease linearly until zero at f_s. The triangle shape is repeated continuously and infinitely. Note that the effective frequency is always between the zero and f_s half or between the zero and π in discrete frequency ω.

Example 2.10

Based on the sampling frequency f_s = 2000 Hz, compute the effective frequencies f_{eff} from following discrete frequencies ω.

$$\omega = \frac{\pi}{4}, \frac{\pi}{2}, \frac{3\pi}{4}, \pi, \frac{5\pi}{4}$$

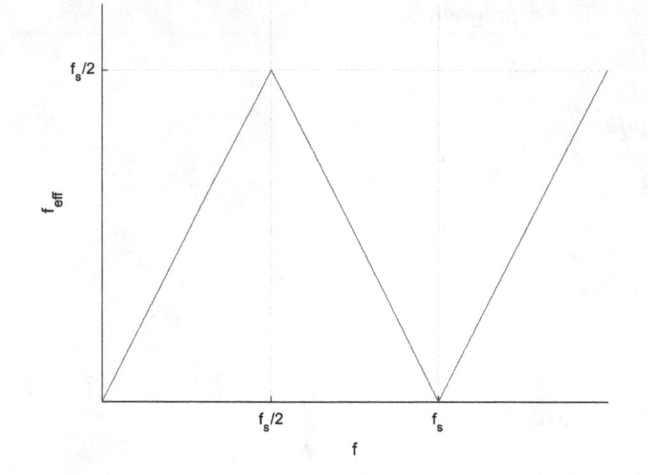

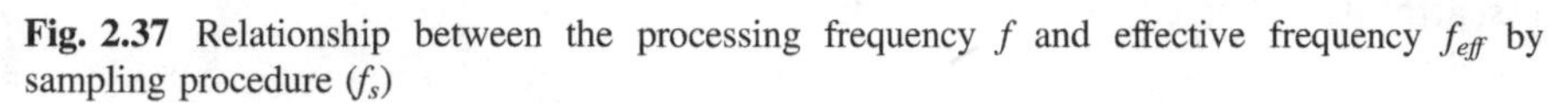

Fig. 2.37 Relationship between the processing frequency f and effective frequency f_{eff} by sampling procedure (f_s)

Solution

The corresponding effective discrete frequencies ω_{eff} are below.

$$\omega_{eff} = \frac{\pi}{4}, \frac{\pi}{2}, \frac{3\pi}{4}, \pi, \frac{3\pi}{4}$$

Therefore, the responding effective frequencies f_{eff} are below.

$$f_{eff} = 250, 500, 750, 1000, 750\,\text{Hz}$$

∎

The discrete frequency also requires the complementary negative frequency to convert into the real number. According to the Euler formula, Eqs. (2.51) and (2.52) are the counterpart of the continuous signal Euler formula for corresponding sinusoid signal. Therefore, the negative frequency similarly contains the conjugated complex magnitude as well.

$$\begin{aligned} x_1[n] &= \frac{A_1 e^{j\omega_1 n} + A_1^* e^{-j\omega_1 n}}{2} = \frac{a_1 e^{j(\omega_1 n + \theta_1)} + a_1 e^{-j(\omega_1 n + \theta_1)}}{2} \\ &= a_1 \cos(\omega_1 n + \theta_1), \quad \text{where } A_1 = a_1 e^{j\theta_1} \text{ and } A_1^* = a_1 e^{-j\theta_1} \end{aligned} \tag{2.51}$$

$$\begin{aligned} x_2[n] &= \frac{B_1 e^{j\omega_1 n} - B_1^* e^{-j\omega_1 n}}{2j} = \frac{b_1 e^{j(\omega_1 n + \theta_1)} - b_1 e^{-j(\omega_1 n + \theta_1)}}{2j} \\ &= b_1 \sin(\omega_1 n + \theta_1), \quad \text{where } B_1 = b_1 e^{j\theta_1} \text{ and } B_1^* = b_1 e^{-j\theta_1} \end{aligned} \tag{2.52}$$

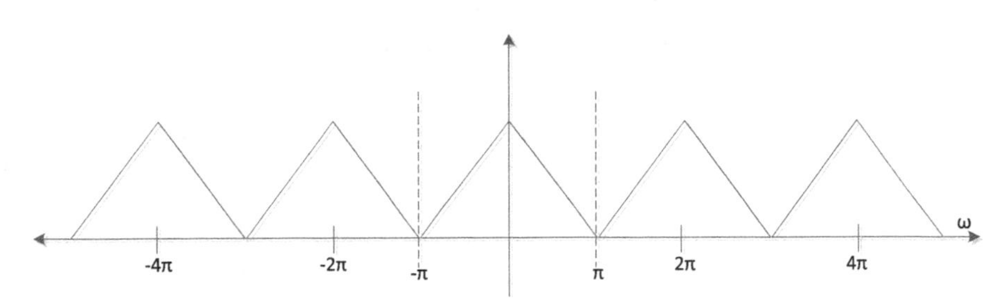

Fig. 2.38 Spectral periodicity in discrete frequency ω

Another characteristic on the frequency distribution for the discrete time signal is periodicity in spectrum. Since the sequence is represented by the integer number, the frequency is repeated every 2π interval as below.

$$\cos(\omega n) = \cos((\omega + 2k\pi)n) = \cos(\omega n + 2k\pi n) \text{ for } n \in \mathbb{Z} \tag{2.53}$$

$$\cos(\Omega t) \neq \cos((\Omega + 2k\pi)t) = \cos(\Omega t + 2k\pi t) \text{ for } t \in \mathbb{R} \tag{2.54}$$

In Eq. (2.53), the ω frequency is equal to the $\omega + 2k\pi$ for any integer number k. The $2k\pi n$ does not cause the radian variation for the given sinusoid. In Eq. (2.54), the $2k\pi t$ could be any radian value due to the real value t; therefore, no spectrum periodicity is reserved for the continuous time signal. Figure 2.38 shows one example of the frequency distribution for the discrete time signal.

The spectrum is repeated for every 2π interval for positive and negative frequency range as well. If the discrete frequency crosses the $\pm\pi$ borders, then spectrum between the neighbors starts to be overlapped and contaminated. The valid discrete frequency distribution from Fig. 2.38 is also meet the Nyquist sampling rate condition. In summary, the discrete frequency is complementary with negative frequency, ranged from $-\pi$ to π, and repeated for every 2π interval.

2.5 Problems

1. The pure Ω frequency signal is represented by complex number as below.

$$x(t) = re^{j\Omega t}$$

The size of the complex number circle r presents the magnitude of the frequency. Derive the magnitude of individual frequencies of following signals

- $r\cos(\Omega t + \theta)$
- $r\sin(\Omega t + \theta)$

2. Find the trigonometric identities of below equations.
   - $\sin\left(\Omega t + \frac{\pi}{2}\right)$
   - $\cos\left(\Omega t + \frac{\pi}{2}\right)$
   - $\sin(\Omega t - \pi)$
   - $\cos\left(\Omega t - \frac{\pi}{2}\right)$
3. Decompose below equation into the non-phased cosine and sine function.
   - $A\sin(\Omega t + \theta)$
4. Find the even and odd part of below signal.
   - $A\sin(\Omega t + \theta)$
5. Using the phasor, prove the equivalence of below equation.
   - $A\cos(\theta)\cos(\Omega \mathrm{t}) - \mathrm{A}\sin(\theta)\sin(\Omega \mathrm{t}) = \mathrm{A}\cos(\Omega \mathrm{t} + \theta)$
6. Equation (2.12) is derived from the following equation.

$$x(t) = \int_0^\infty \frac{a(\Omega) + jb(\Omega)}{2} e^{j\Omega \mathrm{t}} d\Omega + \int_0^\infty \frac{a(\Omega) - jb(\Omega)}{2} e^{-j\Omega \mathrm{t}} d\Omega$$

   To obtain the real $x(t)$, determine the condition of $a(\Omega)$ and $b(\Omega)$. Why?
7. Derive the following equations.
   - $\int_{-\infty}^{\infty} e^{j\Omega \mathrm{t}} dt$
   - $\frac{1}{2\pi} \int_{-\infty}^{\infty} e^{j\left(100\pi t + \frac{\pi}{4}\right)} e^{-j\Omega \mathrm{t}} dt$
8. Compute and draw $X(\Omega)$ of below signal.

$$x(t) = \begin{cases} 1 & \text{for} \quad 0 \le t \le 1 \\ 0 & \text{Otherwise} \end{cases}$$

9. Find the minimum sampling rate for following signal.

$$x(t) = \cos(100\pi t)\sin(200\pi t)$$

10. Find the value of following equations.

$$\begin{cases} \delta[0] = \\ \sum_{n=-\infty}^{\infty} \delta[n] = \\ \sum_{n=-\infty}^{\infty} n\delta[n-3] = \end{cases}$$

$$\begin{cases} \delta(0) = \\ \int_{-\infty}^{\infty} \delta(t)dt = \\ \int_{-\infty}^{\infty} t\delta(t-3)dt = \end{cases}$$

11. Using the unit-step function, modify the equation to extract the one period of following signal.

 - $x_1[n] = \cos\left(\frac{2\pi(n-3)}{6}\right)$
 - $x_2[n] = \cos\left(\frac{2\pi(n-3)}{6}\right) + 2\cos\left(\frac{2\pi}{4}n\right)$

12. Draw the $x[-2(n-1)]$ of given $x[n]$.

$$x[n] = r[n+2](u[n+3] - u[n-4])$$

13. From the given impulse response $h[n]$ as below, determine the LTI system output of shown input $x[n]$.

$$h[n] = \delta[n] - \delta[n-1]$$

$$x[n] = u[n]$$

14. Decide the energy and power existence of below signal.

$$x_1[n] = 0.5^{|n|}$$

$$x_2[n] = \frac{1}{n}(u[n-1] + u[-n+1])$$

15. Based on the sampling frequency f_s = 4000 Hz, compute the effective discrete frequencies ω_{eff} from following frequencies f.

$$f = 500, 1000, 1500, 2000, 2500, 3000, 3500, 4000\,\text{Hz}$$

References

1. Taylor, A.E.: L'Hospital's Rule. Am. Math. Mon. **59**(1), 20–24 (1952). https://doi.org/10.1080/00029890.1952.11988058
2. Bracewell, R.N.: The Fourier Transform and Its Applications, 3rd ed. McGraw Hill (1999)
3. Fourier, J.B.J.: The Analytical Theory of Heat. Cambridge Library Collection—Mathematics. Cambridge University Press, Cambridge (2009)

4. Olver, F.W.J., Standards, N.I.o., Technology, Lozier, D.W., Boisvert, R.F., Clark, C.W.: NIST Handbook of Mathematical Functions Hardback and CD-ROM. Cambridge University Press (2010)
5. Nyquist, H.: Certain topics in telegraph transmission theory. Trans. Am. Inst. Electri. Eng. **47** (2), 617–644 (1928). https://doi.org/10.1109/T-AIEE.1928.5055024
6. Chapter 2—Cartesian Tensors. In: Kundu, P.K., Cohen, I.M., Dowling, D.R. (eds.) Fluid Mechanics. 5th edn, pp. 39–64. Academic Press, Boston (2012)

Chapter 3
Fourier Analysis

The convolution sum and discrete sequence implement the basic digital filter in time domain. In previous chapter, the discrete frequency ω is defined with continuous frequency Ω. Now, we understand the discrete time signal as well as the discrete sinusoid signal with frequency ω. The specific frequency signal can be generated by the linear combination of sinusoids. What if we need to figure out the frequency distribution from a given arbitrary signal? How we can derive the frequency distribution from the discrete time domain? Often frequency analysis is required since the filter is designed based on the frequency information. The Fourier analysis is the powerful tool to transform the information between the frequency and time. The analysis and synthesis equation find the frequency and time information correspondingly. The Fourier analysis is further generalized by the Z-transform to analyze and design the filter in Chap. 5.

3.1 How We Can Find the Magnitude of the Specific Frequency?

You may easily find the magnitude of the single frequency signal. The problem is complicated when you come up against the linearly added multiple sinusoid signal since you have to separate as well as estimate the frequency and magnitude simultaneously. Intuitive solution for the single sinusoid cannot be expanded for the multiple sinusoid. The answer is the exploring the periodicity and orthogonality of the sinusoid signal. Let's consider following instance.

$$\sum_{n=0}^{N-1} e^{j\frac{2\pi}{N}n} = \sum_{n=0}^{N-1} e^{j\omega n} = 0 \text{ for } \forall N \in \mathbb{N} \text{ except } N = 1 \tag{3.1}$$

K. Kim, *Conceptual Digital Signal Processing with MATLAB*,
Signals and Communication Technology,
https://doi.org/10.1007/978-981-15-2584-1_3

Equation (3.1) is derived from the signal periodicity with N period. You can obtain the identical outcome by using the equivalent counterpart as $\cos(2\pi n/N) + j\sin(2\pi n/N)$. To find some insight of the equation, use the illustration in Fig. 3.1 for $N = 6$.

The complex exponential in Fig. 3.1 demonstrates unit size vectors in every $\pi/3$ angular distance. The sum of the vectors indicates the origin of the coordinates; hence, the output is equal to the zero for Eq. (3.1).

The period N is the natural number for discrete time signal; therefore, Eq. (3.1) definition for the discrete frequency ω provides a limited resolution such as $2\pi/2, 2\pi/3, 2\pi/4$, etc. The general definition of discrete frequency ω is given in Eq. (3.2) for dense description of frequency. The multiple k rotations are performed to find the same pattern with N sequence difference.

$$\frac{2\pi k}{N} = \omega = 2\pi \frac{f}{f_s} \text{ for } k,\ N \in \mathbb{N}, \frac{f}{f_s} \in \mathbb{Q}, f, f_s \in \mathbb{R} \tag{3.2}$$

From the given f and f_s, the period N can be calculated by finding the minimum natural number k which produces the natural number from kf_s/f. The equation is given in Eq. (3.3).

$$N = \min\left\{ k\frac{f_s}{f} \in \mathbb{N} \middle| k \in \mathbb{N} \text{ and } \frac{f}{f_s} \in \mathbb{Q} \right\} \tag{3.3}$$

The periodicity is verified by Eq. (3.4).

$$e^{j\omega n} = e^{j\frac{2\pi k}{N}n} = e^{j\frac{2\pi k}{N}(n+N)} = e^{j\frac{2\pi k}{N}n + \frac{2\pi k}{N}N} = e^{j\frac{2\pi k}{N}n} e^{j\frac{2\pi k}{N}N} = e^{j\frac{2\pi k}{N}n} \tag{3.4}$$

As Eq. (3.1), the k cycle summation of the unit magnitude vectors with $2\pi k/N$ angular distance delivers the zero-output due to the symmetric vector sums over the coordinate origin.

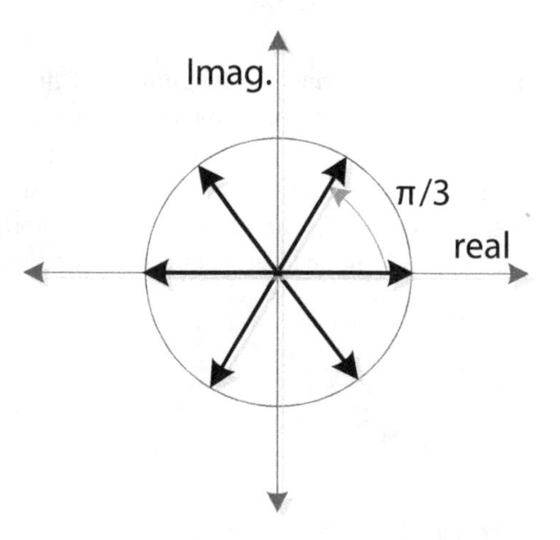

Fig. 3.1 The $e^{j2\pi n/6}$ in complex number domain

Example 3.1
Find the period N of following discrete frequencies.

$$\omega = \frac{5\pi}{3} \text{ and} \frac{4\pi}{3}$$

Solution
The calculation N from the given ω is below.

$$\frac{5\pi}{3} = \frac{2\times 5\pi}{2\times 3} = \frac{10\pi}{6} = \frac{2\pi k}{N} \rightarrow N = 6$$
$$\frac{4\pi}{3} = \frac{2\pi k}{N} \rightarrow N = 3$$

The $5\pi/3$ and $4\pi/3$ require the five and two rotations, correspondingly, to locate another pattern match according to the k value in above equation. ■

Figure 2.34 presents the cyclic motion of $5\pi/3$ and $4\pi/3$ frequency with six and three sample period N, respectively. The inverse of the period N does not represent the discrete frequency ω. We must know the N and k as shown in Eq. (3.2) to calculate the frequency. The general discrete sinusoid function is demonstrated in Eq. (3.5) with discrete frequency ω_1.

$$\begin{aligned} x[n] &= a_1 \cos(\omega_1 n + \theta_1) = a_1 \frac{e^{j(\omega_1 n + \theta_1)} + e^{-j(\omega_1 n + \theta_1)}}{2} \\ &= \frac{a_1 e^{j\theta_1}}{2} e^{j\omega_1 n} + \frac{a_1 e^{-j\theta_1}}{2} e^{-j\omega_1 n} \end{aligned} \tag{3.5}$$

The magnitude for the given signal is $a_1 e^{j\theta_1}/2$ and its complement. In order to find the magnitude, apply the Eq. (3.6).

$$\begin{aligned} X(e^{j\omega}) &= \sum_{n=-\infty}^{\infty} x[n] e^{-j\omega n} = \sum_{n=-\infty}^{\infty} a_1 \cos(\omega_1 n + \theta_1) e^{-j\omega n} \\ &= \frac{a_1 e^{j\theta_1}}{2} \sum_{n=-\infty}^{\infty} e^{j\omega_1 n} e^{-j\omega n} + \frac{a_1 e^{-j\theta_1}}{2} \sum_{n=-\infty}^{\infty} e^{-j\omega_1 n} e^{-j\omega n} \end{aligned} \tag{3.6}$$

To solve Eq. (3.6), we need to understand the sum with complex exponential. Consider below Eq. (3.7).

$$\sum_{n=-\infty}^{\infty} e^{j\omega_1 n} e^{-j\omega n} = \sum_{n=-\infty}^{\infty} e^{j(\omega_1-\omega)n} = \left\{ \begin{array}{ll} \lim\limits_{\substack{k \to \infty \\ k \in \mathbb{N}}} k \sum\limits_{n=0}^{N-1} e^{j(\omega_1-\omega)n} = 0 & \text{for } \omega_1 \neq \omega \\ \sum\limits_{n=0}^{N-1} e^{j(\omega_1-\omega)n} = N & \text{for } \omega_1 = \omega \end{array} \right\}$$

$$= 2\pi\delta(\omega_1 - \omega) \text{ where } N = \begin{cases} \min\left\{ \left(\frac{2\pi k}{(\omega_1-\omega)} \right) \in \mathbb{N} | k \in \mathbb{N} \right\} & \text{for } \omega_1 \neq \omega \\ \frac{2\pi}{(\omega_1-\omega)} = \frac{2\pi}{0} = \infty & \text{for } \omega_1 = \omega \end{cases} \tag{3.7}$$

The real part of the $e^{j(\omega_1-\omega)n}$ is shown in Fig. 3.2 for 0, $2\pi/5$ (1.2566 radian/sample), and $6\pi/10$ (1.8850 radian/sample). For $\omega_1 \neq \omega$ case, the infinite length summation for real part is represented by the product between the period sum and nearly ∞. The individual period sum is zero; hence, the overall sum is zero always. For $\omega_1 = \omega$ case, the $e^{j(\omega_1-\omega)n}$ is constant one for infinite length N which is $2\pi/0$. Therefore, the $\sum_{n=-\infty}^{\infty} e^{j(\omega_1-\omega)n}$ is $2\pi/0$ as ∞ for $\omega_1 = \omega$ situation.

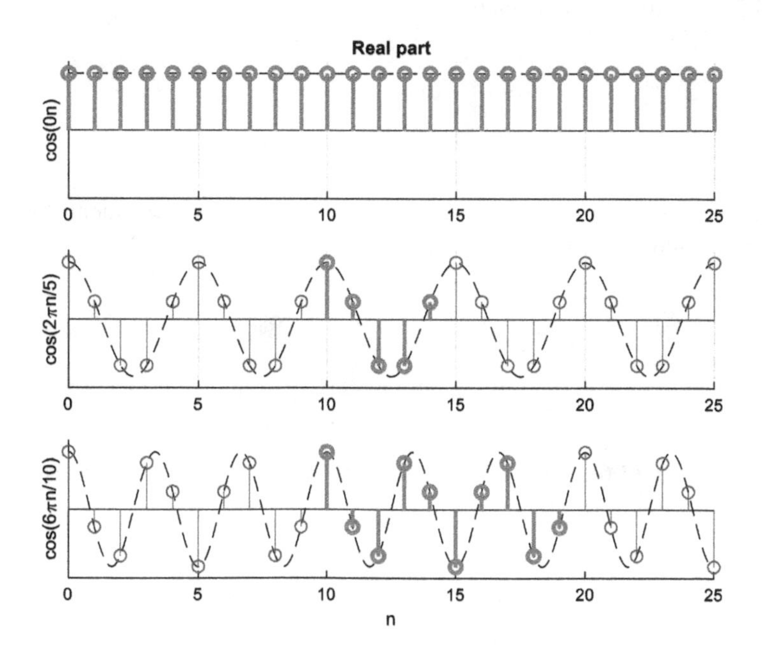

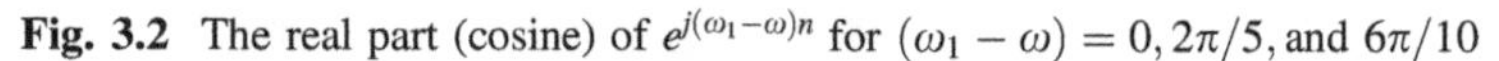

Fig. 3.2 The real part (cosine) of $e^{j(\omega_1-\omega)n}$ for $(\omega_1 - \omega) = 0, 2\pi/5$, and $6\pi/10$

Prog. 3.1 MATLAB program for Fig. 3.2

```
N = 25;
N1 = 5;
k1 = 1;
N2 = 10;
k2 = 3;

nn = 0:1:N;
nn0 = 0:1/100:N;
x1 = cos(2*pi*nn*k1/N1);
x2 = cos(2*pi*nn*k2/N2);
x3 = cos(2*pi*nn);
x01 = cos(2*pi*nn0*k1/N1);
x02 = cos(2*pi*nn0*k2/N2);
x03 = ones(1,length(nn0));

figure,
subplot(311), plot(nn0,x03,'k--'), grid
hold on
stem(nn,x3,'LineWidth',2,'Color',[0.4 0.6 0.7])
hold off
subplot(312), plot(nn0,x01,'k--'), grid
hold on
stem(nn,x1)
stem([10:1:14],cos(2*pi*k1*(10:1:14)/N1),'LineWidth',2,'Color',[0.4 0.6 0.7])
hold off
subplot(313),    plot(nn0,x02,'k--'), grid
hold on
stem(nn,x2), grid
stem([10:1:19],cos(2*pi*k2*(10:1:19)/N2),'LineWidth',2,'Color',[0.4 0.6 0.7])
hold off
```

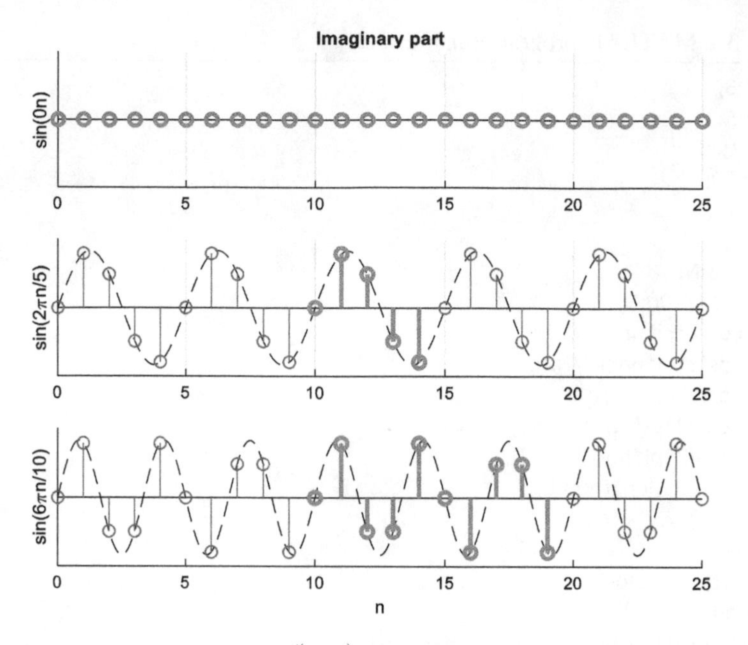

Fig. 3.3 The imaginary part (sine) of $e^{j(\omega_1-\omega)n}$ for $(\omega_1-\omega)=0, 2\pi/5,$ and $6\pi/10$

The imaginary part of the $e^{j(\omega_1-\omega)n}$ is shown in Fig. 3.3 for 0, $2\pi/5$ (1.2566 radian/sample), and $6\pi/10$ (1.8850 radian/sample). The analysis is identical to the real part counterparts except $\omega_1=\omega$ case which delivers the constant zero output.

Prog. 3.2 MATLAB program for Fig. 3.3

```
N = 25;
N1 = 5;
k1 = 1;
N2 = 10;
k2 = 3;

nn = 0:1:N;
nn0 = 0:1/100:N;

y1 = sin(2*pi*nn*k1/N1);
y2 = sin(2*pi*nn*k2/N2);
y3 = sin(2*pi*nn);
y01 = sin(2*pi*nn0*k1/N1);
y02 = sin(2*pi*nn0*k2/N2);
y03 = zeros(1,length(nn0));

figure,
subplot(311), plot(nn0,y03,'k--'), grid
hold on
stem(nn,y3,'LineWidth',2,'Color',[0.4 0.6 0.7])
hold off
ylim([-1.2 1.2])
subplot(312), plot(nn0,y01,'k--'), grid
hold on
stem(nn,y1)
stem([10:1:14],sin(2*pi*k1*(10:1:14)/N1),'LineWidth',2,'Color',[0.4 0.6 0.7])
hold off
subplot(313),   plot(nn0,y02,'k--'), grid
hold on
stem(nn,y2), grid
stem([10:1:19],sin(2*pi*k2*(10:1:19)/N2),'LineWidth',2,'Color',[0.4 0.6 0.7])
hold off
```

The consolidated outputs of the $\sum_{n=-\infty}^{\infty} e^{j(\omega_1-\omega)n}$ are demonstrated in Fig. 3.4. The real and imaginary part of the $e^{j(\omega_1-\omega)n}$ are illustrated on the complex domain with vectors in Fig. 3.4.

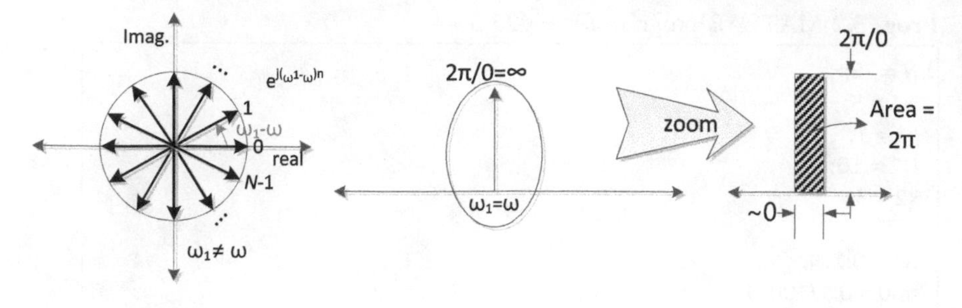

Fig. 3.4 The $\sum_{n=-\infty}^{\infty} e^{j(\omega_1-\omega)n}$ for $\omega_1 \neq \omega$ and $\omega_1 = \omega$

If the ω_1 is not equal to ω then the complex exponential is periodic signal with $(\omega_1 - \omega)$ radian frequency. The sum can be translated to the addition of the infinite repetition of new periods. Since the addition of one period is zero, the whole summation presents zero output. For the $\omega_1 = \omega$ condition, the $e^{j(\omega_1-\omega)n}$ is e^{j0n} that is constant one; therefore, the summation indicates the N with infinite value. Observe that the period of the zero-radian frequency is infinite. The summation with complex exponential follows the property of delta function that presents the infinite outcome on the short point. The area indicated by the given delta function is 2π as shown in Fig. 3.4; hence, the delta function is scaled by the 2π. Therefore, the complete solution is Eq. (3.8).

$$\begin{aligned} X\left(e^{j\omega}\right) &= \frac{a_1 e^{j\theta_1}}{2} \sum_{n=-\infty}^{\infty} e^{j\omega_1 n} e^{-j\omega n} + \frac{a_1 e^{-j\theta_1}}{2} \sum_{n=-\infty}^{\infty} e^{-j\omega_1 n} e^{-j\omega n} \\ &= \frac{a_1 e^{j\theta_1}}{2} 2\pi\delta(\omega_1 - \omega) + \frac{a_1 e^{-j\theta_1}}{2} 2\pi\delta(\omega_1 + \omega) \end{aligned} \quad (3.8)$$

The $X(e^{j\omega})$ argument is $e^{j\omega}$ to imply the 2π periodicity in the spectrum due to the sampling process. The frequency distribution for the $a_1 \cos(\omega_1 n + \theta_1)$ is shown in Fig. 3.5.

The other property used above computation is the orthogonality between the adjacent frequencies. Except the ω_1 frequency, the $\sum_{n=-\infty}^{\infty} x[n]e^{-j\omega n}$ provides the zero output for the given $a_1 \cos(\omega_1 n + \theta_1)$ signal and the $\omega_1 = \omega$ generates the magnitude related output.

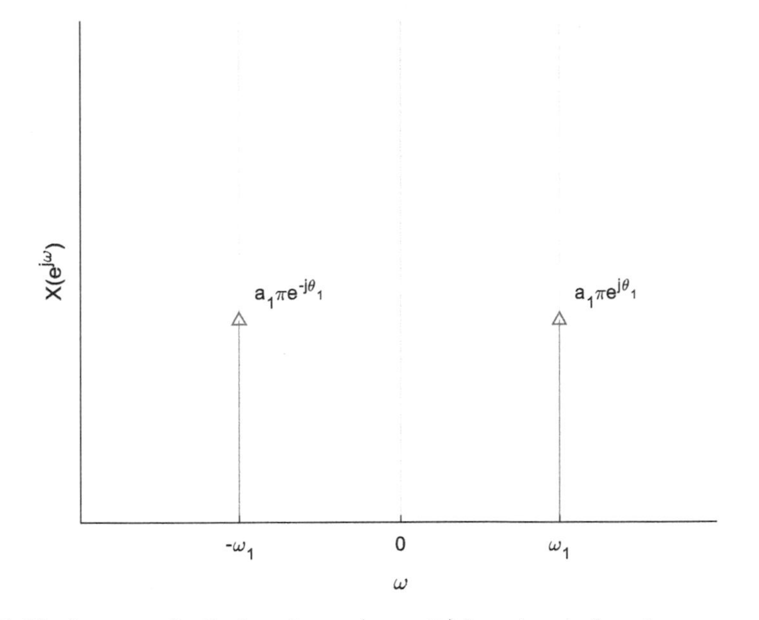

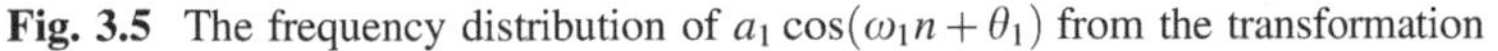

Fig. 3.5 The frequency distribution of $a_1\cos(\omega_1 n+\theta_1)$ from the transformation

$$x[n]=\frac{1}{2\pi}\int_{-\pi}^{\pi}X(e^{j\omega})e^{j\omega n}d\omega$$

$$=\frac{1}{2\pi}\int_{-\pi}^{\pi}\{a_1\pi\, e^{j\theta_1}\delta(\omega_1-\omega)+a_1\pi\, e^{-j\theta_1}\delta(\omega_1+\omega)\}e^{j\omega n}d\omega$$

$$=\frac{1}{2\pi}\int_{-\pi}^{\pi}a_1\pi\, e^{j\theta_1}\delta(\omega_1-\omega)e^{j\omega n}d\omega+\frac{1}{2\pi}\int_{-\pi}^{\pi}a_1\pi\, e^{-j\theta_1}\delta(\omega_1+\omega)e^{j\omega n}d\omega$$

$$\frac{a_1\pi\, e^{j\theta_1}}{2\pi}e^{j\omega_1 n}\int_{-\pi}^{\pi}\delta(\omega_1-\omega)d\omega+\frac{a_1\pi\, e^{-j\theta_1}}{2\pi}e^{-j\omega_1 n}\int_{-\pi}^{\pi}\delta(\omega_1+\omega)d\omega$$

$$=\frac{a_1\pi\, e^{j\theta_1}}{2\pi}e^{j\omega_1 n}+\frac{a_1\pi\, e^{-j\theta_1}}{2\pi}e^{-j\omega_1 n}$$

$$\frac{a_1}{2}e^{j(\omega_1 n+\theta_1)}+\frac{a_1}{2}e^{-j(\omega_1 n+\theta_1)}=a_1\cos(\omega_1 n+\theta_1) \tag{3.9}$$

In order to synthesis the discrete time domain signal, consider the above equations. The synthesis equation is the linear combination of complex exponential with ω radian frequency for the corresponding complex magnitude $X(e^{j\omega})$. Since the digital frequency is ranged from $-\pi$ to π and the frequency is continuous domain, the integral is applied for the given range. Scale down by 2π is required because the analysis equation computes the magnitude with 2π multiplication. The big assumption is that all signal can be decomposed into the infinite or finite combination of the sinusoid signal. We can find each frequency magnitudes by analysis equation and derive the discrete time signal via synthesis equation.

Example 3.2
Compute the frequency distribution of following signals.

$$x_1[n] = a^n u[n] \text{ for } |a|<1$$

$$x_2[n] = u[n+N] - u[n-(N+1)]$$

$$x_3[n] = \delta[n]$$

$$x_4[n] = 3$$

$$x_5[n] = \cos(\omega_1 n)$$

Solution

$$X_1\left(e^{j\omega}\right) = \sum_{n=-\infty}^{\infty} x_1[n]e^{-j\omega n} = \sum_{n=-\infty}^{\infty} a^n u[n]e^{-j\omega n} = \sum_{n=0}^{\infty} a^n e^{-j\omega n}$$

$$= \sum_{n=0}^{\infty} \left(ae^{-j\omega}\right)^n = \frac{1}{1-ae^{-j\omega}} \because \left|ae^{-j\omega}\right| < 1$$

$$X_2\left(e^{j\omega}\right) = \sum_{n=-\infty}^{\infty} (u[n+N] - u[n-(N+1)])e^{-j\omega n} = \sum_{n=-N}^{N} e^{-j\omega n}$$

$$= e^{j\omega N}\left(\frac{1-e^{-j\omega(2N+1)}}{1-e^{-j\omega}}\right) = e^{j\omega N}\frac{e^{-j\frac{\omega(2N+1)}{2}}}{e^{-j\frac{\omega}{2}}}\left(\frac{e^{j\frac{\omega(2N+1)}{2}} - e^{-j\frac{\omega(2N+1)}{2}}}{e^{j\frac{\omega}{2}} - e^{-j\frac{\omega}{2}}}\right)$$

$$= \left(\frac{e^{j\frac{\omega(2N+1)}{2}} - e^{-j\frac{\omega(2N+1)}{2}}}{e^{j\frac{\omega}{2}} - e^{-j\frac{\omega}{2}}}\right) = \frac{\sin\left(\frac{\omega(2N+1)}{2}\right)}{\sin\left(\frac{\omega}{2}\right)}$$

$$X_3\left(e^{j\omega}\right) = \sum_{n=-\infty}^{\infty} \delta[n]e^{-j\omega n} = 1$$

$$X_4\left(e^{j\omega}\right) = \sum_{n=-\infty}^{\infty} 3e^{-j\omega n} = 6\pi\delta(\omega)$$

$$X_5\left(e^{j\omega}\right) = \sum_{n=-\infty}^{\infty} \cos(\omega_1 n)e^{-j\omega n} = \frac{1}{2}\sum_{n=-\infty}^{\infty}\left(e^{-j(\omega-\omega_1)n} + e^{-j(\omega+\omega_1)n}\right)$$

$$= \pi\delta(\omega-\omega_1) + \pi\delta(\omega+\omega_1)$$

Note that above frequency information $X(e^{j\omega})$s represent the distribution over the range between $-\pi$ to π. The distribution is repeated in every 2π. ■

3.2 Signal Property and Frequency Distribution

The analysis equation finds the continuous frequency distribution $X(e^{j\omega})$ of the given discrete time signal $x[n]$. The transfer function converts the discrete domain $x[n]$ into the continuous domain $X(e^{j\omega})$. Certain cases such as pure sinusoid $a_1 \cos(\omega_1 n + \theta_1)$ could present the single frequency distribution. However, the $X(e^{j\omega})$ is the continuous function in general unless the signal is the linear combination of limited discrete frequencies with infinite length to the positive and negative sample sequence. What condition dose the $X(e^{j\omega})$ make discrete distribution? Let's find some general situations which provide the discrete spectral distribution.

Usually the discrete frequency ω denotes the radian frequency (radian/sample) based on the π; therefore, it is difficult to make the ω as integer number. Instead, the ω discretization can be performed by the specifying the spectral distance between the adjacent discrete frequencies as fundamental frequency ω_0. We assume that the new digital frequency is only existed on the integer multiple of fundamental frequency ω_0. The $e^{j\omega_0 kn}$ is the primitive component of the frequency distribution denoting the spectral location at $k\omega_0$ with k integer in Eq. (3.10). Also, the $\delta(\omega - k\omega_0)$ is the equivalent representation for spectral location as shown Eq. (3.11) and Fig. 3.6.

$$\begin{aligned} x[n] = \ldots + a_3^* e^{-j3\omega_0 n} + a_2^* e^{-j2\omega_0 n} + a_1^* e^{-j\omega_0 n} + a_0 e^{j0\omega_0 n} + a_1 e^{j\omega_0 n} \\ + a_2 e^{j2\omega_0 n} + a_3 e^{j3\omega_0 n} + \ldots \end{aligned} \tag{3.10}$$

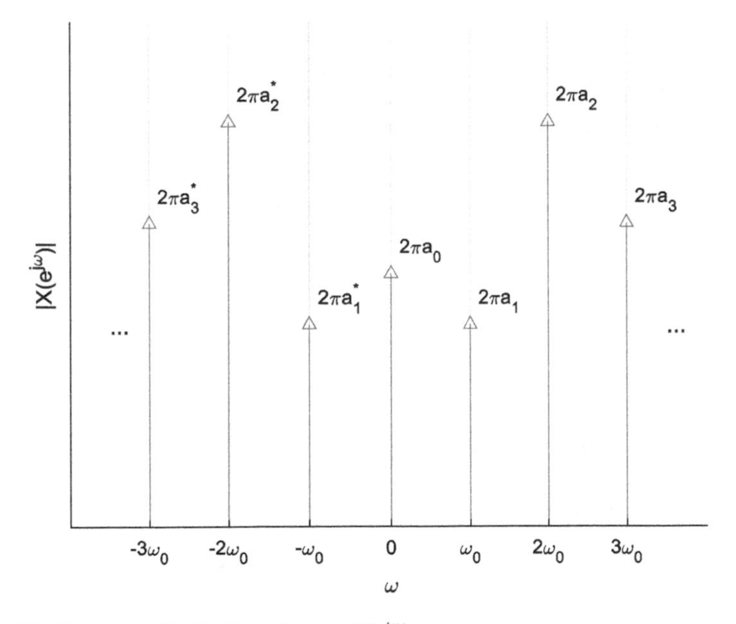

Fig. 3.6 The frequency distribution of given $X(e^{j\omega})$

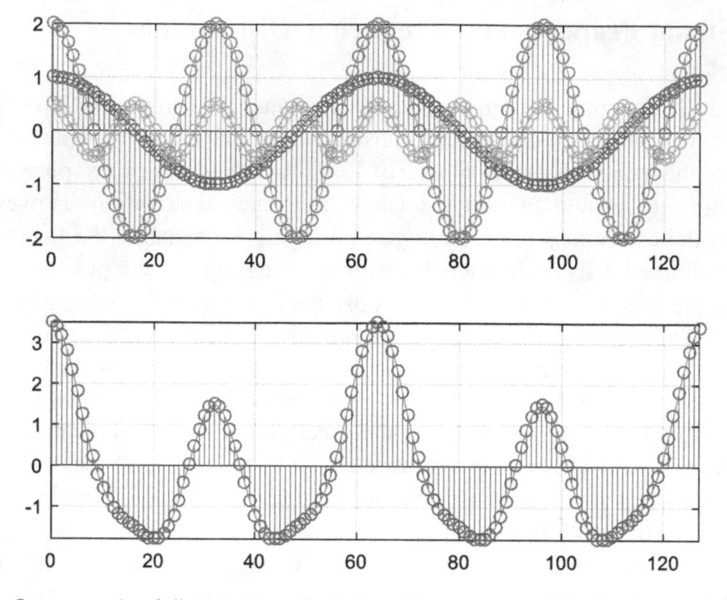

Fig. 3.7 One example of discrete time signal from discrete spectral distribution (ω_0 is $2\pi/64$)

$$\begin{aligned} X(e^{j\omega}) = \ldots &+ 2\pi a_3^* \delta(\omega + 3\omega_0) + 2\pi a_2^* \delta(\omega + 2\omega_0) + 2\pi a_1^* \delta(\omega + \omega_0) \\ &+ 2\pi a_0 \delta(\omega) + 2\pi a_1 \delta(\omega - \omega_0) + 2\pi a_2 \delta(\omega - 2\omega_0) \\ &+ 2\pi a_3 \delta(\omega - 3\omega_0) + \ldots \end{aligned} \tag{3.11}$$

Note that the conversion between the above $x[n]$ and $X(e^{j\omega})$ is performed by the analysis Eq. (3.6) and synthesis Eq. (3.9) equation given previously. The spectral resolution $\Delta\omega$ is equal to the fundamental frequency ω_0 since the discrete frequency is appeared in every ω_0 interval in frequency domain. With this condition, what the signal looks like? Let's see the illustrated signal in Fig. 3.7.

Prog. 3.3 MATLAB program for Fig. 3.7

```
N = 128;
nn = 0:1:N;
x1 = 2*cos(2*pi*nn/32);
x2 = cos(2*pi*nn/64);
x3 = 0.5*cos(2*pi*nn/16);

figure,
subplot(211), stem(nn,[x1' x2' x3']), grid
xlim([0 128])
subplot(212), stem(nn,x1+x2+x3), grid
xlim([0 128])
```

The first figure on Fig. 3.7 shows the three sinusoids with different periods. The blue, red, and yellow plot present the 64, 32, and 16 period respectively. The sum of the sinusoids demonstrated in the second figure indicates periodic signal with 64 sample period since the integer multiple of the 32 and 16 period is 64 period. In the frequency viewpoint, the 64, 32, and 16 sample period is the $2\pi/64$, $2\pi/32$, and $2\pi/16$ discrete frequency correspondingly. Observe that the $2\pi/32$ and $2\pi/16$ are twice and quadruple of the $2\pi/64$. Therefore, the fundamental frequency ω_0 is $2\pi/64$ with 1, 2, and 4 for k values. We can derive the conclusion that the integer multiple combinations of the fundamental frequency provide the periodic signal with period as $2\pi/\omega_0$. Also, we can state that the periodic signal with N period can be decomposed into the fundamental frequency $2\pi/N$ and its integer multiples. If there is any contamination by non-integer multiple of fundamental frequency, the signal establishes new fundamental frequency based on the given frequency distribution to fulfil the integer multiple condition. The signal periodicity decides the frequency discretization property. The non-periodic signal can be considered as the infinite period; hence, the fundamental frequency approaches to zero for continuous frequency distribution.

Example 3.3
Find the new period N and possible frequency components of signal $x[n]$.

$$x_1[n] \text{with 12 sample period}$$

$$x_2[n] \text{with 18 sample period}$$

$$x[n] = x_1[n] + x_2[n]$$

Solution
Least common multiple of periods presents the new period N.

$$\text{LCM}(12, 18) = 36 \text{ samples}$$

The possible frequencies are

$$\omega = \frac{2\pi}{36} k \text{ for } 0 \leq k \leq 18 \text{ and } k \in \mathbb{Z}$$

■

3.3 Discrete-Time Fourier Transform (DTFT)

In the previous sections, the fundamentals of the Discrete-time Fourier transform (DTFT) [1] is explained already. Equations (3.12) and (3.13) are DTFT equations for analysis and synthesis, respectively.

$$X(e^{j\omega}) = \sum_{n=-\infty}^{\infty} x[n]e^{-j\omega n} \tag{3.12}$$

$$x[n] = \frac{1}{2\pi} \int_{-\pi}^{\pi} X(e^{j\omega})e^{j\omega n} d\omega \tag{3.13}$$

Let's do one example to use the above equations.

Example 3.4
The below is the given discrete time signal. Find the frequency distribution.

$$x[n] = \begin{cases} 1 & \text{for} -3 \leq n \leq 3 \\ 0 & \text{Otherwise} \end{cases}$$

Solution
The frequency distribution of the given signal is below.

$$\begin{aligned} X\left(e^{j\omega}\right) &= \sum_{n=-\infty}^{\infty} x[n]e^{-j\omega n} = \sum_{n=-3}^{3} e^{-j\omega n} \\ &= 1 + \left\{e^{-j\omega} + e^{j\omega}\right\} + \left\{e^{-j2\omega} + e^{j2\omega}\right\} + \left\{e^{-j3\omega} + e^{j3\omega}\right\} \\ &= 1 + 2\cos(\omega) + 2\cos(2\omega) + 2\cos(3\omega) \end{aligned}$$

The $X(e^{j\omega})$ distribution is displayed in Fig. 3.9. Note that the given discrete time signal $x[n]$ is not periodic; hence, the $X(e^{j\omega})$ is continuous according to the previous section result.

Prog. 3.4 MATLAB program for Figs. 3.8 and 3.9

```
nn = -4:1:4;
w = -pi:pi/100:pi;
xn = [0 1 1 1 1 1 1 1 0];
Xw = 1+2*cos(w)+2*cos(2*w)+2*cos(3*w);

figure,
stem(nn,xn), grid
xlabel('n')
ylabel('x[n]')

figure,
plot(w,Xw), grid
xlabel('\omega')
ylabel('X(e^{j\omega})')
```

■

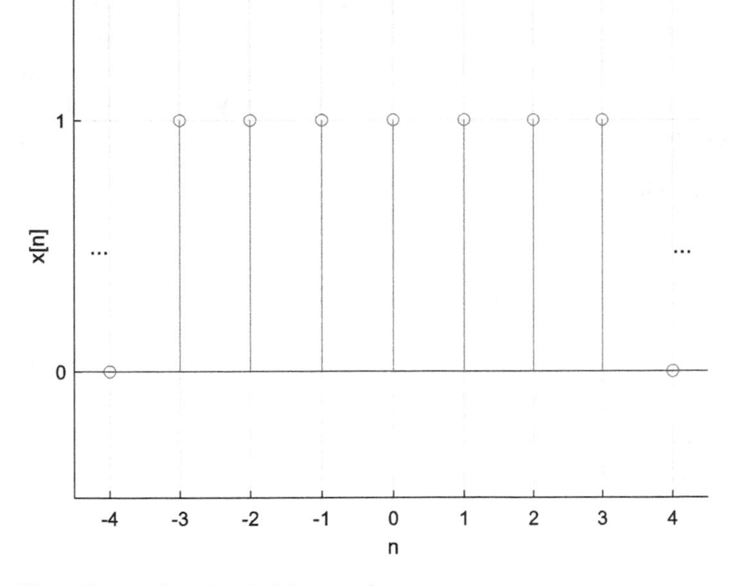

Fig. 3.8 Given discrete time signal $x[n]$ example

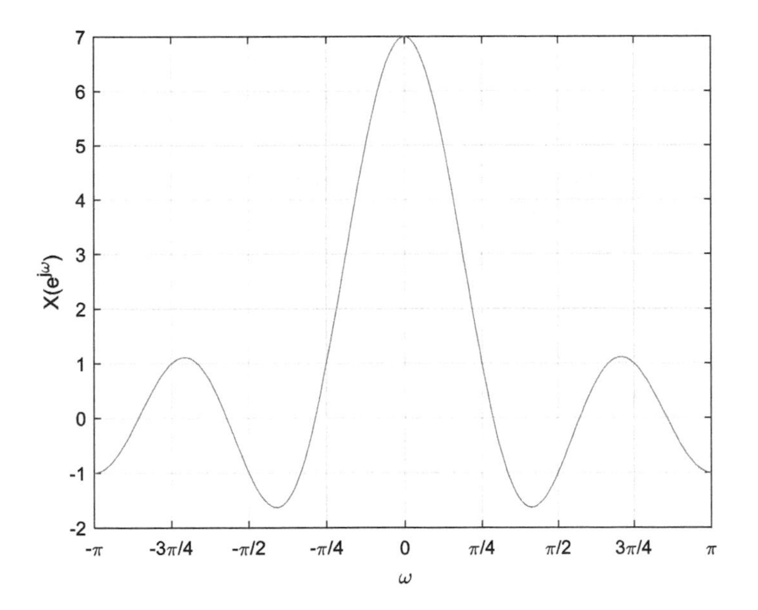

Fig. 3.9 The frequency distribution of given $x[n]$ in Fig. 3.8

The computed $X(e^{j\omega})$ is illustrated at Fig. 3.9. Generally, the $X(e^{j\omega})$ is complex number with magnitude as well as the phase component. The negative value on Fig. 3.9 should be translated as the positive magnitude with π phase because the -1 is equal to the $1e^{j\pi}$. According to the frequency property of the general discrete time signal, the above distribution is repeated in every 2π interval as shown below.

$$\begin{aligned} X\left(e^{j(\omega+2k\pi)}\right) &= 1+2\cos(\omega+2k\pi) \\ &+2\cos(2(\omega+2k\pi))+2\cos(3(\omega+2k\pi)) \\ &= 1+2\cos(\omega)+2\cos(2\omega)+2\cos(3\omega) = X\left(e^{j\omega}\right) \quad k\in\mathbb{Z} \end{aligned}$$

Using the synthesis equation, we can obtain the discrete time sequence from the $X(e^{j\omega})$ as below.

Example 3.5
The below is the given frequency distribution. Find the corresponding discrete-time domain signal $x[n]$.

$$X\left(e^{j\omega}\right) = 1+2\cos(\omega)+2\cos(2\omega)+2\cos(3\omega)$$

Solution

$$\begin{aligned} x[n] &= \frac{1}{2\pi}\int_{-\pi}^{\pi} X\left(e^{j\omega}\right)e^{j\omega n}d\omega \\ &= \frac{1}{2\pi}\int_{-\pi}^{\pi}\{1+2\cos(\omega)+2\cos(2\omega)+2\cos(3\omega)\}e^{j\omega n}d\omega \\ &= \frac{1}{2\pi}\int_{-\pi}^{\pi}\{e^{-j3\omega}+e^{-j2\omega}+e^{-j\omega}+1+e^{j\omega}+e^{j2\omega}+e^{j3\omega}\}e^{j\omega n}d\omega \\ &= \frac{1}{2\pi}\{\int_{-\pi}^{\pi} e^{j(n-3)\omega}d\omega+\int_{-\pi}^{\pi} e^{j(n-2)\omega}d\omega+\int_{-\pi}^{\pi} e^{j(n-1)\omega}d\omega+\int_{-\pi}^{\pi} e^{jn\omega}d\omega \\ &+\int_{-\pi}^{\pi} e^{j(n+1)\omega}d\omega+\int_{-\pi}^{\pi} e^{j(n+2)\omega}d\omega+\int_{-\pi}^{\pi} e^{j(n+3)\omega}d\omega\} \\ &= \delta[n-3]+\delta[n-2]+\delta[n-1]+\delta[n]+\delta[n+1] \\ &+\delta[n+2]+\delta[n+3] = x[n] \end{aligned}$$

■

The integral of the complex exponential in above indicates the discrete delta function as shown in Eq. (3.14).

$$\int_{-\pi}^{\pi} e^{j(n-3)\omega}\, d\omega = \begin{Bmatrix} 0 & \text{for } n \neq 3 \\ 2\pi & \text{for } n = 3 \end{Bmatrix} = 2\pi\delta[n-3] \text{ where } n \in \mathbb{Z} \tag{3.14}$$

If the $(n-3)$ is not equal to zero, the integral accumulates the complex vectors on unit circle over the multiple complete turns specified by $(n-3)$ for zero output. The integral generates non-zero output only at the one specific location that follows the property of the discrete delta function. Therefore, the synthesis equation successfully derives the original given sequence $x[n]$ from the frequency distribution $X(e^{j\omega})$.

The DTFT is the linear operator as below.

$$z[n] = \alpha x[n] + \beta y[n]$$

$$\begin{aligned} Z(e^{j\omega}) &= \sum_{n=-\infty}^{\infty} \{\alpha x[n] + \beta y[n]\} e^{-j\omega n} \\ &= \alpha \sum_{n=-\infty}^{\infty} x[n] e^{-j\omega n} + \beta \sum_{n=-\infty}^{\infty} y[n] e^{-j\omega n} = \alpha X(e^{j\omega}) + \beta Y(e^{j\omega}) \end{aligned} \tag{3.15}$$

Also, synthesis equation is satisfying the linearity as following.

$$Z(e^{j\omega}) = \alpha X(e^{j\omega}) + \beta Y(e^{j\omega})$$

$$\begin{aligned} z[n] &= \frac{1}{2\pi} \int_{-\pi}^{\pi} \{\alpha X(e^{j\omega}) + \beta Y(e^{j\omega})\} e^{j\omega n} d\omega \\ &= \frac{\alpha}{2\pi} \int_{-\pi}^{\pi} X(e^{j\omega}) e^{j\omega n} d\omega + \frac{\beta}{2\pi} \int_{-\pi}^{\pi} Y(e^{j\omega}) e^{j\omega n} d\omega = \alpha x[n] + \beta y[n] \end{aligned} \tag{3.16}$$

The time shift can be represented in the frequency domain as below.

$$\begin{aligned} \sum_{n=-\infty}^{\infty} x[n-d] e^{-j\omega n} &= \sum_{m+d=-\infty}^{\infty} x[m] e^{-j\omega(m+d)} \\ &= \sum_{m=-\infty}^{\infty} x[m] e^{-j\omega m} e^{-j\omega d} = X(e^{j\omega}) e^{-j\omega d} \end{aligned} \tag{3.17}$$

The delay in the time domain is denoted by the complex multiplication in frequency domain. The shift in the frequency domain is illustrated as below.

$$\frac{1}{2\pi} \int_{-\pi}^{\pi} X\left(e^{j(\omega-\omega_s)}\right) e^{j\omega n} d\omega = \frac{1}{2\pi} \int_{-\pi}^{\pi} X(e^{j\varphi}) e^{j(\varphi+\omega_s)n} d(\varphi + \omega_s)$$

$$= \frac{1}{2\pi} \int_{-\pi}^{\pi} X(e^{j\varphi}) e^{j\varphi n} d\varphi e^{j\omega_s n} = x[n] e^{j\omega_s n} \tag{3.18}$$

As previously mentioned, the given real sequence must generate conjugated magnitudes for the positive and negative frequency pair.

$$X(e^{j\omega}) = X^*(e^{-j\omega}) \tag{3.19}$$

The reversal in time provides the reversal in frequency as well.

$$\sum_{n=-\infty}^{\infty} x[-n]e^{-j\omega n} = \sum_{m=-\infty}^{\infty} x[m]e^{j\omega m} = X(e^{-j\omega}) \tag{3.20}$$

According to the Parseval's theorem [2], the energy in time domain is identical to the energy in frequency domain as below.

$$E_x = \sum_{n=-\infty}^{\infty} |x[n]|^2 = \frac{1}{2\pi}\int_{-\pi}^{\pi} |X(e^{j\omega})|^2 \, d\omega \tag{3.21}$$

Example 3.6
Apply the Parseval's theorem on below example.

$$x[n] = \begin{cases} 1 & \text{for} - 3 \leq n \leq 3 \\ 0 & \text{Otherwise} \end{cases}$$

Solution
The time domain energy is below.

$$E_x = \sum_{n=-\infty}^{\infty} |x[n]|^2 = 7$$

The frequency domain distribution and energy are below from Example 3.4.

$$X\left(e^{j\omega}\right) = 1 + 2\cos(\omega) + 2\cos(2\omega) + 2\cos(3\omega)$$

$$\begin{aligned}
E_x &= \frac{1}{2\pi}\int_{-\pi}^{\pi} |1 + 2\cos(\omega) + 2\cos(2\omega) + 2\cos(3\omega)|^2 d\omega \\
&= \frac{1}{2\pi}\left\{\int_{-\pi}^{\pi} 1 d\omega + \int_{-\pi}^{\pi} 4\cos^2(\omega) d\omega + \int_{-\pi}^{\pi} 4\cos^2(2\omega) d\omega + \int_{-\pi}^{\pi} 4\cos^2(3\omega) d\omega\right\} \\
&= \frac{1}{2\pi}\{2\pi + 4\pi + 4\pi + 4\pi\} = 7
\end{aligned}$$

Observe that the cross terms in frequency domain energy computation are zero. For example, shown as below.

$$\frac{1}{2\pi}\int_{-\pi}^{\pi}\cos(\omega)\cos(2\omega)d\omega = 0 \quad \frac{1}{2\pi}\int_{-\pi}^{\pi}\cos(2\omega)\cos(3\omega)d\omega = 0$$

Therefore, energy can be computed in either domain for your computational convenience. ■

3.4 Discrete Fourier Transform (DFT)

Unless you use the symbolic mathematics for the computer, the continuous domain cannot be handled by the digital computer directly. Discrete data or sampled signal can be processed by the digital bits with binary logics specified by the computer software. For versatile processing, the signal is managed in the time domain as well as frequency domain with discrete information handling. We mentioned that the sampling process provides the discrete time sequence that generates the spectrum periodicity in every 2π interval. Also, the signal periodicity presents the discrete frequency distribution with the $2\pi/N$ spectral distance known as fundamental frequency. The discrete Fourier Transform (DFT) [1] is the transformation for the discrete time and discrete frequency distribution; therefore, the conventional computer performs the numerical operations for the transformation. The discrete periodicity and orthogonality provide the important clue to the transformation equations similar to the DTFT derivation. Let's explore the DFT by defining the discrete periodic signal as Eq. (3.22). The signal with period N contains following periods as well.

$$\begin{aligned} x_N[n] &= \alpha_1 s_N[n] + \alpha_2 s_{N/2}[n] + \alpha_3 s_{N/3}[n] + \alpha_4 s_{N/4}[n] + \cdots + \alpha_N s_{N/N}[n] \\ &= x_N[n + mN] \; m \in \mathbb{Z} \end{aligned} \tag{3.22}$$

The $s_N[n]$ is the N period sinusoid with unit magnitude and the α_1 is the constant magnitude for the $s_N[n]$. The $x_N[n]$ only includes the N period as well as its integer divisions such as $N/2$, $N/3$, $N/4$, etc. The integer division of the any given periods can generate N period; in other words, all the sub-period signals complete its period at N sequence location to initiate new $x_N[n]$ period. The frequency representation of the signal is in Eq. (3.23).

$$\begin{aligned} x_N[n] &= \alpha_1 e^{j\frac{2\pi}{N}n} + \alpha_2 e^{j\frac{2\pi}{N}2n} + \alpha_3 e^{j\frac{2\pi}{N}3n} + \alpha_4 e^{j\frac{2\pi}{N}4n} + \cdots + \alpha_N e^{j\frac{2\pi}{N}Nn} \\ &= \sum_{k=0}^{N-1} X[k] e^{j\frac{2\pi}{N}kn} \text{ where } e^{j\frac{2\pi}{N}Nn} = e^{j\frac{2\pi}{N}0n} \end{aligned} \tag{3.23}$$

The signal $x_N[n]$ only has the fundamental frequency $2\pi/N$ and its integer multiples. The $X[k]$ is the complex magnitude for the kth harmonics. Is the $x_N[n]$ the real number? In order to use the Euler equation, the negative frequency is required but not shown in Eq. (3.23). Find the condition to make the real sequence.

$$\begin{aligned}
x_N[n] &= \sum_{k=0}^{N-1} X[k]e^{j\frac{2\pi}{N}kn} \\
&= X[0]e^{j\frac{2\pi}{N}0n} + X[1]e^{j\frac{2\pi}{N}n} + X[2]e^{j\frac{2\pi}{N}2n} + \ldots \\
&+ X[N-2]e^{j\frac{2\pi}{N}(N-2)n} + X[N-1]e^{j\frac{2\pi}{N}(N-1)n} \\
&= X[0]e^{j\frac{2\pi}{N}0n} + X[1]e^{j\frac{2\pi}{N}n} + X[2]e^{j\frac{2\pi}{N}2n} + \ldots \\
&+ X[N-2]e^{j\frac{2\pi}{N}Nn}e^{-j\frac{2\pi}{N}2n} + X[N-1]e^{j\frac{2\pi}{N}Nn}e^{-j\frac{2\pi}{N}n} \\
&= X[0]e^{j\frac{2\pi}{N}0n} + X[1]e^{j\frac{2\pi}{N}n} + X[2]e^{j\frac{2\pi}{N}2n} + \ldots \\
&+ X[N-2]e^{-j\frac{2\pi}{N}2n} + X[N-1]e^{-j\frac{2\pi}{N}n} \\
&= X[0]e^{j\frac{2\pi}{N}0n} + \left\{X[1]e^{j\frac{2\pi}{N}n} + X[N-1]e^{-j\frac{2\pi}{N}n}\right\} \\
&+ \left\{X[2]e^{j\frac{2\pi}{N}2n} + X[N-2]e^{-j\frac{2\pi}{N}2n}\right\} + \cdots \\
&+ \begin{cases} X\left[\frac{N}{2}\right]e^{j\frac{2\pi}{N}\frac{N}{2}n} & \text{for even } N \\ X\left[\frac{N-1}{2}\right]e^{j\frac{2\pi}{N}\frac{N-1}{2}n} + X\left[\frac{N+1}{2}\right]e^{-j\frac{2\pi}{N}\frac{N-1}{2}n} & \text{for odd } N \end{cases}
\end{aligned} \tag{3.24}$$

By using the angular equivalence, the north and south hemisphere of the complex plane shown Fig. 3.10 is assigned to the positive and negative frequency, respectively. The Euler formula pair transforms the complex exponentials to the cosine functions for real number sequence shown in Eq. (3.24).

Therefore, following condition is required for real number sequence.

$$X[k] = X^*[N-k] \tag{3.25}$$

The signal with N period can be generated by the linear combination of complex exponentials as below.

$$x_N[n] = \frac{1}{N}\sum_{k=0}^{N-1} X[k]e^{j\frac{2\pi}{N}kn} \tag{3.26}$$

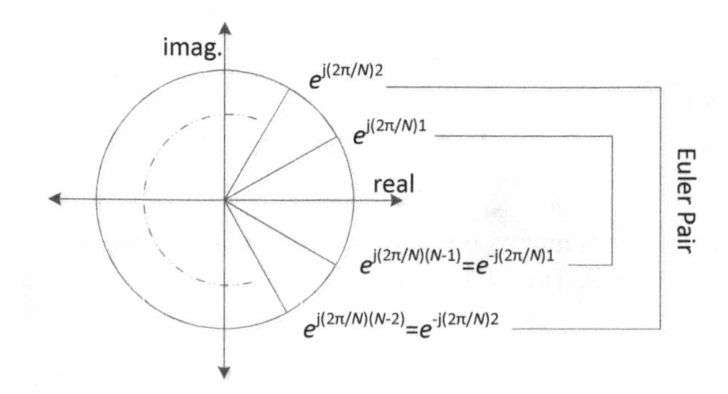

Fig. 3.10 Euler formula pair relationship

From the given periodic sequence, how we can find the complex magnitude for each frequency components? Use the periodicity and orthogonality like DTFT equations.

$$\sum_{n=0}^{N-1} x_N[n]e^{-j\frac{2\pi}{N}kn} = \sum_{n=0}^{N-1}\left(\frac{1}{N}\sum_{l=0}^{N-1} X[l]e^{j\frac{2\pi}{N}ln}\right)e^{-j\frac{2\pi}{N}kn} = \sum_{n=0}^{N-1}\frac{1}{N}\sum_{l=0}^{N-1} X[l]e^{j\frac{2\pi}{N}(l-k)n}$$
$$= \left\{\begin{array}{l} \frac{1}{N}\sum_{l=0}^{N-1} X[l]\sum_{n=0}^{N-1} e^{j\frac{2\pi}{N}(l-k)n} = 0 \text{ for } l \neq k \\ \frac{1}{N}\sum_{l=0}^{N-1} X[l]\sum_{n=0}^{N-1} e^{j\frac{2\pi}{N}(l-k)n} = X[k] \text{ for } l = k \end{array}\right\} = \sum_{l=0}^{N-1} X[l]\delta[k-l] = X[k] \quad (3.27)$$

Therefore,

$$\sum_{n=0}^{N-1} x_N[n]e^{-j\frac{2\pi}{N}kn} = X[k]$$

Finally, DFT analysis and synthesis equation are organized as Eqs. (3.28) and (3.29), respectively.

$$X[k] = \sum_{n=0}^{N-1} x_N[n]e^{-j\frac{2\pi}{N}kn} \quad (3.28)$$

$$x_N[n] = \frac{1}{N}\sum_{k=0}^{N-1} X[k]e^{j\frac{2\pi}{N}kn} \quad (3.29)$$

One example of the discrete periodic signal $x_N[n]$ and corresponding frequency distribution $X[k]$ are illustrated below.

$$\begin{aligned} x_N[n] = \frac{1}{N}\Big\{&X[0]e^{j\frac{2\pi}{N}0n} + X[1]e^{j\frac{2\pi}{N}n} + X[2]e^{j\frac{2\pi}{N}2n} + \ldots \\ &+ X[N-2]e^{j\frac{2\pi}{N}(N-2)n} + X[N-1]e^{j\frac{2\pi}{N}(N-1)n}\Big\} \end{aligned} \quad (3.30)$$

$$\begin{aligned} X[k] = X[0]\delta[k] + X[1]\delta[k-1] + X[2]\delta[k-2] + \ldots \\ + X[N-2]\delta[k-(N-2)] + X[N-1]\delta[k-(N-1)] \end{aligned} \quad (3.31)$$

The $x_N[n]$ and $X[k]$ present the periodic property due to the sampling and discretization in time and frequency domain, respectively. Figure 3.11 shows the one example $x_N[n]$ and $X[k]$ for 16 sample period N. The digital frequency k includes the symmetric pattern distribution for real $x_N[n]$ signal.

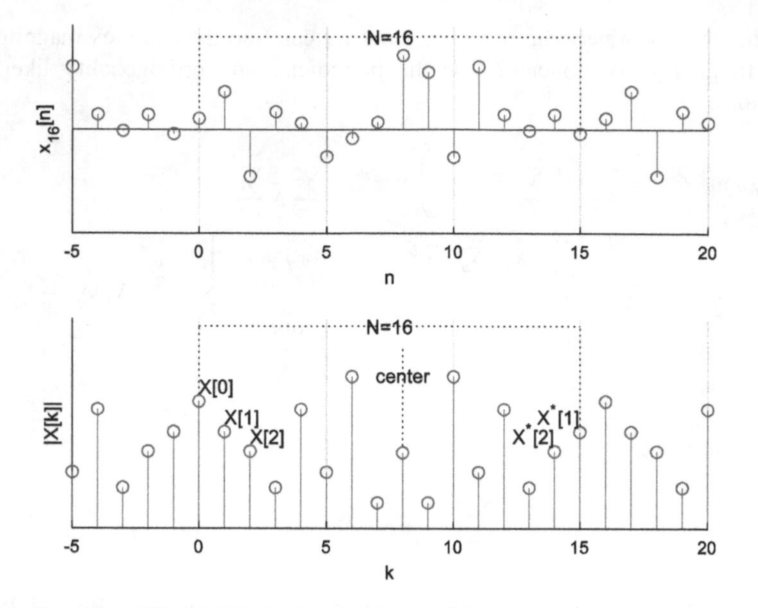

Fig. 3.11 Example $x_N[n]$ and $X[k]$ for 16 sample period N

Prog. 3.5 MATLAB program for Fig. 3.11

```
N = 16;
rng('default')
xn = randn(1,N);
xne = [xn xn xn];
Xk = fft(xn);
Xke = abs([Xk Xk Xk]);
nn = -N:1:2*N-1;
kk = -N:1:2*N-1;

figure,
subplot(211), stem(nn,xne), grid
xlim([-5 20])
ylim([-5 5])
subplot(212), stem(kk,Xke), grid
xlim([-5 20])
ylim([0 15])
```

Example 3.7
Compute the DFT of following signals with length $N > 8$.

$$x_1[n] = a^n(u[n] - u[n-N]) \text{ for } a \in \mathbb{R}$$

$$x_2[n] = u[n] - u[n-N]$$

$$x_3[n] = \delta[n]$$

$$x_4[n] = \cos\left(\frac{2\pi}{N}3n\right)(u[n] - u[n-N])$$

Solution

$$X_1[k] = \sum_{n=0}^{N-1} x_1[n]e^{-j\frac{2\pi}{N}kn} = \sum_{n=0}^{N-1} a^n e^{-j\frac{2\pi}{N}kn} = \sum_{n=0}^{N-1}\left(ae^{-j\frac{2\pi}{N}k}\right)^n$$
$$= \frac{1-\left(ae^{-j\frac{2\pi}{N}k}\right)^N}{1-ae^{-j\frac{2\pi}{N}k}}$$

$$X_2[k] = \sum_{n=0}^{N-1}(u[n] - u[n-N])e^{-j\frac{2\pi}{N}kn} = \sum_{n=0}^{N-1} e^{-j\frac{2\pi}{N}kn}$$
$$= \frac{1-e^{-j\frac{2\pi}{N}kN}}{1-e^{-j\frac{2\pi}{N}k}} = \frac{e^{-j\frac{\pi}{N}kN}}{e^{-j\frac{\pi}{N}k}}\left(\frac{e^{j\frac{\pi}{N}kN} - e^{-j\frac{\pi}{N}kN}}{e^{j\frac{\pi}{N}k} - e^{-j\frac{\pi}{N}k}}\right)$$
$$= e^{-j\frac{\pi}{N}k(N-1)}\frac{\sin(\pi k)}{\sin\left(\frac{\pi}{N}k\right)} = N\delta[k]$$

$$\because \frac{\sin(\pi k)}{\sin\left(\frac{\pi}{N}k\right)} = \begin{cases} 0 & \text{for } 0<k<N \\ N & \text{for } k=0 \end{cases}$$

$$X_3[k] = \sum_{n=0}^{N-1}\delta[n]e^{-j\frac{2\pi}{N}kn} = 1$$

$$X_4[k] = \sum_{n=0}^{N-1}\left(\cos\left(\frac{2\pi}{N}3n\right)(u[n] - u[n-N])\right)e^{-j\frac{2\pi}{N}kn}$$
$$= \frac{1}{2}\sum_{n=0}^{N-1}\left(e^{j\frac{2\pi}{N}3n} + e^{-j\frac{2\pi}{N}3n}\right)e^{-j\frac{2\pi}{N}kn} = \frac{1}{2}\sum_{n=0}^{N-1}\left(e^{-j\frac{2\pi}{N}(k-3)n} + e^{-j\frac{2\pi}{N}(k+3)n}\right)$$
$$= \frac{N}{2}\delta[k-3] + \frac{N}{2}\delta[k+3]$$

Note that above frequency information $X[k]$s represent the distribution over the range between 0 to $N-1$. The distribution is repeated in every N. ■

Relationship between the DTFT and DFT for the periodic signal is below.

$$\begin{aligned}
X(e^{j\omega}) &= \sum_{n=-\infty}^{\infty} x[n]e^{-j\omega n} = \sum_{n=-\infty}^{\infty} x_N[n]e^{-j\omega n} \\
&= \sum_{n=-\infty}^{\infty} \frac{1}{N}\sum_{k=0}^{N-1} X[k]e^{j\frac{2\pi}{N}kn}e^{-j\omega n} = \frac{1}{N}\sum_{k=0}^{N-1} X[k] \sum_{n=-\infty}^{\infty} e^{-j\left(\omega-\frac{2\pi}{N}k\right)n} \\
&= \frac{1}{N}\sum_{k=0}^{N-1} X[k] \sum_{n=-\infty}^{\infty} e^{-j\left(\omega-\frac{2\pi}{N}(k+lN)\right)n} \text{ for } l \in \mathbb{Z} \\
&= \frac{1}{N}\sum_{k=0}^{N-1} X[k]2\pi\delta\left(\omega - \tfrac{2\pi}{N}(k+lN)\right) = \frac{2\pi}{N}\sum_{k=-\infty}^{\infty} X[k]\delta\left(\omega - \tfrac{2\pi}{N}k\right)
\end{aligned} \tag{3.32}$$

$$\begin{aligned}
x[n] &= \frac{1}{2\pi}\int_{-\pi}^{\pi} X(e^{j\omega})e^{j\omega n}d\omega = \frac{1}{2\pi}\int_{-\pi}^{\pi} \frac{2\pi}{N}\sum_{k=-\infty}^{\infty} X[k]\delta\left(\omega - \frac{2\pi}{N}k\right)e^{j\omega n}d\omega \\
&= \frac{1}{N}\sum_{k=-\infty}^{\infty} X[k]e^{j\frac{2\pi}{N}kn}\int_{-\pi}^{\pi}\delta\left(\omega - \frac{2\pi}{N}k\right)d\omega \\
&= \frac{1}{N}\sum_{k=0}^{N-1} X[k]e^{j\frac{2\pi}{N}kn} = x_N[n]
\end{aligned} \tag{3.33}$$

$$\because \int_{-\pi}^{\pi}\delta\left(\omega - \frac{2\pi}{N}k\right)d\omega \neq 0 \text{ for one complete } 2\pi \text{ equivalently } N \text{ sample period}$$

Summarize the relation between the DTFT and DFT below.

$$x[n] = \frac{1}{N}\sum_{k=0}^{N-1} X[k]e^{j\frac{2\pi}{N}kn} = \frac{1}{2\pi}\sum_{k=0}^{N-1} X(e^{jk\omega_0})e^{j\omega_0 kn} = x_N[n] \tag{3.34}$$

$$X(e^{j\omega}) = \frac{2\pi}{N}\sum_{k=-\infty}^{\infty} X[k]\delta\left(\omega - \frac{2\pi}{N}k\right) \tag{3.35}$$

Let's do one example as below.

Example 3.8
The given signal is periodic impulse train with 6 sample period. Compute the DFT and DTFT of the given signal.

$$x_6[n] = \delta_6[n] = \sum_{m=-\infty}^{\infty} \delta[n-6m]$$

Solution

$$X[k] = \sum_{n=0}^{5} \delta[n] e^{-j\frac{2\pi}{6}kn} = 1$$

$$X\left(e^{j\omega}\right) = \frac{2\pi}{6} \sum_{k=-\infty}^{\infty} 1\delta\left(\omega - \frac{2\pi}{6}k\right)$$

The inverse DFT derives the original time domain signal as below.

$$x_6[n] = \frac{1}{6} \sum_{k=0}^{5} 1 e^{jk\frac{2\pi}{6}n} = \sum_{m=-\infty}^{\infty} \delta[n - 6m]$$

Prog. 3.6 MATLAB program for Fig. 3.12

```
N = 6;
xn = [1 0 0 0 0 0];
xne = [xn xn xn];
Xk = fft(xn);
Xke = abs([Xk Xk Xk]);
nn = -N:1:2*N-1;
kk = -N:1:2*N-1;

figure,
subplot(311), stem(nn,xne), grid
xlim([-6 11])
ylim([0 2])
ylabel('x_{6}[n]')
subplot(312), stem(kk,Xke), grid
xlim([-6 11])
ylim([0 2])
ylabel('X[k]')
subplot(313), stem(kk,Xke,'Marker','^'), grid
xlim([-6 11])
ylim([0 2])
ylabel('X(e^{j\omega})')
xticks([-1 0 1 2 3 4 5 6])
xticklabels({'...','0','\pi/3','2\pi/3','\pi','4\pi/3','5\pi/3','...'})
xtickangle(90)
```

■

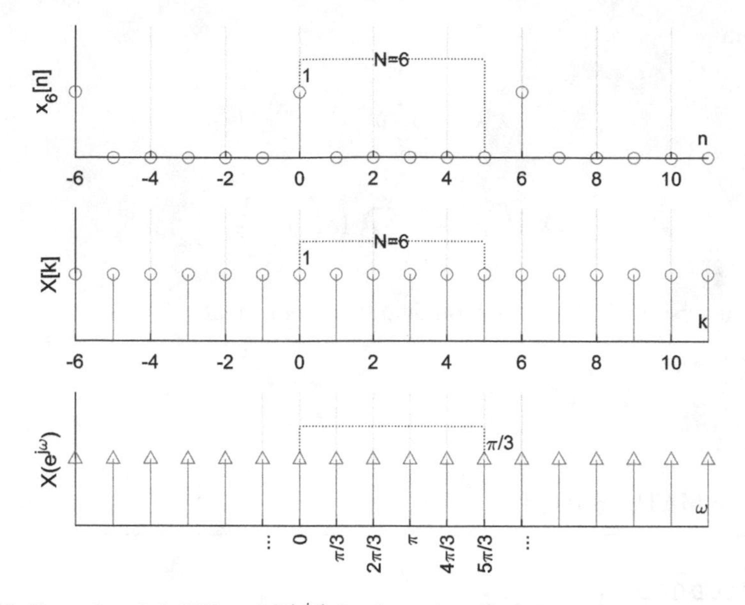

Fig. 3.12 Example $x_N[n]$, $X[k]$, and $X(e^{j\omega})$ for 6 sample period N

The impulse train with six sample period is transformed to the frequency domain by using the DFT and DTFT as shown in Fig. 3.12. The DFT frequency is the discrete domain in spectrum with unit magnitude and the DTFT frequency is the continuous domain in spectrum with $\pi/3$ magnitude for impulse. Note that the DTFT can represent the any arbitrary signal but the DFT can handle the periodic signal only. What if we have non-periodic signal and want to operate with DFT? Let's consider following problem.

$$x[n] = \begin{cases} 1 & \text{for} - 3 \leq n \leq 3 \\ 0 & \text{Otherwise} \end{cases} \tag{3.36}$$

The below is the procedure to find the frequency distribution of the given signal Eq. (3.36). Note that the signa $x[n]$ is illustrated at Fig. 3.8. From the previous example, the DTFT of the given signal is below.

$$X(e^{j\omega}) = 1 + 2\cos(\omega) + 2\cos(2\omega) + 2\cos(3\omega) \tag{3.37}$$

The DFT is following.

$$\begin{aligned} X[k] &= \sum_{n=0}^{N-1} x_N[n]e^{-j\frac{2\pi}{N}kn} = \sum_{n=-3}^{3} e^{-j\frac{2\pi}{7}kn} = 1 + \left\{e^{-j\frac{2\pi}{7}k} + e^{j\frac{2\pi}{7}k}\right\} + \left\{e^{-j\frac{2\pi}{7}2k} + e^{j\frac{2\pi}{7}2k}\right\} \\ &+ \left\{e^{-j\frac{2\pi}{7}3k} + e^{j\frac{2\pi}{7}3k}\right\} \quad = 1 + 2\cos\left(\tfrac{2\pi}{7}k\right) + 2\cos\left(\tfrac{2\pi}{7}2k\right) + 2\cos\left(\tfrac{2\pi}{7}3k\right) \end{aligned} \tag{3.38}$$

The inverse of the DFT is below.

$$
\begin{aligned}
x[n] &= \frac{1}{N}\sum_{k=0}^{N-1} X[k]e^{j\frac{2\pi}{N}kn} = \frac{1}{7}\sum_{k=0}^{6}\left[1+\left\{e^{-j\frac{2\pi}{7}k}+e^{j\frac{2\pi}{7}k}\right\}+\left\{e^{-j\frac{2\pi}{7}2k}+e^{j\frac{2\pi}{7}2k}\right\}\right. \\
&\quad \left.+\left\{e^{-j\frac{2\pi}{7}3k}+e^{j\frac{2\pi}{7}3k}\right\}\right]e^{j\frac{2\pi}{7}kn} \\
&= \frac{1}{7}\sum_{k=0}^{6}\left[e^{j\frac{2\pi}{7}kn}+\left\{e^{-j\frac{2\pi}{7}k(1-n)}+e^{j\frac{2\pi}{7}k(1+n)}\right\}+\left\{e^{-j\frac{2\pi}{7}k(2-n)}+e^{j\frac{2\pi}{7}k(2+n)}\right\}\right. \\
&\qquad \left.+\left\{e^{-j\frac{2\pi}{7}k(3-n)}+e^{j\frac{2\pi}{7}k(3+n)}\right\}\right] \\
&= \delta[n]+\delta[n-1]+\delta[n+1]+\delta[n-2] \\
&\qquad +\delta[n+2]+\delta[n-3]+\delta[n+3]
\end{aligned}
\tag{3.39}
$$

Prog. 3.7 MATLAB program for Fig. 3.13

```
N1 = 7;
w = 0:pi/100:2*pi;
k1 = 0:1:(N1-1);
Xw1 = 1+2*cos(w)+2*cos(2*w)+2*cos(3*w);
Xk1 = 1+2*cos(2*pi*k1/N1)+2*cos(2*pi*2*k1/N1)+2*cos(2*pi*3*k1/N1);

figure,
subplot(211), plot(w,Xw1), grid
ylabel('X(e^{j\omega})')
ylim([-2 8])
hold on
yyaxis right
stem(2*pi*k1/N1,Xk1)
ylabel('X[k]')
ylim([-2 8])
hold off
xlim([0 2*pi])
xticks(2*pi*(0:1:7)/N1)
xticklabels({'2\pi0/7','2\pi1/7','2\pi2/7','2\pi3/7','2\pi4/7','2\pi5/7','2\pi6/7','
2\pi7/7'})
xlabel('\omega')
subplot(212), stem(n1,ones(size(n1))), grid
xlabel('n')
ylabel('x[n]')
ylim([0 2])
```

In this problem, the signal and DFT length is seven; therefore, the seven ones are repeated forever to show the all ones for entire sequence range. The frequency

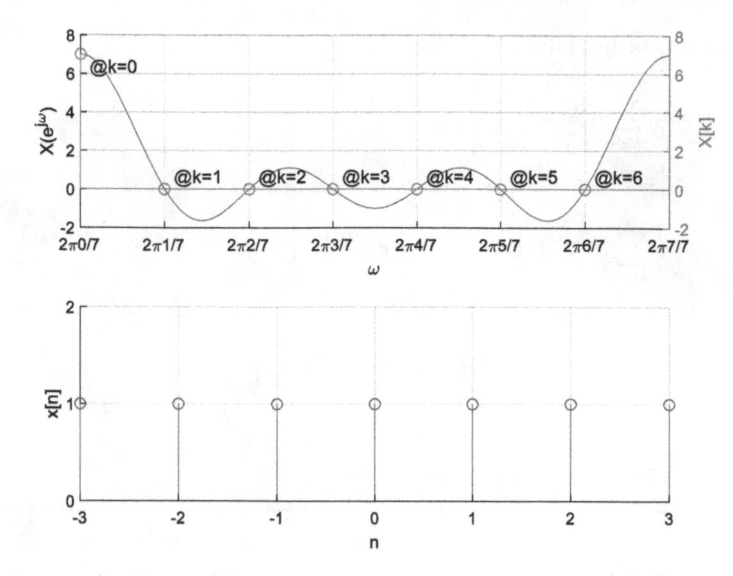

Fig. 3.13 DTFT and DFT of window function Eq. (3.36) and inverse DFT for $x[n]$ for $N = 7$

distribution in Fig. 3.13 from the DFT delivers the single value at the zero-frequency location that represents no fluctuation (constant) signal in time domain. The DTFT result in Fig. 3.13 shown in solid line generates non-periodic signal distribution which denotes the original signal Eq. (3.36) from the given problem. The DFT samples the DTFT continuous outcome in every $2\pi/7$ for digital frequency distribution in terms of k. The sampling is performed improperly due to the short DFT length; therefore, the reconstructed signal does not correspond to the original given signal accurately.

Let's increase the DFT length for 9 as below.

$$
\begin{aligned}
X[k] &= \sum_{n=-4}^{4} e^{-j\frac{2\pi}{9}kn} \\
&= 1 + \left\{ e^{-j\frac{2\pi}{9}k} + e^{j\frac{2\pi}{9}k} \right\} + \left\{ e^{-j\frac{2\pi}{9}2k} + e^{j\frac{2\pi}{9}2k} \right\} + \left\{ e^{-j\frac{2\pi}{9}3k} + e^{j\frac{2\pi}{9}3k} \right\} \\
&= 1 + 2\cos\left(\tfrac{2\pi}{9}k\right) + 2\cos\left(\tfrac{2\pi}{9}2k\right) + 2\cos\left(\tfrac{2\pi}{9}3k\right)
\end{aligned}
\tag{3.40}
$$

The inverse of the DFT is below.

$$\begin{aligned} x[n] &= \frac{1}{N}\sum_{k=0}^{N-1} X[k]e^{j\frac{2\pi}{N}kn} = \frac{1}{9}\sum_{k=0}^{8}\left[1+\left\{e^{-j\frac{2\pi}{9}k}+e^{j\frac{2\pi}{9}k}\right\}+\left\{e^{-j\frac{2\pi}{9}2k}+e^{j\frac{2\pi}{9}2k}\right\}\right. \\ &\quad \left.+\left\{e^{-j\frac{2\pi}{9}3k}+e^{j\frac{2\pi}{9}3k}\right\}\right]e^{j\frac{2\pi}{9}kn} \\ &= \frac{1}{9}\sum_{k=0}^{8}\left[e^{j\frac{2\pi}{9}kn}+\left\{e^{-j\frac{2\pi}{9}k(1-n)}+e^{j\frac{2\pi}{9}k(1+n)}\right\}+\left\{e^{-j\frac{2\pi}{9}k(2-n)}+e^{j\frac{2\pi}{9}k(2+n)}\right\}\right. \\ &\quad \left.+\left\{e^{-j\frac{2\pi}{9}k(3-n)}+e^{j\frac{2\pi}{9}k(3+n)}\right\}\right] \\ &= \delta[n]+\delta[n-1]+\delta[n+1]+\delta[n-2] \\ &\quad +\delta[n+2]+\delta[n-3]+\delta[n+3] \end{aligned} \tag{3.41}$$

Prog. 3.8 MATLAB program for Fig. 3.14

```
N2 = 9;
w = 0:pi/100:2*pi;
k2 = 0:1:(N2-1);
n2 = -4:1:4;
Xw1 = 1+2*cos(w)+2*cos(2*w)+2*cos(3*w);
Xk2 = 1+2*cos(2*pi*k2/N2)+2*cos(2*pi*2*k2/N2)+2*cos(2*pi*3*k2/N2);

figure,
subplot(211), plot(w,Xw1), grid
ylabel('X(e^{j\omega})')
ylim([-2 8])
hold on
yyaxis right
stem(2*pi*k2/N2,Xk2)
ylabel('X[k]')
ylim([-2 8])
hold off
xlim([0 2*pi])
xticks(2*pi*(0:1:9)/N2)
xticklabels({'2\pi0/9','2\pi1/9','2\pi2/9','2\pi3/9','2\pi4/9','2\pi5/9','2\pi6/9','
2\pi7/9','2\pi8/9','2\pi9/9'})
xlabel('\omega')
subplot(212), stem(n2,[0 ones(size(n1)) 0]), grid
xlabel('n')
ylabel('x[n]')
ylim([0 2])
```

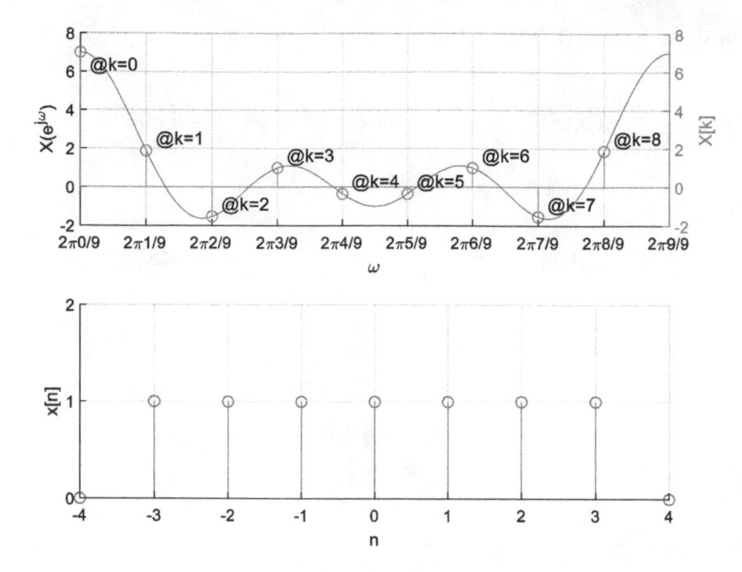

Fig. 3.14 DTFT and DFT of window function Eq. (3.36) and inverse DFT for $x[n]$ for $N = 9$

Now, we can see the certain magnitudes in other than the zero-frequency location in Fig. 3.14. The DTF samples the DTFT outcome in every $2\pi/9$ and the reconstructed signal illustrates the seven ones and two zeros for one period set. Let's increase the DFT length further for 99.

$$\begin{aligned} X[k] &= \sum_{n=-4}^{4} e^{-j\frac{2\pi}{99}kn} \\ &= 1 + \left\{ e^{-j\frac{2\pi}{99}k} + e^{j\frac{2\pi}{99}k} \right\} + \left\{ e^{-j\frac{2\pi}{99}2k} + e^{j\frac{2\pi}{99}2k} \right\} + \left\{ e^{-j\frac{2\pi}{99}3k} + e^{j\frac{2\pi}{99}3k} \right\} \\ &= 1 + 2\cos\left(\tfrac{2\pi}{99}k\right) + 2\cos\left(\tfrac{2\pi}{99}2k\right) + 2\cos\left(\tfrac{2\pi}{99}3k\right) \end{aligned} \tag{3.42}$$

The inverse of the DFT is below.

$$x[n] = \frac{1}{N}\sum_{k=0}^{N-1} X[k]e^{j\frac{2\pi}{N}kn}$$

$$\begin{aligned}
x[n] &= \frac{1}{N}\sum_{k=0}^{N-1} X[k]e^{j\frac{2\pi}{N}kn} \\
&= \frac{1}{99}\sum_{k=0}^{98}\Big[1 + \left\{e^{-j\frac{2\pi}{99}k} + e^{j\frac{2\pi}{99}k}\right\} + \left\{e^{-j\frac{2\pi}{99}2k} + e^{j\frac{2\pi}{99}2k}\right\} \\
&\quad + \left\{e^{-j\frac{2\pi}{99}3k} + e^{j\frac{2\pi}{99}3k}\right\}\Big]e^{j\frac{2\pi}{99}kn} \\
&= \frac{1}{99}\sum_{k=0}^{98}\Big[e^{j\frac{2\pi}{99}kn} + \left\{e^{-j\frac{2\pi}{99}k(1-n)} + e^{j\frac{2\pi}{99}k(1+n)}\right\} + \left\{e^{-j\frac{2\pi}{99}k(2-n)} + e^{j\frac{2\pi}{99}k(2+n)}\right\} \\
&\quad + \left\{e^{-j\frac{2\pi}{99}k(3-n)} + e^{j\frac{2\pi}{99}k(3+n)}\right\}\Big] \\
&= \delta[n] + \delta[n-1] + \delta[n+1] + \delta[n-2] \\
&\qquad + \delta[n+2] + \delta[n-3] + \delta[n+3]
\end{aligned} \tag{3.43}$$

Prog. 3.9 MATLAB program for Fig. 3.15

```
N3 = 99;
w = 0:pi/100:2*pi;
k3 = 0:1:(N3-1);
n3 = -49:1:49;
Xw1 = 1+2*cos(w)+2*cos(2*w)+2*cos(3*w);
Xk3 = 1+2*cos(2*pi*k3/N3)+2*cos(2*pi*2*k3/N3)+2*cos(2*pi*3*k3/N3);

figure,
subplot(211), plot(w,Xw1), grid
ylabel('X(e^{j\omega})')
ylim([-2 8])
hold on
yyaxis right
stem(2*pi*k3/N3,Xk3,'Marker','.')
ylabel('X[k]')
ylim([-2 8])
hold off
xlim([0 2*pi])
xticks([0 pi/4 2*pi/4 3*pi/4 pi 5*pi/4 6*pi/4 7*pi/4 2*pi])
xticklabels({'0','\pi/4','2\pi/4','3\pi/4','\pi','5\pi/4','6\pi/4','7\pi/4','2\pi'})
xlabel('\omega')
subplot(212), stem(n3,[zeros(1,46) ones(size(n1)) zeros(1,46)],'Marker','.'),
grid
xlabel('n')
ylabel('x[n]')
ylim([0 2])
```

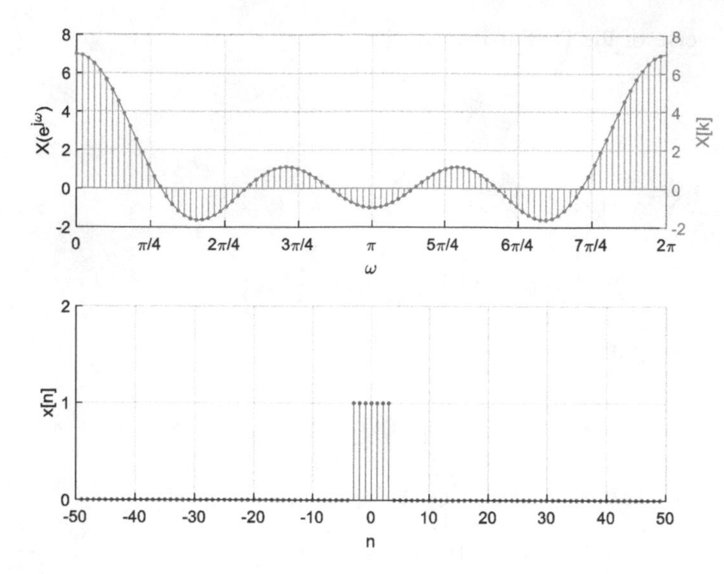

Fig. 3.15 DTFT and DFT of window function Eq. (3.36) and inverse DFT for $x[n]$ for $N = 99$

The DFT from 99 length precisely follows the DTFT continuous line with $2\pi/99$ radian interval in Fig. 3.15. Only seven ones are located in the 99-length period from the reconstruction by inverse DFT at Fig. 3.15. The further increasing in the DFT length expects to show the accurate depiction of the spectrum; however, the periodicity cannot be avoided unless you use the DTFT instead of DFT. For computer processing, the longer DFT length is recommended for the non-periodic signal. To increase the length of the non-periodic signal, zero padding is performed by placing the zero values after the signal portion. The zero padding improves the spectral representation by increasing the frequency resolution as shown above problem. The resolution in radian and cyclic frequency is given as below.

$$\Delta\omega = \omega_0 = \tfrac{2\pi}{N} \quad \text{or} \quad \Delta f = \tfrac{f_s}{N} \tag{3.44}$$

The equivalence location between the DFT and DTFT is subsequence.

$$X[k] \equiv X\left(\tfrac{2\pi}{N}k\right) \quad \text{for } 0 \leq k \leq N-1 \tag{3.45}$$

The length related matter still includes the periodic signal in DFT. The unknown frequency signal is roughly framed by intuitive length in order to convert for the spectrum distribution. Provided that the desired frequency dose not corresponds to the discrete DFT frequencies, the DFT representation loses the accuracy. Let's consider the following signal. The below signal is the discrete cosine wave with 8 period sample.

$$x[n] = \cos\left(\frac{2\pi}{8}n\right) \qquad \text{for } 0 \le n \le 7 \tag{3.46}$$

The DFT of the signal is below. Note that $-j2\pi(k+1)n/8$ term in Eq. (3.47) presents the equal values at $k = -1$, 7, 15, etc. due to the circular property of complex exponential.

$$\begin{aligned} X[k] &= \sum_{n=0}^{7} \cos\left(\frac{2\pi}{8}n\right) e^{-j\frac{2\pi}{8}kn} = \frac{1}{2}\sum_{n=0}^{7} \left(e^{j\frac{2\pi}{8}n} + e^{-j\frac{2\pi}{8}n}\right) e^{-j\frac{2\pi}{8}kn} \\ &= \frac{1}{2}\sum_{n=0}^{7} \left(e^{-j\frac{2\pi}{8}(k-1)n} + e^{-j\frac{2\pi}{8}(k+1)n}\right) = 4\delta[k-1] + 4\delta[k-7] \end{aligned} \tag{3.47}$$

Prog. 3.10 MATLAB program for Fig. 3.16

```
syms w n;
N1 = 8;              % Length
nn1 = 0:1:(N1-1);
nn1e = -N1:2*N1-1;
ww1 = 0:pi/100:2*pi;
xn1 = cos(2*pi*nn1/8);
xn1e = [xn1 xn1 xn1];
bb1 = symsum(cos(2*pi*n/8)*exp(-1j*w*n),n,0,N1-1);
cc1 = simplify(bb1,'Steps',100);
dd1 = subs(cc1,w,ww1);
ee1 = abs(eval(dd1));
Xk1 = fft(xn1);

figure,
subplot(211)
stem(nn1e,xn1e), grid
xlabel('n')
ylabel('x[n]')
subplot(212)
plot(ww1,ee1), grid
ylabel('|X(e^{j\omega})|')
ylim([0 10])
hold on
yyaxis right
stem(nn1*2*pi/N1,abs(Xk1))
ylabel('|X[k]|')
ylim([0 10])
hold off
xlim([0 2*pi])
xticks(2*pi*(0:1:N1-1)/N1)
xticklabels({'0','1','2','3','4','5','6','7'})
xlabel('k')
```

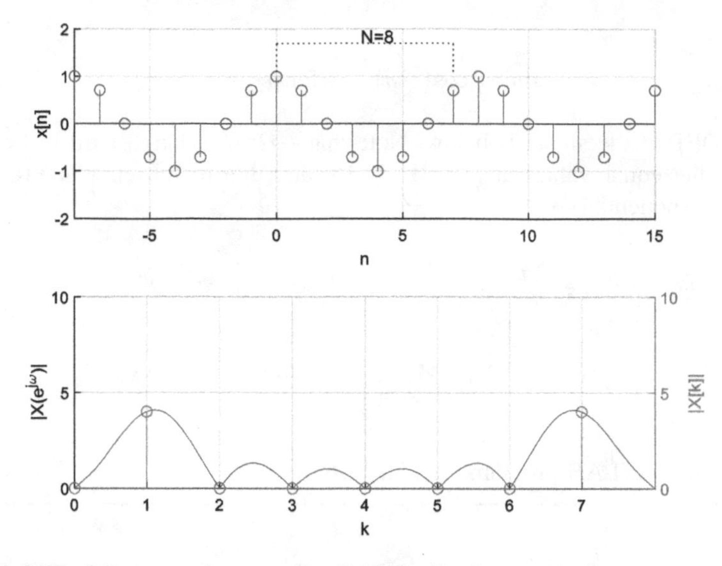

Fig. 3.16 DFT of discrete cosine wave Eq. (3.46) for $N = 8$

The DFT demonstrates the values only at 1 and 7 in k positions as shown in Fig. 3.16. The solid line in the magnitude plot represents the DTFT of the given 8 sample cosine function. The $k = 1$ denotes the $2\pi/8$ radian frequency which specifies the 8-sample period signal. Also, the $k = 7$ renders the $2\pi 7/8$ radian frequency that is equivalent to the $-2\pi/8$. The time sequence of the original signal is well expressed by the DFT since we can observe the clear division between the desired and undesired signal in frequency domain.

Now, we increase the signal length as 9.

$$X[k] = \sum_{n=0}^{8} \cos\left(\tfrac{2\pi}{8} n\right) e^{-j\frac{2\pi}{9}kn} \tag{3.48}$$

The DFT computation generates complex value results. In Fig. 3.17, certain insignificant values are observed along with prominent outcomes due to the frequency mismatch between the signal frequency and DFT frequencies. The given signal frequency is $2\pi/8$ located between the 1 and 2 for k value from DFT with length 9 in Eq. (3.49) and Fig. 3.17.

$$\tfrac{2\pi}{9}(k = 1) < \tfrac{2\pi}{8} < \tfrac{2\pi}{9} 2(k = 2) \tag{3.49}$$

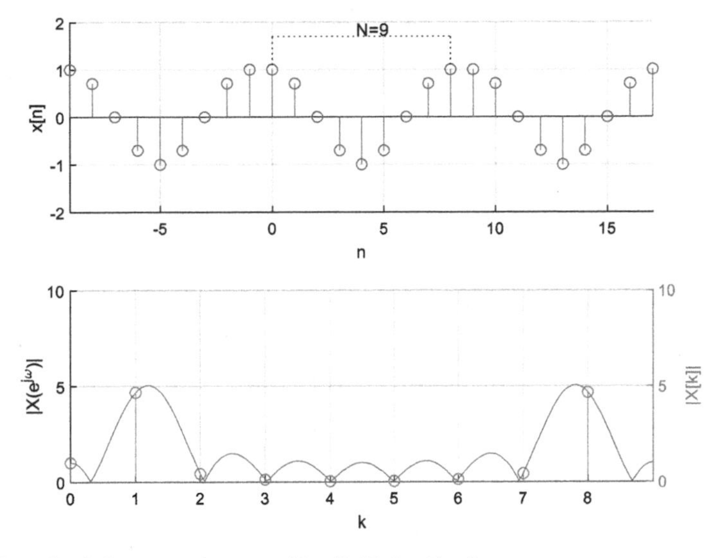

Fig. 3.17 DFT of discrete cosine wave Eq. (3.47) for $N = 9$

The figure matrix in Fig. 3.18 shows the DFT magnitude for length from 10 to 15. The solid line in the magnitude plot is the DTFT result. For length 9 outcome, the DTFT peaks present the actual signal frequencies but the DFT locates the depature frequencies due to the insufficient frequency sampling. This phenomenon delivers the worse frequency estimation as approaching the length 12 in terms of the ratio between the maximum and minimum of DFT magnitude. The 12/8 (1.5) of the k value corresponds to the $2\pi/8$ frequency for length 12 DFT. Further increasing the DFT length improves the frequency representation and the length 16 demonstrates the bestperformance as shown in Fig. 3.19. Note that the length 16 is the complete 2 period of the given cosine signal and the value 2 of the k presents the $2\pi/8$ frequency exactly.

Prog. 3.11 MATLAB program for Fig. 3.17. Identical to **Prog. 3.10** except $N1$ value

```
syms w n;
N1 = 9;            % Length
...
```

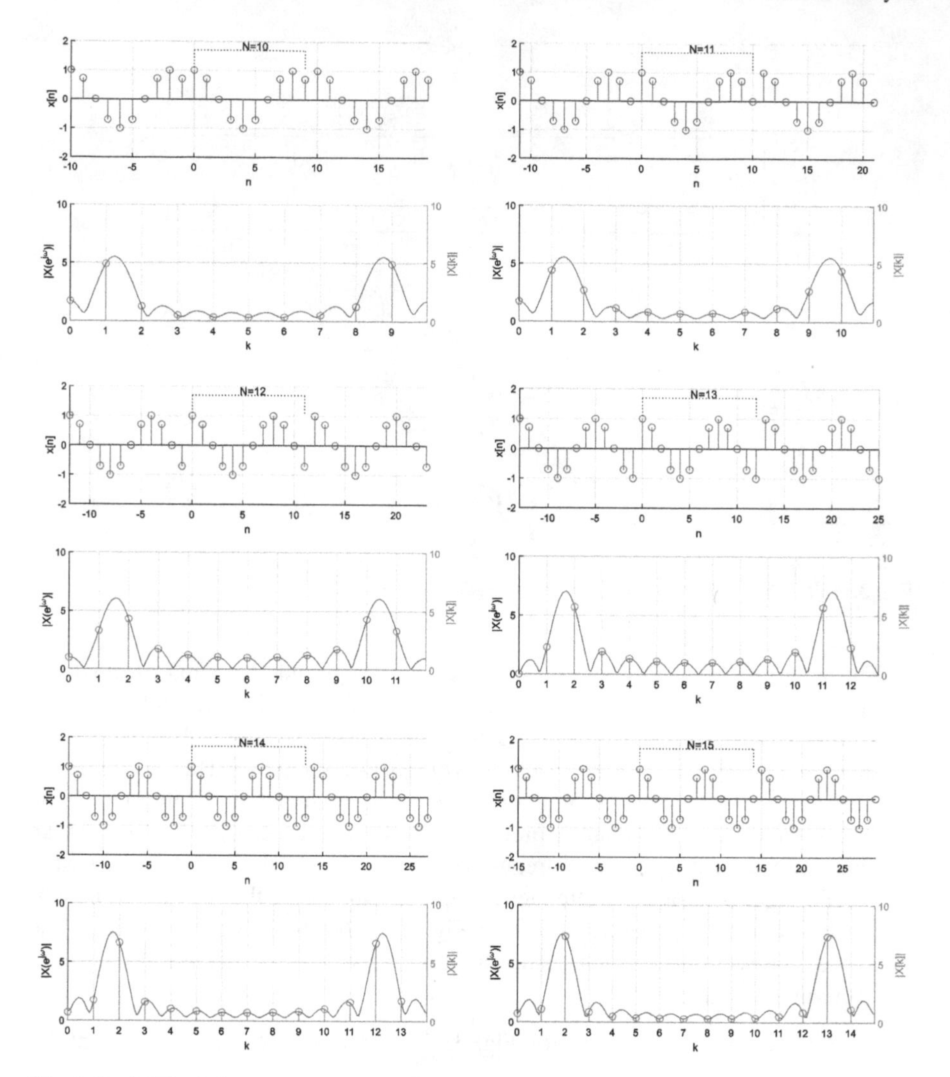

Fig. 3.18 DFT of discrete cosine wave Eq. (3.46) for N = 10,11,12,13,14, and 15 (Solid line indicates the DTFT). MATLAB program is realized by modifying the $N1$ value from Prog. 3.10

For the periodic signal, the longer DFT length is benefit to estimate the precise signal frequency because of the fine grain frequency resoultion by $2\pi/N$. The example below performs the 260 ($8 \times 32 + 4$) length DFT for the cosine function with period 8. The given case is the identically worst case as 12 length from Fig. 3.18 since the frame is finished at the period center. The $2\pi/8$ radian frequency corresponds to the 32.5 of k value as below in Fig. 3.20.

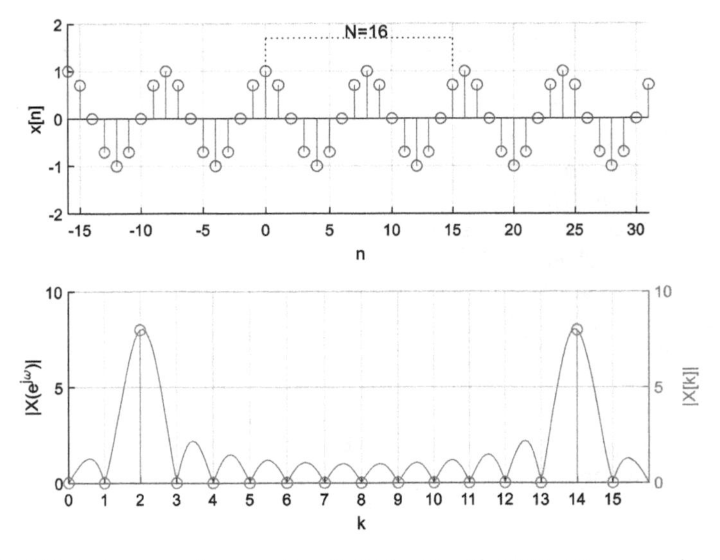

Fig. 3.19 DFT of discrete cosine wave Eq. (3.46) for $N = 16$. MATLAB program is realized by modifying the $N1$ value from Prog. 3.10

$$\tfrac{2\pi}{260}32(k=32) < \tfrac{2\pi}{8} < \tfrac{2\pi}{260}33(k=33) \tag{3.50}$$

Figure 3.20 shows the DFT magnitude for the 260 length signal to represent the $2\pi/8$ radian frequency. The DTFT result is denoted by the solid line and the DTFT peak is located at the center of the k values between 32 and 33. Compare to the DFT with length 12, longer length DFT provides the improved prominence for the signal representation.

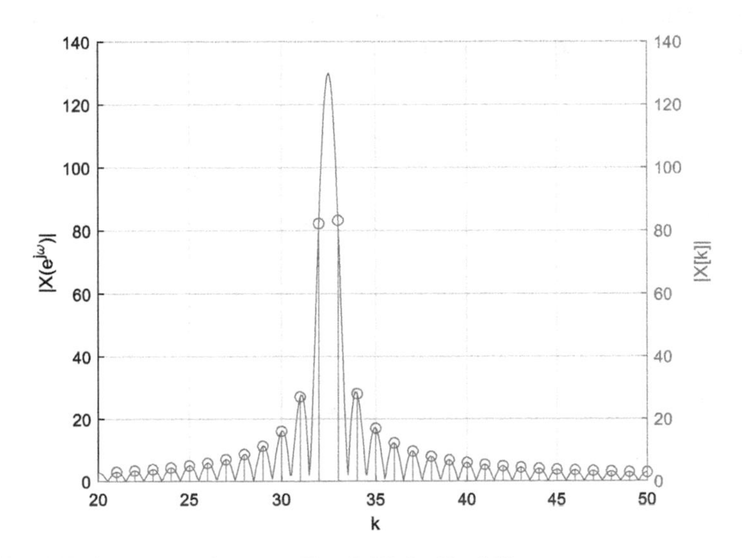

Fig. 3.20 DFT of discrete cosine wave Eq. (3.46) for $N = 260$

Prog. 3.12 MATLAB program for Fig. 3.20

```
syms w n;
N1 = 260;       % Length
nn1 = 0:1:(N1-1);
ww1 = 0:pi/2048:2*pi;
xn1 = cos(2*pi*nn1/8);
bb1 = symsum(cos(2*pi*n/8)*exp(-1j*w*n),n,0,N1-1);
cc1 = simplify(bb1,'Steps',100);
dd1 = subs(cc1,w,ww1);
ee1 = abs(eval(dd1));
Xk1 = fft(xn1);

figure,
plot(ww1,ee1), grid
ylabel('|X(e^{j\omega})|')
ylim([0 140])
hold on
yyaxis right
stem(nn1*2*pi/N1,abs(Xk1))
ylabel('|X[k]|')
ylim([0 140])
hold off
xlim([2*pi*20/N1 2*pi*50/N1])           % From 20 ~ 50 k value
xticks(2*pi*(0:5:N1)/N1)
xticklabels({'0','5','10','15','20','25','30','35','40','45','50','55','60','65','70'});
xlabel('k')
```

If you cannot have the longer periodic signal, zero padding improves the visual spectral resolution by $2\pi/N$. The N includes signal length as well as the zero padding length. The padded zeros are located after the signal sequence as shown in Fig. 3.21.

The attached zeros cannot change the DTFT computations since the only non-zero first signal portion is multiplied with complex exponential as below

$$X(e^{j\omega}) = \sum_{n=-\infty}^{\infty} x[n]e^{-j\omega n} \tag{3.51}$$

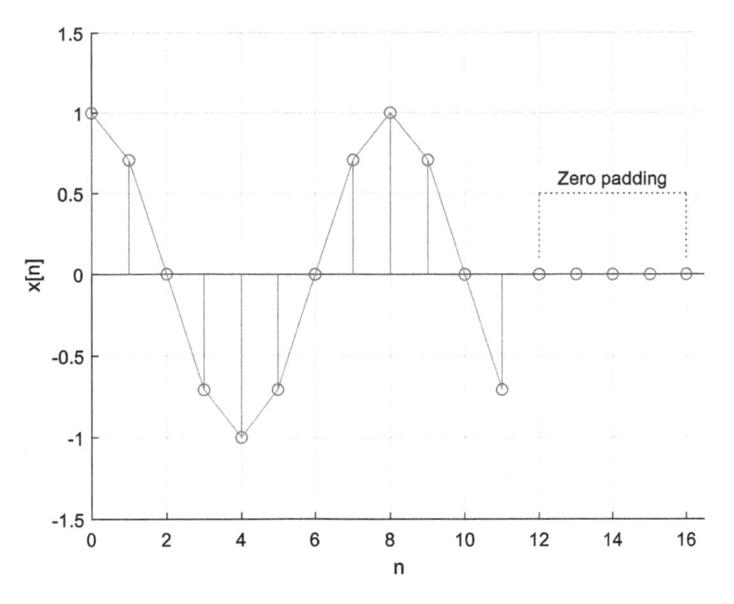

Fig. 3.21 Zero padding of discrete cosine wave Eq. (3.47) for $N = 17$

However, the DFT provides the finer resolution due to the increased overall signal length by zero padding as Eq. (3.52). The multiplication with complex exponential is identical to the DTFT computation but the exponential value is changed by the N for better representation.

$$X[k] = \sum_{n=0}^{N-1} x[n]e^{-j\frac{2\pi}{N}kn} \tag{3.52}$$

Figure 3.22 illustrates the DFT (stem plot) and DTFT (solid plot) results from cosine function for 8 period and 12 length with various zero padding conditions. Note that DTFT outcomes are invariant for all zero padding situations given below and the first plot is the no zero padding case for reference. The augmented signal size by zeros enhances the spetral representation randomly but persistently in length-wise manner. Provided that the system knows the signal or estimation frequency a priori, the designer intelligently selects the DFT parameters for precise detection and estimation. Otherwise, the longer is better in general.

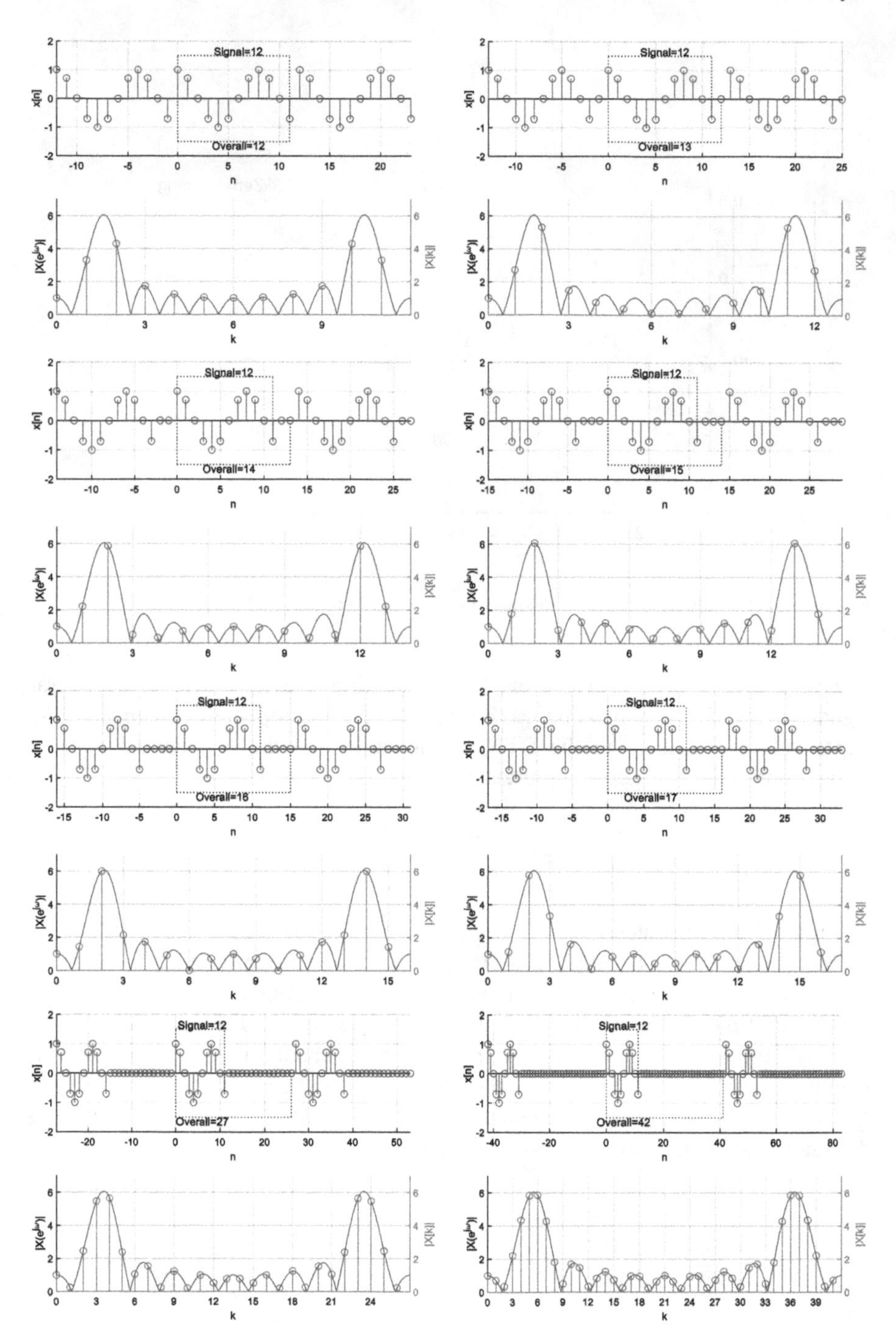

Fig. 3.22 DFT of discrete cosine wave Eq. (3.46) for 12 sample signal length ($0 \leq n \leq 11$) with various zero padding length (Solid line indicates the DTFT)

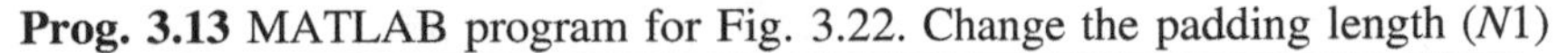

Prog. 3.13 MATLAB program for Fig. 3.22. Change the padding length ($N1$)

```
syms w n;
N = 12;                          % Signal length
N1 = 0;        % Padding size. Choose one [0 1 2 3 4 5 15 30]
nn1 = 0:1:(N-1);
ww1 = 0:pi/256:2*pi;
xn1 = cos(2*pi*nn1/8);
xn1e = [xn1 zeros(1,N1)];
N2 = length(xn1e);          % Length of Signal and padding
xn2 = [xn1e xn1e xn1e];
nn2 = -N2:1:N2*2-1;
nn3 = 0:1:N2-1;
bb1 = symsum(cos(2*pi*n/8)*exp(-1j*w*n),n,0,N-1);
cc1 = simplify(bb1,'Steps',100);
dd1 = subs(cc1,w,ww1);
ee1 = abs(eval(dd1));
Xk1 = fft(xn1e);

figure,
subplot(211)
stem(nn2,xn2), grid
xlabel('n')
ylabel('x[n]')
xlim([-N2 N2*2-1])
ylim([-2 2])
subplot(212)
plot(ww1,ee1), grid
ylabel('|X(e^{j\omega})|')
ylim([0 7])
hold on
yyaxis right
stem(nn3*2*pi/N2,abs(Xk1))
ylabel('|X[k]|')
ylim([0 7])
hold off
xlim([0 2*pi])
xticks(2*pi*(0:3:N2-1)/N2)
xticklabels({'0','3','6','9','12','15','18','21','24','27','30','33','36','39','42'})
xlabel('k')
```

The DFT exhibits unique shift pattern because of the signal periodicity. Let's consider the DFT computation for the triangular signal with length 8. The signal $x[n]$ and exponential $e^{j2\pi kn/8}$ component are decomposed in Table 3.1 for DFT computation.

Table 3.1 Triangular signal $x[n]$ with length 8 and DFT exponential part

n	−1	0	1	2	3	4	5	6	7	8
$x[n]$	0	1	2	3	4	5	4	3	2	0
$e^{-j\frac{2\pi}{8}kn}$	$e^{j\frac{2\pi}{8}k1}$	$e^{-j\frac{2\pi}{8}k0}$	$e^{-j\frac{2\pi}{8}k1}$	$e^{-j\frac{2\pi}{8}k2}$	$e^{-j\frac{2\pi}{8}k3}$	$e^{-j\frac{2\pi}{8}k4}$	$e^{-j\frac{2\pi}{8}k5}$	$e^{-j\frac{2\pi}{8}k6}$	$e^{-j\frac{2\pi}{8}k7}$	$e^{-j\frac{2\pi}{8}k8}$

$$X_1[k] = \sum_{n=0}^{7} x[n]e^{-j\frac{2\pi}{8}kn} \tag{3.53}$$

For the time being, $x[n]$ is not a periodic signal; therefore, the 8 sample sequence leaves the empty values beyond signal length in Table 3.1. The signal is shifted to the right by one sample as $x[n-1]$ and the DFT range follows the sequence span from 1 to 8 as shown in Eq. (3.54).

$$X_2[k] = \sum_{n=1}^{8} x[n-1]e^{-j\frac{2\pi}{8}kn} \tag{3.54}$$

In the Table 3.2, the last sample on the shifted $x[n]$ is multiplied by $e^{j2\pi k8/8}$ that is equivalen to the $e^{j2\pi k0/8}$. The last sample is relocated to the corresponding location as zero index in Table 3.3 and the DFT range is rearranged for signal span in Eq. (3.55).

$$X_2[k] = \sum_{n=0}^{7} x_8[n-1]e^{-j\frac{2\pi}{8}kn} \tag{3.55}$$

Table 3.2 Shifted signal $x[n-1]$ and DFT exponential part

c	−1	0	1	2	3	4	5	6	7	8
x[n−1]	0	0	1	2	3	4	5	4	3	2
$e^{-j\frac{2\pi}{8}kn}$	$e^{j\frac{2\pi}{8}k1}$	$e^{-j\frac{2\pi}{8}k0}$	$e^{-j\frac{2\pi}{8}k1}$	$e^{-j\frac{2\pi}{8}k2}$	$e^{-j\frac{2\pi}{8}k3}$	$e^{-j\frac{2\pi}{8}k4}$	$e^{-j\frac{2\pi}{8}k5}$	$e^{-j\frac{2\pi}{8}k6}$	$e^{-j\frac{2\pi}{8}k7}$	$e^{-j\frac{2\pi}{8}k8}$

Table 3.3 Relocated and shifted signal $x[n-1]$ and DFT exponential part

n	−1	0	1	2	3	4	5	6	7	8
$x[n-1]$	0	2	1	2	3	4	5	4	3	0
$e^{-j\frac{2\pi}{8}kn}$	$e^{j\frac{2\pi}{8}k1}$	$e^{-j\frac{2\pi}{8}k0}$	$e^{-j\frac{2\pi}{8}k1}$	$e^{-j\frac{2\pi}{8}k2}$	$e^{-j\frac{2\pi}{8}k3}$	$e^{-j\frac{2\pi}{8}k4}$	$e^{-j\frac{2\pi}{8}k5}$	$e^{-j\frac{2\pi}{8}k6}$	$e^{-j\frac{2\pi}{8}k7}$	$e^{-j\frac{2\pi}{8}k8}$

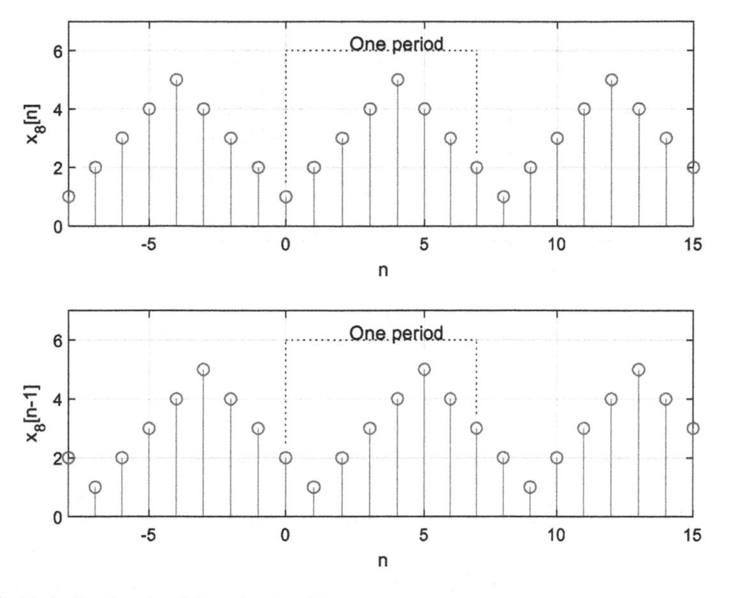

Fig. 3.23 Periodic signal $x_8[n]$ and $x_8[n-1]$

In the DFT computation, the signal is processed and assumed to be periodic with DFT length N. The edge of the signal is moved to the one end and appeared to the other end sequentially. The given $x_8[n]$ and $x_8[n-1]$ are illustrated for the wider range of sequnce in Fig. 3.23.

The most of the DTFT properties are satisfited with DFT. The DFT is the linear operator as below.

$$z[n] = \alpha x[n] + \beta y[n]$$

$$Z[k] = \sum_{n=0}^{N-1} \{\alpha x[n] + \beta y[n]\} e^{-j\frac{2\pi}{N}kn} = \alpha X[k] + \beta Y[k] \tag{3.56}$$

Also, synthesis equation is satisfying the linearity as following.

$$Z[k] = \alpha X[k] + \beta Y[k]$$

$$z[n] = \frac{1}{N} \sum_{n=0}^{N-1} \{\alpha X[k] + \beta Y[k]\} e^{j\frac{2\pi}{N}kn} = \alpha x[n] + \beta y[n] \tag{3.57}$$

The time shift can be represented in the frequency domain as below.

$$\sum_{n=0}^{N-1} x_N[n-d] e^{-j\frac{2\pi}{N}kn} = \sum_{n=0}^{N-1} x_N[n] e^{-j\frac{2\pi}{N}kn} e^{-j\frac{2\pi}{N}kd} = X[k] e^{-j\frac{2\pi}{N}kd} \tag{3.58}$$

Table 3.4 DFT shift property by example

	n	0	1	2	3	4	5	6	7
DFT$\{x_8[n]\}$	$x_8[n]$	$x_8[0]$	$x_8[1]$	$x_8[2]$	$x_8[3]$	$x_8[4]$	$x_8[5]$	$x_8[6]$	$x_8[7]$
	$e^{-j\frac{2\pi}{8}kn}$	$e^{-j\frac{2\pi}{8}k0}$	$e^{-j\frac{2\pi}{8}k1}$	$e^{-j\frac{2\pi}{8}k2}$	$e^{-j\frac{2\pi}{8}k3}$	$e^{-j\frac{2\pi}{8}k4}$	$e^{-j\frac{2\pi}{8}k5}$	$e^{-j\frac{2\pi}{8}k6}$	$e^{-j\frac{2\pi}{8}k7}$
DFT$\{x_8[n-2]\}$	$x_8[n-2]$	$x_8[6]$	$x_8[7]$	$x_8[0]$	$x_8[1]$	$x_8[2]$	$x_8[3]$	$x_8[4]$	$x_8[5]$
	$e^{-j\frac{2\pi}{8}kn}$	$e^{-j\frac{2\pi}{8}k0}$	$e^{-j\frac{2\pi}{8}k1}$	$e^{-j\frac{2\pi}{8}k2}$	$e^{-j\frac{2\pi}{8}k3}$	$e^{-j\frac{2\pi}{8}k4}$	$e^{-j\frac{2\pi}{8}k5}$	$e^{-j\frac{2\pi}{8}k6}$	$e^{-j\frac{2\pi}{8}k7}$
DFT$\{x_8[n-2]\}$	$x_8[n]$	$x_8[0]$	$x_8[1]$	$x_8[2]$	$x_8[3]$	$x_8[4]$	$x_8[5]$	$x_8[6]$	$x_8[7]$
	$e^{-j\frac{2\pi}{8}kn}e^{-j\frac{2\pi}{8}k2}$	$e^{-j\frac{2\pi}{8}k2}$	$e^{-j\frac{2\pi}{8}k3}$	$e^{-j\frac{2\pi}{8}k4}$	$e^{-j\frac{2\pi}{8}k5}$	$e^{-j\frac{2\pi}{8}k6}$	$e^{-j\frac{2\pi}{8}k7}$	$e^{-j\frac{2\pi}{8}k8}$ $e^{-j\frac{2\pi}{8}k0}$	$e^{-j\frac{2\pi}{8}k9}$ $e^{-j\frac{2\pi}{8}k1}$

The explanation of the DFT shift property is presented by the example with table method. The $x[n]$ value and corresponding complex exponential are denoted in each column of Table 3.4. The middle rows provide the DFT of the shifted signal by 2 samples. The bottom rows deliver the shifted DFT by exponential multiplication without relocating the signal according to the Eq. (3.58). Note that the complex exponential equivalence is indicated by the colored circles in Table 3.4.

The shift in frequency domain is shown in Eq. (3.59). The $X[k]$ is also the periodic distribution with N sequence and the description for frequency shift can be realized by the identical table method shown in Table 3.4.

$$\frac{1}{N}\sum_{n=0}^{N-1} X[k-l]e^{j\frac{2\pi}{N}kn} = \frac{1}{N}\sum_{n=0}^{N-1} X[k]e^{j\frac{2\pi}{N}kn}e^{j\frac{2\pi}{N}ln} = x_N[n]e^{j\frac{2\pi}{N}nl} \tag{3.59}$$

Previously stated that the given real sequence requires the conjugated complex numbers for the positive and negative frequency pair according to the Euler equation for cosine function.

$$X[k] = X^*[-k] \quad \text{or} \quad X[k] = X^*[N-k] \tag{3.60}$$

Example 3.9

Find the number property as complex or real number for following signal's DFT. $N = 8$.

$$\begin{aligned} x_1[n] = \delta[n] + 2\delta[n-1] + 3\delta[n-2] + 4\delta[n-3] \\ + 5\delta[n-4] + 4\delta[n-5] + 3\delta[n-6] + 2\delta[n-7] \end{aligned}$$

$$x_2[n] = x_1[n-2]$$

Solution

$$\begin{aligned}
X_1[k] &= \sum_{n=0}^{7} x_1[n]e^{-j\frac{2\pi}{8}kn} \\
&= 1 + 2e^{-j\frac{2\pi}{8}k} + 3e^{-j\frac{2\pi}{8}k2} + 4e^{-j\frac{2\pi}{8}k3} + 5e^{-j\frac{2\pi}{8}k4} \\
&\quad + 4e^{-j\frac{2\pi}{8}k5} + 3e^{-j\frac{2\pi}{8}k6} + 2e^{-j\frac{2\pi}{8}k7} \\
&= 1 + 2e^{-j\frac{2\pi}{8}k} + 3e^{-j\frac{2\pi}{8}k2} + 4e^{-j\frac{2\pi}{8}k3} + 5e^{-j\frac{2\pi}{8}k4} \\
&\quad + 4e^{-j\frac{2\pi}{8}k(8-3)} + 3e^{-j\frac{2\pi}{8}k(8-2)} + 2e^{-j\frac{2\pi}{8}k(8-1)} \\
&= 1 + 2e^{-j\frac{2\pi}{8}k} + 3e^{-j\frac{2\pi}{8}k2} + 4e^{-j\frac{2\pi}{8}k3} + 5e^{-j\frac{2\pi}{8}k4} \\
&\quad + 4e^{j\frac{2\pi}{8}k3} + 3e^{j\frac{2\pi}{8}k2} + 2e^{j\frac{2\pi}{8}k} \\
&= 1 + 2\left(e^{j\frac{2\pi}{8}k} + e^{-j\frac{2\pi}{8}k}\right) + 3\left(e^{j\frac{2\pi}{8}k2} + e^{-j\frac{2\pi}{8}k2}\right) \\
&\quad + 4\left(e^{j\frac{2\pi}{8}k3} + e^{-j\frac{2\pi}{8}k3}\right) + 5e^{-j\frac{2\pi}{8}k4} \\
&= 1 + 4\cos\left(\frac{2\pi}{8}k\right) + 6\cos\left(\frac{2\pi}{8}2k\right) + 8\cos\left(\frac{2\pi}{8}3k\right) + 5(-1)^k \in \mathbb{R}
\end{aligned}$$

The $X_1[k]$ is real number because $x_1[n]$ is symmetric as below.

$$x_1[n] = x_1[8-n]$$

Note that any even symmetric distribution provides the real number on the other domain. Using the property of delay, $X_2[k]$ is below.

$$x_N[n-d] \overset{\text{DFT}}{\leftrightarrow} X[k]e^{-j\frac{2\pi}{N}kd}$$

$$x_2[n] = x_1[n-2] \overset{\text{DFT}}{\leftrightarrow}$$

$$X_2[k] = \left(1 + 4\cos\left(\frac{2\pi}{8}k\right) + 6\cos\left(\frac{2\pi}{8}2k\right) + 8\cos\left(\frac{2\pi}{8}3k\right) + 5(-1)^k\right)e^{-j\frac{2\pi}{8}k2}$$

The $X_2[k]$ is complex number because of $e^{-j\frac{2\pi}{8}k2}$ term which is induced by delay.■

The reversal in time provides the reversal in frequency as well. See Table 3.5.

$$\begin{aligned}
\sum_{n=0}^{N-1} x[-n]e^{-j\frac{2\pi}{N}kn} &= \sum_{n=0}^{N-1} x[N-n]e^{-j\frac{2\pi}{N}kn} = \sum_{n=0}^{N-1} x[n]e^{-j\frac{2\pi}{N}(N-k)n} \\
&= \sum_{n=0}^{N-1} x[n]e^{j\frac{2\pi}{N}(-k)n} = X[-k] - X[N-k]
\end{aligned} \tag{3.61}$$

According to the Parseval's theorem, the energy in time domain is identical to the energy in frequency domain.

Table 3.5 Time reversal example in time domain for DFT

	n	0	1	2	3	4	5	6	7
$\mathrm{DFT}\{x_8[n]\}$	$x_8[n]$	$x_8[0]$	$x_8[1]$	$x_8[2]$	$x_8[3]$	$x_8[4]$	$x_8[5]$	$x_8[6]$	$x_8[7]$
	$e^{-j\frac{2\pi}{8}kn}$	$e^{-j\frac{2\pi}{8}k0}$	$e^{-j\frac{2\pi}{8}k1}$	$e^{-j\frac{2\pi}{8}k2}$	$e^{-j\frac{2\pi}{8}k3}$	$e^{-j\frac{2\pi}{8}k4}$	$e^{-j\frac{2\pi}{8}k5}$	$e^{-j\frac{2\pi}{8}k6}$	$e^{-j\frac{2\pi}{8}k7}$
$\mathrm{DFT}\{x_8[-n]\}$	$x_8[-n]$	$x_8[0]$	$x_8[-1]$	$x_8[-2]$	$x_8[-3]$	$x_8[-4]$	$x_8[-5]$	$x_8[-6]$	$x_8[-7]$
		$x_8[0]$	$x_8[7]$	$x_8[6]$	$x_8[5]$	$x_8[4]$	$x_8[3]$	$x_8[2]$	$x_8[1]$
	$e^{-j\frac{2\pi}{8}kn}$	$e^{-j\frac{2\pi}{8}k0}$	$e^{-j\frac{2\pi}{8}k1}$	$e^{-j\frac{2\pi}{8}k2}$	$e^{-j\frac{2\pi}{8}k3}$	$e^{-j\frac{2\pi}{8}k4}$	$e^{-j\frac{2\pi}{8}k5}$	$e^{-j\frac{2\pi}{8}k6}$	$e^{-j\frac{2\pi}{8}k7}$
$\mathrm{DFT}\{x_8[-n]\}$	$x_8[n]$	$x_8[0]$	$x_8[1]$	$x_8[2]$	$x_8[3]$	$x_8[4]$	$x_8[5]$	$x_8[6]$	$x_8[7]$
	$e^{-j\frac{2\pi}{8}kn}e^{j\frac{2\pi}{8}k8}$	$e^{-j\frac{2\pi}{8}k8}$	$e^{-j\frac{2\pi}{8}k7}$	$e^{-j\frac{2\pi}{8}k6}$	$e^{-j\frac{2\pi}{8}k5}$	$e^{-j\frac{2\pi}{8}k4}$	$e^{-j\frac{2\pi}{8}k3}$	$e^{-j\frac{2\pi}{8}k2}$	$e^{-j\frac{2\pi}{8}k1}$
	$e^{-j\frac{2\pi}{8}(8-k)n}$	$e^{-j\frac{2\pi}{8}k0}$							

$$E_x = \sum_{n=0}^{N-1} |x[n]|^2 = \frac{1}{N}\sum_{k=0}^{N-1} |X[k]|^2$$

Let's place one example to this theorem.

Example 3.10
Apply the Parseval's theorem on below example.

$$x[n] = \cos\left(\frac{2\pi}{4}n\right) \quad \text{for } N = 4$$

Solution The time domain energy is below.

$$E_x = \sum_{n=0}^{3} |x[n]|^2 = 1 + 0 + 1 + 0 = 2$$

The frequency domain distribution and energy are below

$$X[k] = \sum_{n=0}^{3} x[n]e^{-j\frac{2\pi}{4}kn} = 1e^{-j\frac{2\pi}{4}k0} + 0e^{-j\frac{2\pi}{4}k1} - 1e^{-j\frac{2\pi}{4}k2} + 0e^{-j\frac{2\pi}{4}k3}$$
$$= 1 - (-1)^k = 2\delta[k-1] + 2\delta[\mathrm{k}-3]$$

$$E_x = \frac{1}{4}\sum_{k=0}^{3} |X[k]|^2 = \frac{1}{4}(4+4) = 2$$

Therefore, energy can be computed in either domain for your computational convenience. ■

3.5 Problems

1. Show that the even and odd function provides the pure real and imaginary number DTFT, respectively. Use the cosine and sine function without any phase delay.
2. Prove following relationship. What condition is required for below relationship?

$$x[n] = \frac{1}{2\pi}\int_{-\pi}^{\pi} X\left(e^{j\omega}\right)e^{j\omega n}d\omega = \frac{1}{2\pi}\int_{0}^{\pi} m(\omega)\cos(\omega n + \theta(\omega))d\omega \in \mathbb{R}$$
$$= \frac{1}{2\pi}\int_{0}^{\pi} a(\omega)\cos(\omega n)d\omega - \frac{1}{2\pi}\int_{0}^{\pi} b(\omega)\sin(\omega n)d\omega$$

3. Compute the DTFT frequency distribution of following signals.
   - $u[n]$
   - $\sum_{k=-\infty}^{\infty} \delta[n-kM]$
   - $(-1)^n$
4. If the given signal $x[n]$ shows the period $N = 16$ samples. List the all possible frequencies of $X(e^{j\omega})$.
5. The below is the given discrete time signal.

$$x[n] = \delta[n+2] + 2\delta[n+1] + 3\delta[n] + 2\delta[n-1] + \delta[n-2]$$

   - $x[n] \xrightarrow{\text{DTFT}} X(e^{j\omega})$
   - $X(e^{j\omega}) \xrightarrow{\text{Inverse DTFT}} x[n]$
   - Apply Parseval's theorem to compute the energy in both domains.
6. Derive the inverse DTFT of $X(e^{j\omega})$ given below.

$$X(e^{j\omega}) = \begin{cases} 1 & \text{for } -\frac{\pi}{2} < \omega < \frac{\pi}{2} \\ 0 & \text{Otherwise} \end{cases}$$

$$X(e^{j\omega}) = \frac{1}{1 - 0.1e^{-j\omega}} \quad \text{Hint : Geometric series with infinite length}$$

7. The $x[n]$ is obtained by the sampling of $x_c(t)$ with sampling period T_s. The DTFT of $x[n]$ is $X(e^{j\omega})$ and the FT of $x_c(t)$ is $X_c(\Omega)$. Find the relation between $X(e^{j\omega})$ and $X_c(\Omega)$. Most of the procedures are shown below. Follow the procedures and explain the relation in last part.

   The sampled signal can be represented by following definition.

$$x[n] = x_c(nT_s)$$

The continuous time signal $x_c(t)$ and its frequency distribution $X_c(\Omega)$ are given below by using Fourier transform.

$$x_c(t) \overset{\text{FT}}{\leftrightarrow} X_c(\Omega)$$

The sampling process is the multiplication between the continuous signal with impulse train.

$$x_s(t) = x_c(t)s(t) \text{ where } s(t) = \sum_{n=-\infty}^{\infty} \delta(t - nT_s)$$

The frequency distribution of impulse train $s(t)$ is known as below based on Fourier series. As discussed in DFT, the periodic signal only contains the spectrum with the integer multiple of fundamental frequency which is the inverse of the period. The impulse train also presents the periodic signal with T_s; therefore, the spectrum of the train denotes the integer multiple of Ω_s $(= 2\pi f_s = 2\pi/T_s)$.

$$s(t) = \sum_{n=-\infty}^{\infty} \delta(t - nT_s) = \sum_{k=-\infty}^{\infty} S_k e^{jk\Omega_s t} \overset{\text{FT}}{\leftrightarrow}$$

$$S(\Omega) = 2\pi \sum_{k=-\infty}^{\infty} S_k \delta(\Omega - k\Omega_s) \text{ where } \Omega_s = \frac{2\pi}{T_s}$$

Using Fourier series, find the S_k coefficient as below.

$$S_k = \tfrac{1}{T_s}\int_0^{T_s} s(t)e^{-jk\Omega_s t}dt = \tfrac{1}{T_s}\int_0^{T_s} \sum_{n=-\infty}^{\infty} \delta(t - nT_s)e^{-jk\Omega_s t}dt$$
$$= \tfrac{1}{T_s}\int_0^{T_s} \delta(t)e^{-jk\Omega_s t}dt = \tfrac{1}{T_s}$$

The frequency distribution of impulse train is demonstrated as below.

$$S(\Omega) = \frac{2\pi}{T_s} \sum_{k=-\infty}^{\infty} \delta(\Omega - k\Omega_s)$$

The time domain multiplication is equivalent to the convolution integral in frequency domain according to the duality.

$$x_c(t)s(t) \overset{\text{FT}}{\leftrightarrow} \frac{1}{2\pi} X_c(\Omega) * S(\Omega) = X_s(\Omega)$$

The convolution integral with delta function relocates the spectrum to the delta function locations as below.

$$X_s(\Omega) = \frac{1}{T_s} \sum_{k=-\infty}^{\infty} X_c(\Omega - k\Omega_s)$$

Directly perform the Fourier transform on $x(t)s(t)$ as below.

$$X_s(\Omega) = \int_{-\infty}^{\infty} x_s(t)e^{-j\Omega t}dt = \int_{-\infty}^{\infty} x_c(t) \sum_{n=-\infty}^{\infty} \delta(t - nT_s)e^{-j\Omega t}dt$$
$$= \sum_{n=-\infty}^{\infty} x_c(nT_s)e^{-j\Omega nT_s} \int_{-\infty}^{\infty} \delta(t - nT_s)dt = \sum_{n=-\infty}^{\infty} x_c(nT_s)e^{-j\Omega nT_s}$$
$$= X\left(e^{j\Omega T_s}\right) = \sum_{n=-\infty}^{\infty} x_c(nT_s)e^{-j\omega n} = X\left(e^{j\omega}\right) \text{ where } \omega = \Omega T_s \text{ and } x[n] = x_c(nT_s)$$

$$\therefore X\left(e^{j\Omega T_s}\right) = \frac{1}{T_s}\sum_{k=-\infty}^{\infty} X_c(\Omega - k\Omega_s)$$

$$\therefore X\left(e^{j\omega}\right) = \frac{1}{T_s}\sum_{k=-\infty}^{\infty} X_c\left(\frac{\omega}{T_s} - k\frac{2\pi}{T_s}\right)$$

Complete following drawings based on the above equations.

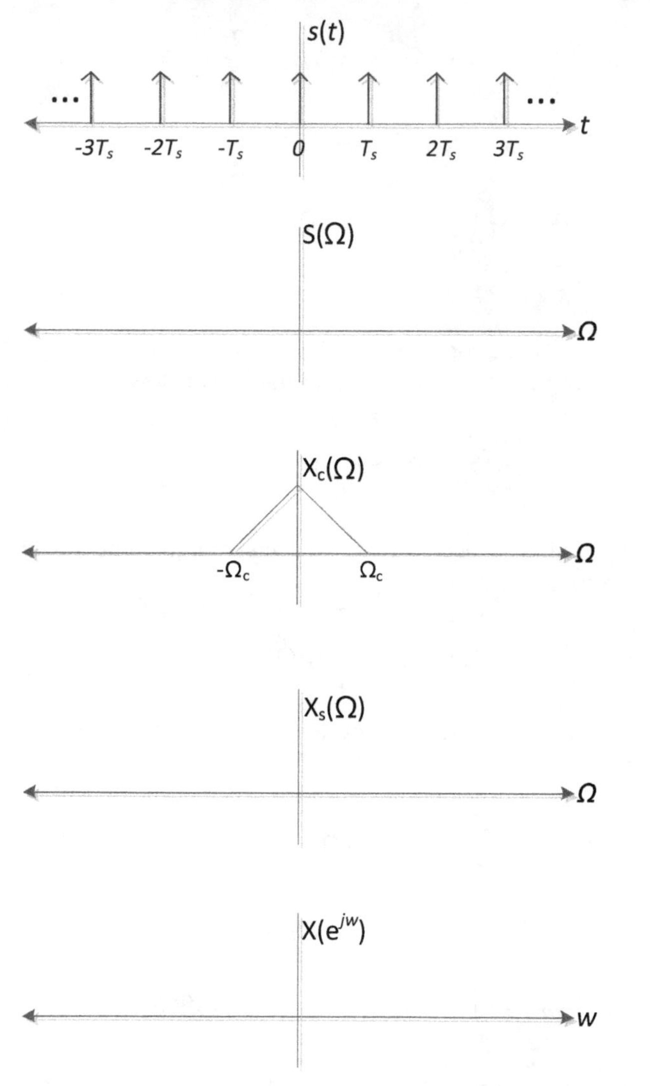

8. Compute following equation.
   $\sum_{n=0}^{N-1} e^{-j\frac{2\pi}{N}kn} =$
9. Find the DFT of following signals.
   - $x_1[n] = \delta[n] + \delta[n-4]$ for $N = 8$
   - $x_2[n] = \cos(\omega n)$ for $\omega \in \mathbb{R}$ and $N \in \mathbb{N}$
   - $x_3[n] = 2\cos^2\left(\frac{2\pi}{8}n\right)$ for $N = 8$
10. In general, the given signal is non-periodic. The only numerical computation for frequency distribution is DFT because of digital frequency k. However, the DFT is designed for applying on the periodic signal. By using the DFT, describe the method to find the approximated DTFT distribution for below two situations. Explain with example. MATLAB code is recommended.
    - $x[n]$ with limited length
    - $x[n]$ with unlimited length
11. Fill out the below DFT table for $x[n-4]$ with $N = 8$.

n	0	1	2	3	4	5	6	7
$x[n]$	1	2	3	4	5	4	3	2
$x[n-3]$								

12. From the Problem 10, derive following.
    - $X[k]$
    - $\mathrm{DFT}(x[n-4])$
    - $\mathrm{DFT}\left(x[n]\cos\left(\frac{2\pi}{8}3n\right)\right)$
13. The given signal is periodic impulse train with 6 sample period. Apply the Parseval's theorem.

$$x_6[n] = \delta_6[n] = \sum_{m=-\infty}^{\infty} \delta[n-6m]$$

14. Derive the inverse DFT of $X[k]$ given below. DFT length is N.

$$X[k] = \begin{cases} 1 & \text{for } 0 \le k \le L \\ 1 & \text{for } N-L \le k < L \\ 0 & \text{Otherwise} \end{cases}$$

Assume that $L \ll N$

15. Determine the inverse DFT of following.

$$x_N[n] \overset{\text{DFT}}{\leftrightarrow} X[k]$$
$$\overset{\text{DFT}}{\leftrightarrow} X^*[k]$$

$$X[k] = \sum_{n=0}^{N-1} x_N[n] e^{-j\frac{2\pi}{N}kn}$$
$$X^*[k] = \left(\sum_{n=0}^{N-1} x_N[n] e^{-j\frac{2\pi}{N}kn}\right)^* = \sum_{n=0}^{N-1} x_N^*[n] e^{j\frac{2\pi}{N}kn} = \sum_{n=0}^{N-1} x_N^*[-n] e^{-j\frac{2\pi}{N}k(-n)}$$
$$x_N^*[-n] = x_N^*[N-n] \overset{\text{DFT}}{\leftrightarrow} X^*[k]$$

References

1. Oppenheim, A.V., Schafer, R.W.: Discrete-time Signal Processing. Prentice Hall (1989)
2. Hazewinkel, M.: Encyclopaedia of Mathematics: Orbit—Rayleigh Equation. Springer, Netherlands (2012)

Chapter 4
Filters in Time Domain

In the first chapter, we discussed about the filter architecture and its fundamental theories without using the any frequency analysis. Basically, the filter generates the output by reducing and amplifying the range of frequencies; therefore, the frequency analysis on the filter is essential part of the filter design. The Chap. 3 provides powerful tools known as Fourier analysis to understand the frequency information for periodic and aperiodic discrete signal. Now it is time to revisit the filter computation and extend the time domain understanding to the frequency domain knowledge. Also, the filter will be further expanded to the recursive filters that uses the output feedback. The simple various filters such as low pass and high pass filter will be derived from the window method as well.

4.1 Digital Filter Revisited

The discrete time signal is the real time signal that provides the signal value in every time interval. Digital filter receives the discrete time signal for modifying the frequency magnitude and delay for desired output. The filter utilizes the recent signal samples by weighting and accumulating as below.

$$y[n] = \sum_{k=-\infty}^{\infty} x[k]h[n-k] \tag{4.1}$$

Equation (4.1) is the convolution sum for discrete time signal. The graphical explanation is shown in Fig. 4.1.

The signal $x[k]$ is the received signal up to the present time n. The future signal values at $n+1$, $n+2$, etc. are shown in Fig. 4.1 but not actually received by the

K. Kim, *Conceptual Digital Signal Processing with MATLAB*,
Signals and Communication Technology,
https://doi.org/10.1007/978-981-15-2584-1_4

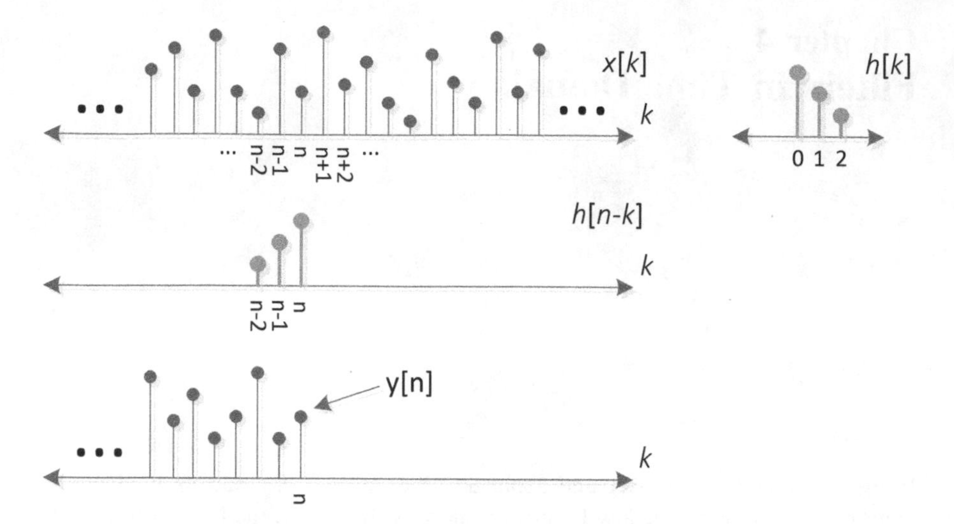

Fig. 4.1 Graphical explanation of convolution sum

filter system. The filter coefficients $h[k]$ is flipped and shifted to the n position as $h[n\text{-}k]$. Note that the shift direction for the $h[n-k]$ is reversed due to the time reversal by $-k$. The multiplication is performed on the input and coefficients for the corresponding location in order to produce the output $y[n]$. The shape and length of the filter coefficients decide the filter function such as low, high, and band pass filter. The output shape follows the profile of the filter coefficients. The filter length determines the frequency focus in inversely proportional manner. The Chap. 1 provided the intuitive filter design stated above. The following sections investigate further for analytical filter design.

4.2 Filter Properties with LTI

Section 1.4 showed that the impulse response is the filter coefficient $h[n]$ if the system meets the linear and time invariant condition. Let's explore the convolution sum equation for linearity condition first. The given signal is below.

$$x[n] = \alpha x_1[n] + \beta x_2[n] \tag{4.2}$$

The convolution sum is performed for the output $y[n]$.

$$
\begin{aligned}
y[n] &= \sum_{k=-\infty}^{\infty} x[k]h[n-k] \\
&= \sum_{k=-\infty}^{\infty} \{\alpha x_1[k] + \beta x_2[k]\}h[n-k] \\
&= \alpha \sum_{k=-\infty}^{\infty} x_1[k]h[n-k] + \beta \sum_{k=-\infty}^{\infty} x_2[k]h[n-k] \\
&= \alpha y_1[n] + \beta y_2[n]
\end{aligned}
\tag{4.3}
$$

$$
\text{where } y_1[n] = \sum_{k=-\infty}^{\infty} x_1[k]h[n-k] \text{ and } y_2[n] = \sum_{k=-\infty}^{\infty} x_2[k]h[n-k]
$$

The linearly combined input produces the output that is equivalent to the identically scaled individual outputs; hence, the convolution sum fulfills the linearity condition. See the next condition time invariance. The delayed signal is below.

$$
x[n] = x_1[n-d] \tag{4.4}
$$

The convolution sum is performed for the output $y[n]$.

$$
\begin{aligned}
y[n] &= \sum_{k=-\infty}^{\infty} x[k]h[n-k] \\
&= \sum_{k=-\infty}^{\infty} x_1[k-d]h[n-k] = \sum_{m+d=-\infty}^{\infty} x_1[m]h[n-m-d] \\
&= \sum_{m=-\infty}^{\infty} x_1[m]h[(n-d)-m] = y_1[n-d]
\end{aligned}
\tag{4.5}
$$

$$
\text{where } y_1[n] = \sum_{k=-\infty}^{\infty} x_1[k]h[n-k]
$$

The input of the delayed signal delivers the equally retarded output signal without modifying output shape. Therefore, the convolution sum satisfies the time invariance condition. Finally, the LTI is completed for the convolution sum. Section 2.3 described the discrete time signal with linear combination of scaled and shifted delta functions as shown in Eq. (4.6).

$$x[n] = \sum_{k=-\infty}^{\infty} x[k]\delta[n-k] \tag{4.6}$$

The decomposed and combined signal is placed on the input of the convolution sum as below.

$$\begin{aligned} y[n] &= \sum_{k=-\infty}^{\infty} x[k]h[n-k] \\ &= \sum_{k=-\infty}^{\infty}\sum_{l=-\infty}^{\infty} x[l]\delta[k-l]h[n-k] = \sum_{l=-\infty}^{\infty} x[l]\sum_{k=-\infty}^{\infty}\delta[k-l]h[n-k] \\ &= \sum_{l=-\infty}^{\infty} x[l]h[n-l]\sum_{k=-\infty}^{\infty}\delta[k-l] = \sum_{l=-\infty}^{\infty} x[l]h[n-l] \end{aligned} \tag{4.7}$$

Note that the impulse signal (delta function) provides the impulse response as $h[n]$. The shifted impulse generates the equally shifted impulse response as below.

$$h[n] = \sum_{k=-\infty}^{\infty} \delta[k]h[n-k] \text{ and } h[n-l] = \sum_{k=-\infty}^{\infty} \delta[k-l]h[n-k] \tag{4.8}$$

Overall, the convolution sum output is the sum of the shifted and scaled version of the impulse response according to the input signal $x[n]$.

Example 4.1
From the given impulse response and $x[n]$, compute the filter output $y[n]$.

$$h[n] = \delta[n] - 2\delta[n-1] + 3\delta[n-2]$$

$$x[n] = -\delta[n] + \delta[n-1] + \delta[n-2]$$

Solution
Figure 4.2 describes the convolution sum and shift & scaled impulse response for given $x[n]$ and $h[n]$. The sum of the $-h[n]$, $h[n-1]$, and $h[n-2]$ are derived from the input signal $x[n] = -\delta[n] + \delta[n-1] + \delta[n-2]$.

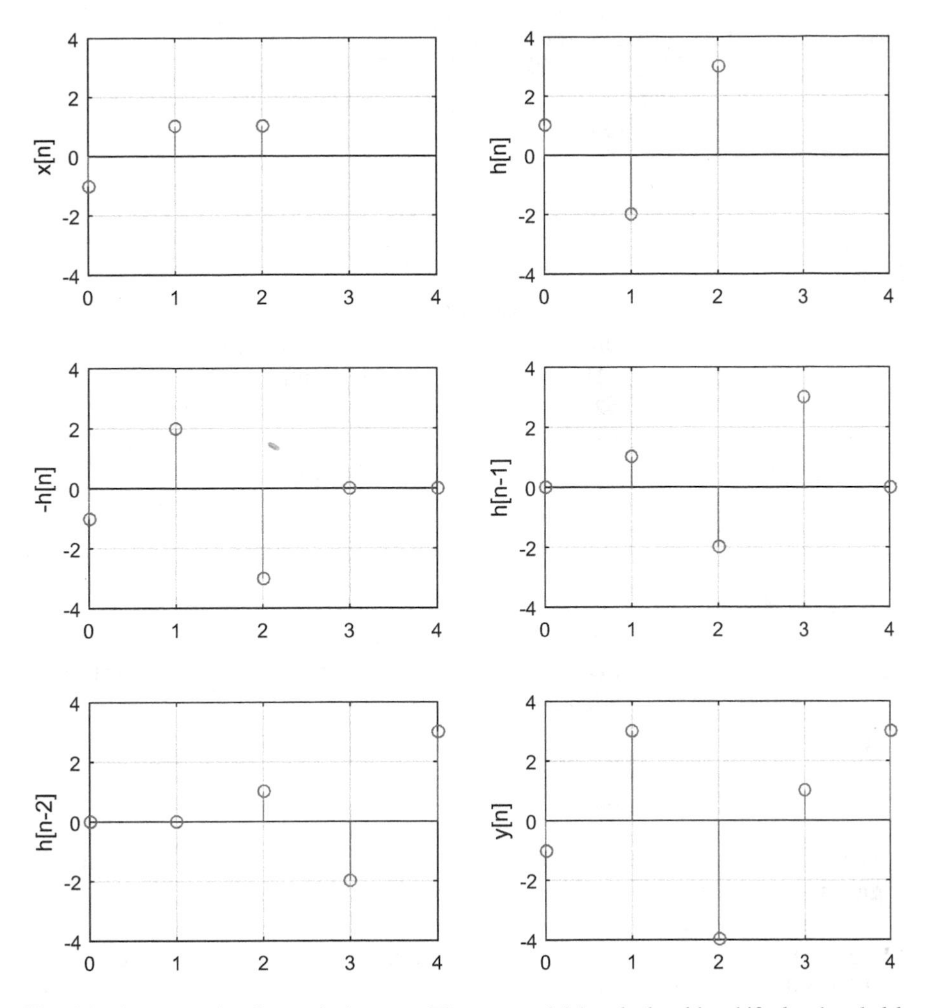

Fig. 4.2 One example of convolution sum. The output $y[n]$ is calculated by shifted and scaled $h[n]$s based on the $x[n]$

Prog. 4.1 MATLAB program for Fig. 4.2

```
hn = [1 -2 3];
xn = [-1 1 1];
y = conv(xn,hn);
y1 = conv(hn,[-1 0 0]);
y2 = conv(hn,[0 1 0]);
y3 = conv(hn,[0 0 1]);
figure,
subplot(321), stem((0:2),xn), grid, xlim([0 4]), ylabel('x[n]'), ylim([-4 4])
subplot(322), stem((0:2),hn), grid, xlim([0 4]), ylabel('h[n]'), ylim([-4 4])
subplot(323), stem((0:4),y1), grid, xlim([0 4]), ylabel('-h[n]'), ylim([-4 4])
subplot(324), stem((0:4),y2), grid, xlim([0 4]), ylabel('h[n-1]'), ylim([-4 4])
subplot(325), stem((0:4),y3), grid, xlim([0 4]), ylabel('h[n-2]'), ylim([-4 4])
subplot(326), stem((0:4),y), grid, xlim([0 4]), ylabel('y[n]'), ylim([-4 4])
```

■

The equation for the convolution sum computes the present output $y[n]$ by using the entire k range as shown in Eq. (4.9). The actual summation takes place only at the overlapped portion between the input signal and impulse response. Every shift by n value changes the overlap area for new output.

$$y[n] = \sum_{k=-\infty}^{\infty} x[k]h[n-k] = \sum_{k=-\infty}^{\infty} h[k]x[n-k] \tag{4.9}$$

The $x[n]$ range is assumed to be infinite in the given convolution sum since the digital filter input is the real time signal. From the past, the data is arrived at the system in every constant time interval continuously. However, the $h[n]$ range is finite from zero time to the positive sequence number. See below for Fig. 4.3.

Figure 4.3 shows two impulse responses with identical but shifted profile. The first $h_1[k]$ starts from the zero index to the positive sequence number. The convolution sum provides the 3 overlapped sequences initiated from the present n to the $n-2$. The second $h_2[k]$ starts from the -1 to the positive sequence number that generates the 3 overlapped sequence initiated from the future $n+1$ to the $n-1$. The future $x[n+1]$ is not available to the system now. Unless the system pauses the one-time sequence, the convolution sum cannot compute the present output $y[n]$. Mathematical representation for the two cases is given in Eqs. (4.10) and (4.11).

$$\begin{gathered} y_1[n] = x[n]h_1[0] + x[n-1]h_1[1] + x[n-2]h_1[2] \\ \text{for } h_1[n] = h_1[0]\delta[n] + h_1[1]\delta[n-1] + h_1[2]\delta[n-2] \end{gathered} \tag{4.10}$$

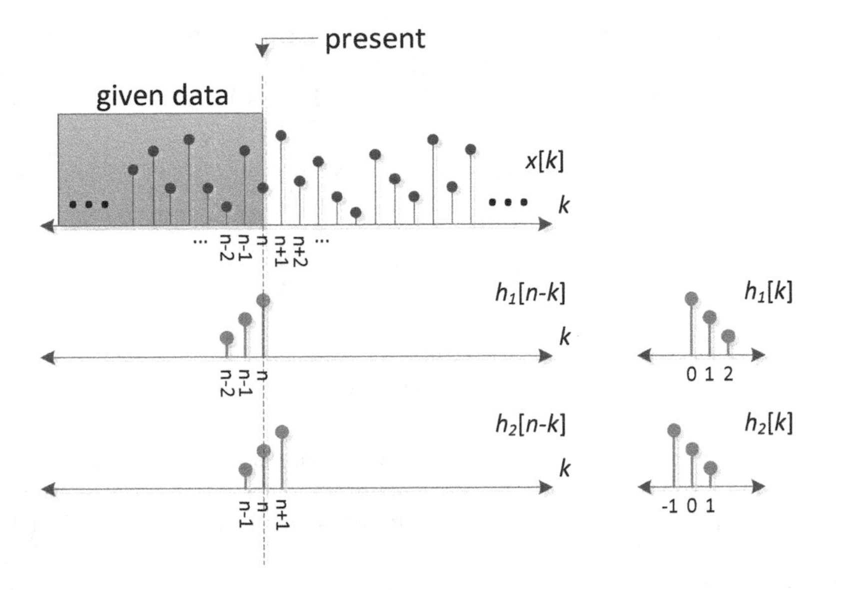

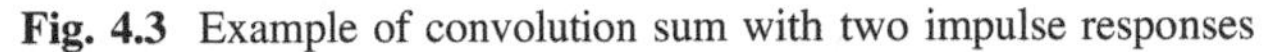

Fig. 4.3 Example of convolution sum with two impulse responses

$$\begin{aligned} y_2[n] &= x[n+1]h_2[-1] + x[n]h_2[0] + x[n-1]h_2[1] \\ \text{for } h_2[n] &= h_2[-1]\delta[n+1] + h_2[0]\delta[n] + h_2[1]\delta[n-1] \end{aligned} \tag{4.11}$$

Observe that $y_2[n]$ is derived by the future input $x[n+1]$. The real time system requires the immediacy by delivering the present output $y[n]$ from present and past input signal. Once the system shows the non-zero $h[n]$ values at negative time indexes, the system cannot meet the real time condition. The system causality is determined by the filter coefficient distribution. The causal system should follow the below condition in Eq. (4.12).

$$h[n] = 0 \quad \text{for } n < 0 \tag{4.12}$$

The non-causal system cannot provide the current output due to the future signal accessibility. However, the delayed output method realizes the system by waiting to receive the future data for convolution sum. After initial waiting, the filter continuously generates the output sequentially.

4.3 FIR and IIR Filters

The length of the impulse response presents significant role in the digital filter performance as explained in Chap. 1. The filter output follows the impulse response shape and concentrates on the shape further for longer impulse response. The convolution sum seems to be the unlimited computation due to the infinite summation range; however, actual computation takes place only at the overlapped range between the input signal and impulse response. For the causal and N length response, the convolution sum is given below as Eqs. (4.13) and (4.14).

$$y[n] = \sum_{k=-\infty}^{\infty} h[k]x[n-k] = \sum_{k=0}^{N-1} h[k]x[n-k] \tag{4.13}$$

$$\text{where } h[n] = 0 \text{ for } \begin{cases} n<0 \\ n \geq N \end{cases}$$

$$\begin{aligned} y[n] = h[0]x[n] + h[1]x[n-1] + h[2]x[n-2] + \ldots \\ + h[N-1]x[n-(N-1)] \end{aligned} \tag{4.14}$$

The required computation for the N length response is N multiplication and $N-1$ accumulation for single output. We can enlarge the filter length by increasing the N number along with increased computational requirement. The filter length is linearly proportional to the computational load. The filter with limited length impulse response is known as the finite impulse response (FIR) filter. For further localization to the specific frequency, can we realize the infinite length filter? By using the conventional convolution sum, it is impossible since the convolution sum needs the infinite multiplications and accumulations for current output as below in Eq. (4.15).

$$y[n] = \lim_{N \to \infty} \sum_{k=0}^{N-1} h[k]x[n-k] \text{ where } h[n] = 0 \text{ for} \begin{cases} n<0 \\ n \geq N \end{cases} \tag{4.15}$$

Let's consider the following structure in Eq. (4.16).

$$y[n] = a_1 y[n-1] + b_0 x[n] \tag{4.16}$$

The current output is computed by the present input $x[n]$ as well as the previous output $y[n-1]$. In order to represent the output in terms of input combinations, the recursive equation is organized as below in Eq. (4.17).

$$y[0] = b_0 x[0] \quad \text{let } x[n] = y[n] = 0 \text{ for } n < 0$$

$$y[1] = a_1 y[0] + b_0 x[1] = a_1 b_0 x[0] + b_0 x[1]$$

$$y[2] = a_1 y[1] + b_0 x[2] = a_1^2 b_0 x[0] + a_1 b_0 x[1] + b_0 x[2]$$

$$y[3] = a_1 y[2] + b_0 x[3] = a_1^3 b_0 x[0] + a_1^2 b_0 x[1] + a_1 b_0 x[2] + b_0 x[3]$$

$$\ldots$$

$$y[n] = b_0 \sum_{k=0}^{n} a_1^{n-k} x[k] \tag{4.17}$$

The output is the accumulated sum of exponentially weighted pervious inputs. Before we derive the impulse response, let's figure out the LTI compliance. The linearity first as below.

$$x[n] = \alpha x_1[n] + \beta x_2[n]$$

$$\begin{aligned} y[n] &= b_0 \sum_{k=0}^{n} a_1^{n-k} x[k] = b_0 \sum_{k=0}^{n} a_1^{n-k} \{\alpha x_1[k] + \beta x_2[k]\} \\ &= \alpha b_0 \sum_{k=0}^{n} a_1^{n-k} x_1[k] + \beta b_0 \sum_{k=0}^{n} a_1^{n-k} x_2[k] = \alpha y_1[n] + \beta y_2[n] \end{aligned} \tag{4.18}$$

The recursive filter successfully meets the linearity condition. The time-invariance is next with time shifted input $x[n-d]$.

$$b_0 \sum_{k=0}^{n} a_1^{n-k} x[k-d] = b_0 \sum_{k=0}^{n} a_1^{n-k} x[m] = *$$

$$\text{let } m = k - d \text{ and } x[n] = 0 \text{ for } n < 0$$

$$* = b_0 \sum_{m+d=0}^{n} a_1^{n-m-d} x[m] = b_0 \sum_{m=0}^{n-d} a_1^{(n-d)-m} x[m] = y[n-d] \tag{4.19}$$

Therefore, the time shifted input provides the relocated output with equal amount of time delay. The given recursive filter safely satisfies the LTI condition. Based on the LTI, we can figure out the impulse response by applying the impulse signal to the input as below in Eqs. (4.20) and (4.21).

$$h[n] = b_0 \sum_{k=0}^{n} a_1^{n-k} \delta[k] = b_0 a_1^n \quad \text{for } n \geq 0 \tag{4.20}$$

$$h[n] = b_0 \{\delta[n] + a_1 \delta[n-1] + a_1^2 \delta[n-2] + a_1^3 \delta[n-3] + \ldots\} \tag{4.21}$$

The impulse response is the exponential function with infinite length. Depending on the base value a_1, the impulse response can be increased or decreased in exponential fashion.

Example 4.2
Find the impulse response $h[n]$ of following difference equations.

$$y_1[n] = 0.75 y_1[n-1] + x[n]$$

$$y_2[n] = 1.25 y_2[n-1] + x[n]$$

Solution
0.75 a_1 less than the magnitude one represents the decreasing exponential. In contrast, the above one magnitude 1.25 illustrates the increasing profile in Fig. 4.4.

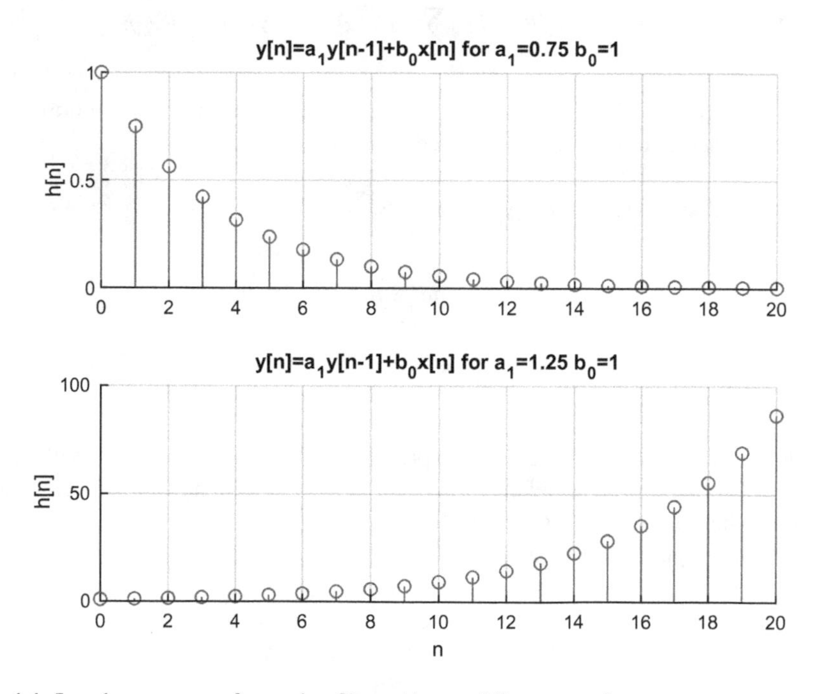

Fig. 4.4 Impulse response of recursive filter with two different a_1 values

$$h_1[n] = 0.75^n \quad \text{for}\, n \geq 0$$

$$h_1[n] = \delta[n] + 0.75\delta[n-1] + 0.75^2\delta[n-2] + 0.75^3\delta[n-3] + \ldots$$

$$h_2[n] = 1.25^n \quad \text{for}\, n \geq 0$$

$$h_1[n] = \delta[n] + 1.25\delta[n-1] + 1.25^2\delta[n-2] + 1.25^3\delta[n-3] + \ldots$$

Prog. 4.2 MATLAB program for Fig. 4.4

```
a1 = 3/4;
a2 = 5/4;
b0 = 1;
n = 0:20;
y1 = b0*(a1.^n);
y2 = b0*(a2.^n);
figure;
subplot(211),stem(n,y1), grid
ylabel('h[n]');
title('y[n]=a_1y[n-1]+b_0x[n] for a_1=0.75 b_0=1')
box off
subplot(212),stem(n,y2), grid
xlabel('n');
ylabel('h[n]');
title('y[n]=a_1y[n-1]+b_0x[n] for a_1=1.25 b_0=1')
box off
```

■

The recursive filter effectively denotes the infinite length impulse response. The convolution sum with finite length impulse response is known as the finite impulse response (FIR) filter. The infinite impulse response (IIR) filter delivers the infinite length impulse response with recursive discrete equation. Does the further recursive filter also provide the exponential impulse response? Let's investigate second order recursive filter in below equations.

$$y[n] = (a_1 + a_2)y[n-1] - a_1a_2y[n-2] + (a_1 - a_2)x[n]$$

$$y[0] = (a_1 - a_2)x[0] \quad \text{let } x[n] = y[n] = 0 \text{ for } n<0$$

$$\begin{aligned} y[1] &= (a_1 + a_2)y[0] + (a_1 - a_2)x[1] \\ &= (a_1 + a_2)(a_1 - a_2)x[0] + (a_1 - a_2)x[1] \\ &= (a_1^2 - a_2^2)x[0] + (a_1 - a_2)x[1] \end{aligned}$$

$$\begin{aligned} y[2] &= (a_1 + a_2)y[1] - a_1a_2y[0] + (a_1 - a_2)x[2] \\ &= (a_1 + a_2)\left(a_1^2 - a_2^2\right)x[0] + (a_1 + a_2)(a_1 - a_2)x[1] \\ &\quad - a_1a_2(a_1 - a_2)x[0] + (a_1 - a_2)x[2] \\ &= \left(a_1^3 - a_2^3\right)x[0] + \left(a_1^2 - a_2^2\right)x[1] + (a_1 - a_2)x[2] \end{aligned}$$

$$\begin{aligned} y[3] &= (a_1 + a_2)y[2] - a_1a_2y[1] + (a_1 - a_2)x[3] \\ &= (a_1 + a_2)\left(a_1^3 - a_2^3\right)x[0] + (a_1 + a_2)\left(a_1^2 - a_2^2\right)x[1] \\ &\quad + (a_1 + a_2)(a_1 - a_2)x[2] - a_1a_2\left(a_1^2 - a_2^2\right)x[0] \\ &\quad - a_1a_2(a_1 - a_2)x[1] + (a_1 - a_2)x[3] \\ &= \left(a_1^4 - a_2^4\right)x[0] + \left(a_1^3 - a_2^3\right)x[1] + \left(a_1^2 - a_2^2\right)x[2] + (a_1 - a_2)x[3] \end{aligned}$$

$$\cdots$$

$$y[n] = \sum_{k=0}^{n}\left(a_1^{(n+1)-k} - a_2^{(n+1)-k}\right)x[k] \tag{4.22}$$

Due to the complexity, the second order recursive filter is exemplified with specific coefficients; however, this case as well as the general cases show the exponential relation between input and output. For conventional IIR filter shown in Eq. (4.23), there are significant limitation to solve the recursive filter by substitution method. The general solution for the IIR filter can be found on Chap. 5 by using the Z-transform. The z domain provides the simple representation of exponential solution with complex numbers. Also, the graphical interpretation of the z domain generates the fast insight on the filter characteristics.

$$\begin{aligned} y[n] &= a_1y[n-1] + a_2y[n-2] + a_3y[n-3] + \ldots \\ &\quad + b_0x[n] + b_1x[n-1] + b_2x[n-2] + \ldots \end{aligned} \tag{4.23}$$

Now, it is time to consider the filter stability. The FIR filter weights the recent N samples of the $x[n]$ based on the impulse response $h[n]$ to compute the current output $y[n]$. If the input $x[n]$ values are bounded; in other words, the incoming signal values are not infinite, the real value impulse response always generates the bounded output. The stable filter always provides the bounded output from the bounded input. This type of stability is known as bounded input and bounded

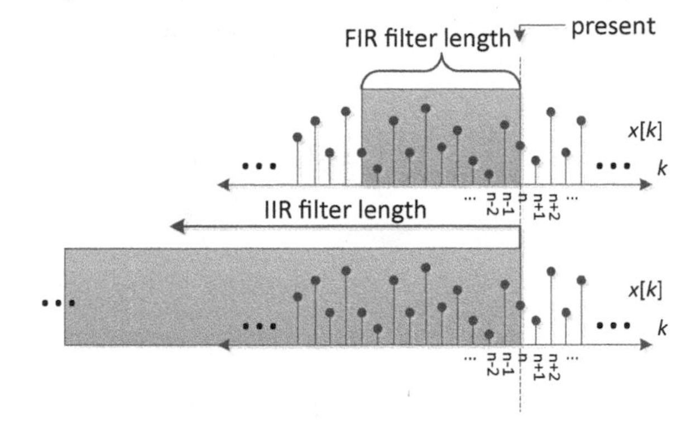

Fig. 4.5 Filter length comparison for FIR and IIR filter

output (BIBO) stability. According to the limited length impulse response, FIR filter satisfies the BIBO stability.

The IIR filter is characterized by the infinite length impulse response that considers the unlimited length of the previous input signal values as shown in Fig. 4.5. The sum of vast products between the input value and impulse response may lead to the infinite output value even for the bounded signal. Let's find the BIBO stability condition for the IIR filter. The infinite length impulse response and its convolution sum is given as Eq. (4.24).

$$y[n] = \lim_{N\to\infty} \sum_{k=0}^{N-1} h[k]x[n-k] = *$$

$$\text{where } h[n] = 0 \text{ for } n < 0 \text{ and } |x[n]| \le M < \infty \forall n$$

$$* \le \left| \lim_{N\to\infty} \sum_{k=0}^{N-1} h[k]x[n-k] \right| \le \lim_{N\to\infty} \sum_{k=0}^{N-1} |h[k]||x[n-k]|$$

$$\le \lim_{N\to\infty} \sum_{k=0}^{N-1} |h[k]|M < \infty \tag{4.24}$$

Since the input is bounded up to M, the bounded output condition is derived below.

$$\sum_{k=0}^{\infty} |h[k]| < \infty \tag{4.25}$$

If the impulse response of the IIR filter is absolute summable, the IIR filter meets the BIBO stability condition. For general order IIR filters, the direct computation of the impulse response and BIBO stability is not feasible in time domain. Further managements of the IIR filter can be easily seen by Z-transform in the next chapter.

Example 4.3

Determine the BIBO stability of following IIR filters.

$$y_1[n] = 0.75y_1[n-1] + x[n]$$

$$y_2[n] = 1.25y_2[n-1] + x[n]$$

Solution

From the Example 4.2, we derived the below impulse responses.

$$h_1[n] = 0.75^n \quad \text{for}\, n \geq 0$$

$$h_2[n] = 1.25^n \quad \text{for}\, n \geq 0$$

$$\sum_{k=0}^{\infty} |h_1[k]| = \sum_{k=0}^{\infty} |0.75^n| = \frac{1}{1-0.75} = 4 < \infty$$

$$\sum_{k=0}^{\infty} |h_2[k]| = \sum_{k=0}^{\infty} |1.25^n| = \infty \because |r| > 1 \,\text{in geometric series}$$

Hence, the filter for $y_1[n]$ is BIBO stable and the filter for $y_2[n]$ is not BIBO stable.

■

4.4 Frequency Response

Chapter 3 provides the frequency analysis tools for periodic and aperiodic signal. The Fourier analysis transforms the signals between the time and frequency domain in order to investigate the spectral distribution of the given signal. The Fourier equations are given below again in Table 4.1.

We can see the frequency profile of the given raw signal and generate the signal with specific frequency combination. Therefore, the input and output of the filter can be individually transformed to the frequency domain for performance

Table 4.1 Fourier transforms for discrete time signal

Discrete-time Fourier transform (DTFT)		Discrete Fourier transform (DFT)	
Analysis	$X(e^{j\omega}) = \sum_{n=-\infty}^{\infty} x[n]e^{-j\omega n}$	Analysis	$X[k] = \sum_{n=0}^{N-1} x_N[n]e^{-j\frac{2\pi}{N}kn}$
Synthesis	$x[n] = \frac{1}{2\pi}\int_{-\pi}^{\pi} X(e^{j\omega})e^{j\omega n}\, d\omega$	Synthesis	$x_N[n] = \frac{1}{N}\sum_{k=0}^{N-1} X[k]e^{j\frac{2\pi}{N}kn}$

verification. What operation is equivalent to the convolution sum in frequency domain? Let's see below for DTFT.

$$\begin{aligned} Y(e^{j\omega}) &= \sum_{n=-\infty}^{\infty} y[n]e^{-j\omega n} = \sum_{n=-\infty}^{\infty} \sum_{k=-\infty}^{\infty} h[k]x[n-k]e^{-j\omega n} \\ &= \sum_{k=-\infty}^{\infty} h[k] \sum_{n=-\infty}^{\infty} x[n-k]e^{-j\omega n} = \sum_{k=-\infty}^{\infty} h[k]X(e^{j\omega})e^{-j\omega k} \\ &= X(e^{j\omega}) \sum_{k=-\infty}^{\infty} h[k]e^{-j\omega k} = X(e^{j\omega})H(e^{j\omega}) \end{aligned} \tag{4.26}$$

The convolution sum in the time domain is represented by the simple multiplication in frequency domain. Due to the multiplication property, the introduced frequency components by input can be reduced or emphasized its magnitude by the frequency response $H(e^{j\omega})$ that is the DTFT of impulse response. However, the filter cannot create arbitrary frequencies from nothing which is equal to zero in magnitude. Therefore, periodic signal input delivers the identical combination of the frequency components in filter output. The DFT expects to have the same relation for the convolution sum. See equations in below.

$$\begin{aligned} Y[k] &= \sum_{n=0}^{N-1} y[n]e^{-j\frac{2\pi}{N}kn} = \sum_{n=0}^{N-1} \sum_{l=-\infty}^{\infty} h[l]x[n-l]e^{-j\frac{2\pi}{N}kn} \\ &= \sum_{l=-\infty}^{\infty} h[l] \sum_{n=0}^{N-1} x[n-l]e^{-j\frac{2\pi}{N}kn} = \sum_{l=-\infty}^{\infty} h[l]X[k]e^{-j\frac{2\pi}{N}kl} \\ &= X[k] \sum_{l=-\infty}^{\infty} h[l]e^{-j\frac{2\pi}{N}kl} = X[k]H[k] \end{aligned} \tag{4.27}$$

According to the relation in Eq. (4.27), some of the frequency components from the input could be removed by weighting from frequency response $H[k]$ which is the DFT of impulse response. Otherwise, the filter passes to the output frequencies with modified magnitude. The frequency response shape decides the filter functionality and performance. Chapter 1 described that the filter output follows the impulse response shape; in other words, the frequency response of the filter is applied over the input spectrum to imitate the shape in time as well as frequency domain.

The DTFT performs the analysis over the whole range for aperiodic signal and the spectrum is represented by the continuous variable ω. Hence, the DTFT length does not matter. You execute the DTFT over the $x[n]$ and $h[n]$ to obtain the $X(e^{j\omega})$ and $H(e^{j\omega})$, respectively. Then, performs the multiplication for convolution sum counterpart in frequency domain and finally develops the convolution sum result by applying the synthesis equation on the multiplication output.

$$\begin{array}{l} x[n] \xrightarrow{\text{DTFT Analysis}} X(e^{j\omega}) \\ h[n] \xrightarrow[\text{DTFT Analysis}]{} H(e^{j\omega}) \end{array} \xrightarrow{\text{Mul.}} X(e^{j\omega})H(e^{j\omega}) \xrightarrow{\text{DTFT Synthesis}} x[n] * h[n] \tag{4.28}$$

Let's do one example as below.

Example 4.4
Compute the filter output of below configuration with DTFT method.

$$x[n] = h[n] = \begin{cases} 1 & \text{for} \quad 0 \leq n \leq 2 \\ 0 & \quad \text{Otherwise} \end{cases}$$

Solution

$$X(e^{j\omega}) = H(e^{j\omega}) = \sum_{n=-\infty}^{\infty} x[n]e^{-j\omega n} = 1 + e^{-j\omega} + e^{-j\omega 2}$$

$$\begin{aligned} X(e^{j\omega})H(e^{j\omega}) &= (1 + e^{-j\omega} + e^{-j\omega 2})(1 + e^{-j\omega} + e^{-j\omega 2}) \\ &= 1 + 2e^{-j\omega} + 3e^{-j\omega 2} + 2e^{-j\omega 3} + e^{-j\omega 4} \end{aligned}$$

$$\begin{aligned} &\frac{1}{2\pi}\int_{-\pi}^{\pi} X(e^{j\omega})\mathrm{H}(e^{j\omega})e^{j\omega n}d\omega \\ &= \frac{1}{2\pi}\int_{-\pi}^{\pi} (1 + 2e^{-j\omega} + 3e^{-j\omega 2} + 2e^{-j\omega 3} + e^{-j\omega 4})e^{j\omega n}d\omega \\ &= \delta[n] + 2\delta[n-1] + 3\delta[n-2] + 2\delta[n-3] + \delta[n-4] = x[n] * h[n] \end{aligned}$$

■

As shown above, the user did not decide the length of the DTFT for the proper convolution sum. The DTFT convolution sum result shows the same outcome from conventional convolution sum process. However, due to continuous variable ω, the general computer cannot process the DTFT computation without using the symbolic mathematics. The DFT is more popular because of the discrete frequency. The DFT also follows the identical process to compute the convolution sum in frequency domain as below in Eq. (4.29).

$$\begin{array}{l} x[n] \xrightarrow{\text{DFT Analysis}} X[k] \\ h[n] \xrightarrow[\text{DFT Analysis}]{} H[k] \end{array} \xrightarrow{\text{Mul.}} X[k]H[k] \xrightarrow{\text{DFT Synthesis}} x[n] * h[n] \tag{4.29}$$

Let's see the DFT convolution sum.

Example 4.5

Compute the filter output of below configuration with DFT method in length 4 and 5.

$$x[n] = h[n] = \begin{cases} 1 & \text{for} \quad 0 \le n \le 2 \\ 0 & \text{Otherwise} \end{cases}$$

Solution

For $N = 4$,

$$X[k] = \mathrm{H}[k] = \sum_{n=0}^{3} x[n]e^{-j\frac{2\pi}{4}kn} = 1 + e^{-j\frac{2\pi}{4}k} + e^{-j\frac{2\pi}{4}k2}$$

$$\begin{aligned} X[k]\mathrm{H}[k] &= \left(1 + e^{-j\frac{2\pi}{4}k} + e^{-j\frac{2\pi}{4}k2}\right)\left(1 + e^{-j\frac{2\pi}{4}k} + e^{-j\frac{2\pi}{4}k2}\right) \\ &= 1 + 2e^{-j\frac{2\pi}{4}k} + 3e^{-j\frac{2\pi}{4}k2} + 2e^{-j\frac{2\pi}{4}k3} + e^{-j\frac{2\pi}{4}k4} \\ &= 2 + 2e^{-j\frac{2\pi}{4}k} + 3e^{-j\frac{2\pi}{4}k2} + 2e^{-j\frac{2\pi}{4}k3} \text{since } e^{-j\frac{2\pi}{4}k4} = 1 \end{aligned}$$

$$\begin{aligned} \frac{1}{4}\sum_{k=0}^{3} X[k]H[k]e^{j\frac{2\pi}{4}kn} &= \frac{1}{4}\sum_{k=0}^{3}\left(2 + 2e^{-j\frac{2\pi}{4}k} + 3e^{-j\frac{2\pi}{4}k2} + 2e^{-j\frac{2\pi}{4}k3}\right)e^{j\frac{2\pi}{4}kn} \\ &= 2\delta[n] + 2\delta[n-1] + 3\delta[n-2] + 2\delta[n-3] \neq x[n] * h[n] \end{aligned}$$

The DFT convolution sum with $N = 4$ does not match with the convolution sum result due to the improper length. The circular property of the DFT wraps the last element $e^{-j2\pi k4/4}$ to the first location e^{-j0}. To avoid this problem, increase the length of the DFT for 5 as below.

For $N = 5$,

$$X[k] = \mathrm{H}[k] = \sum_{n=0}^{4} x[n]e^{-j\frac{2\pi}{5}kn} = 1 + e^{-j\frac{2\pi}{5}k} + e^{-j\frac{2\pi}{5}k2}$$

$$\begin{aligned} X[k]\mathrm{H}[k] &= \left(1 + e^{-j\frac{2\pi}{5}k} + e^{-j\frac{2\pi}{5}k2}\right)\left(1 + e^{-j\frac{2\pi}{5}k} + e^{-j\frac{2\pi}{5}k2}\right) \\ &= 1 + 2e^{-j\frac{2\pi}{5}k} + 3e^{-j\frac{2\pi}{5}k2} + 2e^{-j\frac{2\pi}{5}k3} + e^{-j\frac{2\pi}{5}k4} \end{aligned}$$

$$\begin{aligned} &\frac{1}{5}\sum_{k=0}^{4} X[k]H[k]e^{j\frac{2\pi}{5}kn} \\ &= \frac{1}{5}\sum_{k=0}^{4}\left(1 + 2e^{-j\frac{2\pi}{5}k} + 3e^{-j\frac{2\pi}{5}k2} + 2e^{-j\frac{2\pi}{5}k3} + e^{-j\frac{2\pi}{5}k4}\right)e^{j\frac{2\pi}{5}kn} \\ &= \delta[n] + 2\delta[n-1] + 3\delta[n-2] + 2\delta[n-3] + \delta[n-4] = x[n] * h[n] \end{aligned}$$

■

For the given sequences with l and m length, the convolution sum provides the outcome with $l + m–1$ length. Therefore, the DFT length must cover the convolution sum length as below in Eq. (4.30).

$$\begin{aligned}
&x[n] : l\,\text{length} \xrightarrow{\text{DFT Analysis} \geq l+m-1\, length} X[k] \\
&\text{Must have equal lengths for DFT processing} \xrightarrow{\text{Multiplication}} X[k]H[k] \\
&h[n] : m\,\text{length} \xrightarrow[\text{DFT Analysis} \geq l+m-1\, length]{} H[k]
\end{aligned} \tag{4.30}$$

$$\xrightarrow{\text{DFT Synthesis} \geq l+m-1\, length} x[n] * h[n]$$

With equal or bigger than to the convolution sum length, DFT processing guarantees the identical convolution sum result over simple multiplication in frequency domain. Therefore, DFT length must be chosen with data length consideration. Otherwise, the above the length elements are penetrates into the first results in circular manner known as circular convolution sum that will be explained in Chap. 6. Another example of improper DFT length in below shows the circular manner.

Example 4.6

Compute the filter output of below configuration with DFT method in length 3.

$$x[n] = h[n] = \begin{cases} 1 & \text{for} \quad 0 \leq n \leq 2 \\ 0 & \quad\quad \text{Otherwise} \end{cases}$$

Solution

For $N = 3$,

$$X[k] = \mathrm{H}[k] = \sum_{n=0}^{2} x[n] e^{-j\frac{2\pi}{3}kn} = 1 + e^{-j\frac{2\pi}{3}k} + e^{-j\frac{2\pi}{3}k2}$$

$$\begin{aligned}
X[k]\mathrm{H}[k] &= \left(1 + e^{-j\frac{2\pi}{3}k} + e^{-j\frac{2\pi}{3}k2}\right)\left(1 + e^{-j\frac{2\pi}{3}k} + e^{-j\frac{2\pi}{3}k2}\right) \\
&= 1 + e^{-j\frac{2\pi}{3}k} + e^{-j\frac{2\pi}{3}k2} + e^{-j\frac{2\pi}{3}k} + e^{-j\frac{2\pi}{3}k2} + e^{-j\frac{2\pi}{3}k3} \\
&\quad + e^{-j\frac{2\pi}{3}k2} + e^{-j\frac{2\pi}{3}k3} + e^{-j\frac{2\pi}{3}k4} \\
&= 1 + 2e^{-j\frac{2\pi}{3}k} + 3e^{-j\frac{2\pi}{3}k2} + 2e^{-j\frac{2\pi}{3}k3} + e^{-j\frac{2\pi}{3}k4} \\
&= \left(1 + 2e^{-j\frac{2\pi}{3}k3}\right) + \left(2e^{-j\frac{2\pi}{3}k} + e^{-j\frac{2\pi}{3}k4}\right) + 3e^{-j\frac{2\pi}{3}k2} \\
&= 3 + 3e^{-j\frac{2\pi}{3}k} + 3e^{-j\frac{2\pi}{3}k2}
\end{aligned}$$

$$\text{since } e^{-j\frac{2\pi}{3}k3} = 1 \text{ and } e^{-j\frac{2\pi}{3}k4} = e^{-j\frac{2\pi}{3}k}$$

$$\frac{1}{3}\sum_{k=0}^{2}X[k]H[k]e^{j\frac{2\pi}{3}kn}=\frac{1}{3}\sum_{k=0}^{2}\left(3+3e^{-j\frac{2\pi}{3}k}+3e^{-j\frac{2\pi}{3}k2}\right)e^{j\frac{2\pi}{3}kn}$$
$$=3\delta[n]+3\delta[n-1]+3\delta[n-2]\neq x[n]*h[n]$$

■

Now we understand that the convolution sum in time domain can be represented by the multiplication in the frequency domain. The DTFT or DFT of the impulse response illustrates the filter frequency response which demonstrates the functionality and performance of the filter. Also, the discrete processing with DFT should choose the length with caution in order to avoid the wrap around in circular manner on convolution sum processing over the DFT multiplication.

The time and frequency domain show the mutuality between domains in relation and function. Usually the function relationship is retained in one direction and vice versa known as duality. The convolution sum to multiplication expects to be conserved as multiplication to convolution sum between domains. Let's see the multiplication in time domain based on the DTFT as below in Eq. (4.31).

$$\sum_{n=-\infty}^{\infty}x[n]h[n]e^{-j\omega n}=\sum_{n=-\infty}^{\infty}x[n]\frac{1}{2\pi}\int_{-\pi}^{\pi}H\left(e^{j\varphi}\right)e^{j\varphi n}d\varphi e^{-j\omega n}$$
$$=\frac{1}{2\pi}\int_{-\pi}^{\pi}H\left(e^{j\varphi}\right)\sum_{n=-\infty}^{\infty}x[n]e^{-j(\omega-\varphi)n}d\varphi=\frac{1}{2\pi}\int_{-\pi}^{\pi}H\left(e^{j\varphi}\right)X\left(e^{j(\omega-\varphi)}\right)d\varphi$$

$$x[n]h[n]\overset{DTFT}{\leftrightarrow}\frac{1}{2\pi}X\left(e^{j\omega}\right)*H\left(e^{j\omega}\right) \tag{4.31}$$

The multiplication in time domain is equivalent to the convolution integral in frequency domain with scale. The identical situation can be observed in DFT as below.

$$\sum_{n=0}^{N-1}x[n]h[n]e^{-j\frac{2\pi}{N}kn}=\sum_{n=0}^{N-1}x[n]\frac{1}{N}\sum_{l=0}^{N-1}H[l]e^{j\frac{2\pi}{N}ln}e^{-j\frac{2\pi}{N}kn}$$
$$=\frac{1}{N}\sum_{l=0}^{N-1}\sum_{n=0}^{N-1}x[n]e^{-j\frac{2\pi}{N}(k-l)n}H[l]=\frac{1}{N}\sum_{l=0}^{N-1}X[k-l]H[l]$$

$$x[n]h[n]\overset{DFT}{\leftrightarrow}\tfrac{1}{N}X[k]*H[k] \tag{4.32}$$

The convolution sum and multiplication are the pair of the operation in duality for time and frequency domain. One domain with the convolution sum (or integral) corresponds to the multiplication in the other domain.

4.5 Simple Filter Design

Beside the intuitive filter design at Chap. 1, it is time to build the filter with mathematical fundamentals. In previous chapters, we investigated the discrete signal, convolution sum, frequency, transformations etc. those are the structural materials to develop the systematical filters. Instead of using trial and error method, the designated passing and blocking frequencies are specified in design phase to derive the impulse response of the convolution sum filter. The method utilizes the simple Fourier transformations and its properties to design the filter. Further sophisticate filters can be found in Chap. 5 after the Z-transformation. Recommend using the filter design tools such as MATLAB since the tool provides much optimized results in terms of performance and computation. Otherwise the introducing method is effective and simple to understand and realize the filter from scratch.

- Simple low pass filter first

The low pass filter (LPF) well passes the low frequency components than high frequency section divided by the one threshold frequency value. There are multiple parameters to specify the LPF properties in detail. For the time being, let's consider the LPF with single frequency value that indicates pass band upper limit. Above the f_c value, the LPF tries to block the frequency component by convolution sum process. Figure 4.6 shows the ideal frequency response of the LPF.

We can choose the frequency domain variable in either set as continuous (DTFT) or discrete (DFT) that is the initial point of the filter design. Let's begin with DTFT.

Example 4.7
Design the LPF for Table 4.2 which is the summary of the filter specifications. Use the DTFT method. Show the impulse response from –20 to 20 as n.

Solution
The given frequency information should be changed to the discrete frequency ω based on the sampling frequency f_s. The sampling frequency corresponds to the 2π; therefore, the below equation computes the cutoff discrete frequency as $\pi/4$.

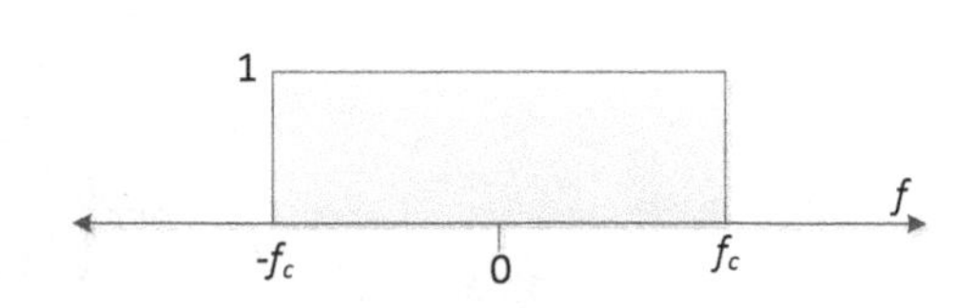

Fig. 4.6 Ideal low pass filter response

Table 4.2 Low pass filter specification for the example

Filter specifications	Value
Filter type	Low pass filter
Filter length	Finite with optimization
Cutoff frequency (f_c)	1000 Hz
Sampling frequency (f_s)	8000 Hz

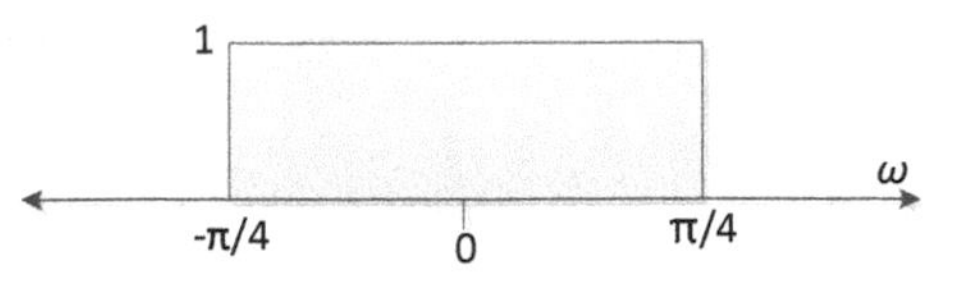

Fig. 4.7 Ideal low pass filter response in discrete frequency

$$\omega_c = \frac{2\pi}{f_s} f_c$$

The frequency response in discrete frequency domain ω is shown in Fig. 4.7. Using the synthesis equation of DTFT, derive the impulse response as below.

$$\begin{aligned} h[n] &= \frac{1}{2\pi} \int_{-\pi}^{\pi} H(e^{j\omega}) e^{j\omega n} d\omega = \frac{1}{2\pi} \int_{-\frac{\pi}{4}}^{\frac{\pi}{4}} 1 e^{j\omega n} d\omega \\ &= \frac{1}{2\pi} \frac{e^{j\omega n}}{jn} \bigg|_{-\frac{\pi}{4}}^{\frac{\pi}{4}} = \frac{1}{2\pi jn} \left(e^{j\frac{\pi}{4}n} - e^{-j\frac{\pi}{4}n}\right) \\ &= \frac{1}{2\pi jn} \left(e^{j\frac{\pi}{4}n} - e^{-j\frac{\pi}{4}n}\right) = \frac{\sin\left(\frac{\pi n}{4}\right)}{\pi n} \end{aligned}$$

$$\lim_{n \to \pm\infty} \frac{\sin\left(\frac{\pi n}{4}\right)}{\pi n} = 0 \text{ and } \lim_{n \to 0} \frac{\sin\left(\frac{\pi n}{4}\right)}{\pi n} = \frac{1}{4}$$

The impulse response shows the infinite length sequence with asymptotes on x axis toward positive and negative infinite. Also, there is one singular point in the impulse response at zero n value and the corresponding outcome is computed by the L'Hôpital's rule [1] as shown in above equations. The impulse response for the 41-length sequence is illustrated in Fig. 4.8.

■

Note that the derived impulse response is the infinite length sequence; therefore, we cannot employ the result directly due to the infinite computation at convolution sum. The proper length should be decided to represent the specified LPF frequency profile. The Parseval's theorem [2] describes the equivalence energy relationship between the time and frequency domain as below.

$$\sum_{n=-\infty}^{\infty} |h[n]|^2 = \frac{1}{2\pi} \int_{-\pi}^{\pi} |H(e^{j\omega})|^2 d\omega \tag{4.33}$$

The energy in frequency domain is below.

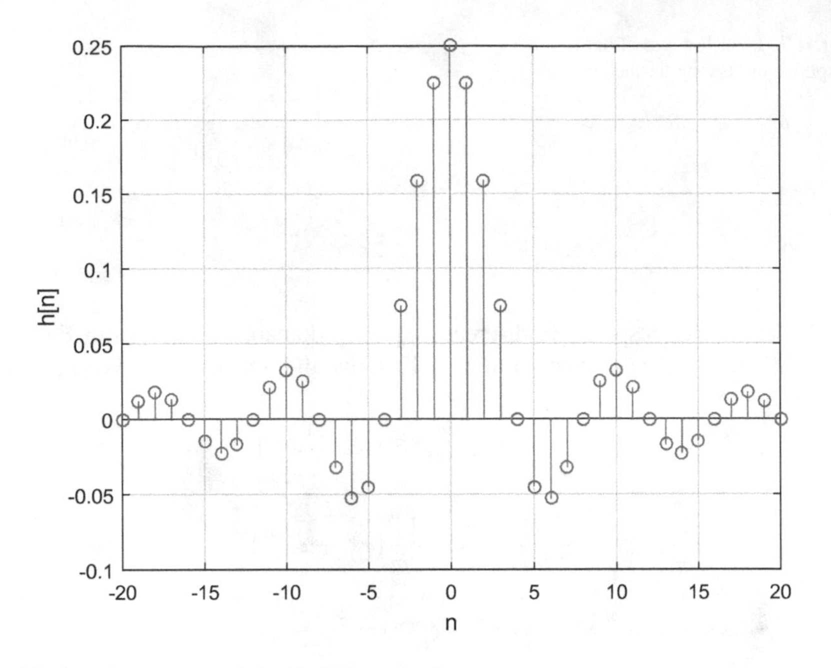

Fig. 4.8 Impulse response of the ideal low pass filter

$$\frac{1}{2\pi}\int_{-\pi}^{\pi} \left|H\left(e^{j\omega}\right)\right|^2 d\omega = \frac{1}{2\pi}\int_{-\frac{\pi}{4}}^{\frac{\pi}{4}} 1^2 d\omega = \frac{\pi/2}{2\pi} = \frac{1}{4} \tag{4.34}$$

Therefore, the time domain energy should be identical to the value $1/4$.

$$\sum_{n=-\infty}^{\infty} \left|\frac{\sin\left(\frac{\pi n}{4}\right)}{\pi n}\right|^2 = \frac{1}{4} \tag{4.35}$$

The energy ratio between the time and frequency domain can be numerically computed by below equation.

$$\frac{100\sum_{n=-l}^{l} \left|\frac{\sin\left(\frac{\pi n}{4}\right)}{\pi n}\right|^2}{0.25} [\%] \tag{4.36}$$

As we increase the impulse response length symmetrically, the time domain is close to the LPF distribution. The figure below depicts the energy ratio defined by Eq. (4.36). The 15 and 83 sample length define the 95 and 99% of energy to the frequency domain, respectively; hence, 83-length impulse response accurately represents the specified LPF property further than the 15-sample counterpart.

Prog. 4.3 MATLAB program for Figs. 4.8 and 4.9

```
syms n w;
nn = (-50:1:50);
f1 = exp(j*w*n);
f2 = int(f1,w,-pi/4,pi/4)/(2*pi);   % Perform IDTFT for h[n]
f3 = simplify(f2);
f4 = subs(f3,n,nn+eps);             % Replace symbol n with real number nn
hn = eval(f4);                      % Evaluate
eee = [0.25^2];
engy = 0;
for kk=0:49
    engy = sum(hn(52:52+kk).^2)*2+0.25^2;
    eee = [eee engy];
end
temp1 = length(eee);
inx1 = 1:temp1;
inx2 = inx1*2-1;

figure,
stem(nn,hn), grid
xlabel('n')
ylabel('h[n]')
xlim([-20 20])
figure,
stem(inx2,eee*100/0.25,':o'), grid
xlim([1 101]);
ylim([50 103]);
xlabel('h[n] length in samples');
ylabel('Energy ratio between time and frequency domain [%]')
```

Each of the frequency distribution can be drawn by the DTFT analysis equation as Eq. (4.37). The corresponding frequency response magnitude is illustrated in Fig. 4.10.

$$
\begin{aligned}
H\left(e^{j\omega}\right) &= \sum_{n=-\infty}^{\infty} h[n]e^{-j\omega n} = \sum_{n=-7\text{ or }-41}^{7\text{ or }41} \frac{\sin\left(\frac{\pi n}{4}\right)}{\pi n} e^{-j\omega n} \\
&= h[0] + \sum_{n=1}^{7\text{ or }41} \frac{\sin\left(\frac{\pi n}{4}\right)}{\pi n} \left(e^{j\omega n} + e^{-j\omega n}\right) \\
&= h[0] + 2\sum_{n=1}^{7\text{ or }41} \frac{\sin\left(\frac{\pi n}{4}\right)}{\pi n} \cos(\omega n)
\end{aligned}
\tag{4.37}
$$

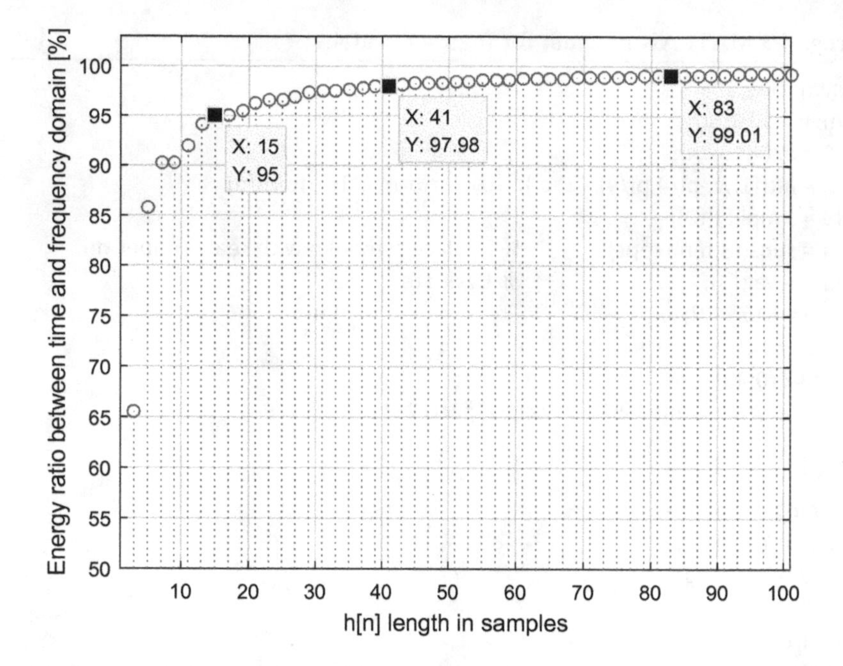

Fig. 4.9 Energy ratio between time and frequency domain for truncated LPF response

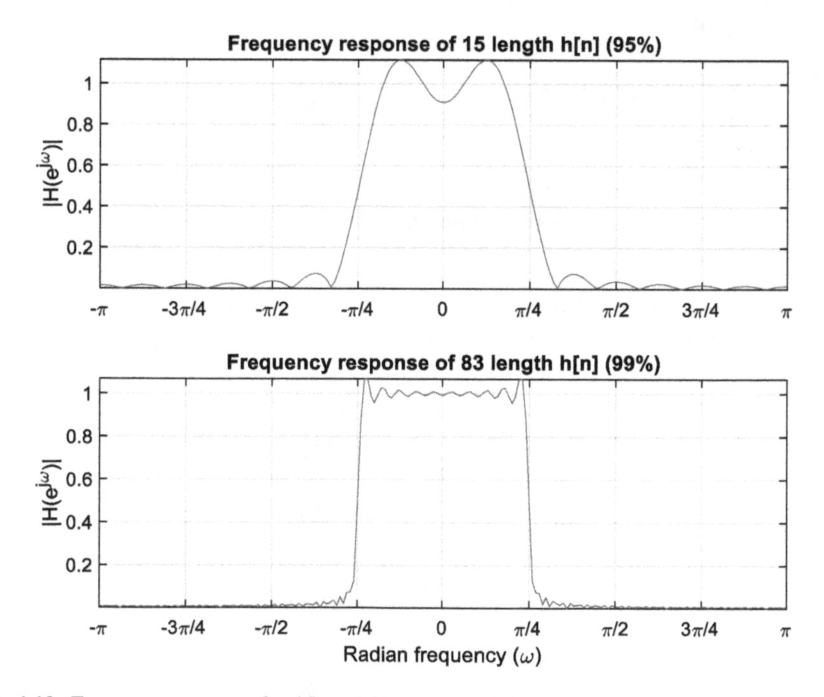

Fig. 4.10 Frequency response for 15- and 83-length LPF $h[n]$

Prog. 4.4 MATLAB program for Fig. 4.10

```
syms w;

n1 = (-7:1:7);
n2 = (-41:1:41);
a1 = (sin(pi*(n1+eps)/4))./(pi*(n1+eps));
a2 = (sin(pi*(n2+eps)/4))./(pi*(n2+eps));
hw1 = sum(a1.*exp(-1i*w*n1));
hw2 = sum(a2.*exp(-1i*w*n2));
w1 = (-pi:pi/100:pi);
ohw1 = eval(subs(hw1,w,w1));
ohw2 = eval(subs(hw2,w,w1));

figure,
subplot(211), plot(w1,abs(ohw1)), grid, axis tight
ax1 = gca;
ax1.XTick = [-pi -3*pi/4 -pi/2 -pi/4 0 pi/4 pi/2 3*pi/4 pi];
ax1.XTickLabel = {'-\pi', '-3\pi/4', '-\pi/2', '-\pi/4', '0', '\pi/4', '\pi/2', '3\pi/4',
'\pi'};
subplot(212), plot(w1,abs(ohw2)), grid, axis tight
ax2 = gca;
ax2.XTick = [-pi -3*pi/4 -pi/2 -pi/4 0 pi/4 pi/2 3*pi/4 pi];
ax2.XTickLabel = {'-\pi', '-3\pi/4', '-\pi/2', '-\pi/4', '0', '\pi/4', '\pi/2', '3\pi/4',
'\pi'};
```

As we expected, the 83-length impulse response accurately follows the given LPF specification. We choose the middle length 41 for the designated impulse response shown in Fig. 4.11 with frequency response as well. Certain overshoots on the edges and sidelobes are observed in the frequency domain and those glitches are intrinsic problem that is the inversely proportional to the impulse response length. Tradeoff should be exercised in terms of numerous factors such as computation, memory, accuracy etc.

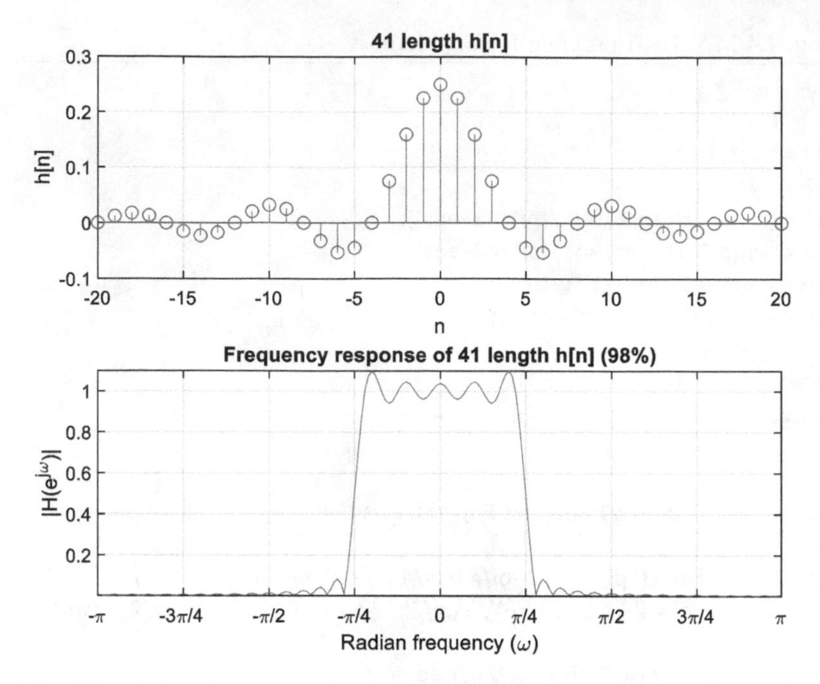

Fig. 4.11 41-length impulse response for given LPF and corresponding frequency response

Prog. 4.5 MATLAB program for Fig. 4.11

```
syms w;
n = (-20:1:20);
hn = sin(pi*(n+eps)/4)./(pi*(n+eps));
hw1 = sum(hn.*exp(-1i*w*n));
w1 = (-pi:pi/100:pi);
ohw1 = eval(subs(hw1,w,w1));
figure,
subplot(211), stem(n,hn), grid
xlabel('n')
ylabel('h[n]')
title('41 length h[n]')
subplot(212), plot(w1,abs(ohw1)), grid, axis tight
ax2 = gca;
ax2.XTick = [-pi -3*pi/4 -pi/2 -pi/4 0 pi/4 pi/2 3*pi/4 pi];
ax2.XTickLabel = {'-\pi', '-3\pi/4', '-\pi/2', '-\pi/4', '0', '\pi/4', '\pi/2', '3\pi/4',
'\pi'};
title('Frequency response of 41 length h[n] (98%)');
xlabel('Digital frequency (\omega)')
ylabel('|H(e^{j\omega})|')
```

This is not the final design since we still have problem with the designed impulse response. Note that the impulse response at negative index refers the future value of the input signal $x[n]$. The given impulse response cannot be realized properly because of the accessibility and causality until the filter obtains the future signals. Equation (4.38) convolution sum shows that $x[n + 20]$, $x[n + 19]$,... and $x[n + 1]$ are the future value of $x[n]$ at the time of n.

$$y[n] = \sum_{k=-20}^{20} h[k]x[n-k] = h[-20]x[n+20] + h[-19]x[n+19] + \ldots \\ + h[0]x[n] + h[1]x[n-1] + \ldots + h[20]x[n-20] \tag{4.38}$$

The solution to this problem is simple. Shift the impulse response to the right until all $h[n]$ values are located at the right-hand side of the plot; therefore, the impulse response is started and placed from the zero n value and higher. As shown in Eq. (4.39), the shift in time domain derives the phase distortion without modifying the magnitude in frequency domain. We specify the frequency response in magnitude; hence, the shifted impulse response also follows the designed requirement in Table 4.2.

$$h[n-d] \xrightarrow{\text{DTFT Analysis}} \sum_{n=-\infty}^{\infty} h[n-d]e^{-j\omega n} \\ = H(e^{j\omega})e^{-j\omega d} = \begin{cases} \text{Magnitude} : |H(e^{j\omega})| \\ \text{Phase} : \angle H(e^{j\omega}) - \omega d \end{cases} \tag{4.39}$$

The final design of the LPF is the $h_{LPF}[n]$, 20 samples delayed version of the $h[n]$, and denoted below with frequency response in magnitude.

$$h_{LPF}[n] = h[n-20] \tag{4.40}$$

Prog. 4.6 MATLAB program for Fig. 4.12

```
syms w;
n = (-20:1:20);
hn = sin(pi*(n+eps)/4)./(pi*(n+eps));
hw1 = sum(hn.*exp(-1i*w*n));
w1 = (-pi:pi/100:pi);
ohw1 = eval(subs(hw1,w,w1));
figure,
subplot(211), stem((0:40),hn), grid
xlabel('n')
ylabel('h_{LPF}[n]')
title('LPF')
subplot(212), plot(w1,abs(ohw1)), grid, axis tight
ax2 = gca;
ax2.XTick = [-pi -3*pi/4 -pi/2 -pi/4 0 pi/4 pi/2 3*pi/4 pi];
ax2.XTickLabel = {'-\pi', '-3\pi/4', '-\pi/2', '-\pi/4', '0', '\pi/4', '\pi/2', '3\pi/4',
'\pi'};
xlabel('\omega')
ylabel('|H_{LPF}(e^{j\omega})|')
```

The shifted impulse response retards the output signal with identical amount of delay as well. The signal shift is represented by the phase in the frequency domain. The phase modification of the impulse response provides the output relocation via the convolution sum process. Equation (4.41) shows the 20 sample in impulse response delay creates the 20-sample output shift to the right. Therefore, the output is delay 20 samples by impulse response shift. The $h_{LPF}[n]$ is final design for the LPF.

$$\begin{aligned} y_{LPF}[n] &= \sum_{k=0}^{40} h_{LPF}[k]x[n-k] = \sum_{k=0}^{40} h[k-20]x[n-k] \\ &= \sum_{m=-20}^{20} h[m]x[(n-20)-m] = y[n-20] \end{aligned} \tag{4.41}$$

Example 4.8

Design the LPF for Table 4.2. Use the DFT method. Show the impulse response from 0 to 99 as n.

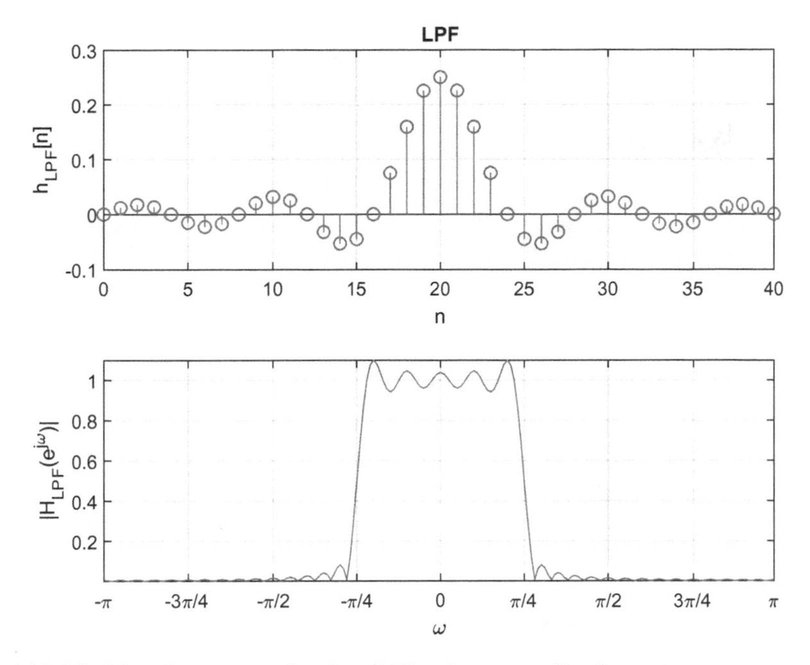

Fig. 4.12 Final impulse response for given LPF and corresponding frequency response

Solution

DFT provides the discrete frequency domain based on the k. The method starts from the DFT frequency domain; hence, we must define the DFT length first for exponential and summation part. Based on the example statement, the DFT length is 100 for N. From the given LPF specification, the equivalent frequency for the given k can be computed by below equation.

$$f_k = \frac{2\pi}{N}\frac{f_s}{2\pi}k = \frac{f_s}{N}k = \Delta_f k$$

The Δ_f is the frequency resolution for the DFT configuration and defined as f_s/N. The derived Δ_f is 80 Hz/sample from 100 N and 8000 Hz sampling frequency. In order to pass the low frequencies until 1000 Hz, the magnitude one should cover up to 12 k value because the 1000/(8000/100) is equal to the 12.5. Note that the real signal requires the conjugated pair on the symmetrical location for the positive and negative k index as shown in below equation and Fig. 4.13. According to the cylindrical and periodical property of DFT, the frequency response for the negative k can be relocated at the other end of given DFT configuration as shown in Fig. 4.13.

$$H[k] = H^*[N - k]$$

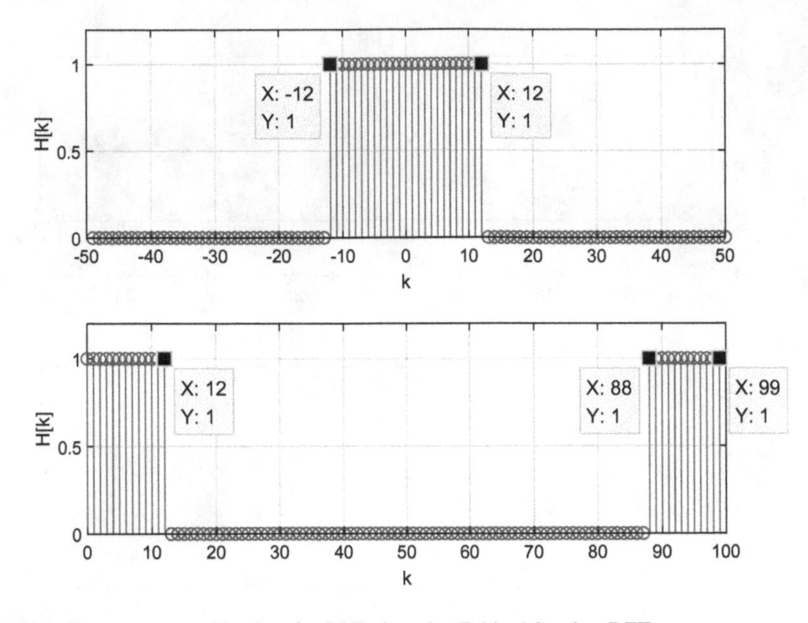

Fig. 4.13 Frequency specification for LPF given by Table 4.2 using DFT

Prog. 4.7 MATLAB program for Fig. 4.13

```
n1 = 0:99;
n2 = -49:1:50;
hk1 = [ones(1,13) zeros(1,100-12-13) ones(1,12)];
hk2 = [zeros(1,37) ones(1,25) zeros(1,38)];
figure,
subplot(211),stem(n2,hk2), grid
xlim([-50 50]);
ylim([0 1.2]);
xlabel('k');
ylabel('H[k]')
subplot(212),stem(n1,hk1), grid
xlim([0 100]);
ylim([0 1.2]);
xlabel('k');
ylabel('H[k]')
```

You can take either DFT configurations as $-49 \sim 50$ or $0 \sim 99$ for k range in LPF design. Based on the $-49 \sim 50$ k range, the DFT synthesis for the LPF impulse response can be written as below.

$$h[n] = \frac{1}{100}\sum_{k=-12}^{12} 1e^{j\frac{2\pi}{100}kn} = \frac{1}{100}\frac{e^{-j\frac{2\pi}{100}12n} - e^{j\frac{2\pi}{100}13n}}{1 - e^{j\frac{2\pi}{100}n}}$$

$$= \frac{1}{100}\frac{\sin\left(\frac{25\pi}{100}n\right)}{\sin\left(\frac{\pi}{100}n\right)} \text{ where } \frac{1}{100}\lim_{n\to 0}\frac{\sin\left(\frac{25\pi}{100}n\right)}{\sin\left(\frac{\pi}{100}n\right)} = \frac{1}{4}$$

The derived impulse response shows 100 sample periodicity due to the numerator and denominator property those represent the 8 and 200 periodicity, respectively. The overall impulse response, numerator, and denominator are illustrated in Fig. 4.14. Note that the singular points at every 100 index delivers finite value 1/4 based on the L'Hôpital's rule [1] shown in above equation.

Prog. 4.8 MATLAB program for Fig. 4.14

```
n1 = 0:99;
hn1 = 0.01*sin(25*pi*(n1+eps)/100)./sin(pi*(n1+eps)/100);
y1 = sin(25*pi*n1/100);
y2 = sin(pi*n1/100);

figure,
subplot(211),stem(n1,hn1), grid
xlim([0 100]);
xlabel('n');
ylabel('h[n]')
subplot(212),
yyaxis left;
plot(n1,0.01*y1)
yyaxis right;
plot(n1,y2), grid
xlim([0 100]);
ylim([-1 1]);
xlabel('n');
```

■

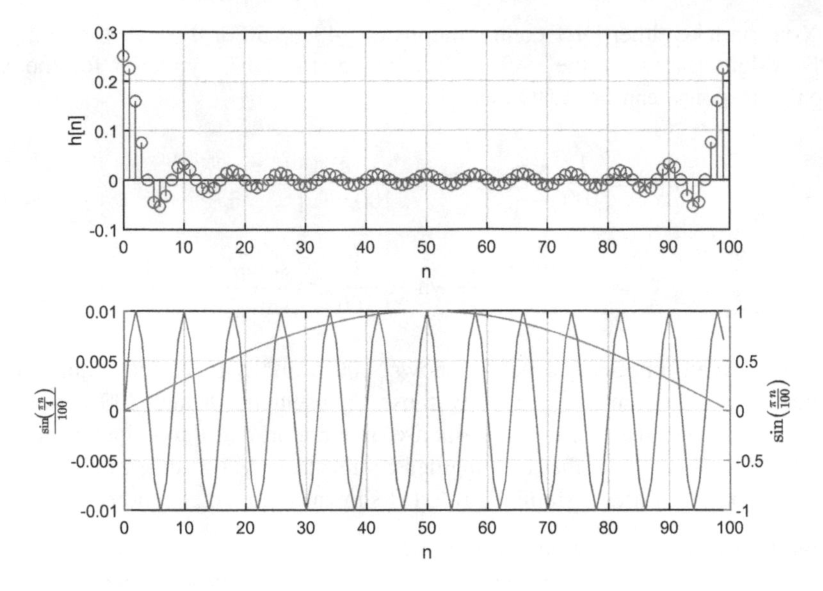

Fig. 4.14 Derived impulse response based on the DFT and its numerator and denominator plot

Likewise, the DFT range, we can observe the either impulse response configurations denoted Fig. 4.15 as $-49 \sim 50$ or $0 \sim 99$ for n range in LPF design. The shape of the impulse response is similar to the previously designed DTFT LPF with periodicity. Does the one period of the impulse response represent final LPF design? Let's see further.

Prog. 4.8 MATLAB program for Fig. 4.15

```
n1 = 0:99;
n2 = -49:1:50;
hn1 = 0.01*sin(25*pi*(n1+eps)/100)./sin(pi*(n1+eps)/100);
hn2 = 0.01*sin(25*pi*(n2+eps)/100)./sin(pi*(n2+eps)/100);
figure,
subplot(211),stem(n2,hn2), grid
xlim([-50 50]);
ylabel('h[n]')
subplot(212),stem(n1,hn1), grid
xlim([0 100]);
xlabel('n');
ylabel('h[n]')
```

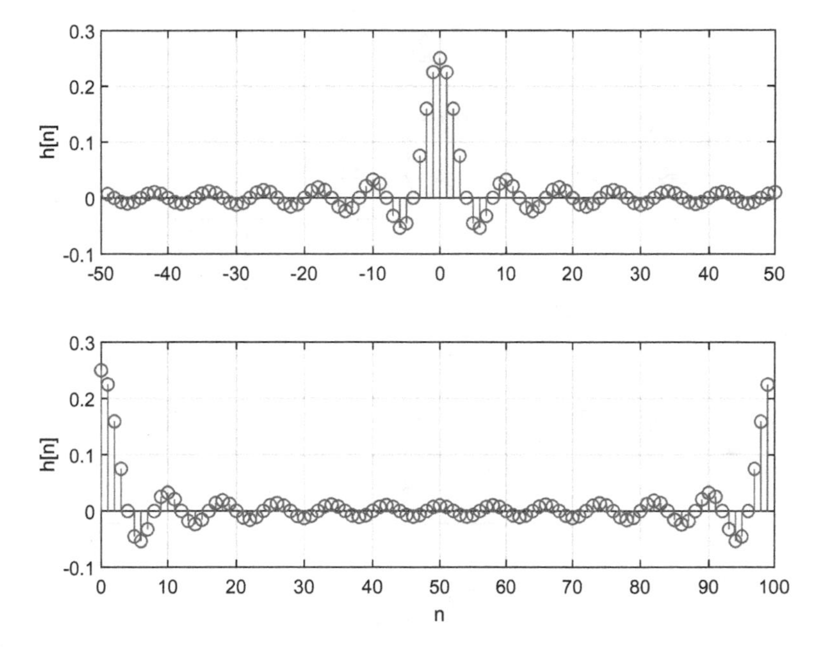

Fig. 4.15 Impulse response configuration as $-49 \sim 50$ and $0 \sim 99$ for LPF with DFT

The convolution sum process cannot handle the infinite length impulse response; therefore, we need to take certain range of the response. In addition, the negative index section cannot be employed due to the accessibility and causality of future input signal. Let's choose the one complete period of impulse response from zero to higher index. The convolution sum between the input signal and selected impulse response provides the output signal. The example in Fig. 4.16 illustrates the input signal with 100 sample period containing the even 100, 50, 25, 20, 10, and 5 sample period signal components. The output for the selected range is presented at the bottom subplot in Fig. 4.16. Note that the output is the part of stream outcome

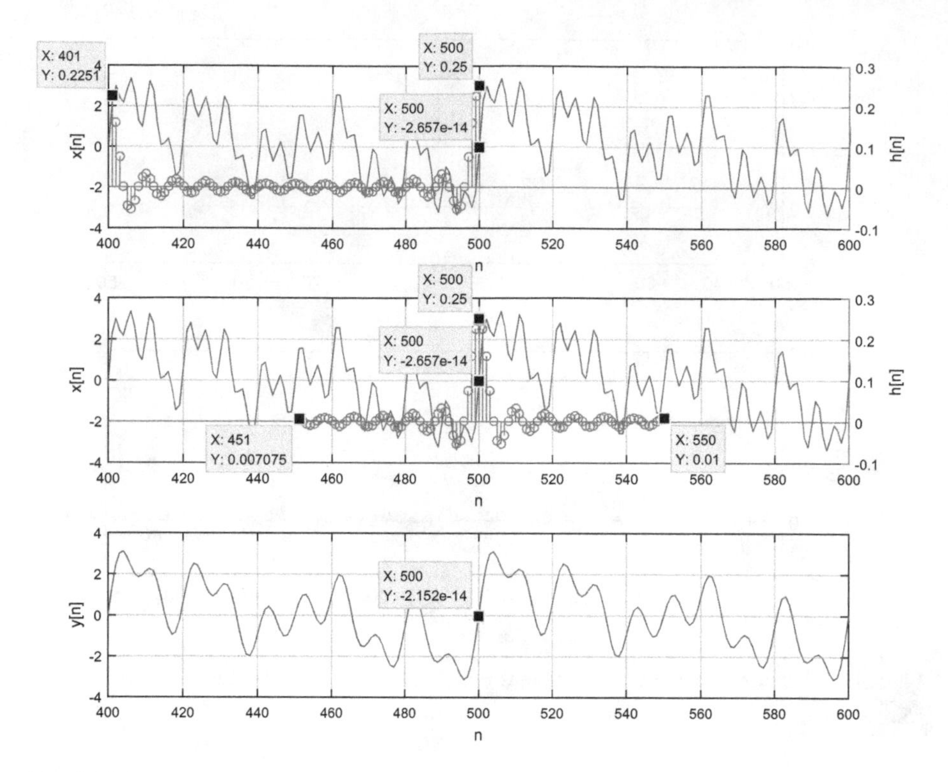

Fig. 4.16 Input signal (100 sample period) with two impulse response configurations. Bottom subplot is the output $y[n]$ from above configurations

without any transition introduced by the incomplete overlapped convolution sum. The transitions generally are appeared at the time edges of input signal such as start and end time. The top of Fig. 4.16 demonstrates the input signal and the impulse response overlapped to compute the output at 500 index as $y[500]$. Provided that the future input values can be accessible, $y[500]$ computation is illustrated in the middle subplot in Fig. 4.16.

Prog. 4.9 MATLAB program for Fig. 4.16

```
n1 = 400:499;
hn1 = 0.01*sin(25*pi*(n1+eps)/100)./sin(pi*(n1+eps)/100);
n3 = 451:550;
hn3 = 0.01*sin(25*pi*(n3+eps)/100)./sin(pi*(n3+eps)/100);
n2 = 0:999;
xn1 =
sin(2*pi*1*n2/100)+sin(2*pi*2*n2/100)+sin(2*pi*4*n2/100)+sin(2*pi*5*n2
/100)+sin(2*pi*10*n2/100)+sin(2*pi*20*n2/100);
yn1 = conv(xn1,hn1);

figure,
yyaxis left;
subplot(311),plot(n2,xn1), grid
ylabel('x[n]')
hold on;
yyaxis right;
stem(n1+1,fliplr(hn1));
ylabel('h[n]')
xlim([400 600]);
xlabel('n');
hold off;

yyaxis left;
subplot(312),plot(n2,xn1), grid
ylabel('x[n]')
hold on;
yyaxis right;
stem(n3,(hn3));
ylabel('h[n]')
xlim([400 600]);
xlabel('n');
hold off;

subplot(313),plot(0:(length(n1)+length(n2)-2),yn1), grid
xlim([400 600]);
ylim([-4 4]);
xlabel('n');
ylabel('y[n]')
```

The impulse response profiles at the top and middle of Fig. 4.16 generate the identical output located at Fig. 4.16 bottom. The signal periodicity derives the future from the past pattern as Eq. (4.42). Therefore, the $x[501]$, $x[502]$, etc. are equivalent to the $x[401]$, $x[402]$, etc., respectively.

$$x[n] = x[n + kN] \text{ where } k \in \mathbb{Z} \text{ and } N \text{ is the signal } x \text{ period} \tag{4.42}$$

The outputs from two impulse responses are only equalized for the periodic signal with same periodic length. Otherwise, the derived impulse response cannot be used directly for the LPF design. Certain truncation and shift are required for the proper convolution sum process like DTFT LPF design. The complete impulse response for given length N is illustrated with frequency response magnitude in Fig. 4.17. Since there is no modification on the impulse response, the LPF response

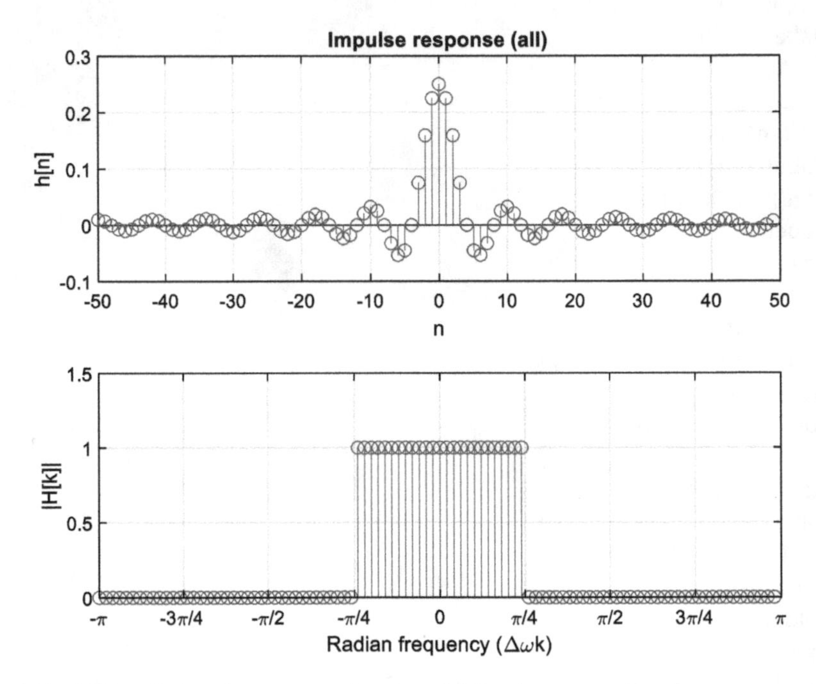

Fig. 4.17 100-length impulse response for given LPF and corresponding frequency response. $\Delta\omega = 2\pi/N$

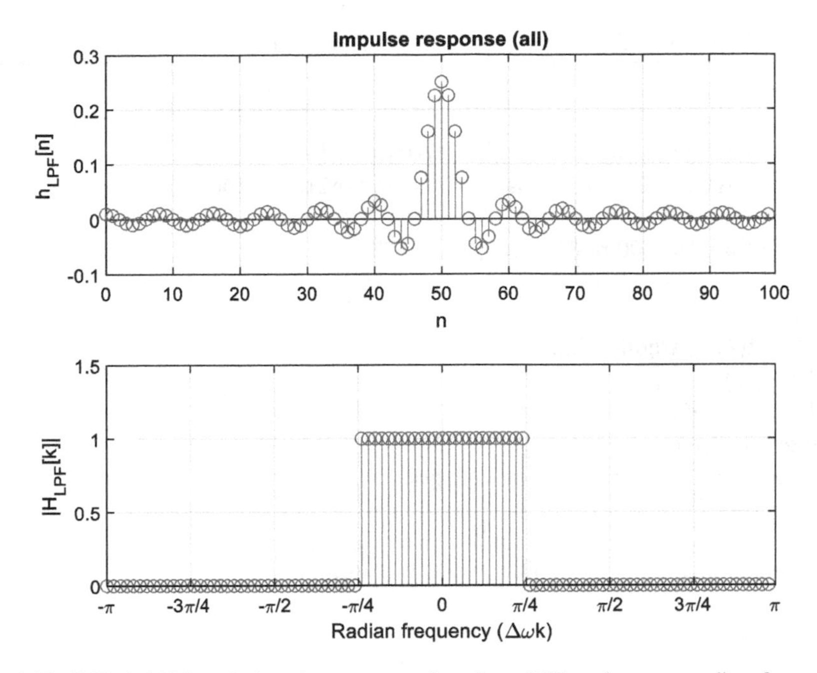

Fig. 4.18 Shifted 100-length impulse response for given LPF and corresponding frequency response. $\Delta\omega = 2\pi/N$

is neat and clean in frequency domain. While the impulse response is periodic and infinite, the frequency response is the DFT of the one period $h[n]$. Note that the $\Delta\omega$ is the radian frequency resolution as $2\pi/N$.

If you want to use the complete impulse response, shift the main lobe to the one period center as Eq. (4.43) and Fig. 4.18. The frequency response is identical to the non-shifted version of the counterpart.

$$h_{LPF}[n] = h\left[n - \frac{N}{2}\right] \tag{4.43}$$

Prog. 4.10 MATLAB program for Figs. 4.17 and 4.18

```
n1 = 0:99;
n2 = -50:1:49;
hn1 = 0.01*sin(25*pi*(n1+eps)/100)./sin(pi*(n1+eps)/100);
hn2 = 0.01*sin(25*pi*(n2+eps)/100)./sin(pi*(n2+eps)/100);
fhn2 = fftshift(fft(hn2));
faxis = (-pi:2*pi/100:pi-2*pi/100);

figure,
subplot(211), stem(n1,hn2), grid
xlabel('n')
ylabel('h_{LPF}[n]')
title('Impulse response (all)')
subplot(212), stem(faxis,abs(fhn2)), grid
ax2 = gca;
ax2.XTick = [-pi -3*pi/4 -pi/2 -pi/4 0 pi/4 pi/2 3*pi/4 pi];
ax2.XTickLabel = {'-\pi', '-3\pi/4', '-\pi/2', '-\pi/4', '0', '\pi/4', '\pi/2', '3\pi/4',
'\pi'};
xlim([-pi pi]);
ylim([0 1.5]);
xlabel('Digital frequency (\Delta\omegak)')
ylabel('|H_{LPF}[k]|')

figure,
subplot(211), stem(n2,hn2), grid
xlabel('n')
ylabel('h[n]')
title('Impulse response (all)')
subplot(212), stem(faxis,abs(fhn2)), grid
ax2 = gca;
ax2.XTick = [-pi -3*pi/4 -pi/2 -pi/4 0 pi/4 pi/2 3*pi/4 pi];
ax2.XTickLabel = {'-\pi', '-3\pi/4', '-\pi/2', '-\pi/4', '0', '\pi/4', '\pi/2', '3\pi/4',
'\pi'};
xlim([-pi pi]);
ylim([0 1.5]);
xlabel('Digital frequency (\Delta\omegak)')
ylabel('|H[k]|')
```

The DFT property describes the time shift as the phase modification without touching the magnitude as Eq. (4.44). Over the time axis, any relocation of the impulse response maintains our design goal of LPF in Table 4.2.

$$h[n-d] \xrightarrow{\text{DFT Analysis}} \sum_{n=0}^{N-1} h[n-d]e^{-j\frac{2\pi}{N}kn}$$
$$= H[k]e^{-j\frac{2\pi}{N}kd} = \begin{cases} \text{Magnitude} : |H[k]| \\ \text{Phase} : \angle H[k] - \frac{2\pi}{N}kd \end{cases} \tag{4.44}$$

The corresponding output is equally delayed by the shifted impulse response at the convolution sum process as Eq. (4.45).

$$y_{LPF}[n] = \sum_{k=0}^{99} h_{LPF}[k]x[n-k] = y\left[n - \tfrac{N}{2}\right] \tag{4.45}$$

The previously shown periodic signal is used to provide the convolution sum output in Fig. 4.19. The overlapped plot between the input signal and impulse response illustrates the $y[550]$ output computation. The overlapped section represents the convolution sum operation with multiplication and accumulation. We

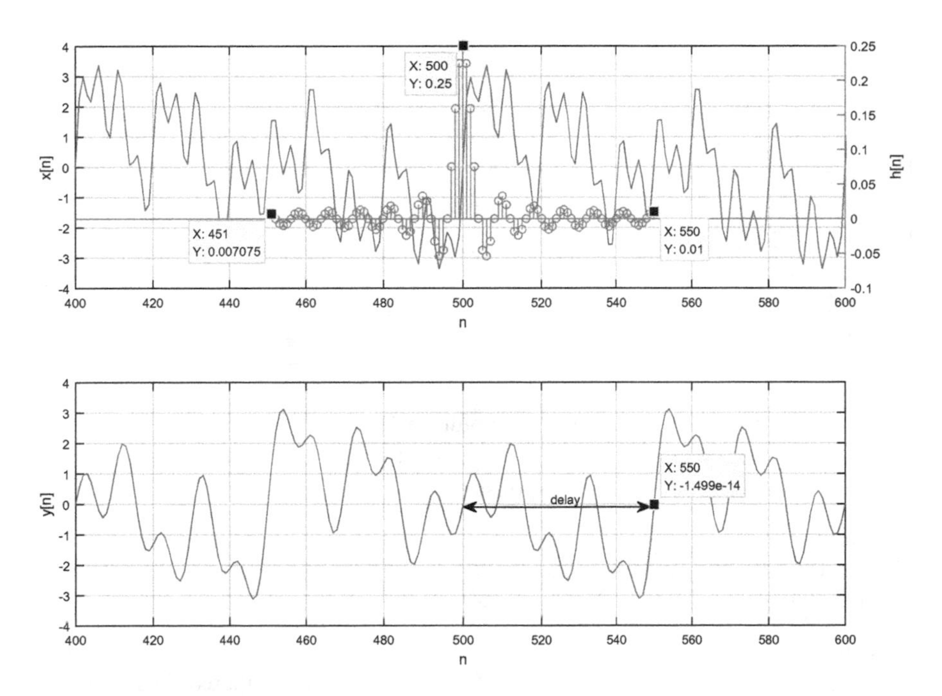

Fig. 4.19 Input signal (100 sample period) with shifted impulse response. Bottom subplot is the corresponding output $y[n]$

assume that the future data is not accessible to maintain the causality of the system in this example. The $x[550]$ and previous 99 data involves to compute the output $y[550]$. Also comparing with the previous instance, the output is delayed for 50 samples due to the impulse response shift. This is one example of LPF based on the DFT method.

Prog. 4.11 MATLAB program for Fig. 4.19

```
n1 = -50:1:49;
hn1 = 0.01*sin(25*pi*(n1+eps)/100)./sin(pi*(n1+eps)/100);
n2 = 0:999;
xn1 =
sin(2*pi*1*n2/100)+sin(2*pi*2*n2/100)+sin(2*pi*4*n2/100)+sin(2*pi*5*n2
/100)+sin(2*pi*10*n2/100)+sin(2*pi*20*n2/100);
yn1 = conv(xn1,hn1);
n3 = 451:550;

yyaxis left;
subplot(211),plot(n2,xn1), grid
ylabel('x[n]')
hold on;
yyaxis right;
stem(n3,fliplr(hn1));
ylabel('h[n]')
xlim([400 600]);
xlabel('n');
hold off;
subplot(212),plot(0:(length(n1)+length(n2)-2),yn1), grid
xlim([400 600]);
ylim([-4 4]);
xlabel('n');
ylabel('y[n]')
```

What if we want to reduce the convolution length? How we can decrease the impulse response size? We can choose from two methods that shrink the DFT length and truncate the impulse response.

Example 4.9
Design the LPF for Table 4.2. Use the DFT method. Show the impulse response from 0 to 19 as n.

Solution
Change the DFT length to design the LPF as 20 N. The computed Δ_f is 400 Hz/sample from 20 N and 8000 Hz sampling frequency. In order to pass the low

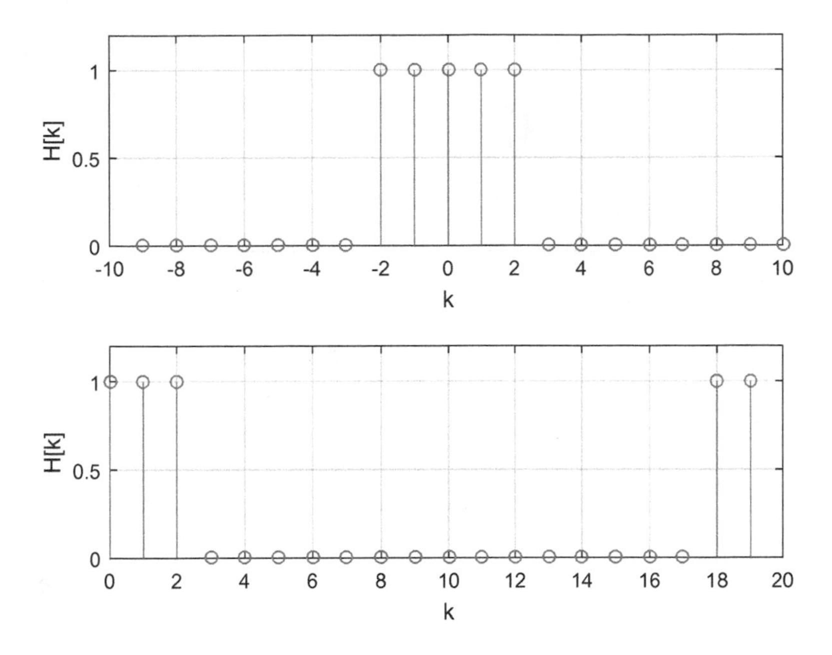

Fig. 4.20 Frequency specification for LPF given by Table 4.2 using DFT

frequencies until 1000 Hz, the magnitude one should cover up to 2 k value because the 1000/(8000/20) is equal to the 2.5. According to the cylindrical and periodical property of DFT, the frequency response for the negative k can be relocated at the other end of given DFT configuration as shown in Fig. 4.20.

Based on the $-9 \sim 10$ k range, the DFT synthesis for the LPF impulse response can be written as below.

$$h[n] = \frac{1}{20}\sum_{k=-2}^{2} 1e^{j\frac{2\pi}{20}kn} = \frac{1}{20}\frac{e^{-j\frac{2\pi}{20}2n} - e^{j\frac{2\pi}{20}3n}}{1 - e^{j\frac{2\pi}{20}n}}$$

$$= \frac{1}{20}\frac{\sin\left(\frac{5\pi}{20}n\right)}{\sin\left(\frac{\pi}{20}n\right)} \text{ where } \frac{1}{20}\lim_{n\to 0}\frac{\sin\left(\frac{5\pi}{20}n\right)}{\sin\left(\frac{\pi}{20}n\right)} = \frac{1}{4}$$

The derived impulse response shows 20 sample periodicity and the singular points at every 20 index delivers finite value 1/4 based on the L'Hôpital's rule [1] shown in above equation. In order to meet the causality of the system, the computed impulse response is shifted to the right for 9 samples as shown in Fig. 4.21.

■

The previously used periodic signal is employed to generate the convolution sum output in Fig. 4.22. The overlapped plot between the input signal and impulse response illustrates the $y[509]$ output computation. The $x[509]$ and previous 19 data

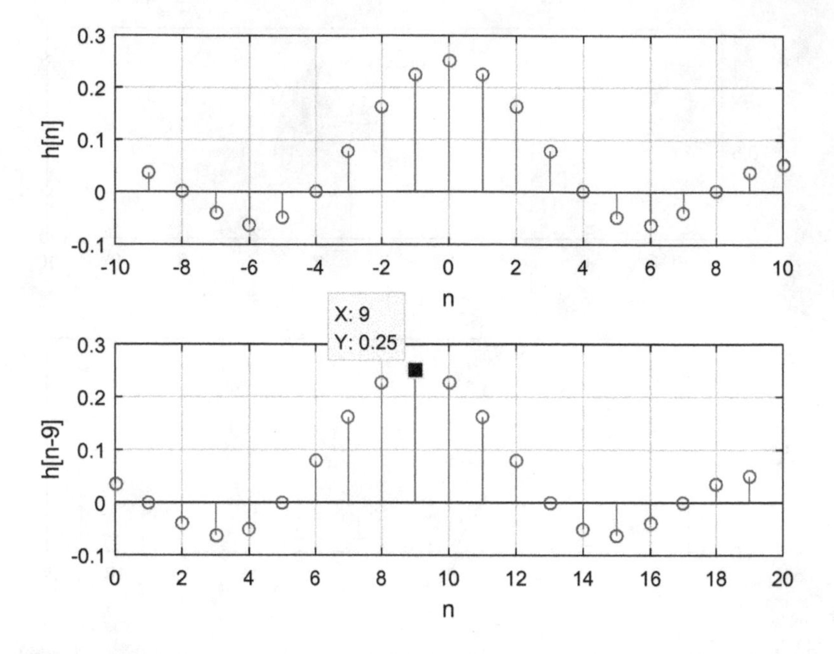

Fig. 4.21 Derived impulse response and shifted impulse response for causal filter system

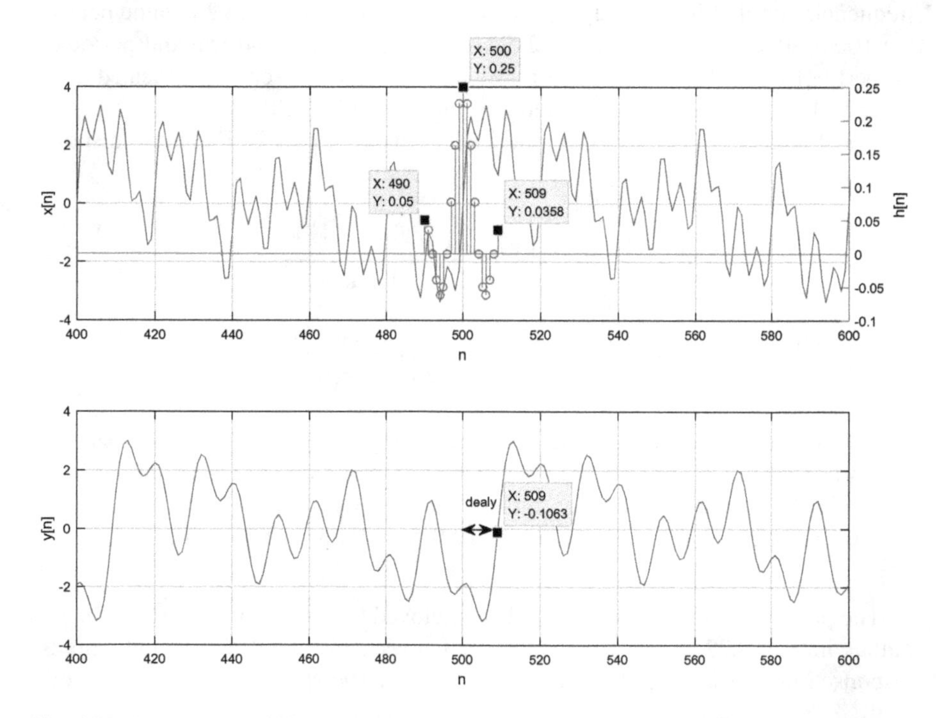

Fig. 4.22 Input signal (100 sample period) with derived impulse response. The overlapped configuration provides for the $y[509]$ computation. Bottom subplot is the corresponding output $y[n]$

involves to compute the output $y[509]$. Also, the output is delayed for 9 samples due to the impulse response shift. This is one example of reduced length LPF based on the DFT method.

Prog. 4.12 MATLAB program for Fig. 4.22

```
n1 = -9:1:10;
hn1 = (1/20)*sin(5*pi*(n1+eps)/20)./sin(pi*(n1+eps)/20);
n2 = 0:999;
xn1 =
sin(2*pi*1*n2/100)+sin(2*pi*2*n2/100)+sin(2*pi*4*n2/100)+sin(2*pi*5*n2
/100)+sin(2*pi*10*n2/100)+sin(2*pi*20*n2/100);
yn1 = conv(xn1,hn1);
n3 = 490:509;

yyaxis left;
subplot(211),plot(n2,xn1), grid
ylabel('x[n]')
hold on;
yyaxis right;
stem(n3,fliplr(hn1));
ylabel('h[n]')
xlim([400 600]);
xlabel('n');
hold off;
subplot(212),plot(0:(length(n1)+length(n2)-2),yn1), grid
xlim([400 600]);
ylim([-4 4]);
xlabel('n');
ylabel('y[n]')
```

Comparing the outputs between the 100 and 20 length DFT for LPF design, the results are identical except the time delay as shown in Fig. 4.23. If the short impulse response works properly, there is no need to increase the impulse response since the length costs the computation and memory of the system. Does the both configurations really provide the same output for all input signals? Let's figure out.

Prog. 4.13 MATLAB program for Fig. 4.23

```
n1 = -9:1:10;
hn1 = (1/20)*sin(5*pi*(n1+eps)/20)./sin(pi*(n1+eps)/20);
n3 = -50:1:49;
hn3 = 0.01*sin(25*pi*(n3+eps)/100)./sin(pi*(n3+eps)/100);
n2 = 0:999;
xn1 =
sin(2*pi*1*n2/100)+sin(2*pi*2*n2/100)+sin(2*pi*4*n2/100)+sin(2*pi*5*n2
/100)+sin(2*pi*10*n2/100)+sin(2*pi*20*n2/100);
yn1 = conv(xn1,hn1);
yn2 = conv(xn1,hn3);

figure,
subplot(211),plot(0:(length(n3)+length(n2)-2),yn2), grid
xlim([400 600]);
ylim([-4 4]);
title('LPF with 100 length (DFT method)')
ylabel('y[n]')
subplot(212),plot(0:(length(n1)+length(n2)-2),yn1), grid
xlim([400 600]);
ylim([-4 4]);
title('LPF with 20 length (DFT method)')
xlabel('n');
ylabel('y[n]')
```

The frequency response of the 20-sample impulse response is illustrated in Fig. 4.24. Below the $\pi/4$ radian frequency components are propagated through the filter without losing the signal magnitude at all. On the other side, above the $\pi/4$ radian frequency components are completely blocked by the filter because of zero magnitude in frequency response. No fluctuations are observed in the passing and blocking frequency band in the frequency response. Similar to the 100-sample impulse response shown in Fig. 4.18, the frequency response demonstrates the ideal response for LPF except low frequency density Δ_f. According to the above filter outputs in Fig. 4.23, short impulse response performs the specified LPF function well.

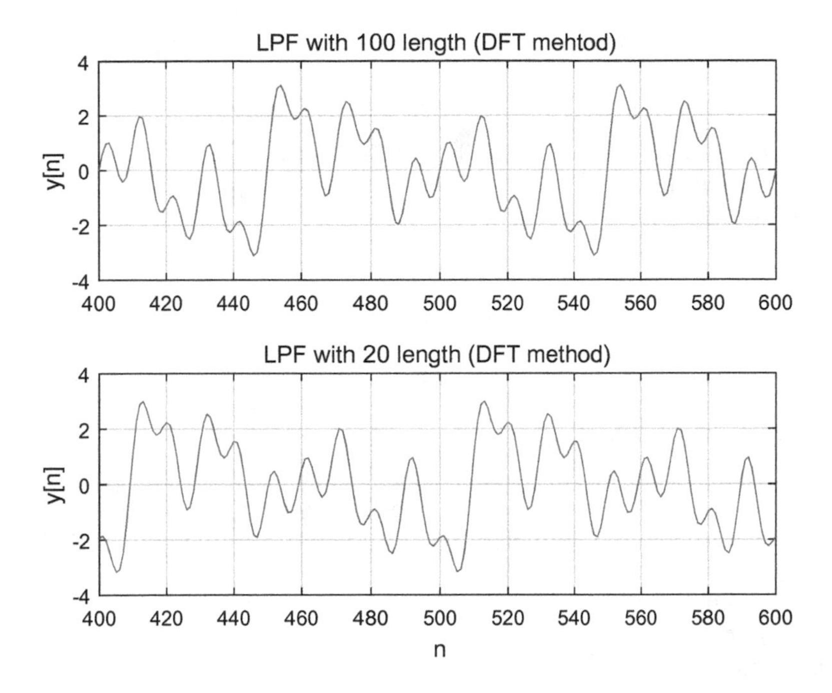

Fig. 4.23 LPF output with 100 and 20 length impulse response

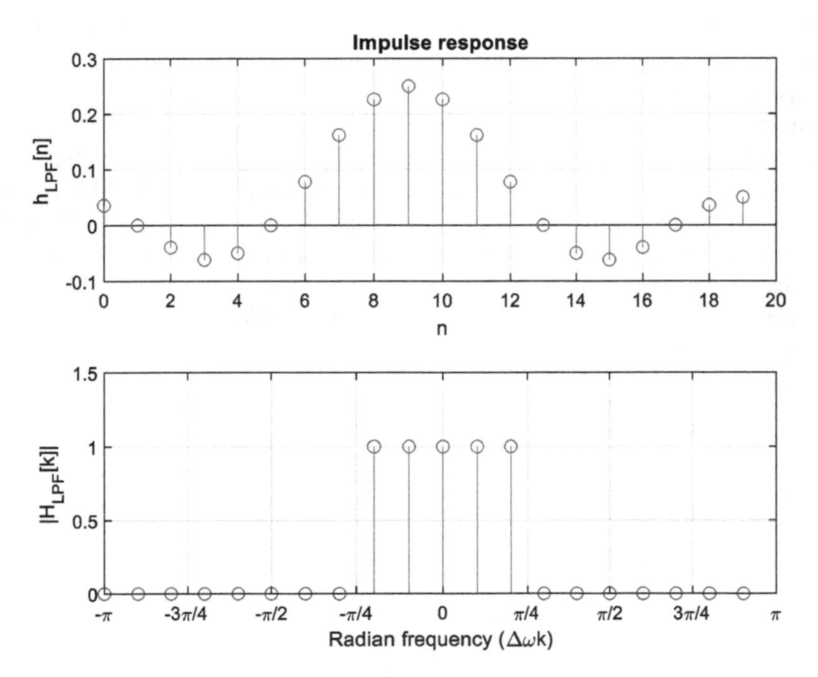

Fig. 4.24 Shifted impulse response and corresponding frequency response in magnitude

Prog. 4.14 MATLAB program for Fig. 4.24

```
n1 = 0:1:19;
n2 = -9:1:10;
hn1 = (1/20)*sin(5*pi*(n2+eps)/20)./sin(pi*(n2+eps)/20);
fhn1 = fftshift(fft(hn1));
faxis = (-pi:2*pi/20:pi-2*pi/20);

figure,
subplot(211), stem(n1,hn1), grid
xlabel('n')
ylabel('h_{LPF}[n]')
title('Impulse response')
subplot(212), stem(faxis,abs(fhn1)), grid
ax2 = gca;
ax2.XTick = [-pi -3*pi/4 -pi/2 -pi/4 0 pi/4 pi/2 3*pi/4 pi];
ax2.XTickLabel = {'-\pi', '-3\pi/4', '-\pi/2', '-\pi/4', '0', '\pi/4', '\pi/2', '3\pi/4',
'\pi'};
xlim([-pi pi]);
ylim([0 1.5]);
xlabel('Digital frequency (\Delta\omegak)')
ylabel('|H_{LPF}[k]|')
```

If the filter performance is insensitive to the filter length, we do not need to choose the longer filter that increases the computation time and output delay. While the discrete frequencies provide the ideal LPF response, the intermediate frequency response is unknown to the filter designer based on the given information. Therefore, the continuous frequency response is required to be investigated for further analysis. Figures 4.25 and 4.26 demonstrate the DTFT of the 20 length LPF designed from DFT method. The continuous and discrete plot represent the DTFT and DFT analysis, respectively. The continuous frequency response from the DTFT denotes the ripples in the passing and blocking band in the frequency domain. The discrete frequency response samples the ideal locations from the continuous counterpart to pretend the perfect response of LPF.

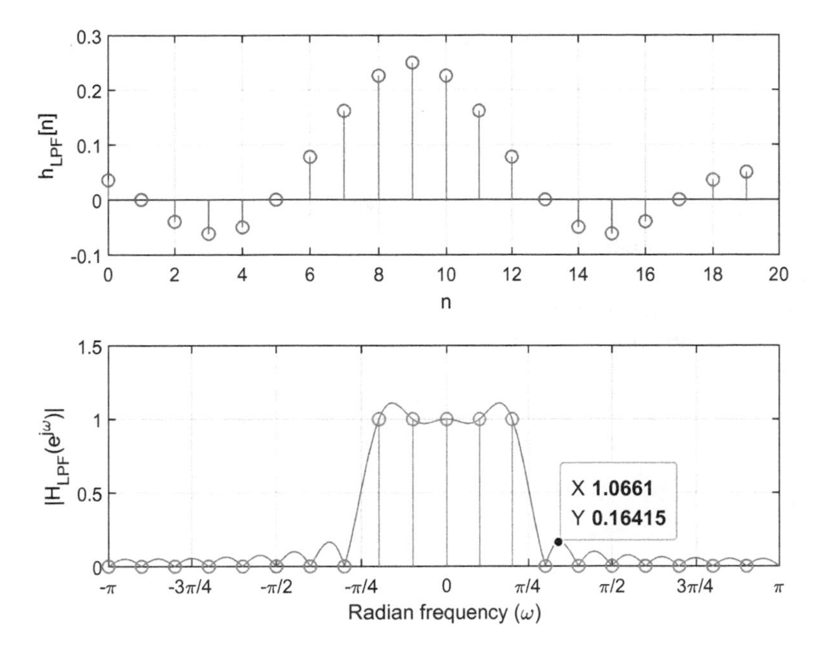

Fig. 4.25 LPF impulse response and corresponding frequency response based on the DFT (discrete) and DTFT (continuous)

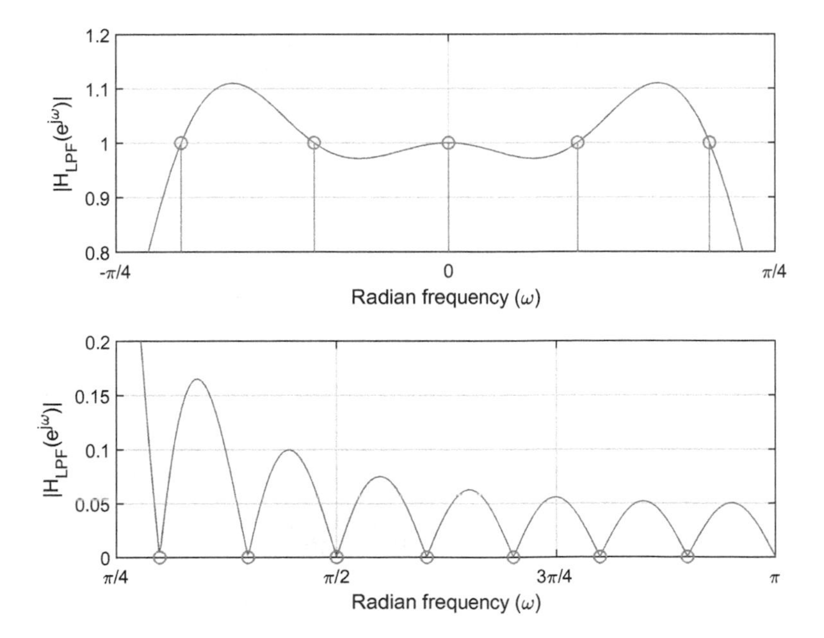

Fig. 4.26 Enlarged passband and stropband of Fig. 4.25 frequency response

Prog. 4.15 MATLAB program for Figs. 4.25 and 4.26

```
syms omega real;
n1 = -9:1:10;
n2 = 0:1:19;
rfreq = linspace(-pi,pi,1000);
hn1 = (1/20)*sin(5*pi*(n1+eps)/20)./sin(pi*(n1+eps)/20);
coeff = exp(1j*omega*n1);
Hint1 = hn1.*coeff;
Homega1 = sum(Hint1);
Homega2 = subs(Homega1,omega,rfreq);
Homega3 = eval(Homega2);
fhn1 = fftshift(fft(hn1));
faxis = (-pi:2*pi/20:pi-2*pi/20);

figure,
subplot(211)
stem(n2,hn1), grid
ylabel('h_{LPF}[n]')
xlabel('n');
subplot(212)
plot(rfreq,abs(Homega3))
hold on;
stem(faxis,abs(fhn1)), grid
ax2 = gca;
ax2.XTick = [-pi -3*pi/4 -pi/2 -pi/4 0 pi/4 pi/2 3*pi/4 pi];
ax2.XTickLabel = {'-\pi', '-3\pi/4', '-\pi/2', '-\pi/4', '0', '\pi/4', '\pi/2', '3\pi/4',
'\pi'};
xlim([-pi pi]);
ylim([0 1.5]);
xlabel('Radian frequency (\omega)')
ylabel('|H_{LPF}(\omega)|')
```

Let's consider the ideal case for the length-20 LPF. The input signal $x[n]$ with 20 sample period shown in Eq. (4.46) contains the five cosine functions and constant value.

$$\begin{aligned} x[n] = 1 + \cos\left(\frac{2\pi}{20}n\right) + \cos\left(\frac{4\pi}{20}n\right) + \cos\left(\frac{8\pi}{20}n\right) \\ + \cos\left(\frac{10\pi}{20}n\right) + \cos\left(\frac{20\pi}{20}n\right) \end{aligned} \tag{4.46}$$

Table 4.3 Input signal $x[n]$ components from Eq. (4.46) for LPF processing

Trigonometric Function	Period	Radian frequency	Cyclic frequency (f_s = 8000 Hz) (Hz)	Propagated?
$\cos\left(\frac{2\pi}{20}n\right)$	20 Samples	$\frac{\pi}{10}$	400	Passed
$\cos\left(\frac{4\pi}{20}n\right)$	10 Samples	$\frac{\pi}{5}$	800	Passed
$\cos\left(\frac{8\pi}{20}n\right)$	5 Samples	$\frac{2\pi}{5}$	1600	Blocked
$\cos\left(\frac{10\pi}{20}n\right)$	4 Samples	$\frac{\pi}{2}$	2000	Blocked
$\cos\left(\frac{20\pi}{20}n\right)$	2 Samples	π	4000	Blocked

Table 4.3 illustrates individual trigonometric functions which denotes the certain integer period. Note that the least common multiple of the all functions is 20 for whole period of the signal. For the given design parameter of the LPF, the signal components up to $\pi/4$ or 1000 Hz are propagated thorough the filter and the higher than the threshold frequency should be blocked by the convolution sum process.

The discrete frequency distributions based on the 20-sample length DFT are presented at Fig. 4.27 for the input and output of the LPF. In addition to the constant value, the input signal retains the all frequency components described in Table 4.3. The output signal from the LPF perfectly maintain or remove the passing and blocking band signal in filter output. Only the zero, $\pi/10$, and $\pi/5$ radian

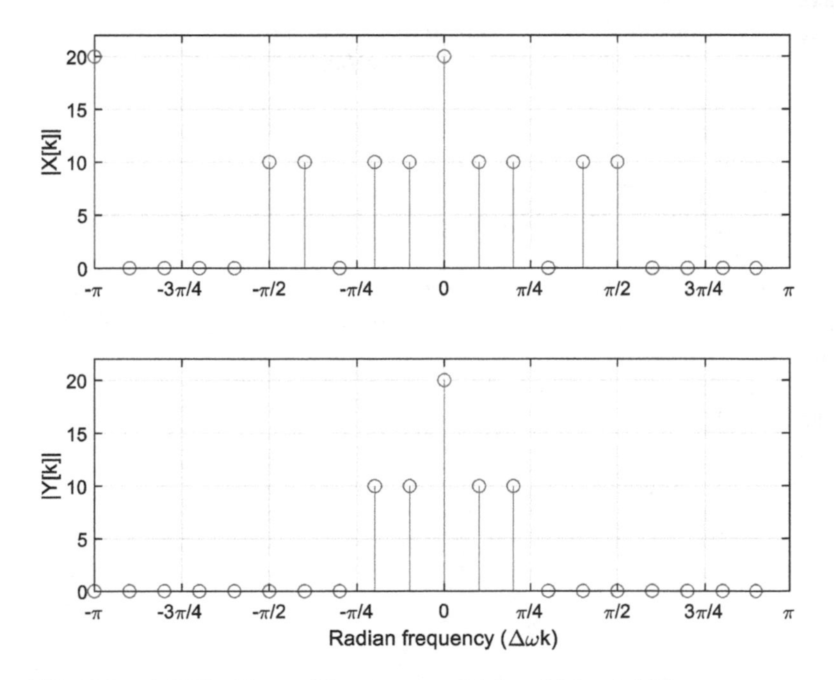

Fig. 4.27 20 length DFT of input $x[n]$ and output $y[n]$ from 20 length LPF

frequencies are survived from the LPF processing without changing the signal magnitude at all. The other frequencies are completely eliminated and barely seen in the output frequency profile. This situation is contributed to the ideal condition that is the signal period is equal to the DFT LPF length.

Prog. 4.16 MATLAB program for Fig. 4.27

```
n1 = -9:1:10;
hn1 = (1/20)*sin(5*pi*(n1+eps)/20)./sin(pi*(n1+eps)/20);
n2 = 0:999;
xn1 =
cos(2*pi*1*n2/20)+cos(2*pi*2*n2/20)+cos(2*pi*4*n2/20)+cos(2*pi*5*n2/2
0)+cos(2*pi*10*n2/20)+1;
yn1 = conv(xn1,hn1);
xn2 = xn1(501:520);
yn2 = yn1(501:520);
fxn2 = fftshift(fft(xn2));
fyn2 = fftshift(fft(yn2));
faxis = (-pi:2*pi/20:pi-2*pi/20);

figure,
subplot(211), stem(faxis,abs(fxn2)), grid
ax2 = gca;
ax2.XTick = [-pi -3*pi/4 -pi/2 -pi/4 0 pi/4 pi/2 3*pi/4 pi];
ax2.XTickLabel = {'-\pi', '-3\pi/4', '-\pi/2', '-\pi/4', '0', '\pi/4', '\pi/2', '3\pi/4',
'\pi'};
xlim([-pi pi]);
ylim([0 22]);
ylabel('|X[k]|');
subplot(212), stem(faxis,abs(fyn2)), grid
ax2 = gca;
ax2.XTick = [-pi -3*pi/4 -pi/2 -pi/4 0 pi/4 pi/2 3*pi/4 pi];
ax2.XTickLabel = {'-\pi', '-3\pi/4', '-\pi/2', '-\pi/4', '0', '\pi/4', '\pi/2', '3\pi/4',
'\pi'};
xlim([-pi pi]);
ylim([0 22]);
ylabel('|Y[k]|')
xlabel('Digital frequency (\Delta\omegak)');
```

In addition to the defined $x[n]$ signal, $7\pi/20$ radian frequency is included to the new input signal in Eq. (4.47). The $7\pi/20$ component presents the 40-sample period and 1400 Hz frequency based on the 8000 Hz sampling frequency. Therefore, the overall period of the new input signal is the 40 instead of 20 samples. The newly inserted frequency should be eliminated by the designed LPF due to the over the limit frequency range.

$$x[n] = 1 + \cos\left(\frac{2\pi}{20}n\right) + \cos\left(\frac{4\pi}{20}n\right) + \cos\left(\frac{7\pi}{20}n\right) + \cos\left(\frac{8\pi}{20}n\right) + \cos\left(\tfrac{10\pi}{20}n\right) + \cos\left(\tfrac{20\pi}{20}n\right) \tag{4.47}$$

The designed LPF computes the output by using the derived 20 length impulse response. While the 20 coefficients for the LPF is generated from the frequency domain, the filter processing is performed by the convolution sum over the complete time domain. The single frame of the infinite input signal is utilized for LPF processing. In order to compare the frequency information between the input and output, the frequency distribution is visualized by the 40 length DFT as shown in Fig. 4.28. The choice of the DFT length is justified by the $7\pi/20(=14\pi/40)$ radian frequency that is the integer multiple of the $1\pi/20(=2\pi/40)$. Once you use the 20 length DFT, the $7\pi/20$ cannot be seen due to the wide stride of frequency distribution.

The filter output shows the modified frequency distribution for low pass property in Fig. 4.29. Above the $\pi/4$ radian frequency (equivalent to the 1000 Hz), most of the frequency components are completely removed from the LPF processing except $7\pi/20$ ($\approx$1.1) location that is not seen on the 20 length DFT design procedure. The DTFT continuous distribution on Figs. 4.25 and 4.26 provides the magnitude 0.1652 at 1.079 rad frequency; therefore, we expect the output magnitude around 3.3 that is the 0.1652×20 at the 1.1 rad frequency. The designed LPF from DFT method guarantees the designated magnitudes on the integer multiple of the fundamental frequency in other words Δf frequency resolution. However, the intermediate continuous frequencies follow the magnitude of DTFT distributions those cannot be observed in the discrete DFT profile.

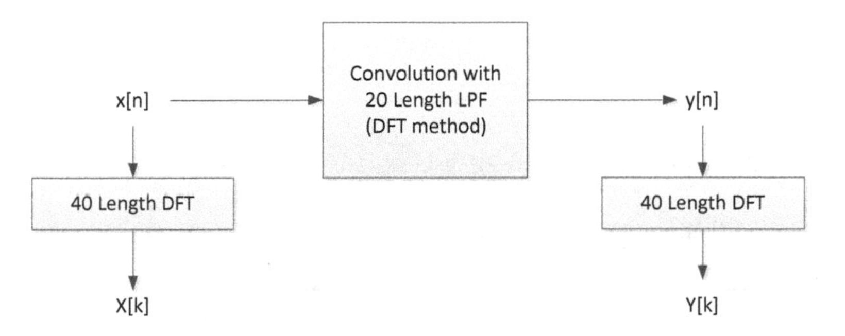

Fig. 4.28 Block diagram to compare the LPF frequency response from given signal

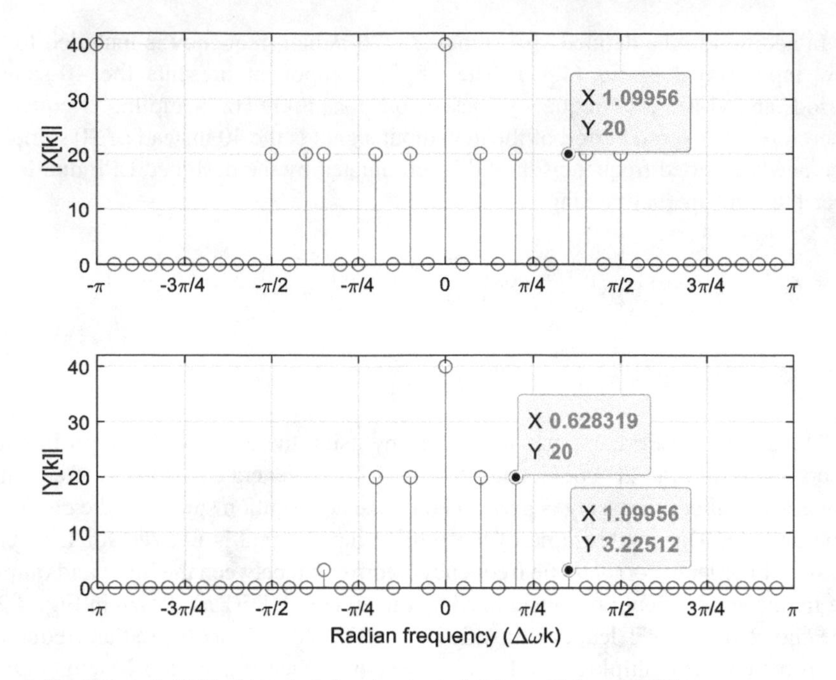

Fig. 4.29 40 length DFT of input $x[n]$ and output $y[n]$ from 20 length LPF

Prog. 4.17 MATLAB program for Figs. 4.29. See **Prog. 4.16** for plot decoration

```
n1 = -9:1:10;
hn1 = (1/20)*sin(5*pi*(n1+eps)/20)./sin(pi*(n1+eps)/20);
n2 = 0:999;
xn1 =
cos(2*pi*1*n2/20)+cos(2*pi*2*n2/20)+cos(2*pi*4*n2/20)+cos(2*pi*5*n2/2
0)+cos(2*pi*10*n2/20)+1+cos(pi*7*n2/20);
yn1 = conv(xn1,hn1);
xn2 = xn1(501:540);
yn2 = yn1(501:540);
fxn2 = fftshift(fft(xn2));
fyn2 = fftshift(fft(yn2));
faxis = (-pi:2*pi/40:pi-2*pi/40);

figure,
subplot(211), stem(faxis,abs(fxn2)), grid
...
subplot(212), stem(faxis,abs(fyn2)), grid
...
```

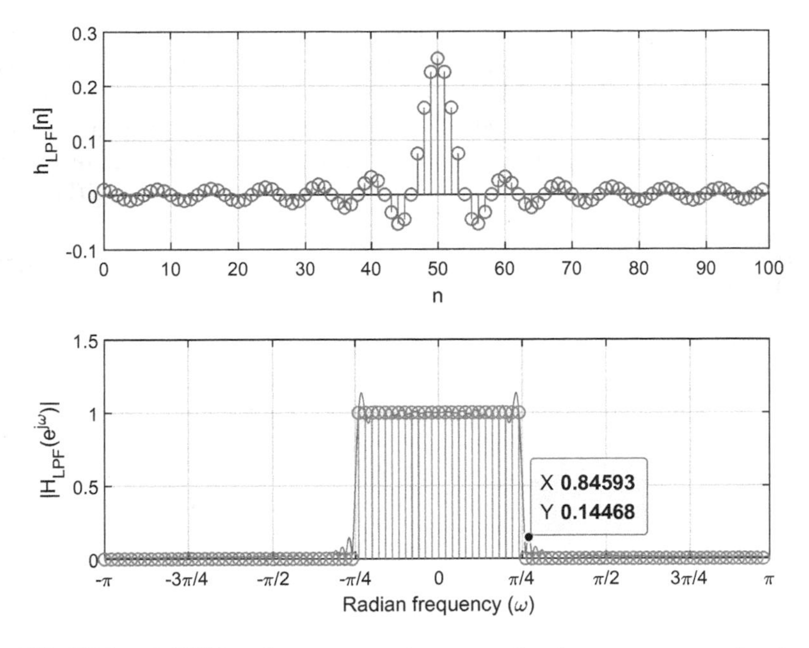

Fig. 4.30 100-length LPF impulse response and corresponding frequency response based on the DFT (discrete) and DTFT (continuous)

The increased DFT length depicts the LPF frequency profile in dense distribution. Figures 4.30 and 4.31 demonstrate the LPF with 100 DFT length. Comparing to the 20 length LPF, the finer grain description is observed in the discrete frequency response. The continuous response from the DTFT still shows the

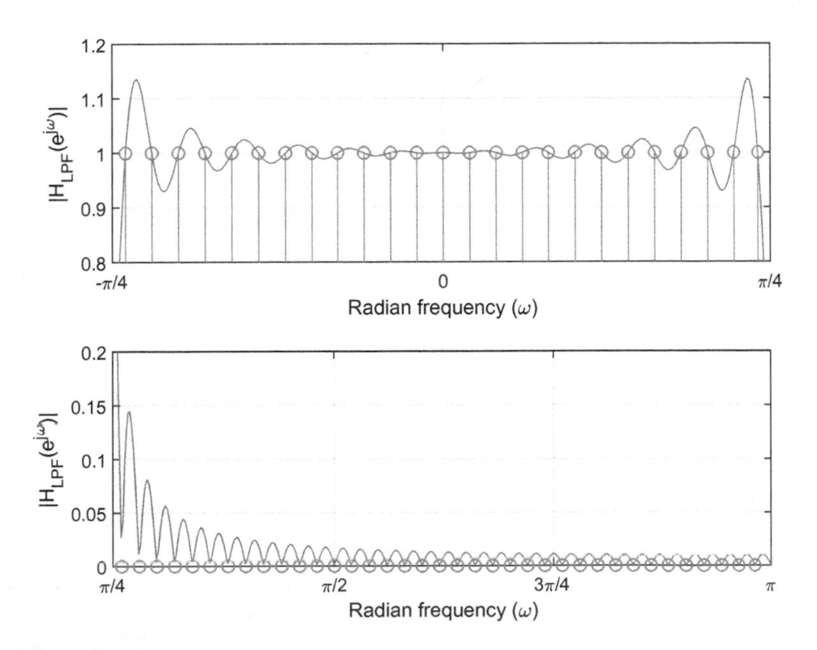

Fig. 4.31 Enlarged passband and stropband of Fig. 4.30 frequency response

fluctuations between the sampled locations in Fig. 4.31; however, the variation height is lower than the 20 length LPF counterpart. The highest blocking band magnitude is the 0.1459 at 0.8439 radian frequency for 100 length LPF and 0.1652 at 1.079 radian frequency for 20 length LPF. Further increasing filter length expects to follow the design parameters closely and accurately with expense of the computation time and output delay as shown in the cases with LPF with DTFT.

Prog. 4.18 MATLAB program for Figs. 4.30 and 4.31

```
syms omega real;
n1 = -50:1:49;
n2 = 0:1:99;
rfreq = linspace(-pi,pi,1000);
hn1 = 0.01*sin(25*pi*(n1+eps)/100)./sin(pi*(n1+eps)/100);
coeff = exp(1j*omega*n1);
Hint1 = hn1.*coeff;
Homega1 = sum(Hint1);
Homega2 = subs(Homega1,omega,rfreq);
Homega3 = eval(Homega2);
fhn1 = fftshift(fft(hn1));
faxis = (-pi:2*pi/100:pi-2*pi/100);

figure,
subplot(211)
stem(n2,hn1), grid
...
subplot(212)
plot(rfreq,abs(Homega3))
hold on;
stem(faxis,abs(fhn1)), grid
hold off;
...
figure,
subplot(211)
plot(rfreq,abs(Homega3))
hold on;
stem(faxis,abs(fhn1)), grid
hold off;
...
subplot(212)
plot(rfreq,abs(Homega3))
hold on;
stem(faxis,abs(fhn1)), grid
hold off;
...
```

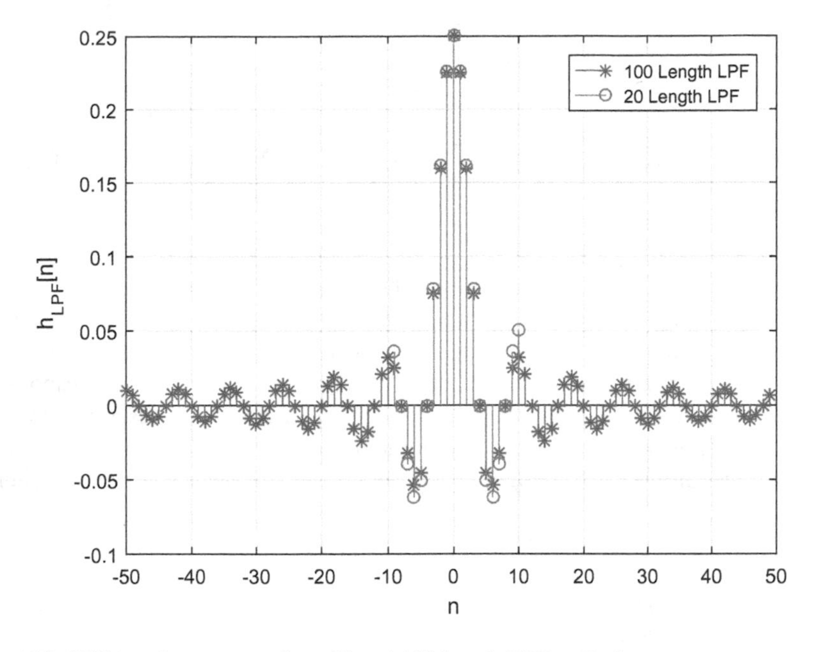

Fig. 4.32 LPF impulse response from 20 and 100 length DFT method

The method to design the LPF with DFT was utilizing the DFT length and specifying the individual frequency requirement on each discrete frequency. If you are using the known periodic discrete signal, this method is perfect for you. Otherwise, the approach suffers from the unspecified frequency responses those can be seen by DTFT analysis. The increased length LPF with DFT improves the description resolution of the frequency response; however, the deviations from desired output are still detected due to the discretization.

In the LPF design with DTFT, we truncate the infinite impulse response to the limited size in order to realize the feasible system. Using the same idea, what is the difference between the truncated and short-length LPF with DFT for the equal size? The previous LPF design for 20 and 100 length will be used for comparison in Fig. 4.32.

Both shows the similar time domain distribution for the overlapped range and the energy for both LPF are identical each other due to the passband ratio that is equal as 1000 Hz/4000 Hz. Note that the 1000 Hz and 4000 Hz is the passband limit and sampling frequency half respectively. According to the Parseval's theorem, frequency domain energy is equal for both length filters as shown in Eq. (4.48).

$$\frac{1}{N}\sum_{k=0}^{N-1}|X[k]|^2 = \frac{1}{20}\sum_{k=-2}^{2}|1|^2 = \frac{1}{100}\sum_{k=-12}^{12}|1|^2$$
$$= \sum_{n=0}^{N-1}|x[n]|^2 = \sum_{n=-9}^{10}|h_{20LPF}[n]|^2 = \sum_{n=-50}^{49}|h_{100LPF}[n]|^2 = 0.25 \quad (4.48)$$

Figure 4.33 demonstrates energy ratio between the time and frequency domain due to the truncation. The frequency domain energy is ideal and constant from the specification and time domain energy is adjusted from the truncation size. The original size of the impulse response is 100 samples and the resized response is changed from 1 to 99 in every odd numbers. The length 20 response contains approximately 96% of the frequency domain energy according to the Fig. 4.33. Note that the untrimmed 20 length LPF includes the 100% energy of the frequency domain.

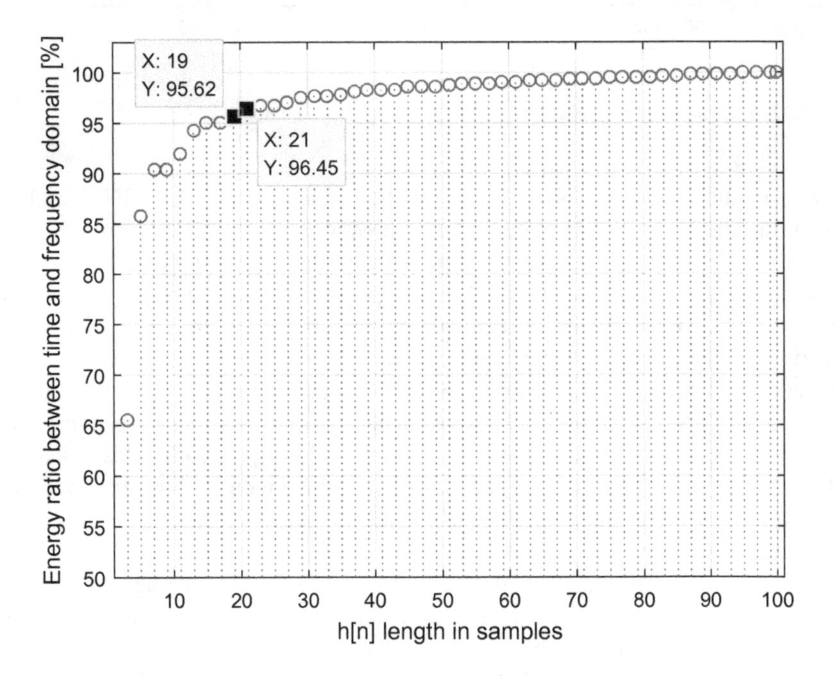

Fig. 4.33 (Time domain energy)/(Frequency domain energy)

Prog. 4.19 MATLAB program for Figs. 4.32 and 4.33

```
n1 = -9:1:10;
hn1 = (1/20)*sin(5*pi*(n1+eps)/20)./sin(pi*(n1+eps)/20);
n2 = -50:1:49;
hn2 = 0.01*sin(25*pi*(n2+eps)/100)./sin(pi*(n2+eps)/100);
n3 = 42:61;
hn3 = hn2(n3);

eee = [0.25^2];
engy = 0;
for kk=0:48
      engy = sum(hn2(52:52+kk).^2)*2+0.25^2;
      eee = [eee engy];
end

engy = sum(hn2(52:100).^2)*2+0.25^2+hn2(1)^2;
eee = [eee engy];
en1 = [(1:2:99) 100];

figure,
stem(n2,hn2,'*'), hold on
stem(n1,hn1), grid, hold off
...
figure,
stem(en1,eee*100/0.25,':o'), grid
...
```

There is subtle difference in time domain distribution for both 20 length LPF designs. The truncated and untrimmed LPF show the proper designated LPF requirement that propagates the frequency components under the 1000 Hz or $\pi/4$ radian frequency. As stated earlier, the untrimmed LPF with DFT guarantees the unit magnitude for the specific periodic components which are represented by the stem plot with o mark in Figs. 4.34 and 4.35. The frequency response for the truncated LPF illustrates slight deviation from the promised magnitude at the passing and periodic frequencies depicted by the stem plot with * mark in Figs. 4.34 and 4.35.

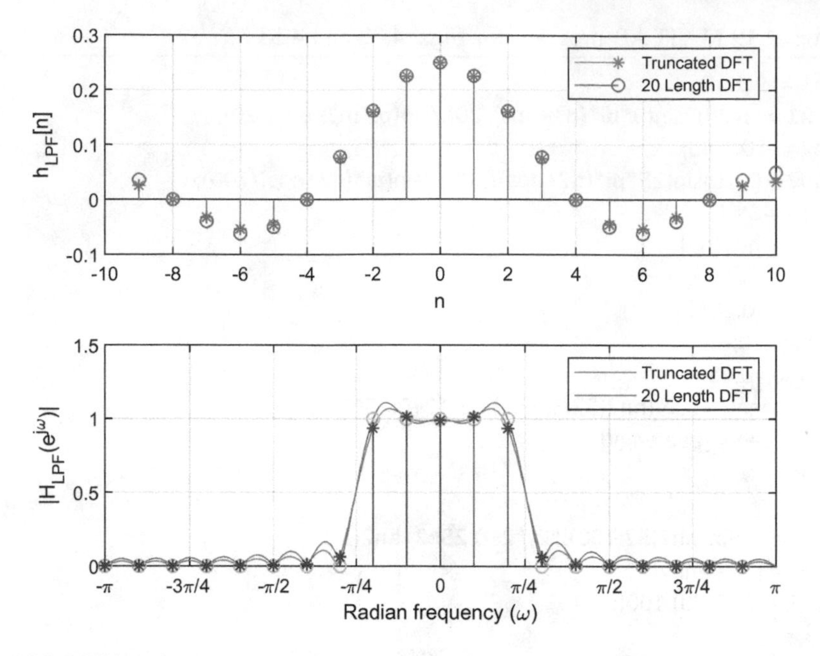

Fig. 4.34 LPF impulse responses and corresponding frequency responses from DFT method. * indicates truncated filter from 100 to 20 and o specifies the untrimmed filter

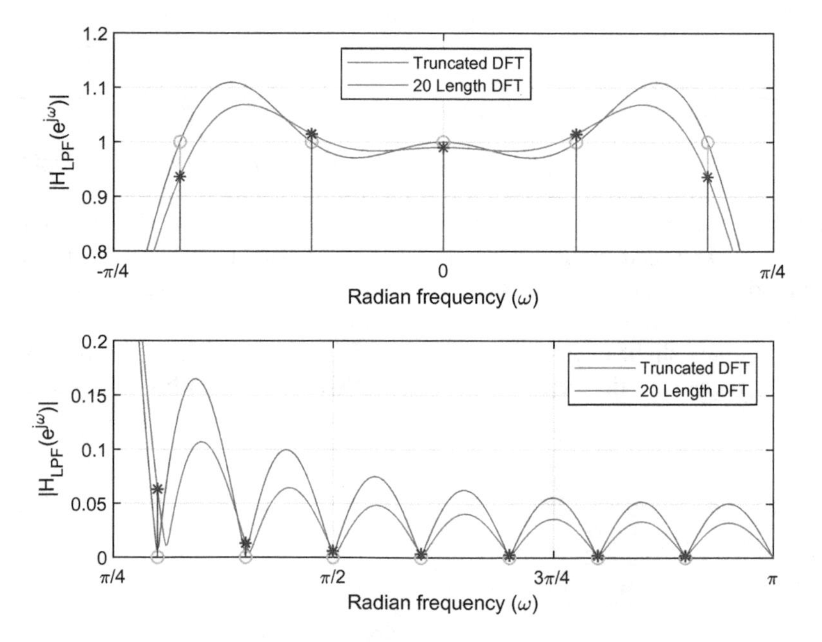

Fig. 4.35 Enlarged passband and stropband of Fig. 4.34 frequency response

Prog. 4.20 MATLAB program for Figs. 4.34 and 4.35

```
syms omega real;
rfreq = linspace(-pi,pi,1000);
n1 = -9:1:10;
hn1 = (1/20)*sin(5*pi*(n1+eps)/20)./sin(pi*(n1+eps)/20);
n2 = -50:1:49;
hn2 = 0.01*sin(25*pi*(n2+eps)/100)./sin(pi*(n2+eps)/100);
n3 = 42:61;
hn3 = hn2(n3);
coeff = exp(1j*omega*n1);
Hint1 = hn1.*coeff;
Hint2 = hn3.*coeff;
Homega1 = sum(Hint1);
Homega2 = sum(Hint2);
Homega3 = subs(Homega1,omega,rfreq);
Homega4 = subs(Homega2,omega,rfreq);
Homega5 = eval(Homega3);
Homega6 = eval(Homega4);
fhn1 = fftshift(fft(hn1));
fhn2 = fftshift(fft(hn3));
faxis = (-pi:2*pi/20:pi-2*pi/20);
figure,
subplot(211)
stem(n1,hn3,'*'), hold on;
stem(n1,hn1), grid, hold off;
subplot(212)
plot(rfreq,abs(Homega6),rfreq,abs(Homega5)), hold on;
stem(faxis,abs(fhn1)),
stem(faxis,abs(fhn2),'*'), grid, hold off;
xlim([-pi pi]), ylim([0 1.5]);
figure,
subplot(211)
plot(rfreq,abs(Homega6),rfreq,abs(Homega5)), hold on;
stem(faxis,abs(fhn1)),
stem(faxis,abs(fhn2),'*'), grid, hold off;
xlim([-pi/4 pi/4]), ylim([0.8 1.2]);
subplot(212)
plot(rfreq,abs(Homega6),rfreq,abs(Homega5)), hold on;
stem(faxis,abs(fhn1)),
stem(faxis,abs(fhn2),'*'), grid, hold off;
xlim([pi/4 pi]), ylim([0 0.2]);
```

In Figs. 4.34 and 4.35, the continuous frequency responses for both LPF provide the opposite performance outcome against the discrete frequency situation. The truncated LPF delivers the less fluctuation in passing and blocking frequency band than untrimmed LPF distribution. Originally, the truncated LPF expects to represent the less variation in the intended band since the filter uses the increased length to describe the given specification. Even the abbreviation of the long length LPF maintains filter performance to the certain point with sacrifice of the designed periodic signal gains. Overall the intermediate frequencies denoted by the continuous distribution are well regulated by the truncated LPF.

We have explored the three methods to design the LPF as DTFT, untrimmed DFT, and truncated DFT. All three methods are compared in Figs. 4.36 and 4.37 with identical length as 20 samples. Note that the truncated DFT method is originally designed for 100 sample length. The discrete plots for the DTFT, untrimmed DFT,

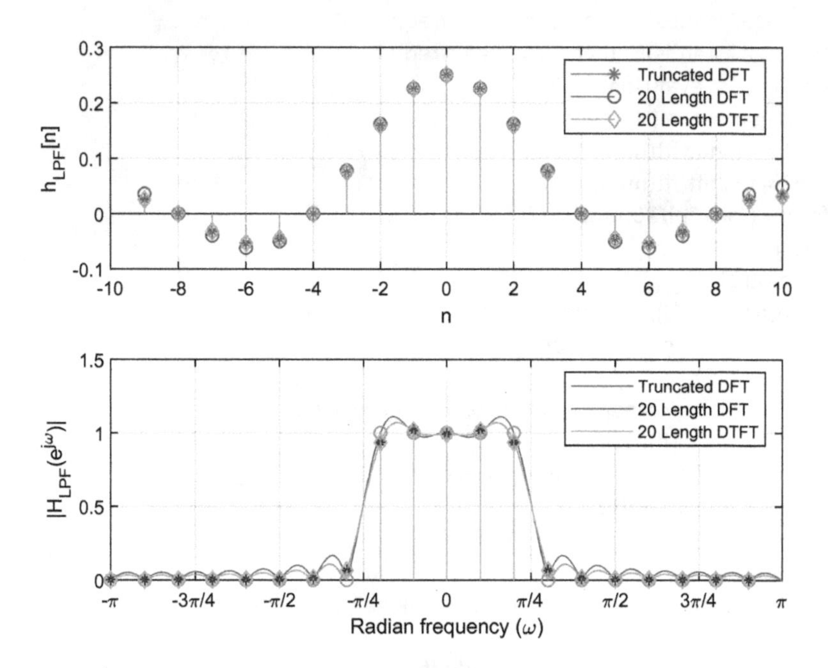

Fig. 4.36 LPF impulse responses and corresponding frequency responses from DFT and DTFT method

Prog. 4.21 MATLAB program for Figs. 4.36 and 4.37. Similar to **Prog. 4.20**. Repeated parts are omitted to save the length

```
syms omega real;
rfreq = linspace(-pi,pi,1000);
n1 = -9:1:10;
...
hn4 = sin(pi*(n1+eps)/4)./(pi*(n1+eps));
coeff = exp(1j*omega*n1);
...
Hint3 = hn4.*coeff;
...
Homega7 = sum(Hint3);
Homega8 = subs(Homega7,omega,rfreq);
Homega9 = eval(Homega8);
...
fhn3 = fftshift(fft(hn4));
faxis = (-pi:2*pi/20:pi-2*pi/20);
figure,
subplot(211)
stem(n1,hn3,'*'), hold on;
stem(n1,hn1,'o');
stem(n1,hn4,'d'), grid, hold off;
subplot(212)
plot(rfreq,abs(Homega6),rfreq,abs(Homega5),rfreq,abs(Homega9)), hold on;
stem(faxis,abs(fhn2),'*'),
stem(faxis,abs(fhn1),'o'),
stem(faxis,abs(fhn3),'d'), grid, hold off;
xlim([-pi pi]), ylim([0 1.5]);
figure,
subplot(211)
plot(rfreq,abs(Homega6),rfreq,abs(Homega5),rfreq,abs(Homega9)), hold on;
stem(faxis,abs(fhn2),'*'),
stem(faxis,abs(fhn1),'o'),
stem(faxis,abs(fhn3),'d'), grid, hold off;
xlim([-pi/4 pi/4]), ylim([0.8 1.2]);
subplot(212)
plot(rfreq,abs(Homega6),rfreq,abs(Homega5),rfreq,abs(Homega9)), hold on;
stem(faxis,abs(fhn2),'*'),
stem(faxis,abs(fhn1),'o'),
stem(faxis,abs(fhn3),'d'), grid, hold off;
xlim([pi/4 pi]), ylim([0 0.2]);
```

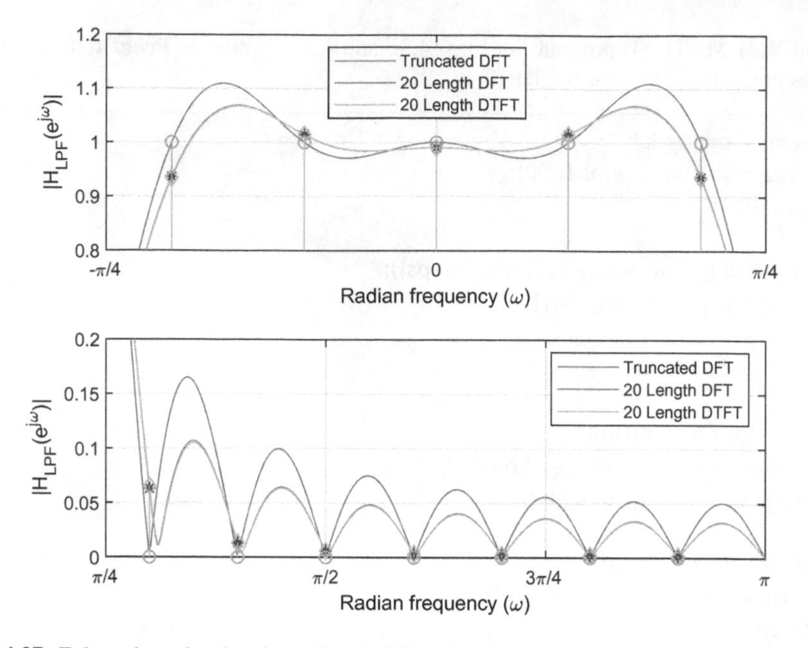

Fig. 4.37 Enlarged passband and stropband of Fig. 4.36 frequency response

and truncated DFT are marked with diamond, circle, and star correspondingly on the figures. The impulse responses show the similar distributions for the three methods. The likeness is higher between the DTFT and truncated DFT method than the untrimmed DFT method especially for the range away from the zero. Observe that DTFT (diamond head) and truncated DFT (star head) distribution are very close and difficult to separate in the figure. Also, the hardly separable condition can be noticed in the frequency domain between the DTFT and truncated DFT method those illustrate the deviation from the designed gains on the discrete frequencies; on the contrary, the untrimmed DFT method holds the intended gains over the frequencies.

The DTFT and truncated DFT method provide comparable performance for the continuous frequency distribution while the truncated DFT employs the increased length for DFT in the initial design stage. The analysis suggests that the DTFT and truncated DFT follow the designed parameters with less variation for any given frequencies. The untrimmed DFT shows the higher fluctuation for the comprehensive frequencies; however, the integer multiples of the frequency resolution maintain the intended gains. Therefore, if the system only controls the periodic signal, the untrimmed DFT is the right choice for the LPF. Figure 4.38 shows the procedure to select the design method for LPF. Though the periodic signal is provided to the system as major input, the lengthy period disturbs to decide the untrimmed DFT method. Note

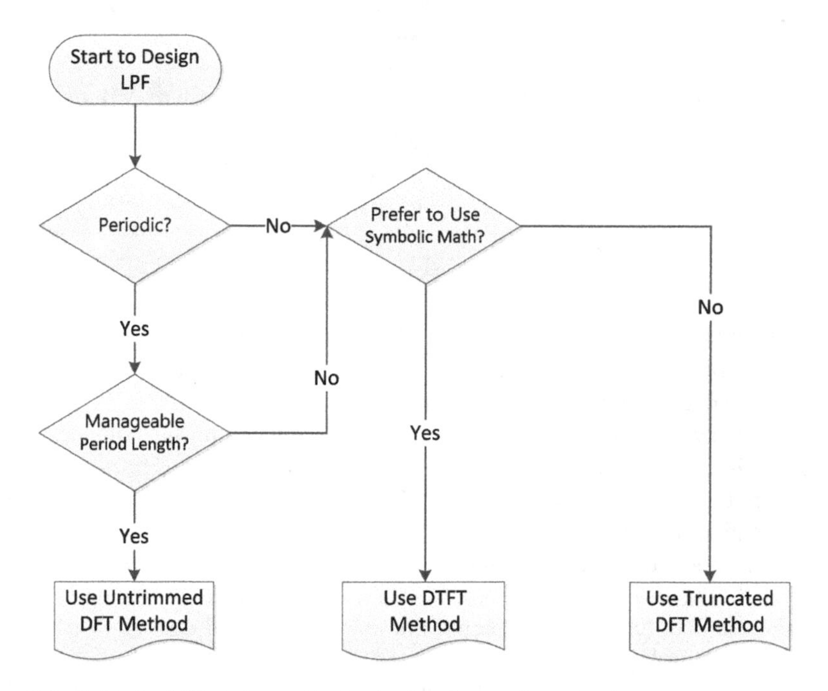

Fig. 4.38 Guide for LPF design method selection from various situations

that the longer impulse response produces the higher computational burden as well as extended delay time. The DTFT method requires to solve the symbolic mathematics that is usually performed by hand or MATLAB symbolic math toolbox. The given example above uses the simple description of the frequency specification but complex shape in the frequency domain may need complicated solution for impulse response. For the untrimmed and truncated DFT method, the computation is straightforward by using the conventional DFT function or program. The individual and discrete frequency specification are placed in the algorithm and the complex number result is computed by the numerical computation.

According to the DTFT and DFT property, the even symmetric in one domain generates the real values only in the other domain. The time or frequency either domain can be applied for the even symmetric as Euler formula in Eq. (4.49).

$$\cos(\omega n) = \frac{e^{j\omega n} + e^{-j\omega n}}{2} \quad \cos\left(\frac{2\pi}{N}kn\right) = \frac{e^{j\frac{2\pi}{N}kn} + e^{-j\frac{2\pi}{N}kn}}{2} \tag{4.49}$$

The frequency specification in even symmetric form provides the impulse response in real value. Subsequently, the symmetric time window frame centered origin is applied over the derived response function for even symmetric impulse response that generates the frequency response in real value. Note that the frequency response without complex number does not modify the input time information; hence, no time shift is observed in the filter output. Table 4.4 shows the

Table 4.4 Frequency and time domain relationship for phase and time shift

DTFT	DFT
$x[n] \overset{DTFT}{\leftrightarrow} X(e^{j\omega})$	$x[n] \overset{DFT}{\leftrightarrow} X[k]$
$x[n-d] \overset{DTFT}{\leftrightarrow} X(e^{j\omega})e^{-j\omega d}$	$x[n-d] \overset{DFT}{\leftrightarrow} X[k]e^{-j\frac{2\pi}{N}kd}$
$Y(e^{j\omega}) = X(e^{j\omega})\lvert H(e^{j\omega})\rvert e^{j\angle H(e^{j\omega})} = \left(X(e^{j\omega})e^{j\angle H(e^{j\omega})}\right)\lvert H(e^{j\omega})\rvert$	$Y[k] = X[k]\lvert H[k]\rvert e^{j\angle H[k]} = \left(X[k]e^{j\angle H[k]}\right)\lvert H[k]\rvert$

frequency and time domain relationship for phase and time shift. The $|H(e^{j\omega})|$ and $|H[k]|$ control the individual frequency magnitude and the $e^{j\angle H(e^{j\omega})}$ and $e^{j\angle H[k]}$ decide each frequency time shift from input signal $x[n]$.

The even symmetric impulse response could be reduced for the compact length; however, the configuration is still maintaining the even symmetric profile. After the truncation, the impulse response is moved to the right until all the response portion in the negative time index is relocated to the zero and higher location. The impulse response positioned at the zero and higher time index assures the system causality. Figure 4.39 illustrates the overall procedure for design the LPF based on the DTFT and DFT method. Observe that the untrimmed DFT method does not perform the trim procedure in Fig. 4.39.

There are number of trade-off between the implementation and performance. The filter designer should choose the right parameters and values to maximize the system performance. This is the end of comprehensive LPF design. The advanced methods for LPF design will be presented after the Z-transform theory in next chapter.

The LPF is the most fundamental structure to design the filter variations such as high pass filter (HPF), band pass filter (BPF), and band stop filter (BSF). The HPF passes the frequency components above the threshold value known as cutoff frequency. The BPF propagates the range of frequency components specified by the lower and upper frequency limits. The BSF suppresses the range of frequency

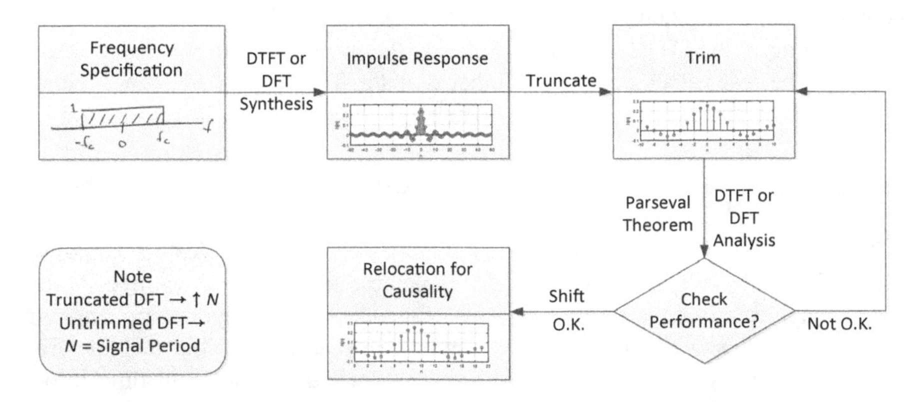

Fig. 4.39 Overall design procedure for LPF based on the DTFT or DFT method

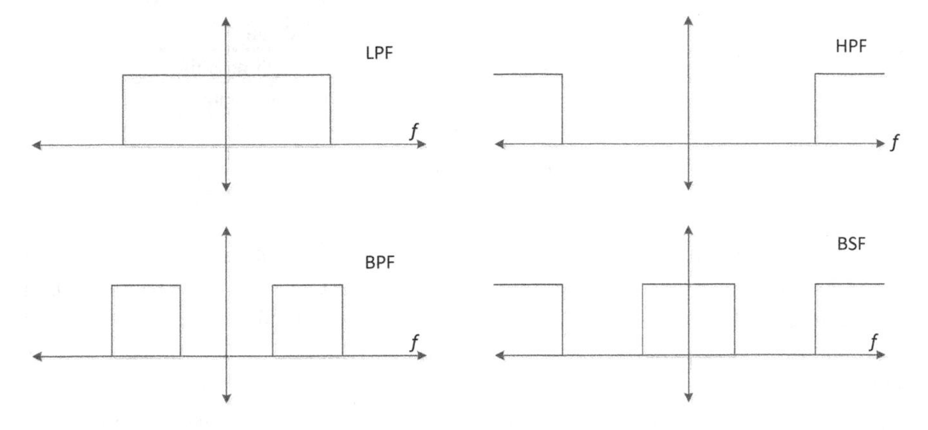

Fig. 4.40 Various filter configurations

components with lower and upper bound. The graphical representation of the filter structures is given in Fig. 4.40.

You can design each filter directly as explained in the LPF design procedure. The frequency distribution is specified and transformed by DTFT or DFT to obtain the raw impulse response. After the truncation and relocation, final impulse response for the given requirement filter is realized for convolution sum process. Since the procedure is identical, leave the direct method examples for HPF, BPF, and BSF to the readers.

According to the DTFT and DFT property, we can consider the design method that relocate and invert the frequency information. The required passing or stopping frequency band is applied to the LPF design which can be shifted and inverted to the desired location in frequency domain. Therefore, the LPF is the basic structure to realize the other filter formats. Table 4.5 is the fundamental equations used for the filter design.

Table 4.5 Fundamental DTFT and DFT properties for the filter transformation

	DTFT		DFT	
	Frequency	Time	Frequency	Time
Frequency Shift	$X\left(e^{j(\omega-\omega_s)}\right)$	$x[n]e^{j\omega_s n}$	$X[k-l]$	$x_N[n]e^{j\frac{2\pi}{N}ln}$
Magnitude inversion	$1-X(e^{j\omega})$	$\delta[n]-x[n]$	$1-X[k]$	$\delta[n]-x_N[n]$
Condition	$X(e^{j\omega})\in\mathbb{R}$ and $\mathrm{Gain}(X(e^{j\omega}))=1$		$X[k]\in\mathbb{R}$ and $\mathrm{Gain}(X[k])=1$	

Table 4.6 High pass filter specification example

Filter specifications	Value
Filter type	High pass filter
Filter length	Finite with optimization
Cutoff frequency (f_c)	3000 Hz
Sampling frequency (f_s)	8000 Hz

You can place the LPF to any location by frequency shift and turn over the magnitude to apply the stop band by magnitude inversion. The shift and inversion in the frequency domain are performed in the time domain via multiplying the computed values with derived impulse response of LPF. Since the impulse response with complex number cannot be used for the convolution sum, the frequency shift utilizes the cosine function instead of the complex exponential based on the Euler relationship. For the proper magnitude inversion, the frequency response should be real number and constant gain in the LPF; otherwise, subtraction from constant value does not invert the frequency response. Let's do one example for HPF as specified in Table 4.6.

Example 4.10
Design the HPF for Table 4.6. Use the LPF from truncated DFT method for 20 sample length.

Solution
Based on the information, the cutoff frequency corresponds to the $3\pi/4$ in discrete frequency. The 2π repeatability with maximum π in discrete frequency domain illustrates Fig. 4.41 frequency distribution for the given HPF. The symmetric structure over the y axis is repeated for every 2π in order to maintain the real number in impulse response. The circle in Fig. 4.41 demonstrates the conjugate relationship between $3\pi/4$ and $5\pi/4$ for Euler formula.

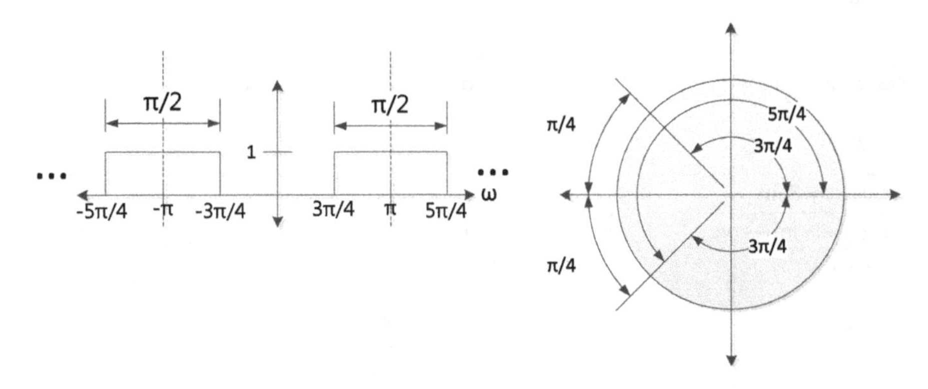

Fig. 4.41 Frequency distribution for desired HPF and its angular profile

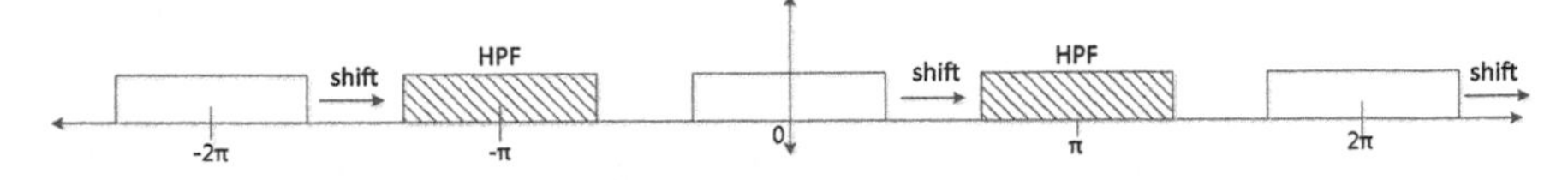

Fig. 4.42 Frequency shift for LPF → HPF

The LPF is devised as identical profile of the HPF located in the $\pm\pi$ center. In this design, the $\pi/2$ bandwidth LPF filter is realized and shifted to the π location as shown in Fig. 4.42. Note that the shift should provide symmetrical (conjugate) profile to preserve the real number impulse response.

Let's assume that the $h_{LPF}[n]$ is implemented by the truncated DFT method for 20 sample length. As stated earlier, the design length of the truncated DFT method is 100 samples. The shift procedure is realized by the complex multiplication and organized as below.

$$h_{HPF}[n] = h_{LPF}[n]e^{j\pi n} = h_{LPF}[n](-1)^n$$

Figures 4.43 and 4.44 demonstrate the preliminary LPF and corresponding shifted HPF. According to the frequency response, the designed HPF properly

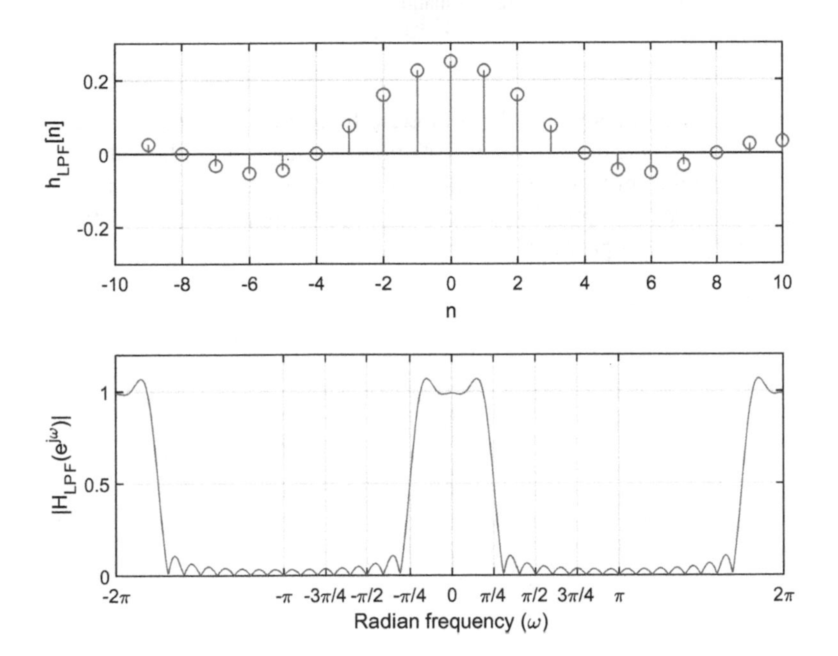

Fig. 4.43 Impulse response and frequency response of baseline LPF

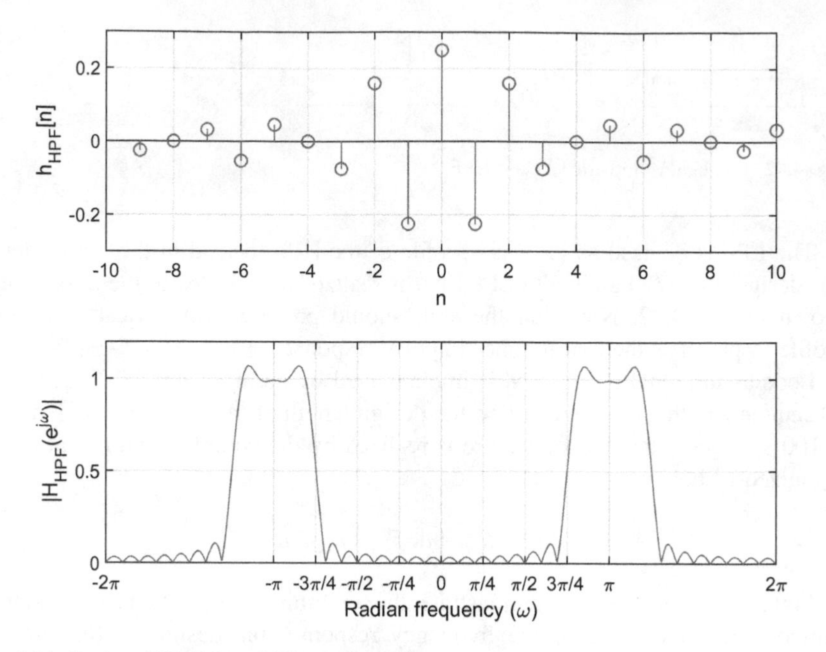

Fig. 4.44 Designed HPF from LPF transformation

propagates the signal above the $3\pi/4$ discrete frequency with real number impulse response. Note that the derived impulse response contains the values at the negative index therefore the shift in time domain is required for system causality. Once the HPF performance is acceptable for the given requirement, this computation is concluding the design procedure.

Prog. 4.22 MATLAB program for Figs. 4.43 and 4.44

```
syms omega real;
rfreq = linspace(-2*pi,2*pi,1000);
n2 = -50:1:49;
hn2 = 0.01*sin(25*pi*(n2+eps)/100)./sin(pi*(n2+eps)/100);
n3 = 42:61;
hn3 = hn2(n3);
n4 = -9:1:10;
hn4 = hn3.*(-1).^n4;
coeff = exp(1j*omega*n4);
Hint2 = hn3.*coeff;
Hint3 = hn4.*coeff;
Homega2 = sum(Hint2);
Homega3 = sum(Hint3);
Homega4 = subs(Homega2,omega,rfreq);
Homega5 = subs(Homega3,omega,rfreq);
Homega6 = eval(Homega4);
Homega7 = eval(Homega5);

figure,
subplot(211)
stem(n4,hn3), grid
ylim([-0.3 0.3]);
...
subplot(212)
plot(rfreq,abs(Homega6)), grid
xlim([-2*pi 2*pi]);
ylim([0 1.2]);
...
figure,
subplot(211)
stem(n4,hn4), grid
ylim([-0.3 0.3]);
...
subplot(212)
plot(rfreq,abs(Homega7)), grid
xlim([-2*pi 2*pi]);
ylim([0 1.2]);
...
```

■

Table 4.7 Band pass filter specification example

Filter specifications	Value
Filter type	Band pass filter
Filter length	Finite with optimization
Lower cutoff frequency (f_l)	1000 Hz
Upper cutoff frequency (f_u)	3000 Hz
Sampling frequency (f_s)	8000 Hz

The succeeding type of the filter is the band pass filter (BPF) which passes the frequency range specified by the lower and upper bound. The design method is identical to the HPF procedure except the variable frequency shift location.

Example 4.11
Design the BPF for Table 4.7. Use the LPF from truncated DFT method for 20 sample length.

Solution
Based on the information, the lower and upper cutoff frequency corresponds to the $\pi/4$ and $3\pi/4$ respectively in discrete frequency. Upon the symmetric distribution constraint, the BPF requirement is illustrated in Fig. 4.45 over discrete frequency domain. The passing band contains $\pi/2$ width and the band midpoint is located at $\pm\pi/2$.

We can consider the BPF design from the previous shift method that relocates the LPF to the right direction. The upper subplot in Fig. 4.46 demonstrates the one direction shift with $\pi/2$ that destroys the symmetric frequency distribution. In this case, the derived impulse response includes the complex number ($h_{LPF}[n]e^{j\frac{\pi}{2}n} = h_{LPF}[n]j^n$) which cannot be realized in the convolution sum. Instead of unidirectional shift, the bidirectional relocation shown in Fig. 4.46 conserves the symmetric profile. Since the bidirectional movement takes place in every 2π locations, the shifted edge cannot cross the discrete frequency boundary that is the $\pm\pi$ of the individual 2πs.

The linear combination of each direction shift is represented by below equation.

$$h_{BPF}[n] = h_{LPF}[n]\left(e^{j\frac{\pi}{2}n} + e^{-j\frac{\pi}{2}n}\right) = 2h_{LPF}[n]\cos\left(\frac{\pi}{2}n\right)$$

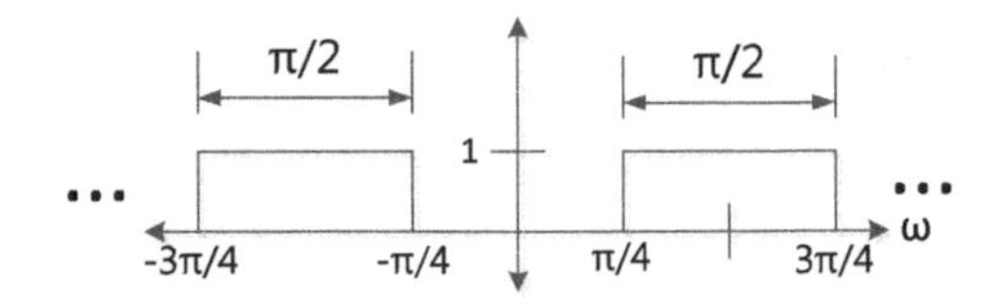

Fig. 4.45 Frequency distribution for desired BPF

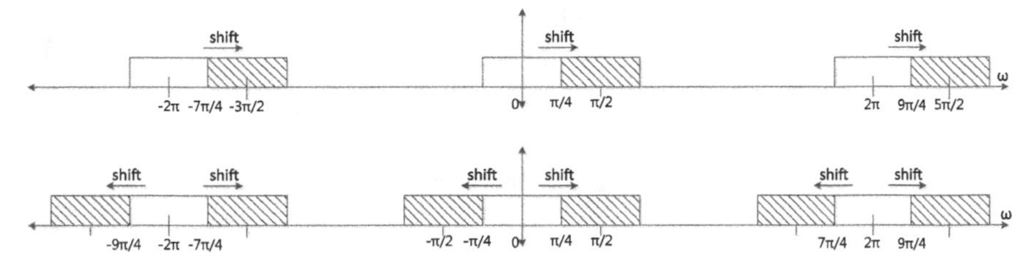

Fig. 4.46 Frequency shift for LPF → BPF

Figures 4.47 and 4.48 show the fundamental LPF and the shifted LPF (BPF). The designed BPF provides the relatively flat gains over the $\pi/4$ and $3\pi/4$ range which meet the initial filter requirement in Table 4.7. If you need the flatter and shaper BPF for better frequency separation, use the longer length in LPF then apply the frequency shift for repositioning.

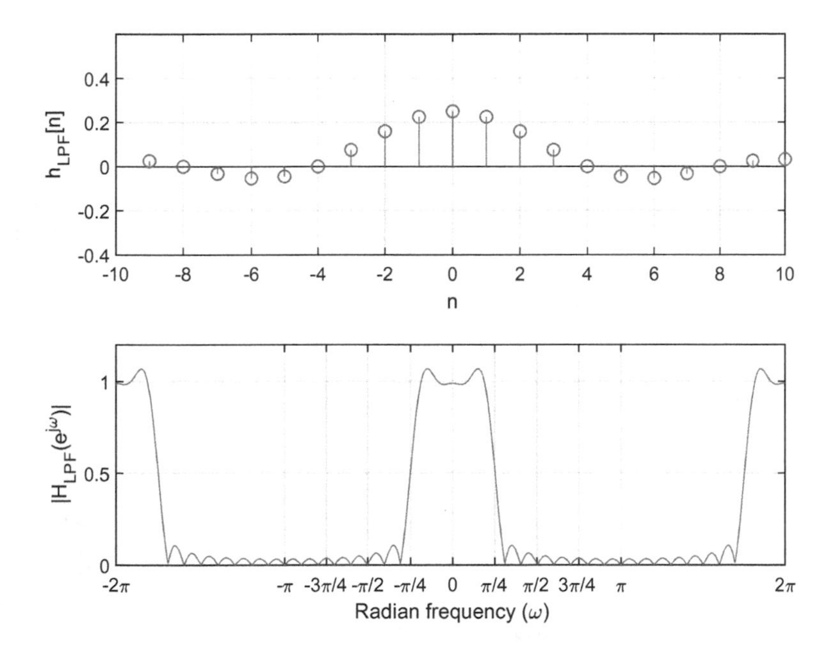

Fig. 4.47 Impulse response and frequency response of baseline LPF

Prog. 4.23 MATLAB program for Figs. 4.47 and 4.48

```
syms omega real;
rfreq = linspace(-2*pi,2*pi,1000);
n2 = -50:1:49;
hn2 = 0.01*sin(25*pi*(n2+eps)/100)./sin(pi*(n2+eps)/100);
n3 = 42:61;
hn3 = hn2(n3);
n4 = -9:1:10;
hn4 = 2*hn3.*cos(pi*n4/2);
coeff = exp(1j*omega*n4);
Hint2 = hn3.*coeff;
Hint3 = hn4.*coeff;
Homega2 = sum(Hint2);
Homega3 = sum(Hint3);
Homega4 = subs(Homega2,omega,rfreq);
Homega5 = subs(Homega3,omega,rfreq);
Homega6 = eval(Homega4);
Homega7 = eval(Homega5);

figure,
subplot(211)
stem(n4,hn3), grid
ylim([-0.4 0.6]);
...
subplot(212)
plot(rfreq,abs(Homega6)), grid
xlim([-2*pi 2*pi]);
ylim([0 1.2]);
...
figure,
subplot(211)
stem(n4,hn4), grid
ylim([-0.4 0.6]);
...
subplot(212)
plot(rfreq,abs(Homega7)), grid
xlim([-2*pi 2*pi]);
ylim([0 1.2]);
...
```

■

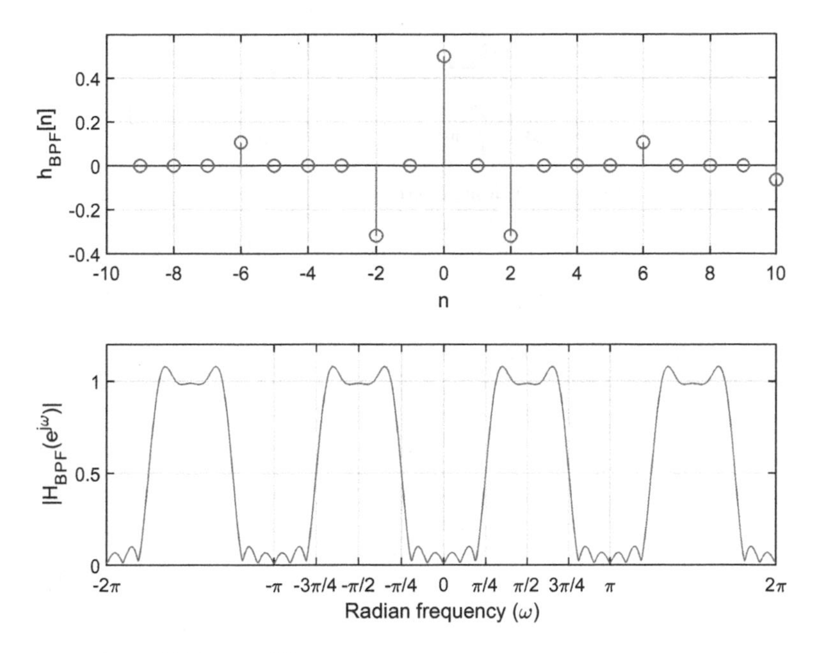

Fig. 4.48 Designed BPF from LPF transformation

The last type of the filter shown in this chapter is the band stop filter (BSF) which blocks the frequency range specified by the lower and upper bound. The design method follows the BPF procedure with additional step that inverts the filter magnitude as low to high and high to low gain.

Example 4.12
Design the BSF for Table 4.8. Use the LPF from truncated DFT method for 20 sample length.

Solution
The above parameters are identical to the BPF design values except the lower and upper frequency specify the blocking instead of passing range. The lower and upper cutoff frequency corresponds to the $\pi/4$ and $3\pi/4$ respectively in discrete frequency. Similar to the flipped version of BPF, the BSF requirement is illustrated below in discrete frequency domain as Fig. 4.49.

Table 4.8 Band stop filter specification example

Filter specifications	Value
Filter type	Band stop filter
Filter length	Finite with optimization
Lower cutoff frequency (f_l)	1000 Hz
Upper cutoff frequency (f_u)	3000 Hz
Sampling frequency (f_s)	8000 Hz

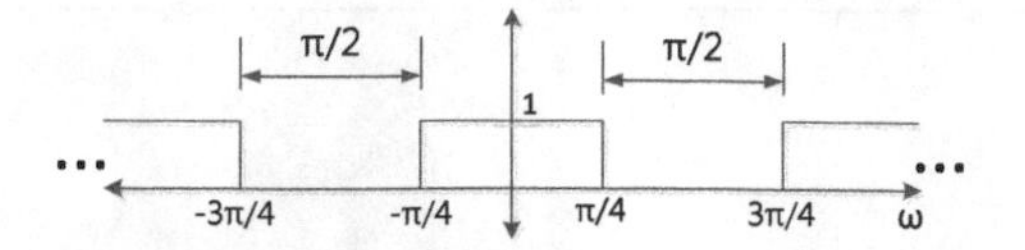

Fig. 4.49 Frequency distribution for desired BSF

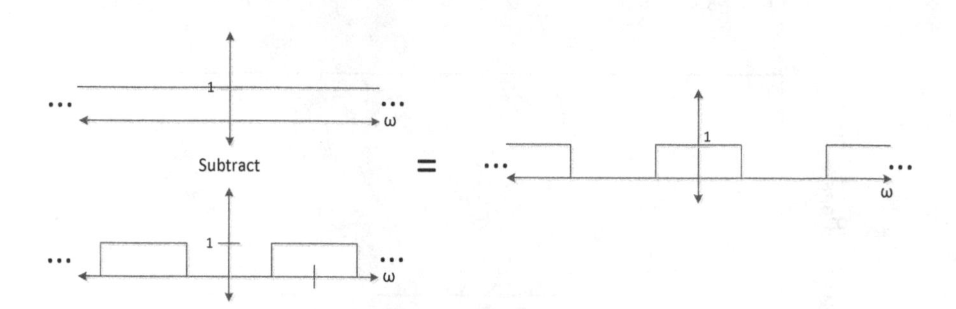

Fig. 4.50 Magnitude inversion for BPF → BSF

The magnitude inversion in frequency domain can be described as the subtracting the frequency response from the designated gain in the BPF as shown in Fig. 4.50. The BPF gain should agree with the subtracting constant and real number; otherwise, the inversion cannot be exercised properly.

The magnitude inversion is mathematically represented in below equation. The subtracting with constant in frequency domain is equivalent to the deduction from the delta function in time domain. According to the linearity in DFT, the individual function is transformed separately for overall operation.

$$\delta[n] - x[n] \overset{\text{DFT}}{\leftrightarrow} 1 - X[k] \, \text{for} \, X[k] \in \mathbb{R} \, \text{and} \, \text{Gain}(X[k]) = 1$$

Figures 4.51 and 4.52 shows the fundamental BPF and the magnitude inverted BPF (BSF). The designed BSF provides the flat low over the $\pi/4$ and $3\pi/4$ and gain one in the other range which meet the initial filter requirement in Table 4.8 If you need the flatter low and shaper edge BSF for better frequency separation, use the longer length in LPF then apply the frequency shift and magnitude inversion sequentially.

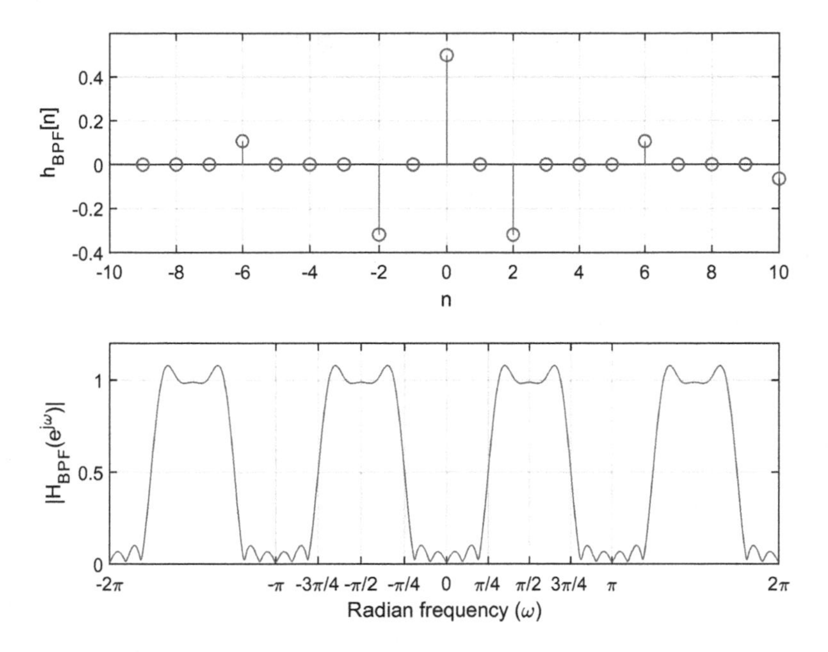

Fig. 4.51 Impulse response and frequency response of baseline BPF

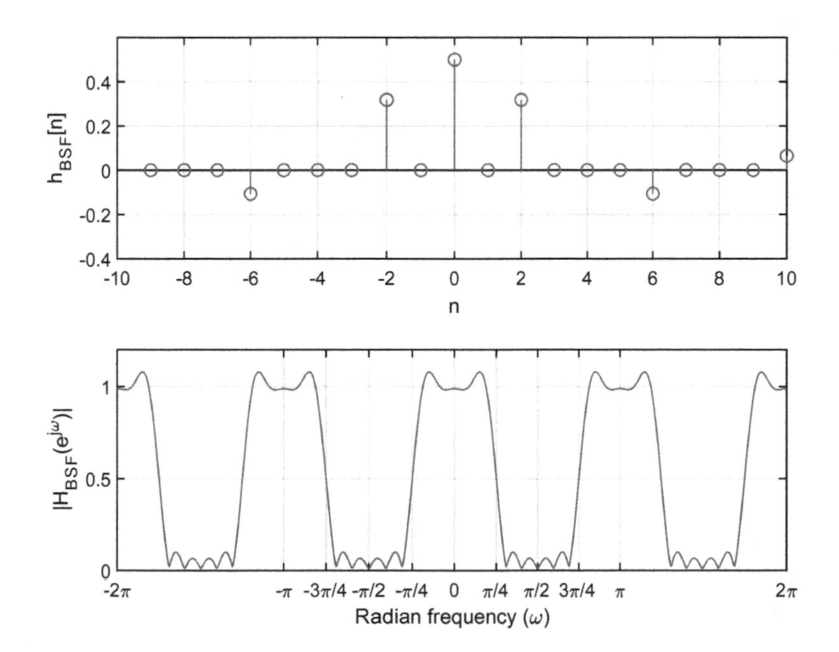

Fig. 4.52 Designed BSF from BPF with magnitude inversion

Prog. 4.24 MATLAB program for Figs. 4.51 and 4.52

```
syms omega real;
rfreq = linspace(-2*pi,2*pi,1000);
n2 = -50:1:49;
hn2 = 0.01*sin(25*pi*(n2+eps)/100)./sin(pi*(n2+eps)/100);
n3 = 42:61;
hn3 = hn2(n3);
n4 = -9:1:10;
hn4 = 2*hn3.*cos(pi*n4/2);
temparr = zeros(size(n4));
temparr(10) = 1;
hn5 = temparr-hn4;
coeff = exp(1j*omega*n4);
Hint2 = hn4.*coeff;
Hint3 = hn5.*coeff;
Homega2 = sum(Hint2);
Homega3 = sum(Hint3);
Homega4 = subs(Homega2,omega,rfreq);
Homega5 = subs(Homega3,omega,rfreq);
Homega6 = eval(Homega4);
Homega7 = eval(Homega5);
figure,
subplot(211)
stem(n4,hn4), grid
ylim([-0.4 0.6]);
...
subplot(212)
plot(rfreq,abs(Homega6)), grid
xlim([-2*pi 2*pi]);
ylim([0 1.2]);
...
figure,
subplot(211)
stem(n4,hn5), grid
ylim([-0.4 0.6]);
...
subplot(212)
plot(rfreq,abs(Homega7)), grid
xlim([-2*pi 2*pi]);
ylim([0 1.2]);
...
```

■

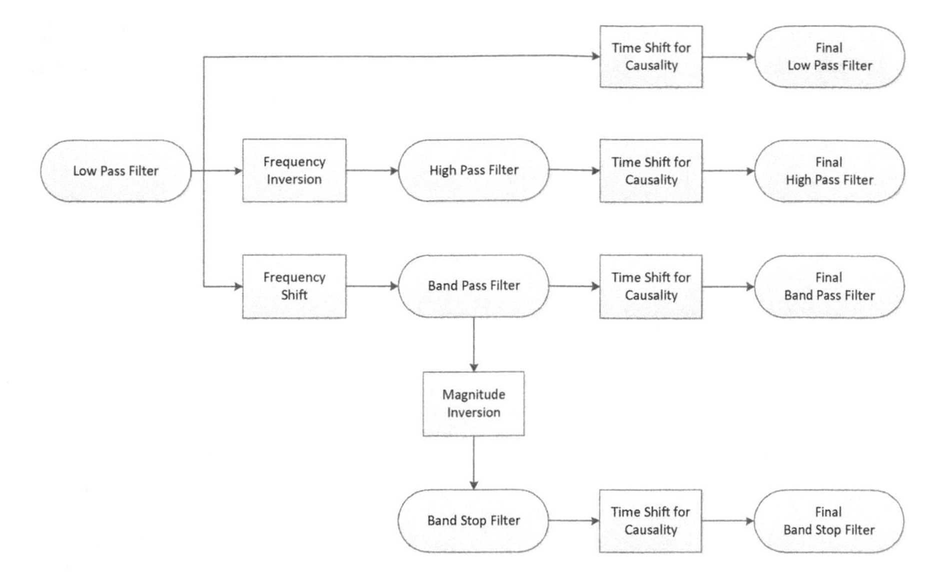

Fig. 4.53 Transformation procedures for various filter types from LPF

Up to now, we have explored the various types of filters as LPF, HPF, BPF, and BSF. The LPF is the building fundamental for the other filter types to specify the designated frequency band for passing or blocking. Further mathematical processing makes the LPF to other types as shown in Fig. 4.53. Note that the frequency shift to the π is described as the frequency inversion in the diagram since the filter profile is completely inverted in term of frequency distribution. At the end of design procedure, time shift is performed for the filter causality.

In this chapter, the filter architecture (convolution sum) shown in the Chap. 1 is extended for the further analysis. The linearity, time-invariance, and causality of the filter are discussed for the convolution sum. The infinite length filter can be realized by the recursive feedback that uses the recent outputs in addition to the input. The Fourier transform simplifies the convolution sum to the multiplication in frequency domain; therefore, impulse response in frequency domain known as frequency response specify the filter property. Based on the understanding above, the various types of the filter are derived and implemented in last part. The next chapter introduces the generalized transform as Z-transform to describe, design and analysis the comprehensive filters that includes the infinite length filter.

4.6 Problems

1. If you have the unit-step response, explain the method to obtain the impulse response from the given information. The unit-step response is the output of the system from the unit-step input.

2. If we have the individual filters as impulse response $h_1[n]$ and $h_2[n]$, derive the overall response of the system in cascade and parallel connection as shown below. Use the convolution sum notation * for final representation.

$$y[n] = \sum_{k=-\infty}^{\infty} x[k]h[n-k] = \sum_{k=-\infty}^{\infty} h[k]x[n-k]$$

$$= x[n] * h[n] = h[n] * x[n]$$

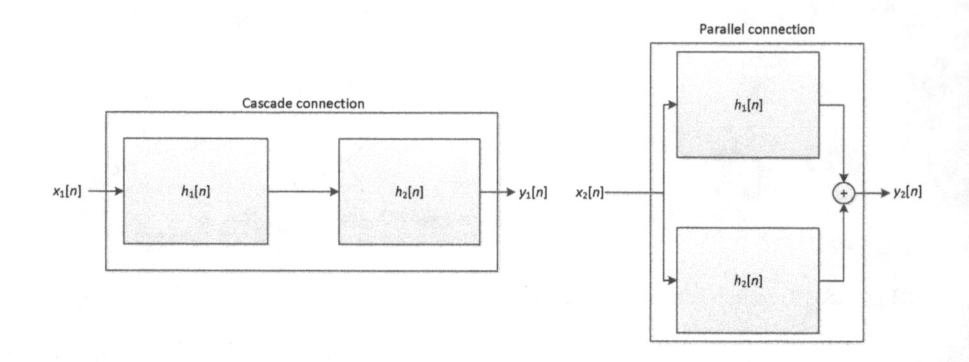

3. Design the digital integrator by using the first order IIR filter which use the $y[n-1]$ with constant coefficient. Note that the integrator performs the accumulation over the all input data.
4. Design the digital differentiator by using the FIR filter. Observe that the differentiator performs the subtraction between the current and previous input. Find the impulse response.
5. The BIBO stability of the system can be tested by using the following equation.

$$\sum_{k=0}^{\infty} |h[k]| < \infty$$

Derive the condition for unstable FIR filter.

6. In order to limit the filter output as below, what should we do? Remember this is the filter.

$$|y[n]| \leq 4x_{max} \forall n$$

7. Compute the filter output of below configuration with DTFT method.

$$x[n] = u[n] - u[n-3]$$

$$h[n] = \delta[n] - \delta[n-1]$$

8. Compute the filter output of below configuration with DFT method. Use the proper length N.

$$x[n] = u[n] - u[n-3]$$

$$h[n] = \delta[n] - \delta[n-1]$$

9. Design the LPF for the below Table. Use the DTFT method. Show the impulse response from –20 to 20 as n.

Filter specifications	Value
Filter type	Low pass filter
Filter length	Finite with optimization
Cutoff frequency (f_c)	1000 Hz
Sampling frequency (f_s)	16000 Hz

10. Draw the frequency response of 95 and 99% for the frequency domain energy from the previous problem.
11. Design the LPF for Problem 9 Table. Use the DFT method. Show the impulse response from 0 to 99 as n.
12. Design the LPF for Problem 9 Table. Use the DFT method. Show the impulse response from 0 to 19 as n.
13. Design the LPF for Problem 9 Table. Use the truncated DFT method for 20 sample length. Design the filter from length 100.
14. Design the HPF for the below Table. Use the LPF from truncated DFT method for 20 sample length. Design the filter from length 100.

Filter specifications	Value
Filter type	High pass filter
Filter length	Finite with optimization
Cutoff frequency (f_c)	7000 Hz
Sampling frequency (f_s)	16000 Hz

15. Design the BPF for the below Table. Use the LPF from truncated DFT method for 20 sample length. Design the filter from length 100.

Filter specifications	Value
Filter type	Band pass filter
Filter length	Finite with optimization
Lower cutoff frequency (f_l)	3000 Hz
Upper cutoff frequency (f_u)	5000 Hz
Sampling frequency (f_s)	16000 Hz

16. Design the BSF for the below Table. Use the LPF from truncated DFT method for 20 sample length. Design the filter from length 100.

Filter specifications	Value
Filter type	Band stop filter
Filter length	Finite with optimization
Lower cutoff frequency (f_l)	3000 Hz
Upper cutoff frequency (f_u)	5000 Hz
Sampling frequency (f_s)	16000 Hz

References

1. Taylor, A.E.: L'Hospital's Rule. Am. Math. Monthly **59**(1), 20–24 (1952). https://doi.org/10.1080/00029890.1952.11988058
2. Hazewinkel, M.: Encyclopaedia of Mathematics: Orbit - Rayleigh Equation. Springer, Netherlands (2012)

Chapter 5
Z-Transform

The fundamentals of the FIR filter design have been explored and investigated so far. The FIR filter designer can find the dedicated methods to control the edges, passing, and stopping band for sophisticated filter design. The essential design procedure of further controlled filter is practically identical to the presented approach with couple of more steps. How about the IIR filter? Upon the best knowledge now, choose the arbitrary coefficients for recursive feedback, derive the impulse response by hand or computer, then obtain the frequency response for filter compliance on the requirement. The convolution sum, DTFT, and DFT barely represent the IIR filter property directly. The Z-transform is the generalized transform for the discrete time domain to describe comprehensive signals and systems. By the Z-transform, the simple rational polynomial function denotes the IIR filter system and pole zero plot visualizes the filter property. The Z-transform extends the design method for the IIR as well as FIR filter.

5.1 Definition and Usage

The Z-transform provides the enhanced tool to understand the signal and system further between time and frequency domain. To understand the meaning of the enhancement, we need to consider the limitation of the DTFT and DFT first. Let's revisit the transforms for the discrete time signal as below in Table 5.1.

All transforms shown above are performed based on the periodic signal $e^{j\omega n}\left(\text{or } e^{j2\pi kn/N}\right)$ that rotates unit circle in the complex domain at $\omega(\text{or } 2\pi k/N)$ rate. The periodic signal magnitude is given by the $X(e^{j\omega})(\text{or } X[k])$ that assumes to be the constant value for the signal length. If we want to use the exponentially increasing or decreasing periodic or non-periodic signal, certain linear combination of the frequency components is required to represent the signal by the transformation. We can simplify this situation by extending the primitive element of the transform as below.

K. Kim, *Conceptual Digital Signal Processing with MATLAB*,
Signals and Communication Technology,
https://doi.org/10.1007/978-981-15-2584-1_5

Table 5.1 Transforms for discrete time signal

Discrete-time Fourier transform (DTFT)		Discrete Fourier transform (DFT)	
Analysis	$X(e^{j\omega}) = \sum_{n=-\infty}^{\infty} x[n]e^{-j\omega n}$	Analysis	$X[k] = \sum_{n=0}^{N-1} x_N[n]e^{-j\frac{2\pi}{N}kn}$
Synthesis	$x[n] = \frac{1}{2\pi}\int_{-\pi}^{\pi} X(e^{j\omega})e^{j\omega n}d\omega$	Synthesis	$x_N[n] = \frac{1}{N}\sum_{k=0}^{N-1} X[k]e^{j\frac{2\pi}{N}kn}$

$$e^{j\omega} \rightarrow z = \left\{re^{j\omega} | r \in \mathbb{R} \text{ and } \omega \in \mathbb{R}\right\} \in \mathbb{C} \tag{5.1}$$

The power of the z represents the $r^n e^{j\omega n}$ which shows the exponential variation with periodicity inside. Plug in the new exponential part to the transform provides the new transformation known as Z-transform [1] below. Note that the z is the continuous complex number domain.

$$X(\mathrm{z}) = \sum_{n=-\infty}^{\infty} x[n]z^{-n} \tag{5.2}$$

What does the Z-transform perform? Similar to the DTFT and DFT, can we expect to obtain the magnitude and phase of the frequency component as well as the exponential part? Let's figure out. The signal model is periodic with exponential part as Eq. (5.3).

$$x_1[n] = \alpha^n e^{j\varphi n} \tag{5.3}$$

Place into the Z-transform as below.

$$\begin{aligned} X_1(z) &= \sum_{n=-\infty}^{\infty} x_1[n]z^{-n} = \sum_{n=-\infty}^{\infty} x_1[n]r^{-n}e^{-j\omega n} \\ &= \sum_{n=-\infty}^{\infty} \alpha^n e^{j\varphi n} r^{-n} e^{-j\omega n} = \sum_{n=-\infty}^{\infty} \left(\frac{\alpha}{r}\right)^n e^{j(\varphi-\omega)n} \end{aligned} \tag{5.4}$$

You may remember the DTFT analysis procedure for orthogonality between the frequency components.

$$\sum_{n=-\infty}^{\infty} e^{j(\omega_1-\omega)n} = 2\pi\delta(\omega_1 - \omega) \tag{5.5}$$

The orthogonality provides the frequency selection to represent the individual spectral complex magnitude. However, the Z-transform cannot guarantee the orthogonality due to the time variant exponential weight with real number α/r. Therefore, the Z-transform does not deliver the individual spectral magnitude in anyway. In different perspectives, we may organize the Z-transform as below.

$$x_2[n] = \alpha^n e^{j\varphi n} u[n] \tag{5.6}$$

$$\begin{aligned} X_2(z) &= \sum_{n=-\infty}^{\infty} x[n]z^{-n} = \sum_{n=0}^{\infty} \alpha^n e^{j\varphi n} z^{-n} = \sum_{n=0}^{\infty} \left(\alpha e^{j\varphi} z^{-1}\right)^n \\ &= \frac{1}{1-\alpha e^{j\varphi} z^{-1}} = \frac{z}{z-\alpha e^{j\varphi}} \text{ for } \left|\alpha e^{j\varphi} z^{-1}\right| < 1 \end{aligned} \tag{5.7}$$

By using the sum of the infinite geometric series, the given signal can be denoted by the rational function with certain condition. The infinite length signal is reduced to the simpler form as rational function. Since the Z-transform is the linear transform as Eq. (5.8), the linear combination of time variant exponential weight periodic signals can be transformed to the corresponding linear combination of rational functions.

$$\sum_{n=-\infty}^{\infty} (\alpha x[n] + \beta y[n]) z^{-n} = \alpha \sum_{n=-\infty}^{\infty} x[n] z^{-n} + \beta \sum_{n=-\infty}^{\infty} y[n] z^{-n} \tag{5.8}$$

Instead of computing the magnitude, the Z-transform converts the infinite geometric series to the straightforward rational function.

$$\alpha^n u[n] \overset{Z}{\leftrightarrow} \frac{1}{1-\alpha z^{-1}} = \frac{z}{z-\alpha} \text{ for } \left|\alpha z^{-1}\right| < 1 \text{ or} |z| > |\alpha| \tag{5.9}$$

The condition to converge the geometric series is controlled by the z values known as the region of convergence (ROC). The signal ratio multiplied with z^{-1} provides the Z-transform common ratio which should be less than one in absolute magnitude. The inequality over the absolute z value specifies the ROC and the corresponding region is indicated by the circular fashion. Equation (5.10) shows the mathematical definition of the ROC.

$$\text{ROC of } Z\{x[n]\} = \{z \,|\, |X(z)| < \infty\} \tag{5.10}$$

The ROC is defined by the signal distribution profile.

Causal signal:

$$x_c[n] = 0 \text{ for n} < 0 \tag{5.11}$$

$$\begin{aligned} &\text{Let} x_c[n] = \alpha^n e^{j\omega n} u[n] \\ &X_c(\text{z}) = \sum_{n=-\infty}^{\infty} x_c[n] z^{-n} = \sum_{n=0}^{\infty} \alpha^n e^{j\omega n} z^{-n} - \sum_{n=0}^{\infty} (\alpha e^{j\omega} z^{-1})^n < \infty \\ &\left|\alpha e^{j\omega} z^{-1}\right| < 1 \rightarrow \left|\alpha e^{j\omega}\right| \left|z^{-1}\right| < 1 \\ &\rightarrow |z| > \left|\alpha e^{j\omega}\right| \rightarrow |z| > |\alpha| \end{aligned} \tag{5.12}$$

Anti-causal signal:

$$x_a[n] = 0 \text{ for } n > 0 \tag{5.13}$$

$$\text{Let } x_a[n] = \beta^n e^{j\omega n} u[-n]$$

$$\begin{aligned} X_a(z) &= \sum_{n=-\infty}^{\infty} x_a[n] z^{-n} = \sum_{n=-\infty}^{0} \beta^n e^{j\omega n} z^{-n} = \sum_{n=0}^{\infty} \beta^{-n} e^{-j\omega n} z^n \\ &= \sum_{n=0}^{\infty} (\beta^{-1} e^{-j\omega} z)^n < \infty \\ &\left|\beta^{-1} e^{-j\omega} z\right| < 1 \rightarrow \left|\beta^{-1} e^{-j\omega}\right| |z| < 1 \\ &\rightarrow |z| < \left|\beta e^{j\omega}\right| \rightarrow |z| < |\beta| \end{aligned} \tag{5.14}$$

Non-causal signal:

$$x_{nc}[n] = x_c[n] + x_a[n] \tag{5.15}$$

$$\text{Let } x_{nc}[n] = \alpha^n e^{j\omega n} u[n] + \beta^n e^{j\omega n} u[-n]$$

$$\begin{aligned} X_{nc}(z) &= \sum_{n=-\infty}^{\infty} x_{nc}[n] z^{-n} = \sum_{n=0}^{\infty} \alpha^n e^{j\omega n} z^{-n} + \sum_{n=-\infty}^{0} \beta^n e^{j\omega n} z^{-n} \\ &= \sum_{n=0}^{\infty} (\alpha e^{j\omega} z^{-1})^n + \sum_{n=0}^{\infty} (\beta^{-1} e^{-j\omega} z)^n \\ &\left|\alpha e^{j\omega} z^{-1}\right| < 1 \text{ and } \left|\beta^{-1} e^{-j\omega} z\right| < 1 \rightarrow |\alpha| < |z| < |\beta| \\ &\text{ROC}\{X_{nc}(z)\} = \text{ROC}\{X_c(z)\} \cap \text{ROC}\{X_a(z)\} \end{aligned} \tag{5.16}$$

Finite-support signal:

$$\begin{aligned} x_f[n] &= x_{nc}[n]\{u[n - N_1] - u[n - N_2]\} \\ &\text{for } -\infty < N_1 < 0 < N_2 < \infty \end{aligned} \tag{5.17}$$

$$\begin{aligned} X_f(z) &= \sum_{n=-\infty}^{\infty} x_f[n] z^{-n} = \sum_{n=N_1}^{N_2} x_f[n] z^{-n} \\ &= x_f[N_1] z^{-N_1} + \ldots + x_f[-1] z^1 + x_f[0] z^0 + x_f[1] z^{-1} \\ &+ \ldots + x_f[N_2] z^{-N_2} \\ &\text{ROC}\{X_f(z)\} = \{z | z \neq 0 \text{ and } z \neq \infty\} \end{aligned} \tag{5.18}$$

Table 5.2 illustrates the Z-transforms for fundamental discrete signals.

Table 5.2 Z-transforms for fundamental discrete time signals

$x[n]$	$X(z)$	ROC
$\delta[n]$	1	Whole z plane
$u[n]$	$\frac{1}{1-z^{-1}} = \frac{z}{z-1}$	$\|z\| > 1$
$\alpha^n u[n]$	$\frac{1}{1-\alpha z^{-1}} = \frac{z}{z-\alpha}$	$\|z\| > \|\alpha\|$
$\cos(\omega_0 n)u[n] = \frac{e^{j\omega_0 n} + e^{-j\omega_0 n}}{2} u[n]$	$\frac{1}{2}\left(\frac{1}{1-e^{j\omega_0}z^{-1}} + \frac{1}{1-e^{-j\omega_0}z^{-1}}\right) = \frac{1-z^{-1}\cos(\omega_0)}{1-2z^{-1}\cos(\omega_0)+z^{-2}}$	$\|z\| > 1$
$\sin(\omega_0 n)u[n] = \frac{e^{j\omega_0 n} - e^{-j\omega_0 n}}{2j} u[n]$	$\frac{1}{2j}\left(\frac{1}{1-e^{j\omega_0}z^{-1}} - \frac{1}{1-e^{-j\omega_0}z^{-1}}\right) = \frac{z^{-1}\sin(\omega_0)}{1-2z^{-1}\cos(\omega_0)+z^{-2}}$	$\|z\| > 1$

Example 5.1

Find the Z-transform of following signals

$$
\begin{aligned}
x_1[n] &= 4\delta[n] - 3\delta[n-1] + 2\delta[n-2] - \delta[n-3] \\
x_2[n] &= \alpha^n \cos(\omega_0 n) u[n] \\
x_3[n] &= u[-n]
\end{aligned}
$$

Solution

$$
\begin{aligned}
X_1(z) &= \sum_{n=-\infty}^{\infty} x_1[n] z^{-n} = 4 - 3z^{-1} + 2z^{-2} - z^{-3} \\
&\mathrm{ROC}\{X_1(\mathrm{z})\} = \{z | z \in \mathbb{C} \text{ and } z \neq 0\} \\
X_2(z) &= \sum_{n=-\infty}^{\infty} x_2[n] z^{-n} = \sum_{n=0}^{\infty} \alpha^n \cos(\omega_0 n) z^{-n} \\
&= \sum_{n=0}^{\infty} \frac{\alpha^n e^{j\omega_0 n} + \alpha^n e^{-j\omega_0 n}}{2} z^{-n} = \frac{1}{2}\sum_{n=0}^{\infty}\left(\alpha e^{j\omega_0} z^{-1}\right)^n + \frac{1}{2}\sum_{n=0}^{\infty}\left(\alpha e^{-j\omega_0} z^{-1}\right)^n \\
&= \frac{1}{2}\left(\frac{1}{1-\alpha e^{j\omega_0} z^{-1}} + \frac{1}{1-\alpha e^{-j\omega_0} z^{-1}}\right) = \frac{1-\alpha z^{-1}\cos(\omega_0)}{1-2\alpha z^{-1}\cos(\omega_0)+\alpha^2 z^{-2}} \\
&\left|\alpha e^{j\omega_0} z^{-1}\right| < 1 \text{ and } \left|\alpha e^{-j\omega_0} z^{-1}\right| < 1 \\
&\rightarrow \left|\alpha z^{-1}\right| < 1 \rightarrow -1 < \alpha z^{-1} < 1 \rightarrow -z < \alpha < z \\
&\rightarrow |z| > |\alpha| \\
&\mathrm{ROC}\{X_2(\mathrm{z})\} = \{z | z \in \mathbb{C} \text{ and } |z| > |\alpha|\}
\end{aligned}
$$

$$\begin{aligned} X_3(z) &= \sum_{n=-\infty}^{\infty} x_3[n]z^{-n} = \sum_{n=-\infty}^{\infty} u[-n]z^{-n} = \sum_{n=-\infty}^{0} z^{-n} \\ &= \sum_{n=0}^{\infty} z^{n} = \frac{1}{1-z} = \frac{-z^{-1}}{1-z^{-1}} \\ \mathrm{ROC}\{X_3(z)\} &= \{z|z \in \mathbb{C} \text{ and } |z| < 1\} \end{aligned}$$

■

The primitive element (eigenfunction) of the DTFT, DFT, and Z-transform is the $e^{j\omega}$, $e^{j2\pi k/N}$, and $z = re^{j\omega}$ respectively. Therefore, the signal is indicated by the corresponding primitive elements those are distributed over the continuous unit circle for DTFT, discrete unit circle for DFT, and whole complex plane for Z-transform as shown in Fig. 5.1.

The Z-transform occupies all type of the signal that contains the continuous frequency periodic signal with subset of the discrete frequency periodic signal. Once the Z-transform is performed, we can convert to the DTFT or DFT by replacing the z with complex exponential as shown in Eq. (5.19).

$$x[n]\begin{cases} \overset{Z}{\leftrightarrow} X(z)|_{z=e^{j\omega}} = X(e^{j\omega}) : \text{DTFT} \\ \overset{Z}{\leftrightarrow} X(z)|_{z=e^{j\frac{2\pi}{N}k}} = X[k] : \text{DFT} \end{cases} \tag{5.19}$$

Basically, the Z-transform is also the transformation with the power of complex number exponential similar to the DTFT and DFT; therefore, the Z-transform follows the most of DTFT and DFT properties. The conventional Z-transform properties are organized in Table 5.3.

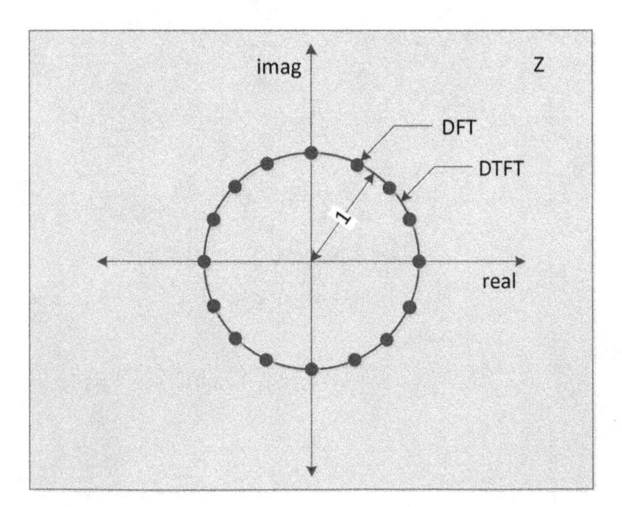

Fig. 5.1 DTFT, DFT, and Z in complex number domain

Table 5.3 Z-transforms properties

Property	Time domain $x[n]$	z domain $X(z)$
Time shift	$x[n-d]$	$X(z)z^{-d}$
Time reversal	$x[-n]$	$X(z^{-1})$
Real signal	$x[n] \in \mathbb{R}$	$X(z) = X^*(z^*)$
Convolution	$x[n] * h[n]$	$X(z)H(z)$
Initial value theorem	$x[n] = x[n]u[n]$	$x[0] = \lim_{z\to\infty} X(z)$
Final value theorem	$x[n] = x[n]u[n]$and $\lim_{n\to\infty} x[n] = x[\infty]$	$x[\infty] = \lim_{z\to 1}(z-1)X(z)$

Example 5.2

Find the Z-transform of following signals

$$x[n] = nu[n]$$

Compute the following values by using the initial value and final value theorem.

$$x[0] =$$
$$x[\infty] =$$

Check the below condition in z domain.

$$x[n] \in \mathbb{R}$$

If the new $x[n]$ is below, check the real value condition again.

$$x[n] = jnu[n]$$

Solution

The given function can be decomposed as below.

$$nu[n] = u[n] * u[n-1]$$
$$u[n] \overset{z}{\leftrightarrow} \frac{1}{1-z^{-1}} u[n-1] \overset{z}{\leftrightarrow} \frac{z^{-1}}{1-z^{-1}}$$
$$u[n] * u[n-1] \overset{z}{\leftrightarrow} \frac{1}{1-z^{-1}} \frac{z^{-1}}{1-z^{-1}} = \frac{z^{-1}}{(1-z^{-1})^2}$$
$$\text{ROC}\{X(\text{z})\} = \{z||z| > 1\}$$

By using the initial value and final value theorem, find $x[0]$ and $x[\infty]$ as below.

$$x[0] = \lim_{z\to\infty} X(z) = \lim_{z\to\infty} \frac{z^{-1}}{(1-z^{-1})^2} = 0$$

$$x[\infty] = \lim_{z\to 1}(z-1)X(z) = \lim_{z\to 1}(z-1)\frac{z^{-1}}{(1-z^{-1})^2}$$

$$= \lim_{z\to 1}\frac{z}{(z-1)} = \infty$$

Determine the real value condition of $x[n]$ by using the following property.

$$X(z) = X^*(z^*)$$

$$X^*(z^*) = \left(\frac{(z^*)^{-1}}{\left(1-(z^*)^{-1}\right)^2}\right)^* = \frac{\left((z^*)^{-1}\right)^*}{\left(1-(z^*)^{-1}\right)^*\left(1-(z^*)^{-1}\right)^*}$$

$$= \frac{z^{-1}}{(1-z^{-1})^2} = X(z)$$

From the new given signal below, check the real value condition.

$$x[n] = jnu[n] \overset{z}{\leftrightarrow} X(z) = \frac{jz^{-1}}{(1-z^{-1})^2}$$

$$X^*(z^*) = \left(\frac{j(z^*)^{-1}}{\left(1-(z^*)^{-1}\right)^2}\right)^* = \frac{-j\left((z^*)^{-1}\right)^*}{\left(1-(z^*)^{-1}\right)^*\left(1-(z^*)^{-1}\right)^*}$$

$$= \frac{-jz^{-1}}{(1-z^{-1})^2} \neq X(z)$$

$$\therefore x[n] \notin \mathbb{R}$$

■

The proof of the above relationships can be easily derived by the DTFT and DFT properties. The DTFT and DFT use the ω and k for the frequency distribution over the one-dimensional axis; however, the Z-transform delivers the information based on the z domain which represents the two-dimensional space with r distance and ω angle. Therefore, the z^{-1} and z^* indicate the $r^{-1}e^{-j\omega}$ and $re^{-j\omega}$ for $z = re^{j\omega}$, respectively as shown in Fig. 5.2.

The real signal property for the Z-transform is shown below.

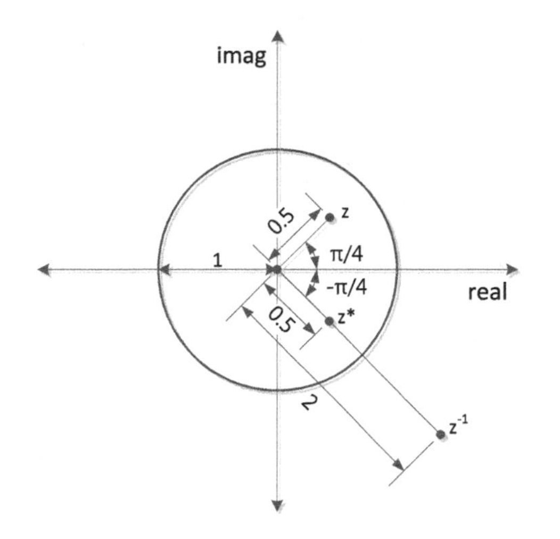

Fig. 5.2 Illustration of $z = 0.5e^{j\pi/4}$, $z^* = 0.5e^{-j\pi/4}$, $z^{-1} = 2e^{-j\pi/4}$

$$\begin{aligned}
&x[n] \in \mathbb{R} \rightarrow x[n] = x^*[n] \\
&X(z) = \sum_{n=-\infty}^{\infty} x[n] z^{-n} \\
&\sum_{n=-\infty}^{\infty} x^*[n] z^{-n} = \sum_{n=-\infty}^{\infty} \{x[n](z^*)^{-n}\}^* = \left\{\sum_{n=-\infty}^{\infty} x[n](z^*)^{-n}\right\}^* = X^*(z^*) \\
&x[n] \in \mathbb{R} \rightarrow x[n] = x^*[n] \rightarrow X(z) = X^*(z^*)
\end{aligned} \tag{5.20}$$

Hence, the Z-transform of the real signal denotes conjugate pair between the one location (z) and symmetric position (z^*) over the real axis on the z plane. The graphical representation of the above condition is illustrated in Fig. 5.3.

We can figure out the initial and final value of the given causal sequence in the z domain as below.

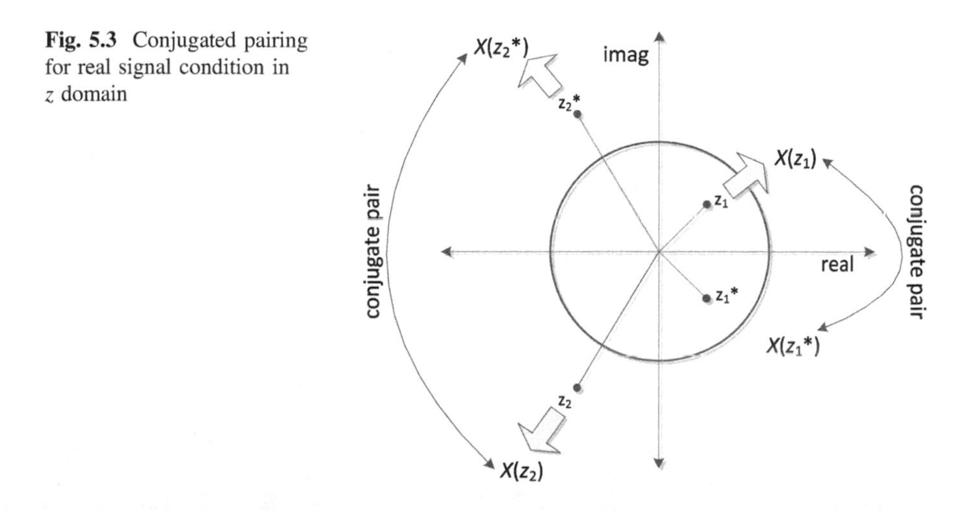

Fig. 5.3 Conjugated pairing for real signal condition in z domain

Initial value theorem: $x[n]$ is causal

$$x[0] = \lim_{z\to\infty} X(z) \tag{5.21}$$

Proof

$$\begin{aligned} x[n] &= x[n]u[n] \\ X(z) &= \sum_{n=0}^{\infty} x[n]z^{-n} = x[0] + x[1]z^{-1} + x[2]z^{-2} + x[3]z^{-3} + \ldots + x[\infty]z^{-\infty} \\ \lim_{z\to\infty} X(z) &= \lim_{z\to\infty} \{x[0] + x[1]z^{-1} + x[2]z^{-2} + x[3]z^{-3} + \ldots + x[\infty]z^{-\infty}\} \\ &= x[0] + x[1]0 + x[2]0 + x[3]0 + \ldots + x[\infty]0 = x[0] \end{aligned}$$

∎

Final value theorem: $x[n]$ is causal

$$x[\infty] = \lim_{z\to 1}(z-1)X(z) \tag{5.22}$$

Proof We assume that the signal asymptotically converges to the constant value $x[\infty]$ as below.

$$x[n] = x[n]u[n] \text{ and } \lim_{n\to\infty} x[n] = x[\infty]$$

And decomposed as below.

$$x[n] = x[\infty]u[n] + w[n] \quad \lim_{n\to\infty} w[n] = 0$$

The example of $w[n]$ is shown in Fig. 5.4.

In order to approach zero, the $w[n]$ common ratio (or pole) should be located within the unit circle in the z plane.

$$X(z) = Z\{x[\infty]u[n] + w[n]\} = x[\infty]\frac{z}{z-1} + W(z)$$

According to the Z-transform above, $X(z)$ contains the $(z-1)$ factor in the denominator. However, $W(z)$ cannot include the $(z-1)$ factor in the denominator because of the common ratio constraint in the $w[n]$.

$$\begin{aligned} (z-1)X(z) &= x[\infty]z + (z-1)W(z) \\ \lim_{z\to 1}(z-1)X(z) &= \lim_{z\to 1} x[\infty]z + \lim_{z\to 1}(z-1)W(z) \end{aligned}$$

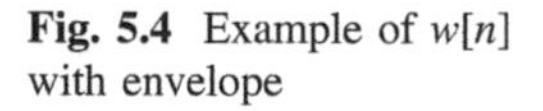

Fig. 5.4 Example of $w[n]$ with envelope

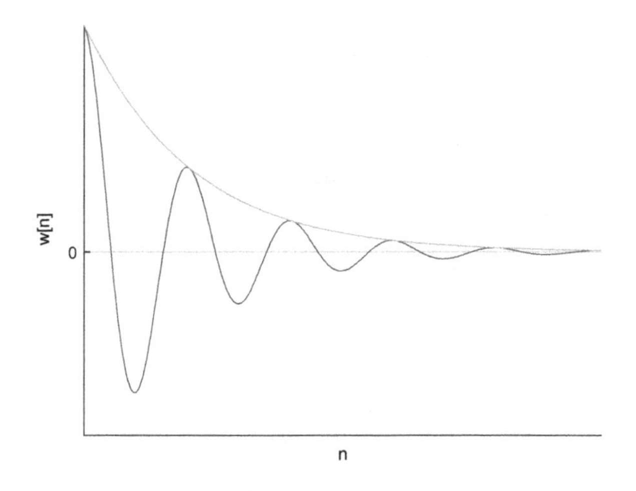

The $(z - 1)$ with $W(z)$ cannot be cancelled out; hence, the last term goes to zero. Finally,

$$\lim_{z \to 1}(z - 1)X(z) = x[\infty]$$

■

The inverse Z-transform can be derived from the DTFT as below.

$$\begin{aligned}
X(z) &= \sum_{n=-\infty}^{\infty} x[n]z^{-n} = \sum_{n=-\infty}^{\infty} (x[n]r^{-n})e^{-j\omega n} = \text{DTFT}\{x[n]r^{-n}\} \\
&x[n]r^{-n} \overset{\text{DTFT}}{\leftrightarrow} X(z) \\
x[n]r^{-n} &= \frac{1}{2\pi}\int_{-\pi}^{\pi} X(\mathrm{z})e^{j\omega n}d\omega \\
x[n] &= \frac{1}{2\pi}\int_{-\pi}^{\pi} X(\mathrm{z})r^n e^{j\omega n}d\omega = \frac{1}{2\pi}\int_{-\pi}^{\pi} X(\mathrm{z})z^n d\omega \\
&\text{where } z = re^{j\omega} \text{ and } d\omega = \frac{1}{j}z^{-1}dz \\
x[n] &= \frac{1}{2\pi j}\oint X(z)z^{n-1}dz
\end{aligned} \tag{5.23}$$

Due to the contour integral over the complex domain, the inverse Z-transform is barely performed to compute the time domain information. By using the Z-transform linearity, the long division and partial fraction method are employed for the inverse Z-transform. Basically, we prefer to build following situation in Eq. (5.24).

$$\begin{aligned} x[n] &= \alpha_0 r_0^n e^{j\omega_0 n} u[n] + \alpha_1 r_1^n e^{j\omega_1 n} u[n] + \alpha_2 r_2^n e^{j\omega_2 n} u[n] + \ldots \\ &\overset{Z}{\leftrightarrow} X(z) = \frac{\alpha_0}{1 - r_0 e^{j\omega_0} z^{-1}} + \frac{\alpha_1}{1 - r_1 e^{j\omega_1} z^{-1}} + \frac{\alpha_2}{1 - r_2 e^{j\omega_2} z^{-1}} + \ldots \end{aligned} \tag{5.24}$$

Once we obtain the linear combination of the first-degree rational functions, the corresponding time domain function can be simply derived by individual replacement based on the Z-transform table. If the given z function is the high order rational function, the long division and partial fraction method are required to decompose the function for the convenient inverse transformation. The general high order rational z function is shown in Eq. (5.25).

$$X(z) = \frac{b_0 + b_1 z^{-1} + b_2 z^{-2} + \ldots + b_n z^{-n}}{a_0 + a_1 z^{-1} + a_2 z^{-2} + \ldots + a_m z^{-m}} \tag{5.25}$$

The numerator and denominator are presented by the n and m degree polynomial, respectively. If the numerator has the higher degree than the denominator, long division is necessary to reduce the numerator degree. Polynomial long division [2] is illustrated in Fig. 5.5.

The example above considers the high order polynomial with two-degree difference between the numerator and denominator. The β_2, β_1, and β_0 are provided to remove the highest negative order at dividend; therefore, each division procedure generates the reduced order at numerator polynomial. Once the remainder order is less than the divisor, stop further division and organize the equation as below.

$$\begin{aligned} X(z) = \frac{b_0 + b_1 z^{-1} + b_2 z^{-2} + \ldots + b_n z^{-n}}{a_0 + a_1 z^{-1} + a_2 z^{-2} + \ldots + a_m z^{-m}} &= \beta_2 z^{-2} + \beta_1 z^{-1} + \beta_0 \\ + \frac{\gamma_0 + \gamma_1 z^{-1} + \ldots + \gamma_{n-3} z^{-(n-3)}}{a_0 + a_1 z^{-1} + a_2 z^{-2} + \ldots + a_m z^{-m}} & \quad \text{for } n = m + 2 \end{aligned} \tag{5.26}$$

The rational function can be further decomposed into the linear combination of first order rational functions as Eq. (5.27). The partial fraction method finds the coefficients of the linearly divided rational functions as ρ_1, ρ_2, and ρ_m. Numerous

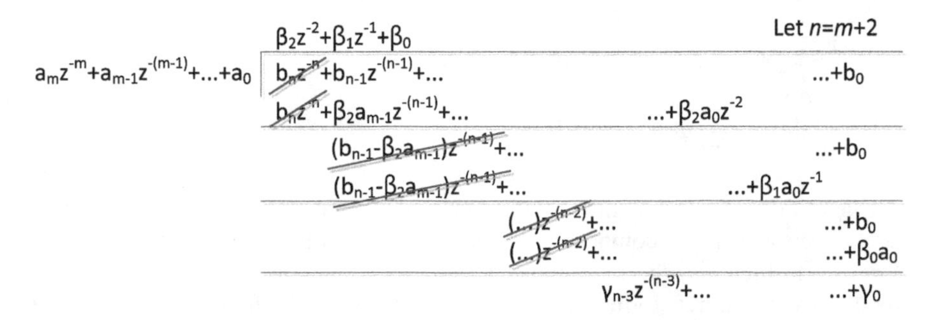

Fig. 5.5 Example of polynomial long division for Eq. (5.25)

approaches are possible to solve the partial fraction method that can be found on the various algebra books [2] and webs [3].

$$
\begin{aligned}
Q(z) &= \frac{\gamma_0 + \gamma_1 z^{-1} + \ldots + \gamma_{n-3} z^{-(n-3)}}{a_0 + a_1 z^{-1} + a_2 z^{-2} + \ldots + a_m z^{-m}} \\
&= \frac{\gamma_0 + \gamma_1 z^{-1} + \ldots + \gamma_{n-3} z^{-(n-3)}}{(1 - \mu_1 z^{-1})(1 - \mu_2 z^{-1})\ldots(1 - \mu_m z^{-1})} \\
&= \frac{\rho_1}{(1 - \mu_1 z^{-1})} + \frac{\rho_2}{(1 - \mu_2 z^{-1})} + \ldots + \frac{\rho_m}{(1 - \mu_m z^{-1})}
\end{aligned} \tag{5.27}
$$

Also, the function can be handled by the positive power of the polynomial as below.

$$
\begin{aligned}
Q(z) &= \frac{\gamma_0 + \gamma_1 z^{-1} + \ldots + \gamma_{n-3} z^{-(n-3)}}{a_0 + a_1 z^{-1} + a_2 z^{-2} + \ldots + a_m z^{-m}} \\
&= \frac{z^m \left(\gamma_0 z^{(n-3)} + \gamma_1 z^{(n-3)-1} + \ldots + \gamma_{n-3}\right)}{z^{(n-3)} \left(a_0 z^m + a_1 z^{m-1} + a_2 z^{m-2} + \ldots + a_m\right)} \\
&= \frac{z\left(\gamma_0 z^{(n-3)} + \gamma_1 z^{(n-3)-1} + \ldots + \gamma_{n-3}\right)}{\left(a_0 z^m + a_1 z^{m-1} + a_2 z^{m-2} + \ldots + a_m\right)} \quad \text{since } n = m + 2 \\
\frac{Q(z)}{z} &= \frac{\left(\gamma_0 z^{(n-3)} + \gamma_1 z^{(n-3)-1} + \ldots + \gamma_{n-3}\right)}{\left(a_0 z^m + a_1 z^{m-1} + a_2 z^{m-2} + \ldots + a_m\right)} \\
&= \frac{\rho_1}{(z - \mu_1)} + \frac{\rho_2}{(z - \mu_2)} + \ldots + \frac{\rho_m}{(z - \mu_m)} \\
Q(z) &= \frac{\rho_1 z}{(z - \mu_1)} + \frac{\rho_2 z}{(z - \mu_2)} + \ldots + \frac{\rho_m z}{(z - \mu_m)}
\end{aligned} \tag{5.28}
$$

Therefore, the final decomposed form of the rational polynomial is given below.

$$
\begin{aligned}
X(z) = \beta_2 z^{-2} + \beta_1 z^{-1} + \beta_0 + \\
\tfrac{\rho_1}{(1 - \mu_1 z^{-1})} + \tfrac{\rho_2}{(1 - \mu_2 z^{-1})} + \ldots + \tfrac{\rho_m}{(1 - \mu_m z^{-1})}
\end{aligned} \tag{5.29}
$$

The corresponding time domain representation is derived by simple substitutions as shown in Eq. (5.30). Assume that the signal is causal.

$$
\begin{aligned}
x[n] = \beta_2 \delta[n-2] + \beta_1 \delta[n-1] + \beta_0 \delta[n] \\
+ \mu_1^n u[n] + \mu_2^n u[n] + \ldots + \mu_m^n u[n]
\end{aligned} \tag{5.30}
$$

The damped (or non-damped) sinusoid can be illustrated by the common ratio form as well for simple Z-transform as below. Further simplification is available and can be found in the conventional Z-transform table [4, 5].

$$
\begin{aligned}
r^n \cos(\omega n + \theta)u[n] &= \frac{r^n e^{j(\omega n+\theta)} + r^n e^{-j(\omega n+\theta)}}{2} u[n] \\
&= \frac{e^{j\theta}}{2}\left(re^{j\omega}\right)^n u[n] + \frac{e^{-j\theta}}{2}\left(re^{-j\omega}\right)^n \\
&\overset{Z}{\leftrightarrow} \frac{e^{j\theta}}{2}\frac{1}{(1-re^{j\omega}z^{-1})} + \frac{e^{-j\theta}}{2}\frac{1}{(1-re^{-j\omega}z^{-1})}
\end{aligned} \tag{5.31}
$$

The Z-transform provides the rational function to represent the geometric series distribution in the time domain signal. The common ratios of the geometric series can be observed in the denominator factor forms even with the combining procedure over the individual first-degree rational functions. The ROC decides the signal time style in terms of length and direction. The conventional causal signal covers the ROC over the outside of the biggest common ratio distance. The inverse process of the Z-transform is performed by the long division and partial fraction method in order to linearly decompose the high order rational function into the first-degree rational functions. Simple replacements of the individual terms denote the corresponding time domain functions.

Example 5.3
Find the inverse Z-transform of following signals. Assume that all signals are causal.

$$
\begin{aligned}
X_1(z) &= 2\left(1-z^{-1}\right)\left(1+2z^{-1}\right)\left(1+0.5z^{-1}\right) \\
X_2(z) &= \frac{3-2z^{-1}+z^{-2}}{1-z^{-1}+z^{-2}-z^{-3}} \\
X_3(z) &= \frac{7-26z^{-1}+17z^{-2}+16z^{-3}-12z^{-4}}{1-6z^{-1}+11z^{-2}-6z^{-3}}
\end{aligned}
$$

Solution

$$
\begin{aligned}
X_1(z) &= 2\left(1-z^{-1}\right)\left(1+2z^{-1}\right)\left(1+0.5z^{-1}\right) \\
&= 2+3z^{-1}-3z^{-2}-2z^{-3} \overset{z}{\leftrightarrow} 2\delta[n]+3\delta[n-1]-3\delta[n-2]-2\delta[n-3] \\
X_2(z) &= \frac{3-2z^{-1}+z^{-2}}{1-z^{-1}+z^{-2}-z^{-3}} = \frac{1}{1-z^{-1}} + \frac{1}{1-jz^{-1}} + \frac{1}{1+jz^{-1}} \\
&\overset{z}{\leftrightarrow} u[n] + e^{-j\frac{\pi}{2}n}u[n] + e^{j\frac{\pi}{2}n}u[n] = u[n] + 2\cos\left(\frac{\pi}{2}n\right)u[n] \\
X_3(z) &= \frac{7-26z^{-1}+17z^{-2}+16z^{-3}-12z^{-4}}{1-6z^{-1}+11z^{-2}-6z^{-3}} \\
&= 1+2z^{-1} + \frac{1}{1-z^{-1}} + \frac{2}{1-2z^{-1}} + \frac{3}{1-3z^{-1}} \\
&\overset{z}{\leftrightarrow} \delta[n] + 2\delta[n-1] + u[n] + 2\cdot 2^n u[n] + 3\cdot 3^n u[n]
\end{aligned}
$$

∎

5.2 Filter and Z-Transform

The variable z in the Z-transform illustrates the two-dimensional space in the complex number domain. Hence, the visualization of the transform requires the three-dimensional illustration. Let's consider the below function.

$$\begin{aligned} X(z) &= \frac{(z+0.8)(z-0.5j)(z+0.5j)}{(z-0.8)(z-0.8e^{j\frac{\pi}{4}})(z-0.8e^{-j\frac{\pi}{4}})} \\ &= \frac{(1+0.8z^{-1})(1-0.5jz^{-1})(1+0.5jz^{-1})}{(1-0.8z^{-1})(1-0.8e^{j\frac{\pi}{4}}z^{-1})(1-0.8e^{-j\frac{\pi}{4}}z^{-1})} \end{aligned} \tag{5.32}$$

The complex number z provides the complex number output $X(z)$ in terms of real/imaginary or magnitude/phase. Figure 5.6 shows the $20\log_{10}|X(z)|$ distribution with unit circle in white. Note that the Z-transform does not deliver the individual spectral magnitude by itself. Therefore, the whole $X(z)$ distribution does not contain the significant information on spectrum.

Prog. 5.1 MATLAB program for Fig. 5.6

```
[X,Y] = meshgrid(-3:0.01:3,-3:0.01:3);
z = X+1j*Y;
XZ = abs((z+0.8).*(z-0.5*1j).*(z+0.5*1j)./((z-0.8).*(z-0.8*exp(1j*pi/4)).*(z-
0.8*exp(-1j*pi/4))));
ii1 = find(XZ==max(max(XZ))); % Remove first pole
XZ(ii1) = 1000;          % Replace max value
ii1 = find(XZ==max(max(XZ))); % Remove second pole
XZ(ii1) = 1000;          % Replace max value
ii1 = find(XZ==max(max(XZ))); % Remove third pole
XZ(ii1) = 1000;          % Replace max value
XZ = XZ+0.001;           % Replace min value
contourf(X,Y,20*log10(XZ)), grid;
hold on
theta = linspace(0,2*pi);
x1 = cos(theta);
y1 = sin(theta);
plot(x1,y1,'w'), axis equal,hold off;     % Unit circle
c = colorbar('Ticks',[-60 60],...
            'TickLabels',{'Zero','Infinite'});
xlabel('Real');
ylabel('Imaginary');
xlim([-2 2]);
ylim([-2 2]);
c.Label.String = '20log_{10}|X(z)|';
```

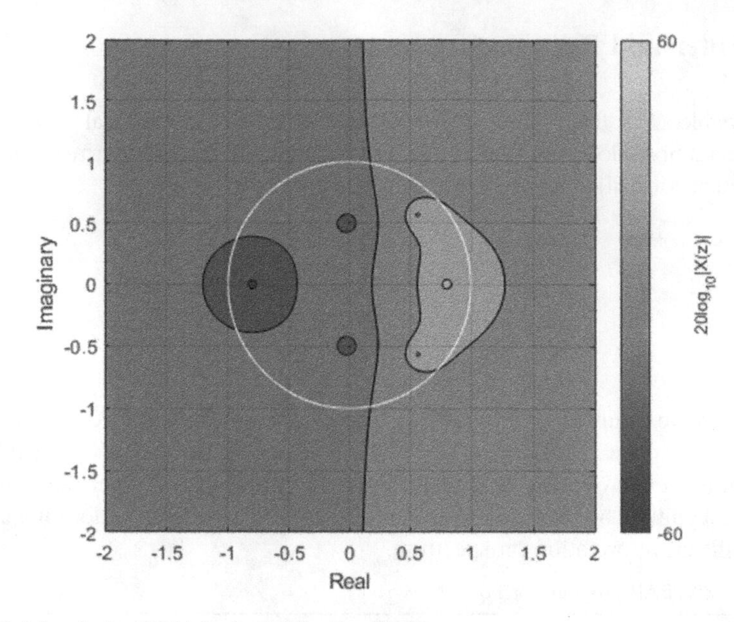

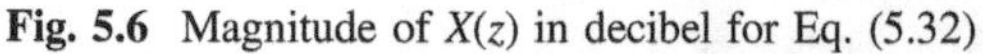

Fig. 5.6 Magnitude of $X(z)$ in decibel for Eq. (5.32)

We discussed that the Z-transform can be converted to the DTFT and DFT by replacing the z with $e^{j\omega}$ and $e^{j2\pi k/N}$, respectively. The DTFT and DFT is the subset of the Z-transform as shown in Fig. 5.7. The $e^{j\omega}$ and $e^{j2\pi k/N}$ represent the complex numbers on the unit circle expressed on Fig. 5.6 with white circle. The Z-transform outcome on the unit circle corresponds to the DTFT and DFT result.

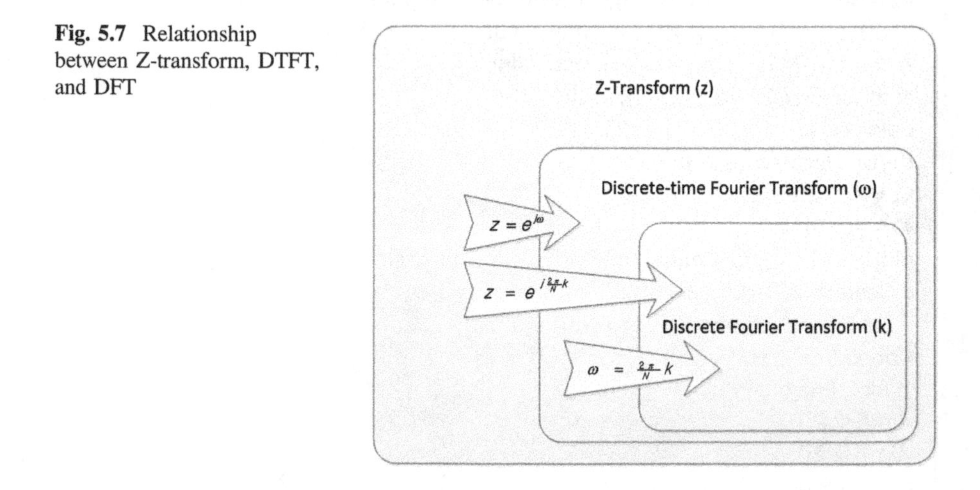

Fig. 5.7 Relationship between Z-transform, DTFT, and DFT

The evaluation on the unit circle from the given Z-transformed function is shown in Fig. 5.8. The corresponding DTFT and DFT of the given function is illustrated in Fig. 5.9. The unfolded version of the unit circle output is identical to the DTFT and DFT result.

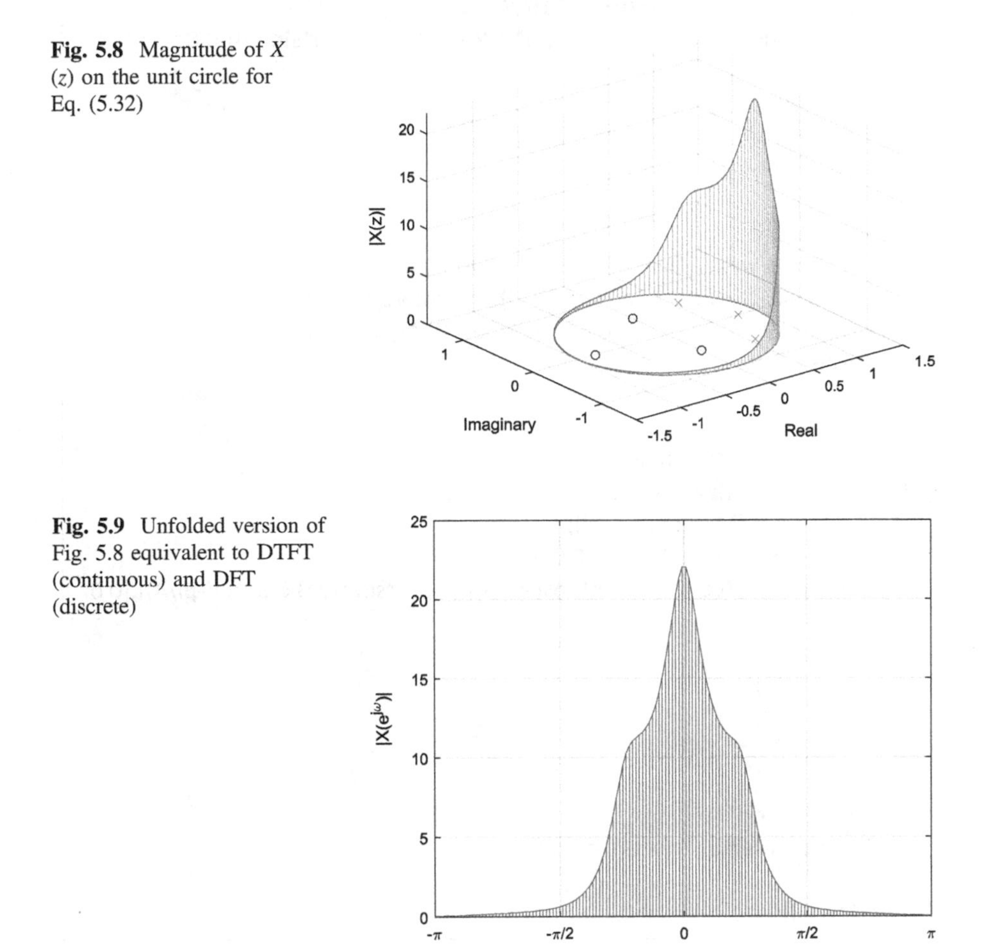

Fig. 5.8 Magnitude of $X(z)$ on the unit circle for Eq. (5.32)

Fig. 5.9 Unfolded version of Fig. 5.8 equivalent to DTFT (continuous) and DFT (discrete)

Prog. 5.2 MATLAB program for Figs. 5.8 and 5.9

```
x1 = @(w)cos(w);
y1 = @(w)sin(w);
z1 = @(w)abs(((cos(w)+1j*sin(w))+0.8).*((cos(w)+1j*sin(w))-
0.5*1j).*((cos(w)+1j*sin(w))+0.5*1j)./(((cos(w)+1j*sin(w))-
0.8).*((cos(w)+1j*sin(w))-0.8*exp(1j*pi/4)).*((cos(w)+1j*sin(w))-0.8*exp(-
1j*pi/4))));
z2 = @(w)w*0;

w1 = linspace(-pi,pi,150);
x2 = cos(w1);
y2 = sin(w1);
z3 = abs(((cos(w1)+1j*sin(w1))+0.8).*((cos(w1)+1j*sin(w1))-
0.5*1j).*((cos(w1)+1j*sin(w1))+0.5*1j)./(((cos(w1)+1j*sin(w1))-
0.8).*((cos(w1)+1j*sin(w1))-0.8*exp(1j*pi/4)).*((cos(w1)+1j*sin(w1))-
0.8*exp(-1j*pi/4))));

figure,
fp1 = fplot3(x1,y1,z1), hold on;
fp2 = fplot3(x1,y1,z2);
stem3(x2,y2,z3,'Marker','none');
scatter3([-0.8 0 0],[0 0.5 -0.5],[0 0 0],'o');
scatter3([0.8 0.8*cos(pi/4) 0.8*cos(-pi/4)],[0 0.8*sin(pi/4) 0.8*sin(-pi/4)],[0 0
0],'x'), hold off;
...
xlim([-1.5 1.5])
ylim([-1.5 1.5])

figure,
plot(w1,z3), grid, hold on;
stem(w1,z3,'Marker','none'), hold off;
...
xlim([-pi pi]);
```

The direct evaluation for spectrum profile requires the computational power due to the complex number arithmetic. Rough estimation is possible by investigating the roots of Z-transformed rational function. The roots of the numerator and denominator polynomial is named zero and pole, respectively. The zero locations deliver the zero output and pole positions represent the infinite output from the z function.

$$\text{zeros} = \{z_z | X(z_z) = 0\} \quad \text{poles} = \{z_p | X(z_p) = \infty\} \tag{5.33}$$

The arbitrary z function is denoted by the rational function and the factors of the numerator and denominator polynomial provides the zeros and poles of the z function as shown in Eq. (5.34).

$$\begin{aligned} X(z) &= \frac{b_0 + b_1 z^{-1} + b_2 z^{-2} + \ldots + b_n z^{-n}}{a_0 + a_1 z^{-1} + a_2 z^{-2} + \ldots + a_m z^{-m}} \\ &= \frac{(1 - \sigma_1 z^{-1})(1 - \sigma_2 z^{-1})\ldots(1 - \sigma_n z^{-1})}{(1 - \mu_1 z^{-1})(1 - \mu_2 z^{-1})\ldots(1 - \mu_m z^{-1})} \\ &\rightarrow \begin{cases} \text{zeros} = \{\sigma_1, \sigma_2, \ldots, \sigma_n\} \\ \text{poles} = \{\mu_1, \mu_2, \ldots, \mu_m\} \end{cases} \end{aligned} \tag{5.34}$$

The given example z function shows the following zeros and poles.

$$\begin{aligned} X(z) &= \frac{(z + 0.8)(z - 0.5j)(z + 0.5j)}{(z - 0.8)(z - 0.8e^{j\frac{\pi}{4}})(z - 0.8e^{-j\frac{\pi}{4}})} \\ &\rightarrow \begin{cases} \text{zeros} = \{-0.8, 0.5j, -0.5j\} : \text{o mark on z} - \text{plane} \\ \text{poles} = \{0.8, 0.8e^{j\frac{\pi}{4}}, 0.8e^{-j\frac{\pi}{4}}\} : \text{x mark on z} - \text{plane} \end{cases} \end{aligned} \tag{5.35}$$

The computed zeros and poles are presented on Fig. 5.8 by o and x mark, individually. We can observe that the radian frequency closer to the poles produce the high magnitude and closer to zeros generate the low magnitude in Fig. 5.8. This geometrical relationship is caused by the Z-transform to DTFT conversion. Equation (5.36) is the conversion for the given z function.

$$X(z)|_{z=e^{j\omega}} = X(e^{j\omega}) = \frac{(e^{j\omega} + 0.8)(e^{j\omega} - 0.5j)(e^{j\omega} + 0.5j)}{(e^{j\omega} - 0.8)(e^{j\omega} - 0.8e^{j\frac{\pi}{4}})(e^{j\omega} - 0.8e^{-j\frac{\pi}{4}})} \tag{5.36}$$

The magnitude of each frequency ω can be evaluated by the Euclidian distance between the poles and zeros to the individual $e^{j\omega}$ positions as below.

$$|X(e^{j\omega})| = \frac{|(e^{j\omega} + 0.8)||(e^{j\omega} - 0.5j)||(e^{j\omega} + 0.5j)|}{|(e^{j\omega} - 0.8)||(e^{j\omega} - 0.8e^{j\frac{\pi}{4}})||(e^{j\omega} - 0.8e^{-j\frac{\pi}{4}})|} \tag{5.37}$$

The product of the pole distances represents the denominator value in the frequency magnitude at ω. In the same manner, the product of the zero distances denotes the numerator value. The final rational number shows the frequency magnitude for the radian frequency ω. The pole-zero plots in Fig. 5.10 show the z function evaluation at the $\pi/6$ and $5\pi/6$ in order to compute the frequency magnitude. The red and

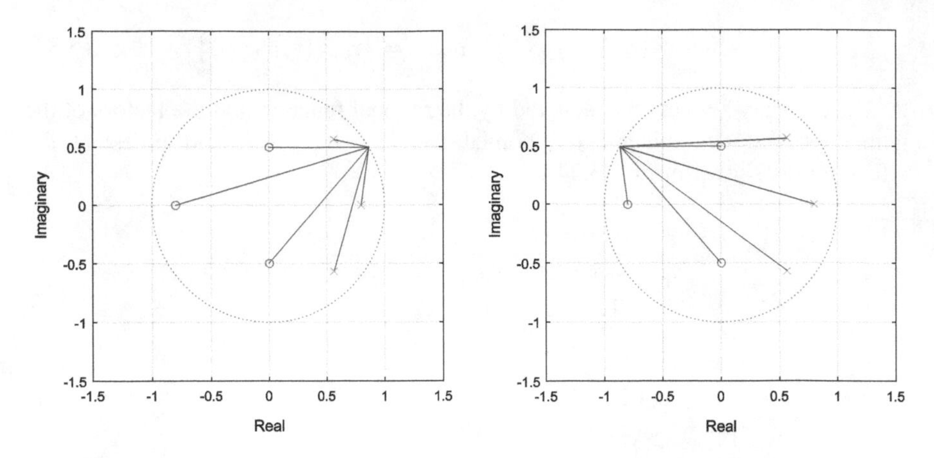

Fig. 5.10 Equation (5.37) evaluation at the $\pi/6$ (left) and $5\pi/6$ (right)

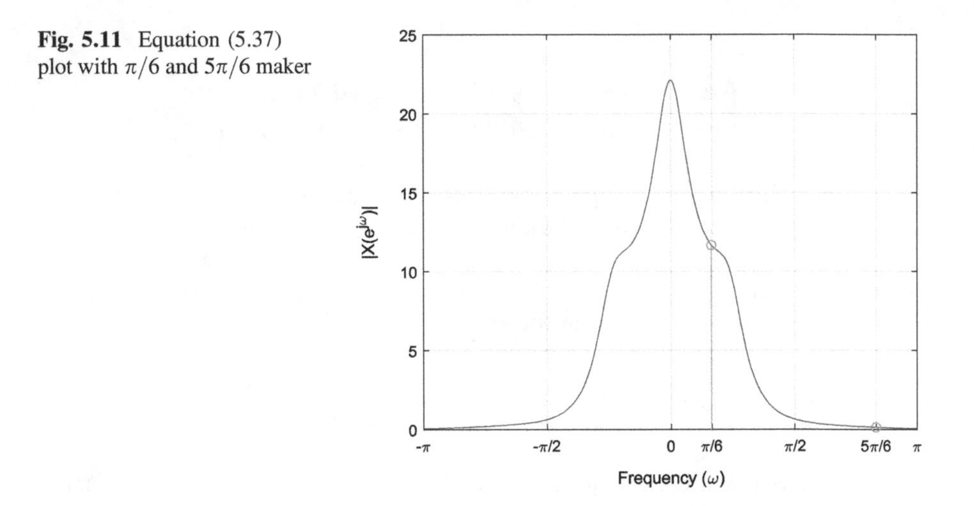

Fig. 5.11 Equation (5.37) plot with $\pi/6$ and $5\pi/6$ maker

blue lines represent the individual denominator and numerator values respectively. The relative pole distances to the $e^{j\pi/6}$ are shorter than the zero's counterparts; therefore, the overall magnitude is high in the frequency response magnitude in Fig. 5.11. The situation for $e^{j5\pi/6}$ is in vice versa. Note that the symmetric locations of poles and zeros produce the even symmetric distribution in frequency response magnitude in Fig. 5.11.

Prog. 5.2 MATLAB program for Fig. 5.11

```
w1 = linspace(-pi,pi,300);
x2 = cos(w1);
y2 = sin(w1);
z3 = abs(((cos(w1)+1j*sin(w1))+0.8).*((cos(w1)+1j*sin(w1))-
0.5*1j).*((cos(w1)+1j*sin(w1))+0.5*1j)./(((cos(w1)+1j*sin(w1))-
0.8).*((cos(w1)+1j*sin(w1))-0.8*exp(1j*pi/4)).*((cos(w1)+1j*sin(w1))-
0.8*exp(-1j*pi/4))));

figure,
plot(w1,z3), grid, hold on;
stem(w1([175 275]),z3([175 275])), hold off;
...
```

Example 5.4
Find the strongest frequency in the following signals.

$$x_1[n] = 0.9^n u[n]$$
$$x_2[n] = 0.9^n \cos\left(\frac{\pi}{2}n\right)u[n]$$
$$x_3[n] = (-0.9)^n u[n]$$

Solution

$$x_1[n] = 0.9^n u[n] \overset{z}{\leftrightarrow} X_1(z) = \frac{1}{1-0.9z^{-1}} = \frac{z}{z-0.9}$$
$$X_1\left(e^{j\omega}\right) = \frac{e^{j\omega}}{e^{j\omega}-0.9}$$

Shortest denominator distance to the unit circle is 0 rad; hence, the strongest frequency is zero radian.

$$x_2[n] = 0.9^n \cos\left(\frac{\pi}{2}n\right)u[n] = \frac{1}{2}0.9^n e^{j\frac{\pi}{2}n}u[n] + \frac{1}{2}0.9^n e^{-j\frac{\pi}{2}n}u[n]$$
$$\overset{z}{\leftrightarrow} X_2(z) = \frac{1/2}{1-0.9e^{j\frac{\pi}{2}}z^{-1}} + \frac{1/2}{1-0.9e^{-j\frac{\pi}{2}}z^{-1}}$$
$$= \frac{z/2}{z-0.9j} + \frac{z/2}{z+0.9j} = \frac{\alpha_2 z^2 + \alpha_1 z}{(z-0.9j)(z+0.9j)}$$
$$X_2\left(e^{j\omega}\right) = \frac{\alpha_2\left(e^{j\omega}\right)^2 + \alpha_1 e^{j\omega}}{\left(e^{j\omega}-0.9j\right)\left(e^{j\omega}+0.9j\right)}$$

Shortest denominator distance to the unit circle is $\pm\frac{\pi}{2}$ radian; hence, the strongest frequency is $\pm\frac{\pi}{2}$ radian.

$$x_3[n] = (-0.9)^n u[n] \overset{z}{\leftrightarrow} X_3(z) = \frac{1}{1+0.9z^{-1}} = \frac{z}{z+0.9}$$
$$X_3(e^{j\omega}) = \frac{e^{j\omega}}{e^{j\omega}+0.9}$$

Shortest denominator distance to the unit circle is π radian; hence, the strongest frequency is π radian.

■

The location of poles and zeros decides the frequency property of the signal as shown above. Let's investigate the poles and zeros location for signal in time domain. The conventional types of the signal are considered as below.

Finite-support and causal signal:

$$\begin{aligned}
x_{fc}[n] &= x_{nc}[n]\{u[n]-u[n-N_1]\} \text{ for } 0<N_1<\infty \\
X_{fc}(z) &= \sum_{n=-\infty}^{\infty} x_{fc}[n]z^{-n} = \sum_{n=0}^{N_1} x_{fc}[n]z^{-n} \\
&= x_{fc}[0]z^0 + x_{fc}[1]z^{-1} + \ldots + x_{fc}[N_2]z^{-N_2} \\
&\text{ROC}\{X_{fc}(z)\} = \{z|z \neq 0\}
\end{aligned} \tag{5.38}$$

Finite-support and non-causal signal:

$$\begin{aligned}
x_{fnc}[n] &= x_{nc}[n]\{u[n-N_1]-u[n-N_2]\} \quad \text{for } -\infty<N_1<0<N_2<\infty \\
X_{fnc}(z) &= \sum_{n=-\infty}^{\infty} x_{fnc}[n]z^{-n} = \sum_{n=N_1}^{N_2} x_{fnc}[n]z^{-n} \\
&= x_{fnc}[N_1]z^{-N_1} + \ldots + x_{fnc}[-1]z^1 \\
&\quad + x_{fnc}[0]z^0 + x_{fnc}[1]z^{-1} + \ldots + x_{fnc}[N_2]z^{-N_2} \\
&\text{ROC}\{X_{fnc}(z)\} = \{z|z \neq 0 \text{ and } z \neq \infty\}
\end{aligned} \tag{5.39}$$

Causal signal:

$$\begin{aligned}
&x_c[n] = 0 \text{ for } n<0 \\
&\text{Let } x_c[n] = \alpha^n e^{j\omega n}u[n] \\
&X_c(z) = \sum_{n=-\infty}^{\infty} x_c[n]z^{-n} = \sum_{n=0}^{\infty} \alpha^n e^{j\omega n} z^{-n} = \sum_{n=0}^{\infty} (\alpha e^{j\omega} z^{-1})^n < \infty \\
&|\alpha e^{j\omega} z^{-1}| < 1 \rightarrow |\alpha e^{j\omega}||z^{-1}| < 1 \rightarrow |z| > |\alpha e^{j\omega}| \rightarrow |z| > |\alpha| \\
&\quad \text{ROC}\{X_c(z)\} = \{z||z| > |\alpha|\}
\end{aligned} \tag{5.40}$$

The typical types of signal and system shows causal property with finite or infinite length. The finite-support signal illustrates the values in zero or positive index for causality otherwise the signal is finite and non-causal signal. Also, the infinite-support signal can be classified as causal and non-causal based on the signal position. See Table 5.4.

Table 5.4 Signal types and corresponding pole locations with examples

Signal type	Pole locations	Example signal	Pole-zero plot
Finite-support and causal signal	$z = 0$	$\sum_{k=0}^{9} 0.5^n \delta[n-k]$	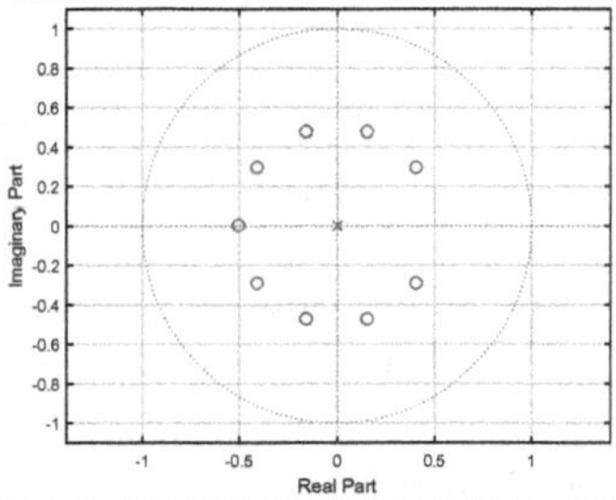
Finite-support and non-causal signal	$z = 0$ and $z = \infty$	$\sum_{k=-1}^{8} 0.5^n \delta[n-k]$	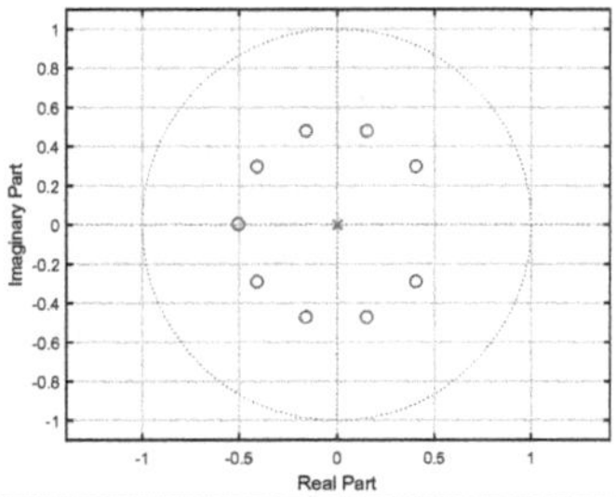
Causal signal (Infinite-support)	Signal dependent but $z \neq \infty$	$0.5^n u[n]$	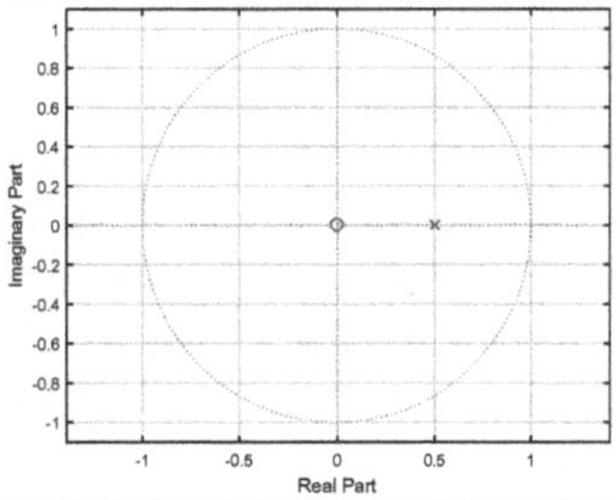
Causal signal + some values in negative index (Infinite-support)	Signal dependent but $z = \infty$	$0.5^n u[n] + 0.5^{-1} \delta[n+1]$	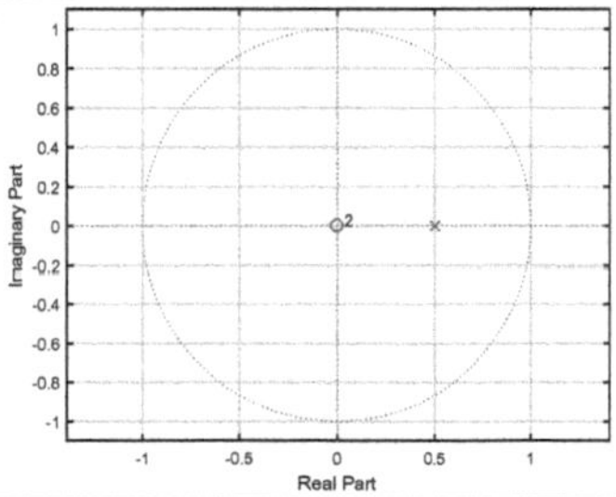

The four classes of the signal are represented for Z-transform analysis with pole distribution. The finite-support signal provides the limited position for the poles and free distribution for the zeros. The causal finite-support signal includes the negative power polynomials for Z-transform as shown $X_1(z)$ in Eq. (5.41); hence, zeros are the solution of the polynomial function and poles are arranged to the origin of the z-plane. The non-causal finite-support signal involves the values at the negative index; therefore, the corresponding Z-transform produces the positive and negative power polynomial as shown $X_2(z)$ in Eq. (5.42). The zeros are allowed to move the z-plane, but the poles are fixed to the origin and infinite.

$$X_1(z) = x_1[0]z^0 + x_1[1]z^{-1} + \ldots + x_1[N]z^{-N} \tag{5.41}$$

$$X_2(z) = x_2[-2]z^2 + x_2[-1]z^1 + x_2[0]z^0 + x_2[1]z^{-1} + \ldots + x_2[N]z^{-N} \tag{5.42}$$

The infinite-support causal signal as shown $X_3(z)$ in Eq. (5.43) can be simplified by the geometric summation for rational function; therefore, the poles freely move on the z domain except infinite value. The negative power polynomials will be disappeared with infinite z value. If there is any signal component at the negative index as $X_4(z)$ in Eq. (5.44), infinite z value is included as the pole position and excluded from the ROC. The positive power polynomials will be amplified to the infinite outcome with infinite z value.

$$X_3(z) = \alpha^0 z^0 + \alpha^1 z^{-1} + \alpha^2 z^{-2} + \ldots + \alpha^{\infty} z^{-\infty} \tag{5.43}$$

$$X_4(z) = \alpha^{-2} z^2 + \alpha^{-1} z^1 + \alpha^0 z^0 + \alpha^1 z^{-1} + \alpha^2 z^{-2} + \ldots + \alpha^{\infty} z^{-\infty} \tag{5.44}$$

If the pole is located within the unit circle, the signal magnitude is gradually decreased. The pole on the unit circle maintains the constant signal magnitude and the pole outside the unit circle provides the diverging signal over time. Therefore, the absolute value of the pole determines the magnitude variation in the time domain signal. The angle of the pole decides variation rate in time domain. The positive real axis represents the zero frequency and negative real axis denotes highest frequency π. The poles on the imaginary axis generate the middle frequency $\pi/2$ based on the given sampling situation. Note that the poles and zeros should be located with conjugated pair manner in z-plane; otherwise, the time domain signal cannot be preserved in real value condition. Three examples of the infinite-support causal signal are shown in Example 5.5 and Table 5.5 with pole-zero plot.

Table 5.5 Examples of the infinite-support causal signal

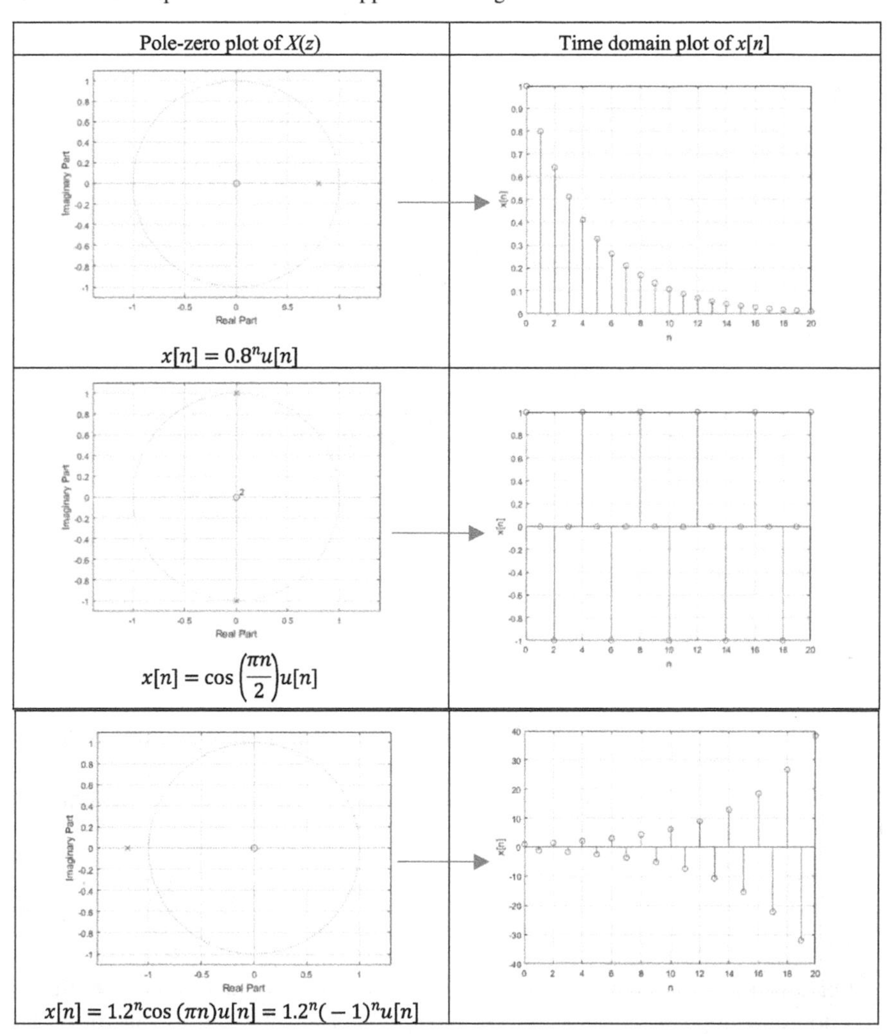

Example 5.5

Draw the pole-zero plot and time domain plot of following signals.

$$x_1[n] = 0.8^n u[n]$$

$$x_2[n] = \cos\left(\frac{\pi}{2}n\right)u[n]$$

$$x_3[n] = 1.2^n \cos(\pi n)u[n]$$

Solution

$$x_1[n] = 0.8^n u[n] \overset{z}{\leftrightarrow} \frac{1}{1-0.8z^{-1}} = \frac{z}{z-0.8}$$
$$\text{zeros} = \{z=0\} \quad \text{poles} = \{z=0.8\}$$
$$x_2[n] = \cos\left(\frac{\pi}{2}n\right)u[n] = \frac{1}{2}e^{j\frac{\pi}{2}n}u[n] + \frac{1}{2}e^{-j\frac{\pi}{2}n}u[n]$$
$$\overset{z}{\leftrightarrow} \frac{1/2}{1-e^{j\frac{\pi}{2}}z^{-1}} + \frac{1/2}{1-e^{-j\frac{\pi}{2}}z^{-1}}$$
$$= \frac{z/2}{z-j} + \frac{z/2}{z+j} = \frac{z}{2}\frac{2z}{(z-j)(z+j)}$$
$$\text{zeros} = \{\text{dual}\, z=0\} \quad \text{poles} = \{z=\pm j\}$$
$$x_3[n] = 1.2^n \cos(\pi n)u[n] = 1.2^n(-1)^n u[n] = (-1.2)^n u[n]$$
$$\overset{z}{\leftrightarrow} \frac{1}{1+1.2z^{-1}} = \frac{z}{z+1.2}$$
$$\text{zeros} = \{z=0\} \quad \text{poles} = \{z=-1.2\}$$

■

Like DTFT and DFT, the Z-transform also provides the convolution sum to multiplication property between the domains as shown in Eq. (5.45).

$$y[n] = x[n] * h[n] \overset{Z}{\leftrightarrow} Y(z) = X(z)H(z) \tag{5.45}$$

This relationship helps to identify the system by finding the rational equation known as the transfer function in Eq. (5.46).

$$H(z) = \frac{Y(z)}{X(z)} \tag{5.46}$$

Once you assign/obtain the proper input/output to/from the system, the transfer function and impulse response can be computed by using the Z-transform. The impulse response is the inverse Z-transform of the derived transfer function as Eq. (5.47).

$$h[n] = \mathrm{Z}^{-1}\{H(z)\} \text{ for given ROC} \tag{5.47}$$

The DTFT and DFT also share the above property; therefore, you can perform the convolution sum by multiplication in the frequency domain as ω and k. However, the signal representation is limited by the transform capability. Certain types of the signal cannot be modelled by the transform effectively; therefore, the signal and system analysis is restricted. The degree of freedom in signal analysis is less in the DTFT and DFT than the Z-transform's counterpart. Let's see the infinite length signal below.

$$x[n] = \delta[n] + r\delta[n-1] + r^2\delta[n-2] + r^3\delta[n-3] + \ldots = r^n u[n] \tag{5.48}$$

DTFT result is below.

$$X\left(e^{j\omega}\right) = 1 + re^{-j\omega} + r^2 e^{-j2\omega} + r^3 e^{-j3\omega} + \ldots \tag{5.49}$$

Z-transform is below.

$$X(z) = 1 + rz^{-1} + r^2 z^{-2} + r^3 z^{-3} + \ldots \tag{5.50}$$

Note that the DFT is not shown above because the transform is for the finite length periodic signal. The DTFT eigenfunction is the $e^{-j\omega}$ and the common ratio for the geometric series is the $re^{-j\omega}$. When the common ratio magnitude is equal to or greater than one, the transform cannot be represented by the rational function according to the geometric summation. However, the Z-transform common ratio is the rz^{-1} and we can control the magnitude by placing the range of the complex number in z known as region of convergence. Even, the infinite length diverging arbitrary signal can be applied for the Z-transform to denote the signal in simple form and further understanding. Equations (5.51) and (5.52) show the corresponding rational functions.

$$\begin{aligned} X\left(e^{j\omega}\right) &= 1 + re^{-j\omega} + r^2 e^{-j2\omega} + r^3 e^{-j3\omega} + \ldots \\ &= \frac{1}{1 - re^{-j\omega}} \text{ only for the given } |r| < 1 \end{aligned} \tag{5.51}$$

$$\begin{aligned} X(z) &= 1 + rz^{-1} + r^2 z^{-2} + r^3 z^{-3} + \ldots \\ &= \frac{1}{1 - rz^{-1}} \text{ for any } r \text{ with ROC } |z| > r \end{aligned} \tag{5.52}$$

The DFT for the finite length geometric series can be shown as below.

$$\begin{aligned} X[k] &= 1 + re^{-j\frac{2\pi}{N}k} + r^2 e^{-j\frac{2\pi}{N}2k} + + r^3 e^{-j\frac{2\pi}{N}3k} + \ldots + r^{(N-1)} e^{-j\frac{2\pi}{N}(N-1)k} \\ &= \frac{1 - r^N e^{-j\frac{2\pi}{N}Nk}}{1 - re^{-j\frac{2\pi}{N}k}} = \frac{\left(r^N - e^{j2\pi k}\right) e^{-j\frac{2\pi}{N}Nk}}{\left(r - e^{j\frac{2\pi}{N}k}\right) e^{-j\frac{2\pi}{N}k}} \\ &= \frac{\prod_{m=0}^{N-1}\left(r - e^{j\frac{2\pi}{N}km}\right)}{\left(r - e^{j\frac{2\pi}{N}k}\right)} e^{-j\frac{2\pi}{N}(N-1)k} \\ &= \left\{\prod_{m=0,2,3,4,5,\ldots,N-1}\left(r - e^{j\frac{2\pi}{N}km}\right)\right\} e^{-j\frac{2\pi}{N}(N-1)k} \end{aligned} \tag{5.53}$$

The finite length geometric series is organized as the polynomial over $e^{-j\frac{2\pi}{N}k}$ without the denominator part; hence, the DFT represents the finite length signal with period N. In conclusion, the infinite length signal can be denoted for limited analysis in DTFT and for no analysis in DFT. The powerful Z-transform successfully handles the any types of signal such as infinite, finite, diverging, converging, and steady in property. The rational function in z domain can be easily applied for the transfer function representation to understand the input and output signal. Excluding the diverging in signal property, the DTFT can perform analysis the signal efficiently. The DFT can manage the finite and periodic signal well. Note that the DFT can approximate the DTFT output by increasing the signal length as shown in the previous discussion. Summary of the transform capability is organized in Fig. 5.12.

One example on the transformation is illustrated in Table 5.6 based on the given infinite length signal for DTFT and Z-transform. The DTFT derives the time domain signal by using the long division; however, the backward analysis is not feasible due to the limited capability of the transform. The Z-transform performs the forward and backward analysis well to compute the time domain and z domain signal.

The conventional causal FIR filter can be represented by the following difference equation. The impulse response or recent input weights are shown as $h[n]$ or b_n in the equation.

$$\begin{aligned} y[n] &= \sum_{k=0}^{L} h[k]x[n-k] = \sum_{k=0}^{L} b_k x[n-k] \\ &= b_0 x[n] + b_1 x[n-1] + b_2 x[n-2] + \ldots + b_L x[n-L] \end{aligned} \tag{5.54}$$

The general form of the filter including the IIR filter is presented below.

$$\begin{aligned} &y[n] + a_1 y[n-1] + a_2 y[n-2] + \ldots + a_M y[n-M] \\ &= b_0 x[n] + b_1 x[n-1] + b_2 x[n-2] + \ldots + b_N x[n-N] \end{aligned} \tag{5.55}$$

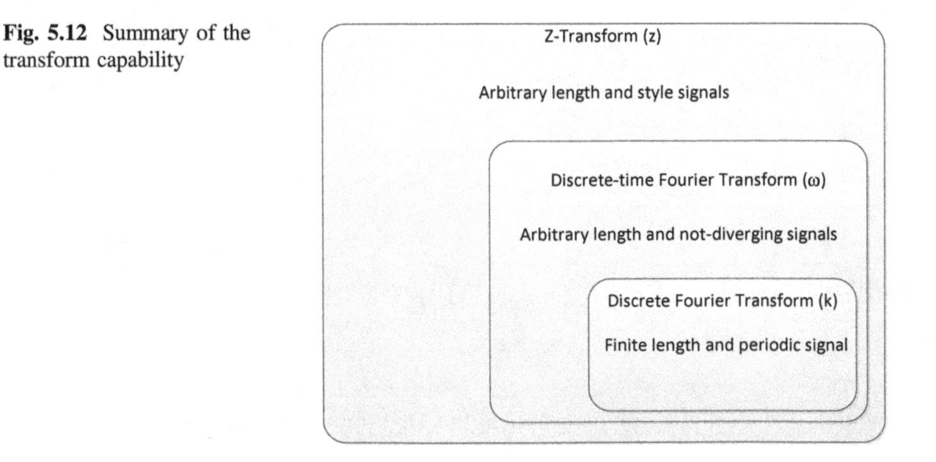

Fig. 5.12 Summary of the transform capability

Table 5.6 Example of transformation to show the Z-trans. and DTFT comparison

DTFT	Z-transform
$X(e^{j\omega}) = \frac{1}{1-e^{-j\omega}}$	$X(z) = \frac{1}{1-z^{-1}}$ ROC : $\lvert z \rvert > 1$
Quotient: $1 + e^{-jw} + e^{-2jw}$ Divisor: $1 - e^{-jw}$, dividend: 1 $1 - e^{-jw}$ e^{-jw} $e^{-jw} - e^{-2jw}$ e^{-2jw} $e^{-2jw} - e^{-3jw}$	Quotient: $1 + z^{-1} + z^{-2}$ Divisor: $1 - z^{-1}$, dividend: 1 $1 - z^{-1}$ z^{-1} $z^{-1} - z^{-2}$ z^{-2} $z^{-2} - z^{-3}$ ....
$X(e^{j\omega}) = \frac{1}{1-e^{-j\omega}} = 1 + e^{-j\omega} + e^{-2j\omega} + \ldots$ $x[n] = \delta[n] + \delta[n-1] + \delta[n-2] + \ldots$ $= u[n]$	$X(z) = \frac{1}{1-z^{-1}} = 1 + z^{-1} + z^{-2} + \ldots$ $x[n] = \delta[n] + \delta[n-1] + \delta[n-2] + \ldots$ $= u[n]$
$x[n] = u[n] = \delta[n] + \delta[n-1] + \delta[n-2] + \ldots$	
$X(e^{j\omega}) = 1 + e^{-j\omega} + e^{-2j\omega} + \ldots$ $\neq \frac{1}{1-e^{-j\omega}}$ Since $\lvert e^{-j\omega} \rvert = 1$	$X(z) = 1 + z^{-1} + z^{-2} + \ldots$ $= \frac{1}{1-z^{-1}}$ ROC : $\lvert z \rvert > 1$

The time delay is represented by the simple multiplication with z power in Z-transform. Before applying the Z-transform to the conventional filter, let's consider the time shift for input $x[n]$ and output $y[n]$. The causal input $x[n]$ to the causal system is assigned by the user as there is no values at the negative time index. Hence the time shift to the right does not create any tails on the left-hand side for example as shown in Eq. (5.56) and Fig. 5.13 top subplot.

$$x[n-2] \overset{Z}{\leftrightarrow} X(z)z^{-2} \tag{5.56}$$

The output $y[n]$ is the system output which possibly contains the initial values. The causal system with causal input initiates the operation from the zero-time index; therefore, the initial values of the output indicate the output prior to the input assignment at negative time index. The initial values are not related to the input and not classified to the output $y[n]$. Figure 5.13 and Eq. (5.57) demonstrate the $y[n]$ with two initial values at $y[-1]$ and $y[-2]$.

$$\begin{aligned} y_{total}[n] &= y[-2]\delta[n+2] + y[-1]\delta[n+1] + y[0]\delta[n] \\ &\quad + y[1]\delta[n-1] + y[2]\delta[n-2] + \ldots \\ y_{total}[n] &= y[-2]\delta[n+2] + y[-1]\delta[n+1] + y[n] \end{aligned} \tag{5.57}$$

The shift to the right by two samples can be denoted as below.

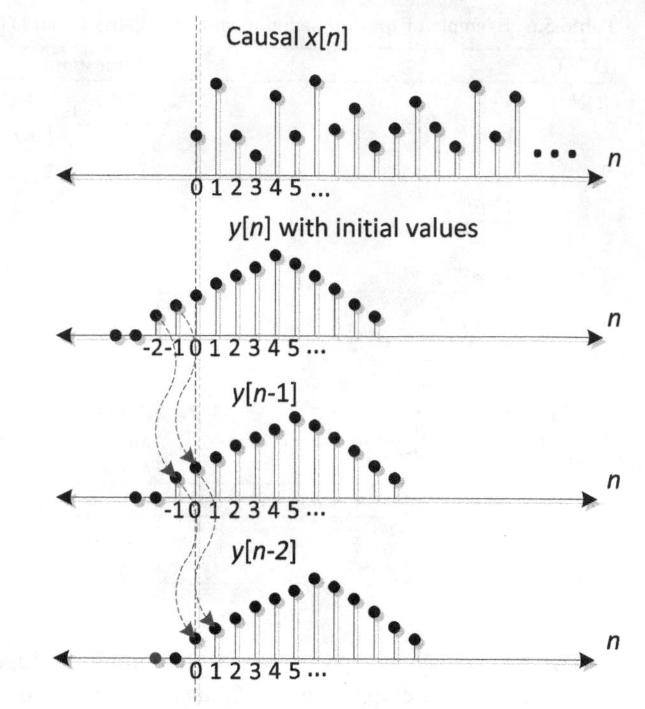

Fig. 5.13 Shift right of $y[n]$ with initial values

$$\begin{aligned} y_{total}[n-2] &= y[-2]\delta[n] + y[-1]\delta[n-1] + y[0]\delta[n-2] \\ &\quad + y[1]\delta[n-3] + y[2]\delta[n-4] + \ldots \\ y_{total}[n-2] &= y[-2]\delta[n] + y[-1]\delta[n-1] + y[n-2] \end{aligned} \tag{5.58}$$

Since the system consider the values from the zero-time index, the shifted tails shown in the working time index should be included to the analysis and transformation as Eq. (5.59).

$$\begin{aligned} & y[n] \overset{Z}{\leftrightarrow} Y(z) \\ & y[n-1] \overset{Z}{\leftrightarrow} y[-1]z^0 + Y(z)z^{-1} \\ & y[n-2] \overset{Z}{\leftrightarrow} y[-2]z^0 + y[-1]z^{-1} + Y(z)z^{-2} \end{aligned} \tag{5.59}$$

Without the initial values on the output, the difference equation for the general filter can be described as Eq. (5.60).

$$\begin{aligned}&y[n]+a_1y[n-1]+a_2y[n-2]+\ldots+a_My[n-M]\\&=b_0x[n]+b_1x[n-1]+b_2x[n-2]+\ldots+b_Nx[n-N]\overset{Z}{\leftrightarrow}\end{aligned}$$

$$\begin{aligned}&Y(z)+a_1Y(z)z^{-1}+a_2Y(z)z^{-2}+\ldots+a_MY(z)z^{-M}\\&=b_0X(z)+b_1X(z)z^{-1}+b_2X(z)z^{-2}+\ldots+b_NX(z)z^{-N}\end{aligned}$$

$$\begin{aligned}&Y(z)\{1+a_1z^{-1}+a_2z^{-2}+\ldots+a_Mz^{-M}\}\\&=X(z)\{b_0+b_1z^{-1}+b_2z^{-2}+\ldots+b_Nz^{-N}\}\end{aligned}$$

$$\frac{Y(z)}{X(z)}=H(z)=\frac{b_0+b_1z^{-1}+b_2z^{-2}+\ldots+b_Nz^{-N}}{1+a_1z^{-1}+a_2z^{-2}+\ldots+a_Mz^{-M}} \tag{5.60}$$

With the initial values on the output, the equation can be depicted as Eq. (5.61).

$$\begin{aligned}&y[n]+a_1y[n-1]+a_2y[n-2]+\ldots+a_My[n-M]\\&=b_0x[n]+b_1x[n-1]+b_2x[n-2]+\ldots+b_Nx[n-N]\overset{Z}{\leftrightarrow}\end{aligned}$$

$$\begin{aligned}&Y(z)+a_1\{y[-1]+Y(z)z^{-1}\}+a_2\{y[-2]+y[-1]z^{-1}+Y(z)z^{-2}\}\\&\quad+\ldots a_M\{y[-M]+y[-(M-1)]z^{-1}+\ldots+Y(z)z^{-M}\}\\&=b_0X(z)+b_1X(z)z^{-1}+b_2X(z)z^{-2}+\ldots+b_NX(z)z^{-N}\end{aligned}$$

$$\begin{aligned}&Y(z)\{1+a_1z^{-1}+a_2z^{-2}+\ldots+a_Mz^{-M}\}\\&\quad+y[-1]\left\{a_1+a_2z^{-1}+a_3z^{-2}+\ldots+a_Mz^{-(M-1)}\right\}\\&\quad+y[-2]\left\{a_2+a_3z^{-1}+a_4z^{-2}+\ldots+a_Mz^{-(M-2)}\right\}+\ldots+y[-M]a_M\\&=X(z)\{b_0+b_1z^{-1}+b_2z^{-2}+\ldots+b_Nz^{-N}\}\end{aligned}$$

$$\begin{aligned}&Y(z)\{1+a_1z^{-1}+a_2z^{-2}+\ldots+a_Mz^{-M}\}=\\&\quad-y[-1]\left\{a_1+a_2z^{-1}+a_3z^{-2}+\ldots+a_Mz^{-(M-1)}\right\}\\&\quad-y[-2]\left\{a_2+a_3z^{-1}+a_4z^{-2}+\ldots+a_Mz^{-(M-2)}\right\}-\ldots-y[-M]a_M\\&\quad+X(z)\{b_0+b_1z^{-1}+b_2z^{-2}+\ldots+b_Nz^{-N}\}\end{aligned}$$

$$\begin{aligned}Y(z)=&\frac{\{b_0+b_1z^{-1}+b_2z^{-2}+\ldots+b_Nz^{-N}\}}{\{1+a_1z^{-1}+a_2z^{-2}+\ldots+a_Mz^{-M}\}}X(z)\\&-\frac{\{a_1+a_2z^{-1}+a_3z^{-2}+\ldots+a_Mz^{-(M-1)}\}}{\{1+a_1z^{-1}+a_2z^{-2}+\ldots+a_Mz^{-M}\}}y[-1]\\&-\frac{\{a_2+a_3z^{-1}+a_4z^{-2}+\ldots+a_Mz^{-(M-2)}\}}{\{1+a_1z^{-1}+a_2z^{-2}+\ldots+a_Mz^{-M}\}}y[-2]-\ldots\\&-\frac{a_M}{\{1+a_1z^{-1}+a_2z^{-2}+\ldots+a_Mz^{-M}\}}y[-M]\end{aligned}$$

$$\begin{aligned} Y(z) = & \frac{\{b_0 + b_1 z^{-1} + b_2 z^{-2} + \ldots + b_N z^{-N}\}}{(1-\alpha_1 z^{-1})(1-\alpha_2 z^{-1})\ldots(1-\alpha_M z^{-1})} X(z) \\ & - \frac{\{a_1 + a_2 z^{-1} + a_3 z^{-2} + \ldots + a_M z^{-(M-1)}\}}{(1-\alpha_1 z^{-1})(1-\alpha_2 z^{-1})\ldots(1-\alpha_M z^{-1})} y[-1] \\ & - \frac{\{a_2 + a_3 z^{-1} + a_4 z^{-2} + \ldots + a_M z^{-(M-2)}\}}{(1-\alpha_1 z^{-1})(1-\alpha_2 z^{-1})\ldots(1-\alpha_M z^{-1})} y[-2] - \ldots \\ & - \frac{a_M}{(1-\alpha_1 z^{-1})(1-\alpha_2 z^{-1})\ldots(1-\alpha_M z^{-1})} y[-M] \end{aligned} \tag{5.61}$$

The output $Y(z)$ is the combination of the rational functions from the $X(z)$, $y[-1]$, $y[-2]$, … and $y[-M]$. The first rational function is the generated by the input signal and the other rational functions are derived from the corresponding initial values. Since all rational functions share the denominator part, zero input response converges to the constant value once the entire poles are located within the unit circle.

The transfer function defines the filter characteristics and property. The continuous frequency response can be calculated by the DTFT as replacing the z with $e^{j\omega}$.

$$H(e^{j\omega}) = H(z)|_{z\to e^{j\omega}} = \frac{b_0 + b_1 e^{-j\omega} + b_2 e^{-j2\omega} + \ldots + b_N e^{-jN\omega}}{1 + a_1 e^{-j\omega} + a_2 e^{-j2\omega} + \ldots + a_M e^{-jM\omega}} \tag{5.62}$$

The discrete frequency response can be obtained by the DFT as exchanging the z with $e^{j2\pi k/N}$.

$$H[k] = H(z)|_{z\to e^{j\frac{2\pi}{N}k}} = \frac{b_0 + b_1 e^{-j\frac{2\pi}{N}k} + b_2 e^{-j\frac{2\pi}{N}2k} + \ldots + b_N e^{-j\frac{2\pi}{N}Nk}}{1 + a_1 e^{-j\frac{2\pi}{N}k} + a_2 e^{-j\frac{2\pi}{N}2k} + \ldots + a_M e^{-j\frac{2\pi}{N}Mk}} \tag{5.63}$$

The filter output in frequency and z domain is the multiplication between the input and filter response in the corresponding domain; therefore, the output follows the profile of the filter response. The higher magnitude in the filter response produces the amplified output in the frequency.

$$\begin{aligned} Y(z) &= X(z)H(z) \\ Y(e^{j\omega}) &= X(e^{j\omega})H(e^{j\omega}) \\ Y[k] &= X[k]H[k] \end{aligned} \tag{5.64}$$

Example 5.6

Derive the impulse response of following IIR filters with and without initial values. The equations are shown previously in Chap. 4.

$$\begin{aligned} y[n] &= a_1 y[n-1] + b_0 x[n] \\ y[n] &= (a_1 + a_2)y[n-1] - a_1 a_2 y[n-2] + (a_1 - a_2)x[n] \end{aligned}$$

Solution

$$
\begin{aligned}
y[n] &= a_1 y[n-1] + b_0 x[n] \text{ with } y[-1] \neq 0 \\
&\overset{Z}{\leftrightarrow} Y(z) = a_1 y[-1] z^0 + a_1 Y(z) z^{-1} + b_0 X(z) \\
Y(z)\{1 - a_1 z^{-1}\} &= a_1 y[-1] + b_0 X(z) \\
Y(z) &= \frac{a_1 y[-1]}{1 - a_1 z^{-1}} + \frac{b_0 X(z)}{1 - a_1 z^{-1}} \\
H(z) &= \frac{a_1 y[-1]}{1 - a_1 z^{-1}} + \frac{b_0}{1 - a_1 z^{-1}} \text{ where } X(z) = 1 \\
h[n] &= \begin{cases} a_1 y[-1] a_1^n u[n] + b_0 a_1^n u[n] & \text{with initial value} \\ b_0 a_1^n u[n] & \text{without initial value} \end{cases}
\end{aligned}
$$

$$
y[n] = (a_1 + a_2) y[n-1] - a_1 a_2 y[n-2] + (a_1 - a_2) x[n] \text{ with } \begin{cases} y[-1] \neq 0 \\ y[-2] \neq 0 \end{cases}
$$

$$
\overset{Z}{\leftrightarrow} Y(z) = (a_1 + a_2) y[-1] z^0 + (a_1 + a_2) Y(z) z^{-1} - a_1 a_2 y[-2] z^0 \\
- a_1 a_2 y[-1] z^{-1} - a_1 a_2 Y(z) z^{-2} + (a_1 - a_2) X(z)
$$

$$
Y(z)\{1 - (a_1 + a_2) z^{-1} + a_1 a_2 z^{-2}\} \\
= (a_1 + a_2) y[-1] - a_1 a_2 y[-2] - a_1 a_2 y[-1] z^{-1} + (a_1 - a_2) X(z)
$$

$$
Y(z)(1 - a_1 z^{-1})(1 - a_2 z^{-1}) \\
= (a_1 + a_2) y[-1] - a_1 a_2 y[-2] - a_1 a_2 y[-1] z^{-1} + (a_1 - a_2) X(z)
$$

$$
\begin{aligned}
Y(z) &= \frac{(a_1 + a_2) y[-1]}{(1 - a_1 z^{-1})(1 - a_2 z^{-1})} - \frac{a_1 a_2 y[-2]}{(1 - a_1 z^{-1})(1 - a_2 z^{-1})} \\
&\quad - \frac{a_1 a_2 y[-1] z^{-1}}{(1 - a_1 z^{-1})(1 - a_2 z^{-1})} + \frac{(a_1 - a_2) X(z)}{(1 - a_1 z^{-1})(1 - a_2 z^{-1})} \\
H(z) &= \frac{(a_1 + a_2) y[-1]}{(1 - a_1 z^{-1})(1 - a_2 z^{-1})} - \frac{a_1 a_2 y[-2]}{(1 - a_1 z^{-1})(1 - a_2 z^{-1})} \\
&\quad - \frac{a_1 a_2 y[-1] z^{-1}}{(1 - a_1 z^{-1})(1 - a_2 z^{-1})} + \frac{(a_1 - a_2)}{(1 - a_1 z^{-1})(1 - a_2 z^{-1})} \text{ where } X(z) = 1 \\
H(z) &= \frac{(a_1 + a_2) y[-1] - a_1 a_2 y[-2]}{(1 - a_1 z^{-1})(1 - a_2 z^{-1})} - \frac{a_1 a_2 y[-1] z^{-1}}{(1 - a_1 z^{-1})(1 - a_2 z^{-1})} \\
&\quad + \frac{(a_1 - a_2)}{(1 - a_1 z^{-1})(1 - a_2 z^{-1})}
\end{aligned}
$$

$$H(z) = \frac{\alpha_1}{(1 - a_1 z^{-1})} + \frac{\alpha_2}{(1 - a_2 z^{-1})} + \frac{\beta_1 z^{-1}}{(1 - a_1 z^{-1})} + \frac{\beta_2 z^{-1}}{(1 - a_2 z^{-1})} + \frac{a_1}{(1 - a_1 z^{-1})} - \frac{a_2}{(1 - a_2 z^{-1})}$$

The corresponding impulse response is below.

$$h[n] = \alpha_1 a_1^n u[n] + \alpha_2 a_2^n u[n] + \beta_1 a_1^{n-1} u[n-1] + \beta_2 a_2^{n-1} u[n-1] + a_1^{n+1} u[n] - a_2^{n+1} u[n]$$

Without the initial values, the last two terms are only survived as below.

$$h[n] = a_1^{n+1} u[n] - a_2^{n+1} u[n]$$

Compare the above result with Eqs. (4.17) and (4.22) which are solved by substitution method.

■

The Z-transform provides the improved method to analyze the filter stability. The stability is defined as the bounded output for the arbitrary bounded input. Hence, the infinite length causal input with limited magnitude should generate the bound output all the time in stable filter system. In the previous discussion, the BIBO stability can be inspected by the impulse response with absolute summability in square manner. In other words, the absolute summable or integrable frequency response in square manner guarantees the BIBO stable filter according to the Parseval's theorem [6] as shown in Eq. (5.65).

$$\sum_{n=-\infty}^{\infty} |h[n]|^2 < \infty \rightarrow \begin{cases} \frac{1}{2\pi} \int_{-\pi}^{\pi} |H(e^{j\omega})|^2 d\omega < \infty \\ \frac{1}{N} \sum_{k=0}^{N-1} |H[k]|^2 < \infty \end{cases} \tag{5.65}$$

The ROC defines the region that the Z-transform exist. If ROC includes the unit circle in the z plane, then the values in the unit circle cannot be diverged to the infinite value. Note that the continuous and discrete unit circle represents the DTFT and DFT for the given signal and system, respectively. The finite values on the unit circle provides the absolute summability and integrability in square fashion; hence, the system deliver the stability in terms of BIBO.

$$h[n] \overset{Z}{\leftrightarrow} H(z); \text{unit circle} \in \text{ROC}; \therefore \text{ BIBO stable} \tag{5.66}$$

The system causality can present further convenience based on the pole distribution. The ROC of the causal system is the region from the outside of the farthest pole from the z plane origin. Once the transfer function includes the all poles within the unit circle, the causal system denotes the BIBO stability as shown in Eq. (5.67).

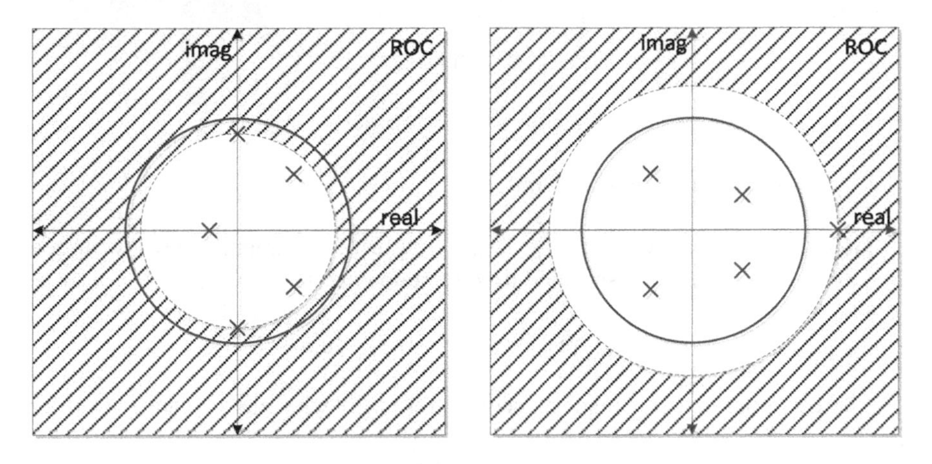

Fig. 5.14 Causal signal example with all poles within the unit circle and not all poles within the unit circle

Figure 5.14 illustrates the two situations as all poles within the unit circle and not all poles in that boundary.

$$h[n] \overset{Z}{\leftrightarrow} H(z); \text{all poles} \in \text{unit cirlce region}; \therefore \text{ BIBO stable} \tag{5.67}$$

Example 5.7

Analyze the difference equation that represents first order IIR filter given as below. We need to explore the impulse response, frequency response, and BIBO stability.

$$y[n] + 0.8y[n-1] = x[n] - x[n-1]$$

Solution

The Z-transform of the filter is shown below.

$$\begin{aligned}
&Y(z) + 0.8\{y[-1] + Y(z)z^{-1}\} = X(z) - X(z)z^{-1} \\
&Y(z)\{1 + 0.8z^{-1}\} + 0.8y[-1] = X(z)\{1 - z^{-1}\} \\
&Y(z)\{1 + 0.8z^{-1}\} = X(z)\{1 - z^{-1}\} - 0.8y[-1] \\
&Y(z) = \frac{X(z)\{1 - z^{-1}\}}{\{1 + 0.8z^{-1}\}} - \frac{0.8y[-1]}{\{1 + 0.8z^{-1}\}} \\
&= \frac{X(z)}{\{1 + 0.8z^{-1}\}} - \frac{X(z)z^{-1}}{\{1 + 0.8z^{-1}\}} - \frac{0.8y[-1]}{\{1 + 0.8z^{-1}\}}
\end{aligned}$$

The transfer function is derived by the input $\delta[n]$ that provides $X(z)$ as 1.

$$H(z) = \frac{1}{\{1+0.8z^{-1}\}} - \frac{z^{-1}}{\{1+0.8z^{-1}\}} - \frac{0.8y[-1]}{\{1+0.8z^{-1}\}}$$

The inverse Z-transform finds the impulse response as below.

$$h[n] = (-0.8)^n u[n] - (-0.8)^{n-1} u[n-1] - 0.8y[-1](-0.8)^n u[n]$$

Generally, the impulse response is the output of the impulse input with zero initial value which can be specified as zero state condition.

$$h[n] = (-0.8)^n u[n] - (-0.8)^{n-1} u[n-1]$$

The zero input response is the output of the zero input with arbitrary initial value on $y[-1]$.

$$y_{zi}[n] = -0.8y[-1](-0.8)^n u[n]$$

The corresponding impulse response and zero input response are illustrated in Figs. 5.15 and 5.16.

The frequency response is calculated by replacing the z with $e^{j\omega}$ in the transfer function without considering the initial values as below.

$$H(e^{j\omega}) = H(z)|_{z\to e^{j\omega}} = \left.\frac{1-z^{-1}}{1+0.8z^{-1}}\right|_{z\to e^{j\omega}} = \frac{1-e^{-j\omega}}{1+0.8e^{-j\omega}}$$

The magnitude and phase of the impulse response is presented as Figs. 5.17 and 5.18.

You can figure out the BIBO stability by inspecting the impulse response as well as pole zero plot. The impulse response develops the sequence in the power of 0.8; hence, the system is expected to be BIBO stable. The pole zero plot shows the BIBO stability by pole position examination. The pole is located at the −0.8 where the unit circle area is; therefore, the system is BIBO stable as shown in Fig. 5.19.

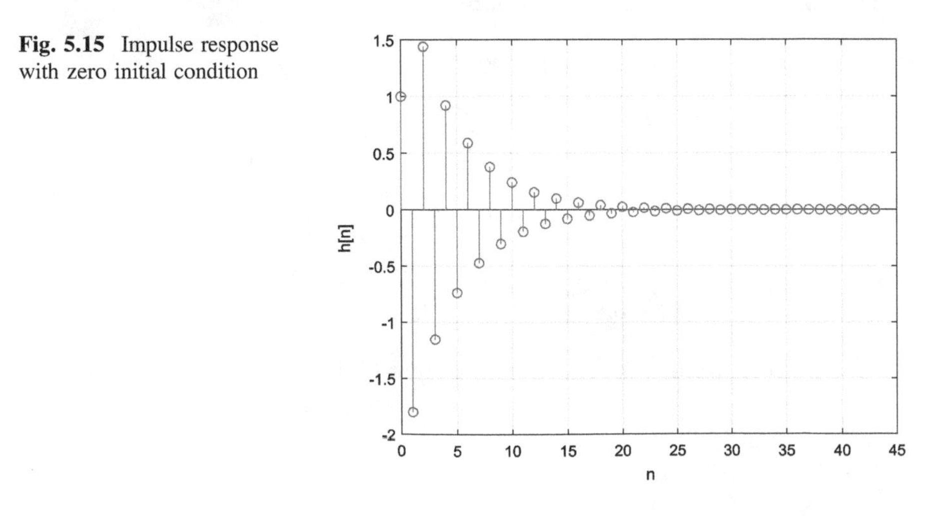

Fig. 5.15 Impulse response with zero initial condition

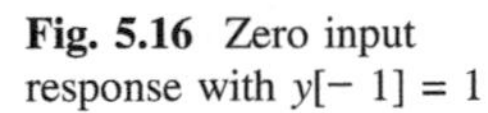

Fig. 5.16 Zero input response with $y[-1] = 1$

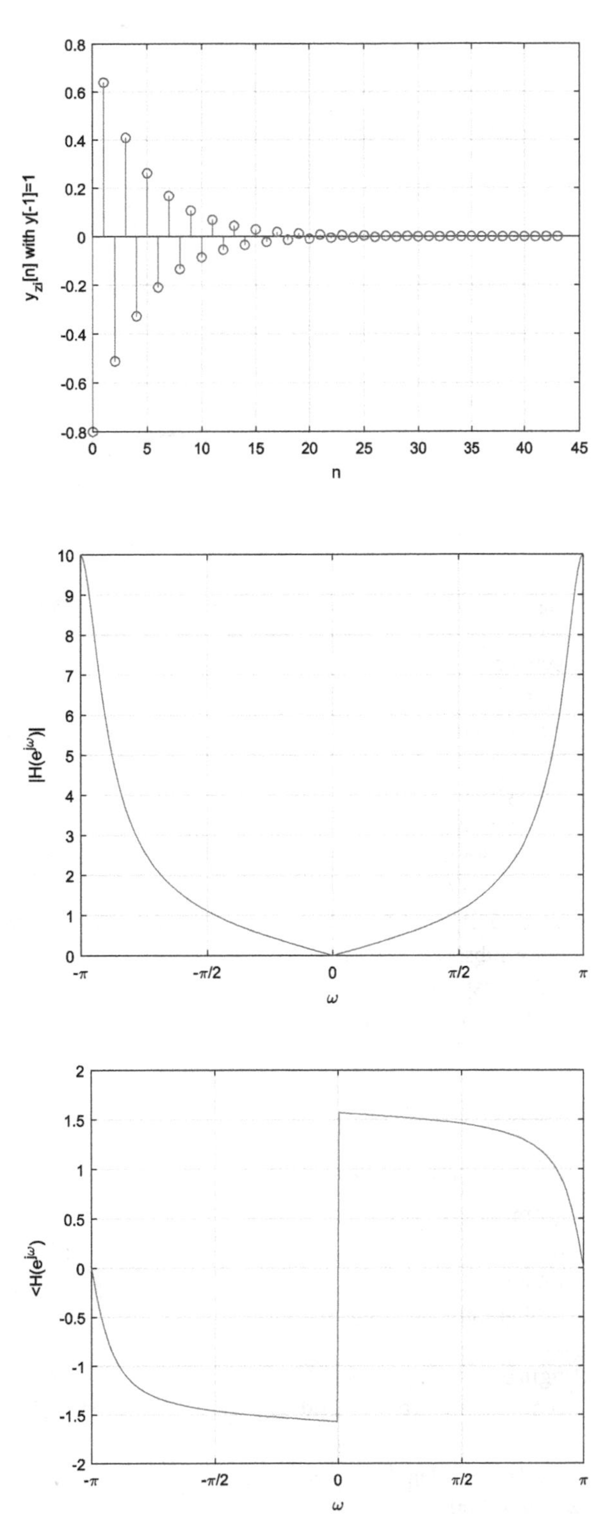

Fig. 5.17 Frequency response in magnitude

Fig. 5.18 Phase response

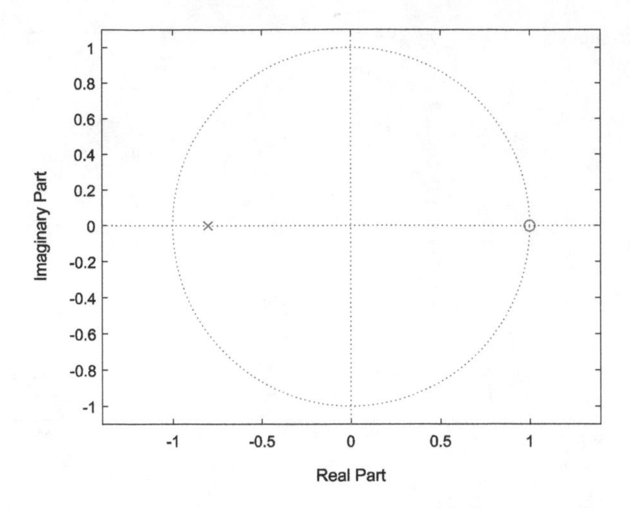

Fig. 5.19 Pole zero plot

Prog. 5.3 MATLAB program for Figs. 5.15, 5.16, 5.17, 5.18 and 5.19

```
syms z w
zy = (1-z^-1)/(1+0.8*z^-1);          % Given equation in Z domain
zn = iztrans(zy);                     % Inverse z transform
[N, D] = numden(zy);
a = sym2poly(D);
b = sym2poly(N);
[H,T] = impz(b,a);
nn = 0:1:43;
yzi = -0.8*(-0.8).^nn;               % Impulse response with initial condition
zw = subs(zy,z,exp(1j*w));
w1 = linspace(-pi,pi,200);
zw1 = eval(subs(zw,w,w1));

figure,    % Fig. 5.15
stem(T,H), grid
...
figure,    % Fig. 5.16
stem(nn,yzi), grid;
...
figure,    % Fig. 5.17
plot(w1,abs(zw1)), grid
...
figure,    % Fig. 5.18
plot(w1,angle(zw1)), grid
...
figure,    % Fig. 5.19
zplane(b,a)
```

The pole zero plot also corresponds to the frequency response. The zero is located at the 0 rad on the unit circle and pole is placed at the -0.8 near the π radian. Therefore, zero frequency cannot be passed, and highest frequency π is amplified by the filter according to the pole zero position.

■

5.3 Intuitive Filter Design

In the previous chapter, we studied the FIR filter design based on the DTFT and DFT. The Z-transform provides the novel power to design IIR filter which cannot be approached by the DTFT and DFT. The proceeding section showed the example of analyzing the difference equation filter to derive the impulse response, frequency response, and filter stability. This section provides the designing the filter equation from the given filter requirement in frequency domain. The intuitive method places the poles and zeros in the z plane to meet the specification based on the pole and zero property. The intuitive iterations optimize the response with specification in terms of order and coefficients over the IIR filter system. Further mathematically optimized methods will be introduced in the next chapter.

The proceeding section stated that the transfer function is the rational function between the input and output in z domain. The frequency response evaluates the transfer function over the unit circle. For the time being, we only consider the magnitude of the frequency response and the phase will be discussed in the next section. The magnitude is computed by the product over the pole and zero distances to the corresponding frequency on the unit circle. The pole close to the unit circle creates the amplified magnitude and the zero near by the unit circle generates the suppressed output for the frequency. Due to the mutual relation, close solution to place the optimized pole zero location requires the mathematical procedure that will be investigated at the next chapter. We only utilize the fundamental knowledges those are the unit circle evaluation, pole effect, and zero effect only.

Single pole:

$$\begin{aligned} y[n] + a_1 y[n-1] &= b_0 x[n] \\ \overset{Z}{\leftrightarrow} Y(z) + a_1 Y(z) z^{-1} &= b_0 X(z) \\ H(z) = \frac{Y(z)}{X(z)} &= \frac{b_0}{1 + a_1 z^{-1}}; |z| > |a_1| \end{aligned} \tag{5.68}$$

The zero-state transfer function above represents the first order IIR filter with pole at $-a_1$ and zero at 0. The single pole should move along with real axis on z plane to preserve the real number coefficient condition at difference equation. Also, the pole must be contained within the unit circle area for stability; therefore, the pole range is above -1 and below 1 in the z domain. As the pole approaches to

1, the zero-frequency component is amplified. The pole near the -1 emphasize the π radian frequency component. With the first order IIR configuration, the LPF and HPF are only accessible. The normalized filter is organized as Eq. (5.69).

$$H(z) = \frac{1 - |a_1|}{1 + a_1 z^{-1}} \tag{5.69}$$

Figures 5.20 and 5.21 show the pole zero plot and corresponding frequency response for range of a_1. Approaching to the unit circle illustrates the strong selective property in the frequency domain. Note that the FIR filter requires the extensive orders to present rapid transition across the requirement edges. With relatively short filter order in IIR configuration, the pole property allows us to design the high discriminating LPF and HPF. The pole closeness to the unit circle is proportional to the filter selectiveness. Do not place the pole on the unit circle; otherwise, the filter stability is destroyed. The impulse response of the first order IIR filter is given in Eq. (5.70).

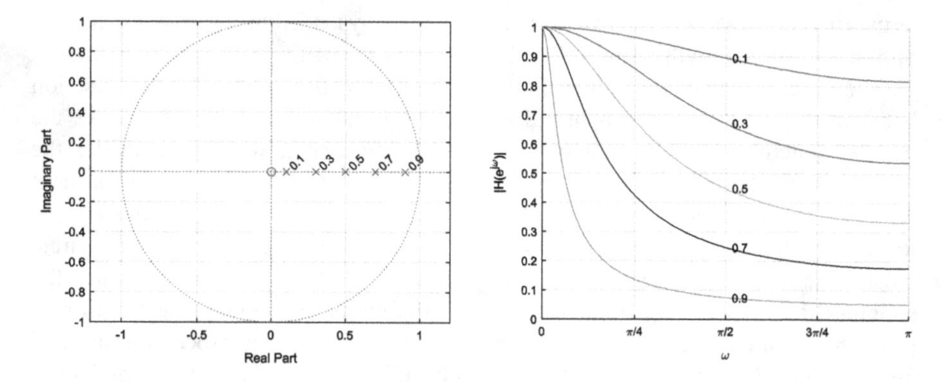

Fig. 5.20 Pole zero plot and frequency response (magnitude) of Eq. (5.69) for positive pole locations

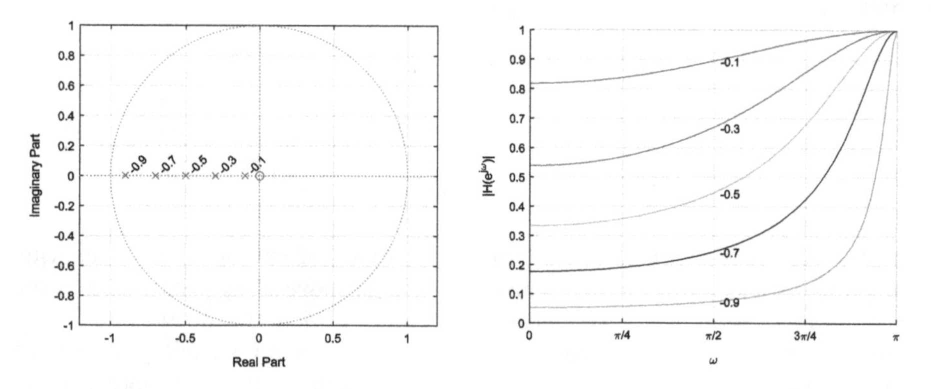

Fig. 5.21 Pole zero plot and frequency response (magnitude) of Eq. (5.69) for negative pole locations

$$h[n] = \{1 - |a_1|\}(-a_1)^n u[n] \tag{5.70}$$

Observe that the b_0 is the $\{1 - |a_1|\}$ in Eq. (5.70) and the maximum gain is limited to the one for all frequency by this modification.

Prog. 5.4 MATLAB program for Figs. 5.20 and 5.21

```
syms z;
w = linspace(0,pi,100);
figure(1), grid, hold on;
figure(2), grid, hold on;

for a1 = -0.1:-0.2:-0.9            % For Fig. 5.21 a1 = 0.1:0.2:0.9
      b0 = 1 - abs(a1);
      hz = b0./(1+a1*z^-1);
      hz1 = subs(hz,z,exp(1j*w));
      hz2 = eval(hz1);
      figure(1), plot(w,abs(hz2));
      str = sprintf('    %0.3g',-a1);
      text(pi/2,abs(hz2(50)),str);
      figure(2), zplane(b0,[1 a1])
      text(-a1,0,str,'rotation',45);
end

figure(1), hold off
figure(2), hold off
figure(1)
ax = gca;
ax.XTick = [0 pi/4 pi/2 3*pi/4 pi];
ax.XTickLabel = {'0','\pi/4','\pi/2','3\pi/4','\pi'};
xlim([0 pi]);
xlabel('\omega');
ylabel('|H(e^{j\omega})|')
figure(2)
xlim([-1.2 1.2]);
ylim([-1.0 1.0]);
```

Dual poles:

$$\begin{aligned} & y[n] + a_1 y[n-1] + a_2 y[n-2] = b_0 x[n] \\ & \overset{Z}{\leftrightarrow} Y(z) + a_1 Y(z) z^{-1} + a_2 Y(z) z^{-2} = b_0 X(z) \\ & H(z) = \frac{Y(z)}{X(z)} = \frac{b_0}{1 + a_1 z^{-1} + a_2 z^{-2}} \end{aligned} \tag{5.71}$$

The IIR filter shown above provides the second order polynomial in denominator and the constant value in numerator. The roots of the polynomial represent the poles that can be freely moved over the z plane. In order to include the real coefficients for the polynomial, the pair of roots should be in the conjugated relationship as $\alpha e^{j\omega}$ and $\alpha e^{-j\omega}$. The BIBO stability condition constrains the complex solution magnitude within the unit circle; hence, the α absolute value should be less than 1. As the poles approach to the $e^{\pm j\omega_0}$, the ω_0 frequency component is amplified. With given second order IIR configuration, the LPF, HPF, and BPF are available. The normalized filter is organized as Eqs. (5.72) and (5.73).

$$\begin{aligned} H(z) &= \frac{(1-\alpha)\sqrt{\alpha^2 - 2\alpha\cos(2\omega) + 1}}{(1 - \alpha e^{j\omega} z^{-1})(1 - \alpha e^{-j\omega} z^{-1})}; |z| > |\alpha| \\ &= \frac{(1-\alpha)\sqrt{\alpha^2 - 2\alpha\cos(2\omega) + 1}}{1 - 2\alpha\cos(\omega) z^{-1} + \alpha^2 z^{-2}} \end{aligned} \tag{5.72}$$

$$\begin{aligned} & y[n] - 2\alpha\cos(\omega) y[n-1] + \alpha^2 y[n-2] \\ & = \left\{(1-\alpha)\sqrt{\alpha^2 - 2\alpha\cos(2\omega) + 1}\right\} x[n] \end{aligned} \tag{5.73}$$

Figures 5.22, 5.23, and 5.24 illustrate the pole zero plots and corresponding frequency responses for range of α with $\pi/2$, $\pi/4$, and $\pi/6$ radian frequency. Similar to the single pole situation, approaching to the unit circle illustrates the strong selective property in the frequency domain as well. The normalization process is exercised for the pole angle that usually provides the highest magnitude in the frequency response. However, the $\pm\pi/4$ and $\pm\pi/6$ pole angles generate the above the one magnitude in the low frequency range at pole away situations from the unit circle. The mutual relation with neighbor pole provides magnitude overlap that produce the higher response at the between pole area. Note that the mutually closer poles cause the higher overlap as shown $\pi/6$ radian frequency in Fig. 5.24. The impulse response of the second order IIR filter is given below.

$$\begin{aligned} H(z) &= \frac{(1-\alpha)\sqrt{\alpha^2 - 2\alpha\cos(2\omega) + 1}}{(1 - \alpha e^{j\omega} z^{-1})(1 - \alpha e^{-j\omega} z^{-1})} \\ &= \frac{\gamma}{(1 - \alpha e^{j\omega} z^{-1})} + \frac{\gamma^*}{(1 - \alpha e^{-j\omega} z^{-1})} \overset{Z}{\leftrightarrow} h[n] = \gamma \alpha^n e^{j\omega n} u[n] \\ &\quad + \gamma^* \alpha^n e^{-j\omega n} u[n] \\ &= 2|\gamma| \alpha^n \cos(\omega n + \measuredangle\gamma) u[n] \end{aligned} \tag{5.74}$$

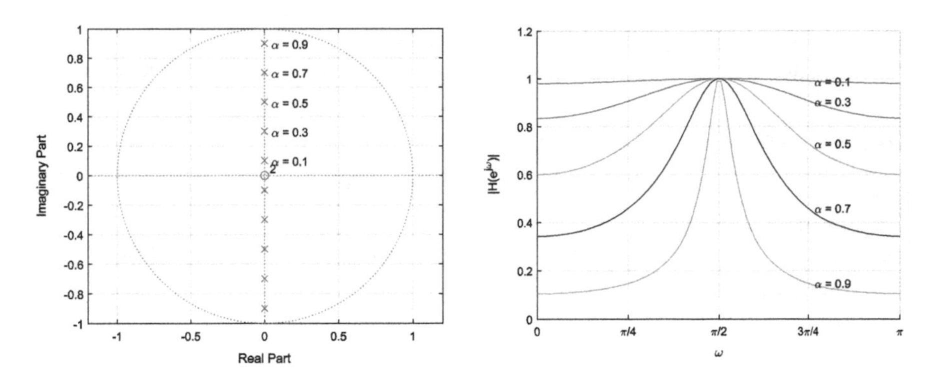

Fig. 5.22 Pole zero plot and frequency response (magnitude) of Eq. (5.72) for $\pi/2$

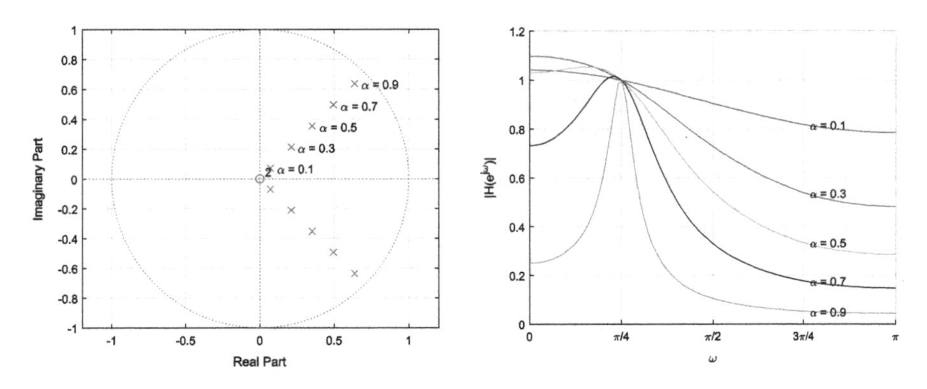

Fig. 5.23 Pole zero plot and frequency response (magnitude) of Eq. (5.72) for $\pi/4$

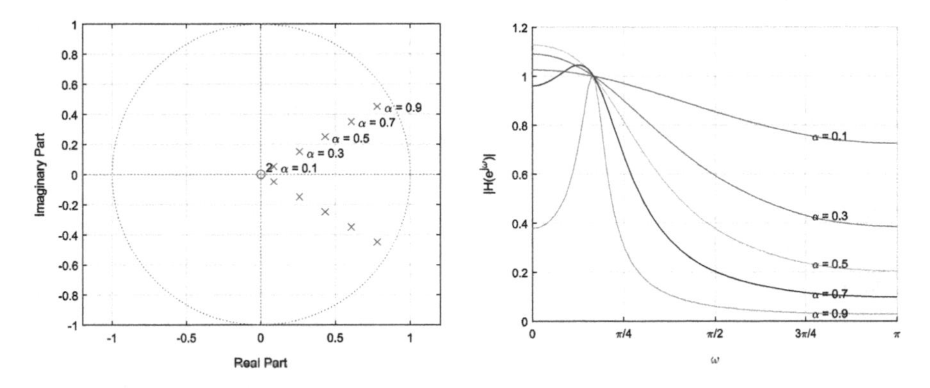

Fig. 5.24 Pole zero plot and frequency response (magnitude) of Eq. (5.72) for $\pi/6$

In Eq. (5.74), the partial fraction computes the γ and its conjugate pair for inverse Z-transform. The conjugated relation preserves the real coefficients for the equation.

Prog. 5.5 MATLAB program for Figs. 5.22, 5.23 and 5.24

```
syms z;
w = linspace(0,pi,100);
an1 = pi/2;            % Fig. 5.23 set an1 = pi/4 and Fig. 5.24 set an1 = pi/6
figure(1), grid, hold on;
figure(2), grid, hold on;

for a1 = 0.1:0.2:0.9
      hz = (1-a1)*sqrt(a1^2-2*cos(2*an1)*a1+1)/(1-2*a1*cos(an1)*z^-
1+a1^2*z^-2);
      hz1 = subs(hz,z,exp(1j*w));
      hz2 = eval(hz1);
      figure(1), plot(w,abs(hz2));
      str = sprintf('   \\alpha = %0.3g',a1);
      text(pi*3/4,abs(hz2(75)),str);
      [N,D] = numden(hz);
      a = sym2poly(D);
      b = sym2poly(N);
      figure(2), zplane(b,a)
      text(a1*cos(an1),a1*sin(an1),str,'rotation',0);
end

figure(1), hold off
ylim([0 1.2]);
figure(2), hold off
figure(1)
ax = gca;
ax.XTick = [0 pi/4 pi/2 3*pi/4 pi];
ax.XTickLabel = {'0','\pi/4','\pi/2','3\pi/4','\pi'};
xlim([0 pi]);
xlabel('\omega');
ylabel('|H(e^{j\omega})|')
figure(2)
xlim([-1.2 1.2]);
ylim([-1.0 1.0]);
```

Single pole with zero:

$$\begin{aligned} y[n] + a_1 y[n-1] &= b_0 x[n] + b_1 x[n-1] \\ &\overset{Z}{\leftrightarrow} Y(z) + a_1 Y(z) z^{-1} = b_0 X(z) + b_1 X(z) z^{-1} \\ H(z) &= \frac{Y(z)}{X(z)} = \frac{b_0 + b_1 z^{-1}}{1 + a_1 z^{-1}} = \frac{b_0\left(1 + \frac{b_1}{b_0} z^{-1}\right)}{1 + a_1 z^{-1}}; |z| > |a_1| \end{aligned} \tag{5.75}$$

The above first order IIR filter contains the pole and zero movement over the z plane; however, the both locations are limited to the real axis due to the real value coefficient condition. The pole location is the $-a_1$ and zero position is the $-b_1/b_0$ with gain control by b_0. Note that the normalization is performed by assigning the proper values at the b_0 and b_1. The study shown below explores the zero-position effect on the frequency response compare to the pole location. Observe that the suggested examples deliver the rough idea on the pole zero location for frequency response. Multiple poles and zeros produce the complicated mutual relation on the response and barely obtainable frequency profile by hand. The first situation is the approaching the zero to the pole. The pole is fixed at the 0.95 and zero is moved toward the pole from 0.1 to 0.9 in every 0.2 stride.

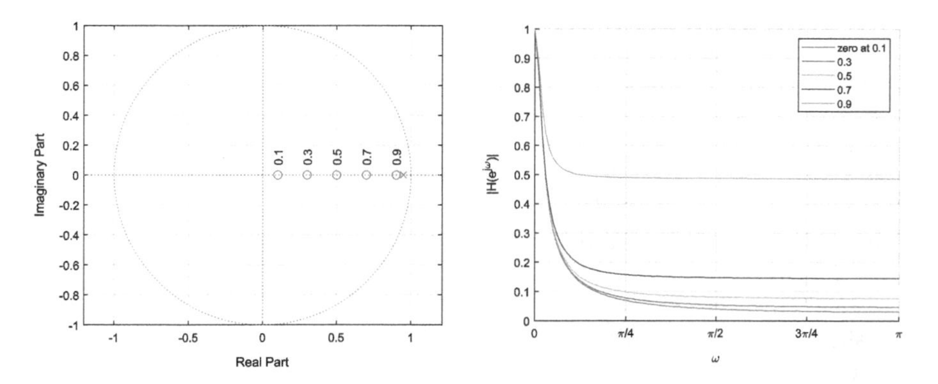

Fig. 5.25 Pole zero plot and frequency response (magnitude) of Eq. (5.75)

Prog. 5.6 MATLAB program for Fig. 5.25

```
syms z;
w = linspace(0,pi,100);
a1 = 0.95;
an1 = 0;
figure(1), grid, hold on;
figure(2), grid, hold on;

for b1 = 0.1:0.2:0.9
	hz = (1-b1*z^-1)*(1-a1)/((1-a1*z^-1)*(1-b1));
	hz1 = subs(hz,z,exp(1j*w));
	hz2 = eval(hz1);
	figure(1), plot(w,abs(hz2));
	str = sprintf('    %0.3g',b1);
	[N,D] = numden(hz);
	a = sym2poly(D);
	b = sym2poly(N);
	figure(2), zplane(b,a)
	text(b1*cos(an1),b1*sin(an1),str,'rotation',90);
end

figure(1), hold off
legend('zero at 0.1','0.3','0.5','0.7','0.9');
figure(2), hold off
figure(1)
ax = gca;
ax.XTick = [0 pi/4 pi/2 3*pi/4 pi];
ax.XTickLabel = {'0','\pi/4','\pi/2','3\pi/4','\pi'};
xlim([0 pi]);
xlabel('\omega');
ylabel('|H(e^{j\omega})|')
figure(2)
xlim([-1.2 1.2]);
ylim([-1.0 1.0]);
```

The 0.95 pole provides the high selective filter in general. As the zero approaches to the pole, the frequency response converges to the higher constant value for the other frequency region. Also, the filter sharpness is slightly blurred for near placed zero. Therefore, the distance between the pole and adjacent zero organizes the convergence value. The next case is a pair of poles zeroes with equidistance moving along the real axis. The distance between the pole and zero is 0.05 with pole closer to unit circle. As the pair advances to the origin, the filter selectiveness is decreased, and convergence value is increased. Therefore, pole zero around the *z* plane origin yields the smooth frequency response.

Prog. 5.7 MATLAB program for Fig. 5.26

```
w = linspace(0,pi,100);
d1 = 0.05;
figure(1), grid, hold on;
figure(2), grid, hold on;

for a1 = 0.1:0.2:0.9
      hz = (1-a1*z^-1)*(1-(a1+d1))/((1-(a1+d1)*z^-1)*(1-a1));
      hz1 = subs(hz,z,exp(1j*w));
      hz2 = eval(hz1);
      figure(1), plot(w,abs(hz2));
      str = sprintf(' %0.3g',a1);
      text(pi/2,abs(hz2(50)),str,'rotation',45);
      [N,D] = numden(hz);
      a = sym2poly(D);
      b = sym2poly(N);
      figure(2), zplane(b,a)
      text(a1,0,str,'rotation',90);
end

figure(1), hold off
figure(2), hold off
figure(1)
ax = gca;
ax.XTick = [0 pi/4 pi/2 3*pi/4 pi];
ax.XTickLabel = {'0','\pi/4','\pi/2','3\pi/4','\pi'};
xlim([0 pi]);
xlabel('\omega');
ylabel('|H(e^{j\omega})|')
figure(2)
xlim([-1.3 1.3]);
ylim([-1.0 1.0]);
```

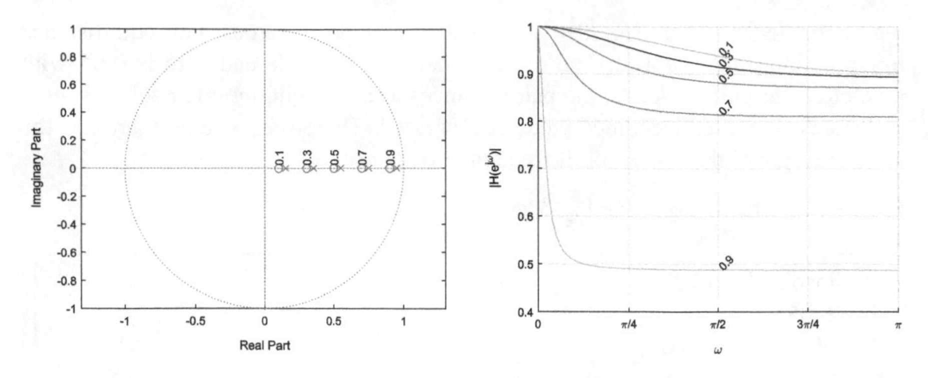

Fig. 5.26 Pole zero plot and frequency response (magnitude) of Eq. (5.75)

The subsequent examination in Fig. 5.27 places the zero above the unit circle since the filter stability does not regulate the zero position. In opposite direction, the pole and zero approach to the unit circle with identical distance and symmetrical manner from 0.75 pole and 1.25 zero. The symmetrical configuration provides relatively flat response and the closer pair presents further uniformity. Hence, the symmetric pole zero pair over the unit circle is useful to construct the flat region in the frequency domain.

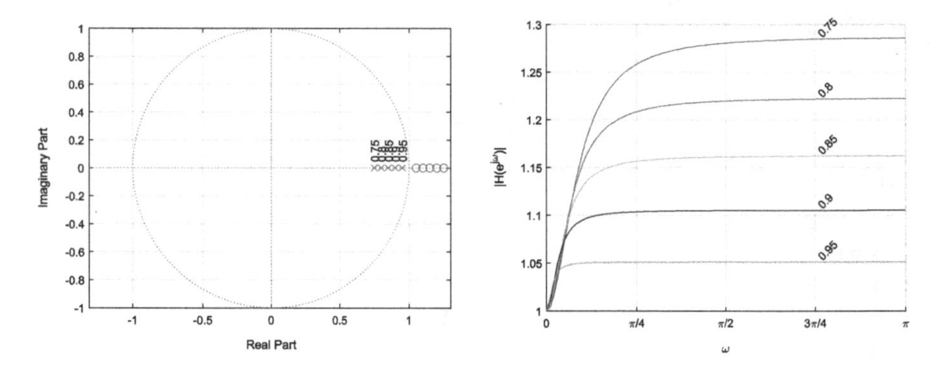

Fig. 5.27 Pole zero plot and frequency response (magnitude) of Eq. (5.75)

Prog. 5.8 MATLAB program for Fig. 5.27

```
w = linspace(0,pi,100);
an1 = 0;
figure(1), grid, hold on;
figure(2), grid, hold on;

for a1 = 0.75:0.05:0.95
    hz = (1-(1+(1-a1))*z^-1)/(1-a1*z^-1);
    hz1 = subs(hz,z,exp(1j*w));
    hz2 = eval(hz1);
    figure(1), plot(w,abs(hz2));
    str = sprintf('   %0.3g',a1);
    text(pi*3/4,abs(hz2(75)),str,'rotation',45);
    [N,D] = numden(hz);
    a = sym2poly(D);
    b = sym2poly(N);
    figure(2), zplane(b,a)
    text(a1*cos(an1),a1*sin(an1),str,'rotation',90);
end

figure(1), hold off
figure(2), hold off
figure(1)
ax = gca;
ax.XTick = [0 pi/4 pi/2 3*pi/4 pi];
ax.XTickLabel = {'0','\pi/4','\pi/2','3\pi/4','\pi'};
xlim([0 pi]);
xlabel('\omega');
ylabel('|H(e^{j\omega})|')
figure(2)
xlim([-1.3 1.3]);
ylim([-1.0 1.0]);
```

The impulse response of the given first order IIR filter is shown below. The polynomial long division computes the μ_0 and γ_1 for inverse Z-transform.

$$\begin{aligned} H(z) &= \frac{b_0 + b_1 z^{-1}}{1 + a_1 z^{-1}} = \mu_0 + \frac{\gamma_1}{1 + a_1 z^{-1}} \\ &\overset{Z}{\leftrightarrow} h[n] = \mu_0 \delta[n] + \gamma_1 (-a_1)^n u[n] \end{aligned} \tag{5.76}$$

Dual poles with zeros:

$$
\begin{aligned}
&y[n] + a_1 y[n-1] + a_2 y[n-2] = b_0 x[n] + b_1 x[n-1] + b_2 x[n-2] \\
&\overset{Z}{\leftrightarrow} Y(z) + a_1 Y(z) z^{-1} + a_2 Y(z) z^{-2} = b_0 X(z) + b_1 X(z) z^{-1} + b_2 X(z) z^{-2} \\
H(z) &= \frac{Y(z)}{X(z)} = \frac{b_0 + b_1 z^{-1} + b_2 z^{-2}}{1 + a_1 z^{-1} + a_2 z^{-2}}
\end{aligned}
\tag{5.77}
$$

The second order IIR filter in Eq. (5.77) contains the dual pole and dual zero movement over the z plane. Both poles and zeros can be placed any locations in the complex domain while the pair positions maintain the conjugate relation. However, the real axis does not limit the pole and zero positions. The single pole and zero experiments are extended to the dual poles and zeros with selective frequencies $\pi/2$, $\pi/4$, and $\pi/6$. Observe that the normalization is performed by dividing the magnitude at the target frequency. The investigation shown in Fig. 5.28 explores the zero-position effect on the frequency response compare to the pole location. The first situation is the approaching the zero to the pole. The pole is fixed at the 0.95 and zero is moved to the pole from 0.1 to 0.9 in every 0.2 stride for the designed frequency.

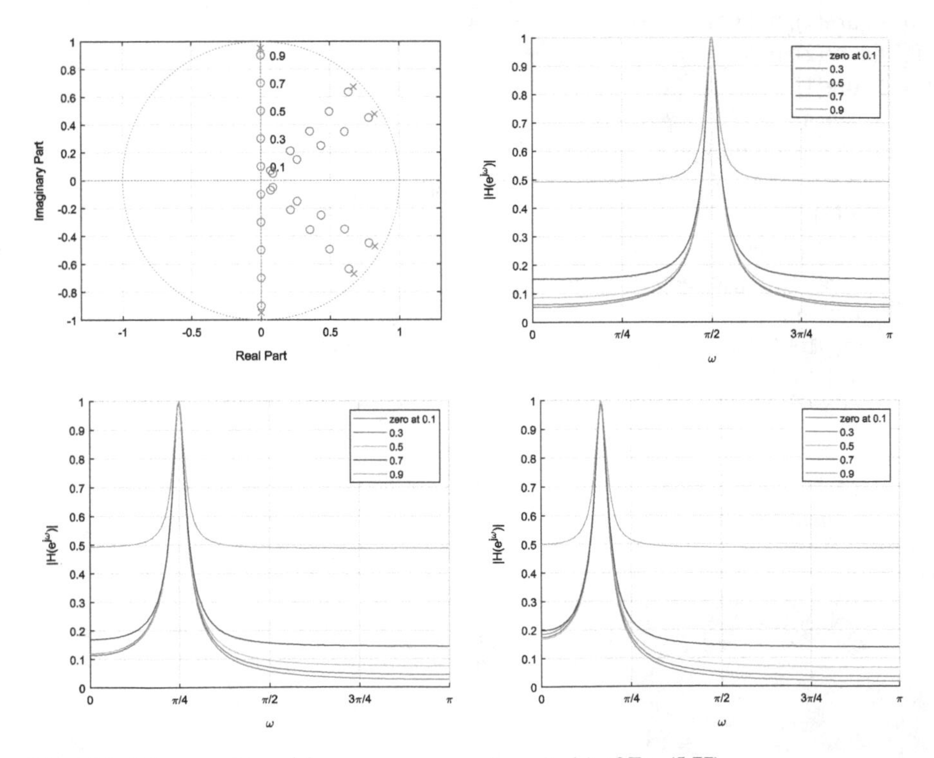

Fig. 5.28 Pole zero plot and frequency response (magnitude) of Eq. (5.77)

Prog. 5.9 MATLAB program for Fig. 5.28

```
syms z;
w = linspace(0,pi,200);
a1 = 0.95;
an1 = [pi/2 pi/4 pi/6];
figure(1), grid, hold on;
figure(2), grid, hold on;
figure(3), grid, hold on;
figure(4), grid, hold on;

for kk = 1:3
    for b1 = 0.1:0.2:0.9
        hz = (1-b1*exp(1j*an1(kk))*z^-1)*(1-b1*exp(-1j*an1(kk))*z^-1)/((1-
a1*exp(1j*an1(kk))*z^-1)*(1-a1*exp(-1j*an1(kk))*z^-1));
        hz1 = subs(hz,z,exp(1j*w));
        hz2 = eval(hz1);
        figure(kk), plot(w,abs(hz2)/max(abs(hz2)));
        str = sprintf('     %0.3g',b1);
        [N,D] = numden(hz);
        a = sym2poly(D);
        b = sym2poly(N);
        figure(4), zplane(b,a)
        if kk == 1
            text((b1*cos(an1(kk))),b1*sin(an1(kk)),str,'rotation',0);
        end
    end
end
figure(1), xlim([0 pi]), hold off
figure(2), xlim([0 pi]), hold off
figure(3), xlim([0 pi]), hold off
figure(4), hold off
```

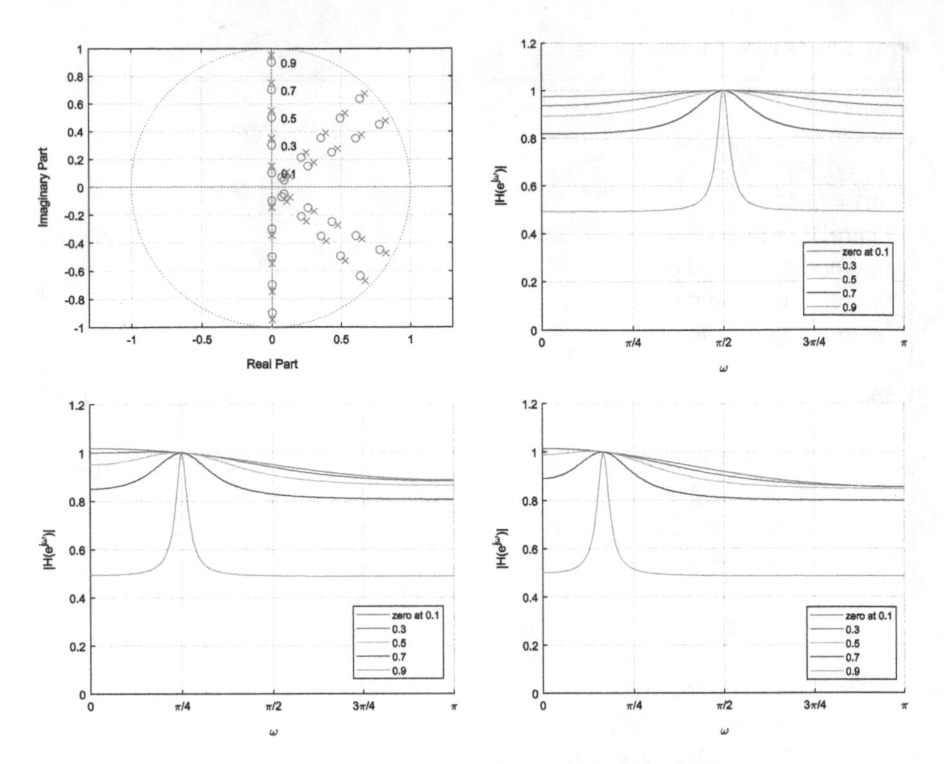

Fig. 5.29 Pole zero plot and frequency response (magnitude) of Eq. (5.77)

Like the single pole and zero, the zero approaching increases the frequency response convergence to the higher constant value. Moreover, the filter sharpness is slightly reduced for near placed zero. Hence, the distance between the pole and adjacent zero decides the convergence value. The case in Fig. 5.29 is a pair of poles and zeroes with equidistance moving along the selective frequency axis. The distance between the pole and zero is 0.05 with pole closer to unit circle. As the pair advances to the origin, the filter selectiveness is decreased, and convergence value is increased. Therefore, pole zero around the z plane origin yields the smooth frequency response.

Prog. 5.10 MATLAB program for Fig. 5.29

```
syms z;
w = linspace(0,pi,200);
a1 = 0.95;
an1 = [pi/2 pi/4 pi/6];
figure(1), grid, hold on;
figure(2), grid, hold on;
figure(3), grid, hold on;
figure(4), grid, hold on;

for kk = 1:3
    for b1 = 0.1:0.2:0.9
        hz = (1-b1*exp(1j*an1(kk))*z^-1)*(1-b1*exp(-1j*an1(kk))*z^-1)/((1-
(b1+0.05)*exp(1j*an1(kk))*z^-1)*(1-(b1+0.05)*exp(-1j*an1(kk))*z^-1));
        hz1 = subs(hz,z,exp(1j*w));
        hz2 = eval(hz1);
        figure(kk), plot(w,abs(hz2)/abs(hz2(round(an1(kk)*200/pi)+1)));
        str = sprintf('    %0.3g',b1);
        [N,D] = numden(hz);
        a = sym2poly(D);
        b = sym2poly(N);
        figure(4), zplane(b,a)
        if kk == 1
            text((b1*cos(an1(kk))),b1*sin(an1(kk)),str,'rotation',0);
        end
    end
end
figure(1), xlim([0 pi]), hold off
figure(2), xlim([0 pi]), hold off
figure(3), xlim([0 pi]), hold off
figure(4), xlim([-1.3 1.3]), ylim([-1.0 1.0]), hold off
```

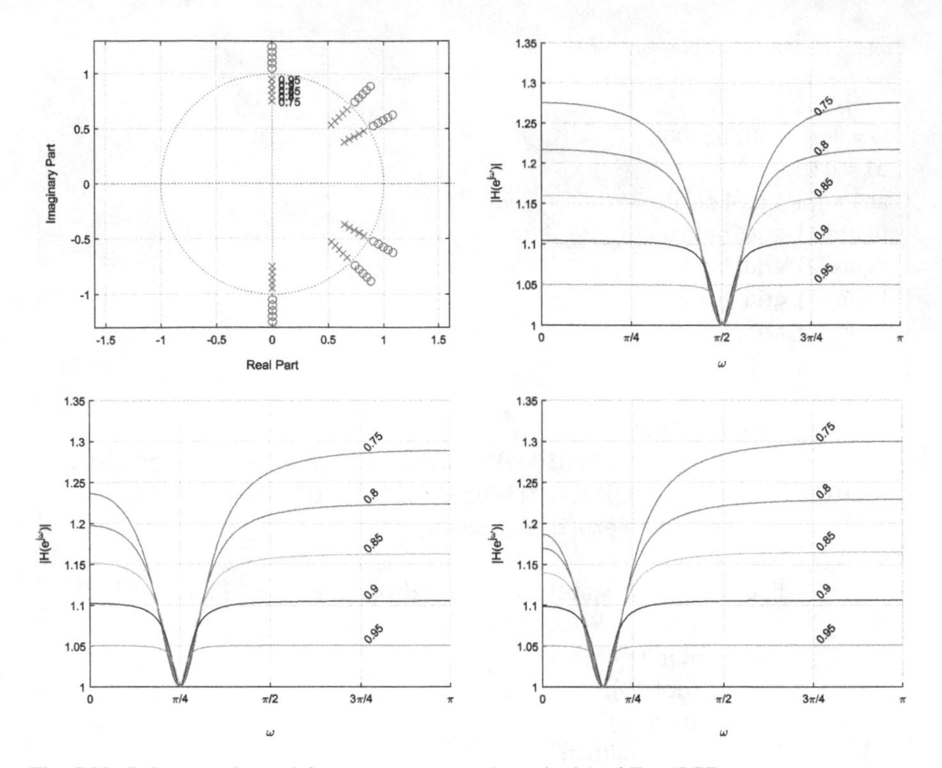

Fig. 5.30 Pole zero plot and frequency response (magnitude) of Eq. (5.77)

The study in Fig. 5.30 locates the zero above the unit circle. In opposite direction, the pole and zero approach to the unit circle with identical distance and symmetrical manner from 0.75 pole and 1.25 zero. Similar to the single pole and zero case, the symmetrical configuration provides relatively flat response and the closer pair presents further uniformity with valley at the designated frequency. Hence, the symmetric pole zero pair over the unit circle is useful to construct the flat region in the frequency domain.

Prog. 5.11 MATLAB program for Fig. 5.30

```
syms z;
w = linspace(0,pi,200);
b1 = 0.1;
an1 = [pi/2 pi/4 pi/6];
figure(1), grid, hold on;
figure(2), grid, hold on;
figure(3), grid, hold on;
figure(4), grid, hold on;

for kk = 1:3
      for a1 = 0.75:0.05:0.95
            hz = (1-(2-a1)*exp(1j*an1(kk))*z^-1)*(1-(2-a1)*exp(-
1j*an1(kk))*z^-1)/((1-a1*exp(1j*an1(kk))*z^-1)*(1-a1*exp(-1j*an1(kk))*z^-
1));
            hz1 = subs(hz,z,exp(1j*w));
            hz2 = eval(hz1);
            figure(kk), plot(w,abs(hz2)/abs(hz2(round(an1(kk)*200/pi)+1)));
            str = sprintf('   %0.3g',a1);
            text(pi*3/4,abs(hz2(200*3/4))/abs(hz2(round(an1(kk)*200/pi)+1)),
str,'rotation',45);
            [N,D] = numden(hz);
            a = sym2poly(D);
            b = sym2poly(N);
            figure(4), zplane(b,a)
            if kk == 1
                  text((a1*cos(an1(kk))),a1*sin(an1(kk)),str,'rotation',0);
            end
      end
end

figure(1), xlim([0 pi]), hold off
figure(2), xlim([0 pi]), hold off
figure(3), xlim([0 pi]), hold off
figure(4), xlim([-1.6 1.6]), ylim([-1.3 1.3]), hold off
```

All of the above investigations, the $\pi/4$ and $\pi/6$ frequencies create the minor asymmetric distribution over the given frequency because the mutual relation with adjacent poles and zeros provides magnitude overlap that produce the higher or lower response at the between the conjugated frequencies. The impulse response of the given second order IIR filter is shown in Eq. (5.78). The polynomial long division computes the μ_0 and γ_1 for inverse Z-transform.

$$\begin{aligned} H(z) &= \frac{b_0 + b_1 z^{-1} + b_2 z^{-2}}{1 + a_1 z^{-1} + a_2 z^{-2}} = \mu_0 + \frac{\gamma_1}{1 - re^{j\omega} z^{-1}} + \frac{\gamma_1^*}{1 - re^{-j\omega} z^{-1}} \\ &\overset{Z}{\leftrightarrow} h[n] = \mu_0 \delta[n] + \gamma_1 r^n e^{j\omega n} u[n] + \gamma_1^* r^n e^{-j\omega n} u[n] \\ &= \mu_0 \delta[n] + 2|\gamma_1| r^n \cos(\omega n + \measuredangle \gamma_1) \end{aligned} \tag{5.78}$$

Multi poles with zeros:

$$\begin{aligned} & y[n] + a_1 y[n-1] + \ldots + a_M y[n-M] \\ &= b_0 x[n] + b_1 x[n-1] + \ldots + b_N x[n-N] \\ &\overset{Z}{\leftrightarrow} Y(z) + a_1 Y(z) z^{-1} + \ldots + a_M Y(z) z^{-M} \\ &= b_0 X(z) + b_1 X(z) z^{-1} + \ldots + b_N X(z) z^{-N} \end{aligned}$$

$$\begin{aligned} H(z) &= \frac{Y(z)}{X(z)} = \frac{b_0 + b_1 z^{-1} + \ldots + b_N z^{-N}}{1 + a_1 z^{-1} + \ldots + a_M z^{-M}} \\ &= \frac{(1-\beta_1 z^{-1})(1-\beta_2 z^{-1})\ldots(1-\beta_N z^{-1})}{(1-\alpha_1 z^{-1})(1-\alpha_2 z^{-1})\ldots(1-\alpha_M z^{-1})} \; |z| > \max\{|\alpha_k|\} \end{aligned} \tag{5.79}$$

The arbitrary number of poles and zeros can contribute to build the designated frequency response. The given equation in Eq. (5.79) contains the M poles and N zeros once there is no cancelations between the poles and zeros. This investigation explores whether the individual known pole-zero placements can illustrate the overall predicted frequency or not. If the realization is true, we can design the IIR filter by adding the poles and zeros on the specific frequency with configurations shown above sections.

Example 5.8
Place the 20 zeros and 20 poles in every $2\pi/20$ radian angles. The 7 pole-zero pairs are located in the low frequency region in symmetric form over unit circle. The 6 pole-zero pairs are placed in the middle frequency range in medium radius and close pairs. The rest of the pole-zero pairs are located at the high frequency section in close poles to the unit circle and far pairs. Inspect the frequency response. Is the overall response close to the individual response combination? Show by MATLAB result.

Solution
With arbitrary parameter values, the transfer function is created as below.

$$H(z) = \frac{\prod_{k=-3}^{3}\left(1 - 1.05e^{j\frac{2\pi}{20}k} z^{-1}\right) \prod_{k=\pm 4}^{\pm 6}\left(1 - 0.7e^{j\frac{2\pi}{20}k} z^{-1}\right) \prod_{k=\pm 7}^{10,-9}\left(1 - 0.7e^{j\frac{2\pi}{20}k} z^{-1}\right)}{\prod_{k=-3}^{3}\left(1 - 0.95e^{j\frac{2\pi}{20}k} z^{-1}\right) \prod_{k=\pm 4}^{\pm 6}\left(1 - 0.75e^{j\frac{2\pi}{20}k} z^{-1}\right) \prod_{k=\pm 7}^{10,-9}\left(1 - 0.95e^{j\frac{2\pi}{20}k} z^{-1}\right)} \; |z| > 0.95$$

The pole zero plot and frequency response are shown in Fig. 5.31. The red lines divide the frequencies between pole-zero configurations in the figure.

Prog. 5.12 MATLAB program for Fig. 5.31

```
syms z k;
w = linspace(0,pi,100);
w1 = linspace(0,pi,10);
wd = 2*pi/20;
a1 = 0.95;
a2 = 0.7;
a3 = 0.7;
HD1 = symprod((1-a1*exp(1j*wd*k)*z^-1),k,-3,3);
HN1 = symprod((1-(2-a1)*exp(1j*wd*k)*z^-1),k,-3,3);
HD2 = symprod((1-(a2+0.05)*exp(1j*wd*k)*z^-1),k,4,6);
HN2 = symprod((1-a2*exp(1j*wd*k)*z^-1),k,4,6);
HD3 = symprod((1-(a2+0.05)*exp(1j*wd*k)*z^-1),k,-6,-4);
HN3 = symprod((1-a2*exp(1j*wd*k)*z^-1),k,-6,-4);
HD4 = symprod((1-0.95*exp(1j*wd*k)*z^-1),k,7,10);
HN4 = symprod((1-a3*exp(1j*wd*k)*z^-1),k,7,10);
HD5 = symprod((1-0.95*exp(1j*wd*k)*z^-1),k,-9,-7);
HN5 = symprod((1-a3*exp(1j*wd*k)*z^-1),k,-9,-7);
H1 = HN1*HN2*HN3*HN4*HN5/(HD1*HD2*HD3*HD4*HD5);
hz1 = subs(H1,z,exp(1j*w));
hz2 = eval(hz1);
[N,D] = numden(H1);
a = sym2poly(D);
b = sym2poly(N);

figure,
plot(w,abs(hz2)), grid
axi1 = axis;
line([2*pi*7/(20*2) 2*pi*7/(20*2)],[axi1(3) axi1(4)],'Color','r')
line([2*pi*13/(20*2) 2*pi*13/(20*2)],[axi1(3) axi1(4)],'Color','r')
xlim([0 pi]);
xlabel('\omega');
ylabel('|H(e^{j\omega})|')
figure,
zplane(b,a), grid
xlim([-1.3 1.3]);
ylim([-1.0 1.0]);
line([0 cos(2*pi*7/(20*2))],[0 sin(2*pi*7/(20*2))],'Color','r')
line([0 cos(2*pi*7/(20*2))],[0 -sin(2*pi*7/(20*2))],'Color','r')
line([0 cos(2*pi*13/(20*2))],[0 sin(2*pi*13/(20*2))],'Color','r')
line([0 cos(2*pi*13/(20*2))],[0 -sin(2*pi*13/(20*2))],'Color','r')
```

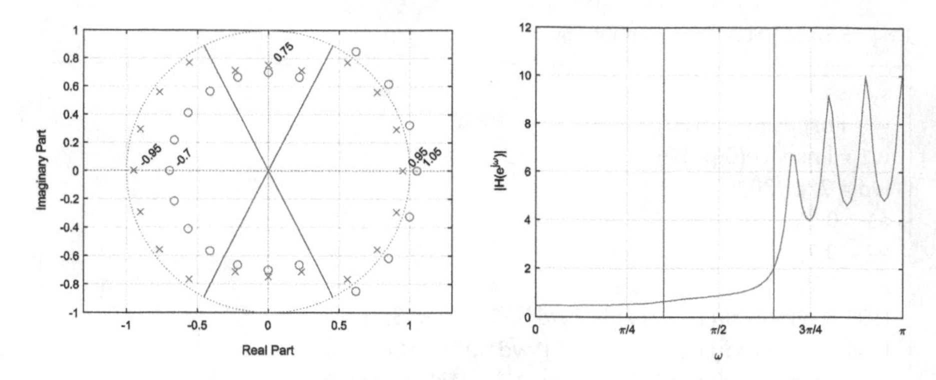

Fig. 5.31 Pole zero plot and frequency response (magnitude) of Example 5.8

The frequency response in Fig. 5.31 shows the overall frequency distribution of the given pole zero configuration. The red lines also represent the section division for individual configuration. While the symmetric pole-zero pairs over the unit circle illustrate the relatively flat frequency response as expected, the rest of the configurations such as medium radius and close pairs and close poles to the unit circle & far pairs barely demonstrate the individual characteristics in the response. The peaky distribution in the high frequency range is derived from the pole zero locations; however, the peak values are not consistent due to the mutual effect from neighbors.

■

Example 5.9

Build the LPF as below. Place the 7 zeros and 7 poles in every $2\pi/20$ radian angles in low frequency section with close symmetric form over unit circle. Locate 13 zeros on the unit circle in every $2\pi/20$ radian for rest of angles. Fix 13 poles on the origin. The desired cutoff frequency of the LPF is the $0.35\,\pi$ from above configuration. Inspect the frequency response. Is the overall response close to the individual response combination? Show by MATLAB result.

Solution

With arbitrary parameter values, the transfer function is created as below.

$$H(z) = \frac{\prod_{k=-3}^{3}\left(1 - 1.01e^{j\frac{2\pi}{20}k}z^{-1}\right)\prod_{k=\pm4}^{10,-9}\left(1 - e^{j\frac{2\pi}{20}k}z^{-1}\right)}{\prod_{k=-3}^{3}\left(1 - 0.99e^{j\frac{2\pi}{20}k}z^{-1}\right)} \quad \text{for } |z| > 0.99$$

The pole zero plot and frequency response are shown in Fig. 5.32. The red line ($0.35\,\pi$) divide the frequencies between pole-zero configurations in the figure.

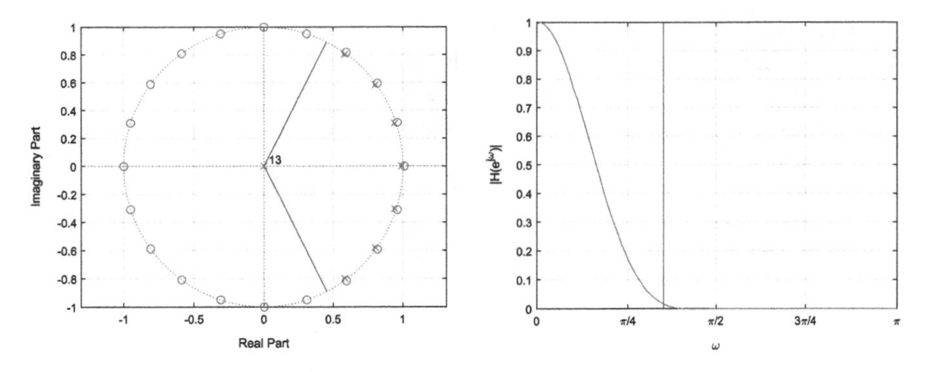

Fig. 5.32 Pole zero plot and frequency response (magnitude) of Example 5.9

Prog. 5.13 MATLAB program for Fig. 5.32

```
syms z k;
w = linspace(0,pi,100);
w1 = linspace(0,pi,10);
wd = 2*pi/20;
a1 = 0.99;
HD1 = symprod((1-a1*exp(1j*wd*k)*z^-1),k,-3,3);
HN1 = symprod((1-(2-a1)*exp(1j*wd*k)*z^-1),k,-3,3);
HN2 = symprod((1-exp(1j*wd*k)*z^-1),k,4,6);
HN3 = symprod((1-exp(1j*wd*k)*z^-1),k,-6,-4);
HN4 = symprod((1-exp(1j*wd*k)*z^-1),k,7,10);
HN5 = symprod((1-exp(1j*wd*k)*z^-1),k,-9,-7);
H1 = HN1*HN2*HN3*HN4*HN5/(HD1);
hz1 = subs(H1,z,exp(1j*w));
hz2 = eval(hz1);
[N,D] = numden(H1);
a = sym2poly(D);
b = sym2poly(N);

figure,
plot(w,abs(hz2)/max(abs(hz2))), grid
axi1 = axis;
line([2*pi*7/(20*2) 2*pi*7/(20*2)],[axi1(3) axi1(4)],'Color','r')
xlim([0 pi]);
xlabel('\omega');
ylabel('|H(e^{j\omega})|')
figure,
zplane(b,a), grid
xlim([-1.3 1.3]);
ylim([-1.0 1.0]);
line([0 cos(2*pi*7/(20*2))],[0 sin(2*pi*7/(20*2))],'Color','r')
line([0 cos(2*pi*7/(20*2))],[0 -sin(2*pi*7/(20*2))],'Color','r')
```

In Fig. 5.32, the low frequency configuration is preserved, and the rest of the frequencies are covered by the zeros on the unit circle in order to create the LPF. The desired cutoff frequency of the LPF is the $0.35\,\pi$ for passing band; however, the magnitude of the LPF is collapsing earlier than expected in the frequency response. The red line in the figure divides the individual pole zero configuration. The zeros on the unit circle shows the strong convergence to the zero magnitude and the unit magnitude is maintained by the symmetric pole zero pairs over the unit circle. Therefore, the given pole zero configuration provides the overall LPF without satisfying the designated specification.

■

Based on the Z-transform, the pole zero placement delivers the IIR filter design method in relaxed manner. We may select the single frequency to be amplified or reduced its magnitude over the IIR filter via approaching the pole or zero to the unit circle in the z plane. The single frequency manipulation is well predictable by the simple calculation in the Z-transform; however, the conventional broadband filters such as LPF, HPF, BPF, and BSF cannot be realized by the simple Z-transform based the method. Between the poles and zeros, the non-linear correlation to the magnitude and phase provides design complexity to the designated IIR filter. In addition, the Z-transform functionality limits the IIR filter design method. Note that the DTFT and DFT directly compute the frequency magnitude and phase; hence, the FIR filter can be designed from the approximated closed method based on the transforms. The Z-transform represents the rational function to figure out the poles/zeros for signal or filter property and dose not deliver the frequency information directly. The advanced methods to design the IIR and FIR filter will be introduced following chapter.

5.4 Relation with Fourier Analysis

Three transforms for discrete time signal are provided previously. The only DFT can be computed by the numerical processing based on the computer and the other transforms can be performed by the symbolic mathematics for analytic purpose. The discrete time signal model and counterpart domain representation are denoted by Table 5.7. The ω, α, and r real value and the n, k, and N are integer value in the equations. As the n increases the integer value, the individual magnitude frequency terms such as $X(e^{j\omega_0})e^{j\omega_0 n}$, $X[k_0]e^{j2\pi k_0 n/N}$, and $\alpha_0 r_0^n e^{j\omega_0 n}$ generate the periodic signals in order to deliver the real value signal. The DTFT and DFT illustrate each frequency component as the delta function signal to represent the magnitude as well as frequency locations in the domain. The Z-transform shows the linear combination of rational functions to indicate the poles and zeros of the signal in the z domain.

Table 5.7 Three transformations with time and counterpart domain

	Discrete time domain	Counterpart domain
DTFT	$x[n] = X(e^{j\omega_0})e^{j\omega_0 n} + X(e^{j\omega_1})e^{j\omega_1 n} + X(e^{j\omega_2})e^{j\omega_2 n} + \ldots$	$X(e^{j\omega}) = 2\pi X(e^{j\omega_0})\delta(\omega - \omega_0) + 2\pi X(e^{j\omega_1})\delta(\omega - \omega_1) + 2\pi X(e^{j\omega_2})\delta(\omega - \omega_2) + \ldots$
DFT	$x[n] = X[k_0]e^{j\frac{2\pi}{N}k_0 n} + X[k_1]e^{j\frac{2\pi}{N}k_1 n} + X[k_2]e^{j\frac{2\pi}{N}k_2 n} + \ldots$	$X[k] = NX[k_0]\delta[\mathrm{k} - k_0] + NX[k_1]\delta[\mathrm{k} - k_1] + NX[k_2]\delta[\mathrm{k} - k_2] + \ldots$
Z-transform	$x[n] = \alpha_0 r_0^n e^{j\omega_0 n} + \alpha_1 r_1^n e^{j\omega_1 n} + \alpha_2 r_2^n e^{j\omega_2 n} + \ldots$	$X(z) = \frac{\alpha_0}{1 - r_0 e^{j\omega_0} z^{-1}} + \frac{\alpha_1}{1 - r_1 e^{j\omega_1} z^{-1}} + \frac{\alpha_2}{1 - r_2 e^{j\omega_2} z^{-1}} + \ldots$

In the previous discussion, we have learned that the Z-transform includes the DTFT and DFT in mathematical viewpoint. The Z-transform can be converted to the DTFT and DFT; however, the Z-transform itself is not the almighty transform to design and analyze the signal and filter. The purpose of individual transform is distinctive. The Z-transform is used to design the IIR filter. The DTFT and DFT is utilized to design the FIR filter. The only DFT is adopted to show the frequency information by using the numerical computer. The transforms are powerful tool to design and analyze the signal and system. Once you use the weapon properly, the information and filter perform what you asked for.

5.5 Problems

1. Find the Z-transform of following signals with associated ROC.
   - $x[n] = u[n] - u[n-10]$
   - $x[n] = 2^n u[-n]$
   - $x[n] = 2^{-n+1} u[n-1]$
   - $x[n] = 3^{-|n|}$
2. Find the Z-transform of following signals with associated ROC.

$$x[n] = n^2 u[n]$$

Use the property below.

$$n^2u[n] = nu[n] * \{u[n] + u[n-1]\}$$

Note that the above property is derived from below.

$$n^2 = \sum_{k=0}^{n} k + \sum_{k=0}^{n} (k-1)$$

Compute the following values by using the initial value and final value theorem.

$$x[0] =$$
$$x[\infty] =$$

Check the below condition in z domain.

$$x[n] \in \mathbb{R}$$

If the new $x[n]$ is below, check the real value condition again.

$$x[n] = jn^2u[n]$$

3. Find the inverse Z-transform of following signals.
   - $X(z) = \frac{1}{1-3^2z^{-2}}$ for $|z| > 3$
   - $X(z) = \frac{1}{1-3^2z^{-2}}$ for $|z| < 3$
   - $X(z) = \frac{1.5-z^{-1}}{1-1.5z^{-1}-z^{-2}}$ for $|z| > 2$
   - $X(z) = \frac{1.5-z^{-1}}{1-1.5z^{-1}-z^{-2}}$ for $0.5 < |z| < 2$
4. Evaluate the frequency magnitude of following signals by using Z-transform.
   - $x[n] = 0.5^n u[n]$ Compute $|H(e^{j\omega})||_{\omega=\frac{\pi}{3}}$
   - $x[n] = 0.7^n \cos\left(\frac{\pi}{3}n\right)u[n]$ Compute $|H(e^{j\omega})||_{\omega=\frac{\pi}{3}}$
   - $x[n] = e^{-n}u[n]$ Compute $|H(e^{j\omega})||_{\omega=\frac{\pi}{2}}$

5. From the given pole-zero plots, derive time domain signals and analyze the stability.

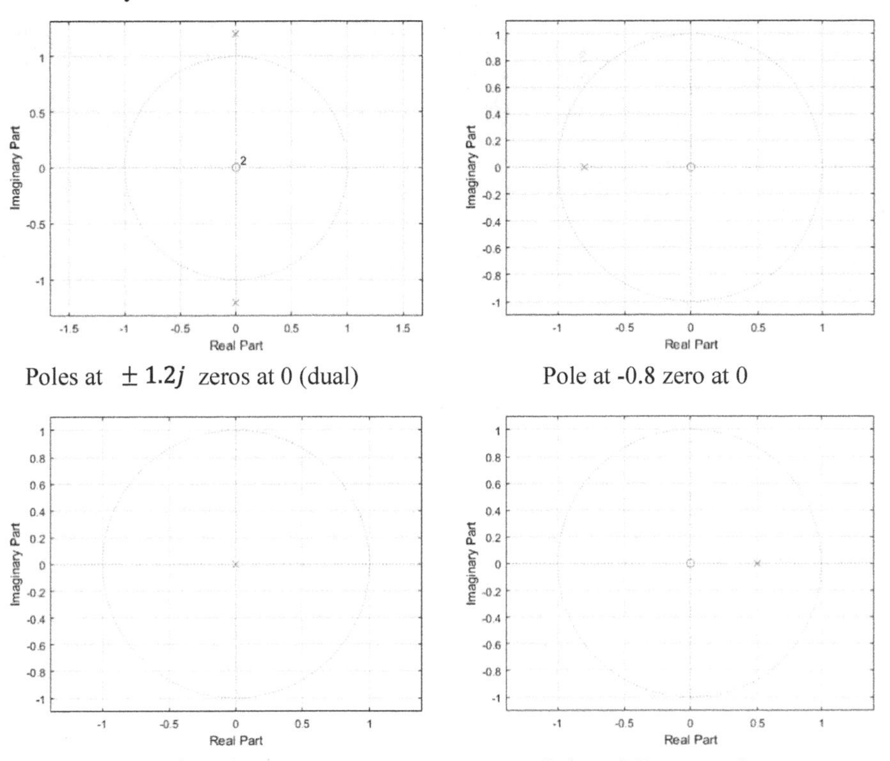

Poles at $\pm 1.2j$ zeros at 0 (dual)

Pole at -0.8 zero at 0

Pole at 0

Pole at 0.5 zeros at 0

6. If the Problem 5 solutions provide the impulse response $h[n]$ of the system, derive the difference equations. Assume that the system is in the initial at rest condition.
7. Design the LPF with IIR structure. Use the minimum order. Zero frequency gain is one and $\pi/4$ frequency gain is less than 0.2. See Fig. 5.20 to determine the order and coefficient. Derive the difference equation.
8. Design the BPF with IIR structure. Use the minimum order. $\pi/2$ frequency gain is one and $\pi/4$ frequency gain is less than 0.2. See Fig. 5.22 to determine the order and coefficient. Derive the difference equation.
9. Design the LPF with IIR structure. Use the minimum order. Approximately maintain the unit magnitude up to $\pi/6$. $\pi/4$ frequency gain is ≈ 0.7 and $\pi/2$ frequency gain is ≈ 0.2. See Fig. 5.24 to determine the order and coefficient. Derive the difference equation.
10. Design the BSF. Use the minimum order. $\pi/2$ frequency gain is zero. Derive the difference equation. Is this IIR or FIR?

References

1. Ragazzini, J.R., Zadeh, L.A.: The analysis of sampled-data systems. Transactions of the American Institute of Electrical Engineers, Part II: Applications and Industry **71**(5), 225–234 (1952). https://doi.org/10.1109/TAI.1952.6371274
2. Larson, R.: Algebra & Trigonometry. Cengage Learning (2016)
3. Wikipedia: Partial fraction decomposition (2020). https://en.wikipedia.org/w/index.php?title=Partial_fraction_decomposition&oldid=954972624
4. Oppenheim, A.V., Schafer, R.W.: Discrete-Time Signal Processing. Prentice Hall (1989)
5. Wikipedia: Z-transform (2020). https://en.wikipedia.org/w/index.php?title=Z-transform&oldid=963886924
6. Hazewinkel, M.: Encyclopaedia of Mathematics: Orbit—Rayleigh Equation. Springer, Netherlands (2012)

Chapter 6
Filter Design

We designed the primitive filters based on the time and frequency domain. The convolution sum operation is used to derive the time domain filter and DTFT (or DFT) is utilized to figure out the FIR filter. The IIR filter is devised from the Z-transform in the previous chapter. We already knew the simple methods to design the basic filter structures. Are these all? Did we miss something? Why we have filter design chapter after all? In general, the filter design is the part of system construction to realize the specific purpose. The target performance is achieved by the dedicated filter operation specified by the sophisticated numbers known as specifications. The previous filter designs only follow personal intuition and ambiguous number for filter implementation; hence, the constructed filter likely shows the unstable performance in overall.

This chapter improves the filter design methods to satisfy the further requirements. We can specify the parameters for transition, ripple, order, etc. in filter shape over frequency domain. The detailed filter requirements can increase the system feasibility by reducing the error propagation from uncertainty. This chapter is initiated from the parameter list and meaning followed by FIR and IIR filter design methods to meet the complex requirements. Although the extensive theories for filter analysis and design are existed in various publications, the specification parameters and design methods in this book are selected from MATLAB contents. The MATLAB already contains the effective and stable design methods in the software toolboxes for successful realization with reduced concerning on the realization complexity. Further implementations beyond the MATLAB territory can be found on the Chap. 7.

K. Kim, *Conceptual Digital Signal Processing with MATLAB*,
Signals and Communication Technology,
https://doi.org/10.1007/978-981-15-2584-1_6

6.1 Filter Specifications

The digital filter requires the specifications in terms of magnitude, frequency, and order. The magnitude in filter specifies the gain for the corresponding frequency. Note that the gain is the ratio between the output amplitude to the input amplitude and the filter magnitude is the function in terms of the frequency. Below is the example to show the filter example in Fig. 6.1.

$$\begin{aligned} x[n] &= a_1 cos(\omega_1 n + \theta_1) = \frac{a_1 e^{j\theta_1}}{2} e^{j\omega_1 n} + \frac{a_1 e^{-j\theta_1}}{2} e^{-j\omega_1 n} \\ y[n] &= a_2 \cos(\omega_1 n + \theta_2) = \frac{a_2 e^{j\theta_2}}{2} e^{j\omega_1 n} + \frac{a_2 e^{-j\theta_2}}{2} e^{-j\omega_1 n} \end{aligned} \tag{6.1}$$

The input and output are the $x[n]$ and $y[n]$ for frequency ω_1, respectively. The given sinusoid amplitude is a_1 and a_2 for input and output, correspondingly. The filter gain can be simply computed as a_2/a_1 in time domain once we have time domain equation for the input and output. Using the Euler equation, we can obtain the frequency domain representation for positive and negative ω_1. The complex scalar portion in front of the exponential frequency presents the complex amplitude for the corresponding frequency. The absolute value of the complex number represents the absolute amplitude such as $a_1/2$ for input and $a_2/2$ for output as shown in Eq. (6.2). Therefore, the filter magnitude or gain is equivalent to the time domain result. If the modelled signal equations are not available, we can use the transformation from the discrete time domain data.

$$\begin{aligned} \left|X\left(e^{j\omega_1}\right)\right| &= \left|X\left(e^{-j\omega_1}\right)\right| = \frac{a_1}{2} \\ \left|Y\left(e^{j\omega_1}\right)\right| &= \left|Y\left(e^{-j\omega_1}\right)\right| = \frac{a_2}{2} \\ \frac{\left|Y\left(e^{j\omega_1}\right)\right|}{\left|X\left(e^{j\omega_1}\right)\right|} &= \frac{\left|Y\left(e^{-j\omega_1}\right)\right|}{\left|X\left(e^{-j\omega_1}\right)\right|} = \frac{a_2}{a_1} \end{aligned} \tag{6.2}$$

The filter selectively passes or suppresses certain frequencies in general; hence, passing band shows unit magnitude and suppressing band exhibits the very small magnitude in designed filter. Therefore, the filter magnitude is represented by the small or very small values with below decimal point. To improve the readability, the filter magnitude is described by the decibel unit instead of absolute scale. The filter magnitude in decibel is shown in Eq. (6.3).

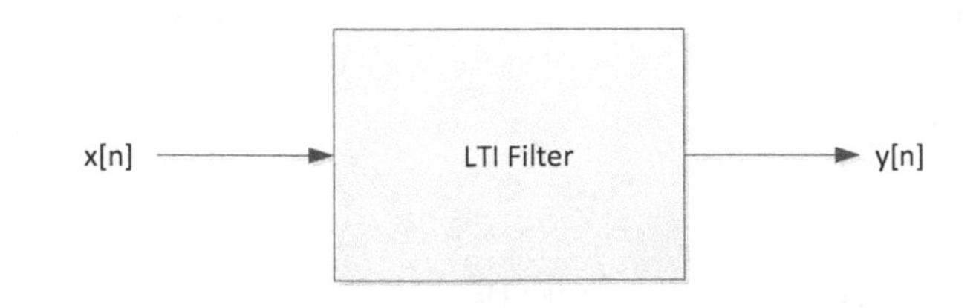

Fig. 6.1 Filter diagram

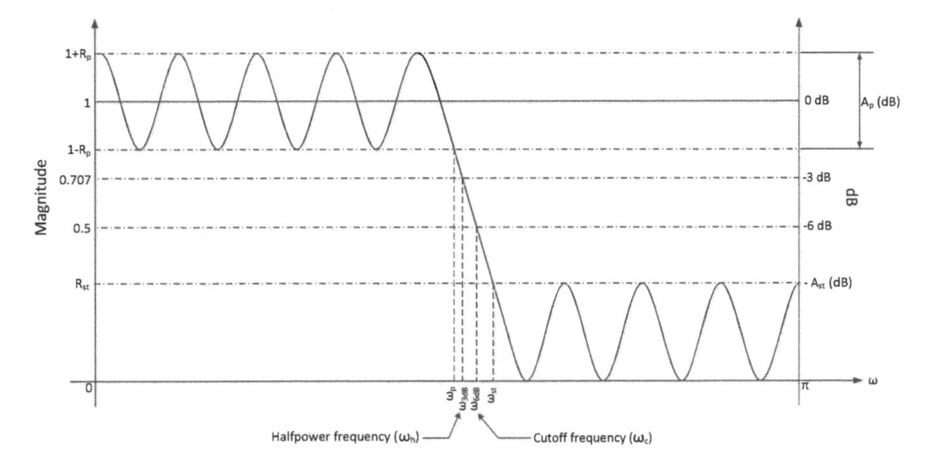

Fig. 6.2 Filter specification with parameters for LPF

$$\text{Magnitude}_{dB}(\omega_1) = 20\log_{10}\frac{|Y(e^{j\omega_1})|}{|X(e^{j\omega_1})|} = 10\log_{10}\frac{|Y(e^{j\omega_1})|^2}{|X(e^{j\omega_1})|^2} \tag{6.3}$$

The beauty of the decibel scale is the compressed and magnified representation range which indicates the huge and tiny numbers by using the relatively manageable numeric. For example, the output amplitude is 0.0001 times less than the input counterpart. The filter magnitude is 0.0001, 10^{-4}, or –80 dB. The 100,000 gain is the 10^5 or 100 dB. Useful decibel numbers are organized as Table 6.1.

Also, due to the logarithm property, the multiplication in the log operator is equivalent to the addition in outside of the operator. Every 3 dB addition or subtraction in magnitude demonstrates the $\sqrt{2}$ multiplication or division in amplitude ratio respectively. Note that the amplitude square ratio presents the multiplication or division by 2 in identical situations. The decibel conversion allows the rapid variations to gradual changes and magnifies the extreme low values to observe the filter performance in detail.

The filter magnitude specification indicates the desired gain in the pass- and stop-band in the filter. Not all filters show the straight magnitude distribution and certain type of the filters creates the magnitude fluctuation over the pass and/or stop frequency band. Therefore, the filter specification denotes the allowed magnitude range in passband and the maximum magnitude in stopband. The variance of the passband gain is described by the upper ($1 + R_p$; in absolute) and lower limit ($1 - R_p$; in absolute) as passband ripple (A_p; in decibel). Note that the passband ripple represents the peak to peak ripple. However, since the lower is the better in stopband magnitude, only upper limit (R_{st}; in absolute) identifies stopband

Table 6.1 Useful decibel numbers and its actual ratios

	60 dB	40 dB	20 dB	6 dB	3 dB	0 dB	−3 dB	−6 dB	−20 dB	−40 dB	−60 dB
$\frac{\lvert Y(e^{j\omega_1})\rvert}{\lvert X(e^{j\omega_1})\rvert}$	1,000	100	10	1.995 ≈ 2	1.413 $\approx \sqrt{2}$	1	0.708 $\approx \sqrt{1/2}$	0.501 $\approx 1/2$	0.1	0.01	0.001
$\frac{\lvert Y(e^{j\omega_1})\rvert^2}{\lvert X(e^{j\omega_1})\rvert^2}$	1,000,000	10,000	100	3.981 ≈ 4	1.995 ≈ 2	1	0.501 $\approx 1/2$	0.251 $\approx 1/4$	0.01	0.0001	0.000001

attenuation (A_{st}; in decibel). The stopband attenuation illustrates the amount of decreased magnitude in decibel from the passband magnitude; therefore, the attenuation is usually represented the negative decibel. By applying the absolute on the actual attenuation, the positive decibel presents the stopband attenuation. Below figure displays the magnitudes for LPF specification.

The passband ripple A_p can be derived from the maximum deviation R_p and vice versa.

$$A_P = 40\log_{10}(1 + R_p) \tag{6.4}$$

$$R_p = 10^{\frac{A_P/2}{20}} - 1 \tag{6.5}$$

The stopband attenuation A_{st} can be computed from the minimum attenuation R_{st} and vice versa.

$$A_{st} = -20\log_{10} R_{st} \tag{6.6}$$

$$R_{st} = 10^{\frac{-A_{st}}{20}} \tag{6.7}$$

Example 6.1
Compute the absolute magnitude of following decibel specifications.

- Passband magnitude: 0 dB
- Passband ripple: 1 dB
- Stopband attenuation: 40 dB

Solution

$$20\log_{10} Mag. = 0\text{dB}; \quad Mag. = 1$$

$$R_p = 10^{\frac{A_P/2}{20}} - 1 = 10^{\frac{1/2}{20}} - 1 = 0.0593$$

$$R_{st} = 10^{\frac{-A_{st}}{20}} = 10^{\frac{-40}{20}} = 0.01$$

The above specifications are organized in Table 6.2 to design the certain filter. We can convert the specifications between the absolute and decibel scale by using the above relations.

∎

Table 6.2 Example of specifications

Specifications	Absolute scale	Decibel scale
Passband magnitude	1	0 dB
Maximum deviation/Passband ripple (peak to peak)	0.0593 (R_p)	1 dB (A_P)
Minimum attenuation/Stopband attenuation	0.01 (R_{st})	40 dB (A_{st})

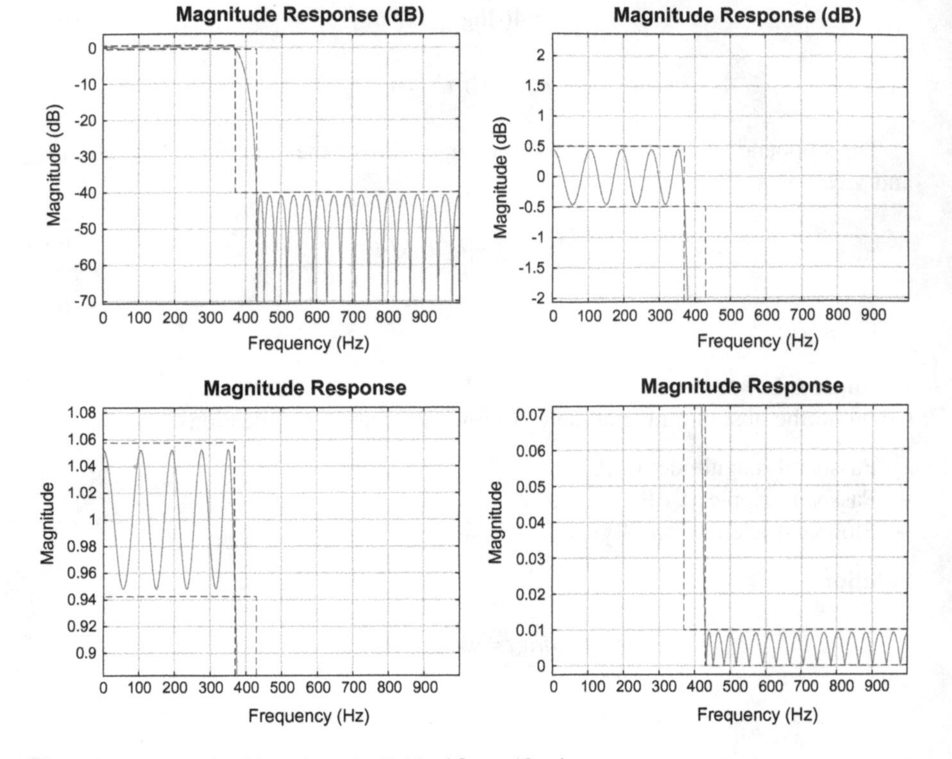

Fig. 6.3 Designed LPF based on the Table 6.2 specifications

By using the MATLAB, the designed LPF is shown at Fig. 6.3 in absolute and decibel scale to meet the Table 6.2 requirement. Note that the frequency specification in LPF should be ignored for the time being. Passband magnitude oscillates between the 0.5 dB (1.0593 in absolute) and –0.5 dB (0.9441 in absolute) to meet the 1 dB pass band ripple. The stopband attenuation demonstrates the below –40 dB (0.01 in absolute) magnitude in Fig. 6.3.

Prog. 6.1 MATLAB program for Fig. 6.3.

```
Fpass = 370;
Fstop = 430;
Ap = 1;
Ast = 40;
Fs = 2000;

d = designfilt('lowpassfir','PassbandFrequency',Fpass,...
   'StopbandFrequency',Fstop,'PassbandRipple',Ap,...
   'StopbandAttenuation',Ast,'SampleRate',Fs);

hfvt = fvtool(d);
```

Next, the frequency values specify the frequency edges for various type filters. The conventional units for the frequency are listed as Table 6.3 and the MATLAB uses the cyclic and normalized frequency. The discrete frequency is used for the mathematical and numerical computation in general. The corresponding frequencies are indicated in Table 6.3 as well.

Example 6.2

Calculate the cyclic, discrete, normalized frequency of following situations.

- $f = 4000$ Hz from $f_s = 10{,}000$ Hz
- $\omega = \frac{\pi}{3}$ from $f_s = 10{,}000$ Hz
- $\frac{1}{6}$ normalized frequency from $f_s = 10{,}000$ Hz

Solution

$$f = 4000 \text{ Hz from } f_s = 10000\text{Hz}$$

$$\omega = 2\pi\frac{f}{f_s} = 2\pi\frac{4000}{10000} = \frac{4\pi}{5}$$

$$\text{Normalized frequency} = \frac{\omega}{\pi} = \frac{4}{5}$$

$$\omega = \frac{\pi}{3} \text{ from } f_s = 10000\text{Hz}$$

$$f = \frac{\omega f_s}{2\pi} = \frac{\pi}{3}\frac{10000}{2\pi} = \frac{5000}{3}\text{Hz}$$

Table 6.3 Frequency units

Frequency type	Unit	Equivalence			Remarks
Cyclic frequency	f in Hz	0	$f_s/4$	$f_s/2$	This type requires the sampling frequency for digital filter
Discrete frequency	ω in radian/sample	0	$\pi/2$	π	$\pm\pi(=\pm\frac{f_s}{2})$ is the maximum digital frequency (in meaningful sense)
Normalized frequency	None	0	1/2	1	The digital frequency is divided by π; therefore, the 1 is the maximum normalized frequency

$$\text{Normalized frequency} = \frac{\omega}{\pi} = \frac{1}{3}$$

$$\frac{1}{6}\text{normalized frequency from } f_s = 10000\,\text{Hz}$$

$$\omega = \pi \text{ Normalized frequency} = \frac{\pi}{6}$$

$$f = \frac{\omega f_s}{2\pi} = \frac{\pi}{6}\frac{10000}{2\pi} = \frac{5000}{6}\ \text{Hz}$$

■

The particular filter type can be described by multiple frequency edges. For example, the LPF requires the one edge to divide the low/high frequency component and the BPF involves at least two edges to allocate the multiple frequency bands. Also, the frequency edge can be defined by the one or two parameters due to the unabrupt transition in the edge. The transition from high to low magnitude is characterized by the passband frequency and stopband frequency parameter. We may specify the edge by using the –3 dB (or –6 dB) below magnitude from the passband magnitude. Table 6.4 summarizes the edge frequency parameters.

The overall LPF specification is illustrated in Fig. 6.2 including the magnitude and frequency requirements. The next specification, the filter order can be identified in time domain and z domain. The highest number of input and output delay in time domain represents the filter order independently. Also, the highest order of the input

Table 6.4 Frequency edge specifications

Frequency edge type	Explanation
Passband frequency	Frequency transition location for $1 - R_p$ magnitude
Stopband frequency	Frequency transition location for R_{st} magnitude
Halfpower frequency for 3 dB point below the passband value	Frequency location for –3 dB below from the desired passband magnitude
Cutoff frequency for 6 dB point below the passband value	Frequency location for –6 dB below from the desired passband magnitude

and output polynomials in z domain presents the filter order equivalently. For example, below filter equation establishes the input N order and output (or feedback) M order. In the same manner, we can specify the system with N numerator order and M denominator order. Note that the FIR filer order demonstrates the arbitrary N input (numerator) order with 1 output (denominator) order due to the feedback absence.

$$\begin{aligned} &y[n]+a_1y[n-1]+a_2y[n-2]+\cdots+a_My[n-M] \\ &=b_0x[n]+b_1x[n-1]+b_2x[n-2]+\cdots+b_Nx[n-N]\overset{Z}{\leftrightarrow} \\ &Y(z)+a_1Y(z)z^{-1}+a_2Y(z)z^{-2}+\cdots+a_MY(z)z^{-M} \\ &=b_0X(z)+b_1X(z)z^{-1}+b_2X(z)z^{-2}+\cdots+b_NX(z)z^{-N} \end{aligned} \tag{6.8}$$

Once we have the N/M order filter, the required number of parameters are $(N + 1)/(M + 1)$ respectively. The numerator involves the b_0, b_1, b_2, ..., b_N values and denominator needs the 1, a_1, a_2, ..., a_M values. Observe that a_0 value should be 1 because of the unit gain for $y[n]$. Now, the digital filter specifications in terms of magnitude, frequency, and order are explained in overall.

6.2 FIR Filters

The FIR filter contains the finite length impulse response to provide the frequency selective filtering on the input signal. Unlike the IIR filter, the FIR filter is dedicated to the discrete-time implementation; therefore, there is no counterpart of FIR filter on continuous-time realization. Note that the classic IIR filters are originated from the continuous-time IIR filters. Due to the finite length, the FIR filters demonstrate following advantages:

- Linear phase (constant delay for whole frequency range)
- Guaranteed stability
- Finite filter transients in time domain
- High feasibility in hardware and software

The linear phase (constant delay) will be explained in the subsequent section. The FIR filter provides the stability because the finite $h[n]$ generates the limited output power based on the convolution sum. The FIR filter transient is finished after complete overlap between the input $x[n]$ and finite $h[n]$ in convolution sum; therefore, the FIR filter shows the finite filter transients in time domain. According to the convolution sum equation, the FIR realization requires the accumulations of weighted and delayed inputs without feedback. Hence, simple multiply and accumulation is the fundamental computation unit for FIR filter computation.

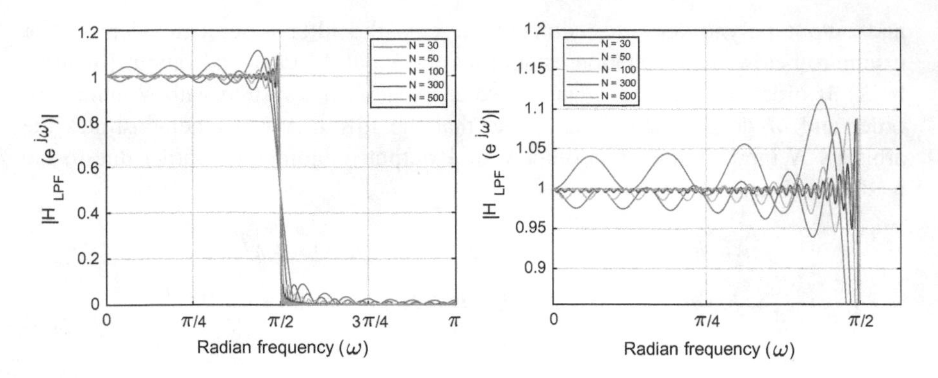

Fig. 6.4 Simple FIR filters for various orders with cutoff frequency $\pi/2$

Previous chapter showed the basic FIR filter design by using the brick wall specification and DTFT and DFT method. We can control the certain parameters of the basic FIR filter such as cutoff and transition frequency by choosing the proper filter length. However, the simple method cannot reduce the pronounced oscillating behavior (a.k.a. Gibbs phenomenon [1]) near the band edges. Figure 6.4 presents the simple FIR filters for order 30, 50, 100, 300, and 500 with cutoff frequency $\pi/2$.

Prog. 6.2 MATLAB program for Fig. 6.4.

```
b1 = fir1(30,0.5,rectwin(31));
b2 = fir1(50,0.5,rectwin(51));
b3 = fir1(100,0.5,rectwin(101));
b4 = fir1(300,0.5,rectwin(301));
b5 = fir1(500,0.5,rectwin(501));
NN = 2^13;
N1 = NN/2+1;
B1 = fft(b1,NN);
B2 = fft(b2,NN);
B3 = fft(b3,NN);
B4 = fft(b4,NN);
B5 = fft(b5,NN);
findex = linspace(0,2*pi,NN);
ff = findex(1:N1);

plot(ff,abs(B1(1:N1)),ff,abs(B2(1:N1)),ff,abs(B3(1:N1)),ff,abs(B4(1:N1)),ff,abs(
B5(1:N1))), grid
...
```

As we increase the order, the FIR filter demonstrates reduced overall ripple in passband/stopband and the sharper response around the cutoff frequency. However, the maximum ripple in the passband and stopband are not manageable. Let's see the LPF in time domain for further investigation.

$$
\begin{aligned}
H_{Ideal\_LPF}\left(e^{j\omega}\right) &= \begin{cases} 1 & 1-\omega_c \leq \omega \leq \omega_c \\ 0 & \text{Otherwise} \end{cases} \\
h_{Ideal\_LPF}[n] &= \frac{1}{2\pi}\int_{-\omega_c}^{\omega_c} 1e^{j\omega n}d\omega = \frac{\sin(\omega_c n)}{\pi n} \qquad (6.9) \\
h_{LPF}[n] &= h_{Ideal_{LPF}}[n]w[n] \text{ where } w[n] = \begin{cases} 1 & 1-n_c \leq n \leq n_c \\ 0 & \text{Otherwise} \end{cases}
\end{aligned}
$$

$$
\begin{aligned}
\overset{\text{DTFT}}{\leftrightarrow} H_{LPF}\left(e^{j\omega}\right) &= \frac{1}{2\pi}H_{Ideal_{LPF}}\left(e^{j\omega}\right) * W\left(e^{j\omega}\right) \\
&= \frac{1}{2\pi}\int_{-\pi}^{\pi} H_{Ideal_{LPF}}\left(e^{j\theta}\right)W\left(e^{j(\omega-\theta)}\right)d\theta \qquad (6.10) \\
&= \frac{1}{2\pi}\int_{-\pi}^{\pi} W\left(e^{j\theta}\right)H_{Ideal_{LPF}}\left(e^{j(\omega-\theta)}\right)d\theta
\end{aligned}
$$

$$
W\left(e^{j\omega}\right) = \sum_{n=-n_c}^{n_c} 1e^{-j\omega n} = \frac{\sin\left(\frac{\omega}{2}(2n_c+1)\right)}{\sin\left(\frac{\omega}{2}\right)} \qquad (6.11)
$$

The impulse response of the FIR LPF can be seen as the multiplication between infinite length sinc function $h_{Ideal\_LPF}[n]$ and finite length window $w[n]$ for finite response length as shown in Eq. (6.9). The frequency domain representation can be understood as the convolution integral between brick wall specification $H_{Ideal\_LPF}\left(e^{j\omega}\right)$ and frequency domain sinc function $W\left(e^{j\omega}\right)$ in Eq. (6.10). Figure 6.5 shows the frequency domain sinc function $W\left(e^{j\omega}\right)$. The increased window size corresponds to the higher and narrower lobes in Eq. (6.11) and Fig. 6.5.

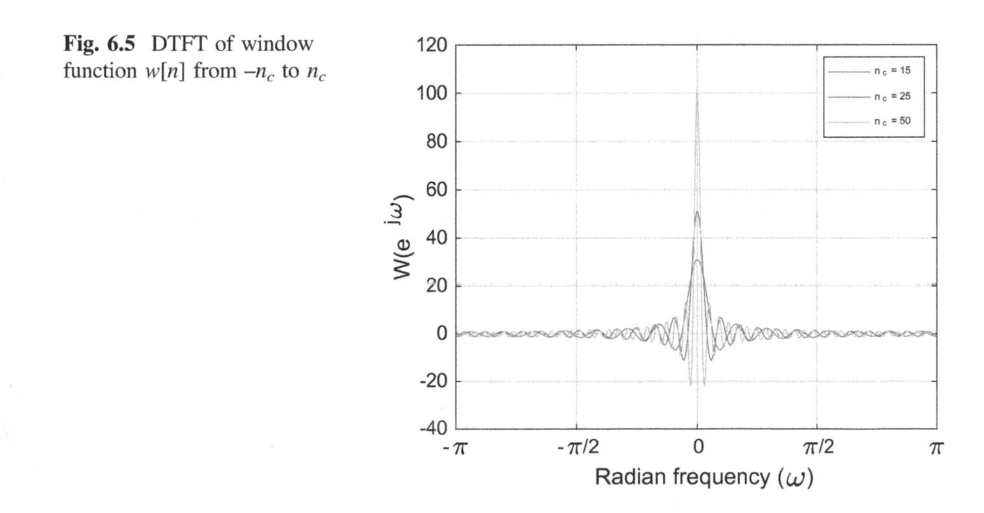

Fig. 6.5 DTFT of window function $w[n]$ from $-n_c$ to n_c

Prog. 6.3 MATLAB program for Fig. 6.5.

```
res = 1001;
NN1 = 30;
NN2 = 50;
NN3 = 100;
nc1 = NN1/2;
nc2 = NN2/2;
nc3 = NN3/2;
ome = linspace(-pi,pi,res);
delw = 2*pi/res;
Ww1 = sin((ome+eps)*(2*nc1+1)/2)./sin((ome+eps)/2);
Ww2 = sin((ome+eps)*(2*nc2+1)/2)./sin((ome+eps)/2);
Ww3 = sin((ome+eps)*(2*nc3+1)/2)./sin((ome+eps)/2);

figure,
plot(ome,Ww1,ome,Ww2,ome,Ww3), grid
...
```

The overall energy in both domains are identical according to the Parseval's theorem [2] as Eq. (6.12). We can approximately calculate the energy in discrete frequency domain ω_k as well.

$$\begin{aligned} \sum_{n=-n_c}^{n_c} 1^2 &= \frac{1}{2\pi}\int_{-\pi}^{\pi}\left|\frac{\sin\left(\frac{\omega}{2}(2n_c+1)\right)}{\sin\left(\frac{\omega}{2}\right)}\right|^2 d\omega \\ &\approx \frac{1}{2\pi}\sum_{k=-k_c}^{k_c}\left|\frac{\sin\left(\frac{\omega_k}{2}(2n_c+1)\right)}{\sin\left(\frac{\omega_k}{2}\right)}\right|^2 \Delta\omega \end{aligned} \tag{6.12}$$

$$\text{where } k_c \gg 1,\ \Delta\omega = \frac{2\pi}{2k_c+1} \text{ and } \omega_k = -\pi + k\Delta\omega$$

The wider window contains the more energy in the frequency domain from Eq. (6.12). Especially for the increased window size, the main lobe illustrates the higher and narrower shape in order to present the neat response after convolution integral in Eq. (6.10). However, the energy ratio between the main lobe and side lobe is almost insensitive to the window size. Note that the main lobe is specified as the area between the first zero crossing on either direction of the origin. We can find the main lobe edges as Eq. (6.13).

$$
\begin{aligned}
W\left(e^{j\omega}\right) &= \frac{\sin\left(\frac{\omega}{2}(2n_c+1)\right)}{\sin\left(\frac{\omega}{2}\right)} = 0 \\
\frac{\omega}{2}(2n_c+1) &= \pm\pi \;\; \omega_m = \pm\frac{2\pi}{2n_c+1}
\end{aligned}
\tag{6.13}
$$

$$
ratio[n_c] = \frac{\text{Energy at main lobe}}{\text{Energy at side lobe}} = \frac{\frac{1}{2\pi}\int_{\omega_m(-)}^{\omega_m(+)} |W(e^{j\omega})|^2 d\omega}{\sum_{n=-n_c}^{n_c} 1^2 - \frac{1}{2\pi}\int_{\omega_m(-)}^{\omega_m(+)} |W(e^{j\omega})|^2 d\omega}
\tag{6.14}
$$

The energy ratio between the main lobe and side lobe is derived at Eq. (6.14) and plotted in Fig. 6.6. The main lobe presents the about 9.3 times energy than the side lobe counterpart and the ratio shows nearly constant in terms of window size. The side lobe in the sinc function contributes to the unwanted spectral leakage which delivers the power to the neighbor frequencies in the convolution integral operation. The shaper main lobe provides the focused convolution integral output as shown in Fig. 6.4. However, we cannot control the side lobe involvement by changing the window length due to the constant energy ratio.

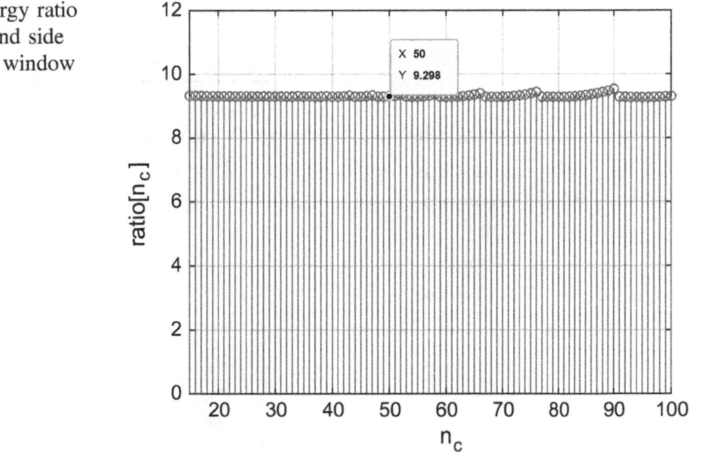

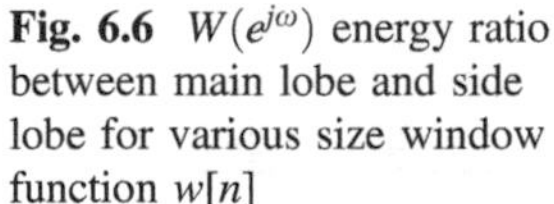

Fig. 6.6 $W(e^{j\omega})$ energy ratio between main lobe and side lobe for various size window function $w[n]$

Prog. 6.4 MATLAB program for Fig. 6.6.

```
res = 1001;
NN = (30:2:200);
nc = NN/2;
ome = linspace(-pi,pi,res);
delw = 2*pi/res;
outWw = [];
zc = 2*pi./(2*nc+1);
zzc = round(zc/delw);
ssrt = ((res-1)/2+1)-zzc;
eend = ((res-1)/2+1)+zzc;
for kk = 1:length(nc)
      Ww = sin((ome+eps)*(2*nc(kk)+1)/2)./sin((ome+eps)/2);
      outWw = [outWw; Ww];
end
mainPower = [];
sidePower = [];
for kk = 1:length(nc)
      temp1 = sum(outWw(kk,ssrt(kk):eend(kk)).^2)*delw/(2*pi);
      temp2 = sum(outWw(kk,1:(ssrt(kk)-1)).^2)*delw/(2*pi);
      temp3 = sum(outWw(kk,(eend(kk)+1):end).^2)*delw/(2*pi);
      mainPower = [mainPower temp1];
      sidePower = [sidePower temp2+temp3];
end
rratio = mainPower./sidePower;
stem(nc,rratio),grid
...
```

Figure 6.7 indicates the maximum ripple situation in the passband and stopband. The convolution integral between the ideal response $H_{Ideal\_LPF}(e^{j\omega})$ and sinc function $W(e^{j\omega})$ is denoted as the integral over the shaded area in both cases. The

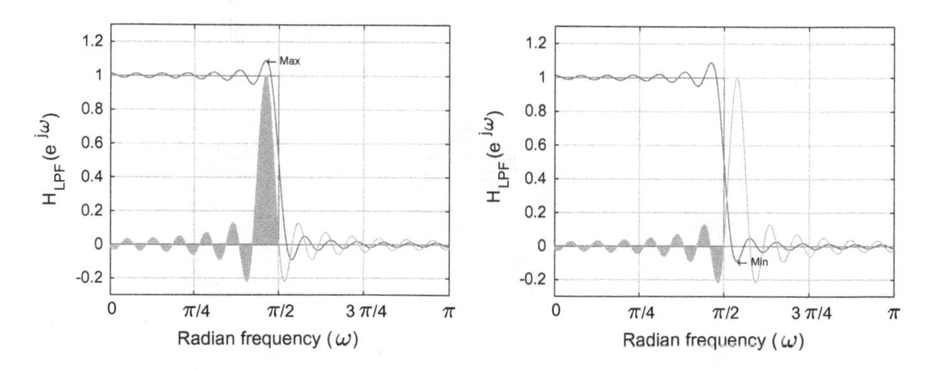

Fig. 6.7 Convolution integral between $H_{Ideal\_LPF}(e^{j\omega})$ & $W(e^{j\omega})$ for maximum ripple situations. Desired filter length is 51 (N) and window length is 25 (n_c)

passband maximum ripple corresponds to the area of main lobe and the side lobe half. The stopband maximum ripple connects to the area of the side lobe half. Since the energy ratio is constant, the maximum ripple in passband and stopband cannot be controllable by changing the window length. The ripples can be narrower but highest magnitude is consistent.

Prog. 6.5 MATLAB program for Fig. 6.7.

```
N = 50;
NN = 2^13;
N1 = NN/2+1;
N2 = N/2;
nnn = (-N2:1:N2);
b2 = sin(pi*(nnn+eps)/2)./(pi*(nnn+eps));
wwindow = [ones(1,N/2+1) zeros(1,NN-(N+1)) ones(1,N/2)];
fwin1 = real(fft(wwindow,NN))/(N+1);
fwin2 = fwin1(1:N1);
b3 = [b2 zeros(1,NN-length(b2))];
b4 = circshift(b3,-25);
B2 = fft(b4,NN);
findex = linspace(0,2*pi,NN);
ff = findex(1:N1);
amp1 = (findex <= pi/2);
[M,I] = max(abs(B2(1:N1)));
fwin3 = circshift(fwin1,I);
fwin4 = fwin3(1:N1);
minloc = max(find(ff<=1.692));
fwin5 = circshift(fwin1,minloc);
fwin6 = fwin5(1:N1);

figure,
plot(ff,real(B2(1:N1)),ff,amp1(1:N1),ff,fwin4), grid, hold on;
stem(ff,[fwin4(1:2048) zeros(1,2049)],'Marker','none','Color',[0.93 0.69
0.13]), hold off;
...
aa = text(ff(I),real(B2(I))+0.005,'\leftarrow Max');
aa.FontSize = 10;

figure,
plot(ff,real(B2(1:N1)),ff,amp1(1:N1),ff,fwin6), grid, hold on;
stem(ff,[fwin6(1:2048) zeros(1,2049)],'Marker','none','Color',[0.93 0.69
0.13]), hold off;
...
aa = text(ff(minloc),real(B2(minloc))+0.005,'\leftarrow Min');
aa.FontSize = 10;
```

The main lobe with one-half side lobe provides the maximum ripple in passband and the one-half side lobe presents the maximum ripple in stopband. The both extreme ripple cases are involved with the one-half side lobe with or without of main lobe presence. Therefore, the area under the one-half side lobe determines the ripple magnitude away from the desired response. Hence, the both ripples indicate the identical deviation from the desired response values. Figure 6.8 shows the boxes and values for the pass-, stop-, and transition band frequency with maximum ripple values. The passband and stopband frequency are decided by the ripple values as the corresponding frequencies to the ripple distance away from the desired magnitude. Each box in the passband and stopband demonstrates the equal distance from the ideal value in magnitude. As we expected, the maximum and minimum ripple exhibits the equal deviation from the ideal filter response.

Prog. 6.6 MATLAB program for Fig. 6.8.

```
N = 50;
NN = 2^13;
N1 = NN/2+1;
N2 = N/2;
nnn = (-N2:1:N2);
b2 = sin(pi*(nnn+eps)/2)./(pi*(nnn+eps));
b3 = [b2 zeros(1,NN-length(b2))];
b4 = circshift(b3,-25);
B2 = real(fft(b4,NN));
findex = linspace(0,2*pi,NN);
ff = findex(1:N1);
amp1 = (findex <= pi/2);
[M1,I1] = max(B2(1:N1));
[M2,I2] = min(B2(1:N1));

figure,
plot(ff,B2(1:N1),ff,amp1(1:N1)), grid
...
straa = sprintf('Max %1.4f',B2(I1));
aa = text(ff(I1),B2(I1)+0.005,['      \leftarrow' straa]);
aa.FontSize = 10;
strbb = sprintf('%1.4f Min',B2(I2));
bb = text(ff(I2)-0.7,B2(I2)+0.005,[strbb '\rightarrow    ']);
bb.FontSize = 10;
cc = min(find(B2<(2-B2(I1))));
dd = max(find(B2(1:NN/2)>-B2(I2)));
...
ee = text(ff(cc)-0.27,0.5,'\Delta\omega \rightarrow ');
ee.FontSize = 10;
gg = text(ff(dd),0.5,'\leftarrow \Delta\omega');
gg.FontSize = 10;
```

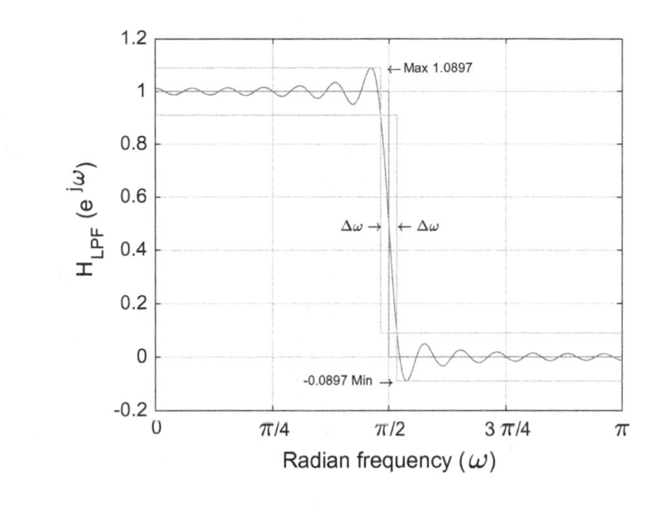

Fig. 6.8 FIR LPF with 51 sample length based on the rectangular window. $\Delta\omega = \omega_{st} - \omega_p$

Above computation for the extreme ripples are derived from the numerical method. We can calculate the ideal ripples from Eqs. (6.15) and (6.16). Equation (6.15) calculates the area of main lobe and one-half side. Equation (6.16) only figures out the one-half side lobe area. The result between the numerical and ideal are almost equal.

$$\max\left(H_{LPF}\left(e^{j\omega}\right)\right) = \int_{-\pi}^{\frac{2\pi}{2n_c+1}} \frac{\sin\left(\frac{\omega}{2}(2n_c+1)\right)}{\sin\left(\frac{\omega}{2}\right)} d\omega \Bigg|_{n_c=25;N=51} \approx 1.0896 \tag{6.15}$$

$$\min\left(H_{LPF}\left(e^{j\omega}\right)\right) = \int_{-\pi}^{-\frac{2\pi}{2n_c+1}} \frac{\sin\left(\frac{\omega}{2}(2n_c+1)\right)}{\sin\left(\frac{\omega}{2}\right)} d\omega \Bigg|_{n_c=25;N=51} \approx -0.0896 \tag{6.16}$$

Due to the equal extreme magnitudes in ripple in pass and stop band, Eqs. (6.4)–(6.7) can be organized to compute the counterpart ripple magnitude in decibel scale.

$$R_p = R_{st}$$

$$10^{\frac{A_p/2}{20}} - 1 = 10^{\frac{-A_{st}}{20}}$$

$$A_{st} = -20\log_{10}\left(10^{\frac{A_p}{40}} - 1\right) \tag{6.17}$$

$$A_p = 40\log_{10}\left(10^{\frac{-A_{st}}{20}} + 1\right) \tag{6.18}$$

The FIR design methods introduced in this chapter provide advanced FIR filters to control the cutoff, transition frequency as well as ripples.

Linear phase FIR filters

The previous filter designs do not consider the phase information. The magnitude of the frequency response is important to provide the selective filter function in terms of the frequency. Based on the determined frequency magnitude by filter, the filter phase value presents the delay location to complete the sinusoidal waveform. Note that the sinusoid as $\alpha\cos(\omega n+\theta)$ is establish by the magnitude, frequency, and phase to localize its position. From the given frequency ω , the filter controls the magnitude and delay by magnitude multiplication as well as phase addition, respectively. The filter retards or advances the propagation by adding or subtracting the phase value to the input individual sinusoid. To maintain the overall shape of the input signal, we prefer to obtain the constant time delay over the entire frequency domain. In order to design the constant time delay filter, let's define the group delay [3] as Eq. (6.19). The argument function delivers the phase angle of given complex number.

$$d(e^{j\omega}) = \mathrm{grd}(\mathrm{H}(\mathrm{e}^{j\omega})) = -\tfrac{\mathrm{d}}{\mathrm{d}\omega}\{\arg(\mathrm{H}(\mathrm{e}^{j\omega}))\} \tag{6.19}$$

To understand the group delay, we assume to have the arbitrary impulse response $h[n]$ with time delay $n_d(\varphi)$ in terms of frequency φ as below.

$$h[n] = \tfrac{1}{2\pi}\int_{-\pi}^{\pi} A(e^{j\varphi})e^{j\varphi(n-n_d(\varphi))}d\varphi \text{ where } A(e^{j\varphi}) \in \mathbb{R} \tag{6.20}$$

The frequency response is the DTFT of $h[n]$ as below.

$$\begin{aligned} H(e^{j\omega}) &= \frac{1}{2\pi}\sum\nolimits_{n=-\infty}^{\infty}\int_{-\pi}^{\pi} A(e^{j\varphi})e^{j\varphi(n-n_d(\varphi))}d\varphi e^{-j\omega n} \\ &= \frac{1}{2\pi}\int_{-\pi}^{\pi} A(e^{j\varphi})e^{-j\varphi n_d(\varphi)}\sum\nolimits_{n=-\infty}^{\infty} e^{-j(\omega-\varphi)n}d\varphi \\ &= \frac{1}{2\pi}\int_{-\pi}^{\pi} A(e^{j\varphi})e^{-j\varphi n_d(\varphi)}2\pi\delta(\omega-\varphi)d\varphi \\ &= A(e^{j\omega})e^{-j\omega n_d(\omega)}\int_{-\pi}^{\pi}\delta(\omega-\varphi)d\varphi = A(e^{j\omega})e^{-j\omega n_d(\omega)} \end{aligned} \tag{6.21}$$

Apply the group delay function to the frequency response as below.

$$\begin{aligned} d(e^{j\omega}) &= -\frac{d}{d\omega}\arg(H(e^{j\omega})) = -\frac{d}{d\omega}(-\omega n_d(\omega)) \\ &= n_d(\omega) + \omega\frac{dn_d(\omega)}{d\omega} \end{aligned} \tag{6.22}$$

$$\text{over } \omega \ (n_d(\omega) = \text{constant})$$

To obtain the constant $d(e^{j\omega})$, the $n_d(\omega)$ should be constant over the entire ω frequency domain. Therefore, the phase information $-\omega n_d(\omega)$ shows the linear variation on ω as shown in Eq. (6.23).

$$d(e^{j\omega}) = \begin{cases} n_d & \arg(H(e^{j\omega})) \rightarrow \text{linear phase} \\ n_d(\omega) + \omega\frac{dn_d(\omega)}{d\omega} & \arg(H(e^{j\omega})) \rightarrow \text{nonlinear} \end{cases} \tag{6.23}$$

The linear phase FIR filter can be designed from the pure real or pure imaginary frequency response. The pure real response indicates the zero or π constant phase and the pure imaginary response specifies the $\pi/2$ or $-\pi/2$ constant phase. The constant phase shows the zero group delay. The corresponding impulse response $h[n]$ should be real.

$$\left.\begin{array}{l} H(e^{j\omega}) \in \mathbb{R} \rightarrow |H(e^{j\omega})|e^{j0} \text{or } |H(e^{j\omega})|e^{j\pi} \\ H(e^{j\omega}) \in j\,\mathbb{R} \rightarrow |H(e^{j\omega})|e^{j\frac{\pi}{2}} \text{or } |H(e^{j\omega})|e^{-j\frac{\pi}{2}} \end{array}\right\} \rightarrow \text{grd}(H(e^{j\omega})) = 0 \tag{6.24}$$

$$H(e^{j\omega}) \overset{\text{DTFT}}{\leftrightarrow} h[n] \in \mathbb{R} \rightarrow \begin{cases} H(e^{j\omega}) \in \mathbb{R} \rightarrow \quad H(e^{j\omega}) = H(e^{-j\omega}) \\ H(e^{j\omega}) \in j\,\mathbb{R} \rightarrow H(e^{j\omega}) = -H(e^{-j\omega}) \end{cases} \tag{6.25}$$

$$H(e^{j\omega}) \in \mathbb{R} \rightarrow h[n] = \frac{1}{2\pi}\int_{-\pi}^{\pi} H(e^{j\omega})e^{j\omega n}d\omega = \frac{1}{\pi}\int_{0}^{\pi} H(e^{j\omega})\cos(\omega n)d\omega$$
$$\rightarrow \text{even symmetric function}$$
$$H(e^{j\omega}) \in j\mathbb{R} \rightarrow h[n] = \frac{1}{2\pi}\int_{-\pi}^{\pi} H(e^{j\omega})e^{j\omega n}d\omega = \frac{1}{\pi}\int_{0}^{\pi} H(e^{j\omega})\sin(\omega n)d\omega$$
$$\rightarrow \text{odd symmetric function} \tag{6.26}$$

According to the above relations, the even and odd symmetric for impulse response $h[n]$ provide the zero group delay FIR filter. The symmetricity and pattern generate the four types of FIR filter with zero group delay as shown below. The type I, II, III, and IV shows the unique symmetric patterns. After designing the zero group delay FIR filter, the $h[n]$ is shifted to right to observe the filter causality. The time shift in the impulse response presents the linear phase property in frequency domain as shown in Eq. (6.27).

$$h[n - n_d] \overset{\text{DTFT}}{\leftrightarrow} H(e^{j\omega})e^{-j\omega n_d} \tag{6.27}$$

The four types of FIR filter for zero group delay are listed as below.

Type I: Even symmetric function with even filter order (odd filter length)

$$h_I[n] = h_I[-n] \tag{6.28}$$

$$\begin{aligned}
h_I[n] &= h_I[0]\delta[n] + h_I[1](\delta[n-1]+\delta[n+1]) + h_I[2](\delta[n-2]+\delta[n+2]) \\
&\quad + \cdots + h_I[K](\delta[n-K]+\delta[n+K]) \\
H_I(e^{j\omega}) &= \sum\nolimits_{n=-\infty}^{\infty} h_I[n]e^{-j\omega n} \\
&= h_I[0] + h_I[1]e^{-j\omega} + h_I[1]e^{j\omega} + h_I[2]e^{-j\omega 2} + h_I[2]e^{j\omega 2} + \cdots \\
&\quad + h_I[K]e^{-j\omega K} + h_I[K]e^{j\omega K} = h_I[0] + 2\sum\nolimits_{n=1}^{K} h_I[n]\cos(\omega n)
\end{aligned} \tag{6.29}$$

Type II: Even symmetric function with odd filter order (even filter length)

$$h_{II}[n] = h_{II}[-n-1] (h_{II}[n] = h_{II}[-n+1] \text{also possible}) \tag{6.30}$$

$$\begin{aligned}
h_{II}[n] &= h_{II}[0](\delta[n]+\delta[n+1]) + h_{II}[1](\delta[n-1]+\delta[n+2]) \\
&\quad + h_{II}[2](\delta[n-2]+\delta[n+3]) + \cdots + h_{II}[K](\delta[n-K]+\delta[n+(K+1)]) \\
H_{II}(e^{j\omega}) &= \sum\nolimits_{n=-\infty}^{\infty} h_{II}[n]e^{-j\omega n} \\
&= h_{II}[0](1+e^{j\omega}) + h_{II}[1](e^{-j\omega}+e^{j\omega 2}) + h_{II}[2](e^{-j\omega 2}+e^{j\omega 3}) + \cdots \\
&\quad + h_{II}[K]\left(e^{-j\omega K} + e^{j\omega(K+1)}\right) \\
&= e^{j\frac{\omega}{2}} h_{II}[0]\left(e^{-j\frac{\omega}{2}} + e^{j\frac{\omega}{2}}\right) + e^{j\frac{\omega}{2}} h_{II}[1]\left(e^{-j\frac{3\omega}{2}} + e^{j\frac{3\omega}{2}}\right) \\
&\quad + e^{j\frac{\omega}{2}} h_{II}[2]\left(e^{-j\frac{5\omega}{2}} + e^{j\frac{5\omega}{2}}\right) + \cdots + e^{j\frac{\omega}{2}} h_{II}[K]\left(e^{-j\frac{(2K+1)\omega}{2}} + e^{j\frac{(2K+1)\omega}{2}}\right) \\
&= 2e^{j\frac{\omega}{2}} \sum\nolimits_{n=0}^{K} h_{II}[n]\cos\left(\tfrac{\omega}{2}(2n+1)\right) \text{ for } h_{II}[n] = h_{II}[-n-1]
\end{aligned} \tag{6.31}$$

$$= 2e^{-j\frac{\omega}{2}} \sum\nolimits_{n=1}^{K} h_{II}[n]\cos\left(\frac{\omega}{2}(2n-1)\right) \text{ for } h_{II}[n] = h_{II}[-n+1]$$

Type III: Odd symmetric function with even filter order (odd filter length)

$$h_{III}[n] = -h_{III}[-n] \ (h_{III}[0] = 0 \text{ due to the odd symmetricity}) \tag{6.32}$$

$$\begin{aligned}
h_{III}[n] &= h_{III}[1](\delta[n-1]-\delta[n+1]) + h_{III}[2](\delta[n-2]-\delta[n+2]) + \cdots \\
&\quad + h_{III}[K](\delta[n-K]-\delta[n+K]) \\
H_{III}(e^{j\omega}) &= \sum\nolimits_{n=-\infty}^{\infty} h_{III}[n]e^{-j\omega n} \\
&= h_{III}[1]e^{-j\omega} - h_{III}[1]e^{j\omega} + h_{III}[2]e^{-j\omega 2} - h_{III}[2]e^{j\omega 2} + \cdots \\
&\quad + h_{III}[K]e^{-j\omega K} - h_{III}[K]e^{j\omega K} \\
&= -2j\sum\nolimits_{n=1}^{K} h_{III}[n]\sin(\omega n)
\end{aligned} \tag{6.33}$$

Type IV: Odd symmetric function with odd filter order (even filter length)

$$h_{IV}[n] = -h_{IV}[-n-1] \;\; (h_{IV}[n] = -h_{IV}[-n+1] \text{ also possible}) \tag{6.34}$$

$$\begin{aligned}
h_{IV}[n] &= h_{IV}[0](\delta[n] - \delta[n+1]) + h_{IV}[1](\delta[n-1] - \delta[n+2]) \\
&\quad + h_{IV}[2](\delta[n-2] - \delta[n+3]) + \cdots + h_{IV}[K](\delta[n-K] - \delta[n+(K+1)]) \\
H_{IV}\left(e^{j\omega}\right) &= \sum\nolimits_{n=-\infty}^{\infty} h_{IV}[n]e^{-j\omega n} \\
&= h_{IV}[0]\left(1 - e^{j\omega}\right) + h_{IV}[1]\left(e^{-j\omega} - e^{j\omega 2}\right) + h_{IV}[2]\left(e^{-j\omega 2} - e^{j\omega 3}\right) + \cdots \\
&\quad + h_{II}[K]\left(e^{-j\omega K} - e^{j\omega(K+1)}\right) \\
&= e^{j\frac{\omega}{2}} h_{IV}[0]\left(e^{-j\frac{\omega}{2}} - e^{j\frac{\omega}{2}}\right) + e^{j\frac{\omega}{2}} h_{IV}[1]\left(e^{-j\frac{3\omega}{2}} - e^{j\frac{3\omega}{2}}\right) \\
&\quad + e^{j\frac{\omega}{2}} h_{IV}[2]\left(e^{-j\frac{5\omega}{2}} - e^{j\frac{5\omega}{2}}\right) + \cdots + e^{j\frac{\omega}{2}} h_{II}[K]\left(e^{-j\frac{(2K+1)\omega}{2}} - e^{j\frac{(2K+1)\omega}{2}}\right) \\
&= -2je^{j\frac{\omega}{2}} \sum\nolimits_{n=0}^{\infty} h_{IV}[n] \sin\left(\frac{\omega}{2}(2n+1)\right) \text{for } h_{IV}[n] = -h_{IV}[-n-1]
\end{aligned} \tag{6.35}$$

$$= -2je^{-j\frac{\omega}{2}} \sum\nolimits_{n=1}^{\infty} h_{IV}[n] \sin\left(\frac{\omega}{2}(2n-1)\right) \text{for } h_{IV}[n] = -h_{IV}[-n+1]$$

Table 6.5 Four types of zero group delay FIR filters (non-causal FIR filters)

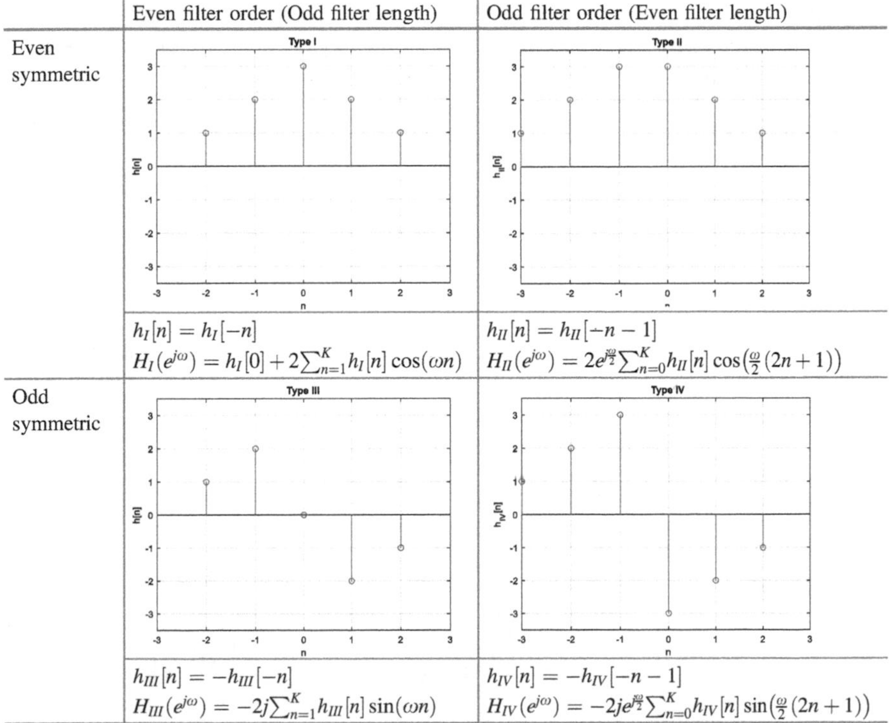

	Even filter order (Odd filter length)	Odd filter order (Even filter length)
Even symmetric	Type I	Type II
	$h_I[n] = h_I[-n]$ $H_I(e^{j\omega}) = h_I[0] + 2\sum_{n=1}^{K} h_I[n]\cos(\omega n)$	$h_{II}[n] = h_{II}[-n-1]$ $H_{II}(e^{j\omega}) = 2e^{j\frac{\omega}{2}}\sum_{n=0}^{K} h_{II}[n]\cos\left(\frac{\omega}{2}(2n+1)\right)$
Odd symmetric	Type III	Type IV
	$h_{III}[n] = -h_{III}[-n]$ $H_{III}(e^{j\omega}) = -2j\sum_{n=1}^{K} h_{III}[n]\sin(\omega n)$	$h_{IV}[n] = -h_{IV}[-n-1]$ $H_{IV}(e^{j\omega}) = -2je^{j\frac{\omega}{2}}\sum_{n=0}^{K} h_{IV}[n]\sin\left(\frac{\omega}{2}(2n+1)\right)$

Table 6.6 Variations of type II and IV

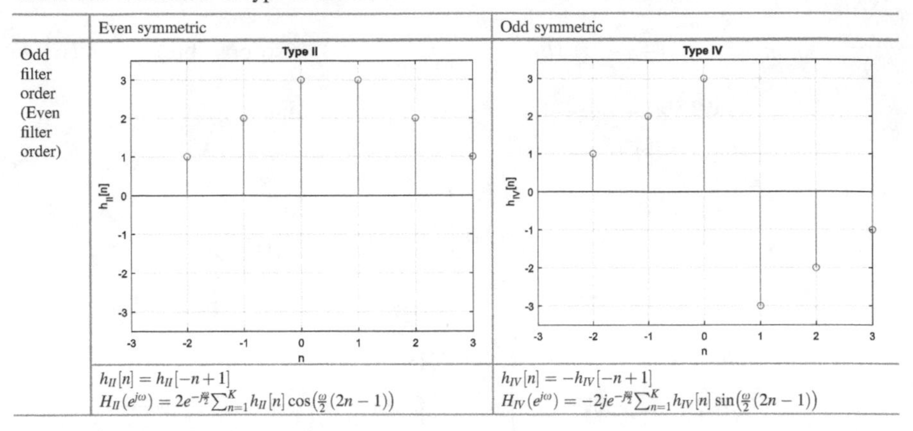

	Even symmetric	Odd symmetric
Odd filter order (Even filter order)	Type II plot	Type IV plot
	$h_{II}[n] = h_{II}[-n+1]$ $H_{II}(e^{j\omega}) = 2e^{-j\frac{\omega}{2}}\sum_{n=1}^{K} h_{II}[n]\cos\left(\frac{\omega}{2}(2n-1)\right)$	$h_{IV}[n] = -h_{IV}[-n+1]$ $H_{IV}(e^{j\omega}) = -2je^{-j\frac{\omega}{2}}\sum_{n=1}^{K} h_{IV}[n]\sin\left(\frac{\omega}{2}(2n-1)\right)$

Four types of zero group delay FIR filter are organized in Table 6.5 with illustrations. The type II and IV variations are listed in Table 6.6 as well. Note that Tables 6.5 and 6.6 filters are all zero group delay filters.

Table 6.7 shows the examples of causal linear phased FIR filters for type I, II, III, and IV. Due to the time shift for causality, the $e^{-j\omega N/2}$ at frequency response provides the linearity in phase; therefore, the *N*/2 group delay is expected for all frequency component ω.

The symmetricity and pattern of the linear phased FIR filters also deliver the restrictions on the filter configuration. Certain frequency style cannot be realized by the linear phase FIR filter type. Below equations analyze the frequency distribution of the filter type. The FIR filter controls the frequency response by placing the zeros in z domain; hence, the Z-transform is applied for analysis.

Type I: Even symmetric function with even filter order (odd filter length)

$$
\begin{aligned}
h_I[n] &= h_I[-n] \\
h_I[n] &= h_I[0]\delta[n] + h_I[1](\delta[n-1] + \delta[n+1]) + h_I[2](\delta[n-2] + \delta[n+2]) \\
&\quad + \cdots + h_I[K](\delta[n-K] + \delta[n+K]) \\
H_I(z) &= \sum_{n=-\infty}^{\infty} h_I[n]z^{-n} \\
&= h_I[0] + h_I[1]z^{-1} + h_I[1]z + h_I[2]z^{-2} + h_I[2]z^2 + \cdots + h_I[K]z^{-K} + h_I[K]z^K \\
&= h_I[0] + h_I[1]\left(z^{-1} + z\right) + h_I[2]\left(z^{-2} + z^2\right) + \cdots \\
&\quad + h_I[K]\left(z^{-K} + z^K\right)
\end{aligned}
\tag{6.36}
$$

Table 6.7 Example of causal linear phased FIR filters for all types

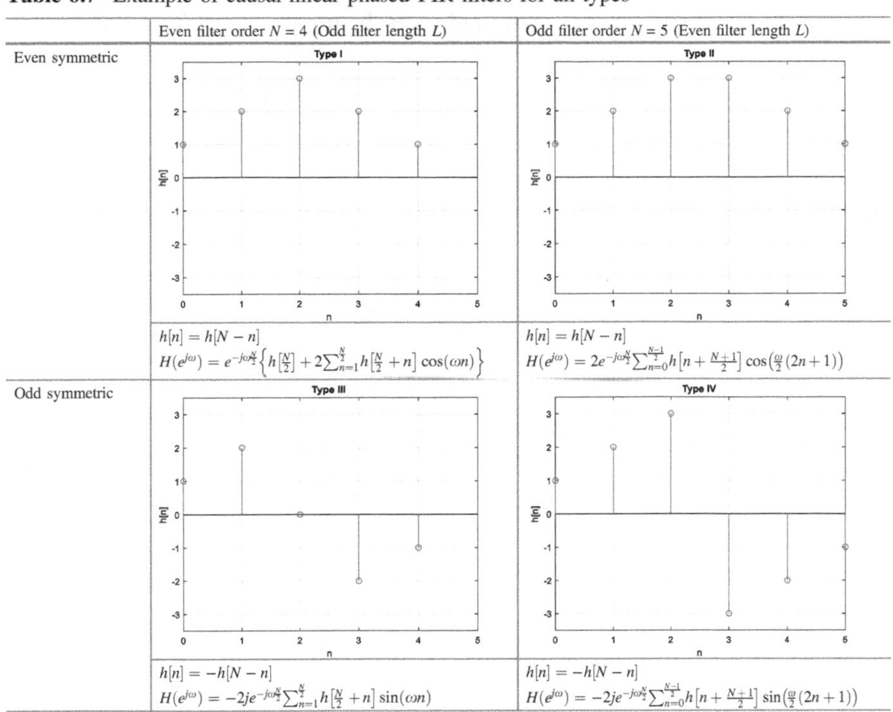

	Even filter order $N = 4$ (Odd filter length L)	Odd filter order $N = 5$ (Even filter length L)
Even symmetric	Type I	Type II
	$h[n] = h[N-n]$ $H(e^{j\omega}) = e^{-j\omega\frac{N}{2}}\left\{h\left[\frac{N}{2}\right] + 2\sum_{n=1}^{\frac{N}{2}} h\left[\frac{N}{2}+n\right]\cos(\omega n)\right\}$	$h[n] = h[N-n]$ $H(e^{j\omega}) = 2e^{-j\omega\frac{N}{2}}\sum_{n=0}^{\frac{N-1}{2}} h\left[n+\frac{N+1}{2}\right]\cos\left(\frac{\omega}{2}(2n+1)\right)$
Odd symmetric	Type III	Type IV
	$h[n] = -h[N-n]$ $H(e^{j\omega}) = -2je^{-j\omega\frac{N}{2}}\sum_{n=1}^{\frac{N}{2}} h\left[\frac{N}{2}+n\right]\sin(\omega n)$	$h[n] = -h[N-n]$ $H(e^{j\omega}) = -2je^{-j\omega\frac{N}{2}}\sum_{n=0}^{\frac{N-1}{2}} h\left[n+\frac{N+1}{2}\right]\sin\left(\frac{\omega}{2}(2n+1)\right)$

There are no specific zeros in z domain. No restrictions mean that all filter types are possible.

Type II: Even symmetric function with odd filter order (even filter length)

$$
\begin{aligned}
h_{II}[n] &= h_{II}[-n-1] \\
h_{II}[n] &= h_{II}[0](\delta[n]+\delta[n+1]) + h_{II}[1](\delta[n-1]+\delta[n+2]) \\
&\quad + h_{II}[2](\delta[n-2]+\delta[n+3]) + \cdots + h_{II}[K](\delta[n-K]+\delta[n+(K+1)]) \\
H_{II}(z) &= \sum_{n=-\infty}^{\infty} h_{II}[n]z^{-n} \\
&= h_{II}[0](1+z) + h_{II}[1]\left(z^{-1}+z^{2}\right) + h_{II}[2]\left(z^{-2}+z^{3}\right) + \cdots \\
&= h_{II}[0](1+z) + z^{-1}h_{II}[1]\left(1+z^{3}\right) + z^{-2}h_{II}[2]\left(1+z^{5}\right) + \cdots \\
&= h_{II}[0](1+z) + z^{-1}h_{II}[1](1+z)\left(1-z+z^{2}\right) \\
&\quad + z^{-2}h_{II}[2](1+z)\left(1-z+z^{2}-z^{3}+z^{4}\right) + \cdots \\
&= (1+z)\left\{h_{II}[0] + z^{-1}h_{II}[1]\left(1-z+z^{2}\right)\right. \\
&\quad \left. + z^{-2}h_{II}[2]\left(1-z+z^{2}-z^{3}+z^{4}\right) + \cdots\right\}
\end{aligned}
\tag{6.37}
$$

Due to the common factor $(1+z)$ in Z-transform, $z = -1$ should be the zero location. The –1 position at z domain indicates the highest frequency in ω domain; therefore, the HPF is not possible.

$$H(-1) = H(e^{j\omega})|_{\omega=\pi} = 0 \tag{6.38}$$

Type III: Odd symmetric function with even filter order (odd filter length)

$$\begin{aligned}
h_{III}[n] &= -h_{III}[-n] \\
h_{III}[n] &= h_{III}[1](\delta[n-1]-\delta[n+1]) + h_{III}[2](\delta[n-2]-\delta[n+2]) + \cdots \\
&\quad + h_{III}[K](\delta[n-K]-\delta[n+K]) \\
H_{III}(z) &= \sum\nolimits_{n=-\infty}^{\infty} h_{III}[n]z^{-n} \\
&= h_{III}[1]z^{-1} - h_{III}[1]z + h_{III}[2]z^{-2} - h_{III}[2]z^2 + \cdots \\
&= h_{III}[1](z^{-1}-z) + h_{III}[2](z^{-2}-z^2) + h_{III}[3](z^{-3}-z^3) + \cdots \\
&= h_{III}[1]z^{-1}(1-z^2) + h_{III}[2]z^{-2}(1-z^4) + h_{III}[3]z^{-3}(1-z^6) + \cdots \\
&= h_{III}[1]z^{-1}(1-z^2) + h_{III}[2]z^{-2}(1-z^2)(1+z^2) \\
&\quad + h_{III}[3]z^{-3}(1-z^2)(1+z^2+z^4) + \cdots \\
&= (1-z^2)\{h_{III}[1]z^{-1} + h_{III}[2]z^{-2}(1+z^2) \\
&\quad + h_{III}[3]z^{-3}(1+z^2+z^4) + \cdots\}
\end{aligned} \tag{6.39}$$

Due to the common factor $(1-z^2)$ in Z-transform, $z = \pm 1$ should be the zero locations. The 1 and –1 position at z domain indicates the lowest and highest frequency in ω domain, respectively; therefore, the LPF and HPF are not possible.

$$\left.\begin{aligned} H(-1) = H(e^{j\omega})|_{\omega=\pi} = 0 \\ H(1) = H(e^{j\omega})|_{\omega=0} = 0 \end{aligned}\right\} \tag{6.40}$$

Type IV: Odd symmetric function with odd filter order (even filter length)

$$\begin{aligned}
h_{IV}[n] &= -h_{IV}[-n-1] \\
h_{IV}[n] &= h_{IV}[0](\delta[n]-\delta[n+1]) + h_{IV}[1](\delta[n-1]-\delta[n+2]) \\
&\quad + h_{IV}[2](\delta[n-2]-\delta[n+3]) + \cdots + h_{IV}[K](\delta[n-K]-\delta[n+(K+1)]) \\
H_{IV}(z) &= \sum\nolimits_{n=-\infty}^{\infty} h_{IV}[n]z^{-n} \\
&= h_{IV}[0](1-z) + h_{IV}[1](z^{-1}-z^2) + h_{IV}[2](z^{-2}-z^3) + \cdots \\
&= h_{IV}[0](1-z) + h_{IV}[1]z^{-1}(1-z^3) + h_{IV}[2]z^{-2}(1-z^5) + \cdots \\
&= h_{IV}[0](1-z) + h_{IV}[1]z^{-1}(1-z)(1+z+z^2) \\
&\quad + h_{IV}[2]z^{-2}(1-z)(1+z+z^2+z^3+z^4) + \cdots \\
&= (1-z)\{h_{IV}[0] + h_{IV}[1]z^{-1}(1+z+z^2) \\
&\quad + h_{IV}[2]z^{-2}(1+z+z^2+z^3+z^4) + \cdots\}
\end{aligned} \tag{6.41}$$

Due to the common factor $(1 - z)$ in Z-transform, $z = 1$ should be the zero location. The 1 position at z domain indicates the lowest frequency in ω domain; therefore, the LPF is not possible. Table 6.8 summarizes the restrictions for each linear phased FIR filter types.

$$H(1) = H(e^{j\omega})|_{\omega=0} = 0 \tag{6.42}$$

Windowing methods for FIR filter

The impulse response of the ideal LPF presents the infinite length in time domain. To realize the FIR filter, the rectangular window is applied over the ideal response. Due to the windowing operation, the LPF frequency response shows the constant level over- and under-shoot around the discontinuity known as Gibb's phenomenon [1]. The inevitable Gibb's phenomenon can be alleviated by choosing the proper window types since the overall frequency response is determined by the convolution integral between the ideal response and window function in frequency domain.

The window function basically emphasizes the impulse response center and underestimates the impulse response edges in time domain. Various types of window are presented in Tables 6.9 and 6.10. Approximate main lobe width represents the radian distance between the first zero crossings around the window center in frequency domain. Relative peak side lobe amplitude demonstrates the ratio between the peak main lobe amplitude and peak side lobe magnitude in decibel scale. The main lobe width corresponds to the frequency transition between the passband and stopband. As shown in Fig. 6.7, the main lobe width denotes the radian distance between the ripple peak in passband and stopband and generally wider than the actual frequency transition. However, the main lobe width is proportional to the frequency transition. The peak side lobe in frequency domain window estimates the approximate ripple magnitude in the frequency response. Exact ripple magnitude can be calculated by the energy in the side lobes based on the distribution.

Table 6.8 Summary of restrictions for linear phased FIR filters

Filter Type	Filter Order N	Filter Length L (Order + 1)	Symmetry of Function (Causal)	Zero Locations $H(z)$	Frequency Response Restriction $H(e^{j\omega})$
Type I	Even	Odd	Even $h[n] = h[N - n]$	No Specific Zeros	No Restriction
Type II	Odd	Even	Even $h[n] = h[N - n]$	$H(-1) = 0$	$H(e^{j\omega})\|_{\omega=\pi} = 0$
Type III	Even	Odd	Odd $h[n] = -h[N - n]$	$H(\pm 1) = 0$	$H(e^{j\omega})\|_{\omega=0} = 0$ $H(e^{j\omega})\|_{\omega=\pi} = 0$
Type IV	Odd	Even	Odd $h[n] = -h[N - n]$	$H(1) = 0$	$H(e^{j\omega})\|_{\omega=0} = 0$

Table 6.9 Conventional windows

Window Type	Defining Equation	Approx. Main Lobe Width (radian)	Relative Peak Side Lobe Amp. (dB)
Rectangular [3]	$w[n] = 1 \; 0 \le n \le N$	$\frac{4\pi}{N+1}$	-13
Bartlett [3]	$w[n] = \begin{cases} \frac{2n}{N} & 0 \le n \le \frac{N}{2} \\ 2 - \frac{2n}{N} & \frac{N}{2} \le n \le N \end{cases}$	$\frac{8\pi}{N}$	-25
Hann [3] (Hanning)	$w[n] = 0.5\left(1 - \cos\left(\frac{2\pi n}{N}\right)\right) \quad 0 \le n \le N$	$\frac{8\pi}{N}$	-31
Hamming [3]	$w[n] = 0.54 - 0.46\cos\left(\frac{2\pi n}{N}\right) \quad 0 \le n \le N$	$\frac{8\pi}{N}$	-41
Blackman [3]	$w[n] = 0.42 - 0.5\cos\left(\frac{2\pi n}{N}\right) + 0.08\cos\left(\frac{4\pi n}{N}\right)$ $0 \le n \le N$	$\frac{12\pi}{N}$	-57
Triangular [3]	$w[n] = \begin{cases} N: \text{odd} = \begin{cases} \frac{2n}{N+1} & 1 \le n \le \frac{N+1}{2} \\ 2 - \frac{2n}{N+1} & \frac{N+1}{2} + 1 \le n \le N \end{cases} \\ N: \text{even} = \begin{cases} \frac{2n-1}{N} & 1 \le n \le \frac{N}{2} \\ 2 - \frac{2n-1}{N} & \frac{N}{2} + 1 \le n \le N \end{cases} \end{cases}$	$\frac{8\pi}{N}$	-25
Kaiser [4]	$w[n] = \frac{I_0\left[\beta\left(1 - \left[\frac{n-N/2}{N/2}\right]^2\right)^{\frac{1}{2}}\right]}{I_0(\beta)} \; 0 \le n \le N$ $I_0(\cdot)$: zeroth-order modified Bessel function of the first kind	$f(N, \beta)$	$f(\beta)$

Along with the conventional windows described in Table 6.9, the MATLAB provides the further windows for various purpose. Certain main lobe width and peak side lobe are illustrated as the function of parameters due to the complexity. However, you can figure out the actual numbers by using the MATLAB Window Designer based on the length and additional parameters. 64-sample length windows are illustrated at Table 6.11 with frequency domain representation.

Below Table 6.12 summarizes window and corresponding MATLAB functions.

FIR filter design example by window method

By using the window method, design the FIR filter for given specification. Following procedure provides the analytical and mathematical approach to realize the designated FIR filter.

Example 6.3

Design the FIR filter for Table 6.13 specification.

Solution

The center of the passband and stopband edge is below. The center frequency is the cutoff frequency which presents half magnitude in the frequency.

$$\omega_c = \omega_p + \frac{\omega_{st} - \omega_p}{2} = \frac{\pi}{4} + \frac{\pi}{8} = \frac{3\pi}{8}$$

Table 6.10 Further windows

Window Type	Defining Equation	~ Main Lobe Width	Peak Side Lobe
Modified Bartlett-Hann [31]	$w[n]=0.62-0.48\left\lvert\left(\frac{n}{N}-0.5\right)\right\rvert+0.38\cos\left(2\pi\left(\frac{n}{N}-0.5\right)\right)$ $0\leq n\leq N$	$f(N)$	–36
Minimum 4-term Blackman-Harris [32]	$w[n]=\begin{cases} a_0-a_1\cos\left(\frac{2\pi n}{N}\right)+a_2\cos\left(\frac{4\pi n}{N}\right)-a_3\cos\left(\frac{6\pi n}{N}\right) & \text{Sym.}\\ a_0-a_1\cos\left(\frac{2\pi n}{N+1}\right)+a_2\cos\left(\frac{4\pi n}{N+1}\right)-a_3\cos\left(\frac{6\pi n}{N+1}\right) & \text{Peri.}\end{cases}$ $0\leq n\leq N$ $a_0=0.35875; a_1=0.48829; a_2=0.14128; a_3=0.01168$	$f(N)$	–92
Bohman [32]	$w(x)=(1-\lvert x\rvert)\cos(\pi\lvert x\rvert)+\frac{1}{\pi}\sin(\pi\lvert x\rvert)-1\leq x\leq 1\ x\in\mathbb{R}$ $w[n]=w(-1+n\Delta x)0\leq n\leq N\quad \Delta x=\frac{2}{N}$	$f(N)$	–46
Chebyshev [33]	Equiripple design method	$f(N,r)$	Define r by User
Flat top weighted [34]	$w[n]=a_0-a_1\cos\left(\frac{2\pi n}{N}\right)+a_2\cos\left(\frac{4\pi n}{N}\right)-a_3\cos\left(\frac{6\pi n}{N}\right)+a_4\cos\left(\frac{8\pi n}{N}\right)$ $0\leq n\leq N$ $a_0=0.21557895; a_1=0.41663158; a_2=0.277263158$ $a_3=0.083578947; a_4=0.006947368$	$f(N)$	$f(N)$
Gaussian [32]	$w[n]=e^{-\frac{1}{2}\left(\frac{\alpha n}{(N-1)/2}\right)^2}\quad -\frac{(N-1)}{2}\leq n\leq\frac{(N-1)}{2}$	$f(N,\alpha)$	$f(N,\alpha)$
Nuttall-defined minimum 4-term Blackman-Harris [35]	$w[n]=\begin{cases} a_0-a_1\cos\left(\frac{2\pi n}{N}\right)+a_2\cos\left(\frac{4\pi n}{N}\right)-a_3\cos\left(\frac{6\pi n}{N}\right) & \text{Sym.}\\ a_0-a_1\cos\left(\frac{2\pi n}{N+1}\right)+a_2\cos\left(\frac{4\pi n}{N+1}\right)-a_3\cos\left(\frac{6\pi n}{N+1}\right) & \text{Peri.}\end{cases}$ $0\leq n\leq N$ $a_0=0.3635819; a_1=0.4891775; a_2=0.1365995; a_3=0.0106411$	$f(N)$	$f(N)$

(continued)

Table 6.10 (continued)

Window Type	Defining Equation	~ Main Lobe Width	Peak Side Lobe
Parzen [32]	$w[n]=\begin{cases}1-6\left(\frac{\lvert n\rvert}{N/2}\right)^2+6\left(\frac{\lvert n\rvert}{N/2}\right)^3 & 0\le\lvert n\rvert\le\frac{N-1}{4}\\ 2\left(1-\frac{\lvert n\rvert}{N/2}\right)^3 & \frac{N-1}{4}<\lvert n\rvert\le\frac{N-1}{2}\end{cases}$ $-\frac{N-1}{2}\le n\le\frac{N-1}{2}$	$f(N)$	–53
Taylor [36]	Smoothed edge version of Chebyshev window at time domain.	$f(N,r,\bar{n})$	Define r by User
Tukey (tapered cosine) [32]	$w(x)=\begin{cases}\frac{1}{2}\left\{1+\cos\left(\frac{2\pi}{r}\left[x-\frac{r}{2}\right]\right)\right\} & 0\le x<\frac{r}{2}\\ 1 & \frac{r}{2}\le x<1-\frac{r}{2}\\ \frac{1}{2}\left\{1+\cos\left(\frac{2\pi}{r}\left[x-1+\frac{r}{2}\right]\right)\right\} & 1-\frac{r}{2}\le x\le 1\end{cases}$ $w[n]=w(n\Delta x)\ 0\le n\le N\quad \Delta x=\frac{1}{N}$	$f(N,r)$	$f(r)$

Table 6.11 Window functions in time and frequency domain (64 length)

Window	Time domain	Frequency domain
Rectangular	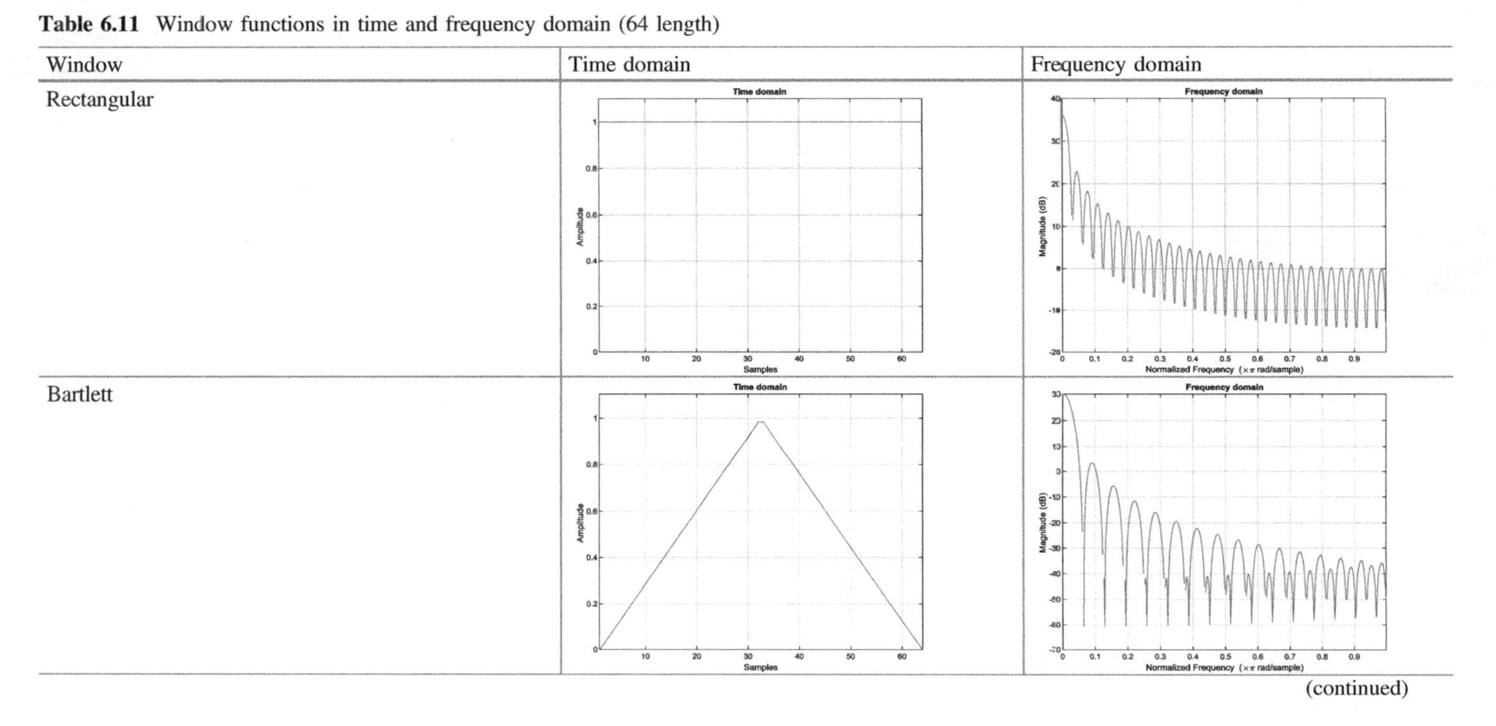	
Bartlett		

(continued)

Table 6.11 (continued)

Window	Time domain	Frequency domain
Hann (Hanning)	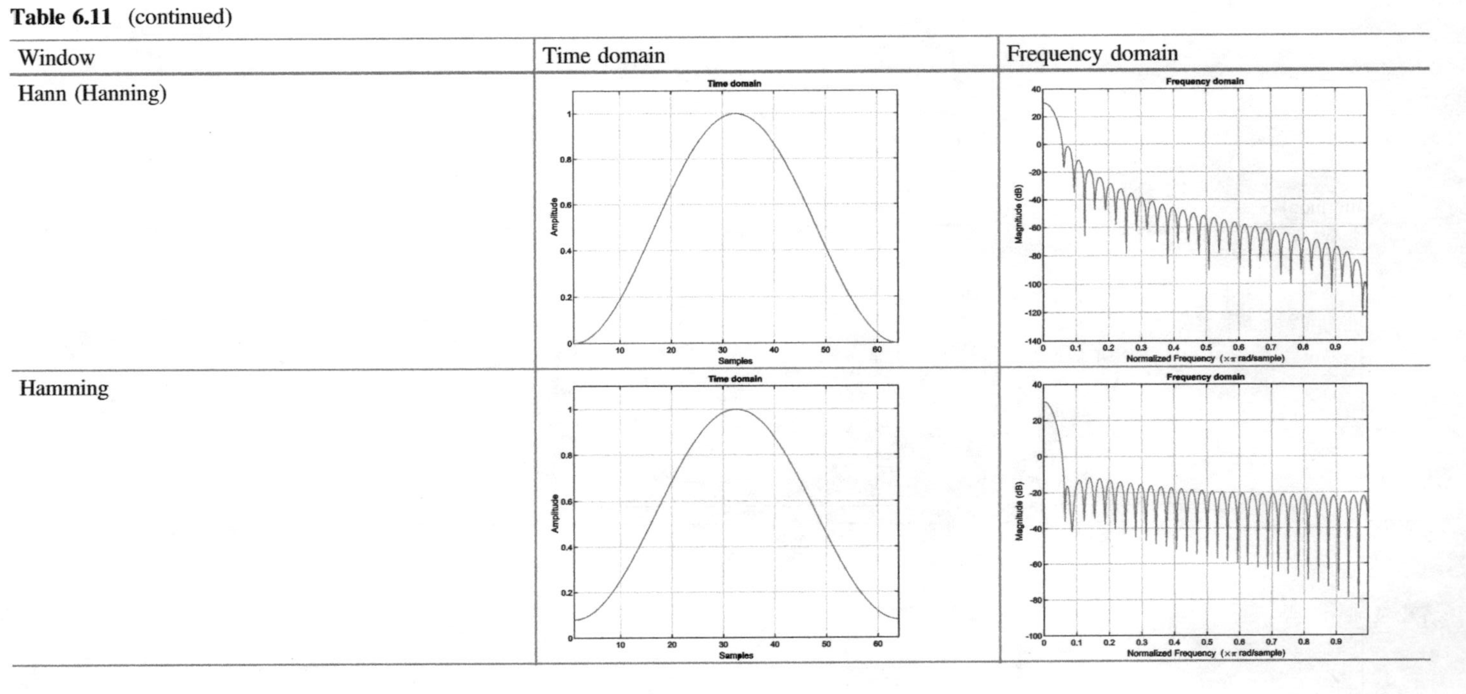	
Hamming		

(continued)

Table 6.11 (continued)

Window	Time domain	Frequency domain
Blackman	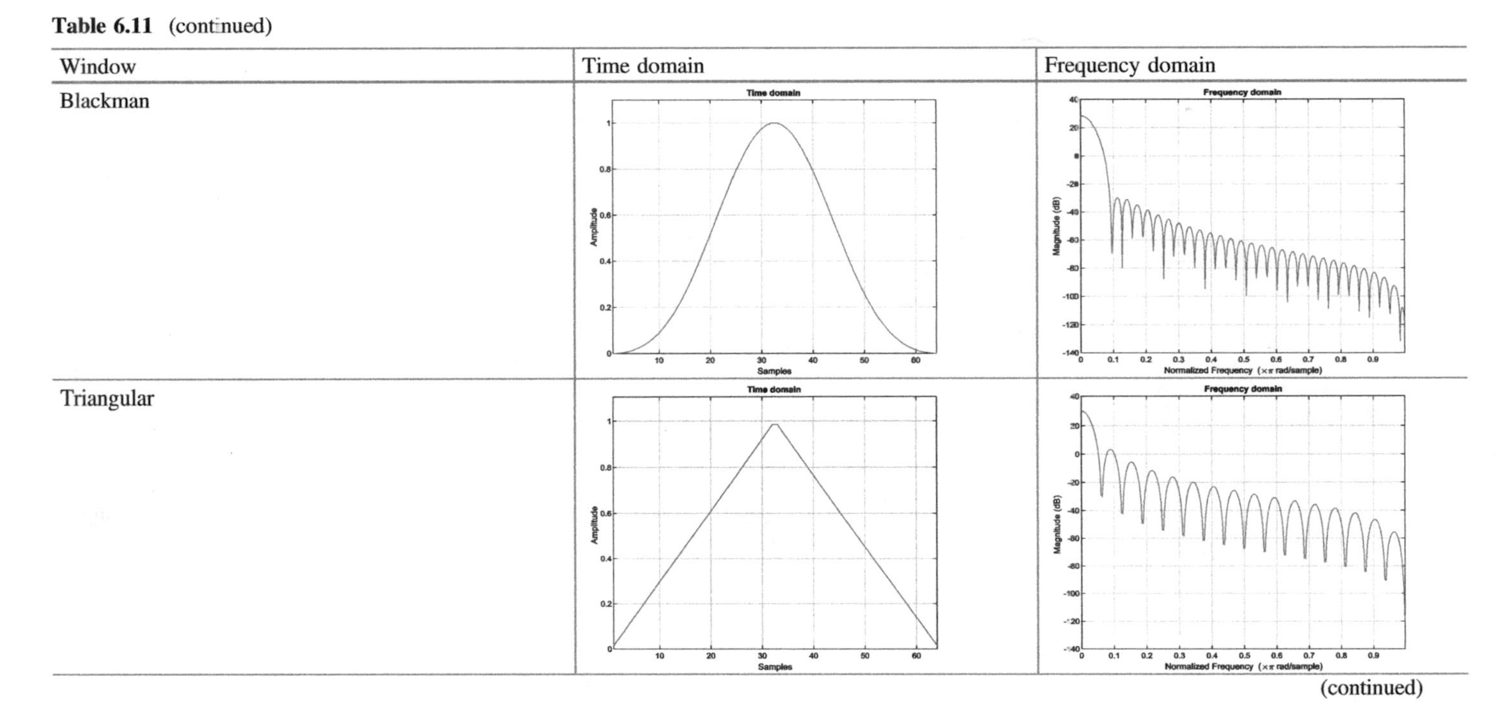	
Triangular		

(continued)

Table 6.11 (continued)

Window	Time domain	Frequency domain
Kaiser β = 0.5	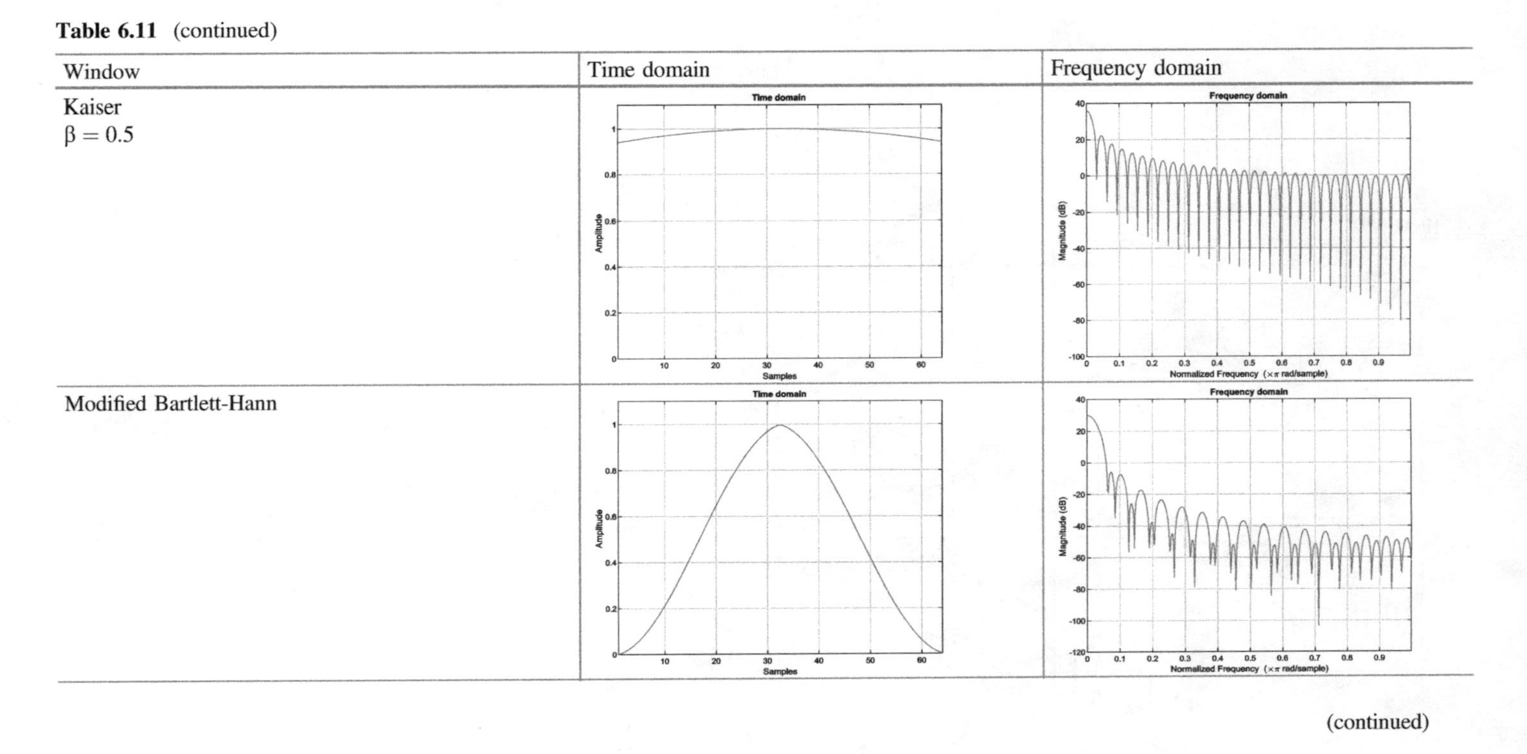	
Modified Bartlett-Hann		

(continued)

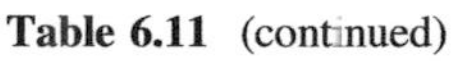

Table 6.11 (continued)

Window	Time domain	Frequency domain
Minimum 4-term Blackman-Harris		
Bohman		

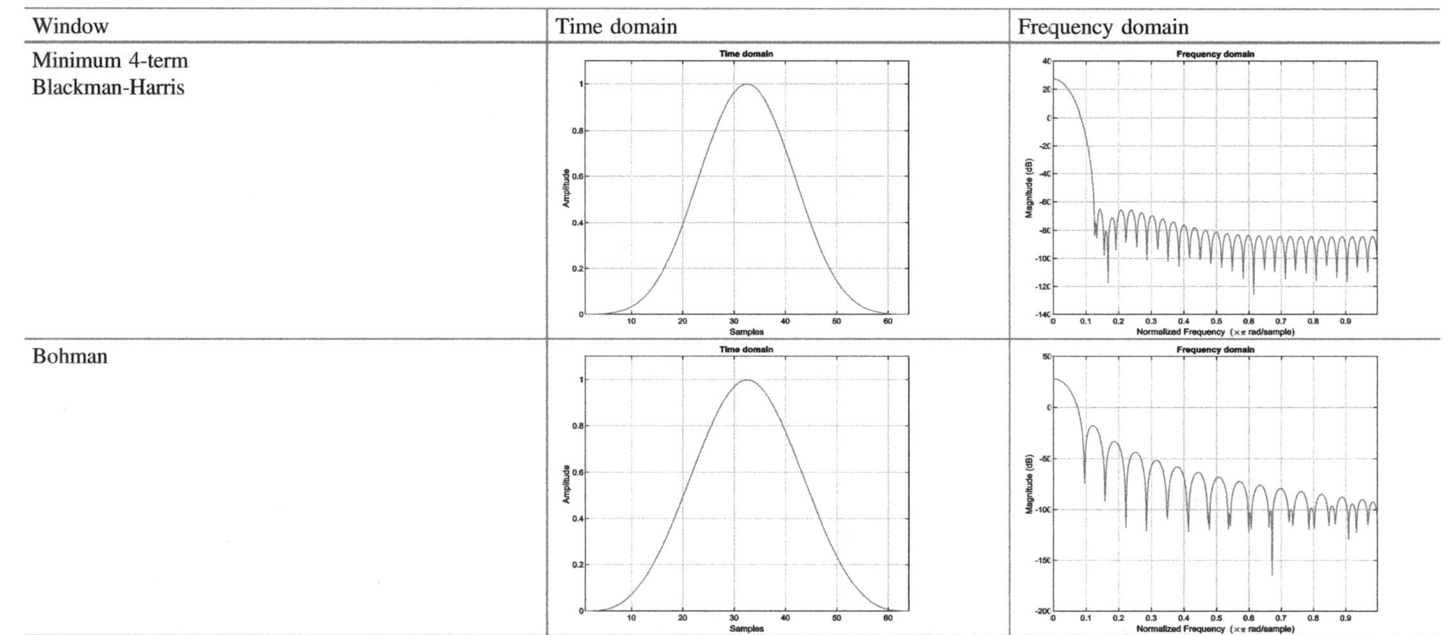

(continued)

Table 6.11 (continued)

Window	Time domain	Frequency domain
$r = 60$	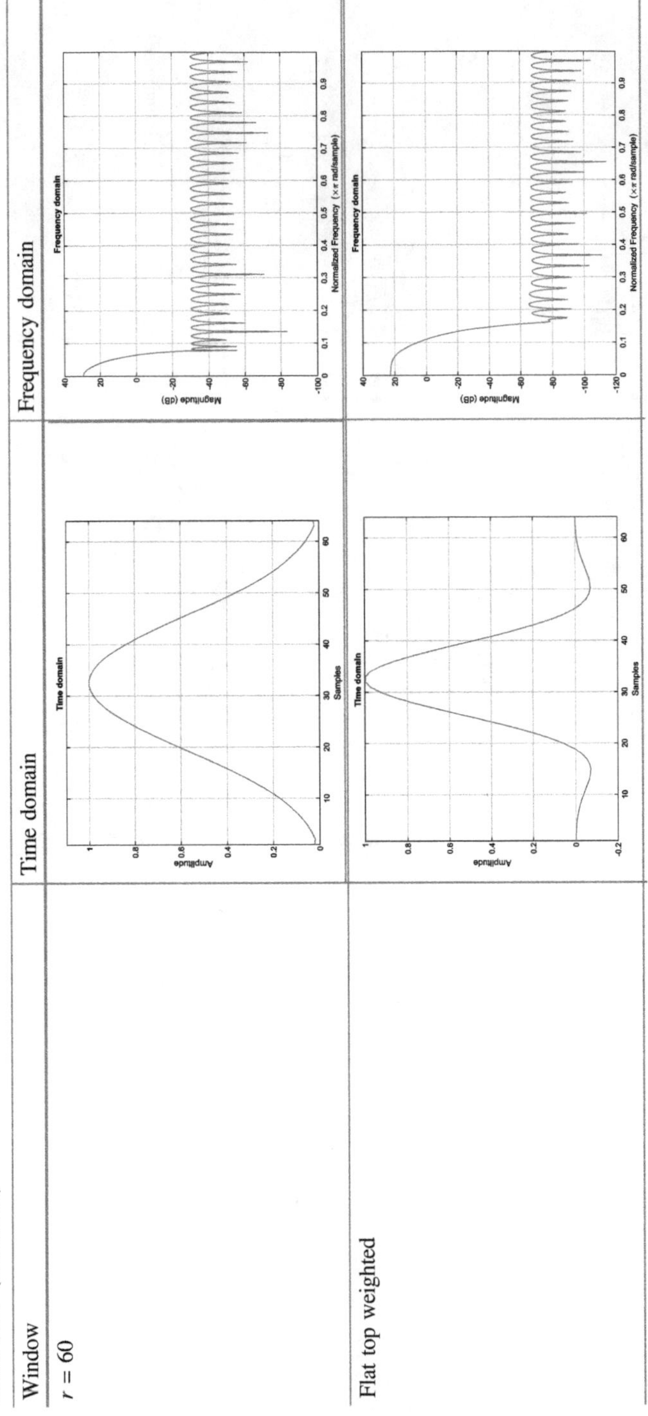	
Flat top weighted		

(continued)

Table 6.11 (continued)

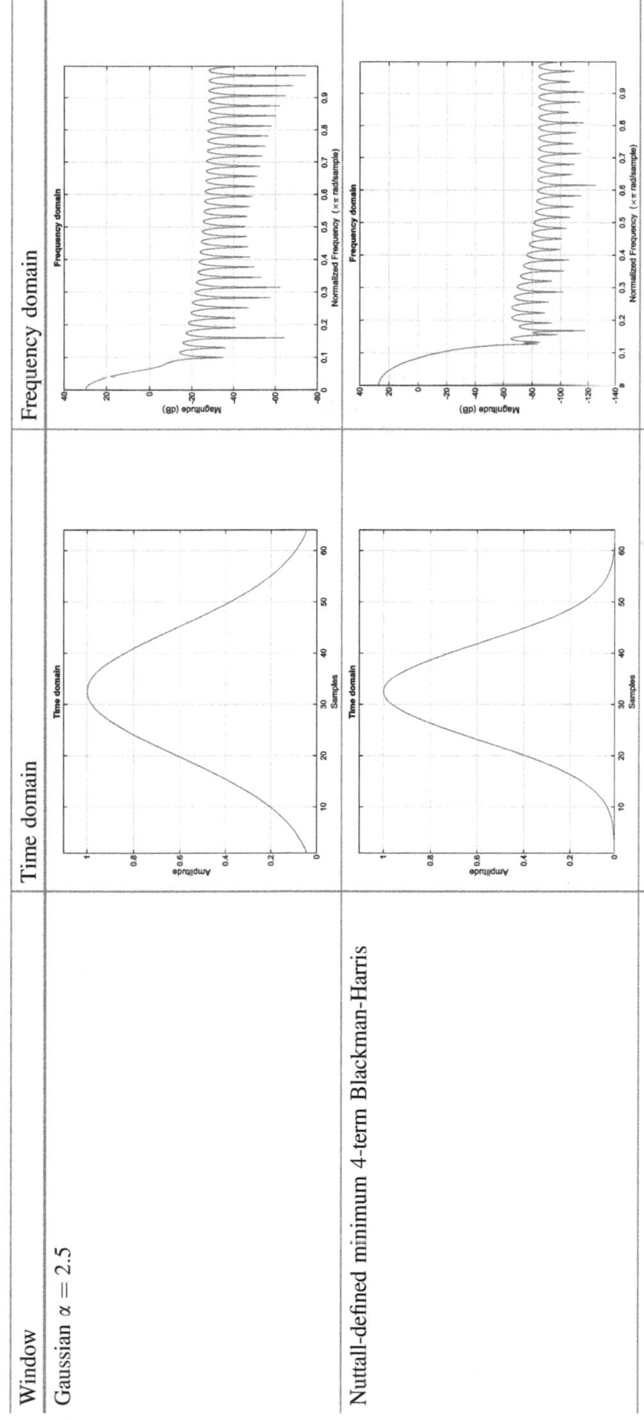

Window	Time domain	Frequency domain
Gaussian $\alpha = 2.5$		
Nuttall-defined minimum 4-term Blackman-Harris		

(continued)

Table 6.11 (continued)

Window	Time domain	Frequency domain
	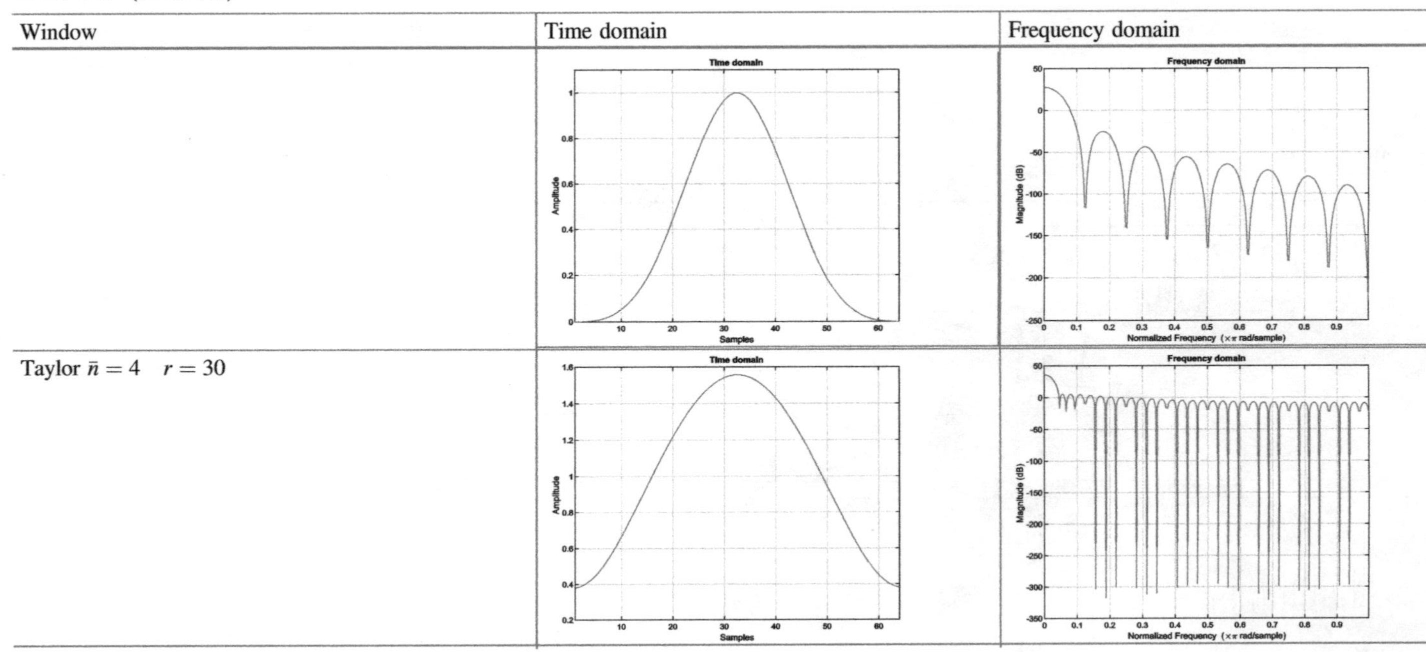	
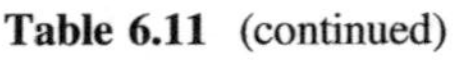 Taylor $\bar{n} = 4 \quad r = 30$		

(continued)

Table 6.11 (continued)

Window	Time domain	Frequency domain
Tukey (tapered cosine) $\alpha = 2.5$	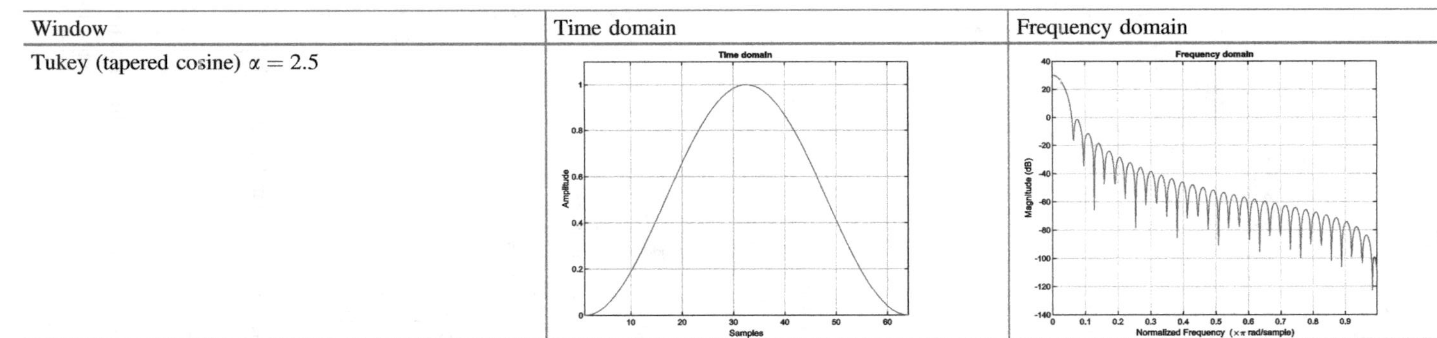	

Table 6.12 Discrete-time window summary and MATLAB functions

MATLAB window functions	Function description
barthannwin	Modified Bartlett-Hann window
bartlett	Bartlett window
blackman	Blackman window
blackmanharris	Minimum four-term Blackman-Harris window
bohmanwin	Bohman window
chebwin	Chebyshev window
enbw	Equivalent noise bandwidth
flattopwin	Flat top weighted window
gausswin	Gaussian window
hamming	Hamming window
hann	Hann (Hanning) window
kaiser	Kaiser window
nuttallwin	Nuttall-defined minimum 4-term Blackman-Harris window
parzenwin	Parzen (de la Vallée Poussin) window
rectwin	Rectangular window
taylorwin	Taylor window
triang	Triangular window
tukeywin	Tukey (tapered cosine) window

Table 6.13 Given filter specification

Filter specifications	Value
Filter type	Low pass filter
Filter realization	FIR with window method
Filter length	N with optimization
Passband frequency (f_p)	1000 Hz; $\omega_p = \pi/4$
Passband magnitude	0 dB
Passband ripple (peak to peak)	1 dB
Stopband frequency (f_{st})	2000 Hz; $\omega_{st} = \pi/2$
Stopband attenuation	40 dB
Sampling frequency (f_s)	8000 Hz

The frequency transition between the passband and stopband edge is below.

$$\Delta\omega = \omega_{st} - \omega_{pa} = \frac{\pi}{2} - \frac{\pi}{4} = \frac{\pi}{4}$$

The main lobe width corresponds to the radian distance between the maximum and minimum ripple; hence, the main lobe width indicates the wider distance than the frequency transition $\Delta\omega$ as shown in Fig. 6.9.

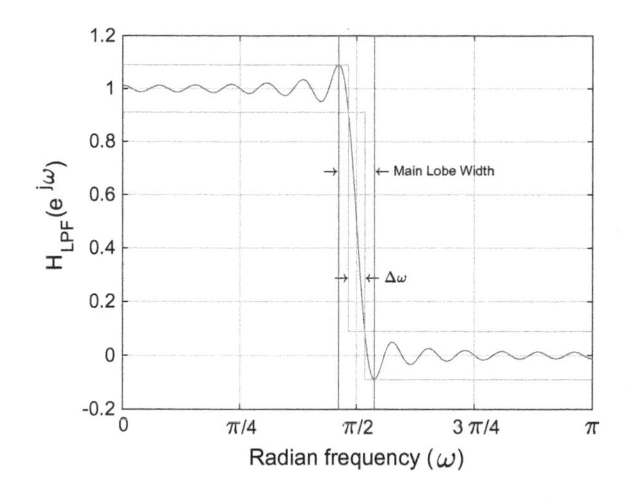

Fig. 6.9 Relationship between main lobe width and Δω. $\omega_{MainLobeWidth} > \Delta\omega$

To meet the specification in wider sense, let the main lobe width considered as frequency transition. Note that the use of the main lobe width designs the filter with sharper transition than the specification. The rectangular window provides below main lobe width in Table 6.9.

$$\frac{4\pi}{N+1} \leq \frac{\pi}{4} N \geq 15$$

We assume that the filter order (*N*) is 16 for Type I filter. The extreme ripples in both bands are identical in terms of deviation from the ideal response magnitude. Also, the deviation is derived from the integral over the one-half side lobe. From Tables 6.9 and 6.10, we can utilize the relative peak side lobe amplitude which is usually higher than the deviation. The rectangular window shows the –13 dB for the relative peak side lobe amplitude and we expect that the filter performance does not meet the requirement. Realization and evaluation are below.

$$h_{Ideal\_LPF}[n] = \frac{1}{2\pi} \int_{-3\pi/8}^{3\pi/8} 1 e^{j\omega n} d\omega = \frac{\sin(3\pi n/8)}{\pi n}$$

$$h_{rect}[n] = \begin{cases} h_{Ideal\_LPF}[n] & \text{for} - 8 \leq n \leq 8 \\ 0 & \text{Otherwise} \end{cases}$$

Prog. 6.7 MATLAB program for Fig. 6.10.

```
N = 16;
NN = 2^13;
N1 = NN/2+1;
N2 = N/2;
nnn = (-N2:1:N2);
b2 = sin(3*pi*(nnn+eps)/8)./(pi*(nnn+eps));
b3 = [b2 zeros(1,NN-length(b2))];
b4 = circshift(b3,-8);
B2 = real(fft(b4,NN));
findex = linspace(0,2*pi,NN);
ff = findex(1:N1);
[M1,I1] = max(B2(1:N1));
[M2,I2] = min(B2(1:N1));

figure,
plot(ff,B2(1:N1)), grid
xlim([0 pi]);
ylim([-0.2 1.2]);
...
straa = sprintf('Max %1.2f dB',40*log10(B2(I1)));
aa = text(ff(I1),B2(I1)+0.005,['      \leftarrow' straa]);
aa.FontSize = 10;
strbb = sprintf('%1.2f dB Min',-20*log10(abs(B2(I2))));
bb = text(ff(I2)-0.7,B2(I2)+0.005,[strbb '\rightarrow    ']);
bb.FontSize = 10;
...
line([ff(I1) ff(I1)],[-0.2 1.2],'Color','magenta');
line([ff(I2) ff(I2)],[-0.2 1.2],'Color','magenta');
hh = text(ff(I1)-0.1,0.7,'\rightarrow ');
hh.FontSize = 10;
ii = text(ff(I2),0.7,'\leftarrow Main Lobe Width');
ii.FontSize = 10;
```

The filter response fails to satisfy the requirement for passband ripple (1 dB) and stopband attenuation (40 dB). Choose the further low side lobe window as hanning. Use the approximation main lobe width of hanning window to determine the filter order (N) as below.

$$\frac{8\pi}{N} \le \frac{\pi}{4} \quad N \ge 32$$

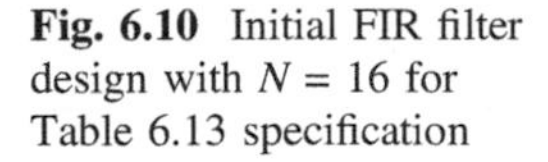
Fig. 6.10 Initial FIR filter design with N = 16 for Table 6.13 specification

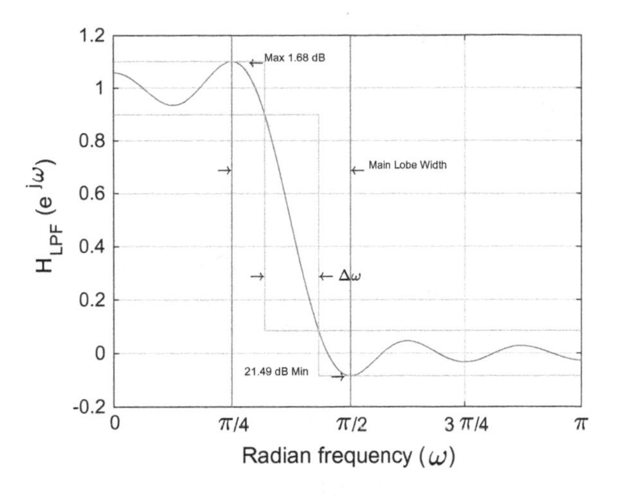

Let the N = 32 for Type I filter. Apply the window as below.

$$h_{hamm.}[n] = \begin{cases} h_{Ideal\_LPF}[n] w_{hamm.}[n] & \text{for } -16 \leq n \leq 16 \\ 0 & \text{Otherwise} \end{cases}$$

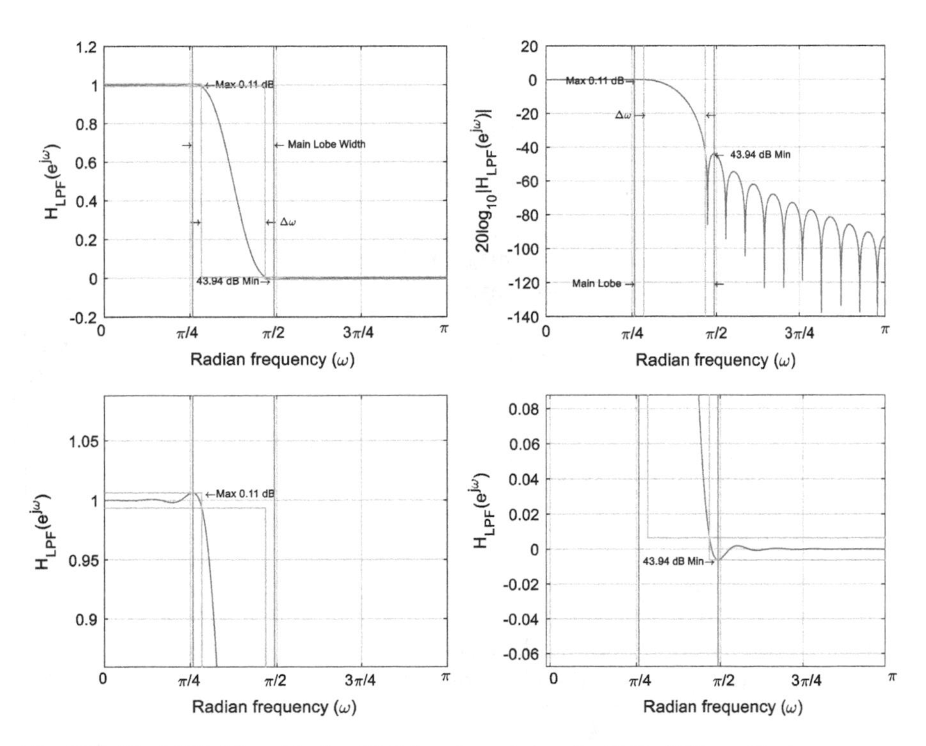

Fig. 6.11 Final FIR filter with N = 32 & hanning window for Table 6.13 specification

Prog. 6.8 MATLAB program for Fig. 6.11.

```
N = 32;
NN = 2^13;
N1 = NN/2+1;
N2 = N/2;
nnn = (-N2:1:N2);
b2 = sin(3*pi*(nnn+eps)/8)./(pi*(nnn+eps));
b2_1 = b2.*(hanning(length(b2)))';
b3 = [b2_1 zeros(1,NN-length(b2_1))];
b4 = circshift(b3,-16);
B2 = real(fft(b4,NN));
findex = linspace(0,2*pi,NN);
ff = findex(1:N1);
[M1,I1] = max(B2(1:N1));
[M2,I2] = min(B2(1:N1));

figure(1)
plot(ff,B2(1:N1)), grid
xlim([0 pi]), ylim([-0.2 1.2]);
straa = sprintf('Max %1.2f dB',40*log10(B2(I1)));
aa = text(ff(I1),B2(I1),['        \leftarrow' straa]);
aa.FontSize = 10;
strbb = sprintf('%1.2f dB Min',-20*log10(abs(B2(I2))));
bb = text(ff(I2)-0.7,B2(I2),[strbb '\rightarrow    ']);
bb.FontSize = 10;
cc = min(find(B2<(2-B2(I1))));
dd = max(find(B2(1:NN/2)>-B2(I2)));
line([ff(I1) ff(I1)],[-0.2 1.2],'Color','magenta');
line([ff(I2) ff(I2)],[-0.2 1.2],'Color','magenta');
figure,
plot(ff,20*log10(abs(B2(1:N1)))), grid
xlim([0 pi]), ylim([-140 20]);
B3 = 20*log10(abs(B2));
straa = sprintf('Max %1.2f dB',40*log10(B2(I1)));
aa = text(ff(I1)-0.63,B3(I1),[ straa '\rightarrow']);
aa.FontSize = 10;
strbb = sprintf('%1.2f dB Min',-20*log10(abs(B2(I2))));
bb = text(ff(I2),B3(I2),['\leftarrow     ' strbb ]);
bb.FontSize = 10;
line([ff(I1) ff(I1)],[-140 20],'Color','magenta');
line([ff(I2) ff(I2)],[-140 20],'Color','magenta');
```

The FIR filter response demonstrates the 0.11 dB maximum passband ripple and 43.94 dB stopband attenuation within filter transition requirement; therefore, the FIR filter is realized properly. The final step is the shifting right to produce the filter causality.

$$h[n] = h_{hamm.}[n - 16]$$

■

The Window Designer from MATLAB provides the various information on the windows. Once the window type, length and parameters are determined, we can observe the time and frequency representation about the specific window. Also following information are presented in Table 6.14.

The described parameters above demonstrate the window information itself. Note that the overall filter response in frequency domain is the convolution integral between the ideal response and window function. Hence, the area under the window function in frequency domain is important. Based on the given values, we cannot predict the exact ripple magnitudes; however, we can roughly estimate the performance by comparing the values between various windows and parameters.

Figure 6.13 illustrates the filter transition from the given 3 dB main lobe width (0.078125; See Fig. 6.12). Note that the main lobe width shows radian distance between the ripple max & min and $\Delta\omega$ is the distance between the locations identified as ω_p and ω_{st}. Usually the $\Delta\omega$ is the transition band based on the filter specification. Once we have passband ripple as 0.11 dB, the stopband ripple can be calculated as below example due to the equal ripple property. The calculations are performed based on the Eqs. (6.17) and (6.18) shown again in below.

Table 6.14 MATLAB Window Designer parameters

Parameter	Explanation	Unit
Leakage factor	Power ratio in the sidelobes to the total window power	Percent
Relative sidelobe attenuation	Height difference between the main lobe peak to the highest side lobe peak	Decibel
Main lobe width (-3 dB)	Main lobe width at 3 dB below ($\approx$0.708) the peak	Normalized Frequency

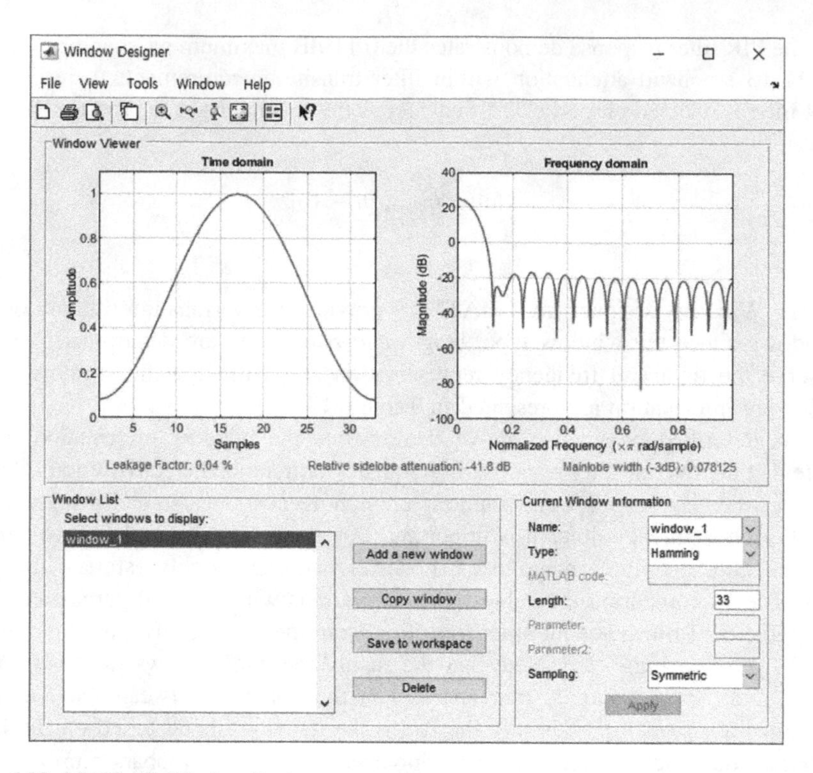

Fig. 6.12 MATLAB Window Designer

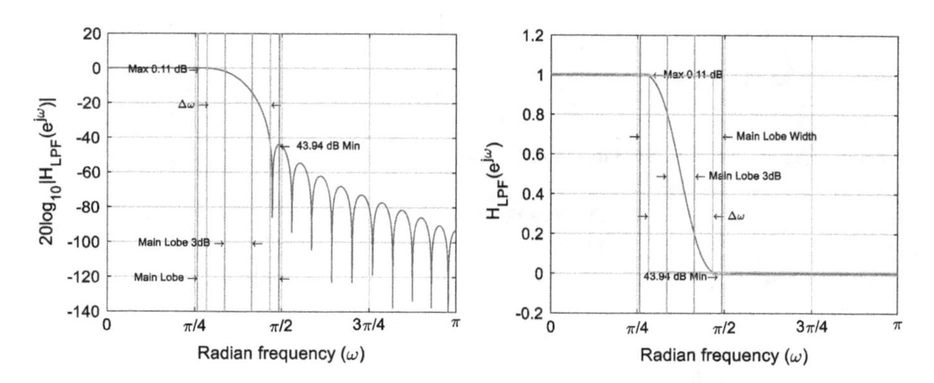

Fig. 6.13 FIR filter (N = 32 & hanning window) for Table 6.13 specification with 3 dB main lobe

Prog. 6.9 MATLAB program for Fig. 6.13.

```
N = 32;
NN = 2^13;
N1 = NN/2+1;
N2 = N/2;
nnn = (-N2:1:N2);
mlwT = 0.078125;    % From window designer hamming L=33
mlw1 = 3*pi/8+mlwT*pi/2;
mlw0 = 3*pi/8-mlwT*pi/2;
b2 = sin(3*pi*(nnn+eps)/8)./(pi*(nnn+eps));
b2_1 = b2.*(hanning(length(b2)))';
b3 = [b2_1 zeros(1,NN-length(b2_1))];
b4 = circshift(b3,-16);
B2 = real(fft(b4,NN));
findex = linspace(0,2*pi,NN);
ff = findex(1:N1);
[M1,I1] = max(B2(1:N1));
[M2,I2] = min(B2(1:N1));

figure,
plot(ff,B2(1:N1)), grid, xlim([0 pi]), ylim([-0.2 1.2]);
line([mlw0 mlw0],[-0.2 1.2],'Color','green');
line([mlw1 mlw1],[-0.2 1.2],'Color','green');
hh = text(mlw0-0.1,0.5,'\rightarrow ');
ii = text(mlw1,0.5,'\leftarrow Main Lobe 3dB');
line([ff(I1) ff(I1)],[-0.2 1.2],'Color','magenta');
line([ff(I2) ff(I2)],[-0.2 1.2],'Color','magenta');
hh = text(ff(I1)-0.1,0.7,'\rightarrow ');
ii = text(ff(I2),0.7,'\leftarrow Main Lobe Width');
figure,
plot(ff,20*log10(abs(B2(1:N1)))), grid, xlim([0 pi]), ylim([-140 20]);
line([mlw0 mlw0],[-140 20],'Color','green');
line([mlw1 mlw1],[-140 20],'Color','green');
hh = text(mlw0-0.78,-100,'Main Lobe 3dB \rightarrow ');
ii = text(mlw1,-100,'\leftarrow');
line([ff(I1) ff(I1)],[-140 20],'Color','magenta');
line([ff(I2) ff(I2)],[-140  20],'Color','magenta');
hh = text(ff(I1)-0.57,-120,'Main Lobe \rightarrow ');
ii = text(ff(I2),-120,'\leftarrow');
```

Example 6.4
Calculate the passband ripple and stopband attenuation from Eqs. (6.17) and (6.18) to verify the equal ripple property. Also compute the parameters in absolute magnitude scale.

Solution

$$43.94\text{dB} = -20\log_{10}\left(10^{\frac{0.11}{40}} - 1\right);\quad A_{st} = -20\log_{10}\left(10^{\frac{A_p}{40}} - 1\right)$$

$$0.11\text{dB} = 40\log_{10}\left(10^{\frac{-43.94}{20}} + 1\right);\quad A_p = 40\log_{10}\left(10^{\frac{-A_{st}}{20}} + 1\right)$$

Compute the ripple magnitude.

$$0.0064 = 10^{\frac{0.11}{40}} - 1;\quad R_p = 10^{\frac{A_p}{40}} - 1$$

$$0.0064 = 10^{\frac{-43.94}{20}};\quad R_{st} = 10^{\frac{-A_{st}}{20}}$$

Find the frequency locations around the cutoff frequency to obtain the $\Delta\omega$.

$$\Delta\omega = \omega_{st} - \omega_p = H^{-1}(0.0064) - H^{-1}(1 - 0.0064) \text{ at first } \omega \text{ around } \omega_c$$

■

Kaiser window method

The Kaiser window [4] is flexible window by changing the order N as well as β value in the algorithm. The experimentally derived equation for estimating the parameters are given as Eqs. (6.43) and (6.44).

$$\Delta\omega = \omega_{st} - \omega_p$$

$$A_{st} = -20\log_{10} R_{st}$$

$$\beta = \begin{cases} 0.1102(A_{st} - 8.7) & A_{st} > 50 \\ 0.5842(A_{st} - 21)^{0.4} + 0.07886(A_{st} - 21) & 21 \le A_{st} \le 50 \\ 0 & A_{st} < 21 \end{cases} \tag{6.43}$$

$$N = \frac{A_{st} - 7.95}{2.285\Delta\omega} \tag{6.44}$$

Table 6.15 Given filter specification

Filter specifications	Value
Filter type	Low pass filter
Filter realization	FIR with Kaiser window
Filter length	N with optimization
Passband frequency (f_p)	1000 Hz; $\omega_p = \pi/4$
Passband magnitude	0 dB
Passband ripple (peak to peak)	1 dB
Stopband frequency (f_{st})	2000 Hz; $\omega_{st} = \pi/2$
Stopband attenuation	40 dB
Sampling frequency (f_s)	8000 Hz

Example 6.5

Design the FIR filter for Table 6.15 specification.

Solution

The frequency transition is calculated as below.

$$\Delta\omega = \omega_{st} - \omega_p = \frac{\pi}{2} - \frac{\pi}{4} = \frac{\pi}{4}$$

The given stopband attenuation and passband ripple are below.

$$A_{st} = 40\text{dB}; \quad A_p = 1\text{dB}$$

Since the linear phase FIR filter presents the equal maximum ripples in both bands, the A_{st} and A_p are mutually related as below.

$$24.55\text{dB} = -20\log_{10}\left(10^{\frac{1}{40}} - 1\right); \quad A_{st} = -20\log_{10}\left(10^{\frac{A_p}{40}} - 1\right)$$

$$0.17\text{dB} = 40\log_{10}\left(10^{\frac{-40}{20}} + 1\right); \quad A_p = 40\log_{10}\left(10^{\frac{-A_{st}}{20}} + 1\right)$$

The $A_p(1\text{dB})$ provides the $24.55\text{dB}(A_{st})$ and the $A_{st}(40\text{dB})$ derives the $0.17\text{dB}\left(A_p\right)$. The lower A_p and higher A_{st} are better in filter performance; therefore, we choose the $40\text{dB}\,A_{st}$ and $0.17\text{dB}\,A_p$ in order to meet the filter requirement. Calculate β and N based on Eqs. (6.43) and (6.44).

$$\beta = 0.5842(A_{st} - 21)^{0.4} + 0.07886(A_{st} - 21) \quad 21 \le A_{st} \le 50$$

$$3.3953 = 0.5842(40 - 21)^{0.4} + 0.07886(40 - 21)$$

$$N = \frac{A_{st} - 7.95}{2.285\Delta\omega}$$

$$17.8588 = \frac{40 - 7.95}{2.285\pi/4}$$

Select filter order N as 18 for type I FIR filter. The center of the passband and stopband edge is below as cutoff frequency.

$$\omega_c = \omega_p + \frac{\omega_{st} - \omega_p}{2} = \frac{\pi}{4} + \frac{\pi}{8} = \frac{3\pi}{8}$$

The impulse response of the ideal LPF is below.

$$h_{Ideal\_LPF}[n] = \frac{1}{2\pi}\int_{-3\pi/8}^{3\pi/8} 1e^{j\omega n} d\omega = \frac{\sin(3\pi n/8)}{\pi n}$$

Apply the window as below.

$$h_{Kaiser}[n] = \begin{cases} h_{Ideal\_LPF}[n] w_{Kaiser}[n] & \text{for } -9 \le n \le 9 \\ 0 & \text{Otherwise} \end{cases}$$

The computed frequency response is shown in Fig. 6.14.

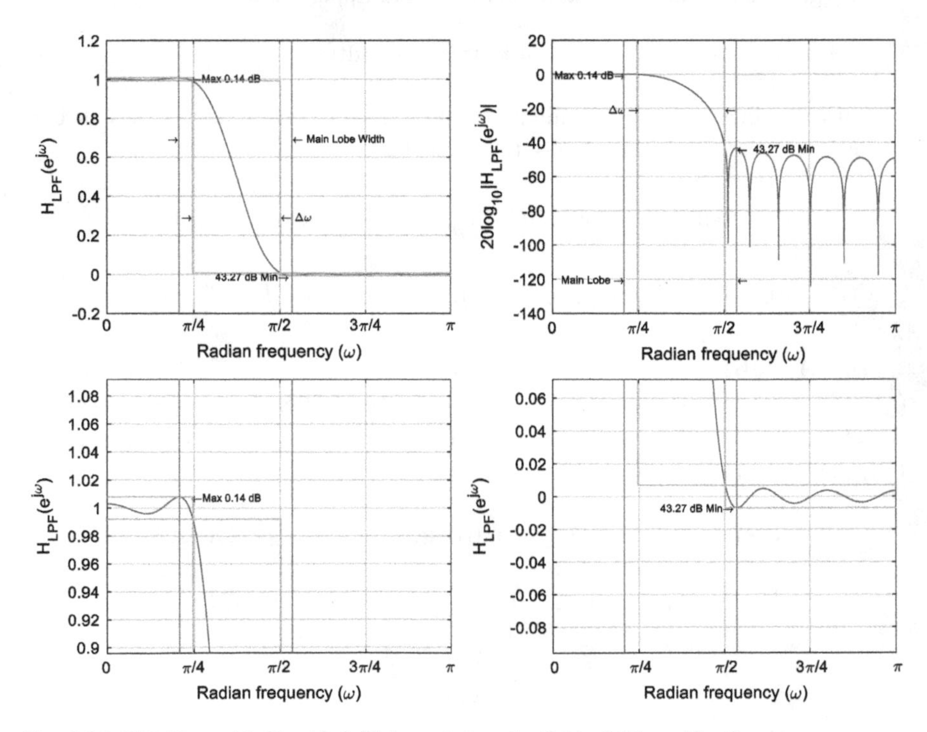

Fig. 6.14 FIR filter with N = 18 & Kaiser window for Table 6.15 specification

Prog. 6.10 MATLAB program for Fig. 6.14.

```
N = 18;
NN = 2^13;
N1 = NN/2+1;
N2 = N/2;
nnn = (-N2:1:N2);
beta = 3.3953;
b2 = sin(3*pi*(nnn+eps)/8)./(pi*(nnn+eps));
b2_1 = b2.*(kaiser(length(b2),beta))';
b3 = [b2_1 zeros(1,NN-length(b2_1))];
b4 = circshift(b3,-N/2);
B2 = real(fft(b4,NN));
findex = linspace(0,2*pi,NN);
ff = findex(1:N1);
[M1,I1] = max(B2(1:N1));
[M2,I2] = min(B2(1:N1));

figure,
plot(ff,B2(1:N1)), grid, xlim([0 pi]), ylim([-0.2 1.2]);
straa = sprintf('Max %1.2f dB',40*log10(B2(I1)));
aa = text(ff(I1),B2(I1),['        \leftarrow' straa]);
aa.FontSize = 10;
strbb = sprintf('%1.2f dB Min',-20*log10(abs(B2(I2))));
bb = text(ff(I2)-0.7,B2(I2),[strbb '\rightarrow    ']);
bb.FontSize = 10;
line([ff(I1) ff(I1)],[-0.2 1.2],'Color','magenta');
line([ff(I2) ff(I2)],[-0.2 1.2],'Color','magenta');
hh = text(ff(I1)-0.1,0.7,'\rightarrow ');
ii = text(ff(I2),0.7,'\leftarrow Main Lobe Width');
figure,
plot(ff,20*log10(abs(B2(1:N1)))), grid, xlim([0 pi]), ylim([-140 20]);
B3 = 20*log10(abs(B2));
straa = sprintf('Max %1.2f dB',40*log10(B2(I1)));
aa = text(ff(I1)-0.63,B3(I1),[ straa '\rightarrow']);
strbb = sprintf('%1.2f dB Min',-20*log10(abs(B2(I2))));
bb = text(ff(I2),B3(I2),['\leftarrow    ' strbb ]);
line([ff(I1) ff(I1)],[-140 20],'Color','magenta');
line([ff(I2) ff(I2)],[-140 20],'Color','magenta');
hh = text(ff(I1)-0.57,-120,'Main Lobe \rightarrow ');
ii = text(ff(I2),-120,'\leftarrow');
```

Calculate the ripple magnitude.

$$0.0081 = 10^{\frac{0.14}{40}} - 1; \quad R_p = 10^{\frac{A_p}{40}} - 1$$

$$0.0069 = 10^{\frac{-43.27}{20}}; \quad R_{st} = 10^{\frac{-A_{st}}{20}}$$

Find the frequency locations around the cutoff frequency to obtain the $\omega_{st}$and ω_p .

$$\omega_{st} = H^{-1}(0.0069) \approx \frac{\pi}{2}; \quad \omega_p = H^{-1}(1 - 0.0081) \approx \frac{\pi}{4}$$

$$\Delta\omega = H^{-1}(0.0069) - H^{-1}(1 - 0.0081) \approx \frac{\pi}{4}$$

The computed $\omega_{st}$and ω_p are illustrated as the vertical yellow lines in Fig. 6.14. All specifications shown in Table 6.15 are satisfied with the filter design. The final step is the shifting right to produce the filter causality.

$$h[n] = h_{Kaiser}[n - 9]$$

■

Type II FIR filter

The Type I linear phase FIR filter is realized based on the real value in frequency domain since the even symmetry function produces the frequency response in real values. For example, the Type I filter before the causality shift is demonstrated below for frequency response.

$$H_I\left(e^{j\omega}\right) = \sum\nolimits_{n=-K}^{K} h_I[n]e^{-j\omega n} = h_I[0] + 2\sum\nolimits_{n=1}^{K} h_I[n]\cos(\omega n) \in \mathbb{R}$$

The Type II is the modified version of the Type I. The modification is made to locate the symmetric center at the between the samples; therefore, the delay component $e^{j\omega/2}$ is shown in the frequency domain as below. Except the $e^{j\omega/2}$, the Type II is pure real response in frequency domain.

$$H_{II}\left(e^{j\omega}\right) = 2e^{j\frac{\omega}{2}}\sum\nolimits_{n=0}^{K} h_{II}[n]\cos\left(\frac{\omega}{2}(2n+1)\right)$$

From the Type II filter, the inverse DTFT provides the impulse response as below

$$
\begin{aligned}
h[n] &= \frac{1}{2\pi}\int_{-\pi}^{\pi} H_{II}\left(e^{j\omega}\right)e^{j\omega n}d\omega \\
&= \frac{1}{2\pi}\int_{-\pi}^{\pi} 2e^{j\frac{\omega}{2}}\sum\nolimits_{m=0}^{K} h_{II}[m]\cos\left(\frac{\omega}{2}(2m+1)\right)e^{j\omega n}d\omega \\
&= \frac{1}{2\pi}\int_{-\pi}^{\pi} 2e^{j\frac{\omega}{2}}\sum\nolimits_{m=0}^{K} h_{II}[m]\frac{e^{-j\omega\frac{2m+1}{2}}+e^{j\omega\frac{2m+1}{2}}}{2}e^{j\omega n}d\omega \\
&= \sum\nolimits_{m=0}^{K} h_{II}[m]\frac{1}{2\pi}\int_{-\pi}^{\pi}\left(e^{j\frac{\omega}{2}}e^{-j\omega\frac{2m+1}{2}}+e^{j\frac{\omega}{2}}e^{j\omega\frac{2m+1}{2}}\right)e^{j\omega n}d\omega \\
&= \sum\nolimits_{m=0}^{K} h_{II}[m]\frac{1}{2\pi}\int_{-\pi}^{\pi}\left(e^{-j\omega m}+e^{j\omega(m+1)}\right)e^{j\omega n}d\omega \\
&= \sum\nolimits_{m=0}^{K} h_{II}[m]\frac{1}{2\pi}\int_{-\pi}^{\pi} e^{-j\omega m}e^{j\omega n}d\omega \\
&\quad+\sum\nolimits_{l=-K-1}^{0} h_{II}[-l-1]\frac{1}{2\pi}\int_{-\pi}^{\pi} e^{-j\omega l}e^{j\omega n}d\omega \quad \text{let } m+1=-l \\
&= \sum\nolimits_{m=0}^{K} h_{II}[m]\frac{1}{2\pi}\int_{-\pi}^{\pi} e^{-j\omega(m-n)}d\omega \\
&\quad+\sum\nolimits_{l=-K-1}^{0} h_{II}[-l-1]\frac{1}{2\pi}\int_{-\pi}^{\pi} e^{-j\omega(l-n)}d\omega \\
&= \sum\nolimits_{m=0}^{K} h_{II}[m]\delta[m-n]+\sum\nolimits_{l=-K-1}^{0} h_{II}[-l-1]\delta[l-n] \\
&= h_{II}[n]\{u[n]-u[n-(K+1)]\} \\
&\quad+h_{II}[-n-1]\{u[n+(K+1)]-u[n]\}
\end{aligned}
\tag{6.45}
$$

where $u[n]$ is unit-step function. The derived $h[n]$ in Eq. (6.45) shows the even symmetricity over the –1/2 sample location. The $h_{II}[n]$ and $h_{II}[-n-1]$ are concatenated mutually for even symmetricity. To implement the Type II filter, Fourier transform is performed as below. Note that the $H_{II}(e^{j\omega})$ is multiplied with $e^{-j\omega/2}$ to remove the –1/2 center location. For convenience purpose, the ω is used for Ω in Fourier transform.

$$
\begin{aligned}
h_{1/2}(t) &= \frac{1}{2\pi}\int_{-\infty}^{\infty} e^{-j\frac{\omega}{2}} H_{II}(\omega) e^{j\omega t} d\omega \\
&= \frac{1}{2\pi}\int_{-\infty}^{\infty} 2\sum\nolimits_{m=0}^{K} h_{II}[m]\cos\left(\frac{\omega}{2}(2m+1)\right) e^{j\omega t} d\omega \\
&= \frac{1}{2\pi}\int_{-\infty}^{\infty} 2\sum\nolimits_{m=0}^{K} h_{II}[m]\frac{e^{-j\omega\frac{2m+1}{2}} + e^{j\omega\frac{2m+1}{2}}}{2} e^{j\omega t} d\omega \\
&= \frac{1}{2\pi}\int_{-\infty}^{\infty} \sum\nolimits_{m=0}^{K} h_{II}[m] e^{-j\omega\frac{2m+1}{2}} e^{j\omega t} d\omega \\
&\quad + \frac{1}{2\pi}\int_{-\infty}^{\infty} \sum\nolimits_{m=0}^{K} h_{II}[m] e^{j\omega\frac{2m+1}{2}} e^{j\omega t} d\omega \\
&= \frac{1}{2\pi}\int_{-\infty}^{\infty} \sum\nolimits_{m=0}^{K} h_{II}[m] e^{-j\omega\frac{2m+1}{2}} e^{j\omega t} d\omega \\
&\quad + \frac{1}{2\pi}\int_{-\infty}^{\infty} \sum\nolimits_{l=-K}^{0} h_{II}[-l] e^{-j\omega\frac{2l-1}{2}} e^{j\omega t} d\omega \\
&= \sum\nolimits_{m=0}^{K} h_{II}[m]\frac{1}{2\pi}\int_{-\infty}^{\infty} e^{-j\omega\left(\frac{2m+1}{2}-t\right)} d\omega \\
&\quad + \sum\nolimits_{l=-K}^{0} h_{II}[-l]\frac{1}{2\pi}\int_{-\infty}^{\infty} e^{-j\omega\left(\frac{2l-1}{2}-t\right)} d\omega \\
&= \sum\nolimits_{m=0}^{K} h_{II}[m]\delta\left(t-\frac{2m+1}{2}\right) + \sum\nolimits_{l=-K}^{0} h_{II}[-l]\delta\left(t-\frac{2l-1}{2}\right) \\
&= h_{II}[K]\delta\left(t+\frac{2K+1}{2}\right) + \cdots + h_{II}[1]\delta\left(t+\frac{3}{2}\right) + h_{II}[0]\delta\left(t+\frac{1}{2}\right) \\
&\quad + h_{II}[0]\delta\left(t-\frac{1}{2}\right) + h_{II}[1]\delta\left(t-\frac{3}{2}\right) + \cdots + h_{II}[K]\delta\left(t-\frac{2K+1}{2}\right)
\end{aligned}
\tag{6.46}
$$

The derived $h_{1/2}(t)$ in Eq. (6.46) shows the even symmetricity over the coordinate origin. To realize the filter, compute the impulse response from the desired frequency response as below.

$$
h(n) = \frac{1}{2\pi}\int_{-\pi}^{\pi} H_{desired}\left(e^{j\omega}\right) e^{j\omega n} d\omega \quad \text{for } H_{desired}\left(e^{j\omega}\right) \in \mathbb{R}
$$

Even symmetricity $h[n] = h[-n]$ is implemented by following conversion for Type II as below.

$$h_{II}[n] = h\left(n + \frac{1}{2}\right) \quad \text{for } n \geq 0$$
$$h_{II}[n] = h_{II}[-n] \tag{6.47}$$

Parseval's theorem is applied for verifying the energy between both domains as Eq. (6.48).

$$\sum_{n=-\infty}^{\infty} |h_{II}[n]|^2 \approx \frac{1}{2\pi} \int_{-\pi}^{\pi} |H_{desired}(e^{j\omega})|^2 d\omega \tag{6.48}$$

Example 6.6
Design the FIR filter for Table 6.16 specification.

Solution
The cutoff frequency and frequency transition are calculated as below.

$$\omega_c = \omega_p + \frac{\omega_{st} - \omega_p}{2} = \frac{\pi}{4} + \frac{\pi}{8} = \frac{3\pi}{8}$$

$$\Delta\omega = \omega_{st} - \omega_p = \frac{\pi}{2} - \frac{\pi}{4} = \frac{\pi}{4}$$

Based on the rectangular window, main lobe width is below.

$$\frac{4\pi}{N+1} \leq \frac{\pi}{4} \; N \geq 15$$

Choose $N = 15$ for Type II filter.

Table 6.16 Given filter specification

Filter specifications	Value
Filter type	Low pass filter
Filter realization	FIR Type II without window method
Filter length	N with optimization
Passband frequency (f_p)	1000 Hz; $\omega_p = \pi/4$
Passband magnitude	0 dB
Stopband frequency (f_{st})	2000 Hz; $\omega_{st} = \pi/2$
Sampling frequency (f_s)	8000 Hz

$$h_{Ideal\_LPF}[n] = \frac{1}{2\pi} \int_{-3\pi/8}^{3\pi/8} 1 e^{j\omega n} d\omega = \frac{\sin(3\pi n/8)}{\pi n}$$

$$\text{Let } h_{Ideal\_LPF}(t) = \frac{\sin(3\pi t/8)}{\pi t}$$

$$\begin{aligned} h[n] &= h_{Ideal_{LPF}}\left(\frac{15}{2}\right)\delta[n] + \cdots + h_{Ideal_{LPF}}\left(\frac{3}{2}\right)\delta[n-6] \\ &+ h_{Ideal_{LPF}}\left(\frac{1}{2}\right)\delta[n-7] + h_{Ideal_{LPF}}\left(\frac{1}{2}\right)\delta[n-8] \\ &+ h_{Ideal\_LPF}\left(\frac{3}{2}\right)\delta[n-9] + \cdots + h_{Ideal\_LPF}\left(\frac{15}{2}\right)\delta[n-15] \end{aligned}$$

The performance of the Type II filter is shown below.

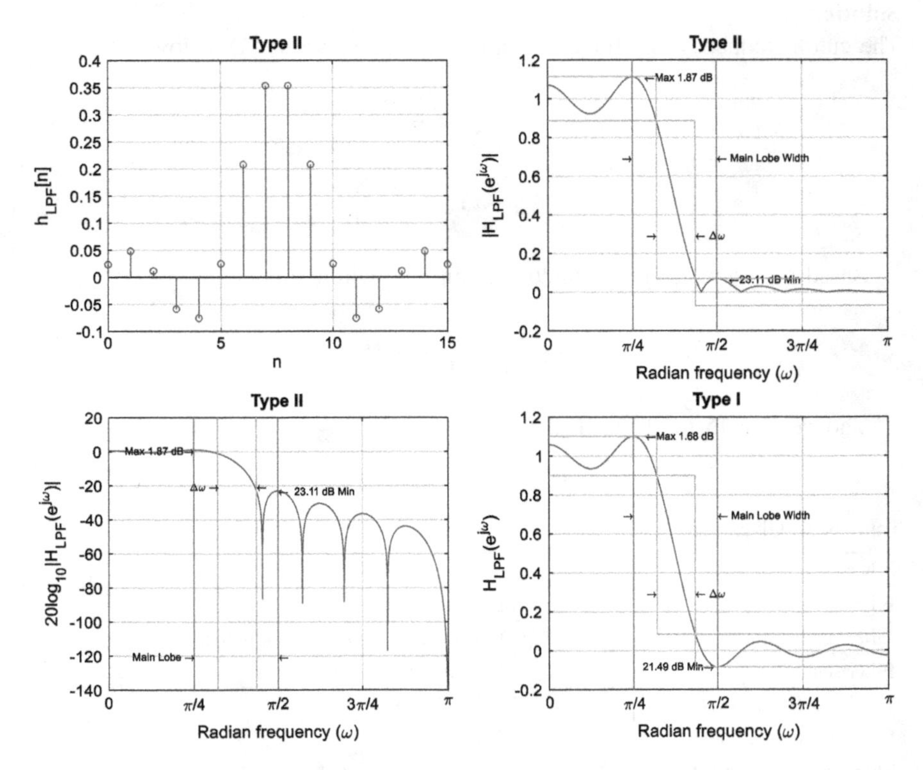

Fig. 6.15 Type II FIR filter with $N = 15$ for Table 6.16 specification. For reference, Type I FIR filter with $N = 16$ is also shown

Prog. 6.11 MATLAB program for Fig. 6.15.

```
N = 15;
NN = 2^13;
N1 = NN/2+1;
findex = linspace(0,2*pi,NN);
ff = findex(1:N1);
t = -15/2:1:15/2;
n = 0:1:(length(t)-1);
ht = sin(3*pi*(t+eps)/8)./(pi*(t+eps));
B2 = abs(fft(ht,NN));
[PKS,LOCS]= findpeaks(B2);
M1 = PKS(1);
I1 = LOCS(1);
M2 = PKS(2);
I2 = LOCS(2);

figure,
stem(n,ht),grid
figure,
plot(ff,B2(1:N1)), grid, xlim([0 pi]), ylim([-0.2 1.2]);
straa = sprintf('Max %1.2f dB',40*log10(B2(I1)));
aa = text(ff(I1),B2(I1),['       \leftarrow' straa]);
strbb = sprintf('%1.2f dB Min',-20*log10(abs(B2(I2))));
bb = text(ff(I2),B2(I2),['       \leftarrow' strbb ]);
line([ff(I1) ff(I1)],[-0.2 1.2],'Color','magenta');
line([ff(I2) ff(I2)],[-0.2 1.2],'Color','magenta');
hh = text(ff(I1)-0.1,0.7,'\rightarrow ');
ii = text(ff(I2),0.7,'\leftarrow Main Lobe Width');
figure,
plot(ff,20*log10(abs(B2(1:N1)))), grid, xlim([0 pi]), ylim([-140 20]);
B3 = 20*log10(abs(B2));
straa = sprintf('Max %1.2f dB',40*log10(B2(I1)));
aa = text(ff(I1)-0.63,B3(I1),[ straa '\rightarrow']);
strbb = sprintf('%1.2f dB Min',-20*log10(abs(B2(I2))));
bb = text(ff(I2),B3(I2),['\leftarrow     ' strbb ]);
line([ff(I1) ff(I1)],[-140 20],'Color','magenta');
line([ff(I2) ff(I2)],[-140 20],'Color','magenta');
hh = text(ff(I1)-0.57,-120,'Main Lobe \rightarrow ');
ii = text(ff(I2),-120,'\leftarrow');
```

For the verification purpose, apply the Parseval's theorem over the designed filter.

$$\frac{1}{2\pi}\int_{-\pi}^{\pi}\left|H\left(e^{j\omega}\right)\right|^2 d\omega = \frac{1}{2\pi}\int_{-3\pi/8}^{3\pi/8} 1^2 d\omega = \frac{1}{2\pi}\frac{3\pi}{4} = \frac{3}{8} = 0.3750$$

$$\sum_{n=0}^{15} |h[n]|^2 = 0.3624 \text{ Energy ratio} = \frac{0.3624}{0.3750} = 0.9664$$

■

Both type filters are illustrated in Fig. 6.16.

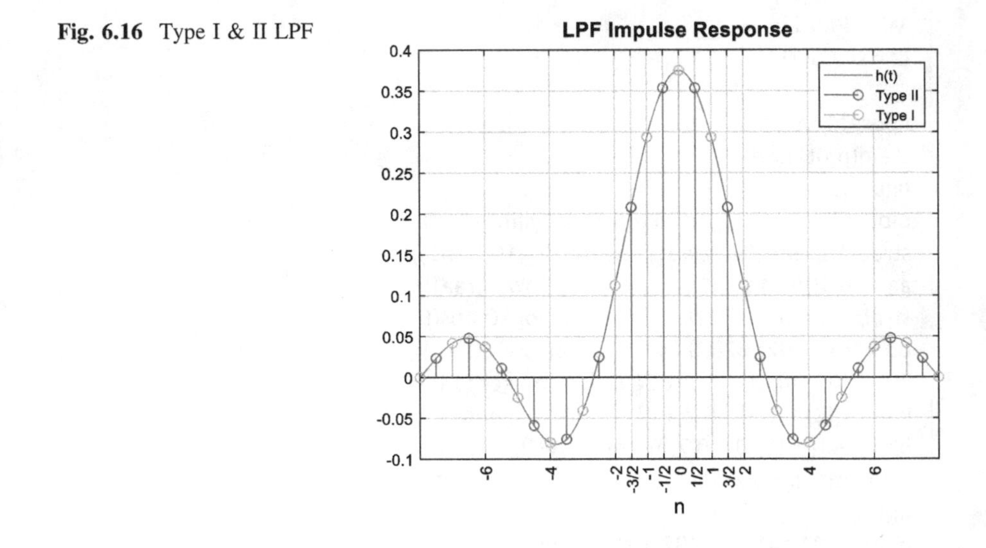

Fig. 6.16 Type I & II LPF

Prog. 6.12 MATLAB program for Fig. 6.16.

```
syms n w;
NN = 2^13;
N = 15;
wc = 3/8;
nn = -N/2:1:N/2;
f1 = linspace(0,2,NN);
nn1 = -(N+1)/2:1:(N+1)/2;
nd = -(N+1)/2:1/100:(N+1)/2;
reff = fir1(N,wc);
aa = exp(1j*w*n);
bb = int(aa,w,-pi*wc,pi*wc)/(2*pi);
cc = simplify(bb);
dd = subs(cc,n,nn);
dd1 = subs(cc,n,nn1+eps);
dd2 = subs(cc,n,nd+eps);
ee = eval(dd);
ee1 = eval(dd1);
ee2 = eval(dd2);
B2 = abs(fft(ee,NN));
B3 = abs(fft(reff,NN));

figure,
plot(nd,ee2), grid, hold on;
stem(nn,ee)
stem(nn1,ee1), hold off;
legend('h(t)','Type II','Type I');
title('LPF Impulse Response')
xlabel('n')
ax2 = gca;
ax2.XTick = [-6 -4 -2 -3/2 -1 -1/2 0 1/2 1 3/2 2 4 6];
ax2.XTickLabel = {'-6','-4','-2','-3/2','-1','-1/2','0','1/2','1','3/2','2','4','6'};
ax2.XTickLabelRotation = 90;
```

According to the Table 6.8, Type II filter cannot be realized as the HPF because of the fixed zero location at –1. For verification purpose, type II HPF example is shown below.

Example 6.7
Design the FIR filter for Table 6.17 specification.

Table 6.17 Given filter specification

Filter specifications	Value
Filter type	High pass filter
Filter realization	FIR Type II without window method
Filter length	N with optimization
Passband frequency (f_p)	1000 Hz; $\omega_p = \pi/2$
Passband magnitude	0 dB
Stopband frequency (f_{st})	2000 Hz; $\omega_{st} = \pi/4$
Sampling frequency (f_s)	8000 Hz

Solution

The center of the passband and stopband edge is below.

$$\omega_c = \omega_{st} + \frac{\omega_p - \omega_{st}}{2} = \frac{\pi}{4} + \frac{\pi}{8} = \frac{3\pi}{8}$$

The frequency transition between the passband and stopband edge is below.

$$\Delta\omega = \omega_p - \omega_{st} = \frac{\pi}{2} - \frac{\pi}{4} = \frac{\pi}{4}$$

The main lobe width for rectangular window presents the frequency transition as previously discussed.

$$\frac{4\pi}{N+1} \leq \frac{\pi}{4} \; N \geq 15$$

Let the N = 15 for Type II filter. Realization and evaluation are below.

$$h_{Ideal_{HPF}}[n] = \frac{1}{2\pi}\int_{-\pi}^{-3\pi/8} 1e^{j\omega n}d\omega + \frac{1}{2\pi}\int_{3\pi/8}^{\pi} 1e^{j\omega n}d\omega$$

$$= \frac{\sin(\pi n)}{\pi n} - \frac{\sin(3\pi n/8)}{\pi n}$$

$$\text{Let } h_{Ideal\_HPF}(t) = \frac{\sin(\pi t)}{\pi t} - \frac{\sin(3\pi t/8)}{\pi t}$$

$$h[n] = h_{Ideal_{HPF}}\left(\frac{15}{2}\right)\delta[n] + \cdots + h_{Ideal_{HPF}}\left(\frac{3}{2}\right)\delta[n-6]$$

$$+ h_{Ideal_{HPF}}\left(\frac{1}{2}\right)\delta[n-7] + h_{Ideal_{HPF}}\left(\frac{1}{2}\right)\delta[n-8] + h_{Ideal_{HPF}}\left(\frac{3}{2}\right)\delta[n-9]$$

$$+ \cdots + h_{Ideal\_HPF}\left(\frac{15}{2}\right)\delta[n-15]$$

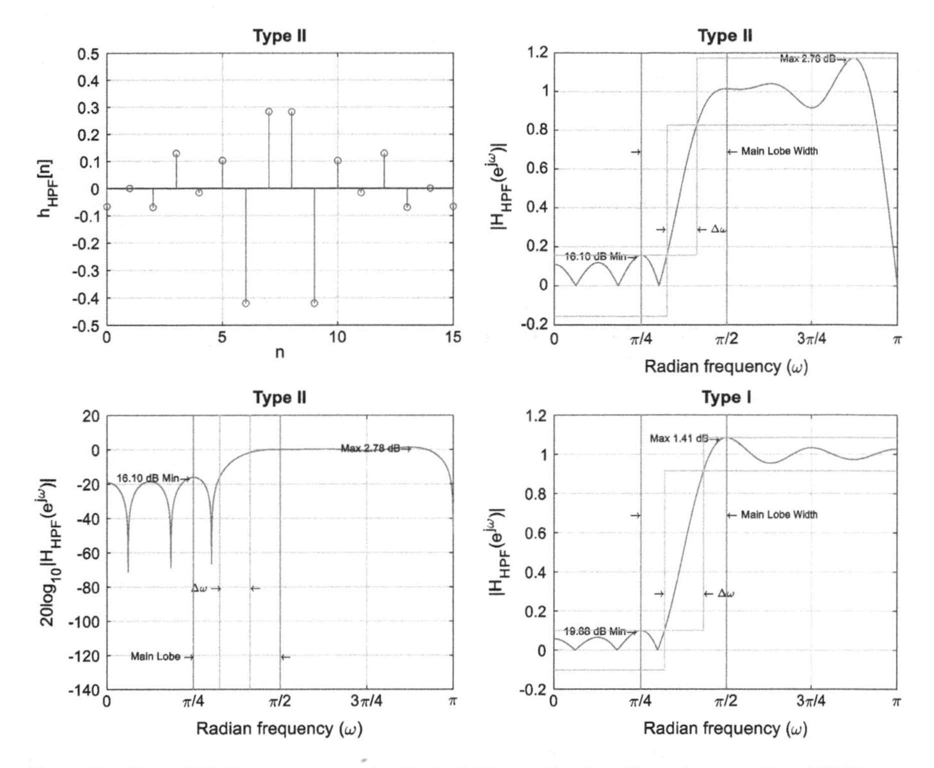

Fig. 6.17 Type II HPF with $N = 15$ for Table 6.17 specification. For reference, Type I HPF with $N = 16$ is also shown

Prog. 6.13 MATLAB program for Fig. 6.17.

```
N = 15;
NN = 2^13;
N1 = NN/2+1;
findex = linspace(0,2*pi,NN);
ff = findex(1:N1);
t = -N/2:1:N/2;
n = 0:1:(length(t)-1);
ht = sin(pi*t)./(pi*t)-sin(3*pi*(t)/8)./(pi*(t));
B2 = abs(fft(ht,NN));
[PKS,LOCS]= findpeaks(B2);
M2 = PKS(2);
I2 = LOCS(2);
M1 = PKS(3);
I1 = LOCS(3);
M3 = PKS(5);
I3 = LOCS(5);

figure,
stem(n,ht),grid, ylim([-0.5 0.5]);
figure,
plot(ff,B2(1:N1)), grid, xlim([0 pi]), ylim([-0.2 1.2]);
straa = sprintf('Max %1.2f dB',40*log10(B2(I3)));
aa = text(ff(I3)-0.7,B2(I3),[straa '\rightarrow    ' ]);
strbb = sprintf('%1.2f dB Min',-20*log10(abs(B2(I2))));
bb = text(ff(I2)-0.7,B2(I2),[strbb '\rightarrow    ']);
line([ff(I1) ff(I1)],[-0.2 1.2],'Color','magenta');
line([ff(I2) ff(I2)],[-0.2 1.2],'Color','magenta');
hh = text(ff(I2)-0.1,0.7,'\rightarrow ');
ii = text(ff(I1),0.7,'\leftarrow Main Lobe Width');
figure,
plot(ff,20*log10(abs(B2(1:N1)))), grid, xlim([0 pi]), ylim([-140 20]);
B3 = 20*log10(abs(B2));
straa = sprintf('Max %1.2f dB',40*log10(B2(I3)));
aa = text(ff(I3)-0.63,B3(I3),[ straa '\rightarrow']);
strbb = sprintf('%1.2f dB Min',-20*log10(abs(B2(I2))));
bb = text(ff(I2)-0.7,B3(I2),[ strbb '\rightarrow']);
line([ff(I1) ff(I1)],[-140 20],'Color','magenta');
line([ff(I2) ff(I2)],[-140 20],'Color','magenta');
hh = text(ff(I2)-0.57,-120,'Main Lobe \rightarrow ');
ii = text(ff(I1),-120,'\leftarrow');
```

For the verification purpose, apply the Parseval's theorem over the designed filter.

$$\frac{1}{2\pi}\int_{-\pi}^{\pi}\left|H\left(e^{j\omega}\right)\right|^2 d\omega = \frac{1}{2\pi}\int_{-\pi}^{-3\pi/8} 1^2 d\omega + \frac{1}{2\pi}\int_{3\pi/8}^{\pi} 1^2 d\omega$$
$$= \frac{1}{2\pi}2\frac{5\pi}{8} = \frac{5}{8} = 0.6250$$
$$\sum\nolimits_{n=0}^{15}\left|h_{TypeIIHPF}[n]\right|^2 = 0.5868$$
$$\text{Energy ratio} = \frac{0.5868}{0.6250} = 0.9389$$

For Type I and $N = 16$, the energy ratio is below.

$$\sum\nolimits_{n=0}^{15}\left|h_{TypeIHPF}[n]\right|^2 = 0.6124$$
$$\text{Energy ratio} = \frac{0.6124}{0.6250} = 0.9799$$

■

Both type filters are illustrated in Fig. 6.18.

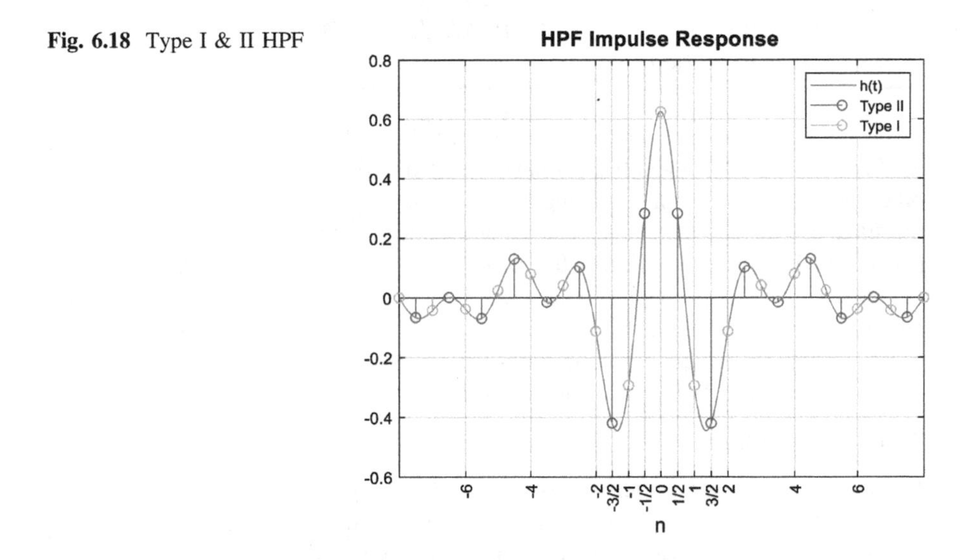

Fig. 6.18 Type I & II HPF

Prog. 6.14 MATLAB program for Fig. 6.18.

```
NN = 2^13;
N = 15;
nn = -N/2:1:N/2;
f1 = linspace(0,2,NN);
nn1 = -(N+1)/2:1:(N+1)/2;
nd = -(N+1)/2:1/100:(N+1)/2;
ht1 = sin(pi*(nn+eps))./(pi*(nn+eps))-sin(3*pi*(nn+eps)/8)./(pi*(nn+eps));
ht2 = sin(pi*(nn1+eps))./(pi*(nn1+eps))-
sin(3*pi*(nn1+eps)/8)./(pi*(nn1+eps));
ht3 = sin(pi*(nd+eps))./(pi*(nd+eps))-sin(3*pi*(nd+eps)/8)./(pi*(nd+eps));
figure,
plot(nd,ht3), grid
hold on
stem(nn,ht1)
stem(nn1,ht2)
legend('h(t)','Type II','Type I')
hold off
title('HPF Impulse Response')
xlabel('n')
ax2 = gca;
ax2.XTick = [-6 -4 -2 -3/2 -1 -1/2 0 1/2 1 3/2 2 4 6];
ax2.XTickLabel = {'-6','-4','-2','-3/2','-1','-1/2','0','1/2','1','3/2','2','4','6'};
ax2.XTickLabelRotation = 90;
```

Due to the generic property of Type II filter, the HPF cannot be realized properly as shown above. The Type II must have the zero at the –1 location in z domain; in other words, the filter gain at the highest frequency is zero. As demonstrated above, the filter magnitude is rapidly decreasing around π radian frequency and highest ripple can be observed nearby the edge. Also, the energy ratio based on the Parseval's theorem presents around 94% which can be compared to the Type I energy ratio 98%. The unsuccessful representation of the spectrum component nearby the highest frequency contributes the low energy ratio in Type II HPF filter. However, the Type I HPF is implemented accurately for $N = 16$ illustrated above.

Type III FIR filter

In contrast, the Type III filter is implemented by using the pure imaginary value in frequency domain because the odd symmetry function generates the frequency response in imaginary values. One example is given below.

$$H_{III}(e^{j\omega}) = -2j\sum_{n=1}^{K} h_{III}[n]\sin(\omega n) \in j\,\mathbb{R}$$

Since the $h_{III}[n]$ is real value, $H_{III}(e^{j\omega})$ is the pure imaginary number. According to the DTFT property, the real $h_{III}[n]$ should provide the following relation.

$$H_{III}(e^{j\omega}) = H_{III}^{*}(e^{-j\omega})$$

Based on the pure imaginary $H_{III}(e^{j\omega})$, the impulse response $h_{III}[n]$ can be derived by the DTFT as Eq. (6.49).

$$h_{III}[n] = \frac{1}{2\pi}\int_{-\omega_{c1}}^{-\omega_{c2}} -jke^{j\omega n}d\omega + \frac{1}{2\pi}\int_{\omega_{c2}}^{\omega_{c1}} jke^{j\omega n}d\omega \quad \text{for } k \in \mathbb{R} \tag{6.49}$$

The k is the constant for the frequency domain gain. The computed impulse response is shown in Eq. (6.50).

$$\begin{aligned} &\frac{1}{2\pi}\int_{-\omega_{c1}}^{-\omega_{c2}} -jke^{j\omega n}d\omega + \frac{1}{2\pi}\int_{\omega_{c2}}^{\omega_{c1}} jke^{j\omega n}d\omega \\ &= \frac{k\cos(n\omega_{c1}) - k\cos(n\omega_{c2})}{n\pi} \end{aligned} \tag{6.50}$$

Because of Type III filter property, the HPF and LPF cannot be realized properly. The Type III must have the zero at –1 and 1 location in z domain; therefore, the filter gain at the highest and lowest frequency are zero. Let's see following example.

Example 6.8
Design the FIR filter for Table 6.18 specification.

Table 6.18 Given filter specification

Filter specifications	Value
Filter type	Band pass filter
Filter realization	FIR Type III without window method
Filter length	$N = 16$
Cutoff frequency 1 (f_{c1})	1000 Hz; $\omega_{c1} = \pi/4$
Passband magnitude	0 dB
Cutoff frequency 2 (f_{c2})	3000 Hz; $\omega_{c2} = 3\pi/4$
Sampling frequency (f_s)	8000 Hz

Solution

$$h_{III}[n] = \frac{1}{2\pi}\int_{-\frac{3\pi}{4}}^{-\frac{\pi}{4}} -je^{j\omega n}d\omega + \frac{1}{2\pi}\int_{\frac{\pi}{4}}^{\frac{3\pi}{4}} je^{j\omega n}d\omega$$

$$= \frac{\cos\left(\frac{3\pi n}{4}\right) - \cos\left(\frac{\pi n}{4}\right)}{n\pi}$$

$$h[n] = h_{III}[-8]\delta[n] + \cdots + h_{III}[0]\delta[n-8] + \cdots + h_{III}[8]\delta[n-16]$$

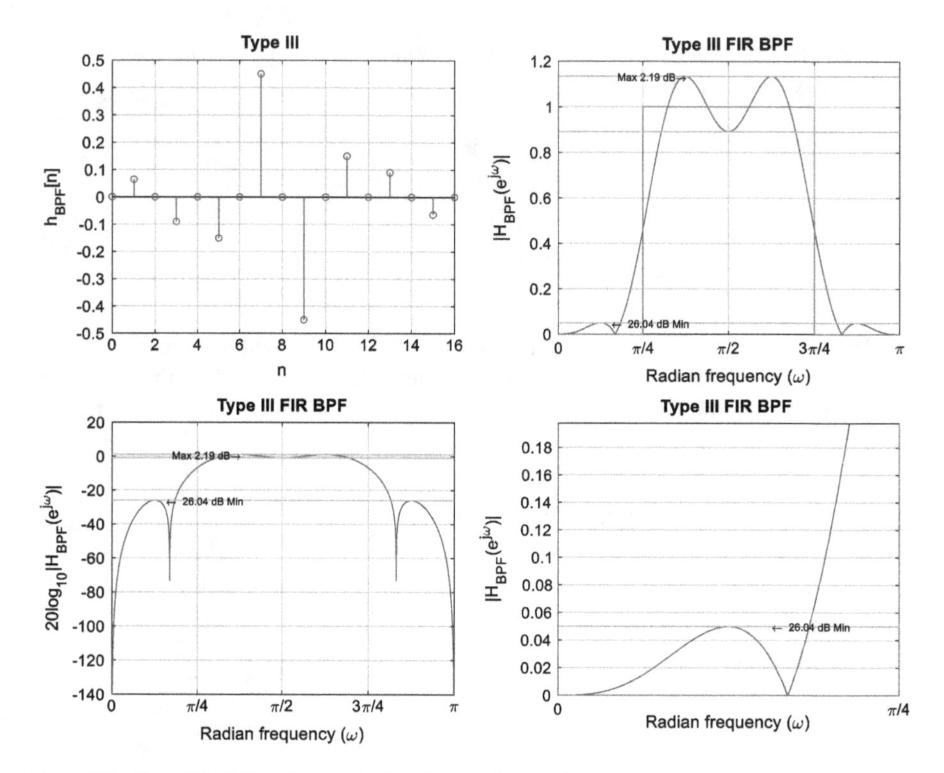

Fig. 6.19 Type III BPF with $N = 16$ for Table 6.18 specification

Prog. 6.15 MATLAB program for Fig. 6.19.

```
N = 16;
NN = 2^13;
nn = (-N/2:1:N/2)+eps;
ff = linspace(0,2*pi,NN);
wc1 = 1/4;
wc2 = 3/4;
dsr = (ff>wc1*pi & ff<wc2*pi);
hn = (cos(wc2*pi*nn)-cos(wc1*pi*nn))./(pi*nn);
Fout = abs(fft(hn,NN));
[PKS,LOCS]= findpeaks(Fout);
M2 = PKS(2);
I2 = LOCS(2);
M1 = PKS(1);
I1 = LOCS(1);
[NPKS,NLOCS]= findpeaks(-Fout);
NM1 = NPKS(2);
NI1 = NLOCS(2);

figure,
stem((0:N),hn), grid
figure,
plot(ff(1:NN/2+1),Fout(1:NN/2+1),ff(1:NN/2+1),dsr(1:NN/2+1)), grid, xlim([0
pi]), ylim([0 1.2]);
straa = sprintf('Max %1.2f dB',40*log10(Fout(I2)));
aa = text(ff(I2)-0.63,Fout(I2),[ straa '\rightarrow']);
strbb = sprintf('%1.2f dB Min',-20*log10(Fout(I1)));
bb = text(ff(I1)+0.1,Fout(I1),['\leftarrow    ' strbb ]);
dBout = 20*log10(Fout);
figure,
plot(ff(1:NN/2+1),dBout(1:NN/2+1)), grid, xlim([0 pi]), ylim([-140 20]);
line([0 pi],[dBout(NI1) dBout(NI1)],'Color',[0.93 0.69 0.13]);
line([0 pi],[dBout(I2) dBout(I2)],'Color',[0.93 0.69 0.13]);
line([0 pi],[dBout(I1) dBout(I1)],'Color',[0.93 0.69 0.13]);
straa = sprintf('Max %1.2f dB',40*log10(Fout(I2)));
aa = text(ff(I2)-0.63,dBout(I2),[ straa '\rightarrow']);
strbb = sprintf('%1.2f dB Min',-20*log10(Fout(I1)));
bb = text(ff(I1)+0.1,dBout(I1),['\leftarrow    ' strbb ]);
```

For the verification purpose, apply the Parseval's theorem over the designed filter.

$$\frac{1}{2\pi}\int_{-\pi}^{\pi}|H(e^{j\omega})|^2 d\omega=\frac{1}{2\pi}\int_{-\pi}^{\pi}H(e^{j\omega})H^*(e^{j\omega})d\omega$$
$$=\frac{1}{2\pi}\int_{-\frac{3\pi}{4}}^{-\frac{\pi}{4}}(-j)(j)e^{j\omega n}d\omega+\frac{1}{2\pi}\int_{\frac{\pi}{4}}^{\frac{3\pi}{4}}(j)(-j)e^{j\omega n}d\omega=\frac{1}{2\pi}2\frac{\pi}{2}=\frac{1}{2}$$
$$\sum_{n=0}^{15}|h[n]|^2=0.4748$$
$$\text{Energy ratio}=\frac{0.4748}{0.5}=0.9496$$

Type IV FIR filter

The Type IV is the modified version of the Type III. The modification is made to locate the symmetric center at the between the samples; therefore, the delay component $e^{j\omega/2}$ is shown in the frequency domain as below. Except the $e^{j\omega/2}$, the Type IV is pure imaginary in frequency response as below.

$$H_{IV}(e^{j\omega})=-2je^{j\frac{\omega}{2}}\sum_{n=0}^{K}h_{IV}[n]\sin\left(\frac{\omega}{2}(2n+1)\right)$$

Similar manner as Type II FIR filter, the inverse DTFT provides the impulse response as below.

$$\begin{aligned}
h[n]&=\frac{1}{2\pi}\int_{-\pi}^{\pi}H_{IV}(e^{j\omega})e^{j\omega n}d\omega\\
&=\frac{1}{2\pi}\int_{-\pi}^{\pi}-2je^{j\frac{\omega}{2}}\sum_{m=0}^{K}h_{IV}[m]\sin\left(\frac{\omega}{2}(2m+1)\right)e^{j\omega n}d\omega\\
&=\frac{1}{2\pi}\int_{-\pi}^{\pi}-2je^{j\frac{\omega}{2}}\sum_{m=0}^{K}h_{IV}[m]\frac{e^{j\omega\frac{2m+1}{2}}-e^{-j\omega\frac{2m+1}{2}}}{2j}e^{j\omega n}d\omega\\
&=\sum_{m=0}^{K}h_{IV}[m]\frac{1}{2\pi}\int_{-\pi}^{\pi}\left(e^{j\frac{\omega}{2}}e^{-j\omega\frac{2m+1}{2}}-e^{j\frac{\omega}{2}}e^{j\omega\frac{2m+1}{2}}\right)e^{j\omega n}d\omega\\
&=\sum_{m=0}^{K}h_{IV}[m]\frac{1}{2\pi}\int_{-\pi}^{\pi}\left(e^{-j\omega m}-e^{j\omega(m+1)}\right)e^{j\omega n}d\omega\\
&=\sum_{m=0}^{K}h_{IV}[m]\frac{1}{2\pi}\int_{-\pi}^{\pi}e^{-j\omega m}e^{j\omega n}d\omega\\
&\quad-\sum_{l=-K-1}^{0}h_{IV}[-l-1]\frac{1}{2\pi}\int_{-\pi}^{\pi}e^{-j\omega l}e^{j\omega n}d\omega\quad \text{let } m+1=-l\\
&=\sum_{m=0}^{K}h_{IV}[m]\frac{1}{2\pi}\int_{-\pi}^{\pi}e^{-j\omega(m-n)}d\omega\\
&\quad-\sum_{l=-K-1}^{0}h_{IV}[-l-1]\frac{1}{2\pi}\int_{-\pi}^{\pi}e^{-j\omega(l-n)}d\omega\\
&=\sum_{m=0}^{K}h_{IV}[m]\delta[m-n]-\sum_{l=-K-1}^{0}h_{IV}[-l-1]\delta[l-n]\\
&=h_{IV}[n]\{u[n]-u[n-(K+1)]\}\\
&\quad-h_{IV}[-n-1]\{u[n+(K+1)]-u[n]\}
\end{aligned}\tag{6.51}$$

where $u[n]$ is unit-step function. The derived $h[n]$ in Eq. (6.51) shows the odd symmetricity over the –1/2 sample location. The $h_{IV}[n]$ and $-h_{IV}[-n-1]$ are concatenated mutually for odd symmetricity. To implement the Type IV filter, Fourier transform is performed as below. Note that the $H_{IV}(e^{j\omega})$ is multiplied by $e^{-j\omega/2}$ to remove the –1/2 center location. For convenience purpose, the ω is used for Ω in Fourier transform.

$$
\begin{aligned}
h_{1/2}(t) &= \frac{1}{2\pi}\int_{-\infty}^{\infty} e^{j\frac{\omega}{2}} H_{IV}(\omega) e^{j\omega n} d\omega \\
&= \frac{1}{2\pi}\int_{-\infty}^{\infty} -2j\sum\nolimits_{m=0}^{K} h_{IV}[m]\sin\left(\frac{\omega}{2}(2m+1)\right) e^{j\omega t} d\omega \\
&= \frac{1}{2\pi}\int_{-\infty}^{\infty} -2j\sum\nolimits_{m=0}^{K} h_{IV}[m]\frac{e^{j\omega\frac{2m+1}{2}} - e^{-j\omega\frac{2m+1}{2}}}{2j} e^{j\omega t} d\omega \\
&= \frac{1}{2\pi}\int_{-\infty}^{\infty} \sum\nolimits_{m=0}^{K} h_{IV}[m] e^{-j\omega\frac{2m+1}{2}} e^{j\omega t} d\omega \\
&- \frac{1}{2\pi}\int_{-\infty}^{\infty} \sum\nolimits_{m=0}^{K} h_{IV}[m] e^{j\omega\frac{2m+1}{2}} e^{j\omega t} d\omega \\
&= \frac{1}{2\pi}\int_{-\infty}^{\infty} \sum\nolimits_{m=0}^{K} h_{IV}[m] e^{-j\omega\frac{2m+1}{2}} e^{j\omega t} d\omega \\
&- \frac{1}{2\pi}\int_{-\infty}^{\infty} \sum\nolimits_{l=-K}^{0} h_{IV}[-l] e^{-j\omega\frac{2l-1}{2}} e^{j\omega t} d\omega \\
&= \sum\nolimits_{m=0}^{K} h_{IV}[m]\frac{1}{2\pi}\int_{-\infty}^{\infty} e^{-j\omega\left(\frac{2m+1}{2}-t\right)} d\omega \\
&- \sum\nolimits_{l=-k}^{0} h_{IV}[-l]\frac{1}{2\pi}\int_{-\infty}^{\infty} e^{-j\omega\left(\frac{2l-1}{2}-t\right)} d\omega \\
&= \sum\nolimits_{m=0}^{K} h_{IV}[m]\delta\left(t-\frac{2m+1}{2}\right) - \sum\nolimits_{l=-K}^{0} h_{IV}[-l]\delta\left(t-\frac{2l-1}{2}\right) \\
&= -h_{IV}[K]\delta\left(t+\frac{2K+1}{2}\right) - \cdots - h_{IV}[1]\delta\left(t+\frac{3}{2}\right) - h_{IV}[0]\delta\left(t+\frac{1}{2}\right) \\
&+ h_{IV}[0]\delta\left(t-\frac{1}{2}\right) + h_{IV}[1]\delta\left(t-\frac{3}{2}\right) + \cdots + h_{IV}[K]\delta\left(t-\frac{2K+1}{2}\right)
\end{aligned}
\tag{6.52}
$$

The derived $h_{1/2}(t)$ in Eq. (6.52) shows the odd symmetricity over the coordinate origin. To realize the filter, compute the impulse response from the desired frequency response as below.

$$
\begin{aligned}
h(n) &= \frac{1}{2\pi}\int_{-\pi}^{\pi} H_{desired}\left(e^{j\omega}\right) e^{j\omega n} d\omega \text{ for } H_{desired}\left(e^{j\omega}\right) \in j\mathbb{R} \\
H_{desired}\left(e^{j\omega}\right) &= -H_{desired}\left(e^{-j\omega}\right)
\end{aligned}
$$

Odd symmetricity $h[n] = -h[-n]$ is implemented by following conversion for Type IV as below.

$$h_{IV}[n] = h\left(n + \frac{1}{2}\right) \quad \text{for } n \geq 0$$
$$h_{IV}[n] = -h_{IV}[-n] \tag{6.53}$$

Parseval's theorem is applied for verifying the energy between both domains as Eq. (6.54).

$$\sum_{n=-\infty}^{\infty} |h_{IV}[n]|^2 \approx \frac{1}{2\pi} \int_{-\pi}^{\pi} H_{desired}(e^{j\omega}) H^*_{desired}(e^{j\omega}) d\omega \tag{6.54}$$

Example 6.9
Design the FIR filter for Table 6.19 specification.

Solution
The center of the passband and stopband edge is below. The center frequency is the cutoff frequency which presents half magnitude in the frequency.

$$\omega_c = \omega_{st} + \frac{\omega_p - \omega_{st}}{2} = \frac{\pi}{4} + \frac{\pi}{8} = \frac{3\pi}{8}$$

The frequency transition between the passband and stopband edge is below

$$\Delta\omega = \omega_p - \omega_{st} = \frac{\pi}{2} - \frac{\pi}{4} = \frac{\pi}{4}$$

The main lobe width for rectangular window presents the frequency transition as previously discussed.

$$\frac{4\pi}{N+1} \leq \frac{\pi}{4} \; N \geq 15$$

Let the $N = 15$ for Type IV filter. Realization and evaluation are below.

Table 6.19 Given filter specification

Filter specifications	Value
Filter type	High pass filter
Filter realization	FIR Type IV without window method
Filter length	N with optimization
Passband frequency (f_p)	2000 Hz; $\omega_p = \pi/2$
Passband magnitude	0 dB
Stopband frequency (f_{st})	1000 Hz; $\omega_{st} = \pi/4$
Sampling frequency (f_s)	8000 Hz

$$h_{Ideal_{HPF}}[n] = \frac{1}{2\pi}\int_{-\pi}^{-3\pi/8} -je^{j\omega n}d\omega + \frac{1}{2\pi}\int_{3\pi/8}^{\pi} je^{j\omega n}d\omega$$

$$= \frac{\cos(\pi n)}{\pi n} - \frac{\cos(3\pi n/8)}{\pi n}$$

$$\text{Let } h_{Ideal\_HPF}(t) = \frac{\cos(\pi t)}{\pi t} - \frac{\cos(3\pi t/8)}{\pi t}$$

$$h[n] = h_{Ideal_{HPF}}\left(\frac{15}{2}\right)\delta[n] + \cdots + h_{Ideal_{HPF}}\left(\frac{3}{2}\right)\delta[n-6]$$

$$+ h_{Ideal_{HPF}}\left(\frac{1}{2}\right)\delta[n-7] + h_{Ideal_{HPF}}\left(\frac{1}{2}\right)\delta[n-8] + h_{Ideal_{HPF}}\left(\frac{3}{2}\right)\delta[n-9]$$

$$+ \cdots + h_{Ideal\_HPF}\left(\frac{15}{2}\right)\delta[n-15]$$

The performance of the Type IV filter is shown in Fig. 6.20.

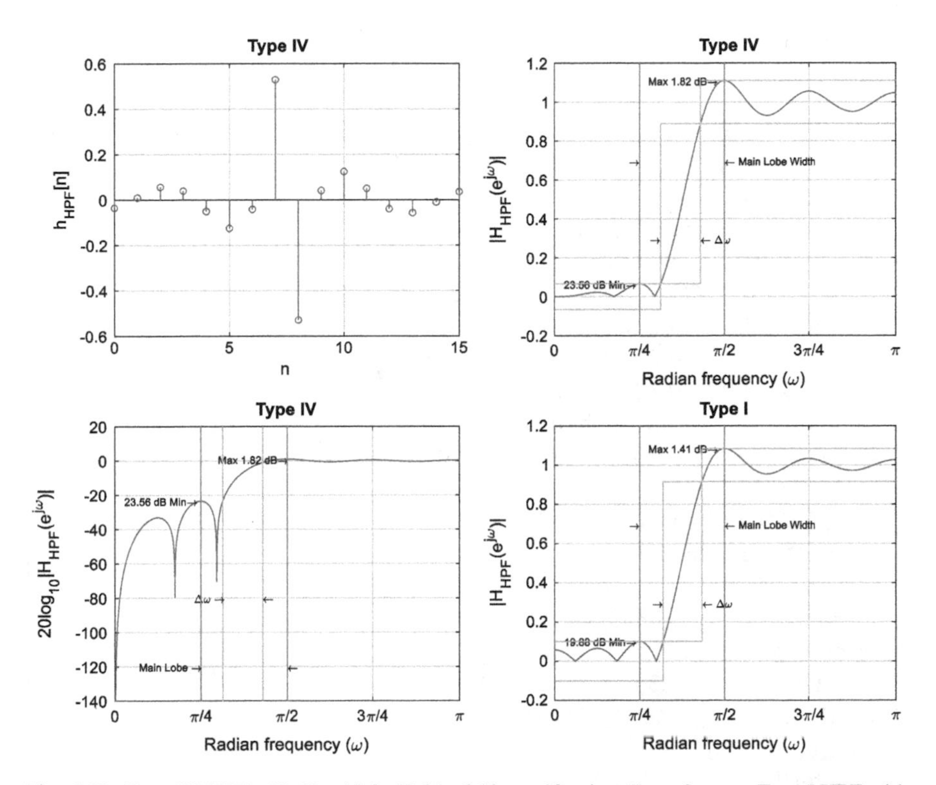

Fig. 6.20 Type IV HPF with $N = 15$ for Table 6.19 specification. For reference, Type I HPF with $N = 16$ is also shown

Prog. 6.16 MATLAB program for Fig. 6.20.

```
N = 15;
NN = 2^13;
N1 = NN/2+1;
t = -N/2:1:N/2;
n = 0:1:(length(t)-1);
ht = cos(pi*t)./(pi*t)-cos(3*pi*(t)/8)./(pi*(t));
B2 = abs(fft(ht,NN));
findex = linspace(0,2*pi,NN);
ff = findex(1:N1);
[PKS,LOCS]= findpeaks(B2);
M2 = PKS(2);
I2 = LOCS(2);
M1 = PKS(3);
I1 = LOCS(3);
M3 = PKS(3);
I3 = LOCS(3);

figure,
stem(n,ht),grid
figure,
plot(ff,B2(1:N1)), grid, xlim([0 pi]), ylim([-0.2 1.2]);
straa = sprintf('Max %1.2f dB',40*log10(B2(I3)));
aa = text(ff(I3)-0.7,B2(I3),[straa '\rightarrow    ' ]);
strbb = sprintf('%1.2f dB Min',-20*log10(abs(B2(I2))));
bb = text(ff(I2)-0.7,B2(I2),[strbb '\rightarrow    ']);
line([ff(I1) ff(I1)],[-0.2 1.2],'Color','magenta');
line([ff(I2) ff(I2)],[-0.2 1.2],'Color','magenta');
hh = text(ff(I2)-0.1,0.7,'\rightarrow ');
ii = text(ff(I1),0.7,'\leftarrow Main Lobe Width');
figure,
plot(ff,20*log10(abs(B2(1:N1)))), grid, xlim([0 pi]), ylim([-140 20]);
B3 = 20*log10(abs(B2));
straa = sprintf('Max %1.2f dB',40*log10(B2(I3)));
aa = text(ff(I3)-0.63,B3(I3),[ straa '\rightarrow']);
strbb = sprintf('%1.2f dB Min',-20*log10(abs(B2(I2))));
bb = text(ff(I2)-0.7,B3(I2),[ strbb '\rightarrow']);
line([ff(I1) ff(I1)],[-140 20],'Color','magenta');
line([ff(I2) ff(I2)],[-140 20],'Color','magenta');
hh = text(ff(I2)-0.57,-120,'Main Lobe \rightarrow ');
ii = text(ff(I1),-120,'\leftarrow');
```

For the verification purpose, apply the Parseval's theorem over the designed filter.

$$\frac{1}{2\pi}\int_{-\pi}^{\pi}\left|H\left(e^{j\omega}\right)\right|^2 d\omega = \frac{1}{2\pi}\int_{-\pi}^{\pi} H\left(e^{j\omega}\right)H^*\left(e^{j\omega}\right)d\omega$$

$$\frac{1}{2\pi}\int_{-\pi}^{-3\pi/8}(-j)jd\omega + \frac{1}{2\pi}\int_{3\pi/8}^{\pi} j(-j)d\omega = \frac{1}{2\pi}2\frac{5\pi}{8} = \frac{5}{8} = 0.6250$$

$$\sum_{n=0}^{15}|h[n]|^2 = 0.6123$$

$$\text{Energy ratio} = \frac{0.6123}{0.6250} = 0.9797$$

■

Advanced example of FIR Type III and Type IV

Differentiator

The differentiator provides the derivative of input signal. The mathematical derivation in strict sense is complicated to design the differentiator but we simplify the process by assuming that the discrete n is real number.

$$\frac{d}{dn}\left(\frac{1}{2\pi}\int_{-\pi}^{\pi} X(e^{j\omega})e^{j\omega n}d\omega\right) = \frac{1}{2\pi}\int_{-\pi}^{\pi} j\omega X(e^{j\omega})e^{j\omega n}d\omega \quad n \in \mathbb{R} \tag{6.55}$$

The corresponding frequency response of the differentiator is shown in Eq. (6.56).

$$H_{diff}(e^{j\omega}) = j\omega \tag{6.56}$$

Due to the odd symmetry in the frequency domain, the FIR Type III and IV are possible.

$$h_{diff}[n] = \frac{1}{2\pi}\int_{-\pi}^{\pi} j\omega e^{j\omega n}d\omega = \frac{\cos(\pi n)}{n} - \frac{\sin(\pi n)}{\pi n^2} \tag{6.57}$$

Let the $N = 16$ for Type III filter.

$$h[n] = h_{diff}[-8]\delta[n] + \cdots + h_{diff}[0]\delta[n-8] + \cdots + h_{diff}[8]\delta[n-16]$$

Let the $N = 15$ for Type IV filter.

$$h[n] = h_{diff}\left(-\frac{15}{2}\right)\delta[n] + \cdots + h_{diff}\left(-\frac{3}{2}\right)\delta[n-6] + h_{diff}\left(-\frac{1}{2}\right)\delta[n-7] + h_{diff}\left(\frac{1}{2}\right)\delta[n-8] + h_{diff}\left(\frac{3}{2}\right)\delta[n-9] + \cdots + h_{diff}\left(\frac{15}{2}\right)\delta[n-15]$$

Realization and evaluation are below.

Prog. 6.17 MATLAB program for Fig. 6.21.

```
syms w n;
NN = 2^13;
N1 = NN/2+1;
findex = linspace(0,2*pi,NN);
f_f = findex(1:N1);
aa = 1j*w*exp(1j*w*n);
bb = int(aa,w,-pi,pi)/(2*pi);
cc = expand(bb);
dd = simplify(cc,'Steps',100);
nn1 = -15/2:1:15/2;
nn2 = -8:1:8;
nn3 = -8:0.001:8;
ee1 = subs(dd,n,nn1);
ee2 = subs(dd,n,nn2+eps);
ee3 = subs(dd,n,nn3+eps);
ff1 = eval(ee1);
ff2 = eval(ee2);
ff3 = eval(ee3);
B1 = abs(fft(ff1,NN));
B2 = abs(fft(ff2,NN));

figure,
plot(nn3,ff3), grid, hold on
stem(nn1,ff1)
stem(nn2,ff2), hold off
figure,
plot(f_f,B1(1:N1),f_f,B2(1:N1),f_f,f_f), grid
xlim([0 pi]);
```

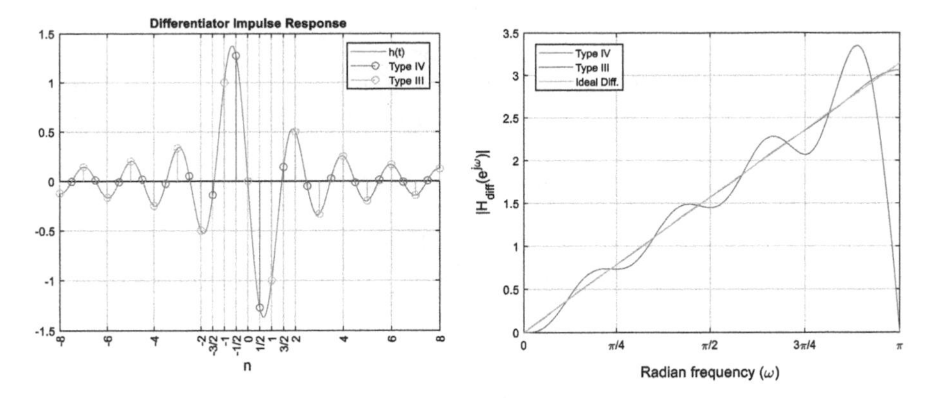

Fig. 6.21 Type III (N = 16) and Type IV (N = 15) differentiator impulse response and frequency response

Hilbert transform

Hilbert transform [3, 5] produces the single sideband (SSB) output in frequency domain. Let's divide $X(e^{j\omega})$ for positive, zero, and negative frequency area as below.

$$x[n] \overset{DTFT}{\leftrightarrow} X(e^{j\omega}) = \begin{cases} X_p(e^{j\omega}) & \text{for } \pi \geq \omega > 0 \\ X_0(e^{j\omega}) & \text{for } \omega = 0 \\ X_n(e^{j\omega}) & \text{for } -\pi < \omega < 0 \end{cases} \tag{6.58}$$

$$X(e^{j\omega}) = X_n(e^{j\omega}) + X_0(e^{j\omega}) + X_p(e^{j\omega})$$

The desired output in frequency domain is zero and positive frequency area only as below.

$$Y(e^{j\omega}) = X_0(e^{j\omega}) + 2X_p(e^{j\omega})$$

Therefore, the positive frequency area and the magnitude inversed negative frequency area are added over the $X(e^{j\omega})$ to provide the desired output as below.

$$X(e^{j\omega}) + \{-X_n(e^{j\omega}) + X_p(e^{j\omega})\} = X_0(e^{j\omega}) + 2X_p(e^{j\omega})$$

The above process can be realized by Eq. (6.59) convolution sum with Eq. (6.60) filter frequency response.

$$\{1 + jH_{Hi}(e^{j\omega})\}X(e^{j\omega}) = X_0(e^{j\omega}) + 2X_p(e^{j\omega})$$

$$\overset{DTFT}{\leftrightarrow} x[n] + jh_{Hi}[n] * x[n] \tag{6.59}$$

$$H_{Hi}(e^{j\omega}) = \begin{cases} -j & \text{for } \pi \geq \omega > 0 \\ 0 & \text{for } \omega = 0 \\ j & \text{for } -\pi < \omega < 0 \end{cases} \tag{6.60}$$

Note that the Eq. (6.59) output is derived by following relation.

$$\begin{gathered} \{1 + jH_{Hi}(e^{j\omega})\}X(e^{j\omega}) \\ = X_n(e^{j\omega}) + X_0(e^{j\omega}) + X_p(e^{j\omega}) + jjX_n(e^{j\omega}) + 0X_0(e^{j\omega}) + j(-j)X_p(e^{j\omega}) \\ = X_0(e^{j\omega}) + 2X_p(e^{j\omega}) \end{gathered}$$

Derive the impulse response by using the DTFT with Eq. (6.60). Due to the odd symmetry in the frequency domain, the FIR Type III and IV are possible.

$$h_{Hi}[n] = \frac{1}{2\pi} \int_{-\pi}^{0} je^{j\omega n} d\omega + \frac{1}{2\pi} \int_{0}^{\pi} -je^{j\omega n} d\omega = \frac{2\{\sin(\frac{\pi n}{2})\}^2}{n\pi}$$

Below is the convolution sum for the Hilbert transform. Be careful with the time delay caused by the causal $h_{Hi}[n]$ filter. The $x[n]$ and $jh_{Hi}[n] * x[n]$ should be in phase to realize the transform; therefore, we should apply the proper delay on the $x[n]$ to synchronize the real and imaginary part of the output.

$$x[n] + jh_{Hi}[n] * x[n] \overset{DTFT}{\leftrightarrow} \{1 + jH_{Hi}(e^{j\omega})\}X(e^{j\omega})$$

We can also use the overall impulse response as below without considering the phase mismatch problem. The below sinc function is equivalent to the delta function in Type III filter but the Type IV shows the different distribution in time domain. Still, the below sinc function deliver the approximate constant magnitude in frequency domain for Type IV.

$$h_{III}[n] \text{or } h_{IV}[n] = \frac{\sin(\pi n)}{\pi n} + jh_{Hi}[n] \overset{DTFT}{\leftrightarrow} 1 + jH_{Hi}(e^{j\omega})$$

Let the $N = 16$ for Type III filter.

$$h[n] = h_{III}[-8]\delta[n] + \cdots + h_{III}[0]\delta[n-8] + \cdots + h_{III}[8]\delta[n-16]$$

Let the $N = 15$ for Type IV filter.

$$\begin{aligned} h[n] = h_{IV}\left(-\frac{15}{2}\right)\delta[n] + \cdots + h_{IV}\left(-\frac{3}{2}\right)\delta[n-6] + h_{IV}\left(-\frac{1}{2}\right)\delta[n-7] \\ + h_{IV}\left(\frac{1}{2}\right)\delta[n-8] + h_{IV}\left(\frac{3}{2}\right)\delta[n-9] + \cdots + h_{IV}\left(\frac{15}{2}\right)\delta[n-15] \end{aligned}$$

Realization and evaluation are below.

Prog. 6.18 MATLAB program for Fig. 6.22.

```
syms w n;
NN = 2^13;
N1 = NN/2+1;
findex = linspace(-pi,pi,NN);
f_f = findex(1:N1);
aa = 1j*exp(1j*w*n);
bb = (int(aa,w,-pi,0)+int(-aa,w,0,pi))/(2*pi);
cc = expand(bb);
dd = simplify(cc,'Steps',100);
nn1 = -15/2:1:15/2;
nn2 = -8:1:8;
nn3 = -8:0.001:8;
ee1 = subs(dd,n,nn1);
ee12 = subs(sin(pi*n)/(n*pi),n,nn1);
ee2 = subs(dd,n,nn2+eps);
ee3 = subs(dd,n,nn3+eps);
ff1 = eval(ee1);
ff12 = eval(ee12);
ff2 = eval(ee2);
ff3 = eval(ee3);
gg1 = ff12+1j*ff1;
gg2 = (nn2==0)+1j*ff2;
B1 = fftshift(abs(fft(gg1,NN)));
B2 = fftshift(abs(fft(gg2,NN)));
B3 = fftshift(abs(fft(ff1,NN)));
B4 = fftshift(abs(fft(ff2,NN)));
B5 = fftshift(abs(fft(ff12,NN)));

figure,
plot(nn3,ff3), grid, hold on
stem(nn1,ff1)
stem(nn2,ff2), hold off
figure,
plot(findex,B1,findex,B2), grid
xlim([-pi pi]);
figure,
plot(findex,B3,findex,B4), grid
xlim([-pi pi]);
```

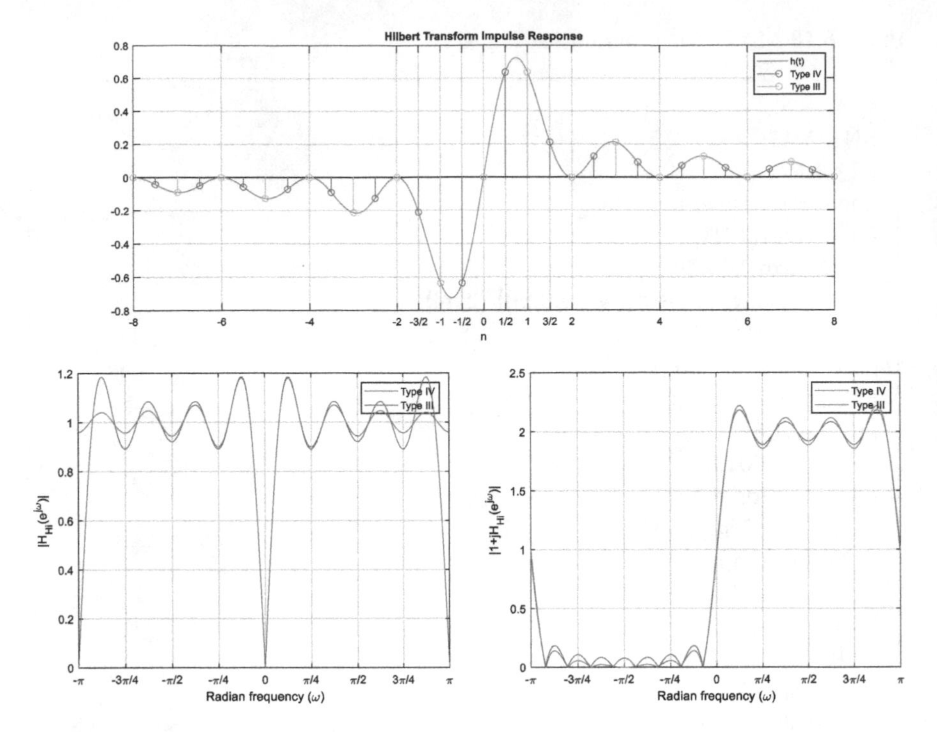

Fig. 6.22 Type III (N = 16) and Type IV (N = 15) Hilbert transform impulse response and frequency response

Remes Exchange Algorithm (Parks and McClellan)

The Remes exchange algorithm [6, 7] presents the equiripple property in passband and stopband by changing the trigonometric problem to polynomial equation. The optimal filter is designed by following the alternation theorem and Remes exchange algorithm. Let's derive the Type I FIR filter for Remes exchange algorithm. Below is the frequency response of the non-causal FIR filter.

$$H(e^{j\omega}) = h\left[\frac{N}{2}\right]e^{j\omega\frac{N}{2}} + \cdots + h[1]e^{j\omega} + h[0] + h[1]e^{-j\omega} + \cdots + h\left[\frac{N}{2}\right]e^{-j\omega\frac{N}{2}}$$

$$H(e^{j\omega}) = h[0] + 2\sum_{n=1}^{N/2} h[n]\cos(\omega n) \tag{6.61}$$

Using the Chebyshev polynomial, the cosine part can be represented by the polynomial as shown in Eq. (6.62).

$$
\begin{aligned}
T_1(\cos(\omega)) &= \cos(\omega) \\
T_2(\cos(\omega)) &= \cos(2\omega) = 2\cos^2(\omega) - 1 \\
T_3(\cos(\omega)) &= \cos(3\omega) = 4\cos^3(\omega) - 3\cos(\omega) \\
&\cdots \\
T_{n+1}(\cos(\omega)) &= 2\cos(\omega)T_n(\cos(\omega)) - T_{n-1}(\cos(\omega))
\end{aligned}
\tag{6.62}
$$

The $c[n]$ in Eq. (6.63) is the Chebyshev polynomial coefficients.

$$
H(e^{j\omega}) = \sum_{n=0}^{N/2} c[n]\cos^n(\omega) \tag{6.63}
$$

Replacing the $\cos(\omega)$ with x completes the polynomial representation as shown in Eq. (6.64).

$$
\text{Let} \quad \cos(\omega) = x, \quad H(x) = \sum_{n=0}^{N/2} c[n]x^n \; 0 \le \omega \le \pi \; -1 \le x \le 1 \tag{6.64}
$$

Maximal error locations show the zero derivative on the position as Eq. (6.65).

$$
\frac{dH(x)}{dx} = 0 \;\; \frac{N}{2} - 1 \text{ maximal error locations} \tag{6.65}
$$

There are additional four boundaries to generate maximal error such as 0, π , and two band edges. Overall number of maximal error locations are $N/2 + 3$ as shown in Fig. 6.23.

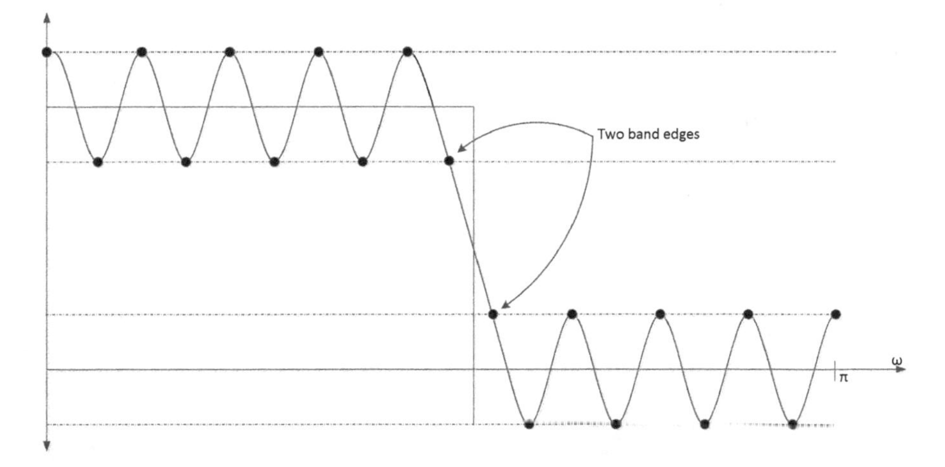

Fig. 6.23 The optimal filter frequency response with maximal error locations for $N = 32$

The error function is defined as Eq. (6.66).

$$E(e^{j\omega}) = W(e^{j\omega})|D(e^{j\omega}) - H(e^{j\omega})| \text{ or } E(x) = W(x)|D(x) - H(x)| \tag{6.66}$$

The $W(e^{j\omega})$ is the positive weight function and $D(e^{j\omega})$ is the desired real-value function. According to the alternation theorem, the $H(e^{j\omega})$ is the unique and best Chebyshev approximation to a given desired function $D(e^{j\omega})$, provided that the error function $E(e^{j\omega})$ shows at least $N/2+2$ maximal error locations. The maximal error values illustrate identical magnitude with alternative sign. The unweighted error function is shown in Eq. (6.67).

$$E(e^{j\omega}) = D(e^{j\omega}) - \sum_{n=0}^{N/2} c[n]\cos^n(\omega) \quad \text{or} \quad E(x) = D(x) - \sum_{n=0}^{N/2} c[n]x^n \tag{6.67}$$

The corresponding maximal error locations are ω_k or x_k where k is 0, 1, 2, …, $N/2$ + 1. Let the maximal error magnitude as Δ. Equation (6.68) presents the maximal error locations.

$$\begin{aligned} D(e^{j\omega_k}) &= \sum_{n=0}^{N/2} c[n]\cos^n(\omega_k) + (-1)^k \Delta \\ \text{or} \quad D(x_k) &= \sum_{n=0}^{N/2} c[n]x_k^n + (-1)^k \Delta \end{aligned} \tag{6.68}$$

Equations (6.69) and (6.70) represent the Eq. (6.68) in matrix form.

$$\begin{bmatrix} 1 & \cos(\omega_0) & \cos^2(\omega_0) & \cdots & \cos^{\frac{N}{2}}(\omega_0) & 1 \\ 1 & \cos(\omega_1) & \cos^2(\omega_1) & \cdots & \cos^{\frac{N}{2}}(\omega_1) & -1 \\ 1 & \cos(\omega_2) & \cos^2(\omega_2) & \cdots & \cos^{\frac{N}{2}}(\omega_2) & 1 \\ 1 & \cos(\omega_3) & \cos^2(\omega_3) & \cdots & \cos^{\frac{N}{2}}(\omega_3) & -1 \\ \vdots & \vdots & \vdots & \ddots & \vdots & \vdots \\ 1 & \cos(\omega_{N/2+1}) & \cos^2(\omega_{N/2+1}) & \cdots & \cos^{\frac{N}{2}}(\omega_{N/2+1}) & (-1)^{N/2+1} \end{bmatrix} \begin{bmatrix} c[0] \\ c[1] \\ c[2] \\ \vdots \\ c\left[\frac{N}{2}\right] \\ \Delta_m \end{bmatrix} = \begin{bmatrix} D(\omega_0) \\ D(\omega_1) \\ D(\omega_2) \\ D(\omega_3) \\ \vdots \\ D(\omega_{N/2+1}) \end{bmatrix} \tag{6.69}$$

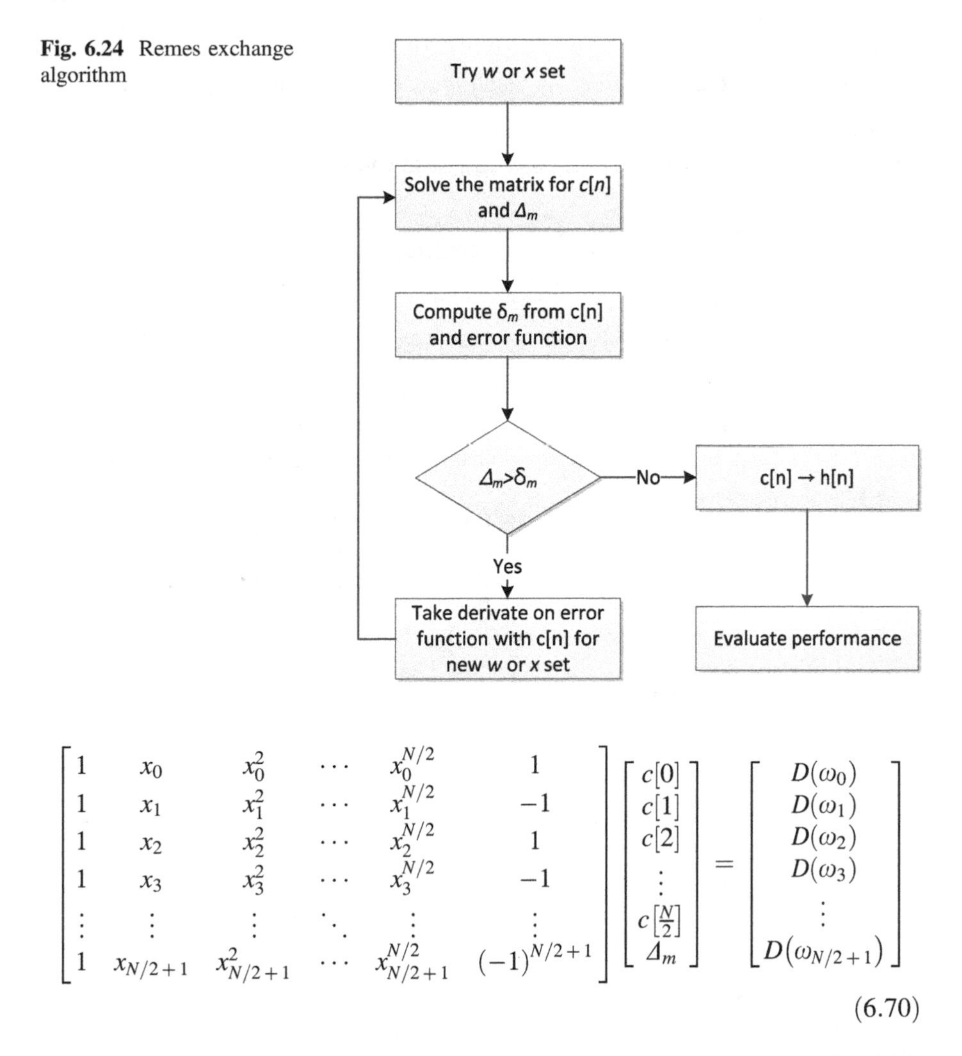

Fig. 6.24 Remes exchange algorithm

$$\begin{bmatrix} 1 & x_0 & x_0^2 & \cdots & x_0^{N/2} & 1 \\ 1 & x_1 & x_1^2 & \cdots & x_1^{N/2} & -1 \\ 1 & x_2 & x_2^2 & \cdots & x_2^{N/2} & 1 \\ 1 & x_3 & x_3^2 & \cdots & x_3^{N/2} & -1 \\ \vdots & \vdots & \vdots & \ddots & \vdots & \vdots \\ 1 & x_{N/2+1} & x_{N/2+1}^2 & \cdots & x_{N/2+1}^{N/2} & (-1)^{N/2+1} \end{bmatrix} \begin{bmatrix} c[0] \\ c[1] \\ c[2] \\ \vdots \\ c\left[\frac{N}{2}\right] \\ \Delta_m \end{bmatrix} = \begin{bmatrix} D(\omega_0) \\ D(\omega_1) \\ D(\omega_2) \\ D(\omega_3) \\ \vdots \\ D(\omega_{N/2+1}) \end{bmatrix} \tag{6.70}$$

The optimal solution is obtained by the iterative approach as shown in Fig. 6.24. Note that Δ_m is the computed Δ value at the m-th iteration. The δ_m is the maximum error value of the given configuration as shown in Eq. (6.71).

$$\begin{aligned} \delta_m &= \max_{\omega \in [0,\pi]} \left| D(\omega) - \sum_{n=0}^{N/2} c[n] \cos^n(\omega) \right| \\ \text{or} \quad \delta_m &= \max_{x \in [-1,1]} \left| D(x) - \sum_{n=0}^{N/2} c[n] x^n \right| \end{aligned} \tag{6.71}$$

Example 6.10

Design the Table 6.20 FIR filter by using the Remes exchange algorithm.

Solution

From the given filter order 6, the filter response can be defined as below.

$$H(e^{j\omega}) = \sum_{n=0}^{3} c[n] \cos^n(\omega)$$

$$D(e^{j\omega_k}) = \sum_{n=0}^{3} c[n] \cos^n(\omega_k) + (-1)^k \Delta$$

The error function $E(e^{j\omega})$ shows at least $N/2+2$ maximal error locations; therefore, $H(e^{j\omega})$ illustrates the 5 peak error locations. The matrix form is shown below.

$$\begin{bmatrix} 1 & \cos(\omega_0) & \cos^2(\omega_0) & \cos^3(\omega_0) & 1 \\ 1 & \cos(\omega_1) & \cos^2(\omega_1) & \cos^3(\omega_1) & -1 \\ 1 & \cos(\omega_2) & \cos^2(\omega_2) & \cos^3(\omega_2) & 1 \\ 1 & \cos(\omega_3) & \cos^2(\omega_3) & \cos^3(\omega_3) & -1 \\ 1 & \cos(\omega_4) & \cos^2(\omega_4) & \cos^3(\omega_4) & 1 \end{bmatrix} \begin{bmatrix} c[0] \\ c[1] \\ c[2] \\ c[3] \\ \Delta_m \end{bmatrix} = \begin{bmatrix} D(e^{j\omega_0}) \\ D(e^{j\omega_1}) \\ D(e^{j\omega_2}) \\ D(e^{j\omega_3}) \\ D(e^{j\omega_4}) \end{bmatrix}$$

First trial of frequency set is below.

$$\begin{aligned} \omega = \{\omega_0, \omega_1, \omega_2, \omega_3, \omega_4\} &= \left\{0, \frac{\pi}{4}, \frac{\pi}{2}, \frac{3\pi}{4}, \pi\right\} \\ &= \{0, 0.7854, 1.5708, 2.3562, 3.1416\} \end{aligned}$$

Table 6.20 Given filter specification

Filter specifications	Value
Filter type	Low pass filter
Filter realization	FIR Type I without window method
Filter length	7 ($N = 6$)
Passband frequency (f_p)	1000 Hz; $\omega_p = \pi/4$
Passband magnitude	0 dB
Stopband frequency (f_{st})	2000 Hz; $\omega_{st} = \pi/2$
Sampling frequency (f_s)	8000 Hz

The corresponding $D(e^{j\omega})$ is below

$$D(e^{j\omega}) = \{D(e^{j\omega_0}), D(e^{j\omega_1}), D(e^{j\omega_2}), D(e^{j\omega_3}), D(e^{j\omega_4})\} = \{1, 1, 0, 0, 0\}$$

Place the ω and $D(e^{j\omega})$ on the matrix form as below.

$$\begin{bmatrix} 1 & \cos(0) & \cos^2(0) & \cos^3(0) & 1 \\ 1 & \cos(\frac{\pi}{4}) & \cos^2(\frac{\pi}{4}) & \cos^3(\frac{\pi}{4}) & -1 \\ 1 & \cos(\frac{\pi}{2}) & \cos^2(\frac{\pi}{2}) & \cos^3(\frac{\pi}{2}) & 1 \\ 1 & \cos(\frac{3\pi}{4}) & \cos^2(\frac{3\pi}{4}) & \cos^3(\frac{3\pi}{4}) & -1 \\ 1 & \cos(\pi) & \cos^2(\pi) & \cos^3(\pi) & 1 \end{bmatrix} \begin{bmatrix} c[0] \\ c[1] \\ c[2] \\ c[3] \\ \Delta_0 \end{bmatrix} = \begin{bmatrix} D(0) \\ D(\frac{\pi}{4}) \\ D(\frac{\pi}{2}) \\ D(\frac{3\pi}{4}) \\ D(\pi) \end{bmatrix}$$

Solution of above matrix is below.

$$\begin{bmatrix} c[0] \\ c[1] \\ c[2] \\ c[3] \\ \Delta_0 \end{bmatrix} = \begin{bmatrix} 0.1250 \\ 0.9142 \\ 0.5000 \\ -0.4142 \\ -0.1250 \end{bmatrix} = \begin{bmatrix} 1 & \cos(0) & \cos^2(0) & \cos^3(0) & 1 \\ 1 & \cos(\frac{\pi}{4}) & \cos^2(\frac{\pi}{4}) & \cos^3(\frac{\pi}{4}) & -1 \\ 1 & \cos(\frac{\pi}{2}) & \cos^2(\frac{\pi}{2}) & \cos^3(\frac{\pi}{2}) & 1 \\ 1 & \cos(\frac{3\pi}{4}) & \cos^2(\frac{3\pi}{4}) & \cos^3(\frac{3\pi}{4}) & -1 \\ 1 & \cos(\pi) & \cos^2(\pi) & \cos^3(\pi) & 1 \end{bmatrix}^{-1} \begin{bmatrix} D(0) \\ D(\frac{\pi}{4}) \\ D(\frac{\pi}{2}) \\ D(\frac{3\pi}{4}) \\ D(\pi) \end{bmatrix}$$

The computed $c[n]$ with Eq. (6.63) generates the first trial frequency response.

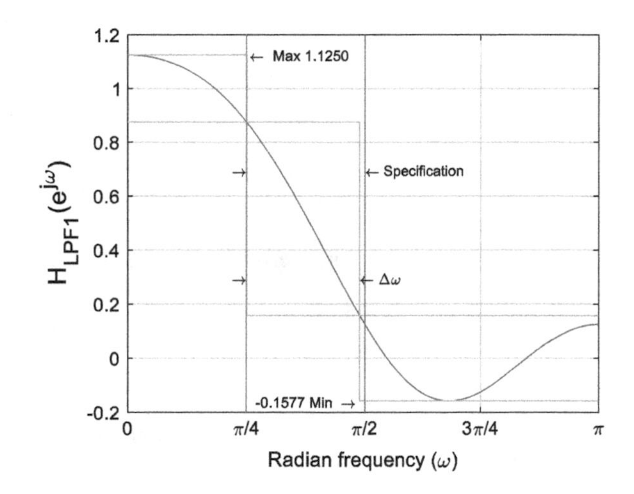

Fig. 6.25 Frequency response of Remes exchange algorithm for first trial

Prog. 6.19 MATLAB program for Fig. 6.25.

```
syms x
ww = linspace(0,pi,1000);
nf1 = [0; 1/4; 1/2; 3/4; 1];
w1 = nf1*pi;
RR1 = [ones(5,1) cos(w1) (cos(w1)).^2 (cos(w1)).^3 [1; -1; 1; -1; 1]];
DD1 = [1; 1; 0; 0; 0];
CS1 = inv(RR1)*DD1;
HW1 = CS1(1)+CS1(2)*cos(ww)+CS1(3)*cos(ww).^2+CS1(4)*cos(ww).^3;
[M1,I1] = min(HW1);
[M2,I2] = max(HW1);
cc1 = max(find(HW1>(-M1)));
cc = min(find(HW1<(2-M2)));

figure
plot(ww,HW1), grid
xlim([0 pi]);
line([0 ww(cc)],[M2 M2],'Color',[0.93 0.69 0.13]);
line([ww(cc1) pi],[M1 M1],'Color',[0.93 0.69 0.13]);
line([pi/4 pi/4],[-0.2 1.2],'Color','magenta');
line([pi/2 pi/2],[-0.2 1.2],'Color','magenta');
straa = sprintf('Max %1.4f',M2);
aa = text(ww(I2)+0.8,M2,['\leftarrow    ' straa]);
strbb = sprintf('%1.4f Min',M1);
bb = text(ww(I1)-1.3,M1,[strbb '    \rightarrow']);
hh = text(pi/4-0.1,0.7,'\rightarrow ');
ii = text(pi/2,0.7,'\leftarrow Specification');
```

The calculated maximum error δ_0 for first trial is below and bigger than Δ_0 .

$$\delta_0 = \max_{\omega\in[0,\pi]} \left| D(\omega) - \sum_{n=0}^{3} c[n]\cos^n(\omega) \right| = 0.1577 > 0.1250 = |\Delta_0|$$

The outcome does not meet the condition ($\delta_m = \Delta_m$); therefore, next iteration is required with new frequency set.

$$H(e^{j\omega}) = \sum_{n=0}^{3} c[n]\cos^n(\omega) \quad \frac{dH(e^{j\omega})}{d\cos(\omega)} = \sum_{n=0}^{2} \tilde{c}[n]\cos^n(\omega) = 0$$

The zero derivative of frequency response provides the new frequency as below.

$$\cos(\omega) = \{1.3498, -0.5451\} \text{ since } |\cos(\omega)| \leq 1$$

$$\cos(\omega) = -0.5451 \text{ and } \omega = 2.1472$$

Second trial of frequency set is below.

$$\begin{aligned}\omega &= \{\omega_0, \omega_1, \omega_2, \omega_3, \omega_4\} = \left\{0, \frac{\pi}{4}, \frac{\pi}{2}, 2.1472, \pi\right\} \\ &= \{0, 0.7854, 1.5708, 2.1472, 3.1416\}\end{aligned}$$

The corresponding $D(e^{j\omega})$ is below

$$D(e^{j\omega}) = \{D(e^{j\omega_0}), D(e^{j\omega_1}), D(e^{j\omega_2}), D(e^{j\omega_3}), D(e^{j\omega_4})\} = \{1, 1, 0, 0, 0\}$$

Place the ω and $D(e^{j\omega})$ on the matrix form as below.

$$\begin{bmatrix} 1 & \cos(0) & \cos^2(0) & \cos^3(0) & 1 \\ 1 & \cos(0.7854) & \cos^2(0.7854) & \cos^3(0.7854) & -1 \\ 1 & \cos(1.5708) & \cos^2(1.5708) & \cos^3(1.5708) & 1 \\ 1 & \cos(2.1472) & \cos^2(2.1472) & \cos^3(2.1472) & -1 \\ 1 & \cos(3.1416) & \cos^2(3.1416) & \cos^3(3.1416) & 1 \end{bmatrix} \begin{bmatrix} c[0] \\ c[1] \\ c[2] \\ c[3] \\ \Delta_1 \end{bmatrix} = \begin{bmatrix} D(0) \\ D(0.7854) \\ D(1.5708) \\ D(2.1472) \\ D(3.1416) \end{bmatrix}$$

Solution of above matrix is below.

$$\begin{bmatrix} 1 & \cos(0) & \cos^2(0) & \cos^3(0) & 1 \\ 1 & \cos(0.7854) & \cos^2(0.7854) & \cos^3(0.7854) & -1 \\ 1 & \cos(1.5708) & \cos^2(1.5708) & \cos^3(1.5708) & 1 \\ 1 & \cos(2.1472) & \cos^2(2.1472) & \cos^3(2.1472) & -1 \\ 1 & \cos(3.1416) & \cos^2(3.1416) & \cos^3(3.1416) & 1 \end{bmatrix}^{-1} \begin{bmatrix} D(0) \\ D(0.7854) \\ D(1.5708) \\ D(2.1472) \\ D(3.1416) \end{bmatrix}$$

$$= \begin{bmatrix} c[0] \\ c[1] \\ c[2] \\ c[3] \\ \Delta_1 \end{bmatrix} = \begin{bmatrix} 0.1328 \\ 0.8699 \\ 0.5000 \\ -0.3699 \\ -0.1328 \end{bmatrix}$$

The computed $c[n]$ with Eq. (6.63) generates the second trial frequency response as shown in Fig. 6.26.

Prog. 6.20 MATLAB program for Fig. 6.26. Continued from Prog. 6.19.

```
% From Prog. 6.19
eq1 = poly2sym(CS1(4:-1:1));
eq2 = diff(eq1);
coeff1 = sym2poly(eq2);
root1 = roots(coeff1);
temp1 = acos(root1);
w2 = w1;
w2(4) = temp1(2);
RR2 = [ones(5,1) cos(w2) (cos(w2)).^2 (cos(w2)).^3 [1; -1; 1; -1; 1]];
DD2 = [1; 1; 0; 0; 0];
CS2 = inv(RR2)*DD2;
HW2 = CS2(1)+CS2(2)*cos(ww)+CS2(3)*cos(ww).^2+CS2(4)*cos(ww).^3;
[M3,I3] = min(HW2);
[M4,I4] = max(HW2);
cc2 = max(find(HW2>(-M3)));
cc3 = min(find(HW2<(2-M4)));

figure,
plot(ww,HW2), grid
xlim([0 pi]);
line([0 ww(cc3)],[M4 M4],'Color',[0.93 0.69 0.13]);
line([ww(cc2) pi],[M3 M3],'Color',[0.93 0.69 0.13]);
line([pi/4 pi/4],[-0.2 1.2],'Color','magenta');
line([pi/2 pi/2],[-0.2 1.2],'Color','magenta');
straa1 = sprintf('Max %1.4f',M4);
aa1 = text(ww(I4)+0.8,M4,['\leftarrow    ' straa1]);
strbb1 = sprintf('%1.4f Min',M3);
bb1 = text(ww(I3)-1.3,M3,[strbb1 '    \rightarrow']);
hh1 = text(pi/4-0.1,0.7,'\rightarrow ');
ii1 = text(pi/2,0.7,'\leftarrow Specification');
```

The calculated maximum error δ_1 for second trial is below and equal to Δ_1 .

$$\delta_1 = \max_{\omega\in[0,\pi]}\left|D(\omega) - \sum_{n=0}^{3} c[n]\cos^n(\omega)\right| = 0.1328 = 0.1328 = |\Delta_1|$$

The outcome satisfies the condition ($\delta_\mathrm{m} = \Delta_m$). Next step provides the impulse response $h[n]$ from $c[n]$. The frequency response from the polynomial is below.

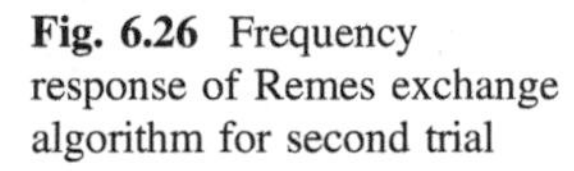

Fig. 6.26 Frequency response of Remes exchange algorithm for second trial

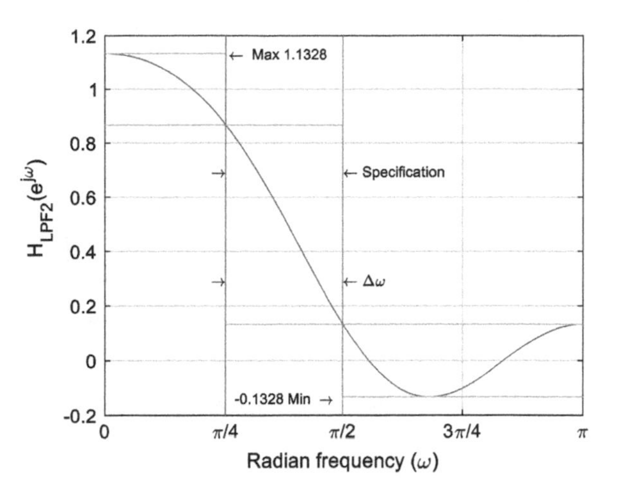

$$H\left(e^{j\omega}\right) = 0.1328 + 0.8699\cos(\omega) + 0.5\cos^2(\omega) - 0.3699\cos^3(\omega)$$

Using the reverse Chebyshev polynomial, the polynomial equation converts to the trigonometric problem as below.

$$\begin{aligned}
\cos^2(\omega) &= (1 + \cos(2\omega))/2 \\
\cos^3(\omega) &= (3\cos(\omega) + \cos(3\omega))/4 \\
H\left(e^{j\omega}\right) &= 0.1328 + 0.8699\cos(\omega) + 0.5\left\{\frac{1 + \cos(2\omega)}{2}\right\} \\
&\quad - 0.3699\left\{\frac{3\cos(\omega) + \cos(3\omega)}{4}\right\} \\
&= 0.3823 + 0.5925\cos(\omega) + 0.25\cos(2\omega) - 0.0925\cos(3\omega) \\
H\left(e^{j\omega}\right) &= 0.3823 + 0.5925\frac{e^{j\omega} + e^{-j\omega}}{2} + 0.25\frac{e^{j2\omega} + e^{-j2\omega}}{2} \\
&\quad - 0.0925\frac{e^{j3\omega} + e^{-j3\omega}}{2} \\
h_{pm}[n] &= -0.0462\delta[n+3] + 0.125\delta[n+2] + 0.2962\delta[n+1] + 0.3823\delta[n] \\
&\quad + 0.2962\delta[n-1] + 0.125\delta[n-2] - 0.0462\delta[n-3] \\
h[n] &= h_{pm}[n-3] = -0.0462\delta[n] + 0.125\delta[n-1] + 0.2962\delta[n-2] \\
&\quad + 0.3823\delta[n-3] + 0.2962\delta[n-4] + 0.125\delta[n-5] - 0.0462\delta[n-6]
\end{aligned}$$

The final impulse response $h[n]$ and corresponding $H(e^{j\omega})$ are illustrated in Fig. 6.27.

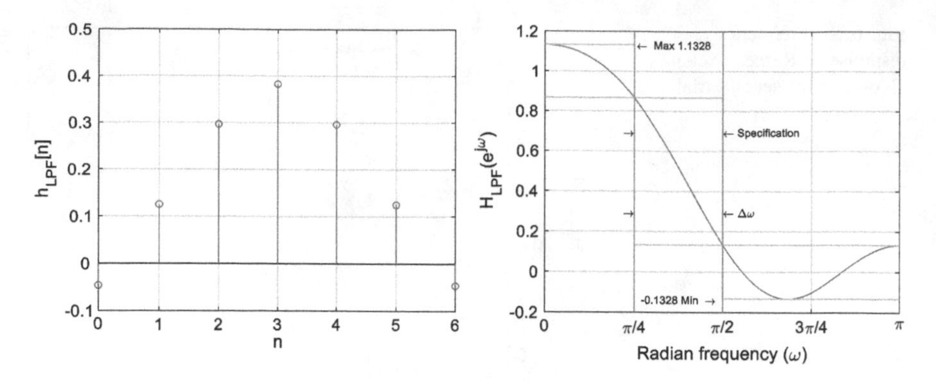

Fig. 6.27 Type I LPF designed by Remes exchange algorithm with $N = 6$ for Table 6.20 specification

Prog. 6.21 MATLAB program for Fig. 6.27. Continued from Prog. 6.20.

```
% From Prog. 6.20
ww = linspace(0,pi,1000);
nn = 0:1:6;
hh = zeros(4,1);
hh(1) = CS2(1)+CS2(3)/2;
hh(2) = CS2(2)+CS2(4)*3/4;
hh(3) = CS2(3)/2;
hh(4) = CS2(4)/4;
hhf = [flipud(hh(2:end))/2; hh(1); hh(2:end)/2];
HW3 = hh(1)+hh(2)*cos(ww)+hh(3)*cos(2*ww)-hh(4)*cos(3*ww);
[M3,I3] = min(HW2);
[M4,I4] = max(HW2);
cc2 = max(find(HW2>(-M3)));
cc3 = min(find(HW2<(2-M4)));

figure,
stem(nn,hhf), grid, ylim([-0.1 0.5]);
figure
plot(ww,HW2), grid, xlim([0 pi]);
line([0 ww(cc3)],[M4 M4],'Color',[0.93 0.69 0.13]);
line([ww(cc2) pi],[M3 M3],'Color',[0.93 0.69 0.13]);
line([pi/4 pi/4],[-0.2 1.2],'Color','magenta');
line([pi/2 pi/2],[-0.2 1.2],'Color','magenta');
straa1 = sprintf('Max %1.4f',M4);
aa1 = text(ww(I4)+0.8,M4,['\leftarrow     ' straa1]);
strbb1 = sprintf('%1.4f Min',M3);
bb1 = text(ww(I3)-1.3,M3,[strbb1 '    \rightarrow']);
hh1 = text(pi/4-0.1,0.7,'\rightarrow ');
ii1 = text(pi/2,0.7,'\leftarrow Specification');
```

The passband ripple A_p can be calculated from the maximum deviation R_p as below.

$$A_P = 40\log_{10}(1+R_p)$$

$$2.1661\text{dB} = 40\log_{10}(1+0.1328)$$

The equivalent stopband attenuation can be calculated as below.

$$A_{st} = -20\log_{10} R_{st}$$

$$17.5360\text{dB} = -20\log_{10} 0.1328$$

■

Frequency sampling design—Type I

The frequency sampling method [7, 8] solves the linear equations between the impulse response and frequency domain amplitude. The equations are given by the DFT. The impulse response length and frequency (amplitude) sampling dimension are identical in order to derive the simple unique solution. The solution provides the coefficients (impulse response) of the desired filter with a frequency response that exactly passes through the sample points specified by the equations. Therefore, the frequency sampling method is an interpolation method and not an approximation approach.

Between the frequency sampling point, there is certain fluctuations. The frequency response exactly passes through the sampling points; however, no response constraint is exercised between the sample points. The desired frequency response $H[k]$ shows the symmetric property as below for real $h_t[n]$ with length L which is $N + 1$ and even N.

$$H[k] = H^*[-k]$$

The corresponding impulse response can be derived as below.

$$\begin{aligned} h_t[n] &= \frac{1}{L}\sum_{k=0}^{L-1} H[k]e^{j\frac{2\pi}{L}kn} = \frac{1}{L}\sum_{k=-N/2}^{N/2} H[k]e^{j\frac{2\pi}{L}kn} \\ &= \frac{1}{L}H[0] + \frac{2}{L}\sum_{k=1}^{N/2} H[k]\cos\left(\frac{2\pi}{L}kn\right) \end{aligned} \tag{6.72}$$

The DFT analysis equation is below to figure out the frequency response for Type I FIR filter.

$$\begin{aligned} H[k] &= \sum_{n=-N/2}^{N/2} h_t[n] e^{-j\frac{2\pi}{L}kn}, \quad H[k] \in \mathbb{R} \\ &= h_t[0] + 2\sum_{n=1}^{N/2} h_t[n] \cos\left(\tfrac{2\pi}{L}kn\right) \end{aligned} \tag{6.73}$$

The matrix form of Eq. (6.73) is below. The exponential part matrix (DFT matrix) is square matrix with $N + 1$ by $N + 1$ size.

$$\begin{bmatrix} e^{j\frac{2\pi}{L}(-\frac{N}{2})\frac{N}{2}} & \cdots & e^{j\frac{2\pi}{L}(-\frac{N}{2})} & 1 & e^{-j\frac{2\pi}{L}(-\frac{N}{2})} & \cdots & e^{-j\frac{2\pi}{L}(-\frac{N}{2})\frac{N}{2}} \\ \vdots & \cdots & \vdots & \vdots & \vdots & \cdots & \vdots \\ e^{j\frac{2\pi}{L}(-1)\frac{N}{2}} & \cdots & e^{j\frac{2\pi}{L}(-1)} & 1 & e^{-j\frac{2\pi}{L}(-1)} & \cdots & e^{-j\frac{2\pi}{L}(-1)\frac{N}{2}} \\ 1 & \cdots & 1 & 1 & 1 & \cdots & 1 \\ e^{j\frac{2\pi}{L}(1)\frac{N}{2}} & \cdots & e^{j\frac{2\pi}{L}(1)} & 1 & e^{-j\frac{2\pi}{L}(1)} & \cdots & e^{-j\frac{2\pi}{L}(1)\frac{N}{2}} \\ \vdots & \vdots & \vdots & \vdots & \vdots & \ddots & \vdots \\ e^{j\frac{2\pi}{L}(\frac{N}{2})\frac{N}{2}} & \cdots & e^{j\frac{2\pi}{L}(\frac{N}{2})} & 1 & e^{-j\frac{2\pi}{L}(\frac{N}{2})} & \cdots & e^{-j\frac{2\pi}{2}(\frac{N}{2})\frac{N}{2}} \end{bmatrix} \begin{bmatrix} h_t[-N/2] \\ \vdots \\ h_t[-1] \\ h_t[0] \\ h_t[1] \\ \vdots \\ h_t[N/2] \end{bmatrix} = \begin{bmatrix} H[-N/2] \\ \vdots \\ H[-1] \\ H[0] \\ H[1] \\ \vdots \\ H[N/2] \end{bmatrix} \tag{6.74}$$

The solution to find the impulse response $h_t[n]$ is derived by using the inverse matrix of exponential part as below.

$$\begin{bmatrix} h_t[-N/2] \\ \vdots \\ h_t[-1] \\ h_t[0] \\ h_t[1] \\ \vdots \\ h_t[N/2] \end{bmatrix} = \begin{bmatrix} e^{j\frac{2\pi}{L}(-\frac{N}{2})\frac{N}{2}} & \cdots & e^{j\frac{2\pi}{L}(-\frac{N}{2})} & 1 & e^{-j\frac{2\pi}{L}(-\frac{N}{2})} & \cdots & e^{-j\frac{2\pi}{L}(-\frac{N}{2})\frac{N}{2}} \\ \vdots & \cdots & \vdots & \vdots & \vdots & \cdots & \vdots \\ e^{j\frac{2\pi}{L}(-1)\frac{N}{2}} & \cdots & e^{j\frac{2\pi}{L}(-1)} & 1 & e^{-j\frac{2\pi}{L}(-1)} & \cdots & e^{-j\frac{2\pi}{L}(-1)\frac{N}{2}} \\ 1 & \cdots & 1 & 1 & 1 & \cdots & 1 \\ e^{j\frac{2\pi}{L}(1)\frac{N}{2}} & \cdots & e^{j\frac{2\pi}{L}(1)} & 1 & e^{-j\frac{2\pi}{L}(1)} & \cdots & e^{-j\frac{2\pi}{L}(1)\frac{N}{2}} \\ \vdots & \vdots & \vdots & \vdots & \vdots & \ddots & \vdots \\ e^{j\frac{2\pi}{L}(\frac{N}{2})\frac{N}{2}} & \cdots & e^{j\frac{2\pi}{L}(\frac{N}{2})} & 1 & e^{-j\frac{2\pi}{L}(\frac{N}{2})} & \cdots & e^{-j\frac{2\pi}{2}(\frac{N}{2})\frac{N}{2}} \end{bmatrix}^{-1} \begin{bmatrix} H[-N/2] \\ \vdots \\ H[-1] \\ H[0] \\ H[1] \\ \vdots \\ H[N/2] \end{bmatrix} \tag{6.75}$$

Note that DFT matrix is unitary matrix [9] since its conjugate transpose (Hermitian conjugate [10]) is the inverse matrix. Also, the DFT matrix represents Vandermonde matrix [11] with the terms of a geometric progression in each row as below.

$$\begin{bmatrix} h_t[-N/2] \\ \vdots \\ h_t[-1] \\ h_t[0] \\ h_t[1] \\ \vdots \\ h_t[N/2] \end{bmatrix} = \frac{1}{L} \begin{bmatrix} e^{-j\frac{2\pi}{L}\left(-\frac{N}{2}\right)\frac{N}{2}} & \cdots & e^{-j\frac{2\pi}{L}\left(-\frac{N}{2}\right)} & 1 & e^{j\frac{2\pi}{L}\left(-\frac{N}{2}\right)} & \cdots & e^{j\frac{2\pi}{L}\left(-\frac{N}{2}\right)\frac{N}{2}} \\ \vdots & \cdots & \vdots & \vdots & \vdots & \cdots & \vdots \\ e^{-j\frac{2\pi}{L}(-1)\frac{N}{2}} & \cdots & e^{-j\frac{2\pi}{L}(-1)} & 1 & e^{j\frac{2\pi}{L}(-1)} & \cdots & e^{j\frac{2\pi}{L}(-1)\frac{N}{2}} \\ 1 & \cdots & 1 & 1 & 1 & \cdots & 1 \\ e^{-j\frac{2\pi}{L}(1)\frac{N}{2}} & \cdots & e^{-j\frac{2\pi}{L}(1)} & 1 & e^{j\frac{2\pi}{L}(1)} & \cdots & e^{j\frac{2\pi}{L}(1)\frac{N}{2}} \\ \vdots & \vdots & \vdots & \vdots & \vdots & \ddots & \vdots \\ e^{-j\frac{2\pi}{L}\left(\frac{N}{2}\right)\frac{N}{2}} & \cdots & e^{-j\frac{2\pi}{L}\left(\frac{N}{2}\right)} & 1 & e^{j\frac{2\pi}{L}\left(\frac{N}{2}\right)} & \cdots & e^{j\frac{2\pi}{2}\left(\frac{N}{2}\right)\frac{N}{2}} \end{bmatrix} \begin{bmatrix} H[-N/2] \\ \vdots \\ H[-1] \\ H[0] \\ H[1] \\ \vdots \\ H[N/2] \end{bmatrix} \tag{6.76}$$

The Euler equation compresses the exponential matrix in short form as below for Type I.

$$\begin{bmatrix} 1 & 2 & 2 & \cdots & 2 \\ 1 & 2\cos\left(\frac{2\pi}{L}(1)1\right) & 2\cos\left(\frac{2\pi}{L}(1)2\right) & \cdots & 2\cos\left(\frac{2\pi}{L}(1)\frac{N}{2}\right) \\ 1 & 2\cos\left(\frac{2\pi}{L}(2)1\right) & 2\cos\left(\frac{2\pi}{L}(2)2\right) & \cdots & 2\cos\left(\frac{2\pi}{L}(2)\frac{N}{2}\right) \\ \vdots & \vdots & \vdots & \cdots & \vdots \\ 1 & 2\cos\left(\frac{2\pi}{L}\left(\frac{N}{2}\right)1\right) & 2\cos\left(\frac{2\pi}{L}\left(\frac{N}{2}\right)2\right) & \cdots & 2\cos\left(\frac{2\pi}{L}\left(\frac{N}{2}\right)\frac{N}{2}\right) \end{bmatrix} \begin{bmatrix} h_t[0] \\ h_t[1] \\ h_t[2] \\ \vdots \\ h_t[N/2] \end{bmatrix} = \begin{bmatrix} H[0] \\ H[1] \\ H[2] \\ \vdots \\ H[N/2] \end{bmatrix} \tag{6.77}$$

$$\begin{bmatrix} h_t[0] \\ h_t[1] \\ h_t[2] \\ \vdots \\ h_t[N/2] \end{bmatrix} = \frac{1}{L} \begin{bmatrix} 1 & 2 & 2 & \cdots & 2 \\ 1 & 2\cos\left(\frac{2\pi}{L}(1)1\right) & 2\cos\left(\frac{2\pi}{L}(1)2\right) & \cdots & 2\cos\left(\frac{2\pi}{L}(1)\frac{N}{2}\right) \\ 1 & 2\cos\left(\frac{2\pi}{L}(2)1\right) & 2\cos\left(\frac{2\pi}{L}(2)2\right) & \cdots & 2\cos\left(\frac{2\pi}{L}(2)\frac{N}{2}\right) \\ \vdots & \vdots & \vdots & \cdots & \vdots \\ 1 & 2\cos\left(\frac{2\pi}{L}\left(\frac{N}{2}\right)1\right) & 2\cos\left(\frac{2\pi}{L}\left(\frac{N}{2}\right)2\right) & \cdots & 2\cos\left(\frac{2\pi}{L}\left(\frac{N}{2}\right)\frac{N}{2}\right) \end{bmatrix} \begin{bmatrix} H[0] \\ H[1] \\ H[2] \\ \vdots \\ H[N/2] \end{bmatrix} \tag{6.78}$$

The complete $h_t[n]$ is below

$$\begin{aligned} h_t[n] = {} & h_t[N/2]\delta[n+N/2] + \cdots + h_t[1]\delta[n+1] + h_t[0]\delta[n] + h_t[1]\delta[n-1] \\ & + \cdots + h_t[N/2]\delta[n-N/2] \end{aligned} \tag{6.79}$$

To make the causal filter, perform the shift operation as below.

$$h[n] = h_t[n - N/2] \tag{6.80}$$

Table 6.21 Given filter specification

Filter specifications	Value
Filter type	Low pass filter
Filter realization	FIR Type I without window method
Filter length	7 ($N = 6$)
Passband frequency (f_p)	1000 Hz; $\omega_p = \pi/4$
Passband magnitude	0 dB
Stopband frequency (f_{st})	2000 Hz; $\omega_{st} = \pi/2$
Sampling frequency (f_s)	8000 Hz

The frequency sampling design method is identical to the DFT method without truncation in Sect. 4.5. The required condition is that the impulse response length is equal to the frequency sampling length. The DFT method output illustrates the equivalent pattern with frequency sampling design method as shown in Figs. 4.25 and 4.26.

Example 6.11
Design the Table 6.21 FIR filter by using the frequency sampling design.

Solution
According to the filter length, 2π radian frequency is divided into 7 segments as below.

$$H[k] \leftrightarrow H\left(\frac{2\pi}{L}k\right) = H\left(\frac{2\pi}{7}k\right)$$

First trial of frequency set is below.

$$\omega = \{\omega_0, \omega_1, \omega_2, \omega_3\} = \left\{0, \frac{2\pi}{7}, 2\frac{2\pi}{7}, 3\frac{2\pi}{7}\right\}$$

The corresponding $H[k]$ is below based on the LPF specification from Table 6.21.

$$\begin{bmatrix} H[0] \\ H[1] \\ H[2] \\ H[3] \end{bmatrix} = \begin{bmatrix} 1 \\ 1 \\ 0 \\ 0 \end{bmatrix}$$

The DFT synthesis equation in short matrix form compute the impulse response half as below.

$$\begin{bmatrix} h_t[0] \\ h_t[1] \\ h_t[2] \\ h_t[3] \end{bmatrix} = \frac{1}{7}\begin{bmatrix} 1 & 2 & 2 & 2 \\ 1 & 2\cos\left(\frac{2\pi}{7}(1)1\right) & 2\cos\left(\frac{2\pi}{7}(1)2\right) & 2\cos\left(\frac{2\pi}{7}(1)3\right) \\ 1 & 2\cos\left(\frac{2\pi}{7}(2)1\right) & 2\cos\left(\frac{2\pi}{7}(2)2\right) & 2\cos\left(\frac{2\pi}{7}(2)3\right) \\ 1 & 2\cos\left(\frac{2\pi}{7}(3)1\right) & 2\cos\left(\frac{2\pi}{7}(3)2\right) & 2\cos\left(\frac{2\pi}{7}(3)3\right) \end{bmatrix}\begin{bmatrix} H[0] \\ H[1] \\ H[2] \\ H[3] \end{bmatrix}$$

$$\begin{bmatrix} 0.4286 \\ 0.3210 \\ 0.0793 \\ -0.1146 \end{bmatrix} = \frac{1}{7}\begin{bmatrix} 1 & 2 & 2 & 2 \\ 1 & 2\cos\left(\frac{2\pi}{7}(1)1\right) & 2\cos\left(\frac{2\pi}{7}(1)2\right) & 2\cos\left(\frac{2\pi}{7}(1)3\right) \\ 1 & 2\cos\left(\frac{2\pi}{7}(2)1\right) & 2\cos\left(\frac{2\pi}{7}(2)2\right) & 2\cos\left(\frac{2\pi}{7}(2)3\right) \\ 1 & 2\cos\left(\frac{2\pi}{7}(3)1\right) & 2\cos\left(\frac{2\pi}{7}(3)2\right) & 2\cos\left(\frac{2\pi}{7}(3)3\right) \end{bmatrix}\begin{bmatrix} 1 \\ 1 \\ 0 \\ 0 \end{bmatrix}$$

The frequency response of first trial is shown in Fig. 6.28. Observe that the stem plot is the design locations. The transition frequency $\Delta\omega(\omega_{st} - \omega_p)$ in Fig. 6.28 occupies further range than the specification; therefore, the first trial should be modified.

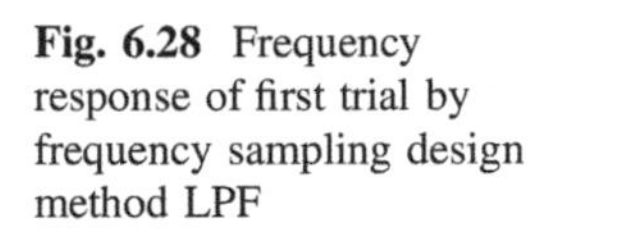

Fig. 6.28 Frequency response of first trial by frequency sampling design method LPF

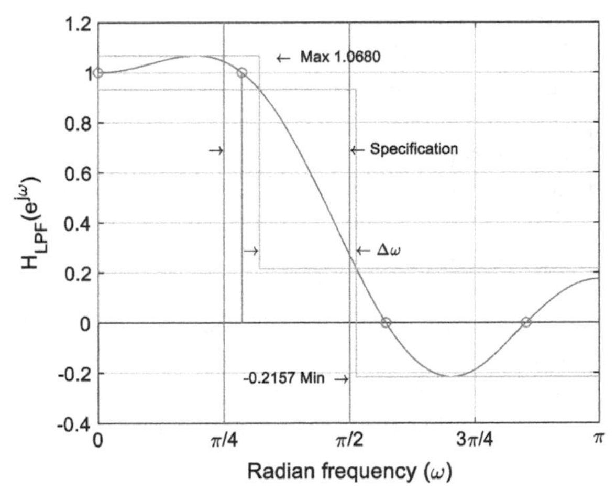

Prog. 6.22 MATLAB program for Fig. 6.28.

```
ww = linspace(0,pi,1000);
nn = (0:6);
RR = [1 2 2 2; ...
    1 2*cos(2*pi/7) 2*cos(2*pi*2/7) 2*cos(2*pi*3/7); ...
    1 2*cos(2*pi*2/7) 2*cos(2*pi*2*2/7) 2*cos(2*pi*2*3/7); ...
    1 2*cos(2*pi*3/7) 2*cos(2*pi*3*2/7) 2*cos(2*pi*3*3/7)];
HH = [1; 1; 0; 0];
ht = RR*HH/7;
HW3 = ht(1)+2*ht(2)*cos(ww)+2*ht(3)*cos(2*ww)+2*ht(4)*cos(3*ww);
hn = [flipud(ht); ht(2:end)];
w0 = [0 2*pi/7 2*2*pi/7 2*3*pi/7];
mag0 = [1 1 0 0];

figure,
plot(ww,HW3), grid, xlim([0 pi]), hold on
stem(w0, mag0), hold off
[M3,I3] = min(HW3);
[M4,I4] = max(HW3);
cc2 = max(find(HW3>(-M3)));
cc3 = min(find(HW3<(2-M4)));
line([0 ww(cc3)],[M4 M4],'Color',[0.93 0.69 0.13]);
line([ww(cc2) pi],[M3 M3],'Color',[0.93 0.69 0.13]);
line([pi/4 pi/4],[-0.4 1.2],'Color','magenta');
line([pi/2 pi/2],[-0.4 1.2],'Color','magenta');
straa1 = sprintf('Max %1.4f',M4);
aa1 = text(ww(I4)+0.5,M4,['\leftarrow    ' straa1]);
strbb1 = sprintf('%1.4f Min',M3);
bb1 = text(ww(I3)-1.3,M3,[strbb1 '    \rightarrow']);
hh1 = text(pi/4-0.1,0.7,'\rightarrow ');
ii1 = text(pi/2,0.7,'\leftarrow Specification');
```

To comply with specification, the new $H[k]$ value is below.

$$\begin{bmatrix} H[0] \\ H[1] \\ H[2] \\ H[3] \end{bmatrix} = \begin{bmatrix} 1 \\ 1 \\ -0.05 \\ 0 \end{bmatrix}$$

The corresponding impulse response is below.

$$\begin{bmatrix} 0.4143 \\ 0.3242 \\ 0.0922 \\ -0.1235 \end{bmatrix} = \frac{1}{7}\begin{bmatrix} 1 & 2 & 2 & 2 \\ 1 & 2\cos\left(\frac{2\pi}{7}(1)1\right) & 2\cos\left(\frac{2\pi}{7}(1)2\right) & 2\cos\left(\frac{2\pi}{7}(1)3\right) \\ 1 & 2\cos\left(\frac{2\pi}{7}(2)1\right) & 2\cos\left(\frac{2\pi}{7}(2)2\right) & 2\cos\left(\frac{2\pi}{7}(2)3\right) \\ 1 & 2\cos\left(\frac{2\pi}{7}(3)1\right) & 2\cos\left(\frac{2\pi}{7}(3)2\right) & 2\cos\left(\frac{2\pi}{7}(3)3\right) \end{bmatrix}\begin{bmatrix} 1 \\ 1 \\ -0.05 \\ 0 \end{bmatrix}$$

The frequency response of second trial is shown in Fig. 6.29. The transition frequency in Fig. 6.29 is located within the specification. You may further evaluate the LPF for better performance.

Prog. 6.23 MATLAB program for Fig. 6.29.

```
ww = linspace(0,pi,1000);
nn = (0:6);
RR = [1 2 2 2; ...
      1 2*cos(2*pi/7) 2*cos(2*pi*2/7) 2*cos(2*pi*3/7); ...
      1 2*cos(2*pi*2/7) 2*cos(2*pi*2*2/7) 2*cos(2*pi*2*3/7); ...
      1 2*cos(2*pi*3/7) 2*cos(2*pi*3*2/7) 2*cos(2*pi*3*3/7)];
HH = [1; 1; -0.05; 0];
ht = RR*HH/7;
HW3 = ht(1)+2*ht(2)*cos(ww)+2*ht(3)*cos(2*ww)+2*ht(4)*cos(3*ww);
hn = [flipud(ht); ht(2:end)];
w0 = [0 2*pi/7 2*2*pi/7 2*3*pi/7];
mag0 = [1 1 -0.05 0];

figure,
plot(ww,HW3), grid, xlim([0 pi]), hold on
stem(w0, mag0), hold off
[M3,I3] = min(HW3);
[M4,I4] = max(HW3);
cc2 = max(find(HW3>(-M3)));
cc3 = min(find(HW3<(2-M4)));
line([0 ww(cc3)],[M4 M4],'Color',[0.93 0.69 0.13]);
line([ww(cc2) pi],[M3 M3],'Color',[0.93 0.69 0.13]);
line([pi/4 pi/4],[-0.4 1.2],'Color','magenta');
line([pi/2 pi/2],[-0.4 1.2],'Color','magenta');
straa1 = sprintf('Max %1.4f',M4);
aa1 = text(ww(I4)+0.5,M4,['\leftarrow    ' straa1]);
strbb1 = sprintf('%1.4f Min',M3);
bb1 = text(ww(I3)-1.3,M3,[strbb1 '    \rightarrow']);
hh1 = text(pi/4-0.1,0.7,'\rightarrow ');
ii1 = text(pi/2,0.7,'\leftarrow Specification');
figure,
stem(nn,hn), grid, ylim([-0.2 0.5])
```

The Type I LPF with non-causality is below. The computed coefficients from the matrix form are the one half of the impulse response.

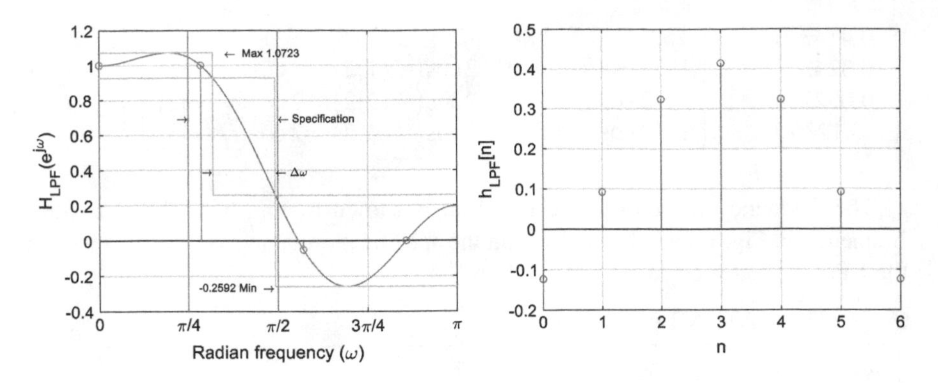

Fig. 6.29 Frequency response of frequency sampling design method LPF Type I

$$h_t[n] = -0.1235\delta[n+3] + 0.0922\delta[n+2] + 0.3242\delta[n+1] + 0.4143\delta[n] \\ + 0.3242\delta[n-1] + 0.0922\delta[n-2] - 0.1235\delta[n-3]$$

Shifting for causality finalizes the filter design.

$$h[n] = h_t[n-3] = -0.1235\delta[n] + 0.0922\delta[n-1] + 0.3242\delta[n-2] \\ + 0.4143\delta[n-3] + 0.3242\delta[n-4] + 0.0922\delta[n-5] - 0.1235\delta[n-6]$$

The passband ripple A_p can be calculated from the maximum deviation R_p as below.

$$A_P = 40\log_{10}(1 + R_p)$$

$$1.2127\text{dB} = 40\log_{10}(1 + 0.0723)$$

The equivalent stopband attenuation can be calculated as below.

$$A_{st} = -20\log_{10} R_{st}$$

$$11.7273\text{dB} = -20\log_{10} 0.2592$$

■

Frequency sampling design—Type II

The desired frequency response $H[k]$ shows the symmetric property as below for real $h_t[n]$ with length L which is N + 1 and odd N.

$$H[k] = H^*[-k]$$

The corresponding impulse response can be derived as below. Note that the Type II filter presents the zero output on π radian frequency as $H[L/2] = 0$.

$$
\begin{aligned}
h_t(n) &= \frac{1}{L}\sum_{k=0}^{L-1} H[k]e^{j\frac{2\pi}{L}kn} = \frac{1}{L}\sum_{k=-(L/2)+1}^{(L/2)-1} H[k]e^{j\frac{2\pi}{L}kn} + \frac{1}{L}H\left[\frac{L}{2}\right]e^{j\frac{2\pi}{L}\frac{L}{2}n}, \quad n \in \mathbb{R} \\
&= \frac{1}{L}H[0] + \frac{2}{L}\sum_{k=1}^{(L/2)-1} H[k]\cos\left(\frac{2\pi}{L}kn\right) + \frac{1}{L}H\left[\frac{L}{2}\right]e^{j\pi n} \\
&= \frac{1}{L}H[0] + \frac{2}{L}\sum_{k=1}^{(L/2)-1} H[k]\cos\left(\frac{2\pi}{L}kn\right), \text{ since } H\left[\frac{L}{2}\right] = 0 \because \text{Type II}
\end{aligned}
\tag{6.81}
$$

The DFT analysis equation is below to figure out the frequency response for Type II FIR filter.

$$
H[k] = 2\sum_{n=0}^{L/2-1} h_t\left(n+\tfrac{1}{2}\right)\cos\left(\tfrac{2\pi}{L}k\left(n+\tfrac{1}{2}\right)\right), \quad H[k] \in \mathbb{R} \tag{6.82}
$$

The matrix form of Eq. (6.75) is below. The cosine matrix (DFT matrix in Euler) is square matrix with $L + 1$ by $L + 1$ size.

$$
\begin{bmatrix} h_t\left(\frac{1}{2}\right) \\ h_t\left(\frac{3}{2}\right) \\ h_t\left(\frac{5}{2}\right) \\ \vdots \\ h_t\left(\frac{N}{2}\right) \end{bmatrix} = \frac{1}{L}
\begin{bmatrix}
1 & 2\cos\left(\frac{2\pi}{L}\left(\frac{1}{2}\right)1\right) & 2\cos\left(\frac{2\pi}{L}\left(\frac{1}{2}\right)2\right) & \cdots & 2\cos\left(\frac{2\pi}{L}\left(\frac{1}{2}\right)\left(\frac{L}{2}-1\right)\right) \\
1 & 2\cos\left(\frac{2\pi}{L}\left(\frac{3}{2}\right)1\right) & 2\cos\left(\frac{2\pi}{L}\left(\frac{3}{2}\right)2\right) & \cdots & 2\cos\left(\frac{2\pi}{L}\left(\frac{3}{2}\right)\left(\frac{L}{2}-1\right)\right) \\
1 & 2\cos\left(\frac{2\pi}{L}\left(\frac{5}{2}\right)1\right) & 2\cos\left(\frac{2\pi}{L}\left(\frac{5}{2}\right)2\right) & \cdots & 2\cos\left(\frac{2\pi}{L}\left(\frac{5}{2}\right)\left(\frac{L}{2}-1\right)\right) \\
\vdots & \vdots & \cdots & \vdots & \\
1 & 2\cos\left(\frac{2\pi}{L}\left(\frac{N}{2}\right)1\right) & 2\cos\left(\frac{2\pi}{L}\left(\frac{N}{2}\right)2\right) & \cdots & 2\cos\left(\frac{2\pi}{L}\left(\frac{N}{2}\right)\left(\frac{L}{2}-1\right)\right)
\end{bmatrix}
\begin{bmatrix} H[0] \\ H[1] \\ H[2] \\ \vdots \\ H[L/2-1] \end{bmatrix}
\tag{6.83}
$$

The complete $h_t[n]$ is below.

$$
\begin{aligned}
h_t[n] &= h_t\left(\frac{N}{2}\right)\delta\left[n+\frac{L}{2}\right] + \cdots + h_t\left(\frac{3}{2}\right)\delta[n+2] + h_t\left(\frac{1}{2}\right)\delta[n+1] \\
&\quad + h_t\left(\frac{1}{2}\right)\delta[n] + h_t\left(\frac{3}{2}\right)\delta[n-1] + \cdots + h_t\left(\frac{N}{2}\right)\delta[n-(L/2-1)]
\end{aligned}
\tag{6.84}
$$

To make the causal filter, perform the shift operation as below.

$$
h[n] = h_t[n - L/2] \tag{6.85}
$$

The Type II frequency sampling design method is identical to the DFT method without truncation in Sect. 4.5 as well. The required condition is that the impulse response length is equal to the frequency sampling length.

Table 6.22 Given filter specification

Filter specifications	Value
Filter type	Low pass filter
Filter realization	FIR Type II without window method
Filter length	8 ($N = 7$)
Passband frequency (f_p)	1000 Hz; $\omega_p = \pi/4$
Passband magnitude	0 dB
Stopband frequency (f_{st})	2000 Hz; $\omega_{st} = \pi/2$
Sampling frequency (f_s)	8000 Hz

Example 6.12

Design the Table 6.22 FIR filter by using the frequency sampling design.

Solution

According to the filter length, 2π radian frequency is divided into 8 segments as below.

$$H[k] \leftrightarrow H\left(\frac{2\pi}{L}k\right) = H\left(\frac{2\pi}{8}k\right)$$

First trial of frequency set is below.

$$\omega = \{\omega_0, \omega_1, \omega_2, \omega_3\} = \left\{0, \frac{2\pi}{8}, 2\frac{2\pi}{8}, 3\frac{2\pi}{8}\right\}$$

The corresponding $H[k]$ is below based on the LPF specification from Table 6.22.

$$\begin{bmatrix} H[0] \\ H[1] \\ H[2] \\ H[3] \end{bmatrix} = \begin{bmatrix} 1 \\ 1 \\ 0 \\ 0 \end{bmatrix}$$

The DFT synthesis equation in short matrix form compute the impulse response half as below.

$$\begin{bmatrix} h_t\left(\frac{1}{2}\right) \\ h_t\left(\frac{3}{2}\right) \\ h_t\left(\frac{5}{2}\right) \\ h_t\left(\frac{7}{2}\right) \end{bmatrix} = \frac{1}{8}\begin{bmatrix} 1 & 2\cos\left(\frac{2\pi}{8}\left(\frac{1}{2}\right)1\right) & 2\cos\left(\frac{2\pi}{8}\left(\frac{1}{2}\right)2\right) & 2\cos\left(\frac{2\pi}{8}\left(\frac{1}{2}\right)3\right) \\ 1 & 2\cos\left(\frac{2\pi}{8}\left(\frac{3}{2}\right)1\right) & 2\cos\left(\frac{2\pi}{8}\left(\frac{3}{2}\right)2\right) & 2\cos\left(\frac{2\pi}{8}\left(\frac{3}{2}\right)3\right) \\ 1 & 2\cos\left(\frac{2\pi}{8}\left(\frac{5}{2}\right)1\right) & 2\cos\left(\frac{2\pi}{8}\left(\frac{5}{2}\right)2\right) & 2\cos\left(\frac{2\pi}{8}\left(\frac{5}{2}\right)3\right) \\ 1 & 2\cos\left(\frac{2\pi}{8}\left(\frac{7}{2}\right)1\right) & 2\cos\left(\frac{2\pi}{8}\left(\frac{7}{2}\right)2\right) & 2\cos\left(\frac{2\pi}{8}\left(\frac{7}{2}\right)3\right) \end{bmatrix}\begin{bmatrix} H[0] \\ H[1] \\ H[2] \\ H[3] \end{bmatrix}$$

$$\begin{bmatrix} 0.3560 \\ 0.2207 \\ 0.0293 \\ -0.1060 \end{bmatrix} = \frac{1}{8}\begin{bmatrix} 1 & 2 & 2 & 2 \\ 1 & 2\cos\left(\frac{2\pi}{8}(1)1\right) & 2\cos\left(\frac{2\pi}{8}(1)2\right) & 2\cos\left(\frac{2\pi}{8}(1)3\right) \\ 1 & 2\cos\left(\frac{2\pi}{8}(2)1\right) & 2\cos\left(\frac{2\pi}{8}(2)2\right) & 2\cos\left(\frac{2\pi}{8}(2)3\right) \\ 1 & 2\cos\left(\frac{2\pi}{8}(3)1\right) & 2\cos\left(\frac{2\pi}{8}(3)2\right) & 2\cos\left(\frac{2\pi}{8}(3)3\right) \end{bmatrix}\begin{bmatrix} 1 \\ 1 \\ 0 \\ 0 \end{bmatrix}$$

Using the following equation, compute the continuous frequency response. Or use the DFT.

$$H\left(e^{j\omega}\right)=2\sum_{n=0}^{L/2-1}h_t\left(n+\frac{1}{2}\right)\cos\left(\omega\left(n+\frac{1}{2}\right)\right)$$

The frequency response of first trial is shown in Fig. 6.30. Note that the stem plot is the design locations. The transition frequency $\Delta\omega\left(\omega_{st}-\omega_p\right)$ in Fig. 6.30 is located within the specification. You may further evaluate the LPF for better performance.

Prog. 6.24 MATLAB program for Fig. 6.30.

```
ww = linspace(0,pi,1000);
nn = (0:7);
RR = [1 2*cos(2*pi*(1/2)/8) 2*cos(2*pi*(1/2)*2/8) 2*cos(2*pi*(1/2)*3/8); ...
      1 2*cos(2*pi*(3/2)/8) 2*cos(2*pi*(3/2)*2/8) 2*cos(2*pi*(3/2)*3/8); ...
      1 2*cos(2*pi*(5/2)/8) 2*cos(2*pi*(5/2)*2/8) 2*cos(2*pi*(5/2)*3/8); ...
      1 2*cos(2*pi*(7/2)/8) 2*cos(2*pi*(7/2)*2/8) 2*cos(2*pi*(7/2)*3/8)];
HH = [1; 1; 0; 0];
ht = RR*HH/8;
HW3 =
2*ht(1)*cos(ww/2)+2*ht(2)*cos((3/2)*ww)+2*ht(3)*cos((5/2)*ww)+2*ht(4)*
cos((7/2)*ww);
hn = [flipud(ht); ht];
w0 = [0 2*pi/8 2*2*pi/8 2*3*pi/8];
mag0 = [1 1 0 0];

figure,
plot(ww,HW3), grid, xlim([0 pi]), hold on;
stem(w0, mag0), hold off
[M3,I3] = min(HW3);
[M4,I4] = max(HW3);
cc2 = max(find(HW3>(-M3)));
cc3 = min(find(HW3<(2-M4)));
line([0 ww(cc3)],[M4 M4],'Color',[0.93 0.69 0.13]);
line([ww(cc2) pi],[M3 M3],'Color',[0.93 0.69 0.13]);
line([pi/4 pi/4],[-0.4 1.2],'Color','magenta');
line([pi/2 pi/2],[-0.4 1.2],'Color','magenta');
straa1 = sprintf('Max %1.4f',M4);
aa1 = text(ww(I4)+0.5,M4,['\leftarrow    ' straa1]);
strbb1 = sprintf('%1.4f Min',M3);
bb1 = text(ww(I3)-1.3,M3,[strbb1 '    \rightarrow']);
hh1 = text(pi/4-0.1,0.7,'\rightarrow ');
ii1 = text(pi/2,0.7,'\leftarrow Specification');
figure,
stem(nn,hn), grid, ylim([-0.2 0.5]);
```

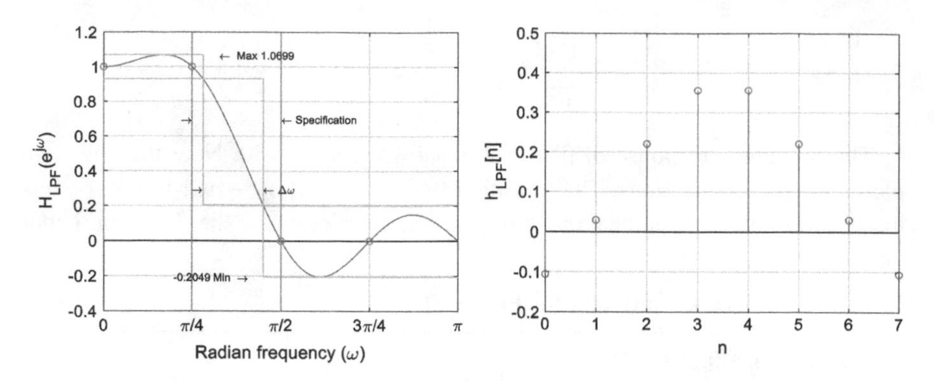

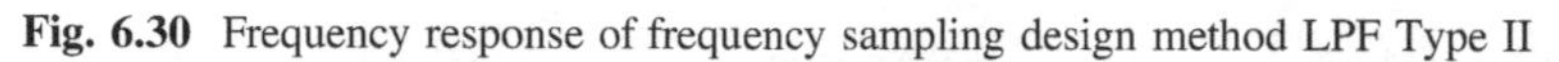

Fig. 6.30 Frequency response of frequency sampling design method LPF Type II

The Type II LPF with causality is below.

$$h[n] = -0.1060\delta[n] + 0.0293\delta[n-1] + 0.2207\delta[n-2] + 0.3560\delta[n-3] \\ + 0.3560\delta[n-4] + 0.2207\delta[n-5] + 0.0293\delta[n-6] - 0.1060\delta[n-7]$$

The passband ripple A_p can be calculated from the maximum deviation R_p as below.

$$A_P = 40\log_{10}(1 + R_p)$$

$$1.1737\text{dB} = 40\log_{10}(1 + 0.0699)$$

The equivalent stopband attenuation can be calculated as below.

$$A_{st} = -20\log_{10} R_{st}$$

$$13.7692\text{dB} = -20\log_{10} 0.2049$$

■

Least squared error frequency domain design—Discrete frequency sampling DFT

Similar to the frequency sampling design, the FIR filter uses a larger number of sampling point than the filter order [7]. The more equations than unknowns indicate the least square solution. The DFT analysis of given non-causal impulse response $h[n]$ Type I is shown in Eq. (6.86). Note that the filter length is L which is $N + 1$.

$$H[k] = h[0] + 2\sum_{n=1}^{N/2} h[n]\cos\left(\tfrac{2\pi}{L}kn\right), 0 \le k \le L-1 \tag{6.86}$$

The DFT synthesis for impulse response $h_t[n]$ can be performed as Eq. (6.87) with L length.

$$
\begin{aligned}
h_t[n] &= \frac{1}{L}\sum_{k=0}^{L-1} H[k]e^{j\frac{2\pi}{L}kn} = \frac{1}{L}\sum_{k=-N/2}^{N/2} H[k]e^{j\frac{2\pi}{L}kn} \\
&= \frac{1}{L}H[0] + \frac{2}{L}\sum_{k=1}^{N/2} H[k]\cos\left(\frac{2\pi}{L}kn\right)
\end{aligned}
\tag{6.87}
$$

The matrix form of Eq. (6.87) is below.

$$
\begin{bmatrix}
e^{j\frac{2\pi}{L}\left(-\frac{N}{2}\right)\frac{N}{2}} & \cdots & e^{j\frac{2\pi}{L}\left(-\frac{N}{2}\right)} & 1 & e^{-j\frac{2\pi}{L}\left(-\frac{N}{2}\right)} & \cdots & e^{-j\frac{2\pi}{L}\left(-\frac{N}{2}\right)\frac{N}{2}} \\
\vdots & \cdots & \vdots & \vdots & \vdots & \cdots & \vdots \\
e^{j\frac{2\pi}{L}(-1)\frac{N}{2}} & \cdots & e^{j\frac{2\pi}{L}(-1)} & 1 & e^{-j\frac{2\pi}{L}(-1)} & \cdots & e^{-j\frac{2\pi}{L}(-1)\frac{N}{2}} \\
1 & \cdots & 1 & 1 & 1 & \cdots & 1 \\
e^{j\frac{2\pi}{L}(1)\frac{N}{2}} & \cdots & e^{j\frac{2\pi}{L}(1)} & 1 & e^{-j\frac{2\pi}{L}(1)} & \cdots & e^{-j\frac{2\pi}{L}(1)\frac{N}{2}} \\
\vdots & \vdots & \vdots & \vdots & \vdots & \ddots & \vdots \\
e^{j\frac{2\pi}{L}\left(\frac{N}{2}\right)\frac{N}{2}} & \cdots & e^{j\frac{2\pi}{L}\left(\frac{N}{2}\right)} & 1 & e^{-j\frac{2\pi}{L}\left(\frac{N}{2}\right)} & \cdots & e^{-j\frac{2\pi}{L}\left(\frac{N}{2}\right)\frac{N}{2}}
\end{bmatrix}
\begin{bmatrix}
h_t[-N/2] \\ \vdots \\ h_t[-1] \\ h_t[0] \\ h_t[1] \\ \vdots \\ h_t[N/2]
\end{bmatrix}
=
\begin{bmatrix}
H[-N/2] \\ \vdots \\ H[-1] \\ H[0] \\ H[1] \\ \vdots \\ H[N/2]
\end{bmatrix}
\tag{6.88}
$$

The Euler equation compresses the exponential matrix in short cosine form as below for Type I.

$$
\begin{bmatrix}
1 & 2 & 2 & \cdots & 2 \\
1 & 2\cos\left(\frac{2\pi}{L}(1)1\right) & 2\cos\left(\frac{2\pi}{L}(1)2\right) & \cdots & 2\cos\left(\frac{2\pi}{L}(1)\frac{N}{2}\right) \\
1 & 2\cos\left(\frac{2\pi}{L}(2)1\right) & 2\cos\left(\frac{2\pi}{L}(2)2\right) & \cdots & 2\cos\left(\frac{2\pi}{L}(2)\frac{N}{2}\right) \\
\vdots & \vdots & \vdots & \cdots & \vdots \\
1 & 2\cos\left(\frac{2\pi}{L}\left(\frac{N}{2}\right)1\right) & 2\cos\left(\frac{2\pi}{L}\left(\frac{N}{2}\right)2\right) & \cdots & 2\cos\left(\frac{2\pi}{L}\left(\frac{N}{2}\right)\frac{N}{2}\right)
\end{bmatrix}
\begin{bmatrix}
h_t[0] \\ h_t[1] \\ h_t[2] \\ \vdots \\ h_t[N/2]
\end{bmatrix}
=
\begin{bmatrix}
H[0] \\ H[1] \\ H[2] \\ \vdots \\ H[N/2]
\end{bmatrix}
\tag{6.89}
$$

Increase number of frequency sampling as $M > L$. Let $P = M - 1$. Again, it is Type I.

$$
\begin{aligned}
&H[k] = h[0] + 2\sum_{n=1}^{N/2} h[n]\cos\left(\frac{2\pi}{M}kn\right) \\
&\text{for } 0 \le k \le M-1 = P \quad \text{and} \quad M > L = N+1
\end{aligned}
\tag{6.90}
$$

The matrix form of Eq. (6.90) is below.

$$\begin{bmatrix} 1 & 2 & 2 & \cdots & 2 \\ 1 & 2\cos\left(\frac{2\pi}{M}(1)1\right) & 2\cos\left(\frac{2\pi}{M}(1)2\right) & \cdots & 2\cos\left(\frac{2\pi}{M}(1)\frac{N}{2}\right) \\ 1 & 2\cos\left(\frac{2\pi}{M}(2)1\right) & 2\cos\left(\frac{2\pi}{M}(2)2\right) & \cdots & 2\cos\left(\frac{2\pi}{M}(2)\frac{N}{2}\right) \\ 1 & 2\cos\left(\frac{2\pi}{M}(3)1\right) & 2\cos\left(\frac{2\pi}{M}(3)2\right) & \cdots & 2\cos\left(\frac{2\pi}{M}(3)\frac{N}{2}\right) \\ \vdots & & & \cdots & \\ 1 & 2\cos\left(\frac{2\pi}{M}\left(\frac{P}{2}\right)(1)\right) & 2\cos\left(\frac{2\pi}{M}\left(\frac{P}{2}\right)(2)\right) & \cdots & 2\cos\left(\frac{2\pi}{M}\left(\frac{P}{2}\right)\left(\frac{N}{2}\right)\right) \end{bmatrix} \begin{bmatrix} h_t[0] \\ h_t[1] \\ h_t[2] \\ \vdots \\ h_t[N/2] \end{bmatrix} = \begin{bmatrix} H[0] \\ H[1] \\ H[2] \\ H[3] \\ \vdots \\ H[P/2] \end{bmatrix} \tag{6.91}$$

The matrix variable representation is below.

$$\mathbf{Rh} = \mathbf{d} \tag{6.92}$$

$$\text{where } \mathbf{R}: \left(\frac{P}{2}+1\right) \times \left(\frac{N}{2}+1\right); \quad \mathbf{h}: \left(\frac{N}{2}+1\right) \times 1; \quad \mathbf{d}: \left(\frac{p}{2}+1\right) \times 1$$

Since the **R** matrix is not square form and shows the overdetermined system. Pseudoinverse [12] is used to compute the impulse response.

$$\left(\mathbf{R}^T\mathbf{R}\right)^{-1}\mathbf{R}^T\mathbf{Rh} = \left(\mathbf{R}^T\mathbf{R}\right)^{-1}\mathbf{R}^T\mathbf{d}$$

$$\mathbf{h} = \left(\mathbf{R}^T\mathbf{R}\right)^{-1}\mathbf{R}^T\mathbf{d} \tag{6.93}$$

The complete $h_t[n]$ is below.

$$\begin{aligned} h_t[n] = {} & h_t[N/2]\delta[n+N/2] + \cdots + h_t[1]\delta[n+1] + h_t[0]\delta[n] + h_t[1]\delta[n-1] \\ & + \cdots + h_t[N/2]\delta[n-N/2] \end{aligned} \tag{6.94}$$

To make the causal filter, perform the shift operation as below.

$$h[n] = h_t[n - N/2] \tag{6.95}$$

The least square error design method provides least square solution based on the overdetermined system which contains the further frequency sampling than the filter length.

Example 6.13
Design the Table 6.23 FIR filter by using the least squared error frequency domain design method.

Solution
The 2π radian frequency is divided into $M = 16$ segments as below.

Table 6.23 Given filter specification

Filter specifications	Value
Filter type	Low pass filter
Filter realization	FIR Type I without window method
Filter length	7 ($N = 6$)
Passband frequency (f_p)	1000 Hz; $\omega_p = \pi/4$
Passband magnitude	0 dB
Stopband frequency (f_{st})	2000 Hz; $\omega_{st} = \pi/2$
Sampling frequency (f_s)	8000 Hz

$$H[k] \leftrightarrow H\left(\frac{2\pi}{M}k\right) = H\left(\frac{2\pi}{16}k\right)$$

First trial of frequency set is below.

$$\begin{aligned}\omega &= \{\omega_0, \omega_1, \omega_2, \omega_3, \omega_4, \omega_5, \omega_6, \omega_7, \omega_8\} \\ &= \left\{0, \frac{2\pi}{16}, 2\frac{2\pi}{16}, 3\frac{2\pi}{16}, 4\frac{2\pi}{16}, 5\frac{2\pi}{16}, 6\frac{2\pi}{16}, 7\frac{2\pi}{16}, 8\frac{2\pi}{16}\right\}\end{aligned}$$

Since the $3 \cdot 2\pi/16$ is located at transition band, we omit the frequency from the designated set. Or, you may choose another value for the transition frequency.

$$\begin{aligned}\omega &= \{\omega_0, \omega_1, \omega_2, \omega_3, \omega_4, \omega_5, \omega_6, \omega_7\} \\ &= \left\{0, \frac{2\pi}{16}, 2\frac{2\pi}{16}, 4\frac{2\pi}{16}, 5\frac{2\pi}{16}, 6\frac{2\pi}{16}, 7\frac{2\pi}{16}, 8\frac{2\pi}{16}\right\}\end{aligned}$$

The corresponding $H[k]$ is below

$$\begin{bmatrix} H[0] \\ H[1] \\ H[2] \\ H[3] \\ H[4] \\ H[5] \\ H[6] \\ H[7] \end{bmatrix} = \begin{bmatrix} 1 \\ 1 \\ 1 \\ 0 \\ 0 \\ 0 \\ 0 \\ 0 \end{bmatrix}$$

The matrix form of Eq. (6.77) is below.

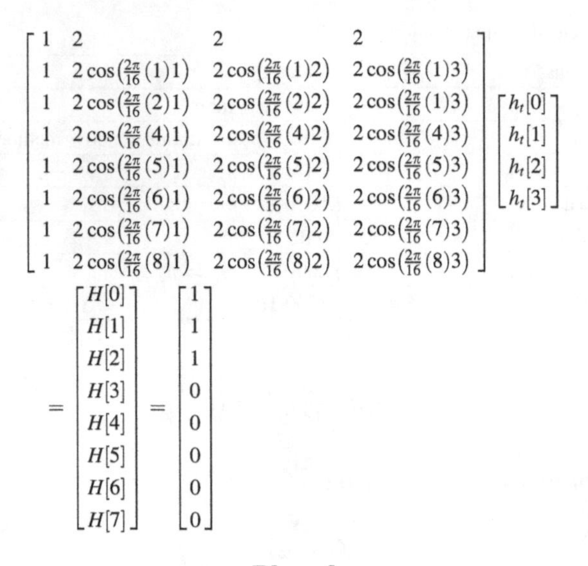

$$\begin{bmatrix} 1 & 2 & 2 & 2 \\ 1 & 2\cos\left(\frac{2\pi}{16}(1)1\right) & 2\cos\left(\frac{2\pi}{16}(1)2\right) & 2\cos\left(\frac{2\pi}{16}(1)3\right) \\ 1 & 2\cos\left(\frac{2\pi}{16}(2)1\right) & 2\cos\left(\frac{2\pi}{16}(2)2\right) & 2\cos\left(\frac{2\pi}{16}(1)3\right) \\ 1 & 2\cos\left(\frac{2\pi}{16}(4)1\right) & 2\cos\left(\frac{2\pi}{16}(4)2\right) & 2\cos\left(\frac{2\pi}{16}(4)3\right) \\ 1 & 2\cos\left(\frac{2\pi}{16}(5)1\right) & 2\cos\left(\frac{2\pi}{16}(5)2\right) & 2\cos\left(\frac{2\pi}{16}(5)3\right) \\ 1 & 2\cos\left(\frac{2\pi}{16}(6)1\right) & 2\cos\left(\frac{2\pi}{16}(6)2\right) & 2\cos\left(\frac{2\pi}{16}(6)3\right) \\ 1 & 2\cos\left(\frac{2\pi}{16}(7)1\right) & 2\cos\left(\frac{2\pi}{16}(7)2\right) & 2\cos\left(\frac{2\pi}{16}(7)3\right) \\ 1 & 2\cos\left(\frac{2\pi}{16}(8)1\right) & 2\cos\left(\frac{2\pi}{16}(8)2\right) & 2\cos\left(\frac{2\pi}{16}(8)3\right) \end{bmatrix} \begin{bmatrix} h_t[0] \\ h_t[1] \\ h_t[2] \\ h_t[3] \end{bmatrix} = \begin{bmatrix} H[0] \\ H[1] \\ H[2] \\ H[3] \\ H[4] \\ H[5] \\ H[6] \\ H[7] \end{bmatrix} = \begin{bmatrix} 1 \\ 1 \\ 1 \\ 0 \\ 0 \\ 0 \\ 0 \\ 0 \end{bmatrix}$$

$$\mathbf{Rh} = \mathbf{d}$$

We can calculate the least square solution by below equation.

$$\mathbf{h} = \left(\mathbf{R}^T\mathbf{R}\right)^{-1}\mathbf{R}^T\mathbf{d} = \begin{bmatrix} 0.3711 \\ 0.2914 \\ 0.0960 \\ -0.0399 \end{bmatrix}$$

In Fig. 6.31 frequency response, observe that the stem plot is the design locations. Unlike to the frequency sampling method, the response does not exactly follow the design locations. In the lease square sense, the response illustrates the

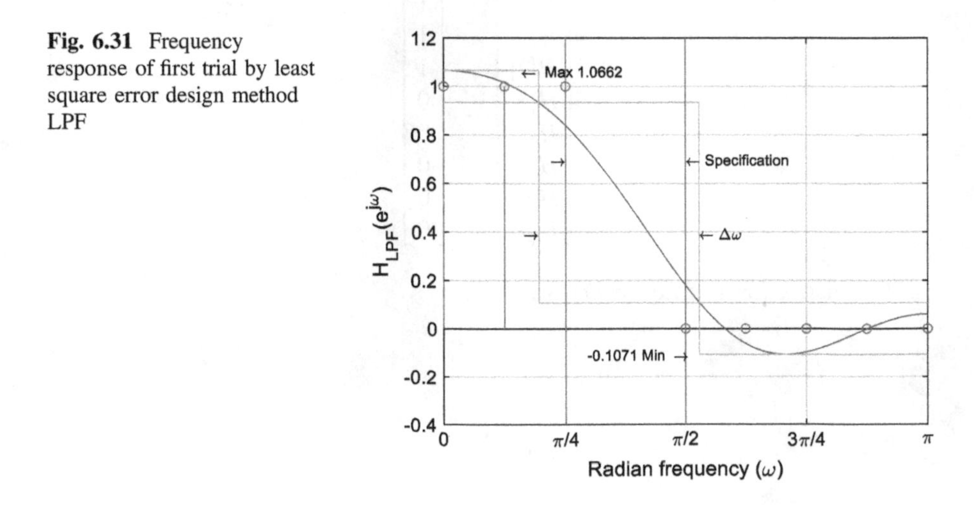

Fig. 6.31 Frequency response of first trial by least square error design method LPF

best curve to represent the specification. The transition frequency $\Delta\omega\left(\omega_{st} - \omega_p\right)$ in Fig. 6.31 occupies further range than the specification; therefore, the first trial should be modified.

Prog. 6.25 MATLAB program for Fig. 6.31.

```
ww = linspace(0,pi,1000);
nn = (0:6);
RR = [1 2 2 2; ...
      1 2*cos(2*pi/16) 2*cos(2*pi*2/16) 2*cos(2*pi*3/16); ...
      1 2*cos(2*pi*2/16) 2*cos(2*pi*2*2/16) 2*cos(2*pi*2*3/16); ...
      1 2*cos(2*pi*4/16) 2*cos(2*pi*4*2/16) 2*cos(2*pi*4*3/16); ...
      1 2*cos(2*pi*5/16) 2*cos(2*pi*5*2/16) 2*cos(2*pi*5*3/16); ...
      1 2*cos(2*pi*6/16) 2*cos(2*pi*6*2/16) 2*cos(2*pi*6*3/16); ...
      1 2*cos(2*pi*7/16) 2*cos(2*pi*7*2/16) 2*cos(2*pi*7*3/16); ...
      1 2*cos(2*pi*8/16) 2*cos(2*pi*8*2/16) 2*cos(2*pi*8*3/16)];
HH = [1; 1; 1; 0; 0; 0; 0; 0];
ht = inv(RR'*RR)*RR'*HH;
HW3 = ht(1)+2*ht(2)*cos(ww)+2*ht(3)*cos(2*ww)+2*ht(4)*cos(3*ww);
hn = [flipud(ht); ht(2:end)];
w0 = [0 2*pi/16 2*2*pi/16 2*4*pi/16 2*5*pi/16 2*6*pi/16 2*7*pi/16
2*8*pi/16];
mag0 = [1 1 1 0 0 0 0 0];

figure,
plot(ww,HW3), grid, xlim([0 pi]), hold on;
stem(w0, mag0), hold off;
[M3,I3] = min(HW3);
[M4,I4] = max(HW3);
line([pi/4 pi/4],[-0.4 1.2],'Color','magenta');
line([pi/2 pi/2],[-0.4 1.2],'Color','magenta');
straa1 = sprintf('Max %1.4f',M4);
aa1 = text(ww(I4)+0.5,M4,['\leftarrow    ' straa1]);
strbb1 = sprintf('%1.4f Min',M3);
bb1 = text(ww(I3)-1.3,M3,[strbb1 '    \rightarrow']);
hh1 = text(pi/4-0.1,0.7,'\rightarrow ');
ii1 = text(pi/2,0.7,'\leftarrow Specification');
figure,
stem(nn,hn), grid, ylim([-0.2 0.5]);
```

To comply with specification, the new $H[k]$ value is below. We adjust certain magnitudes to reduce the transition by trial and error.

$$\begin{bmatrix} H[0] \\ H[1] \\ H[2] \\ H[3] \\ H[4] \\ H[5] \\ H[6] \\ H[7] \end{bmatrix} = \begin{bmatrix} 1 \\ 1.2 \\ 1 \\ 0 \\ -0.2 \\ 0 \\ 0 \\ 0 \end{bmatrix}$$

The corresponding impulse response is below.

$$\mathbf{h} = \left(\mathbf{R}^T\mathbf{R}\right)^{-1}\mathbf{R}^T\mathbf{d} = \begin{bmatrix} 0.3663 \\ 0.3216 \\ 0.1239 \\ -0.0579 \end{bmatrix}$$

The frequency response of second trial is shown in Fig. 6.32. The transition frequency in Fig. 6.32 is located within the specification. You may further evaluate the LPF for better performance.

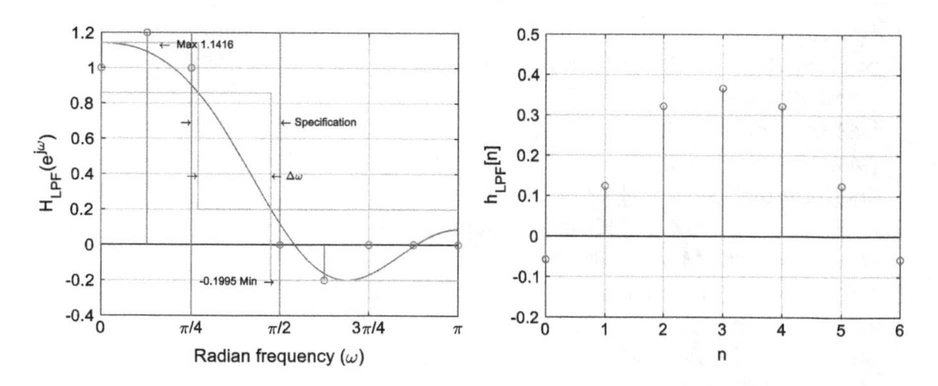

Fig. 6.32 Frequency response of least squared error frequency domain design method LPF Type I

Prog. 6.26 MATLAB program for Fig. 6.32.

```
ww = linspace(0,pi,1000);
nn = (0:6);
RR = [1 2 2 2; ...
      1 2*cos(2*pi/16) 2*cos(2*pi*2/16) 2*cos(2*pi*3/16); ...
      1 2*cos(2*pi*2/16) 2*cos(2*pi*2*2/16) 2*cos(2*pi*2*3/16); ...
      1 2*cos(2*pi*4/16) 2*cos(2*pi*4*2/16) 2*cos(2*pi*4*3/16); ...
      1 2*cos(2*pi*5/16) 2*cos(2*pi*5*2/16) 2*cos(2*pi*5*3/16); ...
      1 2*cos(2*pi*6/16) 2*cos(2*pi*6*2/16) 2*cos(2*pi*6*3/16); ...
      1 2*cos(2*pi*7/16) 2*cos(2*pi*7*2/16) 2*cos(2*pi*7*3/16); ...
      1 2*cos(2*pi*8/16) 2*cos(2*pi*8*2/16) 2*cos(2*pi*8*3/16)];
HH = [1; 1.2; 1; 0; -0.2; 0; 0; 0];
ht = inv(RR'*RR)*RR'*HH;
HW3 = ht(1)+2*ht(2)*cos(ww)+2*ht(3)*cos(2*ww)+2*ht(4)*cos(3*ww);
hn = [flipud(ht); ht(2:end)];
w0 = [0 2*pi/16 2*2*pi/16 2*4*pi/16 2*5*pi/16 2*6*pi/16 2*7*pi/16
2*8*pi/16];
mag0 = [1 1.2 1 0 -0.2 0 0 0];

figure,
plot(ww,HW3), grid, xlim([0 pi]), hold on;
stem(w0, mag0), hold off;
[M3,I3] = min(HW3);
[M4,I4] = max(HW3);
line([pi/4 pi/4],[-0.4 1.2],'Color','magenta');
line([pi/2 pi/2],[-0.4 1.2],'Color','magenta');
straa1 = sprintf('Max %1.4f',M4);
aa1 = text(ww(I4)+0.5,M4,['\leftarrow    ' straa1]);
strbb1 = sprintf('%1.4f Min',M3);
bb1 = text(ww(I3)-1.3,M3,[strbb1 '    \rightarrow']);
hh1 = text(pi/4-0.1,0.7,'\rightarrow ');
ii1 = text(pi/2,0.7,'\leftarrow Specification');
figure,
stem(nn,hn), grid, ylim([-0.2 0.5]);
```

The Type I LPF with non-causality is below. The computed coefficients from the matrix form are the one half of the impulse response.

$$h_t[n] = -0.0579\delta[n+3] + 0.1239\delta[n+2] + 0.3216\delta[n+1] + 0.3663\delta[n] \\ + 0.3216\delta[n-1] + 0.1239\delta[n-2] - 0.0579\delta[n-3]$$

Shifting for causality finalizes the filter design.

$$h[n] = h_t[n-3] = -0.0579\delta[n] + 0.1239\delta[n-1] + 0.3216\delta[n-2] \\ + 0.3663\delta[n-3] + 0.3216\delta[n-4] + 0.1239\delta[n-5] - 0.0579\delta[n-6]$$

The passband ripple A_p can be calculated from the maximum deviation R_p as below.

$$A_P = 40\log_{10}(1 + R_p)$$

$$2.3006\text{dB} = 40\log_{10}(1 + 0.1416)$$

The equivalent stopband attenuation can be calculated as below.

$$A_{st} = -20\log_{10} R_{st}$$

$$14.0011\text{dB} = -20\log_{10} 0.1995$$

■

Least squared error frequency domain design—Integral squared error approximation criterion

The least squared error method in frequency domain from the discrete sampling presents the least squared solution for the non-continuous frequency sampling. For the least squared error method in continuous domain [7], the error is defined as squared L_2-norm shown in Eq. (6.96).

$$\|E(e^{j\omega})\|_2^2 = \frac{1}{2\pi}\int_{-\pi}^{\pi} |H(e^{j\omega}) - H_d(e^{j\omega})|^2 d\omega \tag{6.96}$$

$$\text{where } H\left(e^{j\omega}\right) \in \mathbb{R} \text{ and } H_d\left(e^{j\omega}\right) \in \mathbb{R}$$

where $H_d(e^{j\omega})$ is the desired frequency response of the filter. To minimize the error in least squared sense, we can describe the error function in discrete time domain according to the Parseval's theorem as Eq. (6.97).

$$\|E(e^{j\omega})\|_2^2 = \sum_{n=-\infty}^{\infty} |h[n] - h_d[n]|^2 \tag{6.97}$$

$$\text{where } H\left(e^{j\omega}\right) \overset{\text{DTFT}}{\leftrightarrow} h[n] \; H_d\left(e^{j\omega}\right) \overset{\text{DTFT}}{\leftrightarrow} h_d[n]$$

The bandlimited desired filter generates the infinite length impulse response $h_d[n]$ and the FIR filter $h[n]$ has the finite length response. The error function can be reorganized as Eq. (6.98). The filter length is $L = N+1$ with Type I and zero phase (non-causal).

$$\|E(e^{j\omega})\|_2^2 = \sum_{n=-N/2}^{N/2} |h[n] - h_d[n]|^2 + \sum_{n=N/2+1}^{\infty} 2|h_d[n]|^2 \tag{6.98}$$

The Type II can be described by the simple modification from above equation. The minimum error can be achieved by the equality between the $h[n]$ and $h_d[n]$ for FIR interval and the overall error is demonstrated by the squared residual response sum.

$$h[n] = h_d[n] \text{ for } -N/2 \leq n \leq N/2 \tag{6.99}$$

The final filter is the shifted version of the $h[n]$ for causality. The procedure shown above is identical to the simple filter design at Sect. 4.6. The truncated impulse response from the DTFT is an optimal approximation in the sense that the integral squared error is minimized in continuous frequency domain. Various type windows can apply on the designed impulse response to control the ripples over passband and stopband. However, the window applying does not provide the least squared error solution for the designed filter.

$$\|E(e^{j\omega})\|_2^2 = \sum_{n=-\infty}^{\infty} |w[n]h[n] - h_d[n]|^2$$

$$= \sum_{n=-N/2}^{N/2} |w[n]h[n] - h_d[n]|^2 + \sum_{n=N/2+1}^{\infty} 2|h_d[n]|^2 \tag{6.100}$$

$$\min_{h[n]\in\mathbb{R}\&\text{FIR}} \|E(e^{j\omega})\|_2^2 = \sum_{n=N/2+1}^{\infty} 2|h_d[n]|^2 \tag{6.101}$$

By using the length limited FIR filter, the minimum achievable residual error energy is derived in Eq. (6.101). The inequality between the desired response $h_d[n]$ and windowed response $w[n]h[n]$ represents the greater residual error energy than the minimum value; therefore, the windowed response is the non-least squared error design.

Example 6.14

Design the Table 6.24 FIR filter by using least squared error method in continuous frequency domain.

Solution

The method is identical to the simple filter design at Sect. 4.5 and the extension based on the window method can be found at FIR filters from Sect. 6.2. The simple revisit to realize above filter is below. The center of the passband and stopband edge is below as cutoff frequency.

Table 6.24 Given filter specification

Filter specifications	Value
Filter type	Low pass filter
Filter realization	FIR Type I without window method
Filter length	Optimal value from window
Passband frequency (f_p)	1000 Hz; $\omega_p = \pi/4$
Passband magnitude	0 dB
Stopband frequency (f_{st})	2000 Hz; $\omega_{st} = \pi/2$
Sampling frequency (f_s)	8000 Hz

$$\omega_c = \omega_p + \frac{\omega_{st} - \omega_p}{2} = \frac{\pi}{4} + \frac{\pi}{8} = \frac{3\pi}{8}$$

The frequency transition between the passband and stopband edge is below

$$\Delta\omega = \omega_{st} - \omega_p = \frac{\pi}{2} - \frac{\pi}{4} = \frac{\pi}{4}$$

The main lobe width corresponds to the radian distance between the maximum and minimum ripple; hence, the use of the main lobe width designs the filter with sharper transition than the specification. Below is the main lobe width for rectangular window.

$$\frac{4\pi}{N+1} \leq \frac{\pi}{4} \quad N \geq 15$$

Let the $N = 16$ for Type I filter. According to the window analysis, the extreme ripples in both bands are identical in terms of deviation from the ideal response magnitude. Realization and evaluation are below.

$$h_d[n] = \frac{1}{2\pi} \int_{-3\pi/8}^{3\pi/8} 1e^{j\omega n} d\omega = \frac{\sin(3\pi n/8)}{\pi n}$$

$$\|E(e^{j\omega})\|_2^2 = \sum_{n=-8}^{8} |h[n] - h_d[n]|^2 + \sum_{n=9}^{\infty} 2|h_d[n]|^2$$

$$h[n] = \begin{cases} h_d[n] & \text{for } -8 \leq n \leq 8 \\ 0 & \text{Otherwise} \end{cases}$$

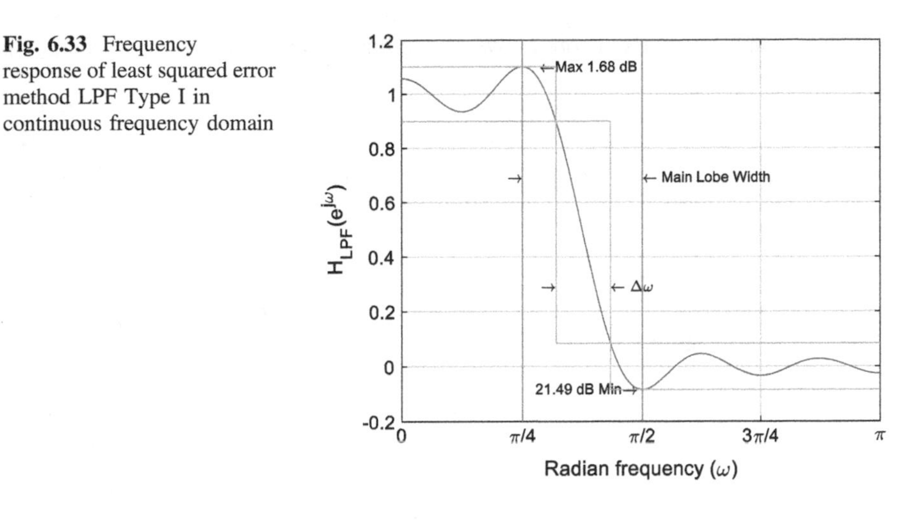

Fig. 6.33 Frequency response of least squared error method LPF Type I in continuous frequency domain

Prog. 6.27 MATLAB program for Fig. 6.33.

```
N = 16;
NN = 2^13;
N1 = NN/2+1;
N2 = N/2;
nnn = (-N2:1:N2);
b2 = sin(3*pi*(nnn+eps)/8)./(pi*(nnn+eps));
b3 = [b2 zeros(1,NN-length(b2))];
b4 = circshift(b3,-8);
B2 = real(fft(b4,NN));
findex = linspace(0,2*pi,NN);
ff = findex(1:N1);
[M1,I1] = max(B2(1:N1));
[M2,I2] = min(B2(1:N1));

figure,
plot(ff,B2(1:N1)), grid, xlim([0 pi]), ylim([-0.2 1.2]);
straa = sprintf('Max %1.2f dB',40*log10(B2(I1)));
aa = text(ff(I1),B2(I1)+0.005,['        \leftarrow' straa]);
strbb = sprintf('%1.2f dB Min',-20*log10(abs(B2(I2))));
bb = text(ff(I2)-0.7,B2(I2)+0.005,[strbb '\rightarrow    ']);
cc = min(find(B2<(2-B2(I1))));
dd = max(find(B2(1:NN/2)>-B2(I2)));
line([ff(I1) ff(I1)],[-0.2 1.2],'Color','magenta');
line([ff(I2) ff(I2)],[-0.2 1.2],'Color','magenta');
hh = text(ff(I1)-0.1,0.7,'\rightarrow ');
ii = text(ff(I2),0.7,'\leftarrow Main Lobe Width');
```

Using the Parseval's theorem, we can compute the residual error energy as below.

$$\|E(e^{j\omega})\|_2^2 = \sum_{n=9}^{\infty} 2|h_d[n]|^2 = \frac{1}{2\pi}\int_{-3\pi/8}^{3\pi/8} 1^2 d\omega - \sum_{n=-8}^{8} |h[n]|^2 = 0.0126$$

Further control of the ripple can be exercised by using the window method on the above filter realization. The final filter is given by the proper shift to preserve the causality. The energy ratio between the desired and calculated filter can be computed by using the Parseval's theorem as well. In the equation below, the numerator and denominator represent the calculated filter energy and desired filter energy, respectively.

$$\frac{\frac{1}{2\pi}\int_{-\pi}^{\pi}|H(e^{j\omega})|^2 d\omega}{\frac{1}{2\pi}\int_{-\pi}^{\pi}|H_d(e^{j\omega})|^2 d\omega} = \frac{\sum_{n=-8}^{8}\left|\frac{\sin(3\pi n/8)}{\pi n}\right|^2}{\frac{1}{2\pi}\int_{-\pi}^{\pi}|H_d(e^{j\omega})|^2 d\omega} = \frac{\sum_{n=-8}^{8}\left|\frac{\sin(3\pi n/8)}{\pi n}\right|^2}{3/8} = 0.9664$$

■

Constrained least square design

The FIR filter can be built by applying the constraint of the upper and lower bound on each band for the frequency response. The constrained least square design method [13] utilizes an iterative least squares algorithm to derive an equiripple response. Let's observe the simplified procedure with L length filter which is $N + 1$ and $M = N/2$ with Type I FIR and zero phase (non-causal) filter.

$$H(e^{j\omega}) = b_0 + \sum_{n=1}^{M} b_n \cos(n\omega) \text{ for } H(e^{j\omega}) \in \mathbb{R} \tag{6.102}$$

$$h[n] = \begin{cases} \frac{1}{2}b_n & \text{for } 1 \le n \le M \\ b_0 & \text{for } n = 0 \\ \frac{1}{2}b_{-n} & \text{for } -M \le n \le -1 \\ 0 & \text{Otherwise} \end{cases}$$

The error is defined as squared L_2-norm shown in Eq. (6.103).

$$\|E(e^{j\omega})\|_2^2 = \frac{1}{\pi}\int_0^{\pi} |H(e^{j\omega}) - H_d(e^{j\omega})|^2 d\omega \tag{6.103}$$

where $H(e^{j\omega}) \in \mathbb{R}$ and $H_d(e^{j\omega}) \in \mathbb{R}$

The equality constraints are employed on the discrete frequency band by upper $U(e^{j\omega})$ and lower $L(e^{j\omega})$ bound as shown in Eq. (6.104).

$$H\left(e^{j\omega}\right) = L\left(e^{j\omega}\right) \text{ for } \omega \in S_l = \left\{\omega_1, \omega_2, \ldots, \omega_q\right\} \in [0, \pi]$$

$$H(e^{j\omega}) = U(e^{j\omega}) \text{ for } \omega \in S_u = \left\{\omega_{q+1}, \omega_{q+2}, \ldots, \omega_r\right\} \in [0, \pi] \tag{6.104}$$

To minimize the squared L_2-norm error subject to above constraints, we use the Lagrangian as shown in Eq. (6.105).

$$\begin{aligned} \mathcal{L} = \left\|E\left(e^{j\omega}\right)\right\|_2^2 - \sum_{i=1}^{q} \mu_i\left(H\left(e^{j\omega_i}\right) - L\left(e^{j\omega_i}\right)\right) \\ + \sum_{i=q+1}^{r} \mu_i\left(H\left(e^{j\omega_i}\right) - U\left(e^{j\omega_i}\right)\right) \end{aligned} \tag{6.105}$$

By placing the derivative of $\mathcal{L}$ with respect to b_k and μ_i to zero, we can obtain the desired solution with the constraint requirements.

$$\frac{\partial \mathcal{L}}{\partial b_k} = \frac{\partial \|E(e^{j\omega})\|_2^2}{\partial b_k} - \sum_{i=1}^{q} \mu_i \frac{\partial H(e^{j\omega_i})}{\partial b_k} + \sum_{i=q+1}^{r} \mu_i \frac{\partial H(e^{j\omega_i})}{\partial b_k} = 0 \tag{6.106}$$

$$\text{for } 0 \leq k \leq M$$

$$H\left(e^{j\omega_i}\right) = L\left(e^{j\omega_i}\right) \text{ for } 1 \leq i \leq q$$

$$H\left(e^{j\omega_i}\right) = U\left(e^{j\omega_i}\right) \text{ for } q+1 \leq i \leq r$$

Based on the Kuhn–Tucker conditions, once the Lagrange multipliers are all nonnegative, the solution from Eq. (6.106) minimize the squared L_2-norm subject to the inequality constraints.

$$H\left(e^{j\omega_i}\right) \geq L\left(e^{j\omega_i}\right) \text{ for } 1 \leq i \leq q$$

$$H(e^{j\omega_i}) \leq U(e^{j\omega_i}) \text{ for } q+1 \leq i \leq r \tag{6.107}$$

The iterative algorithm to solve above equations provides the equiripple response. The algorithm utilizes the multiple exchange algorithm that solves Lagrange multipliers [14] with Kuhn–Tucker conditions [15] on each iteration. The further solution can be found on the paper [13].

Example 6.15
The design example by Table 6.25 utilizes the MATLAB function for the constrained least square design method.

Table 6.25 Given filter specification

Filter specifications	Value
Filter type	Low pass filter
Filter realization	FIR Type I without window method
Filter length	Optimal value from window
Passband frequency (f_p)	1000 Hz; $\omega_p = \pi/4$
Passband magnitude	0 dB
Passband ripple (peak to peak)	1 dB
Stopband frequency (f_{st})	2000 Hz; $\omega_{st} = \pi/2$
Stopband attenuation	40 dB
Sampling frequency (f_s)	8000 Hz

Solution

The center of the passband and stopband edge is below for cutoff frequency.

$$\omega_c = \omega_p + \frac{\omega_{st} - \omega_p}{2} = \frac{\pi}{4} + \frac{\pi}{8} = \frac{3\pi}{8}$$

The frequency transition is below.

$$\Delta\omega = \omega_{st} - \omega_p = \frac{\pi}{2} - \frac{\pi}{4} = \frac{\pi}{4}$$

Below is the main lobe width for rectangular window to determine the filter order N.

$$\frac{4\pi}{N+1} \leq \frac{\pi}{4} \quad N \geq 15$$

Let the $N = 16$ for Type I filter. The maximum deviation R_p can be derived as below.

$$R_p = 10^{\frac{Ap/2}{20}} - 1$$

$$10^{\frac{1/2}{20}} - 1 = 0.0593$$

The minimum attenuation R_{st} can be computed as below.

$$R_{st} = 10^{\frac{-A_{st}}{20}}$$

$$10^{\frac{-40}{20}} = 0.01$$

MATLAB code is below.

Prog. 6.28 MATLAB program for Fig. 6.34.

```
N = 16;                         % Filter order
NN = 2^13;
N1 = NN/2+1;
N2 = N/2;
nnn = (-N2:1:N2);
f = [0 3/8 1];                  % Cutoff frequency = 3/8
a = [1 0];                      % Desired amplitude
up = [1.0593 0.01];             % 1+0.0593 0+0.01
lo = [0.9407 -0.01];            % 1-0.0593 0-0.01
b2 = fircls(N,f,a,up,lo);       % Design the constrained least square FIR filter
b3 = [b2 zeros(1,NN-length(b2))];
b4 = circshift(b3,-8);
B2 = real(fft(b4,NN));
findex = linspace(0,2*pi,NN);
ff = findex(1:N1);
[M1,I1] = max(B2(1:N1));
[M2,I2] = min(B2(1:N1-10));

figure,
plot(ff,B2(1:N1)), grid, xlim([0 pi]), ylim([-0.2 1.2])
straa = sprintf('Max %1.2f dB',40*log10(B2(I1)));
aa = text(ff(I1),B2(I1)+0.005,['       \leftarrow' straa]);
strbb = sprintf('%1.2f dB Min',-20*log10(abs(B2(I2))));
bb = text(ff(I2)-0.7,B2(I2)+0.005,[strbb '\rightarrow     ']);
cc = min(find(B2<(2-B2(I1))));
dd = max(find(B2(1:NN/2-1000)>-B2(I2)));
figure,
LB2 = 20*log10(abs(B2));
plot(ff,LB2(1:N1)), grid, xlim([0 pi]), ylim([-100 20]);
straa = sprintf('Max %1.2f dB',40*log10(B2(I1)));
aa = text(ff(I1),LB2(I1)+1,['       \leftarrow' straa]);
strbb = sprintf('%1.2f dB Min',-20*log10(abs(B2(I2))));
bb = text(ff(I2)-0.7,LB2(I2)+1,[strbb '\rightarrow     ']);
ee = text(ff(cc)-0.1,-20,'\rightarrow ');
gg = text(ff(dd),-20,'\leftarrow \Delta\omega');
line([ff(cc) ff(cc)],[-100 20],'Color',[0.93 0.69 0.13]);
line([ff(dd) ff(dd)],[-100 20],'Color',[0.93 0.69 0.13]);
```

■

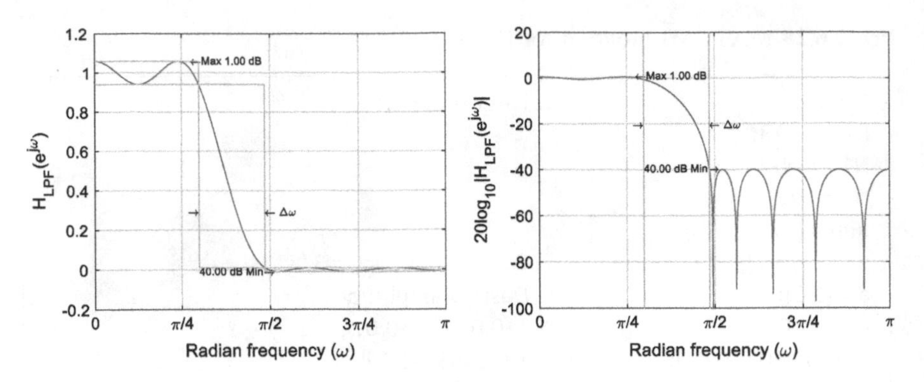

Fig. 6.34 Frequency response of constrained least square design method LPF Type I

Complex and nonlinear-phase equiripple design

Via defining the complex error function, the Remez exchange algorithm [6, 7] can be extended to compute the optimal complex Chebyshev approximation. This generalization of the Parks-McClellan algorithm provides approximation of arbitrary magnitude and phase responses for the FIR filter with asymmetric spectral distribution for positive and negative frequencies. The further information can be found on the paper [16].

Example 6.16
The design example by Fig. 6.35 utilizes the MATLAB function for the complex and nonlinear-phase equiripple design method.

Solution
Let the $N = 16$ for Type I filter. Approximately estimates transition by using the main lobe width of rectangular window as below.

$$\frac{4\pi}{N+1} = \frac{4\pi}{16+1} \approx \frac{\pi}{4} \qquad N = 16$$

Stopband and passband for left frequency edge are below.

$$\omega_{lst} = -\frac{\pi}{2} - \frac{1}{2}\frac{\pi}{4} = -\frac{5\pi}{8}$$

$$\omega_{lp} = -\frac{\pi}{2} + \frac{1}{2}\frac{\pi}{4} = -\frac{3\pi}{8}$$

Stopband and passband for right frequency edge are below.

$$\omega_{rp} = \frac{\pi}{4} - \frac{1}{2}\frac{\pi}{4} = \frac{\pi}{8}$$

$$\omega_{rst} = \frac{\pi}{4} + \frac{1}{2}\frac{\pi}{4} = \frac{3\pi}{8}$$

Prog. 6.29 MATLAB program for Fig. 6.36.

```
N = 16; % Filter order
NN = 2^13;
N1 = NN/2+1;
N2 = N/2;
nnn = (-N2:1:N2);
f = [-1 -5/8 -3/8 1/8 3/8 1]; % f_l±Δ, f_r±Δ
b2 = cfirpm(N,f,@lowpass); % Complex and nonlinear-phase equiripple filter
b3 = [b2 zeros(1,NN-length(b2))];
b4 = circshift(b3,-8);
B2 = real(fftshift(fft(b4,NN)));
LB2 = 20*log10(abs(B2));
findex = linspace(-pi,pi,NN);
ff = findex;
[M1,I1] = max(B2);, [M2,I2] = min(B2);

figure, plot(findex,B2), grid, xlim([-pi pi]), ylim([-0.2 1.2]);
straa = sprintf('Max %1.2f dB',40*log10(B2(I1)));
aa = text(ff(I1),B2(I1)+0.005,['\leftarrow' straa]);
strbb = sprintf('%1.2f dB Min',-20*log10(abs(B2(I2))));
bb = text(ff(I2),B2(I2)+0.005,[strbb
'\rightarrow'],'HorizontalAlignment','right');
cc = min(find(B2<(2-B2(I1))));
dd = max(find(B2(1:NN/2-1000)>-B2(I2)));
line([-5*pi/8 -5*pi/8],[-0.2 1.2],'Color',[0.93 0.69 0.13]);
line([-3*pi/8 -3*pi/8],[-0.2 1.2],'Color',[0.93 0.69 0.13]);
line([pi/8 pi/8],[-0.2 1.2],'Color',[0.93 0.69 0.13]);
line([3*pi/8 3*pi/8],[-0.2 1.2],'Color',[0.93 0.69 0.13]);
line([-pi/2 -pi/2],[-0.2 1.2],'Color','magenta');
line([pi/4 pi/4],[-0.2 1.2],'Color','magenta');
figure, plot(ff,LB2), grid, xlim([-pi pi]), ylim([-120 20]);
straa = sprintf('Max %1.2f dB',40*log10(B2(I1)));
aa = text(ff(I1),LB2(I1)+1,['\leftarrow' straa]);
strbb = sprintf('%1.2f dB Min',-20*log10(abs(B2(I2))));
bb = text(ff(I2),LB2(I2)+1,[strbb '\rightarrow'],'HorizontalAlignment','right');
line([-5*pi/8 -5*pi/8],[-120 20],'Color',[0.93 0.69 0.13]);
line([-3*pi/8 -3*pi/8],[-120 20],'Color',[0.93 0.69 0.13]);
line([pi/8 pi/8],[-120 20],'Color',[0.93 0.69 0.13]);
line([3*pi/8 3*pi/8],[-120 20],'Color',[0.93 0.69 0.13]);
line([-pi/2 -pi/2],[-120 20],'Color','magenta');
line([pi/4 pi/4],[-120 20],'Color','magenta');
```

The frequency response of designed FIR filter is illustrated in Fig. 6.36.

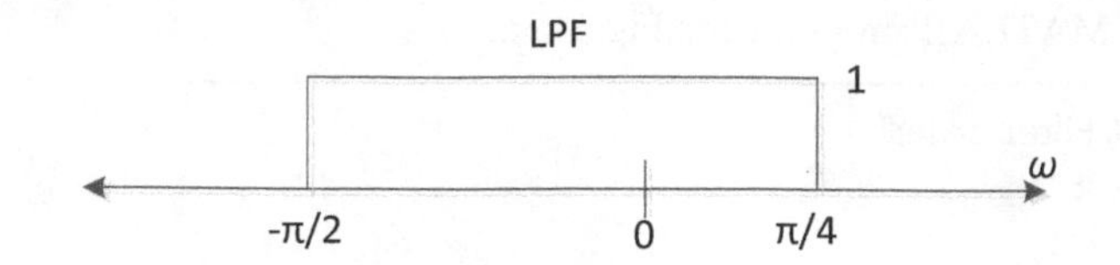

Fig. 6.35 LPF specification

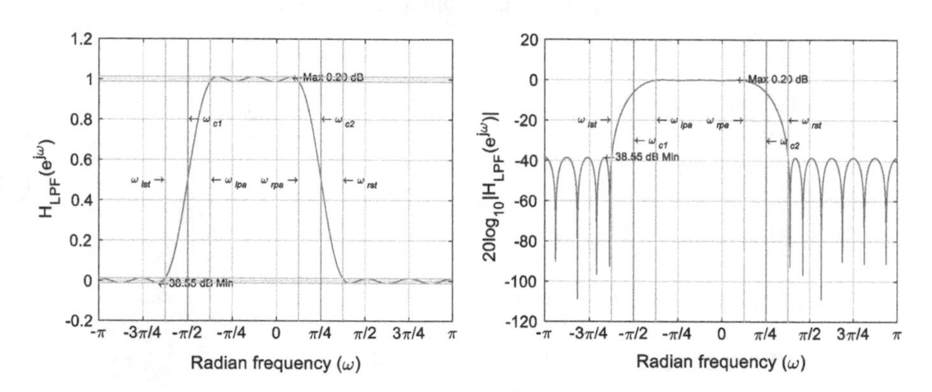

Fig. 6.36 Frequency response of complex and nonlinear-phase equiripple design method LPF Type I

We can convert the LPF to HPF by specifying the parameter in MATLAB function as below.

■

Prog. 6.30 MATLAB program for Fig. 6.37. Use the **Prog. 6.29**.

```
...
b2 = cfirpm(N,f,@highpass); % Complex and nonlinear-phase equiripple filter
...
```

The frequency response of designed HPF filter is presented in Fig. 6.37.

Minimum phase filter design (for FIR and IIR filters)

In general, the filter is designed from the magnitude response. The magnitude specification is decided by the multiple parameters to control the output precisely in terms of the frequency. Once the filter is determined, the question is whether the designed filter is unique or not to the determined magnitude response. To consider this matter, let's revisit the filter transfer function as below.

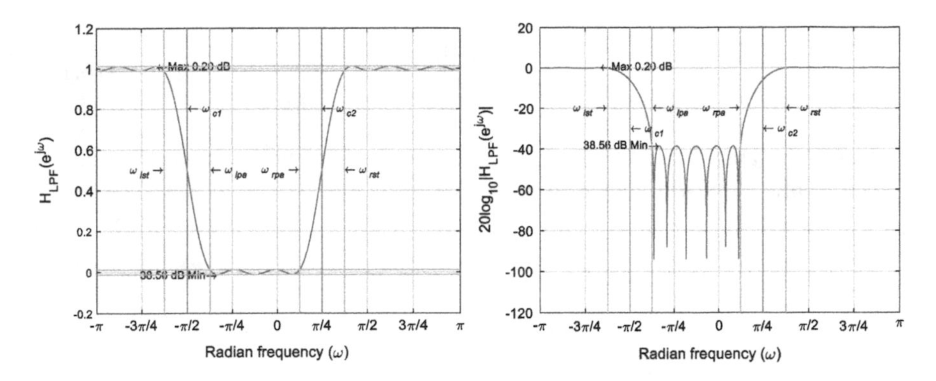

Fig. 6.37 Frequency response of complex and nonlinear-phase equiripple design method HPF Type I

$$H(z) = \frac{Y(z)}{X(z)} \tag{6.108}$$

The frequency response of the filter is computed by the DTFT or DFT which can be derived from the transfer function as below.

$$H\left(e^{j\omega}\right) = H(z)|_{z\to e^{j\omega}}$$
$$H[k] = H(z)|_{z\to e^{\frac{2\pi}{N}k}} \tag{6.109}$$

Therefore, the pole and zero locations from the transfer function play important role to decide the filter properties. The transfer function is the rational function based on the complex or real polynomial based on the z. Let's consider simple polynomial in Eq. (6.110)

$$1 - c_0 z^{-1} \tag{6.110}$$

Note that the negative power is for the system causality. Due to the complex number property, the number conjugate shows the identical magnitude.

$$\left|1 - c_0 z^{-1}\right| = \left|\left(1 - c_0 z^{-1}\right)^*\right| = \left|1 - c_0^*(z^*)^{-1}\right| \tag{6.111}$$

The given polynomial is evaluated over the unit circle in the complex domain to obtain the frequency response. The conjugate of $e^{j\omega}$ is $e^{-j\omega}$.

$$\left\{1 - c_0^*(z^*)^{-1}\right\}\Big|_{z=e^{j\omega}} = \left\{1 - c_0^*\left(z^{-1}\right)^{-1}\right\}\Big|_{z=e^{j\omega}} = \left\{z\left(z^{-1} - c_0^*\right)\right\}\big|_{z=e^{j\omega}} \tag{6.112}$$

The magnitude of $e^{j\omega}$ is one.

$$\begin{aligned} \left|1 - c_0 z^{-1}\right|\big|_{z=e^{j\omega}} &= \left|z\left(z^{-1} - c_0^*\right)\right|\big|_{z=e^{j\omega}} \\ &= |z|\left|z^{-1} - c_0^*\right|\big|_{z=e^{j\omega}} = \left|z^{-1} - c_0^*\right|\big|_{z=e^{j\omega}} \end{aligned} \tag{6.113}$$

Using the above derivation, we can define the filter which passes the all frequency component with unit gain as below.

$$H_{ap}(z) = \frac{z^{-1} - c_0^*}{1 - c_0 z^{-1}} \tag{6.114}$$

The magnitude response of the $H_{ap}(z)$ is one as below.

$$\left|H_{ap}(\omega)\right| = \left|\frac{z^{-1} - c_0^*}{1 - c_0 z^{-1}}\right|\Bigg|_{z=e^{j\omega}} = 1 \tag{6.115}$$

The derived system is named allpass filter [3, 7] since the system passes all frequency components without modifying the magnitude. The transfer function of the allpass filter provides the following pole and zero location.

$$H_{ap}(z) = \frac{z^{-1} - c_0^*}{1 - c_0 z^{-1}} \text{ and pole at } z = c_0 \text{ \& zero at } z = \frac{1}{c_0^*} \tag{6.116}$$

For general complex domain, the pole and zero in the polar coordinates can be described as below.

$$z \in \mathbb{C} \text{ and let } c_0 = re^{j\theta} \text{ then } \frac{1}{c_0^*} = \frac{1}{re^{-j\theta}} = \frac{1}{r}e^{j\theta} \tag{6.117}$$

In the allpass filter, the zero is reflected from the pole as the conjugate reciprocal position. In other words, the pole and zero are placed at the same angle but the magnitude is inversed in polar coordinates. According to the system stability, the poles should be located within the unit circle area; hence, the zeros are placed at the outside of the unit circle. The multiple poles and zeros produce the higher order rational function as below.

$$\begin{aligned} H_{ap}(z) &= \frac{\left(z^{-1} - c_0^*\right)\left(z^{-1} - c_1^*\right)\ldots\left(z^{-1} - c_{m-1}^*\right)}{\left(1 - c_0 z^{-1}\right)\left(1 - c_1 z^{-1}\right)\ldots\left(1 - c_{m-1} z^{-1}\right)} \\ &= \frac{b_0 + b_1 z^{-1} + b_2 z^{-2} + \cdots + b_m z^{-m}}{a_0 + a_1 z^{-1} + a_2 z^{-2} + \cdots + a_m z^{-m}} \end{aligned} \tag{6.118}$$

One example of the allpass filter is given below.

$$H_{ap}(z) = \frac{\left(z^{-1} - 0.5e^{-j\pi/4}\right)\left(z^{-1} - 0.5e^{j\pi/4}\right)\left(z^{-1} + 0.8\right)}{\left(1 - 0.5e^{j\pi/4}z^{-1}\right)\left(1 - 0.5e^{-j\pi/4}z^{-1}\right)\left(1 + 0.8z^{-1}\right)}$$

The corresponding poles and zeros locations are below.

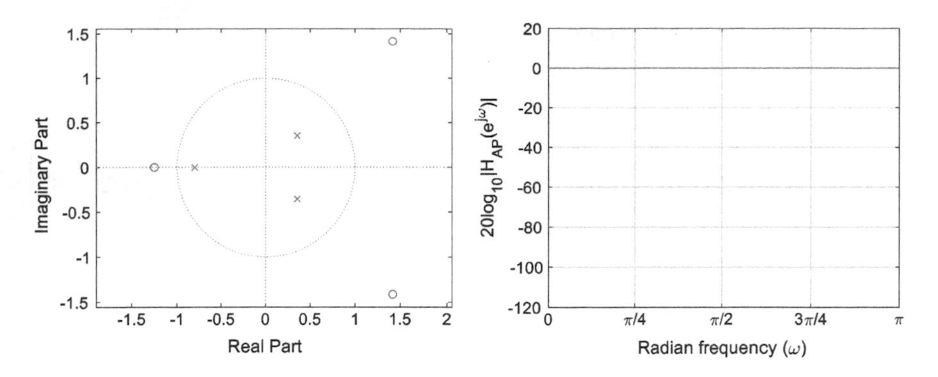

Fig. 6.38 Pole zero plot and frequency response in magnitude for given transfer function

$$\text{Poles}: z = 0.5e^{j\pi/4}, z = 0.5e^{-j\pi/4}, \text{ and } z = -0.8$$

$$\text{zeros}: z = 2e^{j\pi/4}, z = 2e^{-j\pi/4}, \text{and } z = -1.25$$

Figure 6.38 shows the poles and zeros location over the z-plane.

The allpass filter delivers the constant one magnitude in frequency response. However, the time delay caused by the allpass filter denotes the non-constant distribution. The group delay presents the time delay in terms of frequency ω . Below group delay equation is displayed again from Eq. (6.19).

$$d(e^{j\omega}) = \text{grd}(H(e^{j\omega})) = -\tfrac{d}{d\omega}\{\arg(H(e^{j\omega}))\} \tag{6.119}$$

The group delay of given transfer function is shown in Fig. 6.39. The distribution represents the non-linear and non-constant shape.

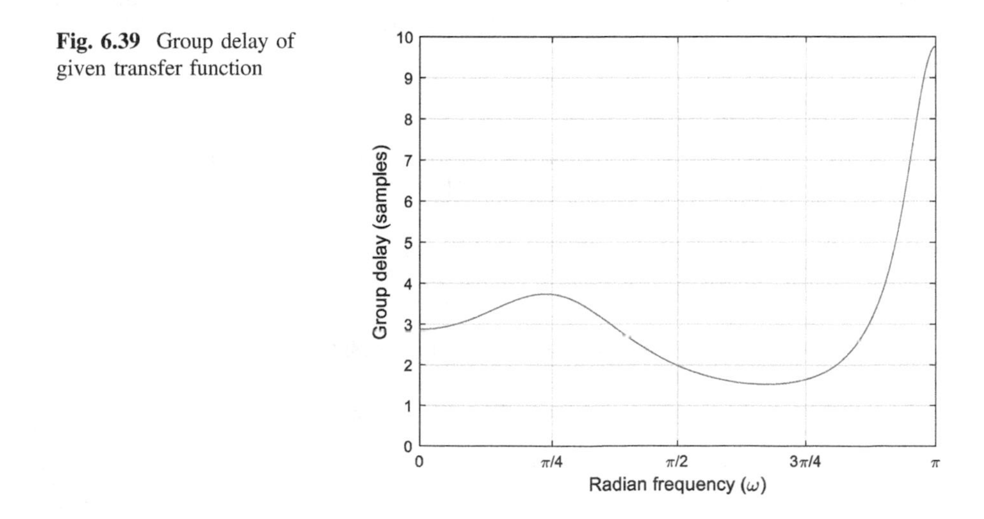

Fig. 6.39 Group delay of given transfer function

Prog. 6.31 MATLAB program for Figs. 6.38 and 6.39.

```
syms z
Hapznum = (z^-1 - 0.5*exp(-j*sym(pi)/4))*(z^-1 - 0.5*exp(j*sym(pi)/4))*(z^-1
+ 0.8);
Hapzden = (1 - 0.5*exp(-j*sym(pi)/4)*z^-1)*(1 - 0.5*exp(j*sym(pi)/4)*z^-
1)*(1 + 0.8*z^-1);
aa = collect(Hapzden,z);
bb = collect(Hapznum,z);
cc = bb/aa;
dd = simplifyFraction(cc);
[num,den] = numden(dd);
ee = sym2poly(collect(num));
ff = sym2poly(collect(den));
[h,w] = freqz(ee,ff,2^13);

figure,
zplane(ee,ff);
figure,
plot(w,20*log10(abs(h))),grid, xlim([0 pi]), ylim([-120 20]);
figure,
[gd,w1] = grpdelay(ee,ff,2^13);
plot(w1,gd),grid, xlim([0 pi]), ylim([0 10]);
```

As shown above, the conjugate reciprocal correspondence between the pole and zero pair provides the unit gain with arbitrary phase response. The transfer function requires the system stability; therefore, the pole should be located within the unit circle and the corresponding zeros is reflected to the conjugate reciprocal position which is the circle outside. If we hope to maintain the magnitude response only then the transfer function can be modified as Eq. (6.120).

$$H(z) = H_0(z)\left(z^{-1} - c_0^*\right) \tag{6.120}$$

The transfer function above contains one zero location $1/c_0^*$ where the position is the unit circle outside as shown in Eq. (6.121).

$$c_0 = re^{j\theta} \text{where} |r| < 1 \text{ hence} \frac{1}{c_0^*} = \frac{1}{r} e^{j\theta} \text{where } \left|\frac{1}{r}\right| > 1 \tag{6.121}$$

By using the allpass filter, we can rewrite the transfer function as below.

$$
\begin{aligned}
H(z) &= H_0(z)\left(z^{-1} - c_0^*\right)\frac{1 - c_0 z^{-1}}{1 - c_0 z^{-1}} \\
&= H_0(z)\left(1 - c_0 z^{-1}\right)\frac{z^{-1} - c_0^*}{1 - c_0 z^{-1}} = H_0(z)\left(1 - c_0 z^{-1}\right)H_{ap}(z)
\end{aligned}
\tag{6.122}
$$

Since the allpass filter demonstrates the unit gain, the zero reflection (conjugate reciprocal) does not change the magnitude of given filter as below.

$$
|H(z)| = \left|H_0(z)\left(z^{-1} - c_0^*\right)\right| = \left|H_0(z)(1 - c_0 z^{-1})\right| \tag{6.123}
$$

Therefore, any zeros on the system can move over the unit circle in conjugate reciprocal fashion without modifying the magnitude response. The general form is shown in Eq. (6.124).

$$
\begin{aligned}
&\left|\frac{1}{\prod_{k=0}^{m-1}\left(1 - \alpha_k z^{-1}\right)}\prod_{k=0}^{n-1}\left(1 - \beta_k z^{-1}\right)\right| \\
&= \left|\frac{1}{\prod_{k=0}^{m-1}\left(1 - \alpha_k z^{-1}\right)}\prod_{k=0}^{n-1}\left(z^{-1} - \beta_k^*\right)\right|
\end{aligned}
\tag{6.124}
$$

Note that the $|\alpha_k|$ is less than one for any k to maintain the system stability. The pole locations are fixed; however, the zeros can be chosen for desired magnitude response. To understand the allpass filter property, we can derive the group delay of the allpass filter as below.

$$
\begin{aligned}
&d(e^{j\omega}) = \mathrm{grd}(H(e^{j\omega})) = -\tfrac{d}{d\omega}\{\arg(H(e^{j\omega}))\} \\
&H(e^{j\omega}) = |H(e^{j\omega})|e^{j\angle H(e^{j\omega})} = |H(e^{j\omega})|e^{j\arg(H(e^{j\omega}))} \\
&\log_e H(e^{j\omega}) = \ln H(e^{j\omega}) \\
&= \ln|H(e^{j\omega})| + j\angle H(e^{j\omega}) = \ln|H(e^{j\omega})| + j\arg(H(e^{j\omega})) \\
&\text{where } H(e^{j\omega}) \neq 0 \text{ and } \angle H(e^{j\omega}) > 0\ \forall\omega \\
&\angle H(e^{j\omega}) = \mathrm{Imag}(\ln H(e^{j\omega})) \\
&d(e^{j\omega}) = -\tfrac{d}{d\omega}\{\angle H(e^{j\omega})\} \\
&= -\tfrac{d}{d\omega}\{\mathrm{Imag}(\ln H(e^{j\omega}))\} = -\mathrm{Imag}\left\{\tfrac{1}{H(e^{j\omega})}\tfrac{dH(e^{j\omega})}{d\omega}\right\}
\end{aligned}
\tag{6.125}
$$

Using the substitution $z = e^{j\omega}(\ln z = j\omega)$.

$$
d(e^{j\omega}) = -\mathrm{Imag}\left\{\frac{jz}{H(z)}\frac{dH(z)}{dz}\right\}\Bigg|_{z=e^{j\omega}} = -\mathrm{Real}\left\{\frac{z}{H(z)}\frac{dH(z)}{dz}\right\}\Bigg|_{z=e^{j\omega}} \tag{6.126}
$$

Based on the above derivation, let's figure out the group delay of allpass filter.

$$
H_{ap}(z) = \tfrac{z^{-1} - c_0^*}{1 - c_0 z^{-1}} = \tfrac{1 - c_0^* z}{z - c_0} \tag{6.127}
$$

$$\frac{dH_{ap}(z)}{dz}=\frac{-c_0^*}{z-c_0}-\frac{1-c_0^*z}{(z-c_0)^2}=\frac{c_0^*c_0-1}{(z-c_0)^2}=\frac{|c_0|^2-1}{(z-c_0)^2} \tag{6.128}$$

$$\begin{aligned}
d\left(e^{j\omega}\right) &= -\mathrm{Real}\left\{\frac{z}{H_{ap}(z)}\frac{dH_{ap}(z)}{dz}\right\}\Bigg|_{z=e^{j\omega}} = -\mathrm{Real}\left\{\frac{z}{H_{ap}(z)}\frac{dH_{ap}(z)}{dz}\right\}\Bigg|_{z=e^{j\omega}} \\
&= -\mathrm{Real}\left\{\frac{z(z-c_0)}{(1-c_0^*z)}\frac{|c_0|^2-1}{(z-c_0)^2}\right\}\Bigg|_{z=e^{j\omega}} = -\mathrm{Real}\left\{\frac{1}{\left(z^{-1}-c_0^*\right)}\frac{|c_0|^2-1}{(z-c_0)}\right\}\Bigg|_{z=e^{j\omega}} \\
&= -\mathrm{Real}\left\{\frac{|c_0|^2-1}{1-c_0z^{-1}-c_0^*z+|c_0|^2}\right\}\Bigg|_{z=e^{j\omega}} \quad \text{Let } c_0=re^{j\theta} \\
&= -\mathrm{Real}\left\{\frac{r^2-1}{1-re^{j\theta}e^{-j\omega}-re^{-j\theta}e^{j\omega}+r^2}\right\} \\
&= -\mathrm{Real}\left\{\frac{r^2-1}{1-r\left(e^{j(\omega-\theta)}+e^{-j(\omega-\theta)}\right)+r^2}\right\} \\
&= -\mathrm{Real}\left\{\frac{r^2-1}{1-2r\cos(\omega-\theta)+r^2}\right\} = \frac{1-r^2}{1-2r\cos(\omega-\theta)+r^2}
\end{aligned} \tag{6.129}$$

Since $|r|<1$ for stable allpass filter

$$d\left(e^{j\omega}\right)=\mathrm{grd}\left(H_{ap}\left(e^{j\omega}\right)\right)=\frac{1-r^2}{1-2r\cos(\omega-\theta)+r^2}>0\ \forall\omega \text{ and } \theta \tag{6.130}$$

The group delay of the allpass filter is always positive because of $|r|<1$ with denominator condition. All possible values of the denominator are presented in Fig. 6.40 for verification.

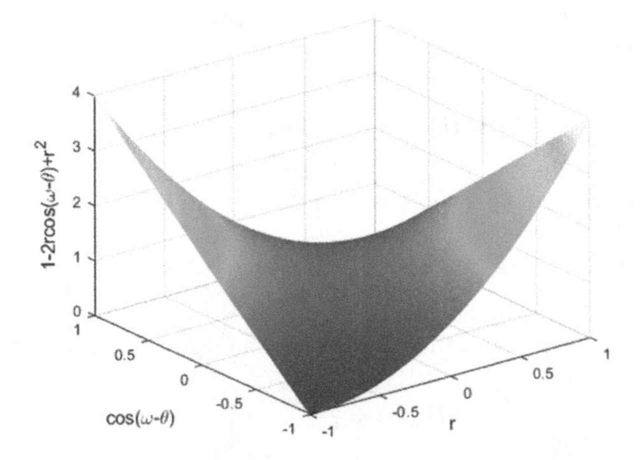

Fig. 6.40 Denominator of allpass filter group delay equation in Eq. (6.130)

Prog. 6.32 MATLAB program for Fig. 6.40.

```
r = -1:0.01:1;
cosw = -1:0.01:1;
[X,Y] = meshgrid(r,cosw);
Z = 1-2*X.*Y+X.^2;
mesh(X,Y,Z);
```

The arbitrary stable transfer function is given as below with m zeros as $1/c_k^*$ located at unit circle outside.

$$\begin{aligned}
H(z) &= H_0(z)\prod_{k=0}^{m-1}\left(z^{-1}-c_k^*\right) = H_0(z)\prod_{k=0}^{m-1}\left\{\left(1-c_kz^{-1}\right)\frac{z^{-1}-c_k^*}{1-c_kz^{-1}}\right\} \\
&= H_0(z)\prod_{k=0}^{m-1}\left(1-c_kz^{-1}\right)\prod_{q=0}^{m-1}\frac{z^{-1}-c_q^*}{1-c_qz^{-1}} = H_0(z)\prod_{k=0}^{m-1}\left(1-c_kz^{-1}\right)H_{ap}(z) \\
&\left|H_0(z)\prod_{k=0}^{m-1}\left(z^{-1}-c_k^*\right)\right| = \left|H_0(z)\prod_{k=0}^{m-1}\left(1-c_kz^{-1}\right)\right| \\
&\mathrm{grd}\left(H(z)|_{z=e^{j\omega}}\right) = \mathrm{grd}\left(\left(H_0(z)\prod_{k=0}^{m-1}\left(1-c_kz^{-1}\right)\right)\Bigg|_{z=e^{j\omega}}\right) \\
&\quad + \mathrm{grd}\left(\left(\prod_{q=0}^{m-1}\frac{z^{-1}-c_q^*}{1-c_qz^{-1}}\right)\Bigg|_{z=e^{j\omega}}\right) \\
&= \mathrm{grd}\left(\left(H_0(z)\prod_{k=0}^{m-1}\left(1-c_kz^{-1}\right)\right)\Bigg|_{z=e^{j\omega}}\right) + \sum_{q=0}^{m-1}\mathrm{grd}\left(\left(\frac{z^{-1}-c_q^*}{1-c_qz^{-1}}\right)\Bigg|_{z=e^{j\omega}}\right) \\
&\mathrm{grd}\left(H(z)|_{z=e^{j\omega}}\right) > \mathrm{grd}\left(\left(H_0(z)\prod_{k=0}^{m-1}\left(1-c_kz^{-1}\right)\right)\Bigg|_{z=e^{j\omega}}\right)
\end{aligned} \tag{6.131}$$

$$\because \mathrm{grd}\left(\left(\frac{z^{-1}-c_q^*}{1-c_qz^{-1}}\right)\Bigg|_{z=e^{j\omega}}\right) > 0 \ \forall q$$

We call the entirely reflected transfer function as minimum phase system [3] since every zero reflection reduces the group delay by allpass filters. The reflection indicates the zero relocations into the unit circle area.

$$H_{min}(z) = H_0(z)\prod_{k=0}^{m-1}\left(1-c_kz^{-1}\right) \tag{6.132}$$

Above derivation provides the minimum group delay property of the minimum phase system. The name minimum phase is originated from the following equation.

$$\arg(H(e^{j\omega})) = \arg(H_{min}(e^{j\omega})) + \arg\left(H_{ap}(e^{j\omega})\right) \tag{6.133}$$

The $\arg(H_{min}(e^{j\omega}))$ presents the minimum phase lag property because of the $\arg(H_{ap}(e^{j\omega}))$. Essential characteristics of the minimum phase system is the least delayed impulse response.

Example 6.17
For the given FIR filter below, reflect the zeros in all possible situations to observe the least delayed impulse response.

$$h_0[n] = \delta[n] - 3\delta[n-4]$$

Solution
The corresponding Z transform is below.

$$H_0(z) = 1 - 3z^{-4}$$

The pole and zero locations are below.

$$H_0(z) = \frac{z^4 - 3}{z^4}$$

$$z^4 = 3e^{j2\pi k}\ z = \sqrt[4]{3}e^{j2\pi k/4} k = 0, 1, 2, \text{and } 3$$

$$z = \pm\sqrt[4]{3}, \pm\sqrt[4]{3}j \text{ for zeros}$$

$$z = 0 \text{ for quartic poles}$$

$$\begin{aligned} H_0(z) &= \left(1 - \sqrt[4]{3}z^{-1}\right)\left(1 + \sqrt[4]{3}z^{-1}\right)\left(1 - \sqrt[4]{3}jz^{-1}\right)\left(1 + \sqrt[4]{3}jz^{-1}\right) \\ &= 1 - 3z^{-4} \end{aligned}$$

Frequency response, pole zero plot, and impulse response of the $H_0(z)$ is shown in Fig. 6.41.

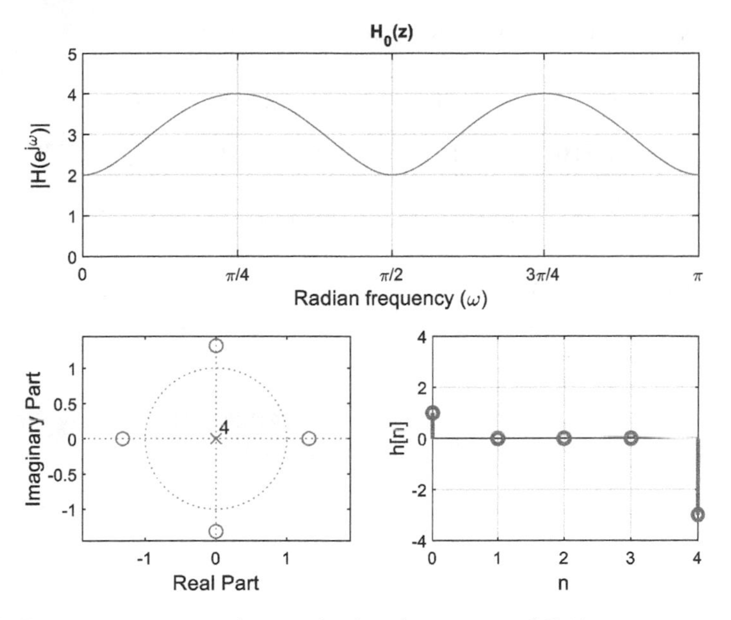

Fig. 6.41 Frequency response, pole zero plot, impulse response of $H_0(z)$

Prog. 6.33 MATLAB program for Fig. 6.41.

```
syms z
NN = 2^13;
N1 = NN/2+1;
findex = linspace(0,2*pi,NN);
f_f = findex(1:N1);
Hznum1 = (1-power(3,1/4)*z^-1)*(1+power(3,1/4)*z^-1)*(1-
1j*power(3,1/4)*z^-1)*(1+1j*power(3,1/4)*z^-1);
aa1 = collect(Hznum1,z);
bb1 = simplifyFraction(aa1);
[num1,den1] = numden(bb1);
cc1 = sym2poly(collect(num1));
dd1 = sym2poly(collect(den1));
ee1 = cc1./dd1(1);
hf1 = abs(fft(ee1,NN));

figure,
subplot(2,2,[1 2]),plot(f_f,hf1(1:N1)), grid, xlim([0 pi]), ylim([0 5]);
subplot(223), zplane(cc1,dd1);
subplot(224),stem((0:4),ee1,'LineWidth',2),grid, xlim([0 4]), ylim([-4 4]);
```

Move one zero into the unit circle area in $H_1(z)$ as below.

$$H_1(z) = \left(z^{-1} - \sqrt[4]{3}\right)\left(1 + \sqrt[4]{3}z^{-1}\right)\left(1 - \sqrt[4]{3}jz^{-1}\right)\left(1 + \sqrt[4]{3}jz^{-1}\right)$$
$$= -1.3161 - 0.7321z^{-1} - 0.9634z^{-2} - 1.2679z^{-3} + 2.2795z^{-4}$$

Frequency response, pole zero plot, and impulse response of the $H_1(z)$ is shown in Fig. 6.42.

Prog. 6.34 MATLAB program for Fig. 6.42. Use the Prog. 6.33.

```
% From Prog. 6.33
...
Hznum1 = (z^-1-power(3,1/4))*(1+power(3,1/4)*z^-1)*(1-
1j*power(3,1/4)*z^-1)*(1+1j*power(3,1/4)*z^-1);
...
```

Move two zeros into the unit circle area in $H_2(z)$ as below.

$$H_2(z) = \left(z^{-1} - \sqrt[4]{3}\right)\left(z^{-1} + \sqrt[4]{3}\right)\left(1 - \sqrt[4]{3}jz^{-1}\right)\left(1 + \sqrt[4]{3}jz^{-1}\right)$$
$$= -1.7321 - 2.0000z^{-2} + 1.7321z^{-4}$$

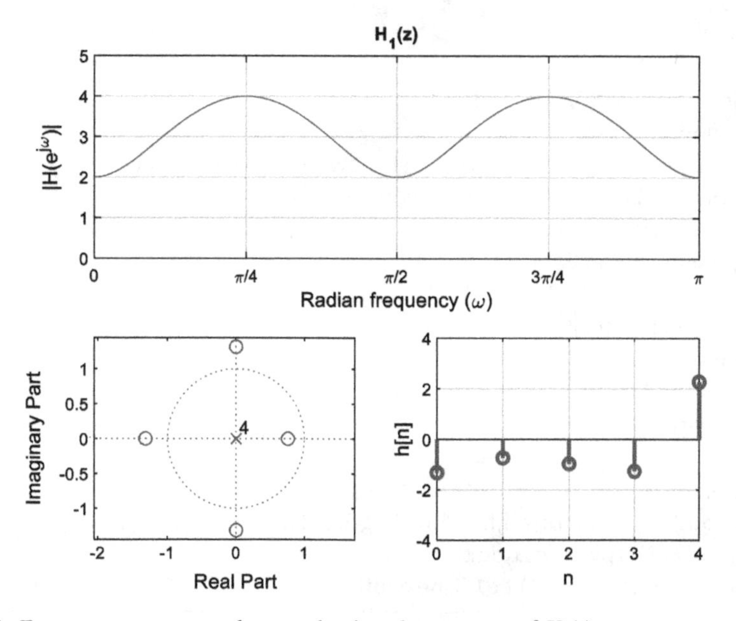

Fig. 6.42 Frequency response, pole zero plot, impulse response of $H_1(z)$

Frequency response, pole zero plot, and impulse response of the $H_1(z)$ is shown in Fig. 6.43.

Prog. 6.35 MATLAB program for Fig. 6.43. Use the Prog. 6.33.

```
% From Prog. 6.33
...
Hznum1 = (z^-1-power(3,1/4))*(z^-1+power(3,1/4))*(1-1j*power(3,1/4)*z^-
1)*(1+1j*power(3,1/4)*z^-1);
...
```

Move all zeros into the unit circle area in $H_{min}(z)$ as below.

$$H_{min}(z) = \left(z^{-1} - \sqrt[4]{3}\right)\left(z^{-1} + \sqrt[4]{3}\right)\left(z^{-1} - \sqrt[4]{3}j\right)\left(z^{-1} + \sqrt[4]{3}j\right) = -3 + z^{-4}$$

Frequency response, pole zero plot, and impulse response of the $H_{min}(z)$ is shown in Fig. 6.44.

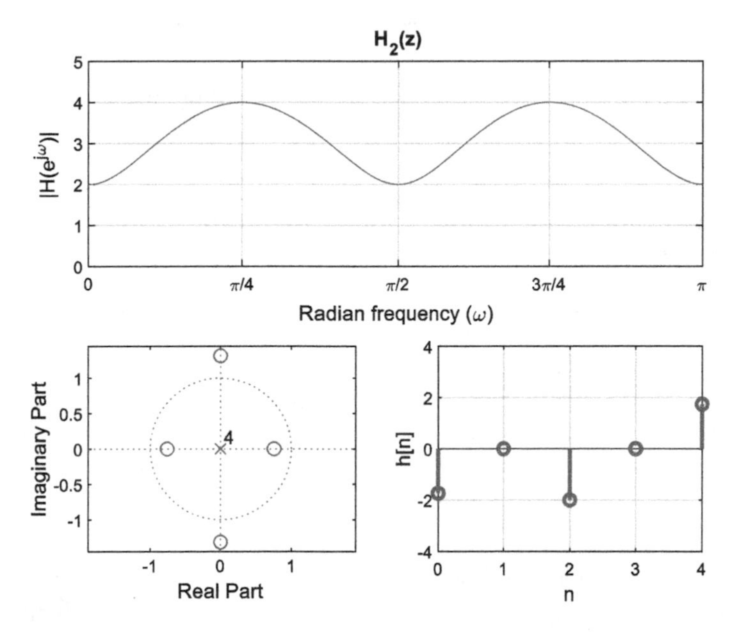

Fig. 6.43 Frequency response, pole zero plot, impulse response of $H_2(z)$

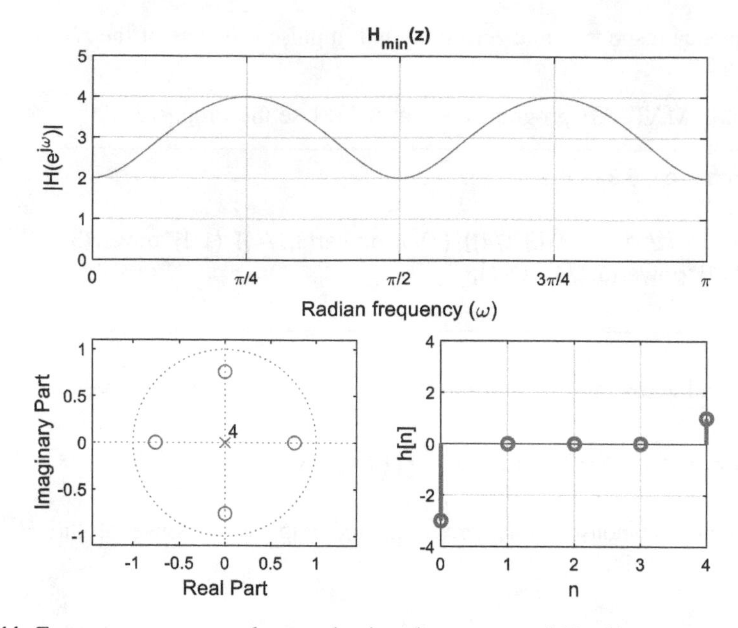

Fig. 6.44 Frequency response, pole zero plot, impulse response of $H_{min}(z)$

Prog. 6.36 MATLAB program for Fig. 6.44. Use the Prog. 6.33.

```
% From Prog. 6.33
...
Hznum1 = (z^-1-power(3,1/4))*(z^-1+power(3,1/4))*(z^-1-
1j*power(3,1/4))*(z^-1+1j*power(3,1/4));
...
```

The reflection of zeros toward unit circle inside does not modify the magnitude response of the designed filters. The group delay of the filter is reduced in the impulse response where the larger samples are located at its left-hand end than other sequences. If we want to design the least energy delay in the filter system, then minimum phase filter is one solution.

■

The group delay discussed above is the averaged phase delay for frequency component combinations. The phase delay is the delay for the single frequency ω_0 as shown in Eq. (6.134).

$$p(e^{j\omega_0}) = \text{phd}(H(e^{j\omega_0})) = -\frac{\arg(H(e^{j\omega_0}))}{\omega_0} \tag{6.134}$$

For example, single frequency sinusoid with envelop function is placed into the LTI filter system. The group delay of the filter indicates the envelop delay between

Table 6.26 Discrete-time FIR filter method summary and MATLAB functions

Filter method	Function description	MATLAB functions
Windowing	Apply window to truncated inverse DFT of brick wall filter specification	**fir1, fir2**
Multiband with transition bands	Equiripple or least squares approach over the frequency range	**firls, firpm**
Constrained least squares	Minimize squared integral error over entire frequency range subject to maximum error constraints	**fircls, fircls1**
Arbitrary response	Arbitrary responses, including nonlinear phase and complex filters	**cfirpm**
Minimum phase	Reflect the zeros into unit circle area	**polystab**

the input and output signal. The phase delay specifies the single frequency sinusoid delay between the signals. Therefore, the group delay presents the overall signal shape delay. Table 6.26 summarizes FIR filter design methods and corresponding MATLAB functions.

6.3 IIR Filters

The IIR filters contain the recursive input from the current and previous outputs known as the feedback. The recursive input is represented by the rational transfer function that is involved with the polynomial in the denominator. The direct IIR filter design is difficult to be realized because of the circular frequency response evaluation based on the exponential function. The IIR filter response is computed by replacing the z variable with the exponential $e^{j\omega}$ as Eq. (6.135). The rational value between the distance products to the poles and zeros decides the complex result of the frequency response. The magnitude and phase are computed from the complex value.

$$H[k] \leftrightarrow H(e^{j\omega})\big|_{\omega=\frac{2\pi}{N}k} \leftrightarrow H(z)\big|_{z=e^{j\omega}} \tag{6.135}$$

Butterworth IIR filter

The previous chapter discussed the mutual correlation between the poles and zeros in the z plane. Since the nonlinear correlation increases the design complexity, the Butterworth IIR filter [3, 17] initiates the design from the linear frequency domain. The linear frequency Ω is the radian frequency from the continuous time signal $x(t)$

which can be decomposed into the infinite linear combination of eigenfunction $e^{j\Omega t}$. The linear property of the frequency is denoted by the $j\Omega$ in the transfer function as $H(j\Omega)$. Note that the circular property of the digital frequency ω is shown by the $e^{j\omega}$ in the transfer function as $H(e^{j\omega})$. Equation (6.136) presents the LPF model for the linear frequency $j\Omega$.

$$|H(j\Omega)|^2 = H(j\Omega)H^*(j\Omega) = H(j\Omega)H(-j\Omega) = \frac{1}{1+\Omega^{2N}} \tag{6.136}$$

$$20\log_{10}|H(j\Omega)| = 10\log_{10}\left(\frac{1}{1+\Omega^{2N}}\right) \approx -20N\log_{10}|\Omega| \text{for } \Omega \gg 1 \tag{6.137}$$

As the frequency Ω is increased, the denominator value becomes the dominant factor to decrease the overall magnitude. Therefore, the frequency response represents the LPF with halfpower frequency $\Omega_h = 1$. The order N illustrates the steepness of the magnitude transition. The higher N value shows the steeper response in the filter as shown in Fig. 6.45.

Since the Eq. (6.136) represents the monotonically decreasing model, the passband edge is defined as the halfpower frequency. Passband peak-to-peak ripple in decibel can be calculated as Eq. (6.138).

$$\begin{aligned}\Delta_p &= 20\log_{10}\left(\frac{\text{Passband maximum magnitude}}{\text{Passband minimum magnitude}}\right) = 10\log_{10}\left(\frac{1}{1/(1+1)}\right) \\ &= 10\log_{10}(1+1) = 3.0103\text{dB for } \Omega_p = \Omega_h\end{aligned} \tag{6.138}$$

The Eq. (6.136) always denotes the one hertz halfpower frequency (also passband frequency). In order to increase the design freedom, the expanding the frequency axis relocates the passband frequency to any frequency by scaling as shown in Eq. (6.139). Figure 6.46 demonstrates the LPF with 100 rad per second Ω_p.

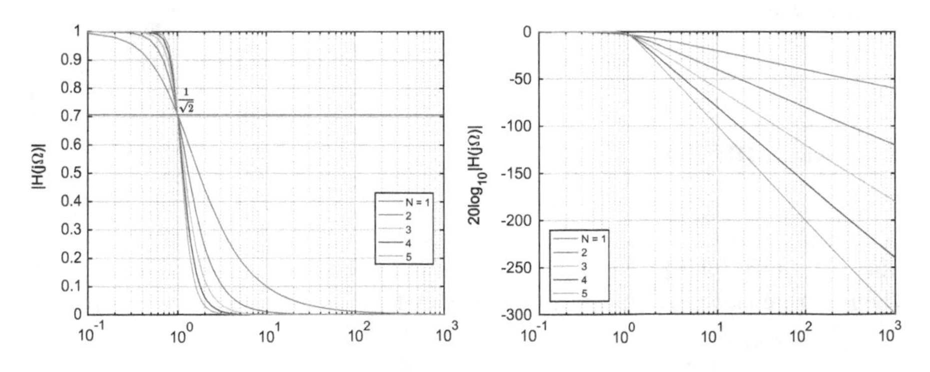

Fig. 6.45 Frequency response of linear frequency model from Eq. (6.136)

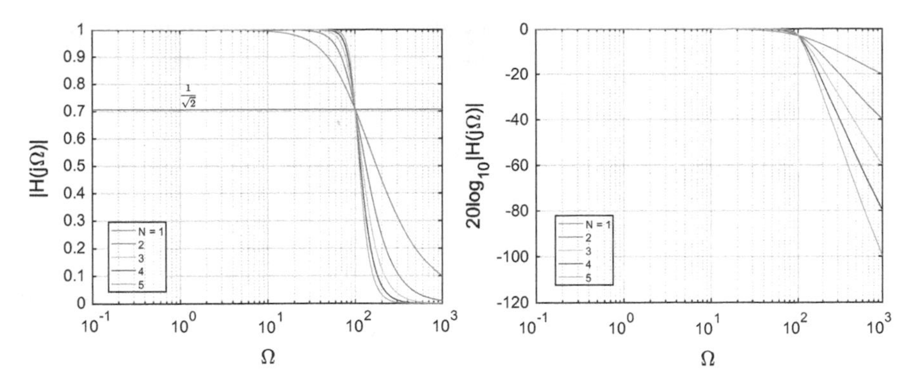

Fig. 6.46 Frequency response of Eq. (6.139) with $\Omega_p = 100$

$$|H(j\Omega)|^2 = \frac{1}{1 + \left(\frac{\Omega}{\Omega_p}\right)^{2N}} \tag{6.139}$$

Prog. 6.37 MATLAB program for Figs. 6.45 and 6.46.

```
ome = 0:0.1:1000;
omec = 100;
N = [2 4 6 8 10]';
hjo = 1./sqrt(1+ome.^N);
lhjo = 20*log10(1./sqrt(1+ome.^N));
hjoc = 1./sqrt(1+(ome/omec).^N);
lhjoc = 20*log10(1./sqrt(1+(ome/omec).^N));

figure,
semilogx(ome,hjo(1,:),ome,hjo(2,:),ome,hjo(3,:),ome,hjo(4,:),ome,hjo(5,:)),
grid
h = line([0.1 1000],[1/sqrt(2) 1/sqrt(2)],'Color',[1 0 0]);
figure,
semilogx(ome,lhjo(1,:),ome,lhjo(2,:),ome,lhjo(3,:),ome,lhjo(4,:),ome,lhjo(5,:))
, grid
figure,
semilogx(ome,hjoc(1,:),ome,hjoc(2,:),ome,hjoc(3,:),ome,hjoc(4,:),ome,hjoc(5,
:)), grid
h = line([0.1 1000],[1/sqrt(2) 1/sqrt(2)],'Color',[1 0 0]);
figure,
semilogx(ome,lhjoc(1,:),ome,lhjoc(2,:),ome,lhjoc(3,:),ome,lhjoc(4,:),ome,lhjo
c(5,:)), grid
```

The stopband is the nearest frequency after the transition due to the monotonic property of the filter. The stopband attenuation in decibels can be computed as Eq. (6.140).

$$\Delta_{\text{st}} = 20\log_{10}\left(\frac{\text{Passband magnitude}}{\text{Stopband magnitude}}\right) = 20\log_{10}\left(\frac{1}{|H(j\Omega_{st})|}\right)$$
$$= 20\log_{10}\left(\frac{1}{\frac{1}{\sqrt{1+\left(\frac{\Omega_{st}}{\Omega_p}\right)^{2N}}}}\right) = 20\log_{10}\left(\sqrt{1+\left(\frac{\Omega_{st}}{\Omega_p}\right)^{2N}}\right) \tag{6.140}$$
$$= 10\log_{10}\left(1+\left(\frac{\Omega_{st}}{\Omega_p}\right)^{2N}\right)$$

The stopband attenuation is proportional to the order selection based on the given stopband and passband frequency. Once we have the frequency specification, the order can be determined by Eq. (6.141).

$$\frac{\Delta_{\text{st}}}{10} = \log_{10}\left(1+\left(\frac{\Omega_{st}}{\Omega_p}\right)^{2N}\right) \rightarrow 10^{\frac{\Delta_{\text{st}}}{10}} = 1+\left(\frac{\Omega_{st}}{\Omega_p}\right)^{2N}$$
$$\rightarrow 10^{\frac{\Delta_{\text{st}}}{10}} - 1 = \left(\frac{\Omega_{st}}{\Omega_p}\right)^{2N} \rightarrow \log_{10}\left(10^{\frac{\Delta_{\text{st}}}{10}} - 1\right) = 2N\log_{10}\left(\frac{\Omega_{st}}{\Omega_p}\right) \tag{6.141}$$
$$N \geq \frac{\log_{10}\left(10^{\frac{\Delta_{\text{st}}}{10}} - 1\right)}{2\log_{10}\left(\frac{\Omega_{st}}{\Omega_p}\right)}$$

Designed filter from the frequency and order specification derives the magnitude equation $|H(j\Omega)|^2$ for the filter. The stable $H(j\Omega)$ filter requires the Laplace transform to choose the proper poles from the combination. Let's apply the Laplace transform on the magnitude filter equation $|H(j\Omega)|^2$ as shown in Eq. (6.142).

$$|H(s)|^2 = H(s)H(-s) = \frac{G}{1+\left(\frac{s}{j\Omega_p}\right)^{2N}} \text{ Let } j\Omega \rightarrow s \tag{6.142}$$

Find poles (denominator solution) of Eq. (6.142) for s variable.

$$|H(s)|^2 = \frac{(-p_0)(-p_1)(-p_2)\ldots(-p_{2N-1})}{(s-p_0)(s-p_1)(s-p_2)\ldots(s-p_{2N-1})} \tag{6.143}$$

$$\text{where } s = \sigma + j\Omega$$

The numerator of the above equation should be normalized by product of whole negative poles. Note that zero s indicates the zero frequency and the zero-frequency gain of the Eq. (6.143) which is one. The poles of Eq. (6.142) can be found as below.

$$\left(\frac{p_k}{j\Omega_p}\right)^{2N} = -1 = e^{j\pi(2k+1)}; \frac{p_k}{\Omega_p} = je^{j\frac{\pi}{2N}(2k+1)} = e^{j\frac{\pi}{2N}(2k+1)+j\frac{\pi}{2}}; \ k \in \mathbb{Z}$$
$$p_k = \Omega_p e^{j\frac{\pi}{2N}(2k+1)+j\frac{\pi}{2}}; \quad k = 0, 1, 2, \ldots, 2N-1 \tag{6.144}$$

On the s plane, the poles are distributed on the circle with Ω_p radius and equal radian distance. According to the Laplace transform and continuous-time filter theory, poles on the left s plane provides the stable filter.

$$\left.\begin{array}{l} p_i^L \in \text{left s plane} \\ p_i^R \in \text{right s plane} \end{array}\right\} i = 1, 2, \ldots, N$$

$$\{p_k\} = \{p_i^L\} \cup \{p_i^R\} \text{ and } \emptyset = \{p_i^L\} \cap \{p_i^R\}$$

Separate the poles into the left s plane as p_i^L and right s plane as p_i^R . Observe that there are no poles on the y axis ($j\Omega$) because of radian offset $e^{j\pi/2}$ in Eq. (6.144).

$$\begin{aligned} H(s) &= \frac{\left(-p_0^L\right)\left(-p_1^L\right)\left(-p_2^L\right)\ldots\left(-p_{N-1}^L\right)}{\left(s-p_0^L\right)\left(s-p_1^L\right)\left(s-p_2^L\right)\ldots\left(s-p_{N-1}^L\right)} \\ &= \frac{\left(-p_0^L\right)\left(-p_1^L\right)\left(-p_2^L\right)\ldots\left(-p_{N-1}^L\right)}{s^N + a_1 s^{N-1} + a_2 s^{N-2} + \cdots + a_{N-2}s^1 + a_{N-1}} \end{aligned} \tag{6.145}$$

Reconstruct the filter equation from the whole p_i^L as above.

$$H(j\Omega) = \frac{\left(-p_0^L\right)\left(-p_1^L\right)\left(-p_2^L\right)\ldots\left(-p_{N-1}^L\right)}{(j\Omega)^N + a_1(j\Omega)^{N-1} + a_2(j\Omega)^{N-2} + \cdots + a_{N-2}(j\Omega)^1 + a_{N-1}} \tag{6.146}$$

$$\text{Let } s \to j\Omega$$

Convert back to $j\Omega$ by replacing with s as Eq. (6.146). The filter with rational function is completed in the linear frequency domain $j\Omega$. The circular frequency $e^{j\omega}$ domain filter requires proper transformation from linear frequency domain filter. The bilinear transformation converts the $j\Omega$ into the z domain which includes the DTFT and DFT domain.

Bilinear transformation

The bilinear transformation [3, 7] is initiated from the z domain definition as below.

$$z = e^{j\omega} = e^{j\frac{2\pi f}{f_s}} = e^{j\frac{\Omega}{f_s}} = e^{j\Omega T_s}$$
$$\log_e z = j\Omega T_s \tag{6.147}$$

The logarithm function cannot directly be applied on the transformation because rational function based on the polynomial is required for final filter equation. Using the natural logarithm series, we can approximate the equation with first order rational function as shown in Eq. (6.148).

$$
\begin{aligned}
j\Omega = \frac{1}{T_s}\log_e z &= \frac{2}{T_s}\left\{\frac{z-1}{z+1} + \frac{1}{3}\left(\frac{z-1}{z+1}\right)^3 + \frac{1}{5}\left(\frac{z-1}{z+1}\right)^5 + \cdots\right\} \approx \frac{2}{T_s}\left(\frac{z-1}{z+1}\right) \\
&= \frac{2}{T_s}\left(\frac{1-z^{-1}}{1+z^{-1}}\right)
\end{aligned}
\tag{6.148}
$$

The following procedure figures out the relationship between Ω and ω in bilinear transformation.

$$
\begin{aligned}
j\Omega = \frac{2}{T_s}\left(\frac{1-z^{-1}}{1+z^{-1}}\right)\bigg|_{z=e^{j\omega}} &= \frac{2}{T_s}\left(\frac{1-e^{-j\omega}}{1+e^{-j\omega}}\right) = \frac{2}{T_s}\left\{\frac{e^{-\frac{j\omega}{2}}\left(e^{\frac{j\omega}{2}} - e^{-\frac{j\omega}{2}}\right)}{e^{-\frac{j\omega}{2}}\left(e^{\frac{j\omega}{2}} + e^{-\frac{j\omega}{2}}\right)}\right\} \\
&= \frac{2}{T_s}\left\{\frac{2je^{-\frac{j\omega}{2}}\left(\frac{e^{\frac{j\omega}{2}} - e^{-\frac{j\omega}{2}}}{2j}\right)}{2e^{-\frac{j\omega}{2}}\left(\frac{e^{\frac{j\omega}{2}} + e^{-\frac{j\omega}{2}}}{2}\right)}\right\} = \frac{2j}{T_s}\frac{\sin\left(\frac{\omega}{2}\right)}{\cos\left(\frac{\omega}{2}\right)} = \frac{2j}{T_s}\tan\left(\frac{\omega}{2}\right)
\end{aligned}
\tag{6.149}
$$

$$
\Omega = \tfrac{2}{T_s}\tan\left(\tfrac{\omega}{2}\right) = 2f_s\tan\left(\tfrac{\omega}{2}\right)
\tag{6.150}
$$

Equation (6.150) represents the exponential relation between the two frequency domains by tangent function. Figure 6.47 illustrates the relationship which indicate the π in ω as the ∞ in Ω . Therefore, the complete Ω frequency is covered by limited length ω frequency.

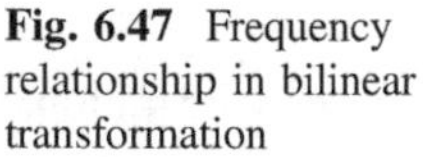

Fig. 6.47 Frequency relationship in bilinear transformation

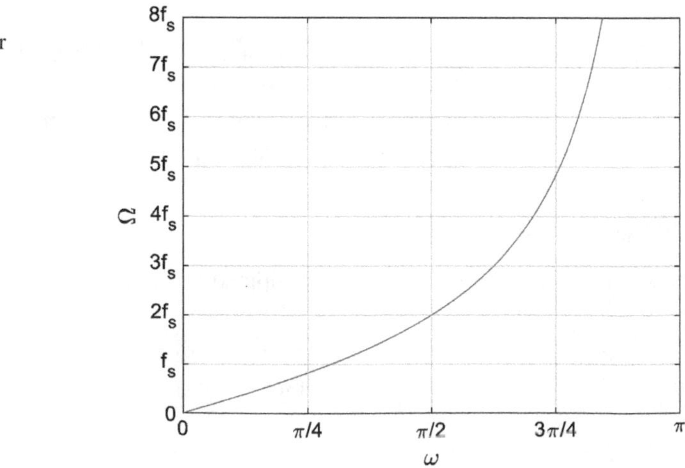

Prog. 6.38 MATLAB program for Fig. 6.47.

```
ome = linspace(0,pi,1000);
fs = 8000;
OM = 2*fs*tan(ome/2);
plot(ome,OM), grid
ylim([0 8000*8]);
xlim([0 pi])
ax = gca;
ax.XTick = [0 pi/4 pi/2 3*pi/4 pi];
ax.XTickLabel = {'0','\pi/4','\pi/2','3\pi/4','\pi'};
ax.YTick = [0 8000 8000*2 8000*3 8000*4 8000*5 8000*6 8000*7 8000*8];
ax.YTickLabel = {'0','f_s','2f_s','3f_s','4f_s','5f_s','6f_s','7f_s','8f_s'};
xlabel('\omega');
ylabel('\Omega')
```

Previous derivation for bilinear transformation performs the projection between $j\Omega$ axis on s plane and unit circle on z plane. Generalization process extends to the whole s and z plane for bilinear transformation as shown in Eq. (6.151).

$$\begin{cases} e^{j\omega} \to z = re^{j\omega} \in \mathbb{C} \\ j\Omega \to s = \sigma + j\Omega \in \mathbb{C} \end{cases}$$

$$z = e^{j\Omega T_s} = e^{\sigma T_s + j\Omega T_s} = e^{\sigma T_s} e^{j\Omega T_s} = re^{j\omega} \text{where} \begin{cases} r = e^{\sigma T_s} \\ \Omega = 2f_s \tan\left(\frac{\omega}{2}\right) \end{cases} \tag{6.151}$$

According to the tangent function, the Ω frequency is ranged from the $-\infty$ to ∞ and also matched to the $-\pi$ to π in ω frequency based on Eq. (6.150). The BIBO stable system must include the unit circle area for ROC in z domain; therefore, the all poles should be located within the unit circle area for causal system in z domain. The unit circle area in the z plane is transformed to the left half of the s plane as shown in Eq. (6.152) and Fig. 6.48.

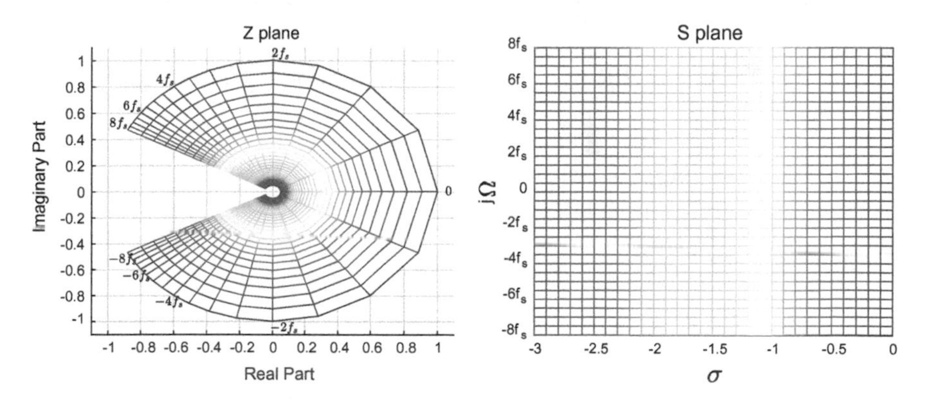

Fig. 6.48 Color coded relation between z and s domain by bilinear transformation

$$r = e^{\sigma T_s} < 1 \rightarrow \sigma < 0$$
$$re^{j\omega} \in \text{Unit circle area} \rightarrow \sigma < 0 \text{ and } -\infty < \Omega < \infty \tag{6.152}$$

Prog. 6.39 MATLAB program for Fig. 6.48.

```
fs = 1;
ts = 1/fs;
aatan = (-4:1/4:4);
of = 2*fs*aatan;
sig = -3:0.1:0;
[of1, sig1] = ndgrid((-4:1/4:4)*2*fs, -3:0.1:0);
s = sig1+1j*of1;
r = exp(sig*ts);
theta = atan(aatan)*2;
z = (r.'*exp(1j*theta))';
figure, mesh(real(s),imag(s),sig1), view(2), colormap(jet), ax = gca;
ax.YTick = [-8*fs -6*fs -4*fs -2*fs 0 2*fs 4*fs 6*fs 8*fs];
ax.YTickLabel = {'-8f_s','-6f_s','-4f_s','-2f_s','0','2f_s','4f_s','6f_s','8f_s',};
xlabel('\sigma');
ylabel('j\Omega')
title('S plane')
figure, mesh(real(z),imag(z),sig1), view(2), colormap(jet);
xlabel('Real Part');
ylabel('Imaginary Part')
title('Z plane')
xlim([-1.1 1.1]);
ylim([-1.1 1.1]);
```

The poles on the left s plane (Laplace transform) are projected on the unit circle area in z domain (Z-transform) for stable IIR filter design. Note that we used the left s plane poles on the Butterworth filter design in the linear frequency domain $j\Omega$.

Example 6.18

Design the Table 6.27 IIR filter by using Butterworth method and bilinear transformation.

Table 6.27 Given filter specification

Filter specifications	Value
Filter type	Low pass filter
Filter realization	Butterworth IIR
Filter length	N with optimization
Passband frequency (f_p)	1000 Hz; $\omega_p = \pi/4$
Passband magnitude	1
Passband peak-to-peak ripple in decibel (Δ_p)	3.0103 dB
Stopband frequency (f_{st})	2000 Hz; $\omega_{st} = \pi/2$
Stopband attenuation in decibels Δ_{st}	$20\log_{10}\left(\frac{1}{0.1}\right) = 20$ dB
Sampling frequency (f_s)	8000 Hz

Solution

Use the bilinear transformation to find the frequency specification in Ω from Eq. (6.150).

$$\Omega_p = 2f_s \tan\left(\frac{\pi}{8}\right) = 6627.4 \text{ rad/s}$$

$$\Omega_{st} = 2f_s \tan\left(\frac{\pi}{4}\right) = 16000 \text{ rad/s}$$

Apply the Eq. (6.141) to compute the filter order as below.

$$N \geq \frac{\log_{10}\left(10^{\frac{20}{10}} - 1\right)}{2\log_{10}\left(\frac{16{,}000}{6627.4}\right)} = 2.6068 \rightarrow 3$$

Create the transfer function in s domain as below.

$$|H(s)|^2 = H(s)H(-s) = \frac{G}{1 + \left(\frac{s}{j\Omega_p}\right)^{2N}} = \frac{G}{1 + \left(\frac{s}{j6627.4}\right)^{6}} \text{ where } s = \sigma + j\Omega$$

Find the poles from the transfer function as below.

$$p_k = 6627.4e^{j\frac{\pi}{6}(2k+1) + j\frac{\pi}{2}} \text{ for } k = -2, -1, 0, 1, 2, 3$$

$$p_k = 6627.4e^{j0\pi}, 6627.4e^{j\frac{\pi}{3}}, 6627.4e^{j\frac{2\pi}{3}}, 6627.4e^{j\pi}, 6627.4e^{j\frac{4\pi}{3}}, 6627.4e^{j\frac{5\pi}{3}}$$

Select the left s plane poles as below.

$$p_i^L = 6627.4e^{j\frac{2\pi}{3}}, 6627.4e^{j\pi}, 6627.4e^{j\frac{4\pi}{3}}$$

Compute the gain G to normalize the response at zero frequency as below.

$$G = \left(-6627.4e^{j\frac{2\pi}{3}}\right)\left(-6627.4e^{j\pi}\right)\left(-6627.4e^{j\frac{4\pi}{3}}\right) = 291091517618.824$$

Assemble the Butterworth equation in s domain as below.

$$H(s) = \frac{291091517618.824}{\left(s - 6627.4e^{j\frac{2\pi}{3}}\right)\left(s - 6627.4e^{j\pi}\right)\left(s - 6627.4e^{j\frac{4\pi}{3}}\right)}$$

Apply the bilinear transformation as below.

$$H(z) = \left.\frac{291091517618.824}{\left(j\Omega - 6627.4e^{j\frac{2\pi}{3}}\right)\left(j\Omega - 6627.4e^{j\pi}\right)\left(j\Omega - 6627.4e^{j\frac{4\pi}{3}}\right)}\right|_{j\Omega = 2f_s\left(\frac{1-z^{-1}}{1+z^{-1}}\right)}$$

The final Butterworth filter in discrete time domain is below.

$$H(z) = \frac{0.0317 + 0.0951z^{-1} + 0.0951z^{-2} + 0.0317z^{-3}}{1 - 1.4590z^{-1} + 0.9104z^{-2} - 0.1978z^{-3}}$$

Figure 6.49 shows the pole zero plot of the design filter. Three poles are located within the unit circle and triple zeros are placed at –1 location.

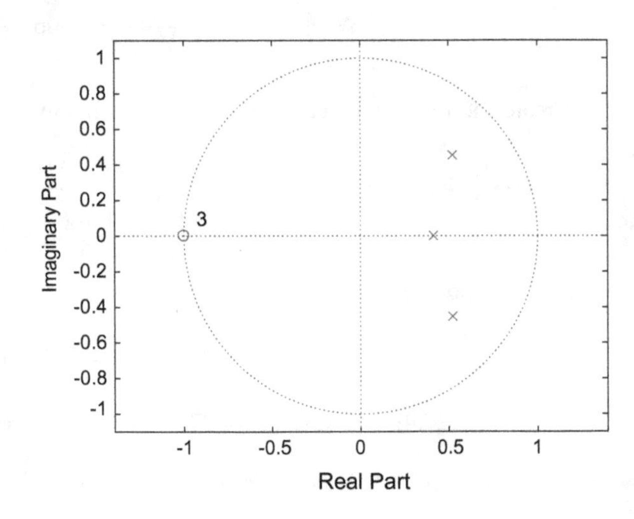

Fig. 6.49 Pole zero plot of designed Butterworth filter for Table 6.27

Prog. 6.40 MATLAB program for Figs. 6.49 and 6.50.

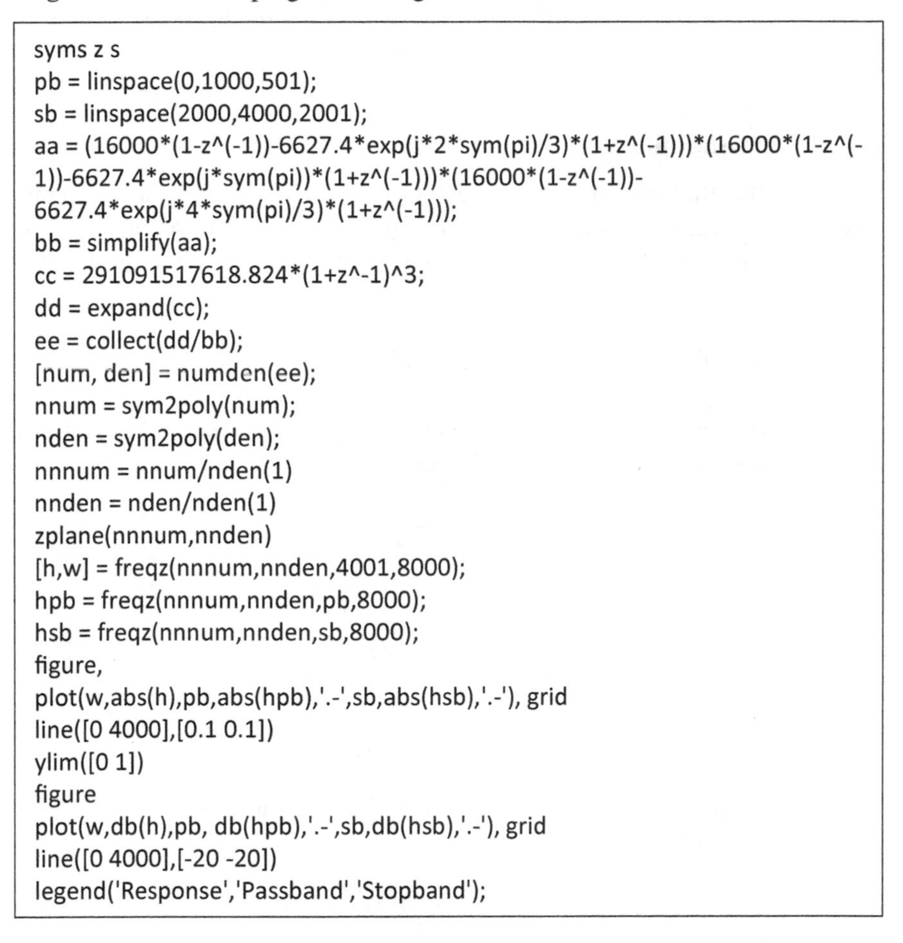

```
syms z s
pb = linspace(0,1000,501);
sb = linspace(2000,4000,2001);
aa = (16000*(1-z^(-1))-6627.4*exp(j*2*sym(pi)/3)*(1+z^(-1)))*(16000*(1-z^(-
1))-6627.4*exp(j*sym(pi))*(1+z^(-1)))*(16000*(1-z^(-1))-
6627.4*exp(j*4*sym(pi)/3)*(1+z^(-1)));
bb = simplify(aa);
cc = 291091517618.824*(1+z^-1)^3;
dd = expand(cc);
ee = collect(dd/bb);
[num, den] = numden(ee);
nnum = sym2poly(num);
nden = sym2poly(den);
nnnum = nnum/nden(1)
nnden = nden/nden(1)
zplane(nnnum,nnden)
[h,w] = freqz(nnnum,nnden,4001,8000);
hpb = freqz(nnnum,nnden,pb,8000);
hsb = freqz(nnnum,nnden,sb,8000);
figure,
plot(w,abs(h),pb,abs(hpb),'.-',sb,abs(hsb),'.-'), grid
line([0 4000],[0.1 0.1])
ylim([0 1])
figure
plot(w,db(h),pb, db(hpb),'.-',sb,db(hsb),'.-'), grid
line([0 4000],[-20 -20])
legend('Response','Passband','Stopband');
```

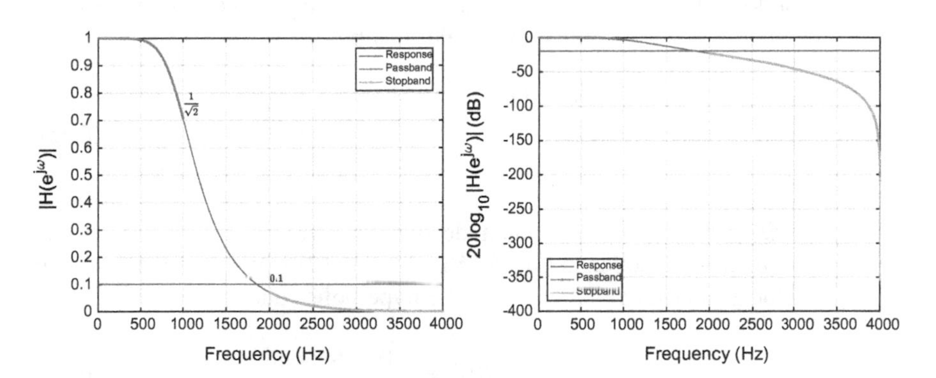

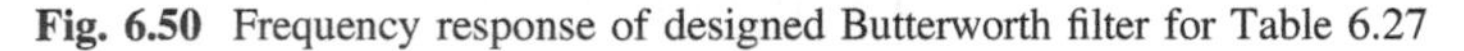

Fig. 6.50 Frequency response of designed Butterworth filter for Table 6.27

Figure 6.50 represents the frequency response of the filter. The passband and stopband satisfy the Table 6.27 specification. Note that the horizontal line in the figures denote the stopband attenuation requirement.

■

Type I Chebyshev filter

Similar to the Butterworth filter, Type I Chebyshev filter [7] uses the rational function with special polynomial as shown in Eq. (6.153). As the $\varepsilon^2 T_N^2(\Omega)$ increases exponentially, the filter magnitude decreases significantly for LPF property.

$$|H(j\Omega)|^2 = H(j\Omega)H^*(j\Omega) = H(j\Omega)H(-j\Omega) = \frac{1}{1+\varepsilon^2 T_N^2(\Omega)} \tag{6.153}$$

The special polynomial is the first kind Chebyshev polynomial $T_N(\Omega)$ [18]. Below show the polynomials up to 7th order.

$$\begin{gathered} T_0(\Omega) = 1 \\ T_1(\Omega) = \Omega \\ T_2(\Omega) = 2\Omega^2 - 1 \\ T_3(\Omega) = 4\Omega^3 - 3\Omega \\ T_4(\Omega) = 8\Omega^4 - 8\Omega^2 + 1 \\ T_5(\Omega) = 16\Omega^5 - 20\Omega^3 + 5\Omega \\ T_6(\Omega) = 32\Omega^6 - 48\Omega^4 + 18\Omega^2 - 1 \\ T_7(\Omega) = 64\Omega^7 - 112\Omega^5 + 56\Omega^3 - 7\Omega \end{gathered} \tag{6.154}$$

The first kind Chebyshev polynomial can be derived from the recursive equation as shown in Eq. (6.155).

$$T_{N+1}(\Omega) = 2\Omega T_N(\Omega) - T_{N-1}(\Omega) \tag{6.155}$$

The first kind Chebyshev polynomial is equivalent to the trigonometric or hyperbolic functions for the range of Ω values as below.

$$T_N(\Omega) = \begin{cases} \cos(N\cos^{-1}(\Omega)) & \text{for} |\Omega| \leq 1 \\ \cosh\left(N\cosh^{-1}(\Omega)\right) & \text{for } \Omega \geq 1 \\ (-1)^N \cosh\left(N\cosh^{-1}(-\Omega)\right) & \text{for } \Omega \leq -1 \end{cases} \tag{6.156}$$

The $T_N(\Omega)$ presents limited magnitude one for less than one Ω in absolute because of the cosine function in Eq. (6.156). Also, the $T_N(\Omega)$ shows rapid growing magnitude for greater than one Ω due to the hyperbolic function.

$$T_N(1) = 1 \text{ and } |T_N(0)| = \begin{cases} 0 & \text{for odd N} \\ 1 & \text{for even N} \end{cases} \tag{6.157}$$

The initial value of the $T_N(\Omega)$ depends on the order parity but the $T_N(1)$ value is fixed as one for all order. Figure 6.51 illustrates the $T_N(\Omega)$ up to 7th order.

Prog. 6.41 MATLAB program for Fig. 6.51.

```
x = linspace(0,2,1000);
t1 = x;
t2 = 2*x.^2-1;
t3 = 4*x.^3-3*x;
t4 = 8*x.^4-8*x.^2+1;
t5 = 16*x.^5-20*x.^3+5*x;
t6 = 32*x.^6-48*x.^4+18*x.^2-1;
t7 = 64*x.^7-112*x.^5+56*x.^3-7*x;
plot(x,t1,x,t2,x,t3,x,t4,x,t5,x,t6,x,t7), grid, xlim([0 2]), ylim([-1.5 12]);
xlabel('\Omega')
ylabel('T_N(\Omega)')
legend('N = 1','2','3','4','5','6','7');
```

As expected, the $T_N(\Omega)$ depicts the fluctuation with limited magnitude up to one Ω and rapidly increasing after the range. The higher order demonstrates the faster monotonic growing in Fig. 6.51. By applying the polynomial to Eq. (6.153), the filter response can be calculated and illustrated as Fig. 6.52. The response is similar to the Butterworth filter; however, we can observe the fluctuation in passband and steeper decreasing in the transition band. In overall, the frequency response represents the LPF with passband frequency $\Omega_p = 1$. Note that the Ω_p and Ω_h are equal in Fig. 6.52 because of unit ε. The order N illustrates the steepness of the magnitude transition. The higher N value shows the steeper response in the filter as shown in Fig. 6.52.

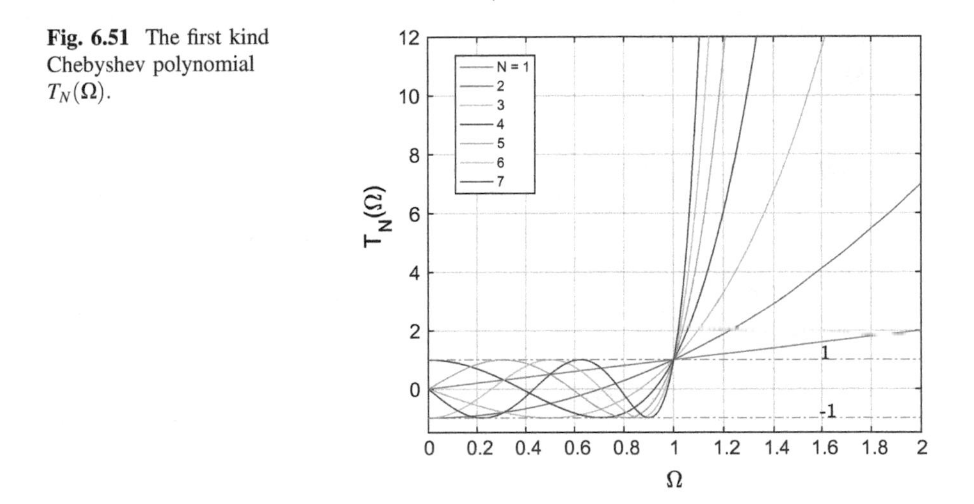

Fig. 6.51 The first kind Chebyshev polynomial $T_N(\Omega)$.

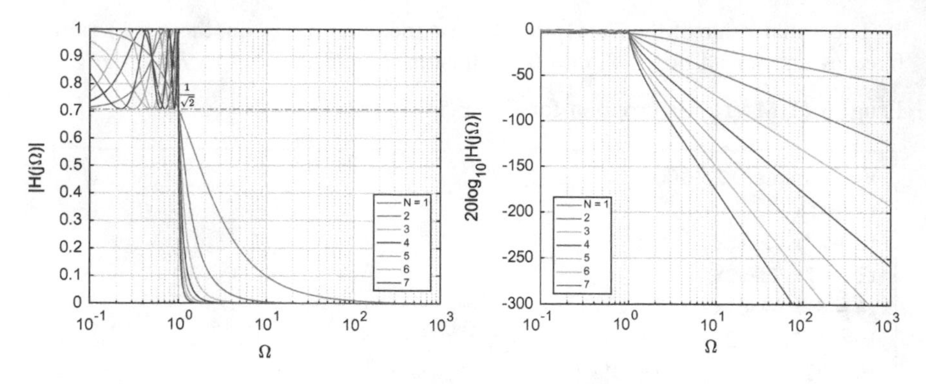

Fig. 6.52 Frequency response of linear frequency model from Eq. (6.153) with $\varepsilon = 1$

Prog. 6.42 MATLAB program for Fig. 6.52.

```
x = linspace(0,1000,100001);
t1 = x;
t2 = 2*x.^2-1;
t3 = 4*x.^3-3*x;
t4 = 8*x.^4-8*x.^2+1;
t5 = 16*x.^5-20*x.^3+5*x;
t6 = 32*x.^6-48*x.^4+18*x.^2-1;
t7 = 64*x.^7-112*x.^5+56*x.^3-7*x;

figure,
semilogx(x,1./sqrt(1+t1.^2),x,1./sqrt(1+t2.^2),x,1./sqrt(1+t3.^2),x,1./sqrt(1+t
4.^2),x,1./sqrt(1+t5.^2),x,1./sqrt(1+t6.^2),x,1./sqrt(1+t7.^2)), grid
h = line([0.1 1000],[1/sqrt(2) 1/sqrt(2)],'LineStyle','-.');
xlim([0.1 1000]);
xlabel('\Omega')
ylabel('|H(j\Omega)|')
legend('N = 1','2','3','4','5','6','7');
figure,
semilogx(x,20*log10(1./sqrt(1+t1.^2)),x,20*log10(1./sqrt(1+t2.^2)),x,20*log1
0(1./sqrt(1+t3.^2)),x,20*log10(1./sqrt(1+t4.^2)),x,20*log10(1./sqrt(1+t5.^2)),
x,20*log10(1./sqrt(1+t6.^2)),x,20*log10(1./sqrt(1+t7.^2))), grid
legend('N = 1','2','3','4','5','6','7');
xlim([0.1 1000]);
ylim([-300 0]);
xlabel('\Omega')
ylabel('20log_{10}|H(j\Omega)|');
```

From the given passband peak-to-peak ripple Δ_p , the ε value can be determined by Eq. (6.158).

$$\Delta_p = 20\log_{10}\left(\frac{1}{1/\sqrt{1+\varepsilon^2}}\right) = 10\log_{10}\left(1+\varepsilon^2\right)$$
$$\varepsilon = \sqrt{10^{\frac{\Delta_p}{10}} - 1} \tag{6.158}$$

Expanding the frequency axis relocates the passband frequency to any frequency by scaling as shown in Eq. (6.159). Figure 6.53 demonstrates the LPF with 100 rad/s Ω_p and 10 dB Δ_p.

$$|H(j\Omega)|^2 = \frac{1}{1+\varepsilon^2 T_N^2\left(\frac{\Omega}{\Omega_p}\right)} \tag{6.159}$$

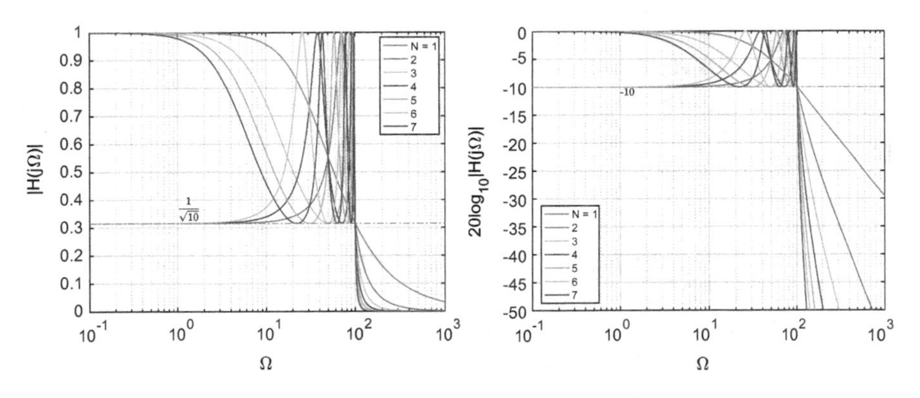

Fig. 6.53 Frequency response of Eq. (6.159) with $\Omega_p = 100$ and $\Delta_p = 10$

Prog. 6.43 MATLAB program for Fig. 6.53.

```
dp = 10;                %passband ripple in decibel
pa = 100;               %passband for LPF
ee = sqrt(10^(dp/10)-1);
x = linspace(0,1000,100001);
t1 = ee*(x/pa);
t2 = ee*(2*(x/pa).^2-1);
t3 = ee*(4*(x/pa).^3-3*(x/pa));
t4 = ee*(8*(x/pa).^4-8*(x/pa).^2+1);
t5 = ee*(16*(x/pa).^5-20*(x/pa).^3+5*(x/pa));
t6 = ee*(32*(x/pa).^6-48*(x/pa).^4+18*(x/pa).^2-1);
t7 = ee*(64*(x/pa).^7-112*(x/pa).^5+56*(x/pa).^3-7*(x/pa));

figure,
semilogx(x,1./sqrt(1+t1.^2),x,1./sqrt(1+t2.^2),x,1./sqrt(1+t3.^2),x,1./sqrt(1+t
4.^2),x,1./sqrt(1+t5.^2),x,1./sqrt(1+t6.^2),x,1./sqrt(1+t7.^2)), grid
h = line([0.1 1000],[1/sqrt(10) 1/sqrt(10)],'LineStyle','-.');
str = '$$\frac{1}{\sqrt{10}}$$';
text(1,1/sqrt(10)+0.05,str,'Interpreter','latex')
xlim([0.1 1000]);
xlabel('\Omega')
ylabel('|H(j\Omega)|')
legend('N = 1','2','3','4','5','6','7');
figure,
semilogx(x,20*log10(1./sqrt(1+t1.^2)),x,20*log10(1./sqrt(1+t2.^2)),x,20*log1
0(1./sqrt(1+t3.^2)),x,20*log10(1./sqrt(1+t4.^2)),x,20*log10(1./sqrt(1+t5.^2)),
x,20*log10(1./sqrt(1+t6.^2)),x,20*log10(1./sqrt(1+t7.^2))), grid
h = line([0.1 1000],[-10 -10],'LineStyle','-.');
str = '-10';
text(1,-10-1,str,'Interpreter','latex')
xlim([0.1 1000]);
ylim([-50 0]);
xlabel('\Omega')
ylabel('20log_{10}|H(j\Omega)|')
legend('N = 1','2','3','4','5','6','7');
```

The stopband is the nearest frequency after the transition due to the monotonic property after passband region. From the given stopband attenuation in decibel Δ_{st}, the required filter order N can be calculated by using Eq. (6.160).

$$\Delta_{st} = 20\log_{10}\sqrt{\frac{1}{|H(j\Omega_{st})|^2}} = 20\log_{10}\left(\frac{1}{1/\sqrt{1+\varepsilon^2 T_N^2\left(\frac{\Omega_{st}}{\Omega_p}\right)}}\right)$$

$$= 10\log_{10}\left(1+\varepsilon^2 T_N^2\left(\frac{\Omega_{st}}{\Omega_p}\right)\right)$$

$$T_N\left(\frac{\Omega_{st}}{\Omega_p}\right) = \sqrt{\frac{10^{\frac{\Delta_{st}}{10}}-1}{\varepsilon^2}}$$

$$T_N\left(\frac{\Omega_{st}}{\Omega_p}\right) = \cosh\left(N\cosh^{-1}\left(\frac{\Omega_{st}}{\Omega_p}\right)\right) = \sqrt{\frac{10^{\frac{\Delta_{st}}{10}}-1}{\varepsilon^2}} \text{ for } \frac{\Omega_{st}}{\Omega_p} > 1$$

$$\cosh x = \frac{e^x+e^{-x}}{2}; \cosh^{-1}x = \log_e\left(x+\sqrt{x^2-1}\right) \text{ for } x \geq 1$$

$$N \geq \frac{\cosh^{-1}\left(\sqrt{\frac{10^{\frac{\Delta_{st}}{10}}-1}{\varepsilon^2}}\right)}{\cosh^{-1}\left(\frac{\Omega_{st}}{\Omega_p}\right)} \tag{6.160}$$

Similar to the Butterworth IIR filter, the stable $H(j\Omega)$ filter requires the Laplace transform to choose the proper poles from the combination. Let's apply the Laplace transform on the magnitude filter equation $|H(j\Omega)|^2$ as shown in Eq. (6.161).

$$|H(s)|^2 = H(s)H(-s) = \frac{G}{1+\varepsilon^2 T_N^2\left(\frac{s}{j\Omega_p}\right)} \text{ Let } j\Omega \to s \tag{6.161}$$

Find poles (denominator solution) of Eq. (6.161) for s variable.

$$|H(s)|^2 = \frac{G}{(s-p_0)(s-p_1)(s-p_2)\ldots(s-p_{2N-1})} \tag{6.162}$$

$$\text{where } s = \sigma + j\Omega$$

The poles of Eq. (6.161) can be found as below.

$$\begin{aligned}\frac{p_k}{\Omega_p} &= \pm\sin\left((2k-1)\frac{\pi}{2N}\right)\sinh\left(\frac{1}{N}\sinh^{-1}\left(\frac{1}{\varepsilon}\right)\right)\\ &\quad + j\cos\left((2k-1)\frac{\pi}{2N}\right)\cosh\left(\frac{1}{N}\sinh^{-1}\left(\frac{1}{\varepsilon}\right)\right)\ \text{for k} = 1,2,\ldots,\text{N}\\ p_k &= \pm\Omega_{pa}\sin\left((2k-1)\frac{\pi}{2N}\right)\sinh\left(\frac{1}{N}\sinh^{-1}\left(\frac{1}{\varepsilon}\right)\right)\\ &\quad + j\Omega_{pa}\cos\left((2k-1)\tfrac{\pi}{2N}\right)\cosh\left(\tfrac{1}{N}\sinh^{-1}\left(\tfrac{1}{\varepsilon}\right)\right)\text{for k} = 1,2,\ldots,\text{N}\end{aligned} \tag{6.163}$$

Separate the poles into the left s plane as p_i^L and right s plane as p_i^R. Observe that poles on the left s plane provides the stable filter. The left and right of the s plane can be divided by the real value. Positive real value indicates right s plane and negative real value specifies left s plane.

$$\left.\begin{array}{l} p_i^L \in \text{left s plane} \\ p_i^R \in \text{right s plane}\end{array}\right\} i = 1,2,\ldots,N$$

$$\begin{aligned}p_k^L &= -\Omega_{pa}\sin\left((2k-1)\frac{\pi}{2N}\right)\sinh\left(\frac{1}{N}\sinh^{-1}\left(\frac{1}{\varepsilon}\right)\right)\\ &\quad + j\Omega_{pa}\cos\left((2k-1)\tfrac{\pi}{2N}\right)\cosh\left(\tfrac{1}{N}\sinh^{-1}\left(\tfrac{1}{\varepsilon}\right)\right)\end{aligned} \tag{6.164}$$

$$\begin{aligned}p_k^R &= \Omega_{pa}\sin\left((2k-1)\frac{\pi}{2N}\right)\sinh\left(\frac{1}{N}\sinh^{-1}\left(\frac{1}{\varepsilon}\right)\right)\\ &\quad + j\Omega_{pa}\cos\left((2k-1)\frac{\pi}{2N}\right)\cosh\left(\frac{1}{N}\sinh^{-1}\left(\frac{1}{\varepsilon}\right)\right)\end{aligned}$$

Reconstruct the filter equation from the whole p_i^L as below.

$$\begin{aligned}H(s) &= \frac{G}{\left(s-p_1^L\right)\left(s-p_2^L\right)\ldots\left(s-p_N^L\right)}\\ &= \frac{G}{s^N + a_1 s^{N-1} + a_2 s^{N-2} + \cdots + a_{N-2}s^1 + a_{N-1}}\end{aligned} \tag{6.165}$$

According to the Chebyshev polynomials, the output for the zero input is different for the even and odd order case as shown in Eq. (6.157); therefore, the zero-frequency gain G is parity sensitive as below. Figure 6.51 also represents the different initial values for the even and odd order number.

$$G = H(s=0) = \begin{cases} 1 & \text{for odd } N \\ 1/\sqrt{1+\varepsilon^2} & \text{for even } N \end{cases}$$

$$H(s=0) = \frac{G}{\left(-p_1^L\right)\left(-p_2^L\right)\ldots\left(-p_N^L\right)} = \begin{cases} 1 & \text{for odd } N \\ 1/\sqrt{1+\varepsilon^2} & \text{for even } N \end{cases}$$

$$G = \begin{cases} \left(-p_1^L\right)\left(-p_2^L\right)\ldots\left(-p_N^L\right) & \text{for odd } N \\ \left(-p_1^L\right)\left(-p_2^L\right)\ldots\left(-p_N^L\right)/\sqrt{1+\varepsilon^2} & \text{for even } N \end{cases} \quad (6.166)$$

Convert back to $j\Omega$ by replacing with s as Eq. (6.167). The filter with rational function is completed in the linear frequency domain $j\Omega$.

$$H(j\Omega) = \frac{G}{(j\Omega)^N + a_1(j\Omega)^{N-1} + a_2(j\Omega)^{N-2} + \cdots + a_{N-2}(j\Omega)^1 + a_{N-1}} \quad (6.167)$$

The bilinear transformation provides the tool to change the linear frequency domain into the z domain for discrete-time filter as shown in Eq. (6.168).

$$H(z) = \left.\frac{G}{(j\Omega)^N + a_1(j\Omega)^{N-1} + \cdots + a_{N-2}(j\Omega)^1 + a_{N-1}}\right|_{j\Omega = 2f_s\left(\frac{1-z^{-1}}{1+z^{-1}}\right)} \quad (6.168)$$

The final Type I Chebyshev filter in discrete time domain is Eq. (6.168).

Example 6.19

Design the Table 6.28 IIR filter by using Type I Chebyshev method and bilinear transformation.

Table 6.28 Given filter specification

Filter specifications	Value
Filter type	Low pass filter
Filter realization	Type I Chebyshev IIR filter
Filter length	N with optimization
Passband frequency (f_p)	1000 Hz; $\omega_p = \pi/4$
Passband magnitude	1
Passband peak-to-peak ripple in decibel (Δ_p)	$20\log_{10}\left(\frac{1}{1/\sqrt{1+1}}\right) = 3.0103\text{dB}$
Stopband frequency (f_{st})	2000 Hz; $\omega_{st} = \pi/2$
Stopband attenuation in decibels Δ_{st}	$20\log_{10}\left(\frac{1}{0.1}\right) = 20$ dB
Sampling frequency (f_s)	8000 Hz

Solution

Use the bilinear transformation to find the frequency specification in Ω from Eq. (6.150).

$$\Omega_p = 2f_s \tan\left(\frac{\pi}{8}\right) = 6627.4\text{rad/s}$$

$$\Omega_{st} = 2f_s \tan\left(\frac{\pi}{4}\right) = 16000\,\text{rad/s}$$

Compute the ε value by using the Eq. (6.158).

$$\varepsilon = \sqrt{10^{\frac{\Delta_p}{10}} - 1} = \sqrt{10^{\frac{3.0103}{10}} - 1} = 1$$

Estimate the proper filter order by using the Eq. (6.160).

$$N \geq \frac{\cosh^{-1}\sqrt{\left(10^{\frac{\Delta_{st}}{10}} - 1\right)/\varepsilon^2}}{\cosh^{-1}\left(\frac{\Omega_{st}}{\Omega_p}\right)} = \frac{\cosh^{-1}\sqrt{\left(10^{\frac{20}{10}} - 1\right)/1^2}}{\cosh^{-1}\left(\frac{16,000}{6627.4}\right)} = 1.9549 \rightarrow 2; \text{even}$$

Derive the left s plane poles from Eq. (6.164).

$$p_k^L = -\Omega_p \sin\left((2k-1)\frac{\pi}{2N}\right)\sinh\left(\frac{1}{N}\sinh^{-1}\left(\frac{1}{\varepsilon}\right)\right)$$
$$+ j\Omega_p \cos\left((2k-1)\frac{\pi}{2N}\right)\cosh\left(\frac{1}{N}\sinh^{-1}\left(\frac{1}{\varepsilon}\right)\right); k = 1, 2$$

$$p_k^L = -2132.7 + j5148.7, -2132.7 - j5148.7$$

Based on the poles and gain equation Eq. (6.166), build the filter equation in s domain as below. Note that the filter order is even.

$$H(s) = \frac{(2132.7 - j\,5148.7)(2132.7 + j\,5148.7)/\sqrt{2}}{(s + 2132.7 - j\,5148.7)(s + 2132.7 + j\,5148.7)}$$
$$= \frac{21960983.69}{s^2 + 4265.4s + 31057520.98}$$

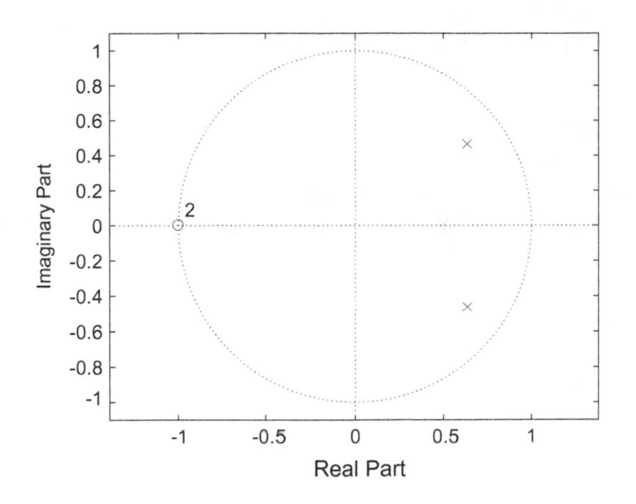

Fig. 6.54 Pole zero plot of designed Type I Chebyshev filter for Table 6.28

Apply the bilinear transformation as below.

$$H(j\Omega) = \frac{(2132.7 - j\,5148.7)(2132.7 + j\,5148.7)/\sqrt{2}}{(j\Omega + 2132.7 - j\,5148.7)(j\Omega + 2132.7 + j\,5148.7)}$$

$$H(z) = \left.\frac{21960983.69}{(j\Omega + 2132.7 - j5148.7)(j\Omega + 2132.7 + j5148.7)}\right|_{j\Omega = 2f_s\left(\frac{1-z^{-1}}{1+z^{-1}}\right)}$$

The final Type I Chebyshev filter in discrete time domain is below.

$$H(z) = \frac{0.0618 + 0.1236z^{-1} + 0.0618z^{-2}}{1 - 1.2662z^{-1} + 0.6158z^{-2}}$$

Figure 6.54 shows the pole zero plot of the design filter. Two poles are located within the unit circle and double zeros are placed at –1 location.

Prog. 6.44 MATLAB program for Figs. 6.54 and 6.55.

```
syms z
pb = linspace(0,1000,501);
sb = linspace(2000,4000,2001);
aa = (21960983.69*(1+z^(-1))^2)/((16000*(1-z^(-1))+(2132.7-
j*5148.7)*(1+z^(-1)))*(16000*(1-z^(-1))+(2132.7+j*5148.7)*(1+z^(-1))));
bb = simplify(aa);
[num den] = numden(bb);
cc = expand(num);
dd = simplify(den);
nnum = sym2poly(cc);
dden = sym2poly(dd);
nnnum = nnum/dden(1);
nnden = dden/dden(1);

figure,
zplane(nnnum,nnden)
[h,w] = freqz(nnnum,nnden,4001,8000);
hpb = freqz(nnnum,nnden,pb,8000);
hsb = freqz(nnnum,nnden,sb,8000);
figure,
plot(w,abs(h),pb,abs(hpb),'.-',sb,abs(hsb),'.-'), grid
line([0 4000],[0.1 0.1])
ylim([0 1])
legend('Response','Passband','Stopband')
ylabel('|H(e^{j\omega})|')
xlabel('Frequency (Hz)')
figure,
plot(w,db(h),pb, db(hpb),'.-',sb,db(hsb),'.-'), grid
line([0 4000],[-20 -20])
legend('Response','Passband','Stopband')
ylabel('20log_{10}|H(e^{j\omega})| (dB)')
xlabel('Frequency (Hz)');
```

Figure 6.55 represents the frequency response of the filter. The passband and stopband satisfy the Table 6.28 specification. Note that the horizontal line in the figures denote the stopband attenuation requirement.

■

Type II Chebyshev filter

Type II Chebyshev filter [7] is originated from the Type I Chebyshev filter. Type II also uses the first kind Chebyshev polynomial $T_N(\Omega)$ [18] as Eq. (6.154).

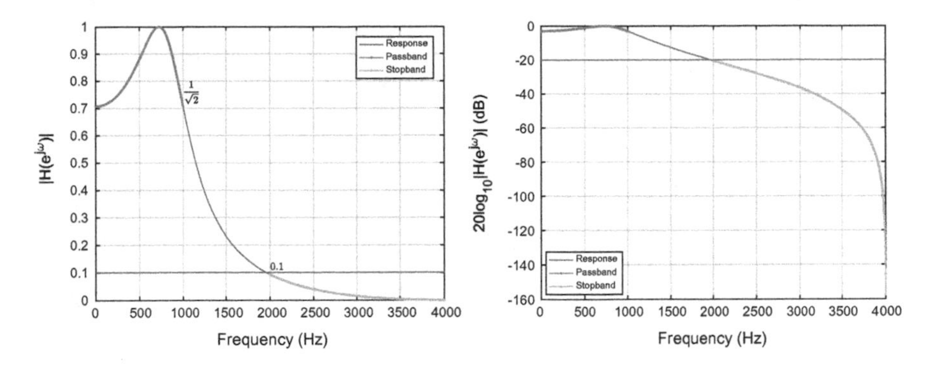

Fig. 6.55 Frequency response of designed Type I Chebyshev filter for Table 6.28

$$|H(j\Omega)|^2 = H(j\Omega)H^*(j\Omega) = H(j\Omega)H(-j\Omega) = \frac{1}{1+\frac{1}{\varepsilon^2 T_N^2\left(\frac{1}{\Omega}\right)}} \tag{6.169}$$

Let's investigate the conceptual derivation procedure of the Type II Chebyshev filter. First, $H_0(j\Omega)$ below is the Type I Chebyshev filter illustrated at Fig. 6.56 with $\varepsilon = 1$.

$$|H_0(j\Omega)| = \sqrt{\frac{1}{1+\varepsilon^2 T_N^2(\Omega)}}; \text{Type I Chebyshev filter} \tag{6.170}$$

The $H_0(j\Omega)$ is inverted in magnitude by placing the $\varepsilon^2 T_N^2(\Omega)$ at numerator as Eq. (6.171). The $H_1(j\Omega)$ is HPF due to the inversion as shown in Fig. 6.57.

$$|H_1(j\Omega)| = \sqrt{\frac{\varepsilon^2 T_N^2(\Omega)}{1+\varepsilon^2 T_N^2(\Omega)}}; \text{Upside down inversion of } |H_0(j\Omega)| \tag{6.171}$$

The $H_1(j\Omega)$ is flipped left right by input variable inversion as $1/\Omega$. The flipping is performed on frequency axis over the $\Omega = 1$ as shown in Fig. 6.58. The inversion and flipping operation make the filter back to the LPF again.

$$\begin{aligned} |H(j\Omega)| &= \sqrt{\frac{\varepsilon^2 T_N^2\left(\frac{1}{\Omega}\right)}{1+\varepsilon^2 T_N^2\left(\frac{1}{\Omega}\right)}} \\ &= \frac{1}{\sqrt{1+\frac{1}{\varepsilon^2 T_N^2\left(\frac{1}{\Omega}\right)}}}; \text{Filp left right of } |H_1(j\Omega)| \text{ over } \Omega = 1 \end{aligned} \tag{6.172}$$

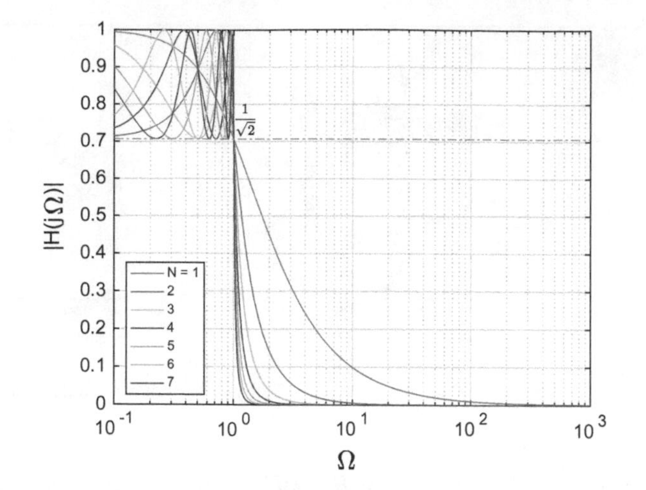

Fig. 6.56 Magnitude response of Eq. (6.170)

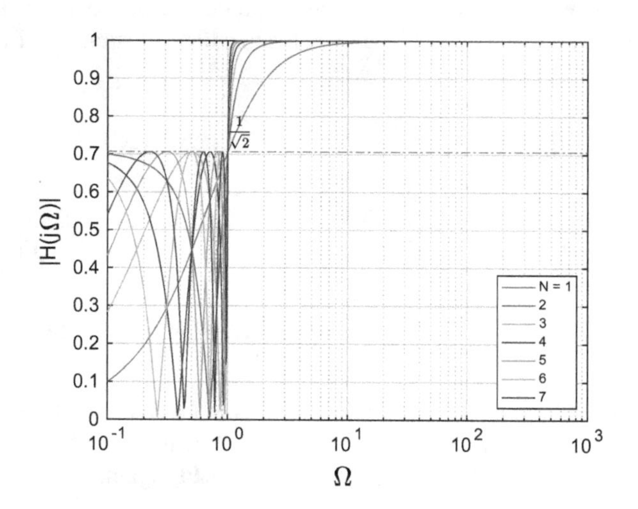

Fig. 6.57 Magnitude response of Eq. (6.171)

To control the ripple in the stopband, Eq. (6.172) is modified as Eq. (6.173).

$$|H(j\Omega)|^2 = \frac{1}{1 + \frac{\varepsilon^2}{T_N^2\left(\frac{\Omega_{st}}{\Omega}\right)}} \tag{6.173}$$

Figure 6.59 depicts Eq. (6.173) for two different parameter values. The increased ε value suppresses the stopband ripples further.

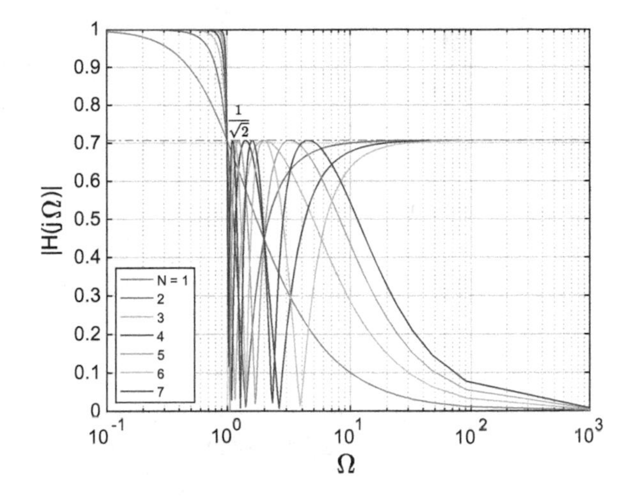

Fig. 6.58 Magnitude response of Eq. (6.172)

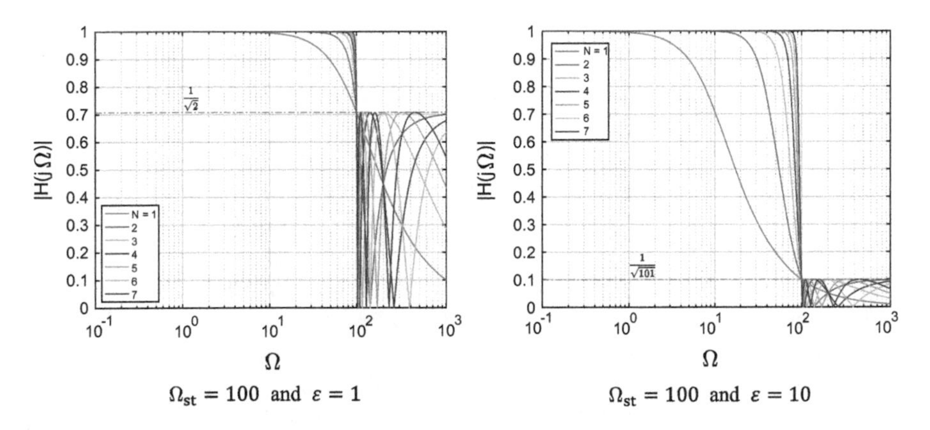

Fig. 6.59 Magnitude response of Eq. (6.173) with different parameters

Prog. 6.45 MATLAB program for Figs. 6.57, 6.58 and 6.59.

```
x = linspace(0.001,1000,100001);
t1 = x;
t2 = 2*x.^2-1;
t3 = 4*x.^3-3*x;
t4 = 8*x.^4-8*x.^2+1;
t5 = 16*x.^5-20*x.^3+5*x;
t6 = 32*x.^6-48*x.^4+18*x.^2-1;
t7 = 64*x.^7-112*x.^5+56*x.^3-7*x;

figure,
semilogx(x,sqrt(t1.^2./(1+t1.^2)),x,sqrt(t2.^2./(1+t2.^2)),x,sqrt(t3.^2./(1+t3.^
2)),x,sqrt(t4.^2./(1+t4.^2)),x,sqrt(t5.^2./(1+t5.^2)),x,sqrt(t6.^2./(1+t6.^2)),x,s
qrt(t7.^2./(1+t7.^2))), grid
h = line([0.1 1000],[1/sqrt(2) 1/sqrt(2)],'LineStyle','-.');
xlim([0.1 1000]), xlabel('\Omega'), ylabel('|H(j\Omega)|'),
legend('N = 1','2','3','4','5','6','7');
x = 1./x;
figure,
semilogx(x,sqrt(t1.^2./(1+t1.^2)),x,sqrt(t2.^2./(1+t2.^2)),x,sqrt(t3.^2./(1+t3.^
2)),x,sqrt(t4.^2./(1+t4.^2)),x,sqrt(t5.^2./(1+t5.^2)),x,sqrt(t6.^2./(1+t6.^2)),x,s
qrt(t7.^2./(1+t7.^2))), grid
h = line([0.1 1000],[1/sqrt(2) 1/sqrt(2)],'LineStyle','-.');
xlim([0.1 1000]), xlabel('\Omega'), ylabel('|H(j\Omega)|');
x = x*100;
figure,
semilogx(x,sqrt(t1.^2./(1+t1.^2)),x,sqrt(t2.^2./(1+t2.^2)),x,sqrt(t3.^2./(1+t3.^
2)),x,sqrt(t4.^2./(1+t4.^2)),x,sqrt(t5.^2./(1+t5.^2)),x,sqrt(t6.^2./(1+t6.^2)),x,s
qrt(t7.^2./(1+t7.^2))), grid
h = line([0.1 1000],[1/sqrt(2) 1/sqrt(2)],'LineStyle','-.');
xlim([0.1 1000]), xlabel('\Omega'), ylabel('|H(j\Omega)|');
figure,
semilogx(x,sqrt(t1.^2./(100+t1.^2)),x,sqrt(t2.^2./(100+t2.^2)),x,sqrt(t3.^2./(1
00+t3.^2)),x,sqrt(t4.^2./(100+t4.^2)),x,sqrt(t5.^2./(100+t5.^2)),x,sqrt(t6.^2./(
100+t6.^2)),x,sqrt(t7.^2./(100+t7.^2))), grid
h = line([0.1 1000],[1/sqrt(101) 1/sqrt(101)],'LineStyle','-.');
xlim([0.1 1000]), xlabel('\Omega'), ylabel('|H(j\Omega)|');
```

Previous IIR filters controls the passband ripple by adjusting the ε value in the rational function. Equation (6.172) is modified as Eq. (6.173) to manage the passband ripple by ε parameter. Note that the passband frequency is required for Eq. (6.174).

$$|H(j\Omega)|^2 = \frac{1}{1+\varepsilon^2 \frac{T_N^2\left(\frac{\Omega_{st}}{\Omega_p}\right)}{T_N^2\left(\frac{\Omega_{st}}{\Omega}\right)}} \tag{6.174}$$

Figure 6.60c demonstrates the controlled passband frequency and ripple by using the Ω_p and ε parameter in Eq. (6.174). The stopband ripple is barely noticeable; however, the fluctuation is observed in decibel plot in Fig. 6.60d.

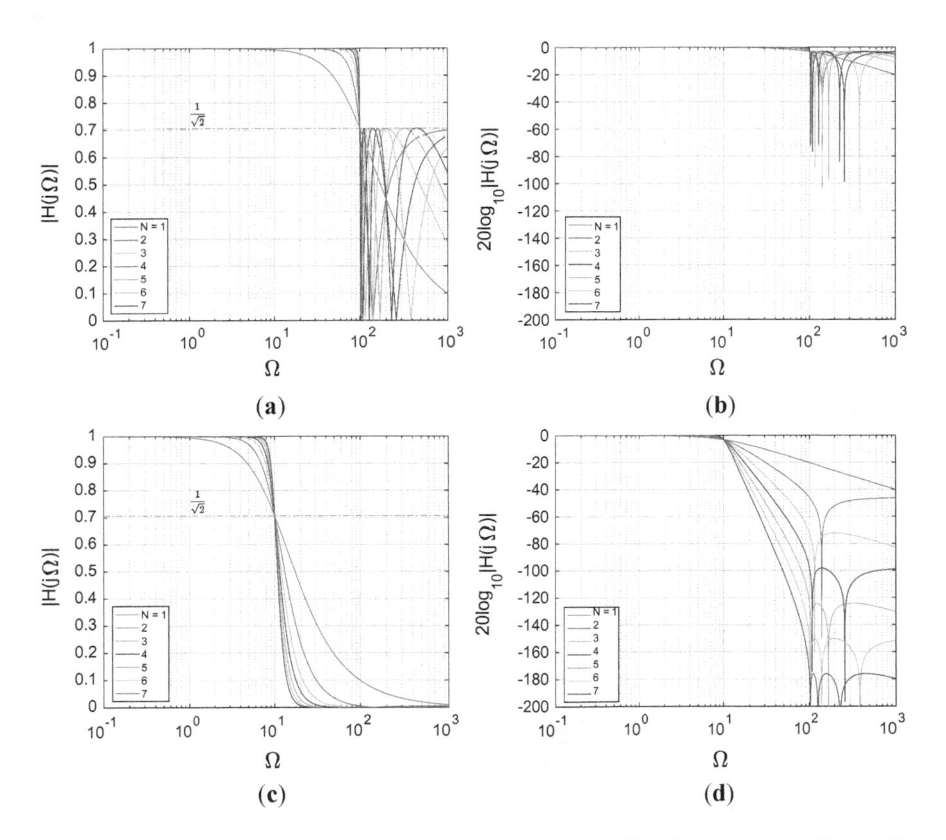

Fig. 6.60 Equation (6.173) response in **a** absolute and **b** decibel for $\varepsilon = 1$ and $\Omega_{st} = 100$. Equation (6.174) response in **c** absolute and **d** decibel for $\varepsilon = 1$, $\Omega_{st} = 100$, and $\Omega_p = 10$

Prog. 6.46 MATLAB program for Fig. 6.60.

```
q = linspace(0.001,1000,100001);
ee = 1;
x = 100./q;
t1 = x;
t2 = 2*x.^2-1;
t3 = 4*x.^3-3*x;
t4 = 8*x.^4-8*x.^2+1;
t5 = 16*x.^5-20*x.^3+5*x;
t6 = 32*x.^6-48*x.^4+18*x.^2-1;
t7 = 64*x.^7-112*x.^5+56*x.^3-7*x;
x = 100./10;
t11 = x;
t21 = 2*x.^2-1;
t31 = 4*x.^3-3*x;
t41 = 8*x.^4-8*x.^2+1;
t51 = 16*x.^5-20*x.^3+5*x;
t61 = 32*x.^6-48*x.^4+18*x.^2-1;
t71 = 64*x.^7-112*x.^5+56*x.^3-7*x;
figure,
semilogx(q,sqrt(t1.^2./(1+t1.^2)),q,sqrt(t2.^2./(1+t2.^2)),q,sqrt(t3.^2./(1+t3.
^2)),q,sqrt(t4.^2./(1+t4.^2)),q,sqrt(t5.^2./(1+t5.^2)),q,sqrt(t6.^2./(1+t6.^2)),q
,sqrt(t7.^2./(1+t7.^2))), grid
h = line([0.1 1000],[1/sqrt(2) 1/sqrt(2)],'LineStyle','-.'), xlim([0.1 1000]);
figure,
semilogx(q,20*log10(sqrt(t1.^2./(1+t1.^2))),q,20*log10(sqrt(t2.^2./(1+t2.^2))
),q,20*log10(sqrt(t3.^2./(1+t3.^2))),q,20*log10(sqrt(t4.^2./(1+t4.^2))),q,20*l
og10(sqrt(t5.^2./(1+t5.^2))),q,20*log10(sqrt(t6.^2./(1+t6.^2))),q,20*log10(sq
rt(t7.^2./(1+t7.^2)))), grid, xlim([0.1 1000]), ylim([-200 0]);
figure,
semilogx(q,sqrt(t1.^2./(ee^2*t11^2+t1.^2)),q,sqrt(t2.^2./(ee^2*t21^2+t2.^2)
),q,sqrt(t3.^2./(ee^2*t31^2+t3.^2)),q,sqrt(t4.^2./(ee^2*t41^2+t4.^2)),q,sqrt(
t5.^2./(ee^2*t51^2+t5.^2)),q,sqrt(t6.^2./(ee^2*t61^2+t6.^2)),q,sqrt(t7.^2./(
ee^2*t71^2+t7.^2))), grid
h = line([0.1 1000],[1/sqrt(2) 1/sqrt(2)],'LineStyle','-.'), xlim([0.1 1000]);
figure,
semilogx(q,20*log10(sqrt(t1.^2./(ee^2*t11^2+t1.^2))),q,20*log10(sqrt(t2.^2.
/(ee^2*t21^2+t2.^2))),q,20*log10(sqrt(t3.^2./(ee^2*t31^2+t3.^2))),q,20*log
10(sqrt(t4.^2./(ee^2*t41^2+t4.^2))),q,20*log10(sqrt(t5.^2./(ee^2*t51^2+t5.
^2))),q,20*log10(sqrt(t6.^2./(ee^2*t61^2+t6.^2))),q,20*log10(sqrt(t7.^2./(ee
^2*t71^2+t7.^2)))), grid, xlim([0.1 1000]), ylim([-200 0]);
```

The passband ripple Δ_p determines the ε parameter as shown in Eq. (6.175). Note that the $|H(j\Omega_p)|$ is $1/\sqrt{1+\varepsilon^2}$ from Eq. (6.174).

$$\Delta_p = 20\log_{10}\left(\frac{1}{1/\sqrt{1+\varepsilon^2}}\right) = 10\log_{10}(1+\varepsilon^2)$$
$$\varepsilon = \sqrt{10^{\frac{\Delta_p}{10}} - 1} \tag{6.175}$$

From the given stopband attenuation in decibel Δ_{st} , the required filter order N can be calculated by using Eq. (6.176). Observe that the $T_N(1)$ is 1, according to the Eq. (6.157).

$$\Delta_{st} = 20\log_{10}\sqrt{\frac{1}{|H(j\Omega_{st})|^2}} = 20\log_{10}\left(\frac{1}{1/\sqrt{1+\varepsilon^2 T_N^2\left(\frac{\Omega_{st}}{\Omega_p}\right)}}\right)$$
$$N \geq \frac{\cosh^{-1}\left(\sqrt{\frac{10^{\frac{\Delta_{st}}{10}}-1}{\varepsilon^2}}\right)}{\cosh^{-1}\left(\frac{\Omega_{st}}{\Omega_p}\right)} \tag{6.176}$$

The stable $H(j\Omega)$ filter requires the Laplace transform to choose the proper poles from the combination. Let's apply the Laplace transform on the magnitude filter equation $|H(j\Omega)|^2$ as shown in Eq. (6.177).

$$|H(s)|^2 = H(s)H(-s) = \frac{1}{1+\varepsilon^2\frac{T_N^2\left(\frac{\Omega_{st}}{\Omega_p}\right)}{T_N^2\left(\frac{j\Omega_{st}}{s}\right)}}$$
$$= \frac{T_N^2\left(\frac{j\Omega_{st}}{s}\right)}{T_N^2\left(\frac{j\Omega_{st}}{s}\right) + \varepsilon^2 T_N^2\left(\frac{\Omega_{st}}{\Omega_p}\right)} \text{Let } j\Omega \to s = \sigma + j\Omega \tag{6.177}$$

The poles of Eq. (6.177) can be found as below.

$$1+\varepsilon^2\frac{T_N^2\left(\frac{\Omega_{st}}{\Omega_p}\right)}{T_N^2\left(\frac{j\Omega_{st}}{s}\right)} = 0 \to 1 + \frac{1}{\varepsilon^2 T_N^2\left(\frac{\Omega_{st}}{\Omega_p}\right)} T_N^2\left(\frac{j\Omega_{st}}{s}\right) = 0$$
$$\to 1 + \gamma^2 T_N^2\left(\frac{j\Omega_{st}}{s}\right) = 0 \text{ where } \gamma^2 = \frac{1}{\varepsilon^2 T_N^2\left(\frac{\Omega_{st}}{\Omega_p}\right)} \tag{6.178}$$

Use the first kind Chebyshev polynomial solution for Type I Chebyshev filter as Eq. (6.179).

$$\frac{\Omega_{st}}{p_k} = \pm\sin\left((2k-1)\frac{\pi}{2N}\right)\sinh\left(\frac{1}{N}\sinh^{-1}\left(\frac{1}{\gamma}\right)\right)$$
$$+j\cos\left((2k-1)\frac{\pi}{2N}\right)\cosh\left(\frac{1}{N}\sinh^{-1}\left(\frac{1}{\gamma}\right)\right)\text{for } k=1,2,\ldots,N$$
$$p_k = \frac{\Omega_{st}}{\pm\sin\left((2k-1)\frac{\pi}{2N}\right)\sinh\left(\frac{1}{N}\sinh^{-1}\left(\frac{1}{\gamma}\right)\right)+j\cos\left((2k-1)\frac{\pi}{2N}\right)\cosh\left(\frac{1}{N}\sinh^{-1}\left(\frac{1}{\gamma}\right)\right)}$$
$$\text{for } k=1,2,\ldots,N \tag{6.179}$$

Separate the poles into the left s plane as p_i^L and right s plane as p_i^R . Note that poles on the left s plane provides the stable filter. The left and right of the s plane can be divided by the real value. Positive real value indicates right s plane and negative real value specifies left s plane.

$$\left.\begin{array}{l} p_i^L \in \text{left s plane} \\ p_i^R \in \text{right s plane} \end{array}\right\} i = 1,2,\ldots,N$$

Zeros of Eq. (6.177) can be found as below.

$$T_N^2\left(\frac{j\Omega_{st}}{s}\right) = 0 \text{ for}\left|\frac{j\Omega_{st}}{s}\right| \le 1 \tag{6.180}$$

Figure 6.61 demonstrates $T_N^2(\Omega)$ distribution and zeros of the function are located between the –1 and 1. Since the vertexes of the curve touch the zero line in Fig. 6.61, the zeros are paired with double.

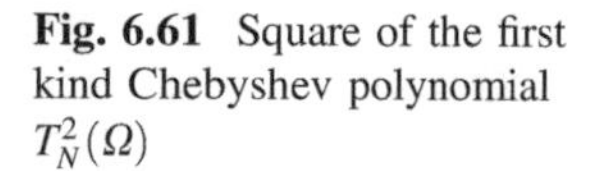
Fig. 6.61 Square of the first kind Chebyshev polynomial $T_N^2(\Omega)$

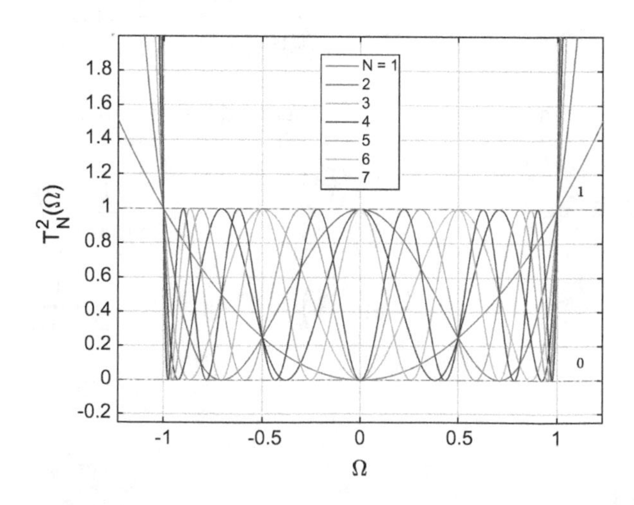

Prog. 6.47 MATLAB program for Fig. 6.61.

```
x = linspace(-2,2,1000);
t1 = (x).^2;
t2 = (2*x.^2-1).^2;
t3 = (4*x.^3-3*x).^2;
t4 = (8*x.^4-8*x.^2+1).^2;
t5 = (16*x.^5-20*x.^3+5*x).^2;
t6 = (32*x.^6-48*x.^4+18*x.^2-1).^2;
t7 = (64*x.^7-112*x.^5+56*x.^3-7*x).^2;
plot(x,t1,x,t2,x,t3,x,t4,x,t5,x,t6,x,t7), grid
line([-2 2],[1 1],'LineStyle','-.')
line([-2 2],[0 0],'LineStyle','-.')
xlim([-2 2]);
ylim([-1.5 12]);
xlabel('\Omega')
ylabel('T^2_N(\Omega)')
axis([-1.23   1.23   -0.25   2.0])
legend('N = 1','2','3','4','5','6','7');
```

The solution of the Eq. (6.180) is below.

$$\cos\left(N\cos^{-1}\left(\tfrac{j\Omega_{st}}{s}\right)\right)\cos\left(N\cos^{-1}\left(\tfrac{j\Omega_{st}}{s}\right)\right) = 0 \qquad (6.181)$$

$$\text{for}\{s \geq j\Omega_{st}\text{or } s \leq -j\Omega_{st}\}$$

Using the trigonometric identities, the above equation is organized as below.

$$\begin{gathered}
\frac{1+\cos\left(2N\cos^{-1}\left(\frac{j\Omega_{st}}{s}\right)\right)}{2} = 0 \\
\cos\left(2N\cos^{-1}\left(\tfrac{j\Omega_{st}}{s}\right)\right) = -1 = \cos((2k-1)\pi) \text{ for } k \in \mathbb{Z} \\
2N\cos^{-1}\left(\tfrac{j\Omega_{st}}{s}\right) = (2k-1)\pi; \cos^{-1}\left(\tfrac{j\Omega_{st}}{s}\right) = (2k-1)\tfrac{\pi}{2N} \\
\tfrac{j\Omega_{st}}{s} = \cos\left((2k-1)\tfrac{\pi}{2N}\right)
\end{gathered} \qquad (6.182)$$

The zeros of Eq. (6.177) are derived as Eq. (6.183).

$$z_k = \frac{j\Omega_{st}}{\cos\left((2k-1)\frac{\pi}{2N}\right)}; k = 1, 2, \ldots, N \qquad (6.183)$$

Collect poles $\left(p_i^{I'}\right)$ and zeros (z_k) to assemble the transfer function in s domain as shown in Eq. (6.184).

$$H(s) = \frac{G(s-z_1)(s-z_2)\ldots(s-z_N)}{\left(s-p_1^L\right)\left(s-p_2^L\right)\ldots\left(s-p_N^L\right)} = \frac{b_0 s^N + b_1 s^{N-1} + b_2 s^{N-2} + \cdots + b_{N-2}s^1 + b_{N-1}}{s^N + a_1 s^{N-1} + a_2 s^{N-2} + \cdots + a_{N-2}s^1 + a_{N-1}} \tag{6.184}$$

The gain G of the transfer function can be computed as below. Equations (6.177) and (6.184) produce the equal value for the zero input to calculate the gain G.

$$H(s = j\Omega = 0) = \frac{G(-z_1)(-z_2)\ldots(-z_N)}{\left(-p_1^L\right)\left(-p_2^L\right)\ldots\left(-p_N^L\right)}$$

$$|H(j\Omega = 0)|^2 = \frac{1}{1+\varepsilon^2 \frac{T_N^2\left(\frac{\Omega_{st}}{\Omega_p}\right)}{T_N^2\left(\frac{\Omega_{st}}{0}\right)}} = \frac{1}{1+\varepsilon^2 \frac{T_N^2\left(\frac{\Omega_{st}}{\Omega_p}\right)}{T_N^2(\infty)}} = 1$$

$$G(-z_1)(-z_2)\ldots(-z_N) = \left(-p_1^L\right)\left(-p_2^L\right)\ldots\left(-p_N^L\right)$$

$$G = \frac{\left(-p_1^L\right)\left(-p_2^L\right)\ldots\left(-p_N^L\right)}{(-z_1)(-z_2)\ldots(-z_N)} \tag{6.185}$$

Convert back to $j\Omega$ by replacing with s as Eq. (6.186). The filter with rational function is completed in the linear frequency domain $j\Omega$.

$$H(j\Omega) = \frac{b_0(j\Omega)^N + b_1(j\Omega)^{N-1} + b_2(j\Omega)^{N-2} + \cdots + b_{N-2}(j\Omega)^1 + b_{N-1}}{(j\Omega)^N + a_1(j\Omega)^{N-1} + a_2(j\Omega)^{N-2} + \cdots + a_{N-2}(j\Omega)^1 + a_{N-1}} = \frac{G(j\Omega-z_1)(j\Omega-z_2)\ldots(j\Omega-z_N)}{\left(j\Omega-p_1^L\right)\left(j\Omega-p_2^L\right)\ldots\left(j\Omega-p_N^L\right)} \tag{6.186}$$

The bilinear transformation provides the tool to change the linear frequency domain into the z domain for discrete-time filter as shown in Eq. (6.187).

$$H(z) = \left.\frac{b_0(j\Omega)^N + b_1(j\Omega)^{N-1} + \cdots + b_{N-2}(j\Omega) + b_{N-1}}{(j\Omega)^N + a_1(j\Omega)^{N-1} + \cdots + a_{N-2}(j\Omega) + a_{N-1}}\right|_{j\Omega = 2f_s\left(\frac{1-z^{-1}}{1+z^{-1}}\right)} \tag{6.187}$$

The final Type II Chebyshev filter in discrete time domain is Eq. (6.187).

Example 6.20

Design the Table 6.29 IIR filter by using Type II Chebyshev method and bilinear transformation.

Solution

Use the bilinear transformation to find the frequency specification in Ω from Eq. (6.150).

Table 6.29 Given filter specification

Filter specifications	Value
Filter type	Low pass filter
Filter realization	Type II Chebyshev IIR filter
Filter length	N with optimization
Passband frequency (f_p)	1000 Hz; $\omega_p = \pi/4$
Passband magnitude	1
Passband peak-to-peak ripple in decibel (Δ_p)	$20\log_{10}\left(\frac{1}{1/\sqrt{1+1}}\right) = 3.0103\text{dB}$
Stopband frequency (f_{st})	2000 Hz; $\omega_{st} = \pi/2$
Stopband attenuation in decibels Δ_{st}	$20\log_{10}\left(\frac{1}{0.1}\right) = 20$ dB
Sampling frequency (f_s)	8000 Hz

$$\Omega_p = 2f_s \tan\left(\frac{\pi}{8}\right) = 6627.4\ \text{rad/s}$$

$$\Omega_{st} = 2f_s \tan\left(\frac{\pi}{4}\right) = 16000\text{rad/s}$$

Compute the ε value by using the Eq. (6.175).

$$\varepsilon = \sqrt{10^{\frac{\Delta_p}{10}} - 1} = \sqrt{10^{\frac{3.0103}{10}} - 1} = 1$$

Estimate the proper filter order by using the Eq. (6.176).

$$N \geq \frac{\cosh^{-1}\sqrt{\left(10^{\frac{\Delta_{st}}{10}} - 1\right)/\varepsilon^2}}{\cosh^{-1}\left(\frac{\Omega_{st}}{\Omega_p}\right)} = \frac{\cosh^{-1}\sqrt{\left(10^{\frac{20}{10}} - 1\right)/1^2}}{\cosh^{-1}\left(\frac{16{,}000}{6627.4}\right)} = 1.9549 \rightarrow 2;\text{even}$$

Derive the whole poles to find left s plane poles from Eq. (6.179).

$$p_k = \frac{\Omega_{st}}{\pm\sin\left((2k-1)\frac{\pi}{2N}\right)\sinh\left(\frac{1}{N}\sinh^{-1}\left(\frac{1}{\gamma}\right)\right) + j\cos\left((2k-1)\frac{\pi}{2N}\right)\cosh\left(\frac{1}{N}\sinh^{-1}\left(\frac{1}{\gamma}\right)\right)}$$

$$\text{where } k = 1, 2 \text{ and } \gamma^2 = \frac{1}{\varepsilon^2 T_N^2\left(\frac{\Omega_{st}}{\Omega_p}\right)}$$

$$\left.\begin{array}{l} p_i^L \in \text{left s plane} \\ p_i^R \in \text{right s plane} \end{array}\right\} i = 1, 2$$

$$p_k = 4656.4 - j5113.8, 4656.4 + j5113.8,$$
$$- 4656.4 - j5113.8, -4656.4 + j5113.8$$
$$p_k^L = -4656.4 - j5113.8, -4656.4 + j5113.8$$

Calculate the zeros by using Eq. (6.183).

$$z_k = \frac{j\Omega_{st}}{\cos\left((2k-1)\frac{\pi}{2N}\right)}; k = 1, 2, \ldots, N$$

$$z_k = j22627.41, -j22627.41$$

Obtain the gain G from Eq. (6.185).

$$G = \frac{(-p_1^L)(-p_2^L)}{(-z_1)(-z_2)} = \frac{(4656.4 + j5113.8)(4656.4 - j5113.8)}{(-j22627.41)(j22627.41)} = 0.09343$$

Follow Eq. (6.186) to build the filter equation in s domain based on the p_i^L , z_k , and G.

$$H(s) = \frac{0.09343(s + j22627.41)(s - j22627.41)}{(s + 4656.4 + j5113.8)(s + 4656.4 - j5113.8)}$$

$$H(s) = \frac{0.09343s^2 + 47833766.89}{s^2 + 9312.8s + 47833011.4}$$

Apply the bilinear transformation as below.

$$H(z) = \left.\frac{0.09343(j\Omega + j22627.41)(j\Omega - j22627.41)}{(j\Omega + 4656.4 + j5113.8)(j\Omega + 4656.4 - j5113.8)}\right|_{j\Omega = 2f_s\left(\frac{1-z^{-1}}{1+z^{-1}}\right)}$$

The final Type II Chebyshev filter in discrete time domain is below.

$$H(z) = \frac{0.15845 + 0.10563z^{-1} + 0.15845z^{-2}}{1 - 0.91939z^{-1} + 0.34191z^{-2}}$$

Figure 6.62 shows the pole zero plot of the design filter. Two poles are located within the unit circle and two zeros are placed on unit circle.

Prog. 6.48 MATLAB program for Figs. 6.62 and 6.63.

```
syms s z;
pb = linspace(0,1000,501);
sb = linspace(2000,4000,2001);
hs = 0.0934253837435988*(s+22627.41j)*(s-
22627.41j)/((s+4656.4+5113.8j)*(s+4656.4-5113.8j));
hz = subs(hs,s,16000*(z-1)/(z+1));
[num1,den1] = numden(hs);
num2 = collect(expand(num1));
den2 = collect(expand(den1));
num3 = sym2poly(num2);
den3 = sym2poly(den2);
num4 = num3./(den3(1));
den4 = den3./(den3(1));
[znum1,zden1] = numden(hz);
znum2 = collect(expand(znum1));
zden2 = collect(expand(zden1));
znum3 = sym2poly(znum2);
zden3 = sym2poly(zden2);
znum4 = znum3./(zden3(1));
zden4 = zden3./(zden3(1));
[h,w] = freqz(znum4,zden4,4001,8000);
hpb = freqz(znum4,zden4,pb,8000);
hsb = freqz(znum4,zden4,sb,8000);

figure,
zplane(znum4,zden4)
figure,
plot(w,abs(h),pb,abs(hpb),'.-',sb,abs(hsb),'.-'), grid
line([0 4000],[0.1 0.1])
ylim([0 1])
legend('Response','Passband','Stopband')
ylabel('|H(e^{j\omega})|')
xlabel('Frequency (Hz)')
figure
plot(w,db(h),pb, db(hpb),'.-',sb,db(hsb),'.-'), grid
line([0 4000],[-20 -20])
legend('Response','Passband','Stopband')
ylabel('20log_{10}|H(e^{j\omega})| (dB)')
xlabel('Frequency (Hz)')
```

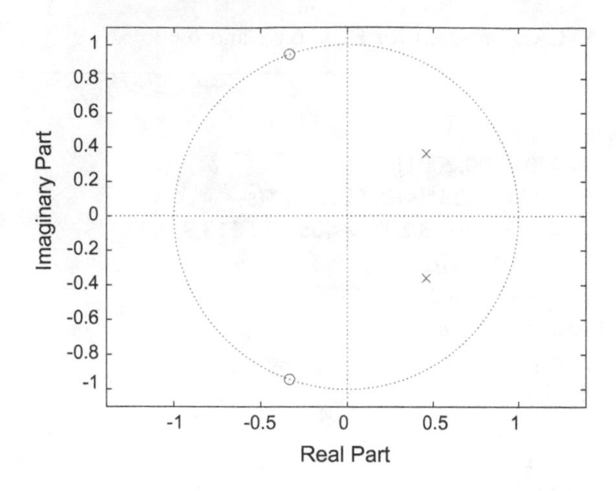

Fig. 6.62 Pole zero plot of designed Type II Chebyshev filter for Table 6.29

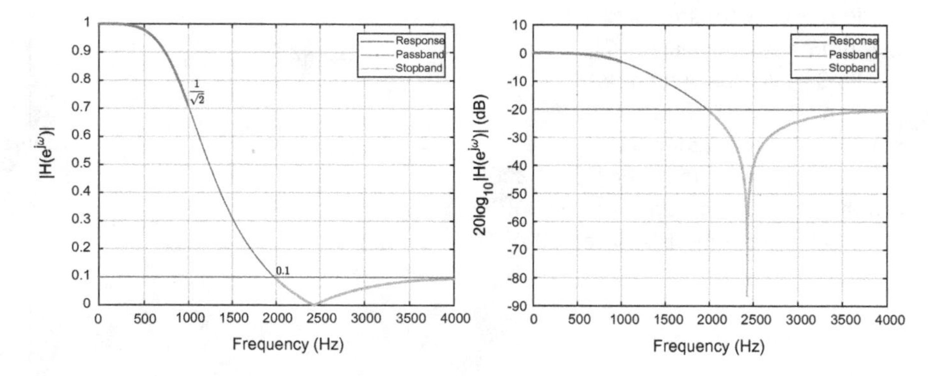

Fig. 6.63 Frequency response of designed Type II Chebyshev filter for Table 6.29

Figure 6.63 represents the frequency response of the filter. The passband and stopband satisfy the Table 6.29 specification. Note that the horizontal line in the figures denote the stopband attenuation requirement. ∎

Elliptic filter

The elliptic filter [7] also utilizes the rational function with special polynomial as shown in Eq. (6.188). As the $\varepsilon^2 R_N^2(\xi, \Omega)$ increases exponentially, the filter magnitude decreases significantly for LPF property.

$$|H(j\Omega)|^2 = H(j\Omega)H^*(j\Omega) = H(j\Omega)H(-j\Omega) = \frac{1}{1+\varepsilon^2 R_N^2(\xi,\Omega)} \quad (6.188)$$

$$\text{where} \begin{cases} \varepsilon : \text{ripple factor} \\ \xi : \text{selectivity factor } (\xi \geq 1) \end{cases}$$

The special function is the elliptic rational function $R_N(\xi, \Omega)$ [19]. Below show the functions up to 4th order.

$$R_1(\xi, \Omega) = \Omega$$

$$R_2(\xi, \Omega) = \frac{(t+1)\Omega^2 - 1}{(t-1)\Omega^2 + 1}$$

$$\text{where } t = \sqrt{1 - (1/\xi^2)}$$

$$R_3(\xi, \Omega) = \Omega \frac{\left(1 - x_p^2\right)\left(\Omega^2 - x_z^2\right)}{\left(1 - x_z^2\right)\left(\Omega^2 - x_p^2\right)}$$

$$\text{where} \begin{cases} x_p^2 = \dfrac{2\xi^2\sqrt{G}}{\sqrt{8\xi^2(\xi^2+1) + 12G\xi^2 - G^3 - \sqrt{G^3}}} \\ x_z^2 = \dfrac{\xi^2}{x_p^2} \\ G = \sqrt{4\xi^2 + \left(4\xi^2(\xi^2-1)\right)^{\frac{2}{3}}} \end{cases}$$

$$R_4(\xi, \Omega) = \frac{(t+1)\left(1+\sqrt{t}\right)^2\Omega^4 - 2(1+t)\left(1+\sqrt{t}\right)\Omega^2 + 1}{(t+1)\left(1-\sqrt{t}\right)^2\Omega^4 - 2(1+t)\left(1-\sqrt{t}\right)\Omega^2 + 1} \quad (6.189)$$

$$\text{where } t = \sqrt{1 - (1/\xi^2)}$$

The properties of the elliptic rational function are below.

$$R_M(R_N(\xi, \xi), R_N(\xi, \Omega)) = R_{M\times N}(\xi, \Omega)$$

$$R_N^2(\xi, \Omega) \leq 1 \text{ for } |\Omega| \leq 1 \quad (6.190)$$

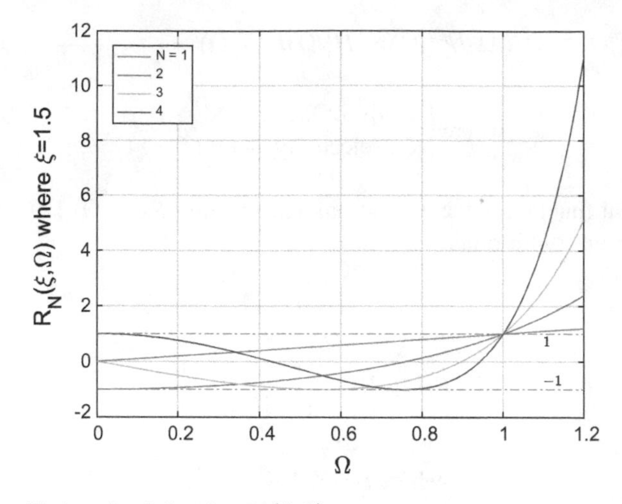

Fig. 6.64 The elliptic rational function $R_N(\xi, \Omega)$

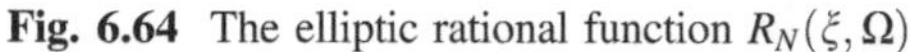

$$R_N^2(\xi, 1) = 1 \tag{6.191}$$

$$R_N^2(\xi, 0)\begin{cases} \neq 0 \text{ for even } N \\ = 0 \quad \text{for odd } N \end{cases} \tag{6.192}$$

$$R_N^2(\xi, \Omega) = R_N^2(\xi, -\Omega) \tag{6.193}$$

$$R_N(\xi, -\Omega) = \begin{cases} R_N(\xi, \Omega) & \text{for } N \text{ even} \\ -R_N(\xi, \Omega) & \text{for } N \text{ odd} \end{cases}$$

$$\lim_{\xi \to \infty} R_N(\xi, \Omega) = T_N(\Omega)$$

The $R_N^2(\xi, \Omega)$ presents limited magnitude one for less than one Ω in absolute according to the property in Eq. (6.190). Also, the $R_N^2(\xi, \Omega)$ shows the even symmetricity over the Ω axis from Eq. (6.193). The initial value of the $R_N^2(\xi, \Omega)$ depends on the order parity but the $R_N^2(\xi, 1)$ value is fixed as one for all orders. Figure 6.64 illustrates the $R_N(\xi, \Omega)$ up to 4th order.

Prog. 6.49 MATLAB program for Fig. 6.64.

```
om = linspace(0,1.2,200);
k = 1.5;
t = sqrt(1-(1/k^2));
G = sqrt(4*k^2+(4*k^2*(k^2-1))^(2/3));
xp2 = 2*k^2*sqrt(G)/(sqrt(8*k^2*(k^2+1)+12*G*k^2-G^3)-sqrt(G^3));
xz2 = k^2/xp2;
R1 = om;
R2 = ((t+1)*om.^2-1)./((t-1)*om.^2+1);
R3 = om.*(1-xp2).*(om.^2-xz2)./((1-xz2).*(om.^2-xp2));
R4 = ((t+1)*(1+sqrt(t))^2*om.^4-2*(1+t)*(1+sqrt(t))*om.^2+1)./((t+1)*(1-
sqrt(t))^2*om.^4-2*(1+t)*(1-sqrt(t))*om.^2+1);

figure,
plot(om,R1,om,R2,om,R3,om,R4), grid
h = line([0 1.2],[1 1],'LineStyle','-.');
h = line([0 1.2],[-1 -1],'LineStyle','-.');
xlabel('\Omega')
ylabel('R_N(\xi,\Omega) where \xi=1.5')
legend('N = 1','2','3','4');
```

The $R_N(\xi, \Omega)$ presents the fluctuation with limited magnitude up to one Ω and rapidly increasing after the range. The higher order illustrates the faster monotonic growing in Fig. 6.64. By applying the function to Eq. (6.188), the filter response can be calculated and depicted as Fig. 6.65. The response is similar to the Type I Chebyshev filter; however, we can observe the fluctuation in stopband and steeper decreasing in the transition band. In overall, the frequency response represents the LPF with passband frequency $\Omega_p = 1$. Note that the Ω_p and Ω_h are equal in Fig. 6.65 because of unit ε. The order N illustrates the steepness of the magnitude transition. The higher N value shows the steeper response in the filter as shown in Fig. 6.65.

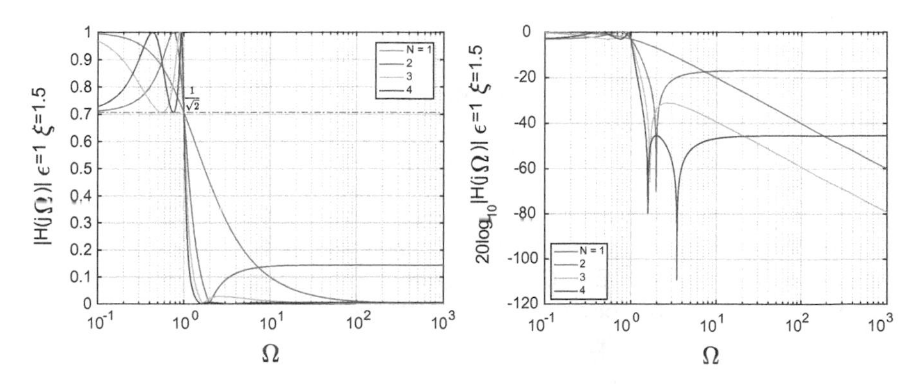

Fig. 6.65 Frequency response of linear frequency model from Eq. (6.188) with $\varepsilon = 1$ and $\xi = 1.5$

Prog. 6.50 MATLAB program for Fig. 6.65.

```
om = linspace(0.1,1000,100000);
k = 1.5;      %Xi
ee = 1;       %Epsilon
t = sqrt(1-(1/k^2));
G = sqrt(4*k^2+(4*k^2*(k^2-1))^(2/3));
xp2 = 2*k^2*sqrt(G)/(sqrt(8*k^2*(k^2+1)+12*G*k^2-G^3)-sqrt(G^3));
xz2 = k^2/xp2;
R1 = om;
R2 = ((t+1)*om.^2-1)./((t-1)*om.^2+1);
R3 = om.*(1-xp2).*(om.^2-xz2)./((1-xz2).*(om.^2-xp2));
R4 = ((t+1)*(1+sqrt(t))^2*om.^4-2*(1+t)*(1+sqrt(t))*om.^2+1)./((t+1)*(1-
sqrt(t))^2*om.^4-2*(1+t)*(1-sqrt(t))*om.^2+1);

figure,
semilogx(om,1./sqrt(1+ee^2*R1.^2),om,1./sqrt(1+ee^2*R2.^2),om,1./sqrt(1
+ee^2*R3.^2),om,1./sqrt(1+ee^2*R4.^2)), grid
h = line([0.1 1000],[1/sqrt(2) 1/sqrt(2)],'LineStyle','-.');
xlim([.1 1000]);
xlabel('\Omega')
ylabel('|H(j\Omega)| \epsilon=1 \xi=1.5')
legend('N = 1','2','3','4')
figure,
semilogx(om,-10*log10(1+ee^2*R1.^2),om,-10*log10(1+ee^2*R2.^2),om,-
10*log10(1+ee^2*R3.^2),om,-10*log10(1+ee^2*R4.^2)), grid
xlim([0.1 1000]);
xlabel('\Omega')
ylabel('20log_{10}|H(j\Omega)| \epsilon=1 \xi=1.5')
legend('N = 1','2','3','4')
```

From the given passband peak-to-peak ripple Δ_p , the ε value can be determined by Eq. (6.194).

$$\Delta_p = 20\log_{10}\left(\frac{1}{1/\sqrt{1+\varepsilon^2}}\right) = 10\log_{10}\left(1+\varepsilon^2\right) \text{ where } R_N^2(\xi,1)=1$$
$$\varepsilon = \sqrt{10^{\frac{\Delta_p}{10}}-1} \tag{6.194}$$

Expanding the frequency axis relocates the passband frequency to any frequency by scaling as shown in Eq. (6.195). Figure 6.66 demonstrates the LPF with 3.0103 dB Δ_p, 1.5 ξ, 66.6667 rad/s Ω_p, and 100 rad/s Ω_{st}.

$$|H(j\Omega)|^2 = \frac{1}{1+\varepsilon^2 R_N^2\left(\xi, \xi\frac{\Omega}{\Omega_{st}}\right)} \text{ where } \xi\frac{\Omega_p}{\Omega_{st}} = 1 \tag{6.195}$$

Prog. 6.51 MATLAB program for Fig. 6.66.

```
x = linspace(0.1,1000,100000);
k = 1.5;      %Xi
om = x*1.5/100;
ee = 1;       %Epsilon
t = sqrt(1-(1/k^2));
G = sqrt(4*k^2+(4*k^2*(k^2-1))^(2/3));
xp2 = 2*k^2*sqrt(G)/(sqrt(8*k^2*(k^2+1)+12*G*k^2-G^3)-sqrt(G^3));
xz2 = k^2/xp2;
R1 = om;
R2 = ((t+1)*om.^2-1)./((t-1)*om.^2+1);
R3 = om.*(1-xp2).*(om.^2-xz2)./((1-xz2).*(om.^2-xp2));
R4 = ((t+1)*(1+sqrt(t))^2*om.^4-2*(1+t)*(1+sqrt(t))*om.^2+1)./((t+1)*(1-
sqrt(t))^2*om.^4-2*(1+t)*(1-sqrt(t))*om.^2+1);

figure,
semilogx(x,1./sqrt(1+ee^2*R1.^2),x,1./sqrt(1+ee^2*R2.^2),x,1./sqrt(1+ee^2*
R3.^2),x,1./sqrt(1+ee^2*R4.^2)), grid
xlim([.1 1000]);
aa = axis;
h = line([0.1 1000],[1/sqrt(2) 1/sqrt(2)],'LineStyle','-.');
h = line([100/1.5 100/1.5],[aa(3) aa(4)],'LineStyle','-.');
xlabel('\Omega')
ylabel('|H(j\Omega)| \epsilon=1 \xi=1.5')
legend('N = 1','2','3','4');

figure,
semilogx(x,-10*log10(1+ee^2*R1.^2),x,-10*log10(1+ee^2*R2.^2),x,-
10*log10(1+ee^2*R3.^2),x,-10*log10(1+ee^2*R4.^2)), grid
xlim([0.1 1000]);
xlabel('\Omega')
ylabel('20log_{10}|H(j\Omega)| \epsilon=1 \xi=1.5')
h = line([100/1.5 100/1.5],[aa(3) aa(4)],'LineStyle','-.');
legend('N = 1','2','3','4')
```

In order to compute stopband attenuation, the discrimination factor is defined as Eq. (6.196). The discrimination factor indicates the minimum value of the $R_N(\xi, \Omega)$ magnitude for $|\Omega| \geq \xi \geq 1$.

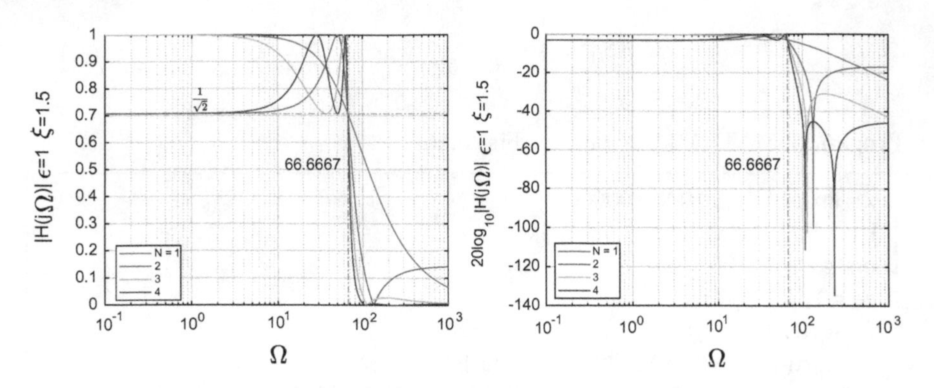

Fig. 6.66 Frequency response of Eq. (6.195) with $\Delta_p = 3.0103$ dB, $\xi = 1.5$, $\Omega_p = 66.6667$ rad/s, and $\Omega_{st} = 100$ rad/s

$$L_N(\xi) = R_N(\xi, \xi) \tag{6.196}$$

Below shows the factor up to 4th order.

$$L_1(\xi) = \xi$$

$$L_2(\xi) = \left(\xi + \sqrt{\xi^2 - 1}\right)^2$$

$$L_3(\xi) = \xi^3 \left(\frac{1 - x_p^2}{\xi^2 - x_p^2}\right)^2$$

$$\text{where} \begin{cases} x_p^2 = \dfrac{2\xi^2\sqrt{G}}{\sqrt{8\xi^2(\xi^2+1) + 12G\xi^2 - G^3} - \sqrt{G^3}} \\ G = \sqrt{4\xi^2 + \left(4\xi^2(\xi^2 - 1)\right)^{\frac{2}{3}}} \end{cases}$$

$$L_4(\xi) = \left(\sqrt{\xi} + (\xi^2 - 1)^{\frac{1}{4}}\right)^4 \left(\xi + \sqrt{\xi^2 - 1}\right)^2 \tag{6.197}$$

Prog. 6.52 MATLAB program for Fig. 6.67.

```
k = linspace(1.001,100,1000);
G = sqrt(4*k.^2+(4*k.^2.*(k.^2-1)).^(2/3));
xp2 = 2.*k.^2.*sqrt(G)./(sqrt(8.*k.^2.*(k.^2+1)+12.*G.*k.^2-G.^3)-
sqrt(G.^3));

L1 = k;
L2 = (k+sqrt(k.^2-1)).^2;
L3 = k.^3.*((1-xp2)./(k.^2-xp2)).^2;
L4 = (sqrt(k)+(k.^2-1).^(1/4)).^4.*((k+sqrt(k.^2-1)).^2);
semilogy(k,L1,k,L2,k,L3,k,L4), grid
legend('N = 1','2','3','4');
xlabel('\xi')
ylabel('L_N(\xi)')
```

From given ε , the selective factor ξ and discrimination factor $L_N(\xi)$ determine the stopband attenuation as shown in Eq. (6.198). Also, we can compute the $L_N(\xi)$ to achieve the target stopband attenuation as shown in Eq. (6.199).

$$\begin{aligned}\Delta_{st} &= 20\log_{10}\sqrt{\frac{1}{|H(j\Omega_{st})|^2}} = 20\log_{10}\left(\frac{1}{1/\sqrt{1+\varepsilon^2 R_N^2\left(\xi, \xi\frac{\Omega_{st}}{\Omega_{st}}\right)}}\right) \\ &= 10\log_{10}\left(1+\varepsilon^2 R_N^2(\xi,\xi)\right) = 10\log_{10}\left(1+\varepsilon^2 L_N^2(\xi)\right)\end{aligned} \tag{6.198}$$

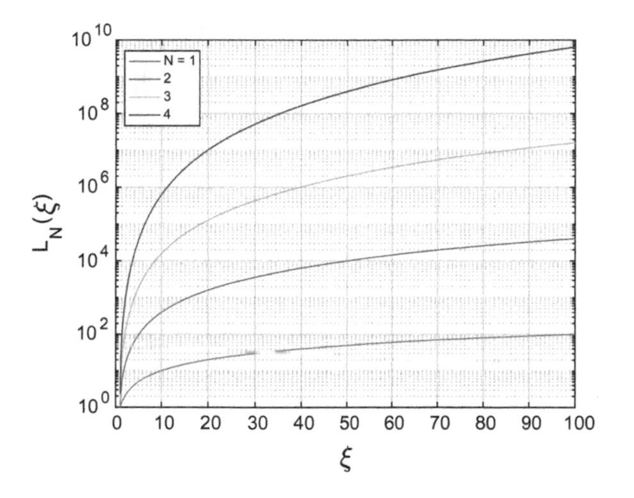

Fig. 6.67 Discrimination factor up to 4th order in terms of selectivity factor ξ

$$L_N(\xi) = \frac{\sqrt{10^{\frac{\Delta_{st}}{10}} - 1}}{\varepsilon} \tag{6.199}$$

For verification purpose, Eq. (6.200) is illustrated at Fig. 6.68 for designated stopband attenuations. Equation (6.201) calculates the selective factor ξ based on the discrimination factor $L_2(\xi)$.

$$|H(j\Omega)|^2 = \frac{1}{1+\varepsilon^2 R_2^2\left(\xi, \xi\frac{\Omega}{\xi}\right)} \tag{6.200}$$

$$L_2(\xi) = \left(\xi + \sqrt{\xi^2 - 1}\right)^2 = \frac{\sqrt{10^{\frac{\Delta_{st}}{10}} - 1}}{\varepsilon} \text{for } \Delta_{st} = 10, 20, 30, \ldots, 100 \tag{6.201}$$

Note that the increased stopband attenuation provides the higher stopband frequency because of Eqs. (6.200) and (6.201). Figure 6.68 well represents the stopband attenuation Δ_{st} for desired value. Once the response passes the stopband frequency, the magnitude draws the valley which asymptotically converges to the designed stopband attenuation.

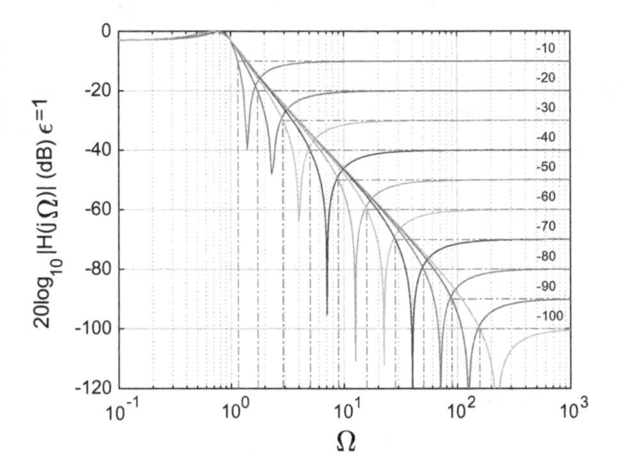

Fig. 6.68 Frequency response of Eq. (6.200) for designated stopband attenuation Δ_{st} and corresponding selective factor ξ

Prog. 6.53 MATLAB program for Fig. 6.68.

```
syms ep;
dst = [10:10:100];
om = linspace(0.1,1000,10000);
ee = 1;
lne = sqrt(10.^(dst/10)-1)./ee;
l2e = (ep+sqrt(ep^2-1))^2;
out = [];
for kk = 1:10
     temp = solve(l2e==lne(kk));
     out = [out eval(temp)];
end
t = sqrt(1-(1./out.^2));
R2 = ((t'+1)*om.^2-1)./((t'-1)*om.^2+1);
semilogx(om,-10*log10(1+ee^2*R2.^2)),grid
ylim([-120 0])
xlabel('\Omega')
ylabel('20log_{10}|H(j\Omega)| (dB) \epsilon=1')
for kk=1:10
     r2 = ((t(kk)+1)*out(kk).^2-1)./((t(kk)-1)*out(kk).^2+1);
     line([out(kk) out(kk)],[-120 -10*log10(1+ee^2*r2.^2)],'LineStyle','-.');
     line([out(kk) 1000],[-10*log10(1+ee^2*r2.^2) -
10*log10(1+ee^2*r2.^2)],'LineStyle','-.');
     str = num2str(-dst(kk));
     text(500,-dst(kk)+3,str);
end
```

After we determine all parameters of the elliptic filter for Eq. (6.202), the poles and zeros of the transfer function should be derived for $H(j\Omega)$. The mathematical solution for $|H(j\Omega)|^2$ with elliptic rational function is beyond this book scope. Instead. we apply the numerical approach to find the numerator and denominator roots of $|H(j\Omega)|^2$ for corresponding zeros and poles, respectively. Below demonstrates the procedure to build the $H(j\Omega)$ from $|H(j\Omega)|^2$

$$|H(j\Omega)|^2 = \frac{1}{1+\varepsilon^2 R_N^2\left(\xi, \xi\frac{\Omega}{\Omega_{st}}\right)} \tag{6.202}$$

$$\text{Let } j\Omega \rightarrow s = \sigma + j\Omega$$

$$\begin{aligned}|H(s)|^2 = H(s)H(-s) &= \frac{1}{1+\varepsilon^2 R_N^2\left(\xi, \xi\frac{s}{j\Omega_{st}}\right)} \\ &= \frac{\beta_0 s^{2N} + \beta_1 s^{2N-1} + \beta_2 s^{2N-2} + \cdots + \beta_{2N-1}s^1 + \beta_{2N}}{s^{2N} + \alpha_1 s^{2N-1} + \alpha_2 s^{2N-2} + \cdots + \alpha_{2N-1}s^1 + \alpha_{2N}}\end{aligned} \quad (6.203)$$

Numerically find the solutions of numerator and denominator for zeros and poles. The factorized polynomials are reconstructed as below.

$$|H(s)|^2 = H(s)H(-s) = \frac{K(s-z_1)(s-z_2)\ldots(s-z_{2N-1})(s-z_{2N})}{(s-p_1)(s-p_2)\ldots(s-p_{2N-1})(s-p_{2N})} \quad (6.204)$$

Separate the all poles into the left and right s plane as below. Zeros can be chosen with various combinations. If we prefer to construct minimum phase filter, select the all zeros from unit circle inside in z domain which corresponds to the left s plane area.

$$\left.\begin{aligned} p_i^L &\in s \text{ plane left} \\ p_i^R &\in s \text{ plane right}\end{aligned}\right\} i = 1, 2, \ldots, N$$

$$z_i \in s \text{ plane left, for } i = 1, 2, \ldots, N$$

Based on the left s plane poles and zeros, the transfer function is assembled as below. Note that G is the zero-input gain.

$$\begin{aligned}H(s) &= \frac{G(s-z_1)(s-z_2)\ldots(s-z_N)}{(s-p_1^L)(s-p_2^L)\ldots(s-p_N^L)} \\ &= \frac{b_0 s^N + b_1 s^{N-1} + b_2 s^{N-2} + \cdots + b_{N-2}s^1 + b_N}{s^N + a_1 s^{N-1} + a_2 s^{N-2} + \cdots + a_{N-2}s^1 + a_N}\end{aligned} \quad (6.205)$$

Convert back to $j\Omega$ by replacing with s as below. The filter with rational function is completed in the linear frequency domain $j\Omega$.

$$\begin{aligned}H(j\Omega) &= \frac{b_0(j\Omega)^N + b_1(j\Omega)^{N-1} + b_2(j\Omega)^{N-2} + \cdots + b_{N-2}(j\Omega)^1 + b_{N-1}}{(j\Omega)^N + a_1(j\Omega)^{N-1} + a_2(j\Omega)^{N-2} + \cdots + a_{N-2}(j\Omega)^1 + a_{N-1}} \\ &= \frac{G(j\Omega - z_1)(j\Omega - z_2)\ldots(j\Omega - z_N)}{(j\Omega - p_1^L)(j\Omega - p_2^L)\ldots(j\Omega - p_N^L)}\end{aligned} \quad (6.206)$$

Table 6.30 Given filter specification

Filter specifications	Value
Filter type	Low pass filter
Filter realization	Elliptic IIR filter
Filter length	N with optimization
Passband frequency (f_p)	1000 Hz; $\omega_p = \pi/4$
Passband magnitude	1
Passband peak-to-peak ripple in decibel (Δ_p)	$20\log_{10}\left(\frac{1}{1/\sqrt{1+1}}\right) = 3.0103\text{dB}$
Stopband frequency (f_{st})	2000 Hz; $\omega_{st} = \pi/2$
Stopband attenuation in decibels Δ_{st}	$20\log_{10}\left(\frac{1}{0.1}\right) = 20\text{dB}$
Sampling frequency (f_s)	8000 Hz

The bilinear transformation provides the tool to change the linear frequency domain into the z domain for discrete-time filter as shown in Eq. (6.207).

$$H(z) = \frac{b_0(j\Omega)^N + b_1(j\Omega)^{N-1} + \cdots + b_{N-2}(j\Omega) + b_{N-1}}{(j\Omega)^N + a_1(j\Omega)^{N-1} + \cdots + a_{N-2}(j\Omega) + a_{N-1}}\Bigg|_{j\Omega = 2f_s\left(\frac{1-z^{-1}}{1+z^{-1}}\right)} \tag{6.207}$$

The final elliptic filter in discrete time domain is Eq. (6.207).

Example 6.21
Design the Table 6.30 IIR filter by using elliptic method and bilinear transformation.

Solution
Use the bilinear transformation to find the frequency specification in Ω from Eq. (6.150).

$$\Omega_p = 2f_s \tan\left(\frac{\pi}{8}\right) = 6627.4\text{rad/s}$$

$$\Omega_{st} = 2f_s \tan\left(\frac{\pi}{4}\right) = 16000\text{rad/s}$$

Compute the ε value by using the Eq. (6.194).

$$\varepsilon = \sqrt{10^{\frac{\Delta_p}{10}} - 1} = \sqrt{10^{\frac{3.0103}{10}} - 1} = 1$$

Let the $N = 2$. Δ_{st} and Ω_{st} are controlled by ξ.

$$\xi = \frac{\Omega_{st}}{\Omega_p} = \frac{16000}{6627.4} = 2.4142 \because \xi \frac{\Omega_p}{\Omega_{st}} = 1$$

Using Eq. (6.198), compute Δ_{st} as below.

$$\begin{aligned}\Delta_{st} &= 10\log_{10}\left(1+\varepsilon^2 L_N^2(\xi)\right) \\ &= 10\log_{10}\left(1+\varepsilon^2\left(\xi+\sqrt{\xi^2-1}\right)^4\right) \text{where } L_2(\xi) = \left(\xi+\sqrt{\xi^2-1}\right)^2\end{aligned}$$

Place the parameter values into the above equation as below. The calculated Δ_{st} is greater than the specification; therefore, we properly selected the order.

$$10\log_{10}\left(1+1^2\left(2.4142+\sqrt{2.4142^2-1}\right)^4\right) = 26.5635 \text{ dB} > 20 \text{ dB} \rightarrow \text{O.K.}$$

The second order elliptic rational function $R_2(\xi, \Omega)$ is below.

$$R_2(\xi,\Omega) = \frac{(t+1)\Omega^2-1}{(t-1)\Omega^2+1} \text{where } t = \sqrt{1-(1/\xi^2)}$$

The t value is below.

$$t = \sqrt{1-(1/2.4142^2)} = 0.9102$$

Build the squared magnitude transfer function with parameters and elliptic rational function as below.

$$\begin{aligned}|H(s)|^2 = H(s)H(-s) &= \frac{1}{1+\varepsilon^2 R_2^2\left(\xi, \xi\frac{s}{j\Omega_{st}}\right)} \\ &= \frac{1}{1+\varepsilon^2\left(\frac{(0.9102+1)\left(\xi\frac{s}{j\Omega_{st}}\right)^2-1}{(0.9102-1)\left(\xi\frac{s}{j\Omega_{st}}\right)^2+1}\right)^2} \\ &= \frac{0.00221s^4+2157165.10916s^2+527558383707008}{s^4+48043766.35091s^2+1055116767414020}\end{aligned}$$

Numerically compute the poles and zeros as below.

$$\begin{aligned} p_k &= 2056.77877 - j5315.28194, 2056.77877 + j5315.28194, \\ &\quad - 2056.77877 + j5315.28194, -2056.77877 - j5315.28194 \end{aligned}$$

$$\begin{aligned} z_k &= 518.39561 - j22097.85371, 518.39561 + j22097.85371, \\ &\quad - 518.39561 + j22097.85371, -518.39561 - j22097.85371 \end{aligned}$$

Select poles and zeros for stable and minimum phase filter. The selection means that the filter only utilizes the poles and zeros from left s plane.

$$\begin{aligned} H(s) &= \frac{G(s - z_1)(s - z_2)\ldots(s - z_N)}{\left(s - p_1^L\right)\left(s - p_2^L\right)\ldots\left(s - p_N^L\right)} \\ &= \frac{G(s + 518.39561 - j22097.85371)(s + 518.39561 + j22097.85371)}{(s + 2056.77877 - j5315.28194)(s + 2056.77877 + j5315.28194)} \end{aligned}$$

Find zero-input gain G as below.

$$H(s = 0) = \frac{1}{\sqrt{1 + \varepsilon^2 R_2^2(\xi, 0)}} = \frac{1}{\sqrt{2}}$$

$$\text{where} |R_2(\xi, 0)| = \left| \frac{(t + 1)0^2 - 1}{(t - 1)0^2 + 1} \right| = 1$$

$$H(s = 0) = \frac{G(518.39561 - j22097.85371)(518.39561 + j22097.85371)}{(2056.77877 - j5315.28194)(2056.77877 + j5315.28194)} = \frac{1}{\sqrt{2}}$$

$$G = 0.04701$$

The filter in s domain is below.

$$H(s) = \frac{0.04701s^2 + 48.74022s + 22968639.16077}{s^2 + 4113.55754s + 32482561.01041}$$

Convert back to $j\Omega$ by replacing with s.

$$H(j\Omega) = \frac{0.04701(j\Omega)^2 + 48.74022j\Omega + 22968639.16077}{(j\Omega)^2 + 4113.55754j\Omega + 32482561.01041}$$

Apply the bilinear transformation as below.

$$H(z) = \left. \frac{0.04701(j\Omega)^2 + 48.74022j\Omega + 22968639.16077}{(j\Omega)^2 + 4113.55754j\Omega + 32482561.01041} \right|_{j\Omega = 2f_s\left(\frac{1-z^{-1}}{1+z^{-1}}\right)}$$

The final elliptic filter in discrete time domain is below.

$$H(z) = \frac{0.10100 + 0.06172z^{-1} + 0.09659z^{-2}}{1 - 1.26174z^{-1} + 0.62847z^{-2}}$$

Figure 6.69 shows the pole zero plot of the design filter. Two poles are located within the unit circle and two zeros are placed nearby unit circle.

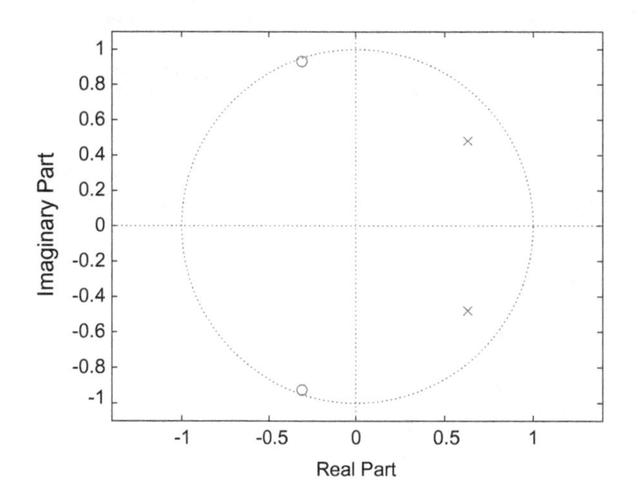

Fig. 6.69 Pole zero plot of designed elliptic filter for Table 6.30

Prog. 6.54 MATLAB program for Figs. 6.69 and 6.70.

```
syms z;
pb = linspace(0,1000,501);
sb = linspace(2000,4000,2001);
hz = (0.10100+0.06172*z^(-1)+0.09659*z^(-2))/(1-1.26174*z^(-
1)+0.62847*z^(-2));
[znum1,zden1] = numden(hz);
znum2 = collect(expand(znum1));
zden2 = collect(expand(zden1));
znum3 = sym2poly(znum2);
zden3 = sym2poly(zden2);
znum4 = znum3./(zden3(1));
zden4 = zden3./(zden3(1));
[h,w] = freqz(znum4,zden4,4001,8000);
hpb = freqz(znum4,zden4,pb,8000);
hsb = freqz(znum4,zden4,sb,8000);

figure,
zplane(znum4,zden4)
figure,
plot(w,abs(h),pb,abs(hpb),'.-',sb,abs(hsb),'.-'), grid
line([0 4000],[0.1 0.1])
ylim([0 1])
legend('Response','Passband','Stopband')
ylabel('|H(e^{j\omega})|')
xlabel('Frequency (Hz)')
figure,
plot(w,db(h),pb, db(hpb),'.-',sb,db(hsb),'.-'), grid
line([0 4000],[-20 -20])
legend('Response','Passband','Stopband')
ylabel('20log_{10}|H(e^{j\omega})| (dB)')
xlabel('Frequency (Hz)')
```

Figure 6.70 represents the frequency response of the filter. The passband and stopband satisfy the Table 6.30 specification. Note that the horizontal line in the figures denote the stopband attenuation requirement.

■

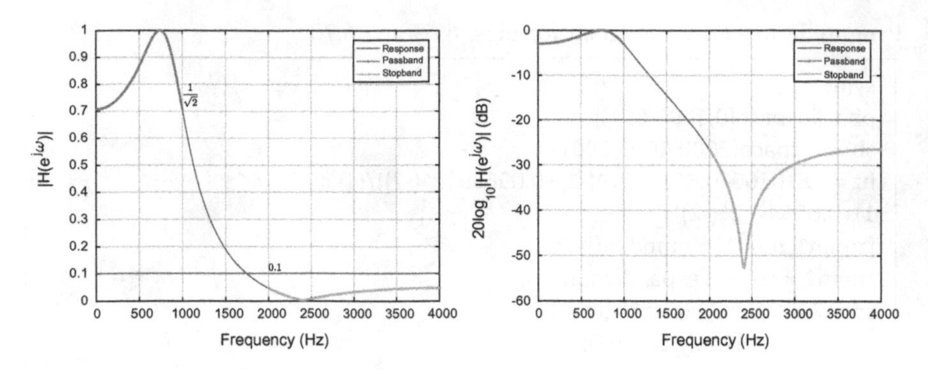

Fig. 6.70 Frequency response of designed elliptic filter for Table 6.30

Parametric methods

The parametric methods design the filter based on the filter models with parameters. Therefore, we have to estimate model as well as parameters in the design procedure. The conventional filter models [20] are below.

Moving average (MA) model:

$$y[n] = b_0x[n] + \cdots + b_{M-1}x[n-(M-1)] + b_Mx[n-M] \tag{6.208}$$

Corresponding frequency response:

$$H(e^{j\omega}) = b_0 + \cdots + b_{M-1}e^{-j\omega(M-1)} + b_Me^{-j\omega M} \tag{6.209}$$

Corresponding transfer function:

$$H(z) = b_0 + \cdots + b_{M-1}z^{-(M-1)} + b_Mz^{-M} \tag{6.210}$$

Autoregressive (AR) model:

$$y[n] + a_1y[n-1] + \cdots + a_Ny[n-N] = b_0x[n] \tag{6.211}$$

Corresponding frequency response:

$$H(e^{j\omega}) = \frac{b_0}{1 + a_1e^{-j\omega} + \cdots + a_Ne^{-j\omega N}} \tag{6.212}$$

Corresponding transfer function:

$$H(z) = \frac{b_0}{1 + a_1z^{-1} + \cdots + a_Nz^{-N}} \tag{6.213}$$

Autoregressive moving average (ARMA) model:

$$\begin{aligned} & y[n]+a_1y[n-1]+\cdots+a_Ny[n-N] \\ & =b_0x[n]+\cdots+b_{M-1}x[n-(M-1)]+b_Mx[n-M] \end{aligned} \tag{6.214}$$

Corresponding frequency response:

$$H(e^{j\omega})=\frac{b_0+\cdots+b_{M-1}e^{-j\omega(M-1)}+b_Me^{-j\omega M}}{1+a_1e^{-j\omega}+\cdots+a_Ne^{-j\omega N}} \tag{6.215}$$

Corresponding transfer function:

$$H(z)=\frac{b_0+\cdots+b_{M-1}z^{-(M-1)}+b_Mz^{-M}}{1+a_1z^{-1}+\cdots+a_Nz^{-N}} \tag{6.216}$$

The filter model is selected by understanding the desired frequency response profile. The MA filter model uses the zeros at z domain and the AR filter model utilizes the poles at z domain. The ARMA filter model employs the zeros and poles at z domain. The zeros close to the unit circle provide the valley shape distribution and the poles nearby the unit circle generate the peaky style response in frequency domain. Note that cluster of poles or zeros could present the flat with slight fluctuation response. Once we choose the filter model, the filter order and coefficients are estimated by the mathematical algorithm. The estimation process requires the statistical signal processing which regards signals as stochastic processes. The statistical signal processing is beyond this book scope; hence, following algorithms only imitate the algorithm for understanding the basic idea only. Note that the statistical signal processing performs the expectation, (auto)covariance, etc. operations based on the stationary white noise.

AR filter estimation (Yule-walker method style) - Approximated

Again, the following AR filter estimation algorithm is not precise as Yule-walker method since we avoid using the techniques from statistical signal processing. The reader can find the exact spectral estimation algorithms on the textbook [20–22]. Only grab the fundamental idea of the algorithm. Below is the conventional AR filter model which is identical to Eq. (6.211).

$$y[n]+a_1y[n-1]+\cdots+a_Ny[n-N]=b_0x[n] \tag{6.217}$$

By placing the impulse signal into the input $x[n]$, we can find the impulse response as below

$$h[n]+a_1h[n-1]+\cdots+a_Nh[n-N]=b_0\delta[n] \tag{6.218}$$

Let's design the 2nd order AR filter as below. For simplicity, we assume that the order is estimated by another algorithm.

$$y[n] + a_1 y[n-1] + a_2 y[n-2] = b_0 x[n] \tag{6.219}$$

The a_1, a_2, and b_0 are required to be estimated. The given impulse signal delivers the impulse response as below.

$$\begin{gathered} h[0] + a_1 h[-1] + a_2 h[-2] = b_0 \delta[0] \\ h[1] + a_1 h[0] + a_2 h[-1] = b_0 \delta[1] \\ h[2] + a_1 h[1] + a_2 h[0] = b_0 \delta[2] \\ h[3] + a_1 h[2] + a_2 h[1] = b_0 \delta[3] \\ h[4] + a_1 h[3] + a_2 h[2] = b_0 \delta[4] \\ \cdots \end{gathered} \tag{6.220}$$

The matrix representation is below for Eq. (6.220). Note that the causal filter shows the zero $h[n]$ for negative n.

$$\begin{bmatrix} h[0] & 0 & 0 \\ h[1] & h[0] & 0 \\ h[2] & h[1] & h[0] \\ h[3] & h[2] & h[1] \\ h[4] & h[3] & h[2] \\ \vdots & \vdots & \vdots \\ h[L-1] & h[L-2] & h[L-3] \end{bmatrix} \begin{bmatrix} 1 \\ a_1 \\ a_2 \end{bmatrix} = \begin{bmatrix} b_0 \\ 0 \\ 0 \\ \vdots \\ 0 \end{bmatrix} \tag{6.221}$$

Segments the matrix as below.

$$\left[\begin{array}{c|cc} h[0] & 0 & 0 \\ \hline h[1] & h[0] & 0 \\ h[2] & h[1] & h[0] \\ h[3] & h[2] & h[1] \\ h[4] & h[3] & h[2] \\ \vdots & \vdots & \vdots \\ h[L-1] & h[L-2] & h[L-3] \end{array}\right] \left[\begin{array}{c} 1 \\ \hline a_1 \\ a_2 \end{array}\right] = \left[\begin{array}{c} b_0 \\ \hline 0 \\ 0 \\ \vdots \\ 0 \end{array}\right] \tag{6.222}$$

Rearrange the matrix as below.

$$\begin{bmatrix} h[1] \\ h[2] \\ h[3] \\ h[4] \\ \vdots \\ h[L-1] \end{bmatrix} + \begin{bmatrix} h[0] & 0 \\ h[1] & h[0] \\ h[2] & h[1] \\ h[3] & h[2] \\ \vdots & \vdots \\ h[L-2] & h[L-3] \end{bmatrix} \begin{bmatrix} a_1 \\ a_2 \end{bmatrix} = \begin{bmatrix} 0 \\ 0 \\ 0 \\ 0 \\ \vdots \\ 0 \end{bmatrix} \tag{6.223}$$

Continuously organize the matrix as below.

$$\begin{bmatrix} h[0] & 0 \\ h[1] & h[0] \\ h[2] & h[1] \\ h[3] & h[2] \\ \vdots & \vdots \\ h[L-2] & h[L-3] \end{bmatrix} \begin{bmatrix} a_1 \\ a_2 \end{bmatrix} = - \begin{bmatrix} h[1] \\ h[2] \\ h[3] \\ h[4] \\ \vdots \\ h[L-1] \end{bmatrix} \tag{6.224}$$

Let's place the name for individual matrix and vectors as below.

$$\boldsymbol{R} = \begin{bmatrix} h[0] & 0 \\ h[1] & h[0] \\ h[2] & h[1] \\ h[3] & h[2] \\ \vdots & \vdots \\ h[L-2] & h[L-3] \end{bmatrix}; \boldsymbol{\alpha} = \begin{bmatrix} a_1 \\ a_2 \end{bmatrix}; \text{and } \boldsymbol{\mu} = \begin{bmatrix} h[1] \\ h[2] \\ h[3] \\ h[4] \\ \vdots \\ h[L-1] \end{bmatrix} \tag{6.225}$$

The solution for α is derived as shown in Eq. (6.226). The apostrophe on variable indicates the matrix transpose.

$$\begin{aligned} \boldsymbol{R\alpha} &= -\boldsymbol{\mu} \\ \boldsymbol{R'R\alpha} &= -\boldsymbol{R'\mu} \\ (\boldsymbol{R'R})^{-1}\boldsymbol{R'R\alpha} &= -(\boldsymbol{R'R})^{-1}\boldsymbol{R'\mu} \\ \boldsymbol{\alpha} &= -(\boldsymbol{R'R})^{-1}\boldsymbol{R'\mu} \end{aligned} \tag{6.226}$$

Table 6.31 Given filter specification

Filter specifications	Value
Filter type	Low pass filter
Filter realization	AR with Yule-walker method style
Filter length	2
Halfpower frequency (f_h)	1000 Hz; $\omega_h = \pi/4$
Passband magnitude	1
Sampling frequency (f_s)	8000 Hz

Once we obtain the a_1 and a_2, the frequency response helps to find b_0 value as below.

$$H(e^{j\omega}) = H(z)\big|_{z=e^{j\omega}} = \frac{b_0}{1 + a_1 e^{-j\omega} + a_2 e^{-j\omega 2}} \tag{6.227}$$

Placing any known $H(e^{j\omega})$ value returns the b_0 as below. This example uses the zero-frequency value.

$$b_0 = \left|H(e^{j0})(1 + a_1 e^{-j0} + a_2 e^{-j02})\right| = \left|H(e^{j0})(1 + a_1 + a_2)\right| \tag{6.228}$$

We may derive the b_0 value from Eq. (6.220) as below.

$$b_0 = h[0] \tag{6.229}$$

The AR filter by using the Yule-walker method requires the impulse response from the unknown filter. The Yule-walker algorithm estimates the coefficients of the unknown AR filter with given order. Therefore, we cannot directly design the AR Yule-walker filter from the specification.

Example 6.22
Design the AR filter first to meet the Table 6.31 requirement and Yule-walker algorithm follows the AR filter to estimate the coefficients. Note that the actual Yule-walker algorithm need the unknown filter output from the white noise input for estimation.

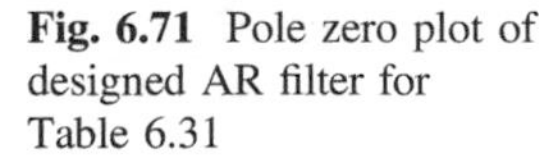

Fig. 6.71 Pole zero plot of designed AR filter for Table 6.31

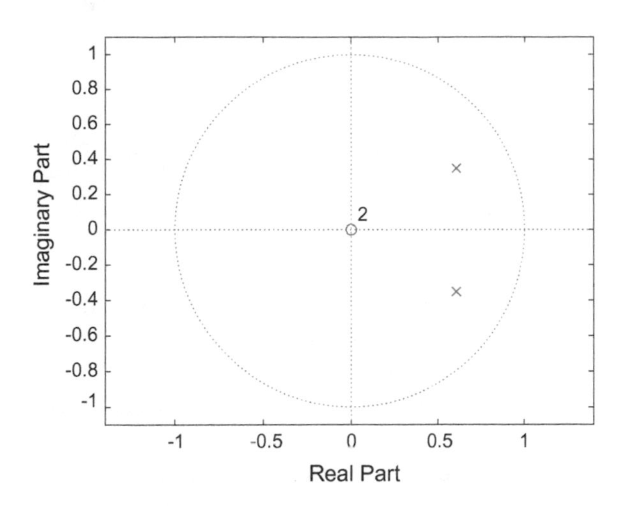

Solution

Figure 5.24 demonstrates the proper LPF for above specification. The pole location $0.7e^{\pm j\pi/6}$ provide the best performance according to Fig. 5.24. Observe that previous IIR filter design methods cannot be used because of model mismatch as ARMA filter model.

$$H_d(z) = \frac{b_0}{(1 - 0.7e^{j\pi/6}z^{-1})(1 - 0.7e^{-j\pi/6}z^{-1})} = \frac{b_0}{1 - 1.2124z^{-1} + 0.49z^{-2}}$$

At halfpower frequency $\pi/4$, the filter magnitude should be $1/\sqrt{2}$ by controlling b_0 value as below.

$$\left|H_d\left(e^{j\pi/4}\right)\right| = \left|\left.\frac{b_0}{1 - 1.2124z^{-1} + 0.49z^{-2}}\right|_{z=e^{j\pi/4}}\right|$$

$$= \left|\frac{b_0}{1 - 1.2124e^{-j\pi/4} + 0.49e^{-j2\pi/4}}\right| = \frac{1}{\sqrt{2}} \rightarrow b_0 = 0.2786$$

The final AR filter for baseline model is shown in below.

$$H_d(z) = \frac{0.2786}{1-1.2124z^{-1} + 0.49z^{-2}}$$

Figure 6.71 shows the pole zero plot of the design AR filter. Two poles are located at $0.7e^{\pm j\pi/6}$ and two zeros are placed at origin.

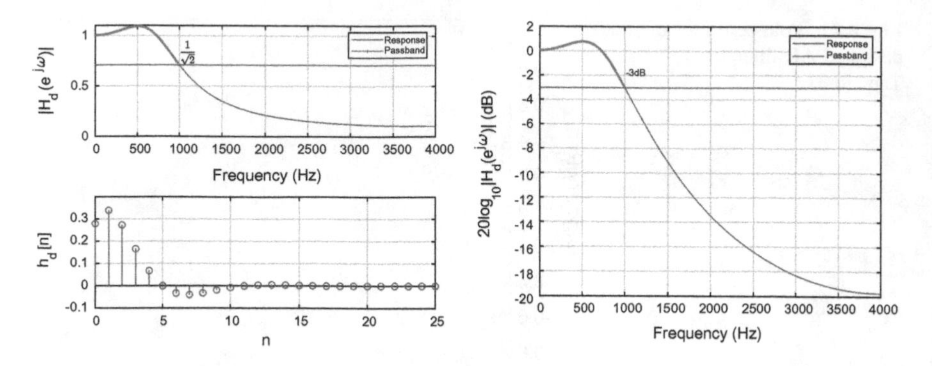

Fig. 6.72 Frequency response and impulse response of designed AR filter for Table 6.31

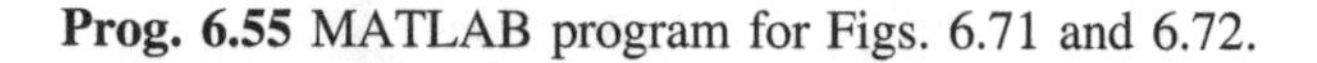

Prog. 6.55 MATLAB program for Figs. 6.71 and 6.72.

```
aangle = pi/6;
alpha = 0.7;
pb = linspace(0,1000,501);
aa = [1 -2*alpha*cos(aangle) alpha^2]
bb = abs(((exp(j*pi/4)-0.7*exp(-j*pi/6))*(exp(j*pi/4)-
0.7*exp(j*pi/6)))/sqrt(2));
[hh0,tt] = impz(bb,aa);
[h0,w0] = freqz(bb,aa,1000,8000);
hpb0 = freqz(bb,aa,pb,8000);

figure,
zplane(bb,aa);
figure,
subplot(211),plot(w0,abs(h0),pb,abs(hpb0),'.-'), grid, ylim([0 1.1]);
xlabel('Frequency (Hz)')
ylabel('|H_d(e^{j\omega})|')
line([0 4000],[1/sqrt(2) 1/sqrt(2)]);
legend('Response','Passband')
subplot(212), stem(tt,hh0), grid, xlim([0 25]), ylim([-0.1 0.4])
xlabel('n')
ylabel('h_d[n]')
figure
plot(w0,db(h0),pb, db(hpb0),'.-'), grid, ylim([-20 2]);
line([0 4000],[db(1/sqrt(2)) db(1/sqrt(2))]);
legend('Response','Passband')
ylabel('20log_{10}|H_d(e^{j\omega})| (dB)')
xlabel('Frequency (Hz)')
```

Figure 6.72 depicts the frequency response and impulse response of the AR filter. The passband satisfies the Table 6.31 specification. Note that the horizontal line in the figures denote the halfpower frequency requirement. Below equation shows the difference equation for $H_d(z)$ in time domain.

$$y_d[n] - 1.2124y_d[n-1] + 0.49y_d[n-2] = 0.2786x_d[n]$$

By placing the delta signal for input, impulse response can be obtained as below.

$$h_d[n] - 1.2124h_d[n-1] + 0.49h_d[n-2] = 0.2786\delta[n]$$

This example employs first six samples of impulse response as below.

$$\begin{aligned} h_d[n] &= 0.2786\delta[n] + 0.3378\delta[n-1] + 0.2731\delta[n-2] \\ &\quad + 0.1655\delta[n-3] + 0.0669\delta[n-4] + 0.0000\delta[n-5] \end{aligned}$$

Build the matrix as below.

$$\begin{bmatrix} h_d[0] & 0 \\ h_d[1] & h_d[0] \\ h_d[2] & h_d[1] \\ h_d[3] & h_d[2] \\ h_d[4] & h_d[3] \end{bmatrix} \begin{bmatrix} a_1 \\ a_2 \end{bmatrix} = - \begin{bmatrix} h_d[1] \\ h_d[2] \\ h_d[3] \\ h_d[4] \\ h_d[5] \end{bmatrix}$$

Actual values are as below.

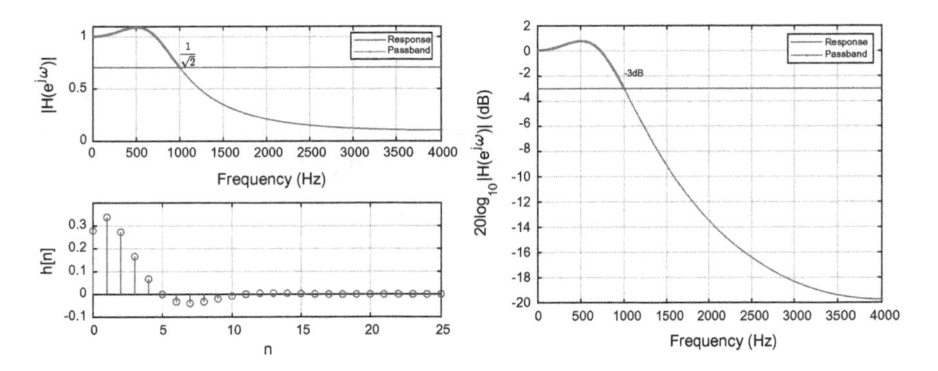

Fig. 6.73 Frequency response and impulse response of designed Yule-walker filter for Table 6.31

$$\begin{bmatrix} 0.2786 & 0 \\ 0.3378 & 0.2786 \\ 0.2731 & 0.3378 \\ 0.1655 & 0.2731 \\ 0.0669 & 0.1655 \end{bmatrix} \begin{bmatrix} a_1 \\ a_2 \end{bmatrix} = - \begin{bmatrix} 0.3378 \\ 0.2731 \\ 0.1655 \\ 0.0669 \\ 0.0000 \end{bmatrix} ; \boldsymbol{R\alpha} = -\boldsymbol{\mu}$$

Solve the matrix for a_1 and a_2 as below.

$$\boldsymbol{\alpha} = -(\boldsymbol{R}'\boldsymbol{R})^{-1}\boldsymbol{R}'\boldsymbol{\mu}$$

$$\boldsymbol{\alpha} = \begin{bmatrix} a_1 \\ a_2 \end{bmatrix} = \begin{bmatrix} -1.2124 \\ 0.4900 \end{bmatrix}$$

Assume that the frequency response $H\left(e^{j\pi/4}\right)$ is $1/\sqrt{2}$ to compute b_0 value.

$$\left|H\left(e^{j\pi/4}\right)\right| = \left|\frac{b_0}{1 - 1.2124z^{-1} + 0.49z^{-2}}\right|_{z=e^{j\pi/4}}$$
$$= \left|\frac{b_0}{1 - 1.2124e^{-j\pi/4} + 0.49e^{-j2\pi/4}}\right| = \frac{1}{\sqrt{2}} \rightarrow b_0 = 0.2786$$

The final Yule-walker AR filter is below.

$$H(z) = \frac{0.2786}{1 - 1.2124z^{-1} + 0.49z^{-2}}$$

The corresponding time domain representation is below.

$$y[n] - 1.2124y[n-1] + 0.49y[n-2] = 0.2786x[n]$$

Figure 6.73 represents the frequency response and impulse of the filter. The passband satisfies the Table 6.31 specification as halfpower frequency requirement. Note that the horizontal line in the figures denote halfpower frequency requirement.

Prog. 6.56 MATLAB program for Fig. 6.73.

```
aangle = pi/6;
alpha = 0.7;
pb = linspace(0,1000,501);
aa = [1 -2*alpha*cos(aangle) alpha^2]
bb = abs(((exp(j*pi/4)-0.7*exp(-j*pi/6))*(exp(j*pi/4)-
0.7*exp(j*pi/6)))/sqrt(2));
[hh0,tt] = impz(bb,aa);
hh = hh0(1:6);
R = [hh(1:end-1) [0; hh(1:end-2)]];
mu = [hh(2:end)];
acoefs = -inv(R'*R)*R'*mu
bcoef = abs((1+acoefs(1)*exp(-j*pi/4)+acoefs(2)*exp(-j*2*pi/4))/sqrt(2));
[hh1,tt1] = impz(bcoef,[1; acoefs]);
[h1,w1] = freqz(bcoef,[1; acoefs],1000,8000);
hpb1 = freqz(bcoef,[1; acoefs],pb,8000);

figure,
subplot(211),plot(w1,abs(h1),pb,abs(hpb1),'.-'), grid, ylim([0 1.1]);
xlabel('Frequency (Hz)')
ylabel('|H(e^{j\omega})|')
line([0 4000],[1/sqrt(2) 1/sqrt(2)]);
legend('Response','Passband')
subplot(212), stem(tt1,hh1), grid, xlim([0 25]), ylim([-0.1 0.4]);
xlabel('n')
ylabel('h[n]')
figure,
plot(w1,db(h1),pb, db(hpb1),'.-'), grid, ylim([-20 2]);
line([0 4000],[db(1/sqrt(2)) db(1/sqrt(2))]);
legend('Response','Passband')
ylabel('20log_{10}|H(e^{j\omega})| (dB)')
xlabel('Frequency (Hz)')
```

The impulse response in above figure also exactly follows $h_d[n]$ in Fig. 6.72. This example exercised the AR Yule-walker method with model and order matching condition. The derived result from the AR Yule-walker method shows the completely identical outcome. Filter model and order mismatch situation could generate the significant performance degradation in frequency response as well as impulse response. Therefore, we need to be aware of matching problem from parametric filter design method.

∎

Maxflat filter

The previously described IIR filters are derived by the rational function with equal or denominator dominant degree of numerator and denominator. Sometimes, it is preferable to design filters having more zeros than poles to achieve an enhanced tradeoff between performance and complexity. The maxflat filter [23] used the unequal order of numerator and denominator on the transfer function for maximum flat frequency response in passband and stopband. The flatness can be explained by using the example in Table 6.32.

Four polynomials y_1, y_2, y_3, and y_4 represents the minimum value 1 at $x = 0$; therefore, the polynomials pass through the identical location $x = 0$ and $y = 1$ as shown in Fig. 6.74. The increase rate after the minimum point is vary based on the order of the polynomial.

Table 6.32 Flatness example with polynomials

$y_1 = x^2 + 1$	$y_2 = x^4 + 1$	$y_3 = x^6 + 1$	$y_4 = x^8 + 1$
$y_1\vert_{x=0} = 1$	$y_2\vert_{x=0} = 1$	$y_3\vert_{x=0} = 1$	$y_4\vert_{x=0} = 1$
$\frac{dy_1}{dx}\Big\vert_{x=0} = 0$	$\frac{dy_2}{dx}\Big\vert_{x=0} = 0$	$\frac{dy_3}{dx}\Big\vert_{x=0} = 0$	$\frac{dy_4}{dx}\Big\vert_{x=0} = 0$
	$\frac{d^2y_2}{dx^2}\Big\vert_{x=0} = 0$	$\frac{d^2y_3}{dx^2}\Big\vert_{x=0} = 0$	$\frac{d^2y_4}{dx^2}\Big\vert_{x=0} = 0$
	$\frac{d^3y_2}{dx^3}\Big\vert_{x=0} = 0$	$\vdots$	$\vdots$
		$\frac{d^5y_3}{dx^5}\Big\vert_{x=0} = 0$	$\frac{d^7y_4}{dx^7}\Big\vert_{x=0} = 0$

⇒ Flatter ⇒

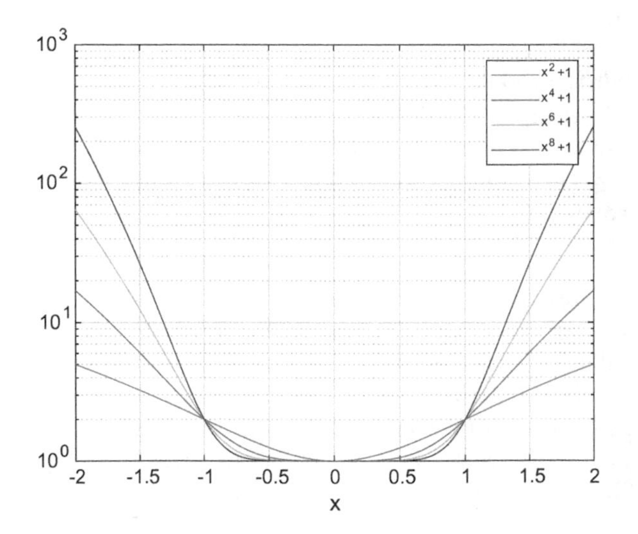

Fig. 6.74 Plot of four polynomials from Table 6.32

Prog. 6.57 MATLAB program for Fig. 6.74.

```
x = linspace(-2,2,10000);
y1 = x.^2+1;
y2 = x.^4+1;
y3 = x.^6+1;
y4 = x.^8+1;
plot(x,y1,x,y2,x,y3,x,y4), grid
semilogy(x,y1,x,y2,x,y3,x,y4), grid
legend('x^2+1','x^4+1','x^6+1','x^8+1')
xlabel('x');
```

The further flatness can be achieved by higher order of derivatives for zero output at the specific input value. At the given example, y_4 shows the maximum flatness because up to the 7th order derivative delivers the zero output for zero input. Figure 6.74 also presents the maximum flatness. Note that Fig. 6.74 utilizes the logarithm scale for y axis to magnify the value around the interesting point. For FIR low pass filter, the flatness is required at the $\omega = 0$ and π . The corresponding representation in mathematics is shown in Eq. (6.230).

$$
\begin{aligned}
&H\left(e^{j\omega}\right) \in \mathbb{R} \\
&H\left(e^{j\omega}\right)\Big|_{\omega=0} = 1, \text{for unit DC gain} \\
&\left.\frac{d^k H\left(e^{j\omega}\right)}{dx^k}\right|_{\omega=0} = 0,\ k = 1, 2, \ldots, L_1 \\
&\left.\frac{d^k H\left(e^{j\omega}\right)}{dx^k}\right|_{\omega=\pi} = 0,\ k = 0, 1, \ldots, L_2
\end{aligned}
\tag{6.230}
$$

Herrmann [24] solved the above problem for FIR filter realization. Selesnick et al. [23] extended FIR filter realization for Butterworth IIR filter with maximum flat response on passband and stopband. The derive procedure and corresponding solution can be found in the papers [23, 24] and will not be presented due to the complexity which is the out of this book scope.

Example 6.23
Using the MATLAB function, fulfill the Table 6.33 maxflat IIR filter.

Table 6.33 Given filter specification

Filter specifications	Value
Filter type	Low pass filter
Filter realization	Maxflat IIR
Filter length	Various $N = 3$ for Butterworth filter (Comparison)
Halfpower frequency (f_c)	1000 Hz; $\omega_h = \pi/4$
Passband magnitude	1
Sampling frequency (f_s)	8000 Hz

Solution

Maxflat MATLAB function receives numerator/denominator order and halfpower frequency argument. The generated numerator and denominator coefficients provide frequency response in magnitude and decibel as shown in Figs. 6.75 and 6.76.

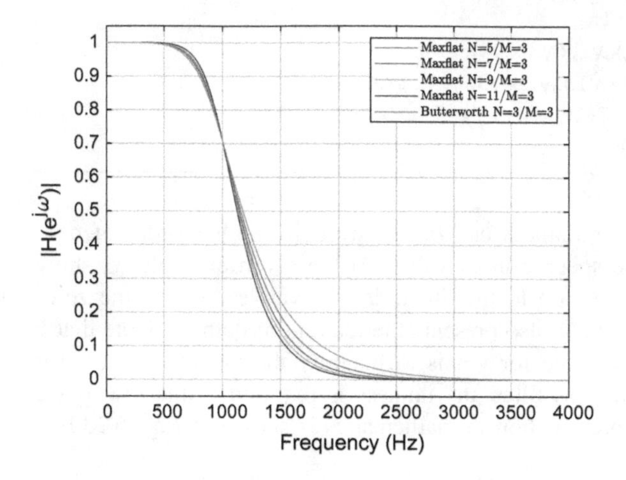

Fig. 6.75 Frequency response in magnitude for various order maxflat filters and Butterworth filter

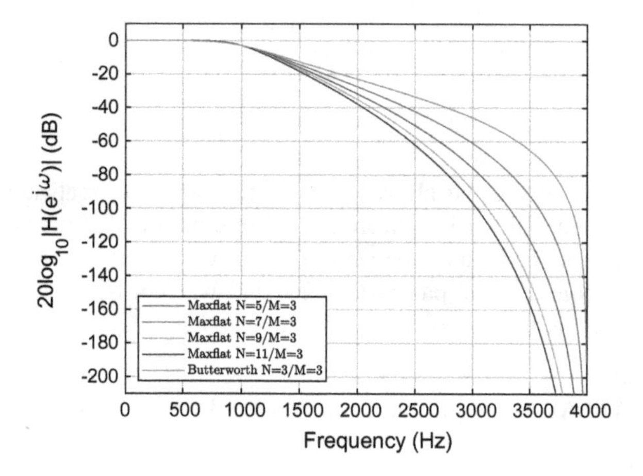

Fig. 6.76 Frequency response in decibel for various order maxflat filters and Butterworth filter

Prog. 6.58 MATLAB program for Figs. 6.75 and 6.76.

```
m = 3;
n1 = 5;
n2 = 7;
n3 = 9;
n4 = 11;
wn = 0.25;
[b1,a1] = maxflat(n1,m,wn);
[b2,a2] = maxflat(n2,m,wn);
[b3,a3] = maxflat(n3,m,wn);
[b4,a4] = maxflat(n4,m,wn);
[b5,a5] = butter(m,wn);
[h1,w1] = freqz(b1,a1,1000);
[h2,w2] = freqz(b2,a2,1000);
[h3,w3] = freqz(b3,a3,1000);
[h4,w4] = freqz(b4,a4,1000);
[h5,w5] = freqz(b5,a5,1000);
ww = w1/pi*4000;

figure,
plot(ww,abs(h1),ww,abs(h2),ww,abs(h3),ww,abs(h4),ww,abs(h5)), grid
h = legend('Maxflat N=5/M=3','Maxflat N=7/M=3','Maxflat
N=9/M=3','Maxflat N=11/M=3','Butterworth N=3/M=3');
set(h,'Interpreter','latex');
xlabel('Frequency (Hz)')
ylabel('|H(e^{j\omega})|')
ylim([-0.05 1.05])
figure,
plot(ww,db(abs(h1)),ww,db(abs(h2)),ww,db(abs(h3)),ww,db(abs(h4)),ww,db(
abs(h5))), grid
h = legend('Maxflat N=5/M=3','Maxflat N=7/M=3','Maxflat
N=9/M=3','Maxflat N=11/M=3','Butterworth N=3/M=3');
set(h,'Interpreter','latex');
xlabel('Frequency (Hz)')
ylabel('20log_{10}|H(e^{j\omega})| (dB)')
ylim([-210 10])
```

In figure legend above, N is numerator order and M is denominator order. As we can see that higher order at the numerator presents the flatter response at passband and stopband. For comparison purpose, Butterworth filter with equal orders for numerator and denominator presents the least flatness in the filter responses. Note that Maxflat filter is the extension of Butterworth filter to deliver the maximum flatness at passband and stopband. ■

Linear prediction filter

The filter can predict the current input by using N previous inputs as shown in Eq. (6.231).

$$\widetilde{x}[n] = -\sum_{k=1}^{N} a_k x[n-k] \tag{6.231}$$

The linear combination of the previous inputs predicts the current input. The estimation error is below.

$$e[n] = x[n] - \widetilde{x}[n] \tag{6.232}$$

With the estimation error, Eq. (6.231) can be organized as below. The $x[n]$ is filter input and $e[n]$ is the filter output.

$$e[n] = x[n] + \sum_{k=1}^{N} a_k x[n-k] \tag{6.233}$$

The z domain representation is below.

$$E(z) = (1 + a_1 z^{-1} + a_2 z^{-2} + \cdots + a_N z^{-N})X(z) \tag{6.234}$$

The estimation error is the residual by removing signal pattern from $x[n]$. The linear combination of the previous inputs generates the best representation of $x[n]$ based on the order N. The optimal predictor minimizes the $e[n]$ in mean square sense for whitening the $e[n]$. Reverse the filter configuration provides the below transfer function.

$$\frac{X(z)}{E(z)} = \frac{1}{(1 + a_1 z^{-1} + a_2 z^{-2} + \cdots + a_N z^{-N})} = H_{AR}(e^{j\omega}) \tag{6.235}$$

The AR model from the $e[n]$ input generates the $x[n]$ output. Since the white noise $e[n]$ has the full and flat spectrum, the output $x[n]$ and the AR filter response are identical in frequency domain as shown in Eq. (6.236).

$$X(e^{j\omega}) = H_{AR}(e^{j\omega})E(e^{j\omega}) \approx H_{AR}(e^{j\omega}) \tag{6.236}$$

The computation of linear prediction filter coefficients in Eq. (6.233) is the estimation process of input signal $x[n]$ based on the AR filter model. Using the coefficients, the linear prediction filter performs the inverse filter operation on $x[n]$ to obtain the white noise $e[n]$ by accurate prediction. The signal generation and linear prediction are shown in Eq. (6.237) in transfer function form.

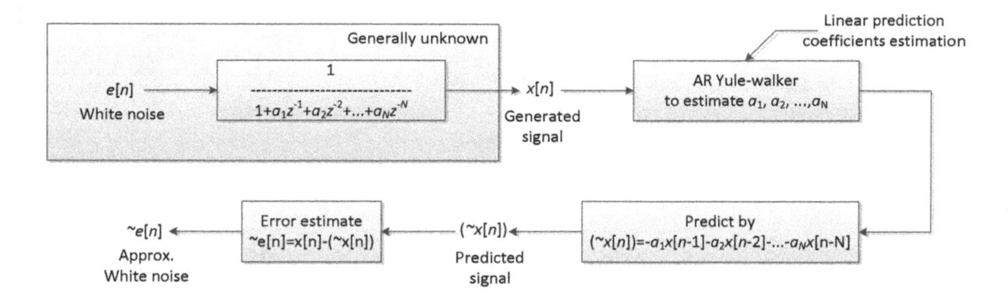

Fig. 6.77 Overall linear prediction filter procedure. Assume order N is known

$$\underbrace{\frac{1}{\left(1+a_1z^{-1}+\cdots+a_Nz^{-N}\right)}}_{\text{Trans.func.for } x[n]}\underbrace{\left(1+a_1z^{-1}+\cdots+a_Nz^{-N}\right)}_{\text{Trans.func. for lin.pred.filter}}\approx 1 \tag{6.237}$$

Therefore, the AR filter generates the $x[n]$ from white noise $e[n]$ and linear prediction filter produces the white noise $e[n]$ from $x[n]$. Figure 6.77 illustrates the overall procedure of the linear prediction filter process.

We assume that the filter order N is known. If the order is required to be estimated as well, we need to exercise the statistical process for optimal order. The linear prediction filter [25] is the FIR filter; however, the linear prediction filter coefficient is frequently used for AR modeling of signal as IIR filter format. The computational procedure and method are identical to the parametric method for AR Yule-walker filter [20–22].

We discussed that the white noise $e[n]$ represents the full and flat spectrum. The squared magnitude in frequency domain corresponds to the autocorrelation operation in time domain as shown in Eq. (6.238).

$$\begin{aligned}\text{constant} &\approx \left|E\left(e^{j\omega}\right)\right|^2 = E\left(e^{j\omega}\right)E^*\left(e^{j\omega}\right) \overset{\text{DTFT}}{\leftrightarrow} e[n]*e[-n] \\ &= \sum_{k=-\infty}^{\infty} e[k]e[k-n] \approx k\delta[n]\end{aligned} \tag{6.238}$$

If the $e[n]$ is the random white noise with independence to adjacent data, the autocorrelation presents the delta function which corresponds to the full and flat spectrum. Further information about the white noise can be found at statistical signal processing books [26, 27].

Prony

The Prony [7, 28] method designs the ARMA IIR filter from time domain response. The difference equation from conventional ARMA model represents the nonlinear property which is difficult to formulize the solution. The Prony method linearizes the ARMA model in matrix description by using Z-transformation. Below is the autoregressive moving average (ARMA) model in difference equation form.

$$\begin{aligned} & y[n] + a_1 y[n-1] + \cdots + a_N y[n-N] \\ & = b_0 x[n] + \cdots + b_{M-1} x[n-(M-1)] + b_M x[n-M] \end{aligned} \tag{6.239}$$

The corresponding transfer function is below.

$$\begin{aligned} \frac{Y(z)}{X(z)} &= \frac{b_0 + \cdots + b_{M-1} z^{-(M-1)} + b_M z^{-M}}{1 + a_1 z^{-1} + \cdots + a_N z^{-N}} = \frac{B(z)}{A(z)} \\ &= H(z) = h[0] + h[1]z^{-1} + h[2]z^{-2} + \cdots \end{aligned} \tag{6.240}$$

The numerator can be denoted by the product between the transfer function and denominator part as below. Observe that the impulse response $h[n]$ length is most likely to be infinite due to the ARMA model.

$$\begin{aligned} B(z) &= H(z)A(z) \\ \left(b_0 + \cdots + b_{M-1} z^{-(M-1)} + b_M z^{-M}\right) \\ &= (h[0] + h[1]z^{-1} + h[2]z^{-2} + \cdots)(1 + a_1 z^{-1} + \cdots + a_N z^{-N}) \end{aligned} \tag{6.241}$$

The above polynomial based on power of z can be represented by matrix form as shown in Eq. (6.242).

$$\begin{bmatrix} b_0 \\ b_1 \\ \vdots \\ b_M \\ 0 \\ \vdots \\ 0 \end{bmatrix} = \begin{bmatrix} h[0] & 0 & 0 & \cdots & 0 \\ h[1] & h[0] & 0 & \cdots & 0 \\ h[2] & h[1] & h[0] & \cdots & 0 \\ \vdots & \vdots & \vdots & \ddots & \vdots \\ h[M] & h[M-1] & h[M-2] & \cdots & \vdots \\ \vdots & \vdots & \vdots & \cdots & \vdots \\ h[L] & h[L-1] & h[L-2] & \cdots & h[L-N] \end{bmatrix} \begin{bmatrix} 1 \\ a_1 \\ a_2 \\ \vdots \\ a_N \end{bmatrix} \tag{6.242}$$

The dimension of above matrix is below.

$$[(L+1) \times 1] = [(L+1) \times (N+1)] \times [(N+1) \times 1]$$

The partitioned matrix is presented as shown in Eq. (6.243).

$$\begin{bmatrix} \boldsymbol{b} \\ \hline \boldsymbol{0} \end{bmatrix} = \begin{bmatrix} \boldsymbol{H}_1 \\ \hline \boldsymbol{h}_1 \quad \boldsymbol{H}_2 \end{bmatrix} \begin{bmatrix} 1 \\ \hline \boldsymbol{a}^{\dagger} \end{bmatrix} \tag{6.243}$$

The submatrix dimension is below.

$$\begin{bmatrix} (M+1) \times 1 \\ \hline (L-M) \times 1 \end{bmatrix} = \begin{bmatrix} (M+1) \times (N+1) \\ \hline (L-M) \times 1 \quad (L-M) \times N \end{bmatrix} \begin{bmatrix} 1 \times 1 \\ \hline N \times 1 \end{bmatrix}$$

Prony algorithm solves the following linear algebra for $\boldsymbol{a}$ and $\boldsymbol{b}$ vector.

$$\boldsymbol{0} = \boldsymbol{h}_1 + \boldsymbol{H}_2\boldsymbol{a}^{\dagger}$$

$$\boldsymbol{b} = \boldsymbol{H}_1\boldsymbol{a} \tag{6.244}$$

Apply the derived a and b coefficients to the transfer function as below.

$$H(z) = \frac{b_0 + \cdots + b_{M-1}z^{-(M-1)} + b_M z^{-M}}{1 + a_1 z^{-1} + \cdots + a_N z^{-N}} \tag{6.245}$$

The final Prony filter in discrete time domain is Eq. (6.245). Similar to the Yule-walker method, the Prony filter based on the ARMA model requires the designed impulse response which satisfies the specification with unknown coefficients. The Prony algorithm estimates the coefficients of the unknown ARMA filter with given order. Therefore, we cannot directly design the ARMA Prony filter from the specification.

Example 6.24
Design the Table 6.34 IIR filter by using Prony method. By using the Butterworth method, example below provides the ARMA filter first to meet the Table 6.34 requirement and Prony algorithm follows the ARMA filter to estimate the coefficients.

Solution
In previous example, the Butterworth method presented the following ARMA IIR filter design to meet the Table 6.34 specification.

$$H_d(z) = \frac{0.0317 + 0.0951z^{-1} + 0.0951z^{-2} + 0.0317z^{-3}}{1 - 1.4590z^{-1} + 0.9104z^{-2} - 0.1978z^{-3}}$$

The pole zero plot of the above filter is demonstrated at Fig. 6.49 and the frequency response and impulse response are illustrated at Fig. 6.78. The passband satisfies the Table 6.34 specification. Note that the horizontal line in the figures denote the halfpower frequency requirement.

Table 6.34 Given filter specification

Filter specifications	Value
Filter type	Low pass filter
Filter realization	Prony IIR
Filter length	N = 3 and M = 3
Passband frequency (f_p)	1000 Hz; $\omega_p = \pi/4$
Passband magnitude	1
Passband peak-to-peak ripple in decibel (Δ_p)	3.0103 dB
Stopband frequency (f_{st})	2000 Hz; $\omega_{st} = \pi/2$
Stopband attenuation in decibels Δ_{st}	$20\log_{10}\left(\frac{1}{0.1}\right) = 20\text{dB}$
Sampling frequency (f_s)	8000 Hz

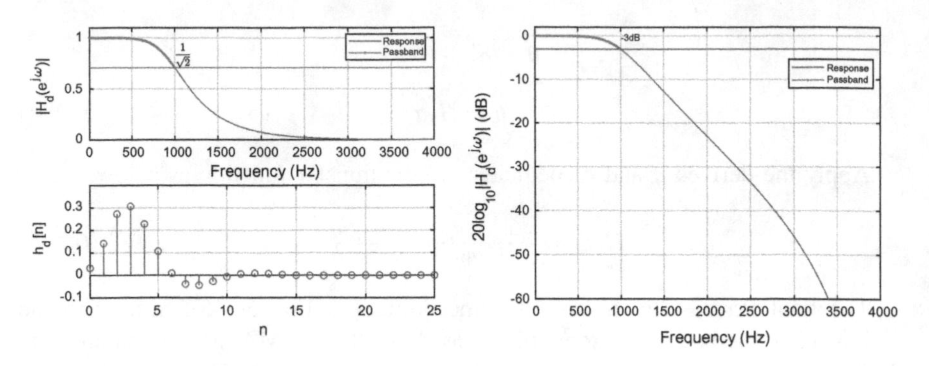

Fig. 6.78 Frequency response and impulse response of designed Butterworth filter for Table 6.34

Prog. 6.59 MATLAB program for Fig. 6.78.

```
pb = linspace(0,1000,501);
sb = linspace(2000,4000,2001);
a0 = [1 -1.4590 0.9104 -0.1978];
b0 = [0.0317 0.0951 0.0951 0.0317];
[h0,w0] = freqz(b0,a0,1000,8000);
[hh0,t0] = impz(b0,a0);
hpb0 = freqz(b0,a0,pb,8000);
hsb0 = freqz(b0,a0,sb,8000);

figure,
subplot(211),plot(w0,abs(h0),pb,abs(hpb0),'.-'), grid, ylim([0 1.1]);
xlabel('Frequency (Hz)');
ylabel('|H_d(e^{j\omega})|');
line([0 4000],[1/sqrt(2) 1/sqrt(2)]);
legend('Response','Passband');
subplot(212), stem(t0,hh0), grid, xlim([0 25]), ylim([-0.1 0.4]);
xlabel('n');
ylabel('h_d[n]');
figure
plot(w0,db(h0),pb, db(hpb0),'.-'), grid, ylim([-60 2]);
line([0 4000],[db(1/sqrt(2)) db(1/sqrt(2))]);
legend('Response','Passband');
ylabel('20log_{10}|H_d(e^{j\omega})| (dB)');
xlabel('Frequency (Hz)');
```

Below equation shows the difference equation for $H_d(z)$ in time domain.

$$\begin{aligned} &y_d[n] - 1.4590y_d[n-1] + 0.9104y_d[n-2] - 01978y_d[n-3] \\ &= 0.0317x_d[n] + 0.0951x_d[n-1] + 0.0951x_d[n-2] + 0.0317x_d[n-3] \end{aligned}$$

By placing the delta signal for input, impulse response can be obtained as below.

$$\begin{aligned} &h_d[n] - 1.4590h_d[n-1] + 0.9104h_d[n-2] - 01978h_d[n-3] \\ &= 0.0317\delta[n] + 0.0951\delta[n-1] + 0.0951\delta[n-2] + 0.0317\delta[n-3] \end{aligned}$$

This example employs first seven samples of impulse response as below.

$$\begin{aligned} h_d[n] = &\ 0.0317\delta[n] + 0.1414\delta[n-1] + 0.2725\delta[n-2] + 0.3068\delta[n-3] \\ &+ 0.2276\delta[n-4] + 0.1066\delta[n-5] + 0.0090\delta[n-6] \end{aligned}$$

Build the matrix for Eq. (6.242) as below.

$$\begin{bmatrix} b_0 \\ b_1 \\ b_2 \\ b_3 \\ 0 \\ 0 \\ 0 \end{bmatrix} = \begin{bmatrix} h_d[0] & 0 & 0 & 0 \\ h_d[1] & h_d[0] & 0 & 0 \\ h_d[2] & h_d[1] & h_d[0] & 0 \\ h_d[3] & h_d[2] & h_d[1] & h_d[0] \\ h_d[4] & h_d[3] & h_d[2] & h_d[1] \\ h_d[5] & h_d[4] & h_d[3] & h_d[2] \\ h_d[6] & h_d[5] & h_d[4] & h_d[3] \end{bmatrix} \begin{bmatrix} 1 \\ a_1 \\ a_2 \\ a_3 \end{bmatrix}$$

Actual values are as below.

$$\begin{bmatrix} b_0 \\ b_1 \\ b_2 \\ b_3 \\ \hline 0 \\ 0 \\ 0 \end{bmatrix} = \left[\begin{array}{c|ccc} 0.0317 & 0 & 0 & 0 \\ 0.1414 & 0.0317 & 0 & 0 \\ 0.2725 & 0.1414 & 0.0317 & 0 \\ 0.3068 & 0.2725 & 0.1414 & 0.0317 \\ \hline 0.2276 & 0.3068 & 0.2725 & 0.1414 \\ 0.1066 & 0.2276 & 0.3068 & 0.2725 \\ 0.0090 & 0.1066 & 0.2276 & 0.3068 \end{array}\right] \begin{bmatrix} 1 \\ \hline a_1 \\ a_2 \\ a_3 \end{bmatrix}$$

Partitioning matrix as below.

$$\begin{bmatrix} \boldsymbol{b} \\ \hline \boldsymbol{0} \end{bmatrix} = \begin{bmatrix} \boldsymbol{H}_1 \\ \hline \boldsymbol{h}_1 \quad \boldsymbol{H}_2 \end{bmatrix} \begin{bmatrix} 1 \\ \hline \boldsymbol{a}^{\dagger} \end{bmatrix}$$

The submatrices are used for $\boldsymbol{a}$ and $\boldsymbol{b}$ coefficient as below. The $\boldsymbol{a}$ coefficient computation is below.

$$\boldsymbol{0} = \boldsymbol{h}_1 + \boldsymbol{H}_2\boldsymbol{a}^{\dagger}$$

$$\begin{bmatrix} 0 \\ 0 \\ 0 \end{bmatrix} = \begin{bmatrix} 0.2276 \\ 0.1066 \\ 0.0090 \end{bmatrix} + \begin{bmatrix} 0.3068 & 0.2725 & 0.1414 \\ 0.2276 & 0.3068 & 0.2725 \\ 0.1066 & 0.2276 & 0.3068 \end{bmatrix} \begin{bmatrix} a_1 \\ a_2 \\ a_3 \end{bmatrix}$$

The $\boldsymbol{b}$ coefficient computation is below.

$$\boldsymbol{b} = \boldsymbol{H}_1 \boldsymbol{a}$$

$$\begin{bmatrix} b_0 \\ b_1 \\ b_2 \\ b_3 \end{bmatrix} = \begin{bmatrix} 0.0317 & 0 & 0 & 0 \\ 0.1414 & 0.0317 & 0 & 0 \\ 0.2725 & 0.1414 & 0.0317 & 0 \\ 0.3068 & 0.2725 & 0.1414 & 0.0317 \end{bmatrix} \begin{bmatrix} 1 \\ a_1 \\ a_2 \\ a_3 \end{bmatrix}$$

Solve the matrix for $\boldsymbol{a}$ as below.

$$\begin{bmatrix} a_1 \\ a_2 \\ a_3 \end{bmatrix} = \begin{bmatrix} -1.4590 \\ 0.9104 \\ -0.1978 \end{bmatrix} = -\begin{bmatrix} 0.3068 & 0.2725 & 0.1414 \\ 0.2276 & 0.3068 & 0.2725 \\ 0.1066 & 0.2276 & 0.3068 \end{bmatrix}^{-1} \begin{bmatrix} 0.2276 \\ 0.1066 \\ 0.0090 \end{bmatrix}$$

Solve the matrix for $\boldsymbol{b}$ as below.

$$\begin{bmatrix} b_0 \\ b_1 \\ b_2 \\ b_3 \end{bmatrix} = \begin{bmatrix} 0.0317 \\ 0.0951 \\ 0.0951 \\ 0.0317 \end{bmatrix} = \begin{bmatrix} 0.0317 & 0 & 0 & 0 \\ 0.1414 & 0.0317 & 0 & 0 \\ 0.2725 & 0.1414 & 0.0317 & 0 \\ 0.3068 & 0.2725 & 0.1414 & 0.0317 \end{bmatrix} \begin{bmatrix} 1 \\ -1.4590 \\ 0.9104 \\ -0.1978 \end{bmatrix}$$

The final Prony ARMA filter is below.

$$H(z) = \frac{0.0317 + 0.0951z^{-1} + 0.0951z^{-2} + 0.0317z^{-3}}{1 - 1.4590z^{-1} + 0.9104z^{-2} - 0.1978z^{-3}}$$

The corresponding time domain representation is below.

$$\begin{aligned} & y[n] - 1.4590y[n-1] + 0.9104y[n-2] - 01978y[n-3] \\ & = 0.0317x[n] + 0.0951x[n-1] + 0.0951x[n-2] + 0.0317x[n-3] \end{aligned}$$

Figure 6.79 represents the frequency response and impulse of the filter. The passband satisfies the Table 6.34 specification as halfpower frequency requirement. Note that the horizontal line in the figures denote halfpower frequency requirement.

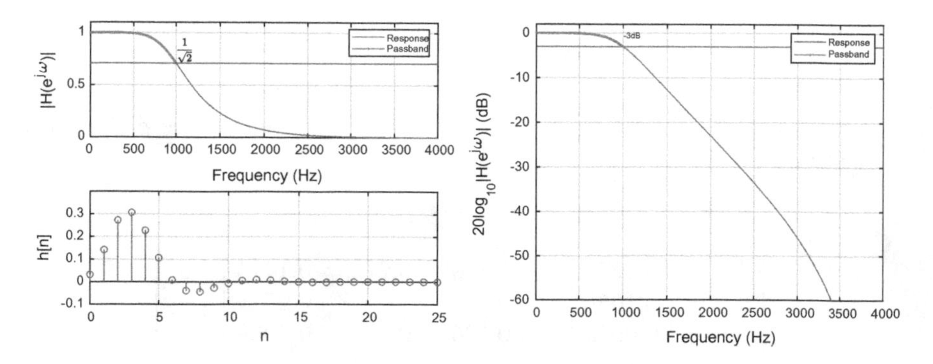

Fig. 6.79 Frequency response and impulse response of designed Prony filter for Table 6.34

Prog. 6.60 MATLAB program for Fig. 6.79.

```
pb = linspace(0,1000,501);
sb = linspace(2000,4000,2001);
a0 = [1 -1.4590 0.9104 -0.1978];
b0 = [0.0317 0.0951 0.0951 0.0317];
[hh0,t0] = impz(b0,a0);
h1 = [hh0(5:7)];
H2 = toeplitz(hh0(4:6),hh0(4:-1:2));
H1 = toeplitz(hh0(1:4),zeros(4,1));
a1 = -inv(H2)*h1;
b1 = H1*[1; a1];
[h3,w3] = freqz(b1,[1; a1],1000,8000);
[hh3,t3] = impz(b1,[1; a1]);
hpb3 = freqz(b1,[1; a1],pb,8000);
hsb3 = freqz(b1,[1; a1],sb,8000);

figure,
subplot(211),plot(w3,abs(h3),pb,abs(hpb3),'.-'), grid, ylim([0 1.1]);
xlabel('Frequency (Hz)');
ylabel('|H(e^{j\omega})|');
line([0 4000],[1/sqrt(2) 1/sqrt(2)]);
legend('Response','Passband');
subplot(212), stem(t3,hh3), grid, xlim([0 25]), ylim([-0.1 0.4]);
xlabel('n');
ylabel('h[n]');
figure,
plot(w3,db(h3),pb, db(hpb3),'.-'), grid, ylim([-60 2]);
line([0 4000],[db(1/sqrt(2)) db(1/sqrt(2))]);
legend('Response','Passband');
ylabel('20log_{10}|H(e^{j\omega})| (dB)');
xlabel('Frequency (Hz)');
```

The impulse response in above figure also exactly follows $h_d[n]$ in Fig. 6.78. This example exercised the Prony ARMA filter with model and order matching condition. The derived result from the Prony method shows the completely identical outcome. Filter model and order mismatch situation could generate the significant performance degradation in frequency response as well as impulse response. Therefore, we need to be aware of matching problem from parametric filter design method.

■

Steiglitz-McBride algorithm

As a part of parametric methods, Steiglitz-McBride algorithm [29] finds an IIR filter coefficients with given desired impulse response. The Steiglitz-McBride iterative method is applied to identify an unknown linear system (or transfer function) by

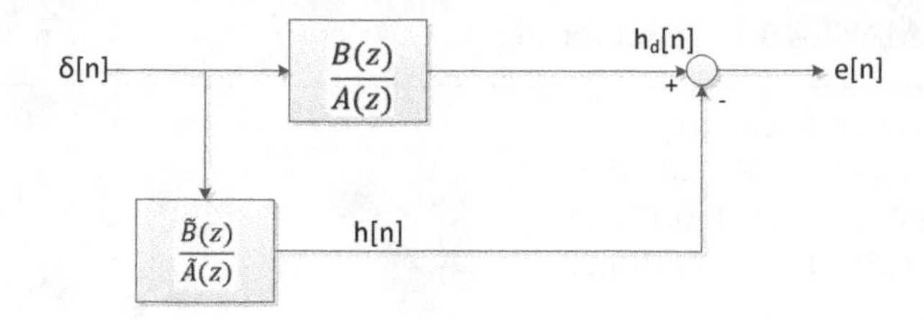

Fig. 6.80 Simplified system model for Stiglitz-McBride algorithm

minimizing the mean square error between the system and model output. Figure 6.80 demonstrates the simplified system model for Stiglitz-McBride algorithm.

The Steiglitz-McBride algorithm figures out the target system $B(z)/A(z)$ by minimizing the energy of the $e[n]$ for equalizing model system $\widetilde{B}(z)/\widetilde{A}(z)$ to the target. The impulse response of the individual system is defined as below.

$$h_d[n] \overset{Z}{\leftrightarrow} \frac{B(z)}{A(z)} \text{ and } h[n] \overset{Z}{\leftrightarrow} \frac{\widetilde{B}(z)}{\widetilde{A}(z)} \tag{6.246}$$

The system error $e[n]$ is represented as below.

$$e[n] = h_d[n] - h[n] \overset{Z}{\leftrightarrow} E(z) = \frac{B(z)}{A(z)} - \frac{\widetilde{B}(z)}{\widetilde{A}(z)} \tag{6.247}$$

The system solution is presented by below equation.

$$\widetilde{\theta} = \underset{\theta}{\operatorname{argmin}} \sum_n |e[n]|^2 \text{where } \theta = \{a_1, \ldots, a_M, b_0, \ldots, b_N\} \text{from} \frac{\widetilde{B}(z)}{\widetilde{A}(z)} \tag{6.248}$$

$$\text{Then } h[n] \approx \widetilde{h}[n] \text{ and } \frac{\widetilde{B}(z)}{\widetilde{A}(z)} \approx \frac{B(z)}{A(z)}.$$

Due to the recursive equation, the solution by Eq. (6.248) indicates the highly nonlinear and intractable property. The system model is modified as below to linearize the solution.

Figure 6.81 shows the iterative method to find the optimal solution for the Fig. 6.80 model. Below equations provide the verification of equivalence.

$$\begin{aligned} \sum_n |e[n]|^2 &= \frac{1}{2\pi j} \oint_c \left| H_d(z) \frac{\widetilde{A}_i(z)}{\widetilde{A}_{i-1}(z)} - \frac{\widetilde{B}_i(z)}{\widetilde{A}_{i-1}(z)} \right|^2 z^{-1} dz \\ &= \frac{1}{2\pi j} \oint_c \left| \frac{B(z)}{A(z)} - \frac{\widetilde{B}_i(z)}{\widetilde{A}_i(z)} \right|^2 \left| \frac{\widetilde{A}_i(z)}{\widetilde{A}_{i-1}(z)} \right|^2 z^{-1} dz \end{aligned} \tag{6.249}$$

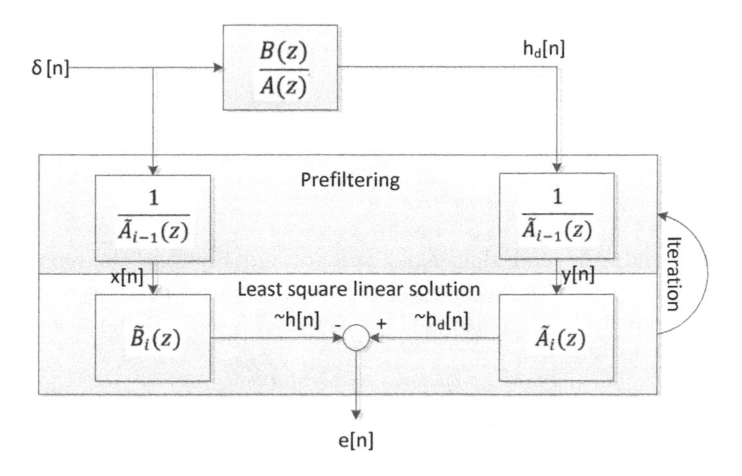

Fig. 6.81 Modified system identification model for Stiglitz-McBride algorithm

$$= \frac{1}{2\pi j}\oint_{c}\left|H_d(z)\widetilde{A}_i(z) - \widetilde{B}_i(z)\right|^2\left|\frac{1}{\widetilde{A}_{i-1}(z)}\right|^2 z^{-1}dz \qquad (6.250)$$

As the error norm approaches to zero, Eq. (6.249) denotes that the target and model system are became identical. Therefore, the optimal numerator and denominator coefficients can be derived by solving Eq. (6.252). The $\widetilde{A}_{i-1}(z)$ and $\widetilde{A}_i(z)$ are related; therefore, the iterative method should be exercised.

$$\widetilde{\theta} = \underset{\theta}{\operatorname{argmin}} \sum_n |e[n]|^2$$

$$= \underset{\theta}{\operatorname{argmin}} \oint_c \left|H_d(z)\widetilde{A}_i(z) - \widetilde{B}_i(z)\right|^2\left|\frac{1}{\widetilde{A}_{i-1}(z)}\right|^2 z^{-1}dz \qquad (6.251)$$

$$= \underset{\theta}{\operatorname{argmin}} \oint_c \left|H_d(z)\widetilde{A}_i(z) - \widetilde{B}_i(z)\right|^2 z^{-1}dz \qquad (6.252)$$

$$= \underset{\theta}{\operatorname{argmin}} \oint_c \left|\frac{B(z)}{A(z)} - \frac{\widetilde{B}_i(z)}{\widetilde{A}_i(z)}\right|^2 z^{-1}dz$$

As shown in Eqs. (6.251) and (6.252), the prefiltering by $1/\widetilde{A}_{i-1}(z)$ converts the equation to the linear problem as $\left|H_d(z)\widetilde{A}_i(z) - \widetilde{A}_i(z)\right|^2$. The prefiltering with delta function and desired response is shown below.

$$x_i[n] = -a_{(i-1)1}x_i[n-1] - a_{(i-1)2}x_i[n-2] - \cdots \\ -a_{(i-1)M}x_i[n-M] + \delta[n] \tag{6.253}$$

$$y_i[n] = -a_{(i-1)1}y_i[n-1] - a_{(i-1)2}y_i[n-2] - \cdots \\ -a_{(i-1)M}y_i[n-M] + h_d[n] \tag{6.254}$$

The left-hand side and right-hand side of the Fig. 6.81 is represented by Eqs. (6.253) and (6.254), respectively. Note that the matrix operation below signifies convolution sum. The matrix size is $L + 1$ rows and $N + 1$ columns (or $M + 1$).

$$\begin{bmatrix} x_i[0] & 0 & 0 & \cdots & 0 \\ x_i[1] & x_i[0] & 0 & \cdots & 0 \\ x_i[2] & x_i[1] & x_i[0] & \cdots & 0 \\ \vdots & \vdots & \vdots & \ddots & \vdots \\ x_i[N] & x_i[N-1] & x_i[N-2] & \cdots & x_i[0] \\ x_i[N+1] & x_i[N] & x_i[N-1] & \cdots & x_i[1] \\ \vdots & \vdots & \vdots & \ddots & \vdots \\ x_i[L] & x_i[L-1] & x_i[L-2] & \cdots & x_i[L-N] \end{bmatrix} \begin{bmatrix} b_{i0} \\ b_{i1} \\ b_{i2} \\ \vdots \\ b_{iN} \end{bmatrix} = \widetilde{h}[n] \tag{6.255}$$

$$\begin{bmatrix} y_i[0] & 0 & 0 & \cdots & 0 \\ y_i[1] & y_i[0] & 0 & \cdots & 0 \\ y_i[2] & y_i[1] & y_i[0] & \cdots & 0 \\ \vdots & \vdots & \vdots & \ddots & \vdots \\ y_i[M] & y_i[M-1] & y_i[M-2] & \cdots & y_i[0] \\ y_i[M+1] & y_i[M] & y_i[M-1] & \cdots & y_i[1] \\ \vdots & \vdots & \vdots & \ddots & \vdots \\ y_i[L] & y_i[L-1] & y_i[L-2] & \cdots & y_i[L-M] \end{bmatrix} \begin{bmatrix} a_{i0} \\ a_{i1} \\ a_{i2} \\ \vdots \\ a_{iM} \end{bmatrix} = \widetilde{h}_d[n] \tag{6.256}$$

With zero error condition, the Eqs. (6.255) and (6.256) should be identical as below.

$$\widetilde{h}[n] = \widetilde{h}_d[n] \text{ for } e[n] = 0 \tag{6.257}$$

Equations (6.255) and (6.256) can be combined for Eq. (6.258) condition as below.

$$\left[\begin{array}{c|cccccccc} -y_i[0] & 0 & \cdots & 0 & x_i[0] & 0 & \cdots & 0 \\ -y_i[1] & -y_i[0] & \cdots & 0 & x_i[1] & x_i[0] & \cdots & 0 \\ \vdots & \vdots & \ddots & \vdots & \vdots & \vdots & \ddots & \vdots \\ -y_i[M] & -y_i[M-1] & \cdots & -y_i[0] & x_i[N] & x_i[N-1] & \cdots & x_i[0] \\ \vdots & \vdots & \ddots & \vdots & \vdots & \vdots & \ddots & \vdots \\ -y_i[L] & -y_i[L-1] & \cdots & -y_i[L-M] & x_i[L] & x_i[L-1] & \cdots & x_i[L-N] \end{array}\right] \times \begin{bmatrix} a_{i0} \\ \hline \vdots \\ a_{iM} \\ b_{i0} \\ \vdots \\ b_{iN} \end{bmatrix} = \begin{bmatrix} 0 \\ 0 \\ 0 \\ 0 \\ \vdots \\ 0 \end{bmatrix} \tag{6.258}$$

The first column of the matrix and first row of the column vector are partitioned as below.

$$\begin{bmatrix} 0 & \cdots & 0 & x_i[0] & 0 & \cdots & 0 \\ -y_i[0] & \cdots & 0 & x_i[1] & x_i[0] & \cdots & 0 \\ \vdots & \ddots & \vdots & \vdots & \vdots & \ddots & \vdots \\ -y_i[M-1] & \cdots & -y_i[0] & x_i[N] & x_i[N-1] & \cdots & x_i[0] \\ \vdots & \ddots & \vdots & \vdots & \vdots & \ddots & \vdots \\ -y_i[L-1] & \cdots & -y_i[L-M] & x_i[L] & x_i[L-1] & \cdots & x_i[L-N] \end{bmatrix} \begin{bmatrix} a_{i1} \\ \vdots \\ a_{iM} \\ b_{i0} \\ \vdots \\ b_{iN} \end{bmatrix} = \begin{bmatrix} y_i[0] \\ y_i[1] \\ \vdots \\ y_i[M] \\ \vdots \\ y_i[L] \end{bmatrix} = \boldsymbol{Hc} = \boldsymbol{y} \tag{6.259}$$

We can solve the Eq. (6.259) with linear solver such as QR solver to find minimum norm residual solution as below.

$$c = \text{Linear Solver}(\boldsymbol{H}, \boldsymbol{y}) \tag{6.260}$$

Repeat the procedures from Eqs. (6.253) and (6.254) for several iterations. The iterative method converges to the optimal solution rapidly. The Prony method [7, 28] provides the good initial guess of $a_{(i-1)}$ coefficients for Eqs. (6.253) and (6.254). The final Steiglitz-McBride filter in discrete time domain is Eq. (6.261).

$$H(z) = \frac{b_0 + \cdots + b_{N-1} z^{-(N-1)} + b_N z^{-N}}{1 + a_1 z^{-1} + \cdots + a_M z^{-M}} \tag{6.261}$$

In summary, the Steiglitz-McBride algorithm tries to compute the least square solution to minimize the error power in time domain based on the initial estimate (usually done by Prony) of the denominator coefficient $\widetilde{A}_1(z)$. The input and output of the unknown system are prefiltered by $1/\widetilde{A}_1(z)$. The filtered signals are used for the least square solution to derive the $\widetilde{A}_2(z)$ and $\widetilde{B}_2(z)$. Repeat the procedures for converged solution.

Similar to the Yule-walker method, the Steiglitz-McBride filter based on the ARMA model requires the designed impulse response which satisfies the specification with unknown coefficients. The Steiglitz-McBride algorithm estimates the coefficients of the unknown ARMA filter with given order. Therefore, we cannot directly design the ARMA Steiglitz-McBride filter from the specification.

Example 6.25

Design the Table 6.35 IIR filter by using Steiglitz-McBride method. By using the Butterworth method, example below provides the ARMA filter first to meet the Table 6.35 requirement and Steiglitz-McBride algorithm follows the ARMA filter to estimate the coefficients.

Solution

In previous example, the Butterworth method presented the following ARMA IIR filter design to meet the Table 6.35 specification.

Table 6.35 Given filter specification

Filter specifications	Value
Filter type	Low pass filter
Filter realization	Steiglitz-McBride IIR
Filter length	$N = 3$ and $M = 3$
Passband frequency (f_p)	1000 Hz; $\omega_p = \pi/4$
Passband magnitude	1
Passband peak-to-peak ripple in decibel (Δ_p)	3.0103 dB
Stopband frequency (f_{st})	2000 Hz; $\omega_{st} = \pi/2$
Stopband attenuation in decibels Δ_{st}	$20\log_{10}\left(\frac{1}{0.1}\right) = 20$ dB
Sampling frequency (f_s)	8000 Hz

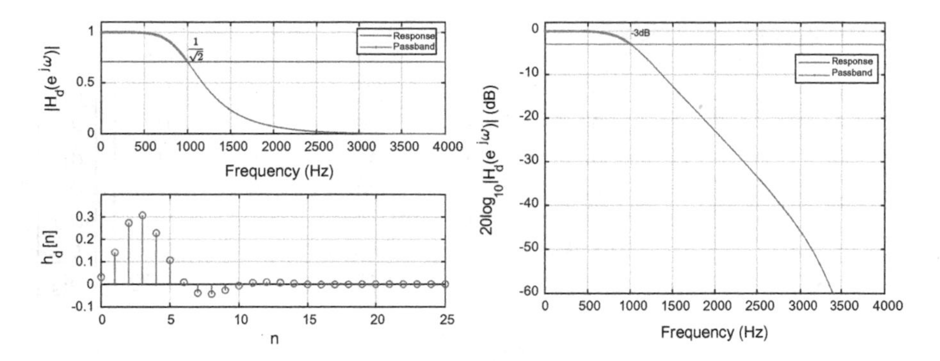

Fig. 6.82 Frequency response and impulse response of designed Butterworth filter for Table 6.35. See Prog. 6.59 for realization

$$H_d(z) = \frac{0.0317 + 0.0951z^{-1} + 0.0951z^{-2} + 0.0317z^{-3}}{1 - 1.4590z^{-1} + 0.9104z^{-2} - 0.1978z^{-3}}$$

The pole zero plot of the above filter is demonstrated at Fig. 6.49 and the frequency response and impulse response are illustrated at Fig. 6.82. The passband satisfies the Table 6.35 specification. Note that the horizontal line in the figures denote the halfpower frequency requirement.

This example employs first seven samples ($L = 6$) of Eq. (6.246) impulse response as below.

$$\begin{aligned} h_d[n] = {} & 0.0317\delta[n] + 0.1414\delta[n-1] + 0.2725\delta[n-2] + 0.3068\delta[n-3] \\ & + 0.2276\delta[n-4] + 0.1066\delta[n-5] + 0.0090\delta[n-6] \end{aligned}$$

Using the Prony AR filter, initial prefilter coefficient is derived as below.

$$a_0 = [1.0000 - 2.3874\ 2.4372 - 1.0675]^T$$

Perform the prefiltering with a_0 as below.

$$x_1[n] = 2.3874x_1[n-1] - 2.4372x_1[n-2] + 1.0675x_1[n-3] + \delta[n]$$

$$y_1[n] = 2.3874y_1[n-1] - 2.4372y_1[n-2] + 1.0675y_1[n-M] + h_d[n]$$

Create Eq. (6.259) matrix as below ($L = 6$).

$$\begin{bmatrix} 0 & 0 & 0 & x_1[0] & 0 & 0 & 0 \\ -y_1[0] & 0 & 0 & x_1[1] & x_1[0] & 0 & 0 \\ -y_1[1] & -y_1[0] & 0 & x_1[2] & x_1[1] & x_1[0] & 0 \\ -y_i[2] & -y_1[1] & -y_1[0] & x_1[3] & x_1[2] & x_1[1] & x_1[0] \\ -y_1[3] & -y_1[2] & -y_1[1] & x_1[4] & x_1[3] & x_1[2] & x_1[1] \\ -y_i[4] & -y_1[3] & -y_1[2] & x_1[5] & x_1[4] & x_1[3] & x_1[2] \\ -y_1[5] & -y_1[4] & -y_1[3] & x_1[6] & x_1[5] & x_1[4] & x_1[3] \end{bmatrix} \begin{bmatrix} a_{11} \\ a_{12} \\ a_{13} \\ b_{10} \\ b_{11} \\ b_{12} \\ b_{13} \end{bmatrix} = \begin{bmatrix} y_1[0] \\ y_1[1] \\ y_1[2] \\ y_1[3] \\ y_1[4] \\ y_1[5] \\ y_1[6] \end{bmatrix}$$

Actual values are below.

$$\begin{bmatrix} 0 & 0 & 0 & 1.0000 & 0 & 0 & 0 \\ -0.0317 & 0 & 0 & 2.3874 & 1.0000 & 0 & 0 \\ -0.2170 & -0.0317 & 0 & 3.2623 & 2.3874 & 1.0000 & 0 \\ -0.7133 & -0.2170 & -0.0317 & 3.0372 & 3.2623 & 2.3874 & 1.0000 \\ -1.5147 & -0.7133 & -0.2170 & 1.8486 & 3.0372 & 3.2623 & 2.3874 \\ -2.3368 & -1.5147 & -0.7133 & 0.4936 & 1.8486 & 3.0372 & 3.2623 \\ -2.7552 & -2.3368 & -1.5147 & -0.0848 & 0.4936 & 1.8486 & 3.0372 \end{bmatrix} \begin{bmatrix} a_{11} \\ a_{12} \\ a_{13} \\ b_{10} \\ b_{11} \\ b_{12} \\ b_{13} \end{bmatrix}$$

$$= \begin{bmatrix} 0.0317 \\ 0.2170 \\ 0.7133 \\ 1.5147 \\ 2.3368 \\ 2.7552 \\ 2.5084 \end{bmatrix} = \boldsymbol{Hc} = \boldsymbol{y}$$

Solve above equation with linear solver to find minimum norm residual solution as below. In this case, the $\boldsymbol{H}$ matrix is square; hence, we can use the matrix inversion.

$$\boldsymbol{c} = \text{Linear Solver}(\boldsymbol{H}, \boldsymbol{y})$$

$$\boldsymbol{c} = \boldsymbol{H}^{-1}\boldsymbol{y}$$

The derived coefficients are presented as below.

$$\boldsymbol{c} = \begin{bmatrix} a_{11} \\ a_{12} \\ a_{13} \\ b_{10} \\ b_{11} \\ b_{12} \\ b_{13} \end{bmatrix} = \begin{bmatrix} -1.4590 \\ 0.9104 \\ -0.1978 \\ 0.0317 \\ 0.0951 \\ 0.0951 \\ 0.0317 \end{bmatrix}$$

Organize the coefficients for denominator and numerator as below. Note that the a_{10} is always one.

$$a_1 = [1.0000 - 1.4590\ 0.9104 - 0.1978]^T$$

$$b_1 = [0.0317\ 0.0951\ 0.0951\ 0.0317]^T$$

If further iterations are required, execute the prefiltering with updated coefficient a_1 and follow the procedures for new a_2 and b_2 coefficient. Perform the iterative procedures until the results converge to the optimal values. In this case, the a_1 and a_2 provide the identical result as $H_d(z)$. The final Steiglitz-McBride ARMA filter is below.

$$H(z) = \frac{0.0317 + 0.0951z^{-1} + 0.0951z^{-2} + 0.0317z^{-3}}{1 - 1.4590z^{-1} + 0.9104z^{-2} - 0.1978z^{-3}}$$

The corresponding time domain representation is below.

$$\begin{aligned} & y[n] - 1.4590y[n-1] + 0.9104y[n-2] - 01978y[n-3] \\ & = 0.0317x[n] + 0.0951x[n-1] + 0.0951x[n-2] + 0.0317x[n-3] \end{aligned}$$

Figure 6.83 represents the frequency response and impulse of the filter. The passband satisfies the Table 6.35 specification as halfpower frequency requirement. Note that the horizontal line in the figures denote halfpower frequency requirement.

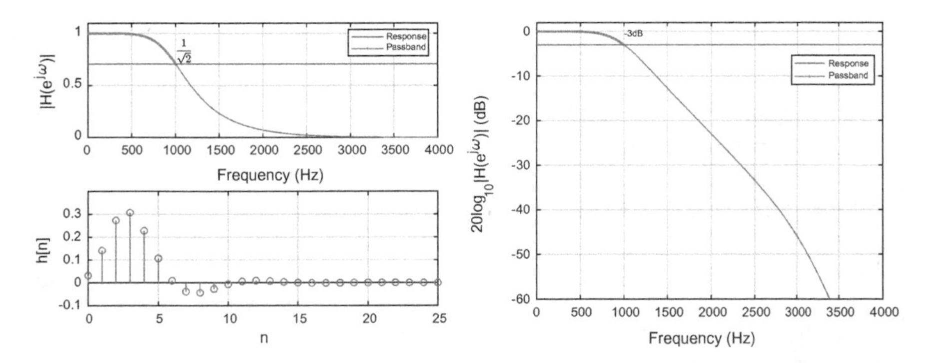

Fig. 6.83 Frequency response and impulse response of designed Steiglitz-McBride method for Table 6.35

Prog. 6.61 MATLAB program for Fig. 6.83.

```
K = 3;, M = 3;, N = 3;, Q = 1;, L = 7;
% K: Given filter order, M: Pole order, N: Zero order, Q:Iteration number
pb = linspace(0,1000,501);
sb = linspace(2000,4000,2001);
a = [1 -1.4590 0.9104 -0.1978];        % Designed Butterworth
b = [0.0317 0.0951 0.0951 0.0317];
[h0,t0] = impz(b,a);                    % Desired impulse response
h = h0(1:L);
[~,a0] = prony(h,0,K);          % Initial A(z) guess
x0 = [1; zeros(L-1,1)];        % Delta input for the system
aa = a0;
for kk = 1:Q
     y = filter(1,aa,h);
     x = filter(1,aa,x0);
     Y1 = toeplitz(y,[y(1); zeros(L-1,1)]);
     X1 = toeplitz(x,[x(1); zeros(L-1,1)]);
     Y2 = Y1(1:L,1:M+1);
     X2 = X1(1:L,1:N+1);
     HH = [-Y2(:,2:end) X2];
     coefs = HH\y;
     aa = [1; coefs(1:M)];
     bb = coefs(M+1:end);
end
[h00,w00] = freqz(bb,aa,1000,8000);
[hh00,t00] = impz(bb,aa);
hpb0 = freqz(bb,aa,pb,8000);
hsb0 = freqz(bb,aa,sb,8000);

figure,
subplot(211),plot(w00,abs(h00),pb,abs(hpb0),'.-'), grid, ylim([0 1.1]);
xlabel('Frequency (Hz)')
line([0 4000],[1/sqrt(2) 1/sqrt(2)]);
legend('Response','Passband')
subplot(212), stem(t00,hh00), grid, xlim([0 25]), ylim([-0.1 0.4]);
xlabel('n')
figure
plot(w00,db(h00),pb, db(hpb0),'.-'), grid, ylim([-60 2])
line([0 4000],[db(1/sqrt(2)) db(1/sqrt(2))]);
legend('Response','Passband')
xlabel('Frequency (Hz)')
```

The impulse response in above figure also exactly follows $h_d[n]$ in Fig. 6.82. This example exercised the Steiglitz-McBride ARMA filter with model and order matching condition. The derived result from the Steiglitz-McBride method shows the completely identical outcome. Filter model and order mismatch situation could generate the significant performance degradation in frequency response as well as impulse response. Therefore, we need to be aware of matching problem from parametric filter design method.

■

Frequency transformation in IIR filter

The previously designed IIR filters such as Butterworth (Maxflat), Type I Chebyshev, Type II Chebyshev, and Elliptic filter are derived for the LPF. Other types of filter cannot be designed from the IIR filter method directly. The frequency transformation changes the filter type from the prototype LPF. The transformation can be performed in s (Laplace transform) or z (z transform) domain. We only consider the z domain transformation in this book. The fundamental transformation form is shown in Eq. (6.262).

$$H_d(z) = H(H_t(z)) \tag{6.262}$$

The desired type transfer function $H_d(z)$ is obtained from the prototype LPF $H(z)$ with replacement transformation function $H_t(z)$ as shown above. The transformation function $H_t(z)$ is the allpass mapping filter which is the extension of allpass filter shown in Eq. (6.114). To determine the transformation function, we have to locate the frequency locations of the selected feature as shown in Fig. 6.84. Halfpower frequencies are chosen for the feature.

Table 6.36 demonstrates the conventional transformation functions with feature frequencies indicated by Fig. 6.84. The z in the prototype filter is replaced with $H_t(z)$ for preferred filter type. The further derivation and extension of the transformation functions can be found on reference [30].

After the transformation, the new type filter shows the below rational polynomial function. The order of numerator and denominator could be increased because of the transformation function order.

$$\begin{aligned} H_d(z) &= \frac{b_0 + b_1 z^{-1} + b_2 z^{-2} + \cdots + b_N z^{-N}}{1 + a_1 z^{-1} + a_2 z^{-2} + \cdots + a_M z^{-M}} \\ &= \frac{b_0\left(1-d_1 e^{j\mu_1} z^{-1}\right)\left(1-d_2 e^{j\mu_2} z^{-1}\right)\ldots\left(1-d_N e^{j\mu_N} z^{-1}\right)}{\left(1-c_1 e^{j\gamma_1} z^{-1}\right)\left(1-c_2 e^{j\gamma_2} z^{-1}\right)\ldots\left(1-c_M e^{j\gamma_M} z^{-1}\right)} \end{aligned} \tag{6.263}$$

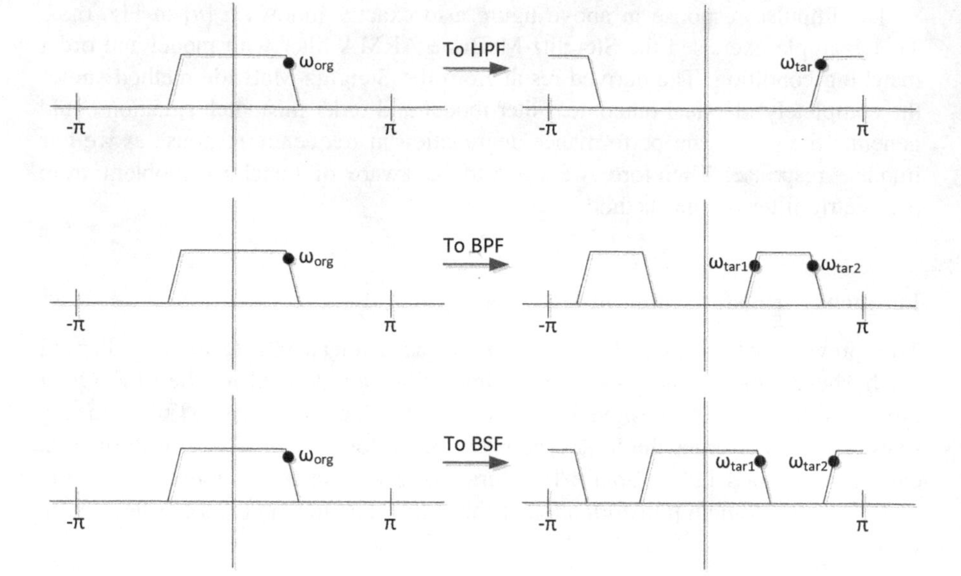

Fig. 6.84 Spectral distribution and feature frequencies of frequency transformation example

Table 6.36 Transformation functions and parameters

Transform property	Transform function $H_t(z)$	Parameters
LPF → HPF	$-\left(\frac{\alpha+z^{-1}}{1+\alpha z^{-1}}\right)$	$\alpha=-\left(\frac{\cos\left(\frac{1}{2}(\omega_{org}+\omega_{tar})\right)}{\cos\left(\frac{1}{2}(\omega_{org}-\omega_{tar})\right)}\right)$
LPF → BPF	$-\left(\frac{\alpha-\beta(1+\alpha)z^{-1}+z^{-2}}{1-\beta(1+\alpha)z^{-1}+\alpha z^{-2}}\right)$	$\omega_{bw}=\lvert\omega_{tar2}-\omega_{tar1}\rvert$ $\alpha=\frac{\sin\left(\frac{1}{2}(\omega_{org}-\omega_{bw})\right)}{\sin\left(\frac{1}{2}(\omega_{org}+\omega_{bw})\right)}$ $\beta=\frac{\cos\left(\frac{1}{2}(\omega_{tar1}+\omega_{tar2})\right)}{\cos\left(\frac{1}{2}\omega_{bw}\right)}$
LPF → BSF	$\left(\frac{\alpha-\beta(1+\alpha)z^{-1}+z^{-2}}{1-\beta(1+\alpha)z^{-1}+\alpha z^{-2}}\right)$	$\omega_{bw}=\lvert\omega_{tar2}-\omega_{tar1}\rvert$ $\alpha=\frac{\cos\left(\frac{1}{2}(\omega_{org}+\omega_{bw})\right)}{\cos\left(\frac{1}{2}(\omega_{org}-\omega_{bw})\right)}$ $\beta=\frac{\cos\left(\frac{1}{2}(\omega_{tar1}+\omega_{tar2})\right)}{\cos\left(\frac{1}{2}\omega_{bw}\right)}$

The transformation could provide the filter instability by placing the poles at unit circle outside. Using the allpass filter property in Eq. (6.122), poles are relocated into the unit circle area. Also, the zeros are moved into the unit circle inside for minimum phase filter as shown in Eq. (6.264).

$$\begin{aligned} H(z) &= \frac{1 - r_1 e^{j\theta_1} z^{-1}}{1 - r_0 e^{j\theta_0} z^{-1}} = \frac{1 - r_1 e^{j\theta_1} z^{-1}}{1 - r_0 e^{j\theta_0} z^{-1}} \frac{z^{-1} - r_0 e^{-j\theta_0}}{z^{-1} - r_0 e^{-j\theta_0}} \frac{z^{-1} - r_1 e^{-j\theta_1}}{z^{-1} - r_1 e^{-j\theta_1}} \\ &= \frac{z^{-1} - r_1 e^{-j\theta_1}}{z^{-1} - r_0 e^{-j\theta_0}} \frac{z^{-1} - r_0 e^{-j\theta_0}}{1 - r_0 e^{j\theta_0} z^{-1}} \frac{1 - r_1 e^{j\theta_1} z^{-1}}{z^{-1} - r_1 e^{-j\theta_1}} = \frac{z^{-1} - r_1 e^{-j\theta_1}}{z^{-1} - r_0 e^{-j\theta_0}} H_{ap1}(z) H_{ap2}(z) \\ \therefore & \left| \frac{1 - r_1 e^{j\theta_1} z^{-1}}{1 - r_0 e^{j\theta_0} z^{-1}} \right| = \left| \frac{z^{-1} - r_1 e^{-j\theta_1}}{z^{-1} - r_0 e^{-j\theta_0}} \right| \text{where} |r_0| > 1 \text{ and} |r_1| > 1 \end{aligned} \tag{6.264}$$

The new rational function is derived as below with gain adjustment parameter w. Note that, even with identical poles and zeros, the filter magnitude presents various distributions with identical shape.

$$\begin{aligned} H_d(z) &= \frac{b_0 + b_1 z^{-1} + b_2 z^{-2} + \cdots + b_N z^{-N}}{1 + a_1 z^{-1} + a_2 z^{-2} + \cdots + a_M z^{-M}} \\ &\xrightarrow{P\&Z \text{ relocations}} \frac{w\left(\beta_0 + \beta_1 z^{-1} + \beta_2 z^{-2} + \cdots + \beta_N z^{-N}\right)}{1 + \alpha_1 z^{-1} + \alpha_2 z^{-2} + \cdots + \alpha_M z^{-M}} = H_{dsta}(z) \end{aligned} \tag{6.265}$$

Once the transformation and pole & zero relocations are performed, the filter gain should be adjusted to obtain identical amplitude at the reference point such as zero frequency. Derive the w parameter based on Eq. (6.266).

$$\left|H_d(e^{j0})\right|^2 = \left|H_{dsta}(e^{j0})\right|^2 \tag{6.266}$$

Example 6.26
In previous example, the Butterworth method presented the following ARMA IIR LPF to meet 1000 Hz halfpower frequency based on the 8000 Hz sampling frequency.

$$H(z) = \frac{0.0317 + 0.0951 z^{-1} + 0.0951 z^{-2} + 0.0317 z^{-3}}{1 - 1.4590 z^{-1} + 0.9104 z^{-2} - 0.1978 z^{-3}}$$

The corresponding pole zero plot is below.
Based on the given LPF, design the following filters.

1. HPF with 3000 Hz halfpower frequency.
2. BPF with passband between 1000 Hz and 3000 Hz as halfpower frequency.
3. BSF with stopband between 1000 Hz and 3000 Hz as halfpower frequency.

Solution

Transformation is performed as below for HPF with 3000 Hz halfpower frequency. The feature frequencies from protype and target filter are shown as well.

$$H_{HPF0}(z) = \left.\frac{0.0317 + 0.0951z_0^{-1} + 0.0951z_0^{-2} + 0.0317z_0^{-3}}{1 - 1.4590z_0^{-1} + 0.9104z_0^{-2} - 0.1978z_0^{-3}}\right|_{z_0=-(z^{-1})}$$

$$\text{where } \omega_{org} = \frac{\pi}{4}, \omega_{tar} = \frac{3\pi}{4}, \text{ and } \alpha = 0$$

The derived HPF is below.

$$H_{HPF0}(z) = \frac{-0.1603 + 0.4808z^{-1} - 0.4808z^{-2} + 0.1603z^{-3}}{1 + 4.6026z^{-1} + 7.3761z^{-2} + 5.0556z^{-3}}$$

The pole zero plot of the above transfer function indicates the unstable filter due to the poles located at unit circle outside as shown in Fig. 6.86. By reflecting the poles into the unit circle, the filter demonstrates the stable property as shown in Fig. 6.86.

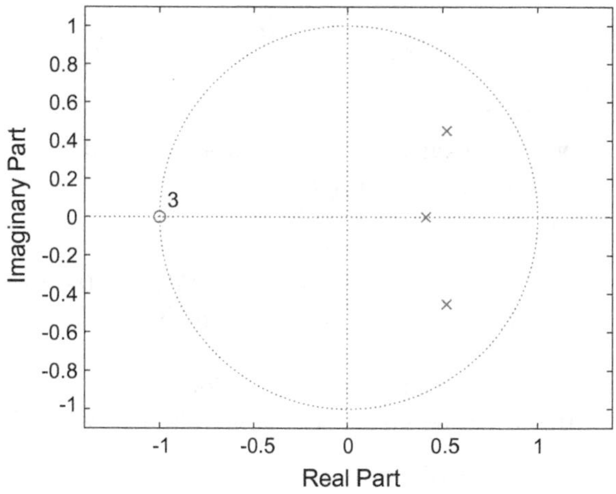

Fig. 6.85 Pole zero plot of Butterworth LPF

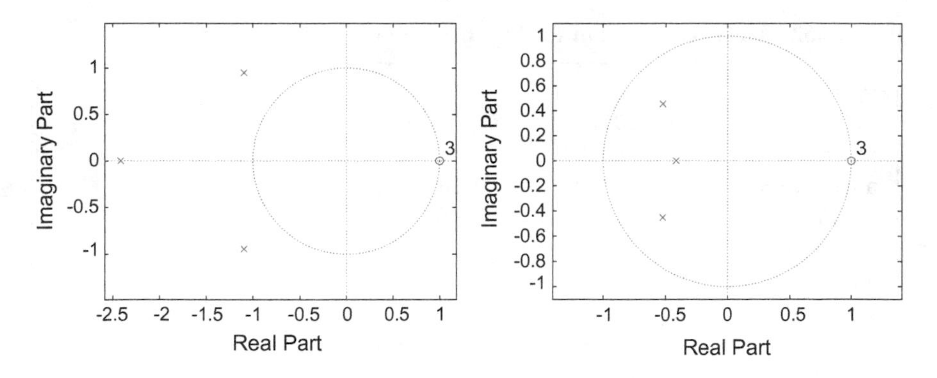

Fig. 6.86 Pole zero plot of HPF from transformation (left) and reflection (right)

The transformed and reflected filter is shown in below.

$$H_{HPF1}(z) = \frac{-0.0317 + 0.0951z^{-1} - 0.0951z^{-2} + 0.0317z^{-3}}{1 + 1.4590z^{-1} + 0.9104z^{-2} + 0.1978z^{-3}}$$

The frequency response of above transfer function is shown in Fig. 6.87.

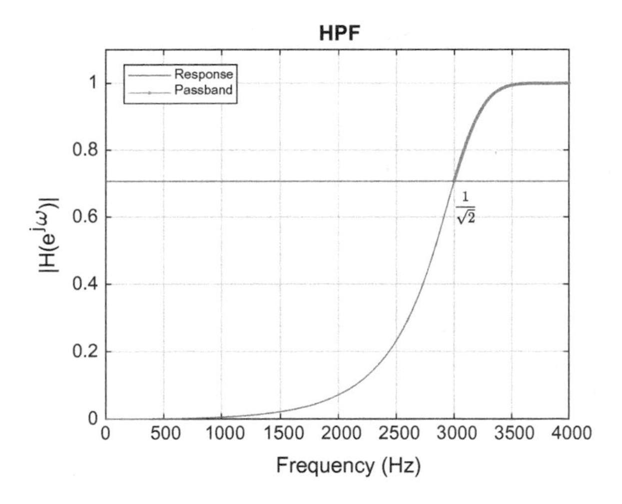

Fig. 6.87 Frequency response of transformed and reflected HPF

Prog. 6.62 MATLAB program for Fig. 6.85 ~ Fig. 6.87.

```
syms s z;
pb0 = linspace(0,1000,501);
pb1 = linspace(3000,4000,501);
a = [1 -1.4590 0.9104 -0.1978];          % Designed Butterworth
b = [0.0317 0.0951 0.0951 0.0317];
figure, zplane(b,a);
worg = sym('pi')/4;
wtar = 3*sym('pi')/4;
alp = -(cos((1/2)*(worg+wtar))/cos((1/2)*(worg-wtar)));
z1 = -(alp+z^-1)/(1+alp*z^-1);
nums = poly2sym(b,s);
dens = poly2sym(a,s);
Hs = nums/dens;
Hz = (subs(Hs,s,z1));
[numz, denz] = numden(Hz);
b0 = sym2poly(numz);
a0 = sym2poly(denz);
figure, zplane(b0,a0);
b1 = polystab(b0)*norm(b0)/norm(polystab(b0));   % Relocate zero for min
phase
a1 = polystab(a0)*norm(a0)/norm(polystab(a0));   % Relocate pole for
stability
figure, zplane(b1,a1);
[h0,w0] = freqz(b,a,1000,8000);
[h1,w1] = freqz(b1,a1,1000,8000);
hpb0 = freqz(b,a,pb0,8000);
hpb1 = freqz(b1,a1,pb1,8000);

figure,
plot(w1,abs(h1),pb1,abs(hpb1),'.-'), grid, ylim([0 1.1])
xlabel('Frequency (Hz)')
ylabel('|H(e^{j\omega})|')
line([0 4000],[1/sqrt(2) 1/sqrt(2)]);
legend('Response','Passband')
title('HPF')
```

Transformation is performed as below for BPF with passband between 1000 Hz and 3000 Hz as halfpower frequency. The feature frequencies from protype and target filter are shown as well.

$$H_{BPF0}(z) = \left. \frac{0.0317 + 0.0951 z_0^{-1} + 0.0951 z_0^{-2} + 0.0317 z_0^{-3}}{1 - 1.4590 z_0^{-1} + 0.9104 z_0^{-2} - 0.1978 z_0^{-3}} \right|_{z_0 = -\left(\frac{1-\sqrt{2}+z^{-2}}{1+(1-\sqrt{2})z^{-2}}\right)}$$

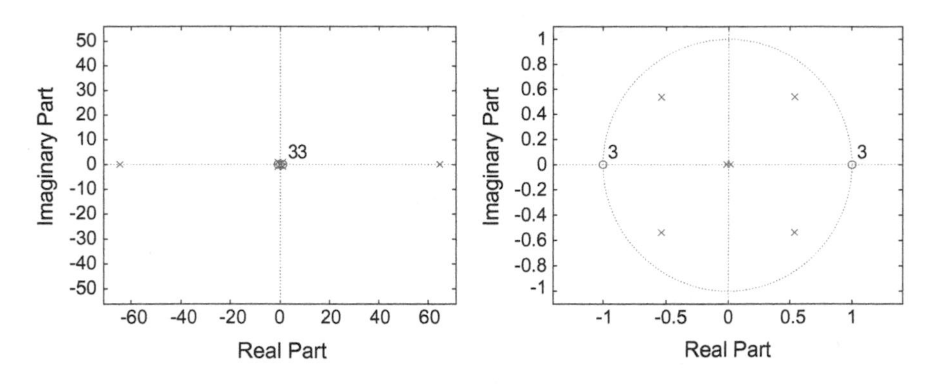

Fig. 6.88 Pole zero plot of BPF from transformation (left) and reflection (right)

$$\text{where } \omega_{org} = \frac{\pi}{4}, \omega_{tar1} = \frac{\pi}{4}, \omega_{tar2} = \frac{3\pi}{4}, \ \alpha = 1 - \sqrt{2}, \text{and } \beta = 0$$

The derived BPF is below.

$$H_{BPF0}(z) = \frac{0.2085 - 0.6253z^{-2} - 0.6253z^{-4} - 0.2085z^{-3}}{1.0000 - 4169.8z^{-2} + 1.8571z^{-4} - 12503z^{-6}}$$

The pole zero plot of the above transfer function indicates the unstable filter due to the poles located at unit circle outside as shown in Fig. 6.88. By reflecting the poles into the unit circle, the filter demonstrates the stable property as shown in Fig. 6.88. The zeros reflections do not change any zero locations since all zeros are on unit circle.

The transformed and reflected filter is shown in below.

$$H_{BPF1}(z) = \frac{\begin{array}{c}0.1667 + 5.3541\times10^{-8}z^{-1} - 0.5002z^{-2} \\ -1.0708\times10^{-7}z^{-3} + 0.5002z^{-4} + 5.3540\times10^{-8}z^{-5} - 0.1667z^{-6}\end{array}}{\begin{array}{c}1.0000 - 1.5543\times10^{-15}z^{-1} - 1.4854\times10^{-4}z^{-2} + 7.2164\times10^{-16}z^{-3} \\ + 3.3351\times10^{-1}z^{-4} + 9.6223\times10^{-19}z^{-5} - 7.9981\times10^{-5}z^{-6}\end{array}}$$

The frequency response of above transfer function is shown in Fig. 6.89.

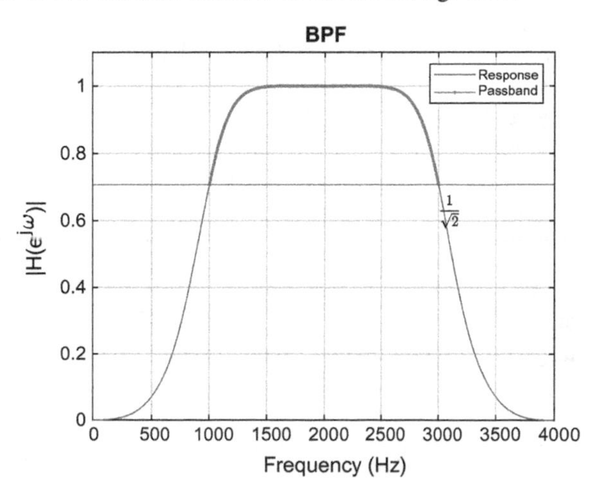

Fig. 6.89 Frequency response of transformed and reflected BPF

Prog. 6.63 MATLAB program for Figs. 6.88 and 6.89.

```
syms s z;
pb0 = linspace(0,1000,501);
pb1 = linspace(1000,3000,501);
a = [1 -1.4590 0.9104 -0.1978];          % Designed Butterworth
b = [0.0317 0.0951 0.0951 0.0317];
worg = sym('pi')/4;
wtar1 = sym('pi')/4;
wtar2 = 3*sym('pi')/4;
wbw = abs(wtar2-wtar1);
alp = (sin((1/2)*(worg-wbw))/sin((1/2)*(worg+wbw)));
bet = cos((1/2)*(wtar1+wtar2))/cos((1/2)*wbw);
z1 = -(alp-bet*(1+alp)*z^-1+z^-2)/(1-bet*(1+alp)*z^-1+alp*z^-2);
nums = poly2sym(b,s);
dens = poly2sym(a,s);
Hs = nums/dens;
Hz = (subs(Hs,s,z1));
[numz, denz] = numden(Hz);
b0 = sym2poly(numz);
a0 = sym2poly(denz);
figure, zplane(b0,a0);
b1 = polystab(b0)*norm(b0)/norm(polystab(b0));    % Relocate zero for min
phase
a1 = polystab(a0)*norm(a0)/norm(polystab(a0));    % Relocate pole for
stability
figure, zplane(b1,a1);
[h0,w0] = freqz(b,a,1000,8000);
[h1,w1] = freqz(b1,a1,1000,8000);
hpb0 = freqz(b,a,pb0,8000);
hpb1 = freqz(b1,a1,pb1,8000);

figure,
plot(w1,abs(h1),pb1,abs(hpb1),'.-'), grid, ylim([0 1.1])
xlabel('Frequency (Hz)')
ylabel('|H(e^{j\omega})|')
line([0 4000],[1/sqrt(2) 1/sqrt(2)]);
legend('Response','Passband')
title('BPF')
```

Transformation is performed as below for BSF with stopband between 1000 Hz and 3000 Hz as halfpower frequency. The feature frequencies from protype and target filter are shown as well.

$$H_{BSF0}(z) = \left. \frac{0.0317 + 0.0951 z_0^{-1} + 0.0951 z_0^{-2} + 0.0317 z_0^{-3}}{1 - 1.4590 z_0^{-1} + 0.9104 z_0^{-2} - 0.1978 z_0^{-3}} \right|_{z_0 = \left(\frac{\sqrt{2} - 1 + z^{-2}}{1 + (\sqrt{2} - 1) z^{-2}} \right)}$$

$$\text{where } \omega_{org} = \frac{\pi}{4}, \omega_{tar1} = \frac{\pi}{4}, \omega_{tar2} = \frac{3\pi}{4}, \; \alpha = \sqrt{2} - 1, \text{and } \beta = 0$$

The derived BSF is below.

$$H_{BSF0}(z) = \frac{2.0845 \times 10^3 + 6.2534 \times 10^3 z^{-2} + 6.2534 \times 10^3 z^{-4} + 2.0845 \times 10^3 z^{-6}}{1.0000 + 4.1698 \times 10^3 z^{-2} + 1.8571 z^{-4} + 1.2503 \times 10^4 z^{-6}}$$

The pole zero plot of the above transfer function indicates the unstable filter due to the poles located at unit circle outside as shown in Fig. 6.90. By reflecting the poles into the unit circle, the filter demonstrates the stable property as shown in Fig. 6.90. The zeros reflections do not change any zero locations since all zeros are on unit circle.

The transformed and reflected filter is shown in below.

$$H_{BSF1}(z) = \frac{\begin{array}{c} 1.6672 \times 10^{-1} + 3.7334 \times 10^{-16} z^{-1} + 5.0015 \times 10^{-1} z^{-2} \\ -2.5183 \times 10^{-16} z^{-3} + 5.0015 \times 10^{-1} z^{-4} - 6.2516 \times 10^{-16} z^{-5} + 1.6671 \times 10^{-1} z^{-6} \end{array}}{\begin{array}{c} 1.0000 + 1.7764 \times 10^{-15} z^{-1} + 1.4854 \times 10^{-4} z^{-2} \\ -7.7716 \times 10^{-16} z^{-3} + 3.3351 \times 10^{-1} z^{-4} - 1.4908 \times 10^{-19} z^{-5} + 7.9981 \times 10^{-5} z^{-6} \end{array}}$$

The frequency response of above transfer function is shown in Fig. 6.91.

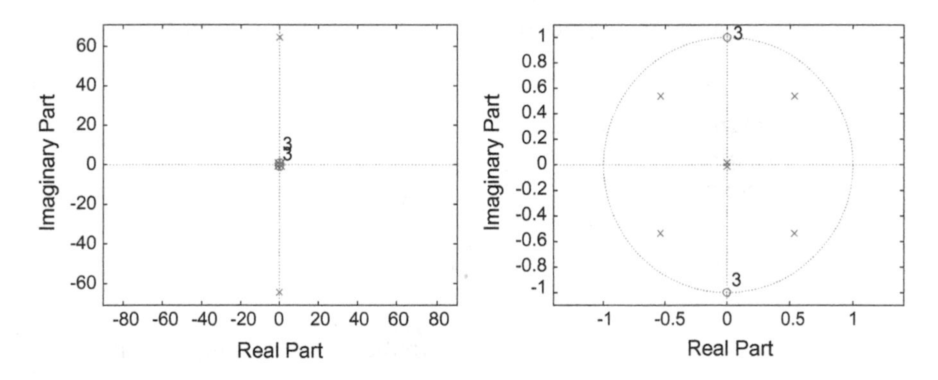

Fig. 6.90 Pole zero plot of BSF from transformation (left) and reflection (right)

Prog. 6.64 MATLAB program for Figs. 6.90 and 6.91.

```
syms s z;
pb0 = linspace(0,1000,501);
pb1 = linspace(0,1000,501);
pb2 = linspace(3000,4000,501);
a = [1 -1.4590 0.9104 -0.1978];          % Designed Butterworth
b = [0.0317 0.0951 0.0951 0.0317];
worg = sym('pi')/4;
wtar1 = sym('pi')/4;
wtar2 = 3*sym('pi')/4;
wbw = abs(wtar2-wtar1);
alp = (cos((1/2)*(worg+wbw))/cos((1/2)*(worg-wbw)));
bet = cos((1/2)*(wtar1+wtar2))/cos((1/2)*wbw);
z1 = (alp-bet*(1+alp)*z^-1+z^-2)/(1-bet*(1+alp)*z^-1+alp*z^-2);
nums = poly2sym(b,s);
dens = poly2sym(a,s);
Hs = nums/dens;
Hz = (subs(Hs,s,z1));
[numz, denz] = numden(Hz);
b0 = sym2poly(numz);
a0 = sym2poly(denz);
figure, zplane(b0,a0);
b1 = polystab(b0)*norm(b0)/norm(polystab(b0));    % Relocate zero for min
phase
a1 = polystab(a0)*norm(a0)/norm(polystab(a0));    % Relocate pole for
stability
figure, zplane(b1,a1);
[h0,w0] = freqz(b,a,1000,8000);
[h1,w1] = freqz(b1,a1,1000,8000);
hpb0 = freqz(b,a,pb0,8000);
hpb1 = freqz(b1,a1,pb1,8000);
hpb2 = freqz(b1,a1,pb2,8000);

figure,
plot(w1,abs(h1),pb1,abs(hpb1),'.-',pb2,abs(hpb2),'.-'), grid, ylim([0 1.1]);
xlabel('Frequency (Hz)');
ylabel('|H(e^{j\omega})|');
line([0 4000],[1/sqrt(2) 1/sqrt(2)]);
legend('Response','Passband','Passband');
title('BSF');
```

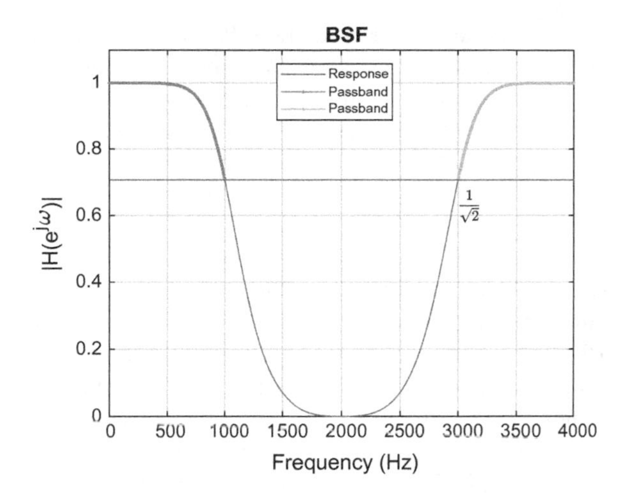

Fig. 6.91 Frequency response of transformed and reflected BSF

Following figures consolidate the prototype and transformed filter responses as summary (Figure 6.92).

∎

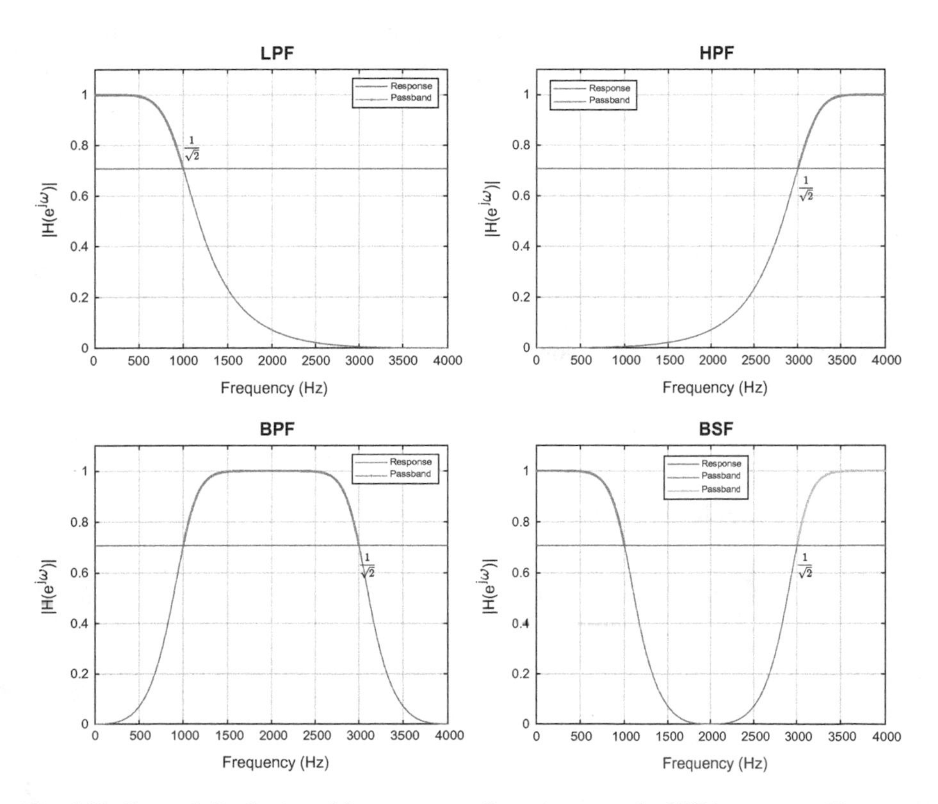

Fig. 6.92 Spectral distribution of frequency transformation example. LPF is prototype filter

Note that the Yule-walker, linear prediction, Prony, and Steiglitz-McBride filter are the parametric methods which follow the impulse response of the unknown filter. Therefore, no direct filter design is feasible from the given specification. However, once the IIR filter is designed from the parametric method, the transformation can be applied for the target filter type as shown in previous examples.

Impulse invariance

The classic s (or $j\Omega$) domain filter design for continuous time is transformed to z domain by using bilinear transformation. The bilinear transformation converts the complete Ω frequency range to the 2π ω frequency range by wrapping the left-side of s plane into the unit circle of z plane. The bilinear transformation is derived from the frequency domain viewpoint. Time domain also provides the method to transform the continuous to discrete time filter. The impulse invariance method [3, 7] samples the designed impulse response of continuous time filter with constant period for discrete time IIR filter. The mathematical representation is below.

$$h_d[n] = T_s h_c(nT_s) \tag{6.267}$$

T_s is sampling period and $h_c(t)$ is the impulse response of continuous time filter. The direct sampling in time domain does not give the rational function for IIR filter. The transfer function in s domain filter is required to derive the IIR filter in z domain. The partial fraction of the Laplace domain transfer function is below.

$$H_c(s) = \sum_{k=1}^{M} \frac{G_k}{s - s_k} \tag{6.268}$$

The time domain impulse response is below.

$$h_c(t) = \sum_{k=1}^{M} G_k e^{s_k t} \tag{6.269}$$

Sampling the response is below.

$$T_s h_c(nT_s) = T_s \sum_{k=1}^{M} G_k e^{s_k n T_s} = T_s \sum_{k=1}^{M} G_k (e^{s_k T_s})^n \tag{6.270}$$

Build the $H(z)$ based on the common ratio $e^{s_k T_s}$ as below.

$$H(z) = \sum_{n=0}^{\infty} \left\{ \sum_{k=1}^{M} T_s G_k (e^{s_k T_s})^n \right\} z^{-n} = \sum_{k=1}^{M} \frac{T_s G_k}{1 - e^{s_k T_s} z^{-1}} \tag{6.271}$$

The Fourier transform of the sampled impulse response is the repeated $H_c(j\Omega)$ in every sampling frequency f_s as below.

$$FT(h_c(nT_s)) = \sum_{k=-\infty}^{\infty} H_c\left(j\Omega + j\frac{2\pi}{T_s}k\right)$$

$$\text{where } H_c(j\Omega) = \int_{-\infty}^{\infty} h_c(t)e^{-j\Omega t}dt \tag{6.272}$$

According to Eq. (2.53), the sampling process provides the spectral duplication in every 2π distance which represents the sampling frequency f_s in cyclic frequency. Figure 6.93 shows the spectral profile of LPF $H_c(j\Omega)$ and sampled time filter in cyclic frequency domain. In general, the continuous time filter cannot be bandlimited; hence, the sampled time filter presents aliasing side effect due to the tails from the consecutive spectral distributions. Beyond the $f_s/2$ frequency, the $|FT(h_c(nT_s))|$ in Fig. 6.93 indicates the overlapping which may causes the performance degradation in IIR filter in discrete time domain.

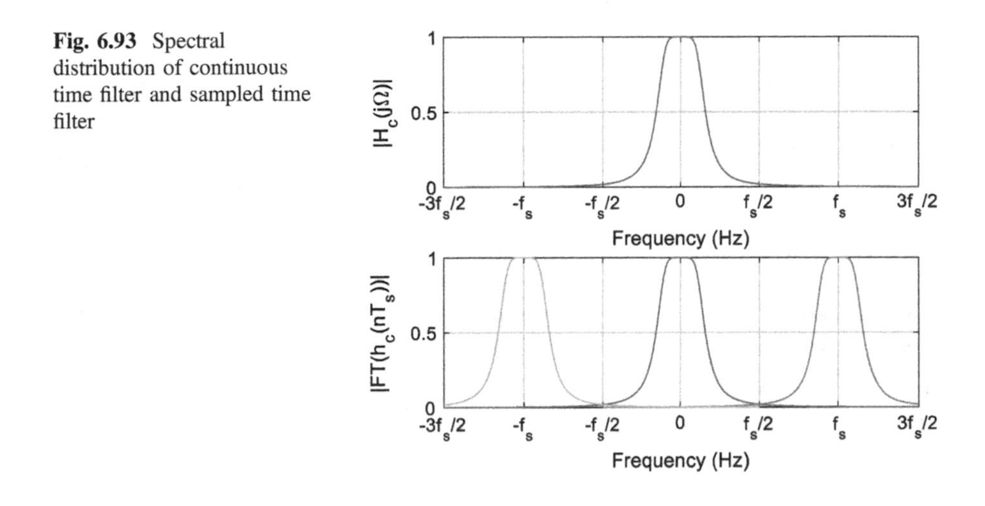

Fig. 6.93 Spectral distribution of continuous time filter and sampled time filter

Prog. 6.65 MATLAB program for Fig. 6.93.

```
p = poly([6627.4*exp(j*2*pi/3) -6627.4 6627.4*exp(-j*2*pi/3)]); %
Butterworth poles
w5 = -12000*2*pi:0.1:12000*2*pi;
h5 = freqs(291091517618.824,p,w5);
ofset = 8000*2*pi;

figure,
subplot(211), plot(w5/(2*pi),abs(h5)),grid
xlim([-12000 12000]);
xticks([-12000 -8000 -4000 0 4000 8000 12000]);
xticklabels({'-3f_s/2','-f_s','-f_s/2','0','f_s/2','f_s','3f_s/2'});
xlabel('Frequency (Hz)');
ylabel('|H_c(j\Omega)|');
subplot(212), plot(w5/(2*pi),abs(h5),(w5+ofset)/(2*pi),abs(h5),(w5-
ofset)/(2*pi),abs(h5)),grid
xlim([-12000 12000]);
xticks([-12000 -8000 -4000 0 4000 8000 12000]);
xticklabels({'-3f_s/2','-f_s','-f_s/2','0','f_s/2','f_s','3f_s/2'});
xlabel('Frequency (Hz)');
ylabel('|FT(h_c(nT_s))|');
```

The sampling frequency f_s (or period T_s) cannot be used to control the aliasing in impulse invariance method since the s domain and z domain poles are mutually related. The pole locations in z domain are $e^{s_k T_s}$ in Eq. (6.271); therefore, the T_s variation should trigger the s_k modification to maintain the desired filter specification. The impulse invariance method is proper to apply for the bandlimited filters only. Otherwise, the designed filter suffers from the significant distortion in frequency response.

Example 6.27

The Butterworth filter (f_p = 1000 Hz, f_{st} = 2000 Hz, and f_s = 8000 Hz) before the bilinear transformation with left s plane poles and the appropriate gain is shown below.

$$H_c(s) = \frac{291091517618.824}{\left(s - 6627.4e^{j\frac{2\pi}{3}}\right)\left(s - 6627.4e^{j\pi}\right)\left(s - 6627.4e^{j\frac{4\pi}{3}}\right)}$$

Compare the impulse invariance with bilinear transformation.

Solution

Derive the partial fraction of the transfer function as below.

$$H_c(s) = \frac{3826.3e^{-j\frac{5\pi}{6}}}{\left(s - 6627.4e^{j\frac{2\pi}{3}}\right)} + \frac{6627.4}{\left(s - 6627.4e^{j\pi}\right)} + \frac{3826.3^{j\frac{5\pi}{6}}}{\left(s - 6627.4e^{j\frac{4\pi}{3}}\right)}$$

The corresponding impulse response $h_c(t)$ is below.

$$\begin{aligned} h_c(t) &= 3826.3e^{-j\frac{5\pi}{6}}e^{6627.4e^{j\frac{2\pi}{3}}t} + 6627.4e^{(-6627.4)t} + 3826.3^{j\frac{5\pi}{6}}e^{6627.4e^{j\frac{4\pi}{3}}t} \\ &= 7652.6e^{(-3313.7)t}\cos\left(5739.5t - \frac{5\pi}{6}\right) + 6627.4e^{(-6627.4)t} \end{aligned}$$

Apply the impulse invariance method on partial fraction $H_c(s)$ as below.

$$\begin{aligned} H_{imp}(z) &= \frac{3826.3e^{-j\frac{5\pi}{6}}}{\left(1 - e^{\frac{6627.4}{8000}e^{j\frac{2\pi}{3}}}z^{-1}\right)} + \frac{6627.4}{\left(1 - e^{-\frac{6627.4}{8000}}z^{-1}\right)} + \frac{3826.3^{j\frac{5\pi}{6}}}{\left(1 - e^{\frac{6627.4}{8000}e^{j\frac{4\pi}{3}}}z^{-1}\right)} \\ &= \frac{-9.09495 \times 10^{-13} + 1256.78z^{-1} + 727.066z^{-2}}{1 - 1.4326z^{-1} + 0.8717z^{-2} - 0.1907z^{-3}} \end{aligned}$$

For reference, the transfer function from bilinear transformation method is below.

$$H_{bil}(z) = \frac{0.0317 + 0.0951z^{-1} + 0.0951z^{-2} + 0.0317z^{-3}}{1 - 1.4590z^{-1} + 0.9104z^{-2} - 0.1978z^{-3}}$$

Prog. 6.66 MATLAB program for Fig. 6.94.

```
Ts = 1/8000;
tt = 0:Ts/100:22/8000;
p = poly([6627.4*exp(j*2*pi/3) -6627.4 6627.4*exp(-j*2*pi/3)]); %
Butterworth poles
[r1,p1,k1] = residue(291091517618.824,p); % Partial fraction
[num,den] = residue(r1,exp(p1*Ts),[]);
[h0,w0] = freqz(num,den,1000,8000);
[h1,t1] = impz(num,den);
h2 = r1.*exp(tt.*p1);
b0 = [0.0317 0.095 0.0951 0.0317];    % Butterworth
a0 = [1 -1.4590 0.9104 -0.1978];
[h4,w4] = freqz(b0,a0,1000,8000);
w5 = -12000*2*pi:0.1:12000*2*pi;
h5 = freqs(291091517618.824,p,w5);
ofset = 8000*2*pi;

figure,
subplot(211), plot(w0,abs(h0)*Ts), grid, hold on
xlabel('Frequency (Hz)')
ylabel('|H(e^{j\omega})|')
plot(w4, abs(h4)), hold off, ylim([0 1.1]);
line([0 4000],[1/sqrt(2) 1/sqrt(2)]);
line([0 4000],[0.1 0.1]);
legend('Impulse Inv.','Bilinear')
subplot(212), stem(t1*Ts,h1*Ts), grid, xlim([0 22/8000]), hold on
ylabel('T_sh_c(t) & h_d[n]')
plot(tt,sum(h2)*Ts), hold off
xlabel('t (second)')
legend('Impulse Inv.','T_sh_c(t)')
figure,
subplot(211), plot(w5/(2*pi),db(abs(h5))),grid, xlim([-12000 12000])
xticks([-12000 -8000 -4000 0 4000 8000 12000]);
xticklabels({'-12000','-8000','-4000','0','4000','8000','12000'});
xlabel('Frequency (Hz)')
ylabel('20log_{10}(|H_c(j\Omega)|)')
subplot(212), plot(w5/(2*pi),db(abs(h5)),(w5+ofset)/(2*pi),db(abs(h5)),(w5-
ofset)/(2*pi),db(abs(h5))),grid, xlim([-12000 12000])
xticks([-12000 -8000 -4000 0 4000 8000 12000]);
xticklabels({'-12000','-8000','-4000','0','4000','8000','12000'});
xlabel('Frequency (Hz)')
ylabel('20log_{10}(|FT(h_c(nT_s))|)')
```

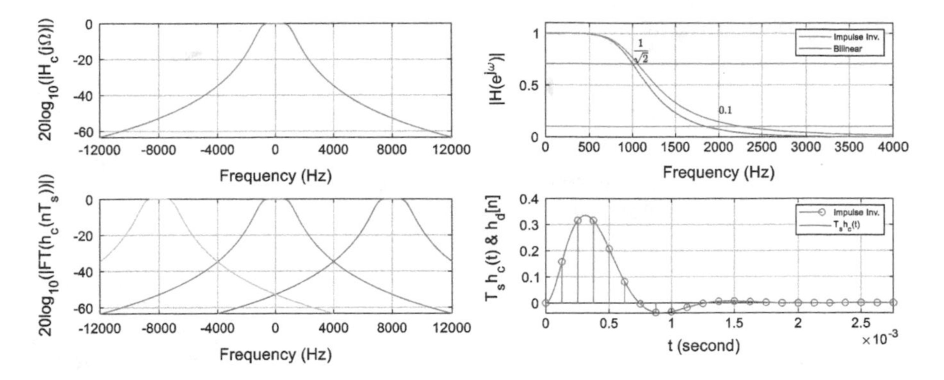

Fig. 6.94 Spectral distribution of continuous time LPF and sampled time LPF on left. Designed LPF frequency and impulse response by impulse invariance and bilinear transformation method on right

Figure 6.94 demonstrates the $|H_c(j\Omega)|$ and $|FT(h_c(nT_s))|$ for designed LPF from Butterworth. Also, the figure shows the frequency response from impulse invariance and bilinear transformation method. The impulse invariance method samples the continuous time filter in every T_s interval on the figure as well. According to the frequency response, the impulse invariance method does not meet the LPF requirement from Table 6.27 (f_p = 1000 Hz, f_{st} = 2000 Hz, and Δ_{st} = 20 dB).

■

Because of the frequency aliasing distorting by neighbor distributions, the impulse invariance presents the higher magnitude in the frequency response. By tightening the LPF requirement, the impulse invariance provides the adequate performance in the design LPF. The impulse invariance method could require the multiple iterations to verify the filter performance. On the other hand, the bilinear transformation reflects the complete left-hand area of s domain into the unit circle area of z domain; hence, one trial of closed-form expression finds the optimal solution in general. Table 6.37 summarizes the IIR filter design methods and corresponding MATLAB functions.

Table 6.37 Discrete-time IIR filter method summary and MATLAB functions

Filter method	Function description	MATLAB functions
Analog prototyping (Based on the classical lowpass prototype filter in *s* domain, obtain a digital filter by filter discretization.)	Complete design functions	**butter, cheby1, cheby2, ellip**
	Lowpass Butterworth filters with more zeros	**maxflat**
	Order estimation functions	**buttord, cheb1ord, cheb2ord, ellipord**
	Frequency transformation functions	**iirlp2hp, iirlp2bp, iirlp2bs**
	Filter discretization functions	**bilinear, impinvar**
Parametric modeling (Derive modelled digital filter which follows a prescribed time domain response.)	Time-domain modeling functions	**aryule(yulewalk), lpc, prony, stmcb**
Minimum phase	Reflect the poles and zeros into unit circle area	**polystab**

6.4 Problems

1. Compute the absolute magnitude of following decibel specifications.
   - Passband magnitude: 0 dB
   - Passband ripple: 0.1 dB
   - Stopband attenuation: 80 dB
2. Calculate the cyclic, discrete, normalized frequency of following situations.
   - $f = 2000$ Hz from $f_s = 10{,}000$ Hz
   - $\omega = \frac{\pi}{4}$ from $f_s = 10{,}000$ Hz
   - $\frac{3}{4}$ normalized frequency from $f_s = 10{,}000$ Hz
3. Design the FIR filter for following specification.

Filter specifications	Value
Filter type	Low pass filter
Filter realization	FIR with hanning window
Filter length	*N* with optimization
Passband frequency (f_p)	2000 Hz; $\omega_p = \pi/2$

(continued)

(continued)

Filter specifications	Value
Passband magnitude	0 dB
Passband ripple (peak to peak)	1 dB
Stopband frequency (f_{st})	3000 Hz; $\omega_{st} = 3\pi/4$
Stopband attenuation	40 dB
Sampling frequency (f_s)	8000 Hz

4. Design the Problem 3 FIR filter with Kaiser window.
5. Design the FIR filter for following specification.

Filter specifications	Value
Filter type	Low pass filter
Filter realization	FIR Type II without window method
Filter length	N with optimization
Passband frequency (f_p)	2000 Hz; $\omega_p = \pi/2$
Passband magnitude	0 dB
Stopband frequency (f_{st})	3000 Hz; $\omega_{st} = 3\pi/4$
Sampling frequency (f_s)	8000 Hz

6. Design the FIR filter for following specification.

Filter specifications	Value
Filter type	Band pass filter
Filter realization	FIR Type III without window method
Filter length	$N = 16$
Cutoff frequency 1 (f_{c1})	4000/3 Hz; $\omega_{c1} = \pi/3$
Passband magnitude	0 dB
Cutoff frequency 2 (f_{c2})	8000/3 Hz; $\omega_{c2} = 2\pi/3$
Sampling frequency (f_s)	8000 Hz

7. Design the FIR filter for following specification.

Filter specifications	Value
Filter type	High pass filter
Filter realization	FIR Type IV without window method
Filter length	*N* with optimization
Passband frequency (f_p)	3000 Hz; $\omega_p = 3\pi/4$
Passband magnitude	0 dB
Stopband frequency (f_{st})	2000 Hz; $\omega_{st} = \pi/2$
Sampling frequency (f_s)	8000 Hz

8. Design the following FIR filter by using the Remes exchange algorithm.

Filter specifications	Value
Filter type	Low pass filter
Filter realization	FIR Type I without windowing method
Filter length	7 ($N = 6$)
Passband frequency (f_p)	2000 Hz; $\omega_p = \pi/2$
Passband magnitude	0 dB
Stopband frequency (f_{st})	3000 Hz; $\omega_{st} = 3\pi/4$
Sampling frequency (f_s)	8000 Hz

9. Design the FIR filter by using the frequency sampling method based on the Problem 8 specification.
10. Design the FIR filter by using the least squared error frequency domain method based on the Problem 8 specification.
11. Design the following FIR filter by using the constrained least square method. MATLAB realization only.

Filter specifications	Value
Filter type	Low pass filter
Filter realization	FIR Type I without window method
Filter length	Optimal value from window
Passband frequency (f_p)	2000 Hz; $\omega_p = \pi/2$
Passband magnitude	0 dB
Passband ripple (peak to peak)	1 dB

(continued)

(continued)

Filter specifications	Value
Stopband frequency (f_{st})	3000 Hz; $\omega_{st} = 3\pi/4$
Stopband attenuation	40 dB
Sampling frequency (f_s)	8000 Hz

12. Design the following FIR filter by using the complex and nonlinear-phase equiripple method. MATLAB realization only.

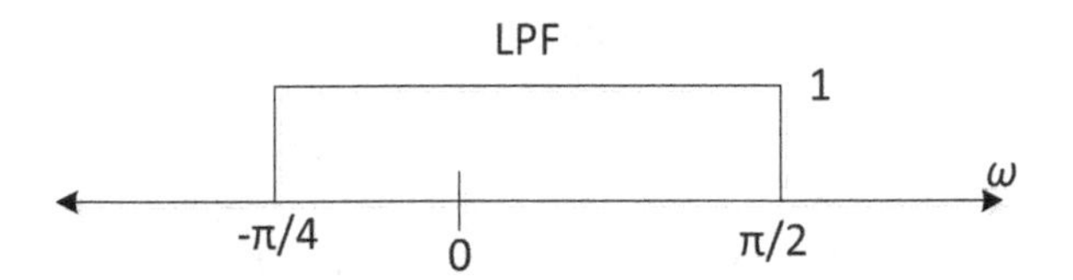

13. For the given filter below, design the minimum phase FIR filter.

$$h[n] = \delta[n] + 4\delta[n-4]$$

14. Design the following IIR filter by using Butterworth method and bilinear transformation.

Filter specifications	Value
Filter type	Low pass filter
Filter realization	Butterworth IIR
Filter length	N with optimization
Passband frequency (f_p)	2000 Hz; $\omega_p = \pi/2$
Passband magnitude	1
Passband peak-to-peak ripple in decibel (Δ_p)	3.0103 dB
Stopband frequency (f_{st})	3000 Hz; $\omega_{st} = 3\pi/4$
Stopband attenuation in decibels Δ_{st}	$20\log_{10}\left(\frac{1}{0.1}\right) = 20$ dB
Sampling frequency (f_s)	8000 Hz

15. Design the following IIR filter by using Type I Chebyshev method and bilinear transformation.

Filter specifications	Value
Filter type	Low pass filter
Filter realization	Type I Chebyshev IIR filter
Filter length	N with optimization
Passband frequency (f_p)	2000 Hz; $\omega_p = \pi/2$
Passband magnitude	1
Passband peak-to-peak ripple in decibel (Δ_p)	$20\log_{10}\left(\frac{1}{1/\sqrt{1+1}}\right) = 3.0103$ dB
Stopband frequency (f_{st})	3000 Hz; $\omega_{st} = 3\pi/4$
Stopband attenuation in decibels Δ_{st}	$20\log_{10}\left(\frac{1}{0.1}\right) = 20$ dB
Sampling frequency (f_s)	8000 Hz

16. Design the following IIR filter by using Type II Chebyshev method and bilinear transformation.

Filter specifications	Value
Filter type	Low pass filter
Filter realization	Type II Chebyshev IIR filter
Filter length	N with optimization
Passband frequency (f_p)	2000 Hz; $\omega_p = \pi/2$
Passband magnitude	1
Passband peak-to-peak ripple in decibel (Δ_p)	$20\log_{10}\left(\frac{1}{1/\sqrt{1+1}}\right) = 3.0103$ dB
Stopband frequency (f_{st})	3000 Hz; $\omega_{st} = 3\pi/4$
Stopband attenuation in decibels Δ_{st}	$20\log_{10}\left(\frac{1}{0.1}\right) = 20$ dB
Sampling frequency (f_s)	8000 Hz

17. Design the following IIR filter by using elliptic method and bilinear transformation.

Filter specifications	Value
Filter type	Low pass filter
Filter realization	Elliptic IIR filter
Filter length	N with optimization
Passband frequency (f_p)	2000 Hz; $\omega_p = \pi/2$
Passband magnitude	1
Passband peak-to-peak ripple in decibel (Δ_p)	$20\log_{10}\left(\frac{1}{1/\sqrt{1+1}}\right) = 3.0103$ dB

(continued)

(continued)

Filter specifications	Value
Stopband frequency (f_{st})	3000 Hz; $\omega_{st} = 3\pi/4$
Stopband attenuation in decibels Δ_{st}	$20\log_{10}\left(\frac{1}{0.1}\right) = 20$ dB
Sampling frequency (f_s)	8000 Hz

18. Design the AR filter based on the Yule-walker algorithm which estimates the coefficients of the unknown AR filter $H_d(z)$ with given 2nd order.

$$H_d(z) = \frac{0.25}{1 - 1.2z^{-1} + 0.45}$$

19. Design the following IIR filter by using maxflat method. MATLAB realization only with maxflat function.

Filter specifications	Value
Filter type	Low pass filter
Filter realization	Maxflat IIR
Filter length	Numerator order = 9 Denominator order = 3
Halfpower frequency (f_h)	2000 Hz; $\omega_h = \pi/2$
Passband magnitude	1
Sampling frequency (f_s)	8000 Hz

20. Design the ARMA filter based on the Prony algorithm which estimates the coefficients of the unknown ARMA filter $H_d(z)$ with given 2nd order.

$$H_d(z) = \frac{0.10100 + 0.06172z^{-1} + 0.09659z^{-2}}{1 - 1.26174z^{-1} + 0.62847z^{-2}}$$

21. Design the ARMA filter based on the Steiglitz-McBride method which estimates the coefficients of the unknown ARMA filter $H_d(z)$ with given 2nd order.

$$H_d(z) = \frac{0.10100 + 0.06172z^{-1} + 0.09659z^{-2}}{1 - 1.26174z^{-1} + 0.62847z^{-2}}$$

22. The Elliptic method presented the following ARMA IIR LPF to meet 1000 Hz halfpower frequency based on the 8000 Hz sampling frequency.

$$H(z) = \frac{0.10100 + 0.06172z^{-1} + 0.09659z^{-2}}{1 - 1.26174z^{-1} + 0.62847z^{-2}}$$

Based on the given LPF, design the following filters by using the transformation functions

- HPF with 3000 Hz halfpower frequency.
- BPF with passband between 1000 Hz and 3000 Hz as halfpower frequency.
- BSF with stopband between 1000 Hz and 3000 Hz as halfpower frequency.

23. The Type I Chebyshev IIR filter (f_p = 1000 Hz, f_{st} = 2000 Hz, and f_s = 8000 Hz) before the bilinear transformation with left s plane poles and the appropriate gain is shown below.

$$H_c(s) = \frac{21960983.69}{s^2 + 4265.4s + 31057520.98}$$

Compare the impulse invariance with bilinear transformation.

References

1. Hewitt, E., Hewitt, R.E.: The Gibbs-Wilbraham phenomenon: An episode in fourier analysis. Arch. Hist. Exact Sci. **21**(2), 129–160 (1979). https://doi.org/10.1007/BF00330404
2. Hazewinkel, M.: Encyclopaedia of Mathematics: Orbit—Rayleigh Equation. Springer, Netherlands (2012)
3. Oppenheim, A.V., Schafer, R.W.: Discrete-Time Signal Processing. Prentice Hall, (1989)
4. Kaise, J.F.: Nonrecursive digital filter design using the I_0-Sinh window function. In: IEEE International Symposium on Circuits and Systems, San Francisco, California, USA, 22–25 April 1974. IEEE
5. Marple, S.L.: Computing the discrete-time 'analytic' signal via FFT. In: Conference Record of the Thirty-First Asilomar Conference on Signals, Systems and Computers (Cat. No.97CB36136), 2–5 November 1997, vol.1322, pp. 1322–1325 (1997)
6. Rabiner, L.R., McClellan, J.H., Parks, T.W.: FIR digital filter design techniques using weighted Chebyshev approximation. Proc. IEEE **63**(4), 595–610 (1975). https://doi.org/10.1109/PROC.1975.9794
7. Parks, T.W., Burrus, C.S.: Digital Filter Design. Wiley, (1987)
8. Jackson, L.B.: Digital Filters and Signal Processing, 3rd edn. Kluwer Academic Publishers, Boston (1996)
9. Wikipedia: Unitary matrix (2020). https://en.wikipedia.org/w/index.php?title=Unitary_matrix&oldid=961644790
10. Wikipedia: Conjugate transpose (2020). https://en.wikipedia.org/w/index.php?title=Conjugate_transpose&oldid=961399224
11. Wikipedia: Vandermonde matrix (2020). https://en.wikipedia.org/w/index.php?title=Vandermonde_matrix&oldid=965588394

12. Wikipedia: Moore–Penrose inverse (2020). https://en.wikipedia.org/w/index.php?title=Moore%E2%80%93Penrose_inverse&oldid=961651512
13. Selesnick, I.W., Lang, M., Burrus, C.S.: Constrained least square design of FIR filters without specified transition bands. IEEE Trans. Signal Process. **44**(8), 1879–1892 (1996). https://doi.org/10.1109/78.533710
14. Wikipedia: Lagrange multiplier (2020). https://en.wikipedia.org/w/index.php?title=Lagrange_multiplier&oldid=964661260
15. Wikipedia: Karush–Kuhn–Tucker conditions (2020). https://en.wikipedia.org/w/index.php?title=Karush%E2%80%93Kuhn%E2%80%93Tucker_conditions&oldid=966216684
16. Karam, L.J., McClellan, J.H.: Complex Chebyshev approximation for FIR filter design. IEEE Trans. Circuits Syst. II: Analog. Digit. Signal Process. **42**(3), 207–216 (1995). https://doi.org/10.1109/82.372870
17. Butterworth, S.: On the theory of filter amplifiers. Exp. Wirel. Wirel. Eng. **7**, 536–541 (1930)
18. Wikipedia: Chebyshev polynomials. https://en.wikipedia.org/w/index.php?title=Chebyshev_polynomials&oldid=962339092 (2020)
19. Wikipedia: Elliptic rational functions (2020). https://en.wikipedia.org/w/index.php?title=Elliptic_rational_functions&oldid=946797692
20. Stoica, P., Moses, R.L.: Introduction to Spectral Analysis. Prentice Hall (1997)
21. Yule, G.U.: On a method of investigating periodicities in disturbed series, with special reference to Wolfer's sunspot numbers. Philos. Trans. R. Soc. Lond. Ser. A (Containing Papers of a Mathematical or Physical Character) **226**, 267–298 (1927)
22. Walker, G.T.: On periodicity in series of related terms. Proc. R. Soc. Lond. Ser. A (Containing Papers of a Mathematical and Physical Character) **131**(818), 518–532 (1931). https://doi.org/10.1098/rspa.1931.0069
23. Selesnick, I.W., Burrus, C.S.: Generalized digital Butterworth filter design. In: 1996 IEEE International Conference on Acoustics, Speech, and Signal Processing Conference Proceedings, 9–9 May 1996, vol. 1363, pp. 1367–1370 (1996)
24. Herrmann, O.: On the approximation problem in nonrecursive digital filter design. IEEE Trans. Circuit Theory **18**(3), 411–413 (1971). https://doi.org/10.1109/TCT.1971.1083275
25. Makhoul, J.: Linear prediction: a tutorial review. Proc. IEEE **63**(4), 561–580 (1975). https://doi.org/10.1109/PROC.1975.9792
26. Papoulis, A.: Probability, Random Variables, and Stochastic Processes. McGraw-Hill (1991)
27. Kay, S.M.: Fundamentals of Statistical Signal Processing. Estimation Theory, vol. 1. Prentice Hall (1993)
28. Hauer, J.F., Demeure, C.J., Scharf, L.L.: Initial results in Prony analysis of power system response signals. IEEE Trans. Power Syst. **5**(1), 80–89 (1990). https://doi.org/10.1109/59.49090
29. Steiglitz, K., McBride, L.: A technique for the identification of linear systems. IEEE Trans. Autom. Control. **10**(4), 461–464 (1965). https://doi.org/10.1109/TAC.1965.1098181
30. Constantinides, A.G.: Spectral transformations for digital filters. Proc. Inst. Electr. Eng. **117**(8), 1585–1590 (1970). https://doi.org/10.1049/piee.1970.0281
31. Yeong Ho, H., Pearce, J.A.: A new window and comparison to standard windows. IEEE Trans. Acoust. Speech Signal Process. **37**(2), 298–301 (1989). https://doi.org/10.1109/29.21693
32. Harris, F.J.: On the use of windows for harmonic analysis with the discrete Fourier transform. Proc. IEEE **66**(1), 51–83 (1978). https://doi.org/10.1109/PROC.1978.10837
33. Elliott, D.F.: Handbook of Digital Signal Processing. Academic, New York (1987)
34. D'Antona, G., Ferrero, A.: Digital Signal Processing for Measurement Systems: Theory and Applications. Springer, US (2006)
35. Nuttall, A.: Some windows with very good sidelobe behavior. IEEE Trans. Acoust. Speech Signal Process. **29**(1), 84–91 (1981). https://doi.org/10.1109/TASSP.1981.1163506
36. Brookner, E.: Practical Phased-array Antenna Systems. Artech House (1991)

Chapter 7
Implementation Matters

The previously explained filter design provided the proper type, method, and order of the filter to meet the desired specification. Once we derived the transfer function as Eq. (7.1), everything seems to be finalized.

$$H(z) = \frac{b_0 + \cdots + b_{N-1}z^{-(N-1)} + b_N z^{-N}}{1 + a_1 z^{-1} + \cdots + a_M z^{-M}} \tag{7.1}$$

From the transfer function above, we find the difference equation for time domain realization. The filter output $y[n]$ is computed by the combination of weighted inputs and outputs as shown in Eq. (7.2). The computation is performed by the digital processors based on the binary number system.

$$y[n] = b_0 x[n] + \cdots + b_M x[n-M] - a_1 y[n-1] - \cdots a_N y[n-N] \tag{7.2}$$

Unfortunately, the computation is impossible to deliver due to the infinite precision of the real numbers. Note that the inputs, outputs, and coefficients are real numbers. The time in the filter is discretized by the sampling process. In the same manner, the inputs, outputs, and coefficients are required to be discretized for storing and processing the values within the digital processor. The discretization process for the magnitude is named as the quantization. Because of the limited precision, the quantization process loses the information; therefore, we expect the performance degradation of the filter. The various filter realization methods demonstrate the distinctive sensitivity to the quantization error. The random nature of the quantization error involves the statistical analysis which will be introduced in rudimentary level.

K. Kim, *Conceptual Digital Signal Processing with MATLAB*,
Signals and Communication Technology,
https://doi.org/10.1007/978-981-15-2584-1_7

7.1 Number Representations

– Binary numbers

In conventional computers and processors, the binary number is used for digital arithmetic. The information is represented by the simple high (1) or low (0) of the electrical voltage; therefore, the individual digit of the binary number contains the one or zero only. According to the translation, the binary number can be converted to the various conventional number systems. Frequently, the integer number is utilized for the corresponding binary number. Equation (7.3) shows the $B + 1$ bit binary number. Note that the digit in the binary number is known as a bit.

$$b_B b_{B-1} \ldots b_2 b_1 b_0 \tag{7.3}$$

The leftmost b_B and rightmost b_0 bit denotes the most significant bit (MSB) and least significant bit (LSB), respectively. For integer conversion, we assume that the binary point is placed after LSB. The weighted sum from sequential power of two presents the unsigned integer value as below.

$$\text{Unsigned integer} = \sum_{k=0}^{B} b_k 2^k \tag{7.4}$$

2's complement method is employed for the signed integer from binary number. The MSB signifies the negative number with highest weight as shown below. The sum of all lower bits below the MSB is less than the single MSB weight; hence, the MSB serves as the sign bit.

$$\text{Signed integer} = -b_B 2^B + \sum_{k=0}^{B-1} b_k 2^k \tag{7.5}$$

The unsigned and signed inter range are given as below. Observe that overall integer range is identical for unsigned and signed representation with different coverage.

$$\begin{gathered} 0 \leq \text{Unsigned integer} \leq 2^{B+1} - 1 \\ -2^B \leq \text{Signed integer} \leq 2^B - 1 \end{gathered} \tag{7.6}$$

Because of the limited size arithmetic in binary number, the overflow contaminates the computation result significantly. The binary addition and subtraction do not break the arithmetic rules in the binary system. However, the carry-out from or to the MSB ruins the decimal representation. Figure 7.1 is the binary arithmetic diagram for 3-bit size. Note that unsigned integer is ranged from 0 to 7 and signed integer is covered from −4 to 3.

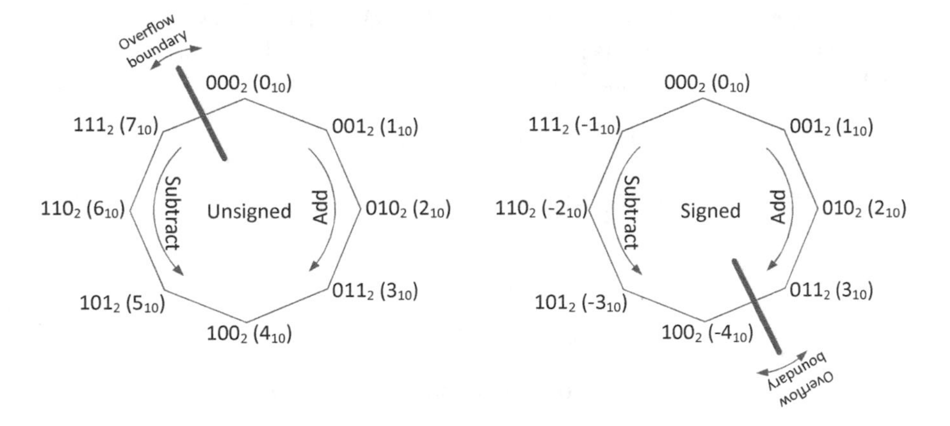

Fig. 7.1 Binary number arithmetic for 3-bit size and 2's complement

The circular property of the binary number generates the abrupt representation in decimal number. The impact of overflow is significant in signal description; therefore, the overflow should be avoided or minimized within the possible limits.

– Floating point numbers

The real number can be represented by the floating-point number which uses the power of the base with signed significand. The exponent and signed significand will be represented by the binary number. The power base is implicitly specified in the floating-point number system. Figure 7.2 is the example of floating-point number with 10 base.

The floating-point number increases the dynamic range of the signal representation by storing the significand and exponent. The digital signal processing is performed by the conventional computer or special processor; therefore, the signal should be denoted by the binary number for execution. By converting the significand and exponent to the binary number, we can use the very large or small magnitude numbers in the system. Equation (7.7) is the small-scale example for floating-point number by 8 bits.

$$\begin{gathered}\underbrace{b_7}_{s}\ \underbrace{b_6b_5b_4b_3}_{e}\ \underbrace{b_2b_1b_0}_{f} \\ \text{Floating point number}_{10}(\text{FP}\#10) = (-1)^s \times (1.f \times 2^{e-7})_{10}\end{gathered} \tag{7.7}$$

The f mantissa and e exponent are represented by the unsigned binary number. Note that the subscript 10 presents the decimal conversion. The biased exponent

Fig. 7.2 Floating-point number example

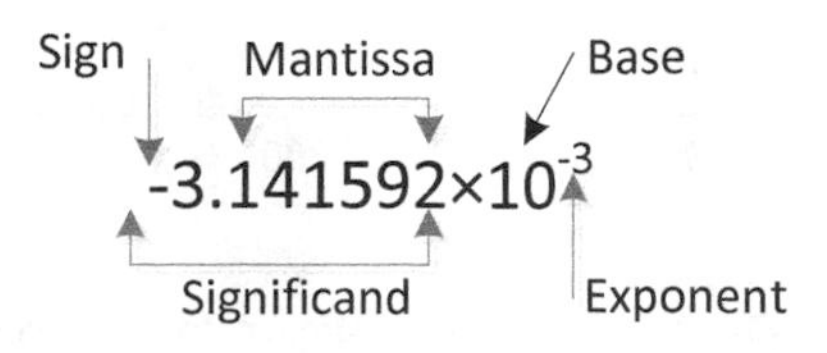

shifts the numeric range to cover the signed number. To demonstrate the proper significand, the binary point is located at the left of f MSB. Also, we assume that the implicit 1 is comprised at the binary point left as shown below.

$$1.b_2b_1b_0 \rightarrow 1 + b_2 2^{-1} + b_1 2^{-2} + b_0 2^{-3} \tag{7.8}$$

Below is the range of significand and exponent based on the above configuration.

$$\begin{aligned} 1 \leq |1.f| \leq 1.8750 \\ -7 \leq e - 7 \leq 8 \end{aligned} \tag{7.9}$$

Therefore, the overall range of floating-point number absolute is below.

$$1 \times 2^{-7} \approx 0.078 \leq |FP\#10| \leq 1.8750 \times 2^8 = 480 \tag{7.10}$$

From the given 8-bit floating-point number, above range is divided into the 128 positive numbers and the corresponding identical negative numbers. Figure 7.3 depicts the 128 positive numbers for the floating-point number. Observe that the vertical grids show the floating-point number positions.

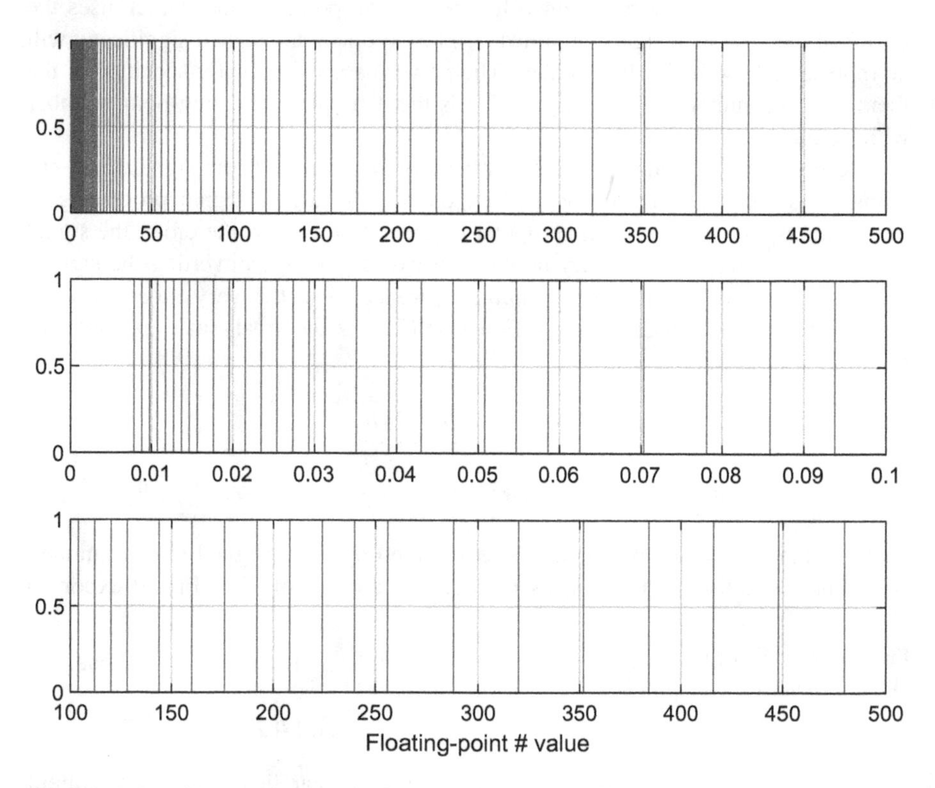

Fig. 7.3 128 positive numbers for the floating-point number with 8-bit configuration on Eq. (7.7)

Prog. 7.1 MATLAB program for Fig. 7.3.

```
ff = 0:1:7;
ee = -7:1:8;
fract = 1+ff*2^-3;
out = [];
for kk=1:length(ee)
    out = [out fract.*2^ee(kk)];
end
figure
subplot(311), h = stem(out,ones(size(out)));, grid
h.Marker = 'none';
subplot(312), h = stem(out,ones(size(out)));, grid
h.Marker = 'none';
xlim([0 0.1])
subplot(313), h = stem(out,ones(size(out)));, grid
h.Marker = 'none';
xlim([100 500]);
xlabel('Floating-point # value');
```

Due to the exponent, the floating-point number shows the non-constant resolution between adjacent numbers. The overall plot demonstrates the dense population nearby the zero and coarse distribution around high. The other subplots magnify the low and high zone to verify the observation.

The problems of the above configuration are zero and special numbers notations. With the implicit 1 at the significant, no methods can be applied to define the zero. The second subplot in Fig. 7.3 demonstrates the significant white region which provides the descriptive void around zero. Also, the conventional computation requires the infinity and not-a-number (NaN; 0/0) representation to describe the unexpected results. The following bit reservation manages the special numbers.

The zero, infinity, and NaN are defined by the 8-bit floating point number as shown in Table 7.1. The zero exponent denormalizes the significand to present the values around zero. Note that the denormalization includes the exponent as the second least (−6 not −7) value as shown in Eq. (7.11).

$$\mathrm{FP\#10} = \begin{cases} (-1)^s \times \left(1.f \times 2^{e-7}\right)_{10} & \text{for } 1 \le e \le 14 \quad \text{(normalized)} \\ (-1)^s \times \left(0.f \times 2^{-6}\right)_{10} & \text{for } e = 0 \quad \text{(denormalized)} \end{cases}. \tag{7.11}$$

Table 7.1 Floating-point number bit configurations for special numbers

s	e	f	Decimal #	Remark
0_2	0000_2	000_2	0	Denormalized
1_2	0000_2	000_2	−0	Denormalized
0_2	1111_2	000_2	∞	
1_2	1111_2	000_2	$-\infty$	
0_2	1111_2	$\neq 000_2$	NaN	
1_2	1111_2	$\neq 000_2$	NaN	

Following is the range of significand and exponent based on the above configuration.

$$\begin{aligned} 1 \le |1.f| \le 1.8750 \text{ for } e \ne 0 \text{ (normalized)} \\ 0 \le |0.f| \le 0.8750 \text{ for } e = 0 \text{ (denormalized)} \\ -6 \le e - 7 \le 7 \text{ exclude } e = 0 \text{ and } 15 \end{aligned} \tag{7.12}$$

Therefore, overall range of floating-point number absolute is shown in Eq. (7.13).

$$0 \le |FP\#10| \le 1.8750 \times 2^7 = 240 \tag{7.13}$$

The denormalized and normalized floating-point representation are depicted in Fig. 7.4 with stems.

Prog. 7.2 MATLAB program for Fig. 7.4.

```
ff = 0:1:7;
ee = -6:1:7;
ee1 = -7:1:7;
fract1 = 1+ff*2^-3;
fract2 = ff*2^-3;
out1 = [];
out2 = [];
out1 = [out1 fract2.*2^-6];
for kk=1:length(ee)
      out1 = [out1 fract1.*2^ee(kk)];
end
for qq=1:length(ee1)
      out2 = [out2 fract1.*2^ee1(qq)];
end

figure,
subplot(311), h = stem(out1,ones(size(out1)));, grid
h.Marker = 'none';
ylabel('Denormalized')
subplot(312), h = stem(out1,ones(size(out1)));, grid
h.Marker = 'none';
xticks([0 1/128 1/64 1/32 1/16])
xticklabels({'0','1/128','1/64','1/32','1/16'})
xlim([0 1/16])
ylabel('Denormalized')
subplot(313), h = stem(out2,ones(size(out2)));, grid
h.Marker = 'none';
xticks([0 1/128 1/64 1/32 1/16])
xticklabels({'0','1/128','1/64','1/32','1/16'})
xlim([0 1/16])
ylabel('Normalized only')
xlabel('Floating-point # value')
```

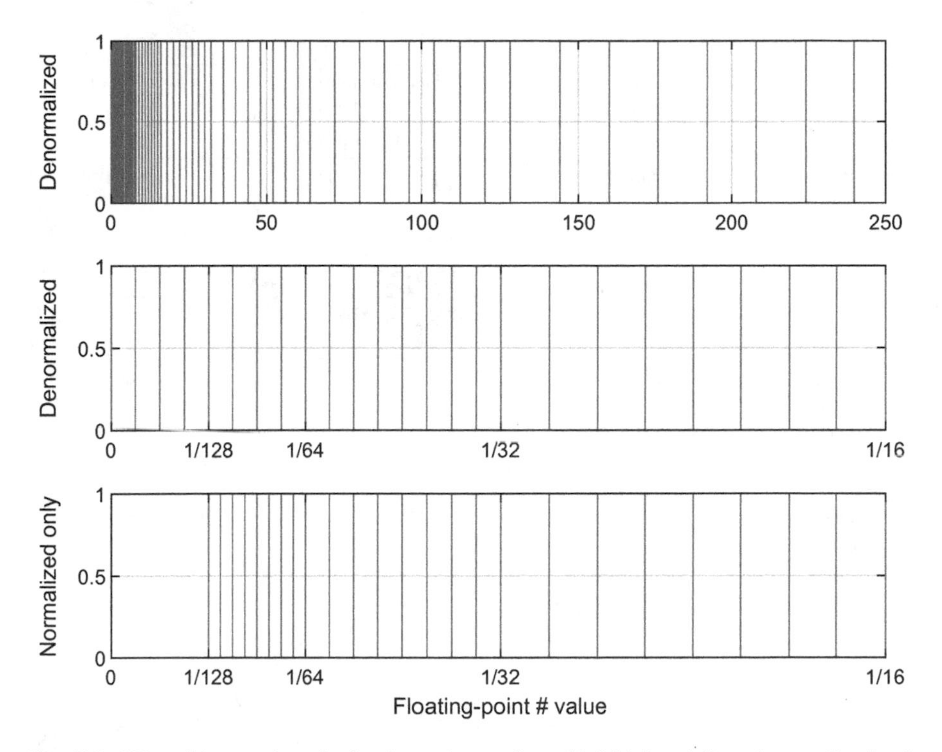

Fig. 7.4 128 positive numbers for floating-point number with 8-bit length from denormalized and normalized configuration

With denormalization, the dynamic range is decreased by the reduced exponent limit as shown in Fig. 7.4. However, the denormalization stuffs the blank around the zero (shown in third subplot of Fig. 7.4) with consistent floating-point numbers (shown in second subplot of Fig. 7.4). Note that the infinities and NaN are not presented in above figure. From the given configuration, the floating-point number specifications can be derived as Table 7.2.

IEEE Standard 754 [1] specifies the general floating-point numbers in single precision and double precision as Eq. (7.14). Note that MATLAB uses the double precision floating-point number as default.

$$\begin{aligned} &\underbrace{b_{31}}_{s}\ \underbrace{b_{30}b_{29}\ldots b_{24}b_{23}}_{e:8\text{ bits}}\ \underbrace{b_{22}b_{21}\ldots b_{1}b_{0}}_{f:23\text{ bits}} : \text{single precision (32 bits)} \\ &\underbrace{b_{63}}_{s}\ \underbrace{b_{62}b_{61}\ldots b_{53}b_{52}}_{e:11\text{ bits}}\ \underbrace{b_{51}b_{50}\ldots b_{1}b_{0}}_{f:52\text{ bits}} : \text{double precision(64 bits)} \end{aligned} \tag{7.14}$$

The corresponding decimal number is derived as Eq. (7.15).

Table 7.2 8-bit floating-point number specifications

Specification	Binary $\underbrace{b_7}_{s}\ \underbrace{b_6b_5b_4b_3}_{e}\ \underbrace{b_2b_1b_0}_{f}$	Decimal number $(-1)^s\times(1.f\times 2^{e-7})_{10}$ $(-1)^s\times(0.f\times 2^{-6})_{10}$	MATLAB
Smallest positive normalized floating-point number	00001000_2	$1\times 2^{-6}\approx 0.0156$	**realmin**
Largest positive floating-point number	01110111_2	$(1+2^{-1}+2^{-2}+2^{-3})\times 2^7=240$	**realmax**
Floating-point relative accuracy (Distance from 1.0 to the next larger number)	00111001_2	$(1+2^{-3})\times 2^0-1=0.1250$	**eps**
Smallest positive denormalized floating-point number	00000001_2	$2^{-3}\times 2^{-6}=2^{-9}\approx 0.0020$	None

$$FP\#10_{\text{single}}=\begin{cases}(-1)^s\times\left(1.f\times 2^{e-127}\right)_{10} & \text{for } 1\le e\le 254 \quad \text{(normalized)}\\ (-1)^s\times\left(0.f\times 2^{-126}\right)_{10} & \text{for } e=0 \quad \text{(denormalized)}\end{cases}. \tag{7.15}$$

$$FP\#10_{\text{double}}=\begin{cases}(-1)^s\times\left(1.f\times 2^{e-1023}\right)_{10} & \text{for } 1\le e\le 2046 \quad \text{(normalized)}\\ (-1)^s\times\left(0.f\times 2^{-1022}\right)_{10} & \text{for } e=0 \quad \text{(denormalized)}\end{cases}.$$

From the given configuration, IEEE 754 single and double precision floating-point number specifications can be derived as Table 7.3.

In the filter design and realization, we have to consider the quantization effects generated from the finite length variable and arithmetic for infinite precision number. The single or double precision floating-point number shown above indicates the high dynamic range with high resolution around zero; therefore, the floating-point number is adequately accurate for the processing. With floating-point

Table 7.3 The single and double precision floating-point number specifications

Specification	Decimal # for single	Decimal # for double
Smallest positive normalized floating-point number	$\approx 1.1755\times 10^{-38}$	$\approx 2.2251\times 10^{-308}$
Largest positive floating-point number	$\approx 3.4028\times 10^{38}$	$\approx 1.7977\times 10^{308}$
Floating-point relative accuracy (Distance from 1.0 to the next larger number)	$\approx 1.1921\times 10^{-7}$	$\approx 2.2204\times 10^{-16}$

number, we can assume that the quantization problems are ignored and restricted to the fixed-point number. Note that the floating-point number arithmetic requires the expensive hardware and/or software to process the number in real time.

– Fixed point numbers

The fixed-point number fixes the binary point at the specific point to represent the corresponding decimal number. Equation (7.5) described binary to integer conversion assumed that the binary point is located at the right of LSB. For the effectiveness of the fixed-point number arithmetic, following configuration is used. The reason will be explained in later section.

$$b_0.b_{-1}b_{-2}...b_{-B+1}b_{-B} \tag{7.16}$$

The $B + 1$ bit configuration places the binary point at the MSB right. The decimal number from the configuration is the signed fraction number based on the 2's complement. The conversion equation is below.

$$\text{Fraction number} = -b_0 + \sum_{k=1}^{B} b_{-k}2^{-k} \tag{7.17}$$

The range of the fraction number is below.

$$-1 \leq \text{Fraction number} \leq 1 - 2^{-B} \tag{7.18}$$

The smallest difference between numbers is below.

$$\Delta = 2^{-B} \tag{7.19}$$

Unlike the floating-point number, the fixed-point number demonstrates the constant distance Δ between the adjacent numbers. The fixed-point number provides the simple arithmetic computations which can be built from the inexpensive processor and uncomplicated algorithm. The computation methods for the fixed-point number will be introduced in later section.

– Signal quantization

Previously, the signal $x[n]$ is sampled from the $x(t)$ with period T. The time is discretized but the $x[n]$ magnitude is still real value which cannot be represented by finite length variable. The signal quantization process converts the real magnitude to the finite precision value such as fixed-point number. The below presents the quantization process for fixed-point number with $B + 1$ bits. Note that the Q_B presents the quantization process.

$$\hat{x} = Q_B(x) \text{where} \, x \in \mathbb{R} \tag{7.20}$$

The quantized values can be derived from the following equation.

$$\hat{x} = X_m\left(-b_0 + \sum_{k=1}^{B} b_{-k}2^{-k}\right) \text{where } X_m = \max_n |x[n]| \tag{7.21}$$

The minimum difference between the quantized numbers is below.

$$\Delta_{QB} = X_m 2^{-B} \tag{7.22}$$

The fraction number by fixed-point binary produces the less than one magnitude; hence, the quantized value $\hat{x}$ cannot be increased further than the signal maximum X_m. The selected approximation policy finds proper binary value to represent the given value $x[n]$. The rounding makes a real value x to the nearest discrete $\hat{x}$ value. A real value x which is located at the exact middle location is approximated to the higher (right) and nearest discrete $\hat{x}$ value. The truncation in 2's complement (in abbreviated form, truncation) transfers a real value x to the nearest discrete $\hat{x}$ value toward negative infinity direction. Note that the truncation is identical to the floor function. For example, one random real x value has the two adjacent $\hat{x}$ values at left and right of the x. The round chooses the nearest distance $\hat{x}$ from two adjacent values and the truncation selects the $\hat{x}$ in the left direction from the two. The transfer equation for round and truncation is below. Note that the $\lfloor\,\rfloor$ denotes floor function.

$$Q_B(x) = \Delta_{QB}\left\lfloor \frac{x}{\Delta_{QB}} + \frac{1}{2} \right\rfloor \text{for rounding} \tag{7.23}$$

$$Q_B(x) = \Delta_{QB}\left\lfloor \frac{x}{\Delta_{QB}} \right\rfloor \text{for truncation} \tag{7.24}$$

The binary representation can be calculated by following equations.

$$b_0 b_{-1} \ldots b_{-B+1} b_{-B} = \left(\left\lfloor \frac{x}{\Delta_{QB}} + \frac{1}{2} \right\rfloor\right)_2 \text{for rounding} \tag{7.25}$$

$$b_0 b_{-1} \ldots b_{-B+1} b_{-B} = \left(\left\lfloor \frac{x}{\Delta_{QB}} \right\rfloor\right)_2 \text{for truncation} \tag{7.26}$$

The example binary conversion is presented below based on the maximum magnitude X_m as 2 and assigned bits B as 2 in Fig. 7.5.

The quantization error is defined as below.

$$e = \hat{x} - x = Q_B(x) - x \tag{7.27}$$

The error range for both approximations are below.

$$-\frac{\Delta_{QB}}{2} < e \leq \frac{\Delta_{QB}}{2} \text{ for rounding} \tag{7.28}$$

$$-\Delta_{QB} < e \leq 0 \text{ for truncation} \tag{7.29}$$

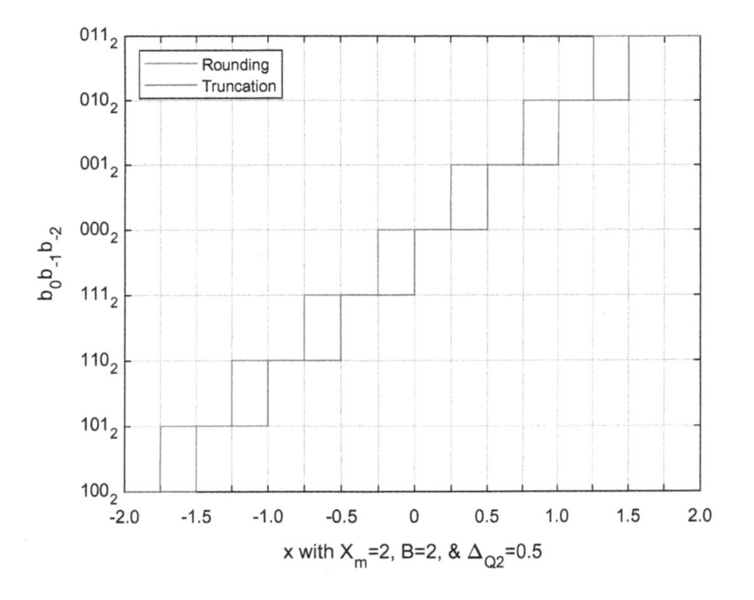

Fig. 7.5 Binary conversion example for rounding and truncation with 2 X_m and 2 B

The overall error distance is Δ_{QB} for both methods with individual initial and final error range.

– Simple statistical analysis

The quantization error e is the continuous unpredictable random value. We describe the e as the random variable E whose possible values e are the random process (quantization in this case) outcomes. Often the random variable is analyzed by the distribution, mean, variance, etc. The expected value μ_E can be calculated by the probability density function $f_E(e)$ or arithmetic mean. The Γ indicates the probability density function range. We assume that the signal is initiated from zero time [0].

$$\mathbb{E}[E] = \mu_E = \int_\Gamma e f_E(e)de = \lim_{N\to\infty}\left(\frac{1}{N}\sum_{k=0}^{N-1} e[k]\right) \tag{7.30}$$

The variance also can be computed by the $f_E(e)$ probability density function or second moment [2] as below.

$$\begin{aligned} Var(E) &= \sigma_E^2 = \mathbb{E}\left[(E - \mu_E)^2\right] = \lim_{N\to\infty}\left(\frac{1}{N}\sum_{k=0}^{N-1}(e[k] - \mu_E)^2\right) \\ &= \mathbb{E}[E^2] - \mathbb{E}[E]^2 = \mathbb{E}[E^2] - \mu_E^2 = \int_\Gamma e^2 f_E(e)de - \mu_E^2 \end{aligned} \tag{7.31}$$

The signal power of $e[n]$ is defined as below.

$$P_e = \frac{1}{N}\sum_{k=0}^{N-1} e^2[k] \approx \mathbb{E}[E^2] \tag{7.32}$$

The variance of random variable E is below.

$$\sigma_E^2 \approx \frac{1}{N}\sum_{k=0}^{N-1}(e[k] - \mu_E)^2 \approx \frac{1}{N}\sum_{k=0}^{N-1} e^2[k] - \mu_E^2 \tag{7.33}$$

Observe that the dividing factor N is replaced with $N-1$ for unbiased estimation of variance. The given signal power and variance show the following relationship.

$$P_e = \sigma_E^2 \text{ for } \mu_E = 0 \tag{7.34}$$

The signal-to-noise ratio (SNR) in decibel is defined as the ratio between the signal power and noise power as below.

$$SNR = 10\log_{10}\left(\frac{P_{signal}}{P_{noise}}\right) = 10\log_{10}\left(\frac{\sigma_{signal}^2}{\sigma_{noise}^2}\right) \text{ for } \mu = 0 \tag{7.35}$$

For zero mean noise and signal, the SNR can be calculated from the variance ratio as above. Once we know the statistical information, the SNR can be derived by the estimated the variances. Following example computes the SNR in rounding quantization.

Example 7.1
Calculate the SNR in round quantization. Assume that the probability density function is uniform as shown in Fig. 7.6.

Solution
The minimum distance between the quantization level is given below.

$$\Delta_{QB} = X_m 2^{-B}$$

The quantization error range and corresponding probability density function are below. Assume that we have uniform distribution.

$$f_E(e) = \frac{1}{\Delta_{QB}} \text{ for } -\frac{\Delta_{QB}}{2} < e \leq \frac{\Delta_{QB}}{2}$$

Note that the total area under the probability density function should be one as below due to the function property.

$$\int_{-\frac{\Delta_{QB}}{2}}^{\frac{\Delta_{QB}}{2}} f_E(e)de = 1$$

The expected value (mean) of the quantization error is below.

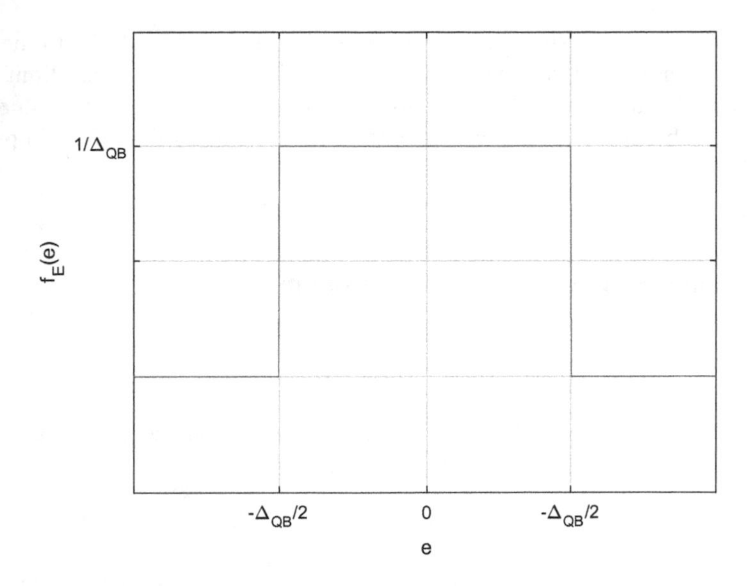

Fig. 7.6 Example probability function for rounding

$$\mu_E = \int_{-\frac{\Delta_{QB}}{2}}^{\frac{\Delta_{QB}}{2}} e f_E(e) de = \frac{1}{\Delta_{QB}} \int_{-\frac{\Delta_{QB}}{2}}^{\frac{\Delta_{QB}}{2}} e de = 0$$

The variance of the quantization error is below.

$$\begin{aligned}\sigma_{noise}^2 &= \int_{-\frac{\Delta_{QB}}{2}}^{\frac{\Delta_{QB}}{2}} e^2 f_E(e) de = \int_{-\frac{\Delta_{QB}}{2}}^{\frac{\Delta_{QB}}{2}} e^2 \frac{1}{\Delta_{QB}} de = \frac{2}{\Delta_{QB}} \int_0^{\frac{\Delta_{QB}}{2}} e^2 de \\ &= \frac{2}{\Delta_{QB}} \left[\frac{e^3}{3}\right]_0^{\frac{\Delta_{QB}}{2}} = \frac{\Delta_{QB}^2}{12} = \frac{X_m^2 2^{-2B}}{12}\end{aligned}$$

Let the maximum signal magnitude is N times signal standard deviation σ_{signal}.below.

$$X_m = N\sigma_{signal}$$

The corresponding signal variance is below.

$$\sigma_{signal}^2 = \frac{X_m^2}{N^2}$$

The SNR is below.

$$\begin{aligned}SNR =& 10\log_{10}\left(\frac{\sigma_{signal}^2}{\upsilon_{noise}^2}\right) = 10\log_{10}\left(\frac{X_m^2/N^2}{X_m^2 2^{-2B}/12}\right) = 10\log_{10}\left(\frac{12}{N^2 2^{-2B}}\right) \\ =& 10\log_{10}(12) - 20\log_{10}(N) + 20B\log_{10}(2) \\ =& 10.7918 - 20\log_{10}(N) + 6.0206B\end{aligned}$$

According to the SNR equation above, adding the additional bits to the quantization increases the 6 dB. Also, as we expand the maximum signal limit X_m by N times of signal standard deviation, the SNR decreases in logarithm fashion. For instance, 16-bit quantization level (B = 15) with three times limit (N = 3) presents 91.5584 dB SNR. ∎

7.2 Filter Implementation Components

– Fundamental blocks

The filter realization can be decomposed into three fundamental blocks in time domain. Below is the conventional filter equation.

$$y[n] = b_0x[n] + \cdots + b_Mx[n-M] - a_1y[n-1] - \cdots a_Ny[n-N] \tag{7.36}$$

For current filter output $y[n]$, the filter performs the multiplication, addition, and delay operations with proper order and coefficients. The filter design procedure determined parameters such as filter type, order, and coefficients in advance. The basic three blocks multiplication, addition, and delay are illustrated as below in Fig. 7.7.

Each single line represents the one signal state and propagation through box, arrow, or circle changes the signal state. The multiplication multiplies the signal with constant a_1. The addition adds the more than two signal inputs toward the circle to create single output. The delay retards the signal time by one sample via using the memory element. The certain combination of the above three blocks provides the filter realization in a specific method.

– Fixed point number computations

According to the basic filter blocks, the filter implementation requires the multiplication and addition as the arithmetic. The binary fractions show the special properties and requirement to perform the arithmetic. Let's assume that we have B + 1 bit binary fraction number as below.

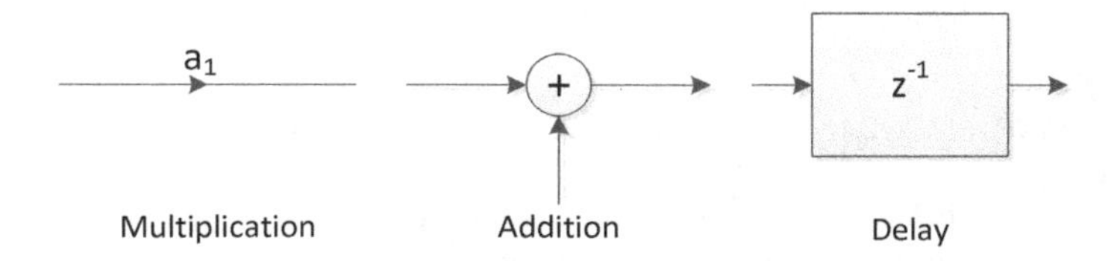

Fig. 7.7 Fundamental filter blocks for time domain realization

$$b_0.b_{-1}b_{-2}\ldots b_{-B+1}b_{-B} \tag{7.37}$$

The x_1 and x_2 are the random binary fraction number with possible range as below.

$$-1 \leq x_1 \leq 1 - 2^{-B} \text{ and } -1 \leq x_2 \leq 1 - 2^{-B} \tag{7.38}$$

The feasible range of $x_1 + x_2$ is below.

$$-2 \leq x_1 + x_2 \leq 2 - 2^{-B+1} \tag{7.39}$$

To avoid the overflow, addition outcome should be stored in the $B + 2$ bit memory as below. Note that additional bit is appended at the left of MSB.

$$b_1 b_0.b_{-1}b_{-2}\ldots b_{-B+1}b_{-B} \tag{7.40}$$

The corresponding decimal fraction number can be computed by weighted sum as below.

$$x_1 + x_2 = -2b_1 + b_0 + \sum_{k=1}^{B} b_{-k}2^{-k} \tag{7.41}$$

For general situation, the accumulated range for N number addition is below.

$$-N \leq \sum_{k=1}^{N} x_k \leq N - N2^{-B} \tag{7.42}$$

The required extra bits to avoid overflow is below. Observe that the $\lceil\rceil$ is the ceiling function which provides the least integer greater than or equal to the argument.

$$D = \lceil \log_2 N \rceil \tag{7.43}$$

Place the D number of bit on the MSB left. The below bit configuration guarantees the proper value representation by N number additions.

$$b_D \ldots b_1 b_0.b_{-1}b_{-2}\ldots b_{-B+1}b_{-B} \tag{7.44}$$

One example of binary addition for 2's complement is below.

$$\begin{array}{r} 0.110_2 \rightarrow 0.750_{10} \\ +0.101_2 \rightarrow 0.625_{10} \\ \hline 01.011_2 \rightarrow 1.375_{10} \end{array} \qquad \begin{array}{r} 1.001_2 \rightarrow -0.875_{10} \\ +1.111_2 \rightarrow -0.125_{10} \\ \hline 11.000_2 \rightarrow -1.000_{10} \end{array}$$

The unified bit format for input and output causes overflow problem in addition operation. See the below example.

$$\begin{array}{rr} 0.110_2 \rightarrow 0.750_{10} & 1.001_2 \rightarrow -0.875_{10} \\ +0.101_2 \rightarrow 0.625_{10} & +1.110_2 \rightarrow -0.250_{10} \\ \hline 1.011_2 \rightarrow -0.625_{10} & 0.111_2 \rightarrow 0.8750_{10} \end{array}$$

The carry-out from the MSB addition is abandoned for limited size output; therefore, the output sign is inverted due to the missing information. The impact of the overflow is substantial because of the circulation in decimal representation. With limited length memory, we cannot completely prevent the overflow from the addition. However, the saturation addition decreases the overflow influence by maintaining the numeric boundary. Once the result is bigger or lower than the upper or lower boundary, the output is the edge value. The output is still incorrect, but the impact is lessened by minimizing error distance. See below for example.

$$\begin{array}{rr} 0.110_2 \rightarrow 0.750_{10} & 1.001_2 \rightarrow -0.875_{10} \\ +0.101_2 \rightarrow 0.625_{10} & +1.110_2 \rightarrow -0.250_{10} \\ \hline 1.011_2 \rightarrow -0.625_{10} & 0.111_2 \rightarrow 0.8750_{10} \\ \rightarrow 0.111_2 \rightarrow 0.875_{10} & \rightarrow 1.000_2 \rightarrow -1.000_{10} \end{array}$$

Followed by overflow detection algorithm, the addition algorithm decides the positive (0.111_2) or negative (1.000_2) edge value as final output in overflow situation. Figure 7.8 shows the non-saturation and saturation arithmetic with 4-bit binary signed integer. If the arithmetic decimal output is out of the 4-bit range, the final output is translated or modified to the proper value based on the saturation policy.

With the number x_1 and x_2, the range of multiplication is below. The product range cannot be increased further than one because of the individual x_1 and x_2 magnitude.

$$0 \leq |x_1 x_2| \leq 1 \tag{7.45}$$

Except zero, the minimum magnitude of the individual and product is below.

$$\begin{aligned} \min|x_1| &= 2^{-B} \text{ for } x_1 \neq 0 \\ \min|x_1 x_2| &= 2^{-2B} \text{ for } x_1 x_2 \neq 0 \end{aligned} \tag{7.46}$$

The product generates the $2B + 2$ bit output as below. Two bits above the binary point indicate the redundant sign bits; hence, the redundancy removal establishes the final output as below.

$$b_1 b_0 . b_{-1} b_{-2} \ldots b_{-2B+1} b_{-2B} \rightarrow b_0 . b_{-1} b_{-2} \ldots b_{-2B+1} b_{-2B} \tag{7.47}$$

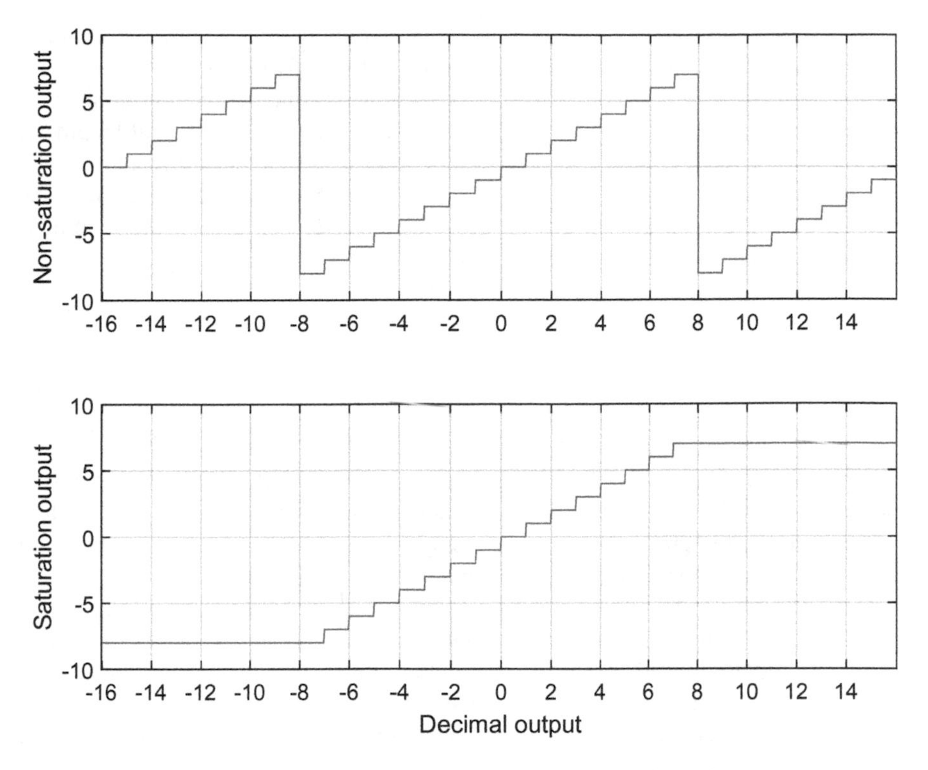

Fig. 7.8 Non-saturation and saturation arithmetic with 4-bit binary signed integer

The range of product outcome is below.

$$-1 \leq b_0.b_{-1}b_{-2}\ldots b_{-2B+1}b_{-2B} \leq 1 - 2^{-2B} \tag{7.48}$$

The conversion equation to the decimal fraction is below.

$$x_1x_2 = -b_0 + \sum_{k=1}^{2B} b_{-k}2^{-k} \tag{7.49}$$

One example of binary multiplication for 2's complement is below.

$$\begin{array}{ll} 0.110_2 \rightarrow 0.750_{10} & 1.001_2 \rightarrow -0.875_{10} \\ \underline{\times 0.101_2 \rightarrow 0.625_{10}} & \underline{\times 1.111_2 \rightarrow -0.125_{10}} \\ 0.000110_2 & 1.111001_2 \\ 0.00000_2 & 1.11001_2 \\ 0.0110_2 & 1.1001_2 \\ \underline{0.000_2} & \underline{0.111_2} \\ 0\underline{0.011110}_2 \rightarrow 0.46875_{10} & 1\underline{0.000111}_2 \rightarrow 0.109375_{10} \end{array}$$

With $2B + 2$ bit configuration, most product outputs are properly symbolized except one. The positive one from the product of negative ones cannot be represented. The situation is the only overflow experienced from the fixed-point number multiplication in binary fraction form. On the other hand, the shorter bit length for the product causes the problem known as underflow. The reduced bit length loose the output resolution; therefore, the outputs are appraised to the nearest value according to the approximation rule. For example, $2B$ bits below binary point is converted to the B bits as shown subsequently.

$$b_1 b_0.b_{-1}b_{-2}\ldots b_{-2B+1}b_{-2B} \rightarrow b_0.b_{-1}b_{-2}\ldots b_{-B+1}b_{-B} \tag{7.50}$$

$$\Delta_{Q2B} = 2^{-2B} \rightarrow \Delta_{QB} = 2^{-B} \text{ for } X_m = 1$$

The corresponding precision is decreased to the 2^{-B} from 2^{-2B}; hence, the values in the 2^{-B} interval should be approximated to the left or right. Below figure presents the $b_0.b_{-1}b_{-2}b_{-3}$ and $b_0.b_{-1}$ quantization positions. If the 3 bits below binary point is the product output, then mapping to the 1 bit below configuration requires the three approximations in every interval as shown in Fig. 7.9.

The discrete values around zero are quantized to the one bit below the binary point number as below. Note that a value x which is located at the exact middle location is approximated to the higher $\hat{x}$ value in rounding method on Table 7.4.

To avoid the underflow problem, the multiplication requires significant bit length. For instance, M times multiplications with $B + 1$ bit length multiplicand and

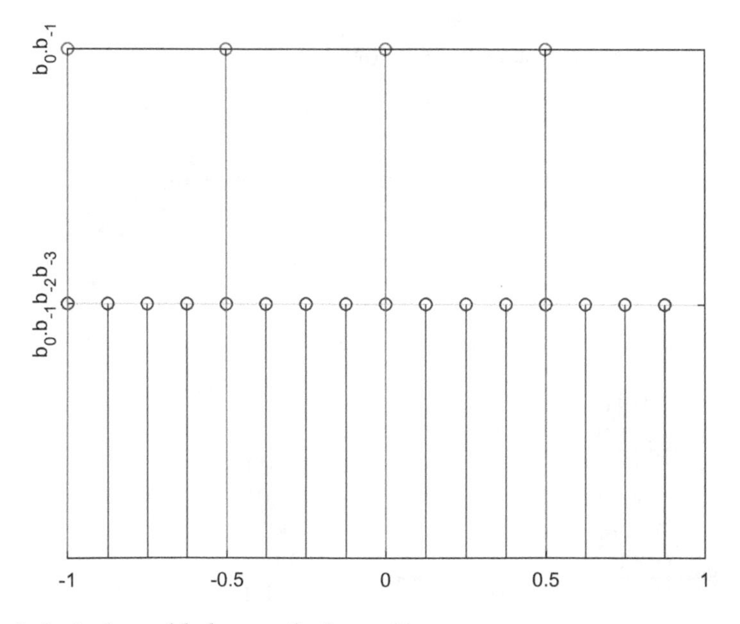

Fig. 7.9 $b_0.b_{-1}b_{-2}b_{-3}$ and $b_0.b_{-1}$ quantization positions

Table 7.4 Conversion from $b_0.b_{-1}b_{-2}b_{-3}$ to $b_0.b_{-1}$

$B = 3$	Fraction	Round	Truncate	$B = 3$	Fraction	Round	Truncate
0.001_2	0.125_{10}	0.0_2	0.0_2	1.111_2	-0.125_{10}	0.0_2	1.1_2
0.010_2	0.250_{10}	0.1_2	0.0_2	1.110_2	-0.250_{10}	0.0_2	1.1_2
0.011_2	0.375_{10}	0.1_2	0.0_2	1.101_2	-0.375_{10}	1.1_2	1.1_2

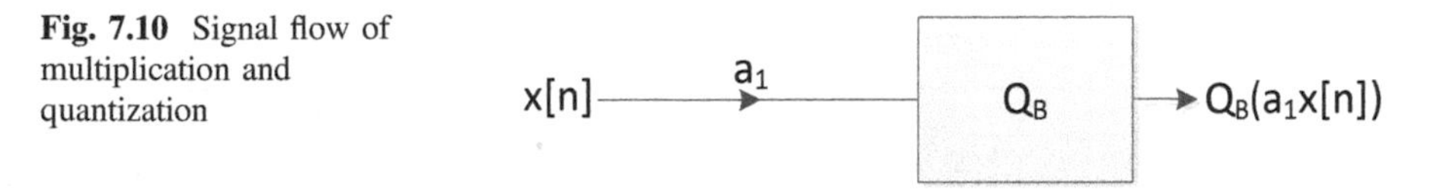

Fig. 7.10 Signal flow of multiplication and quantization

multiplier involve $M \times (B+1)$ bit length to keep the product correctly. Normally the linear increasement of the memory size is not feasible; therefore, the product is placed in the shrunk memory size. The underflow can be seen as the quantization problem which is mapping the higher resolution information to the lower precision dimension. The statistical analysis is identical to the quantization problem.

Figure 7.10 depicts the signal multiplication with constant a_1 followed by the quantization with $B + 1$ bits. We assume that $x[n]$ and a_1 are binary fraction number with $B + 1$ bits individually. The noise variance from the quantization is approximated below. Observe that the binary fraction $a_1x[n]$ has the maximum magnitude X_m as 1.

$$\sigma^2_{noise} = \frac{2^{-2B}}{12} \tag{7.51}$$

While the above quantization error presents the discrete probability distribution, we can approximate to the continuous probability density function based on the dense distribution of $2B + 2$ bit quantization levels. Overall 2^{-2B} factor is identical with minor constant variation. Further assignment of bits to the quantization process reduces the noise variance as well as the power.

– Coefficient quantization

One famous example of coefficient quantization error is Wilkinson's polynomial [3–5]. Below Wilkinson example demonstrates the 20 order FIR filter with linearly increasing zero locations. The other FIR filter $H_2(z)$ shows the identical transfer function as $H_1(z)$ except error in z^{-1} degree. The error is negligible; however, the impact is substantial as shown in Fig. 7.11 for zero plot in z-plane.

$$H_1(z) = \prod_{k=1}^{20} (1 - kz^{-1}) \tag{7.52}$$

$$H_2(z) = \prod_{k=1}^{20} (1 - kz^{-1}) - 2^{-23}z^{-1} \tag{7.53}$$

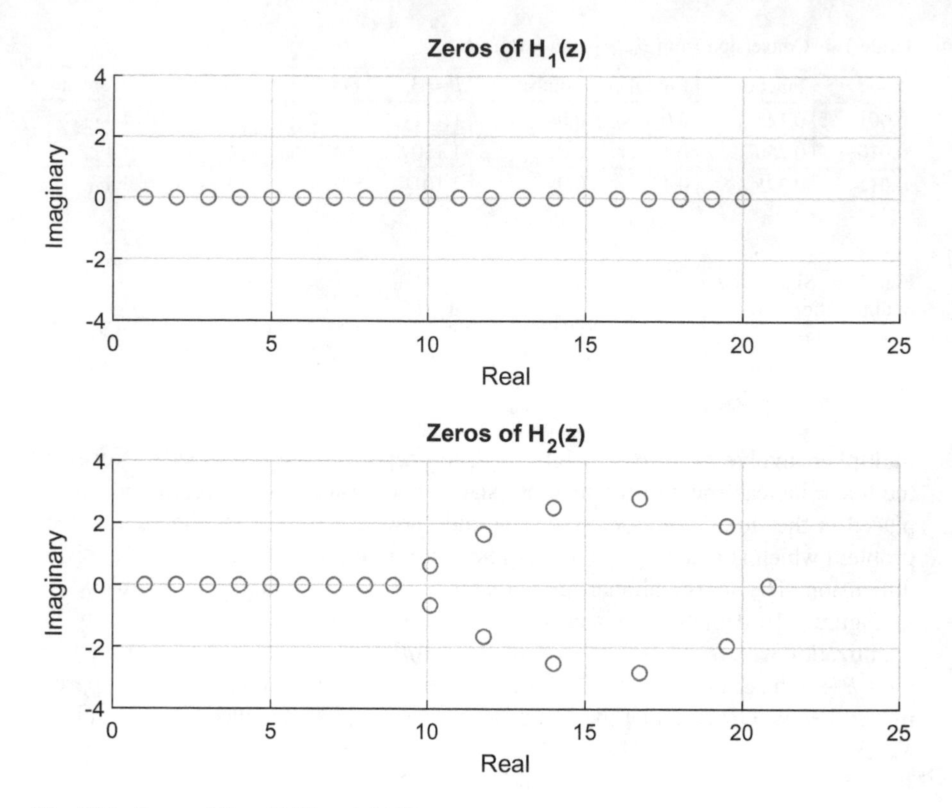

Fig. 7.11 Zeros of Eqs. (7.52) and (7.53)

Prog. 7.3 MATLAB program for Fig. 7.11.

```
syms z k

y1 = symprod(1 - k*z^-1, k, 1, 20);
y2 = expand(y1) - 2^-23*z^-1;
y3 = collect(y2);
y4 = solve(y3);
y5 = solve(y1);
double(y4);

figure,
subplot(211), scatter(real(y5),imag(y5)),grid
axis([ 0     25     -4      4]);
title('Zeros of H_1(z)')
xlabel('Real')
ylabel('Imaginary')
subplot(212), scatter(real(y4),imag(y4)), grid
axis([ 0     25     -4      4]);
title('Zeros of H_2(z)')
xlabel('Real')
ylabel('Imaginary')
```

Around the 10 real zero location in above figure, the zeros start to diverge and converge to the elliptical shape; therefore, the filter property is expected to be deviated from the design specification. Usually, the higher order polynomial is vulnerable to the tiny quantization errors which generate ill-condition state as significant property modification. Let's analyze the sensitivity of individual zero location with respect to filter coefficients. Rewrite the filter equation as below.

$$H_1(z) = \prod_{k=1}^{20}(1 - kz^{-1}) = \prod_{k=1}^{20}(1 - z_k z^{-1}) = 1 + \sum_{k=1}^{20} b_k z^{-k} \tag{7.54}$$

$$\text{where } z_k = k$$

The sensitivity of zero z_j to coefficient b_k is defined and written as below. Used chain rule.

$$\frac{\partial z_j}{\partial b_k} = \left.\frac{\partial z_j}{\partial H_1(z)}\right|_{z=z_j} \left.\frac{\partial H_1(z)}{\partial b_k}\right|_{z=z_j} = \frac{\left(\partial H_1(z)/\partial b_k\right)_{z=z_j}}{\left(\partial H_1(z)/\partial z_j\right)_{z=z_j}} \tag{7.55}$$

The numerator and denominator partial derivatives are derived as below.

$$\left.\frac{\partial H_1(z)}{\partial b_k}\right|_{z=z_j} = z_j^{-k} \tag{7.56}$$

$$\left.\frac{\partial H_1(z)}{\partial z_j}\right|_{z=z_j} = -z_j^{-20} \prod_{i=1, i\neq j}^{20} \left(z_j - z_i\right) \tag{7.57}$$

$$\text{where } H_1(z) = z^{-20} \prod_{k=1}^{20} (z - z_k)$$

The derived sensitivity is below.

$$\frac{\partial z_j}{\partial b_k} = \frac{z_j^{-k}}{-z_j^{-20} \prod_{i=1, i\neq j}^{20} (z_j - z_i)} = -\frac{z_j^{20-k}}{\prod_{i=1, i\neq j}^{20} (z_j - z_i)} \tag{7.58}$$

The jth zero location error can be described by the coefficient errors as below.

$$\Delta z_j = \sum_{k=1}^{20} \frac{\partial z_j}{\partial b_k} \Delta b_k \tag{7.59}$$

In the Wilkinson's polynomial example, we modify the z^{-1} order coefficient only.

$$\Delta z_j = \frac{\partial z_j}{\partial b_1} \Delta b_1 \tag{7.60}$$

The sensitivity with respect to b_1 coefficient is below.

$$\frac{\partial z_j}{\partial b_1} = -\frac{z_j^{20-1}}{\prod_{i=1, i\neq j}^{20} (z_j - z_i)} \tag{7.61}$$

The zero location z_{10} (10 in z-plane) sensitivity to b_1 is below.

$$\frac{\partial z_{10}}{\partial b_1} = -\frac{z_{10}^{20-1}}{\prod_{i=1,i\neq 10}^{20}(z_{10}-z_i)} = -\frac{10^{19}}{\prod_{i=1,i\neq 10}^{20}(10-z_i)} = -\frac{10^{19}}{9!10!} \tag{7.62}$$

The deviation of z_{10} is below. Note that b_1 error is -2^{-23}.

$$\Delta z_{10} = \frac{\partial z_{10}}{\partial b_1}\Delta b_1 = -\frac{10^{19}}{9!10!}\left(-2^{-23}\right) = 0.9053 \tag{7.63}$$

The individual zero sensitivity to b_1 error is below according to the above equation. Also, logarithm sensitivity base ten is illustrated below to magnify the visual dynamic range.

Prog. 7.4 MATLAB program for Fig. 7.12.

```
syms z k
kk = 1:20;
sensi = zeros(1,20);
sensi(1) = 1^19/(factorial(0)*factorial(19));
sensi(2) = 2^19/(factorial(1)*factorial(18));
sensi(3) = 3^19/(factorial(2)*factorial(17));
sensi(4) = 4^19/(factorial(3)*factorial(16));
sensi(5) = 5^19/(factorial(4)*factorial(15));
sensi(6) = 6^19/(factorial(5)*factorial(14));
sensi(7) = 7^19/(factorial(6)*factorial(13));
sensi(8) = 8^19/(factorial(7)*factorial(12));
sensi(9) = 9^19/(factorial(8)*factorial(11));
sensi(10) = 10^19/(factorial(9)*factorial(10));
sensi(11) = 11^19/(factorial(10)*factorial(9));
sensi(12) = 12^19/(factorial(11)*factorial(8));
sensi(13) = 13^19/(factorial(12)*factorial(7));
sensi(14) = 14^19/(factorial(13)*factorial(6));
sensi(15) = 15^19/(factorial(14)*factorial(5));
sensi(16) = 16^19/(factorial(15)*factorial(4));
sensi(17) = 17^19/(factorial(16)*factorial(3));
sensi(18) = 17^19/(factorial(17)*factorial(2));
sensi(19) = 19^19/(factorial(18)*factorial(1));
sensi(20) = 20^19/(factorial(19)*factorial(0));

figure,
subplot(211),bar(kk,(sensi)), grid
ylabel('|¡Óz_j/¡Ób_1|')
xlabel('j')
subplot(212),bar(kk,log10(sensi)), grid
ylabel('log_{10}(|¡Óz_j/¡Ób_1|)')
xlabel('j')
```

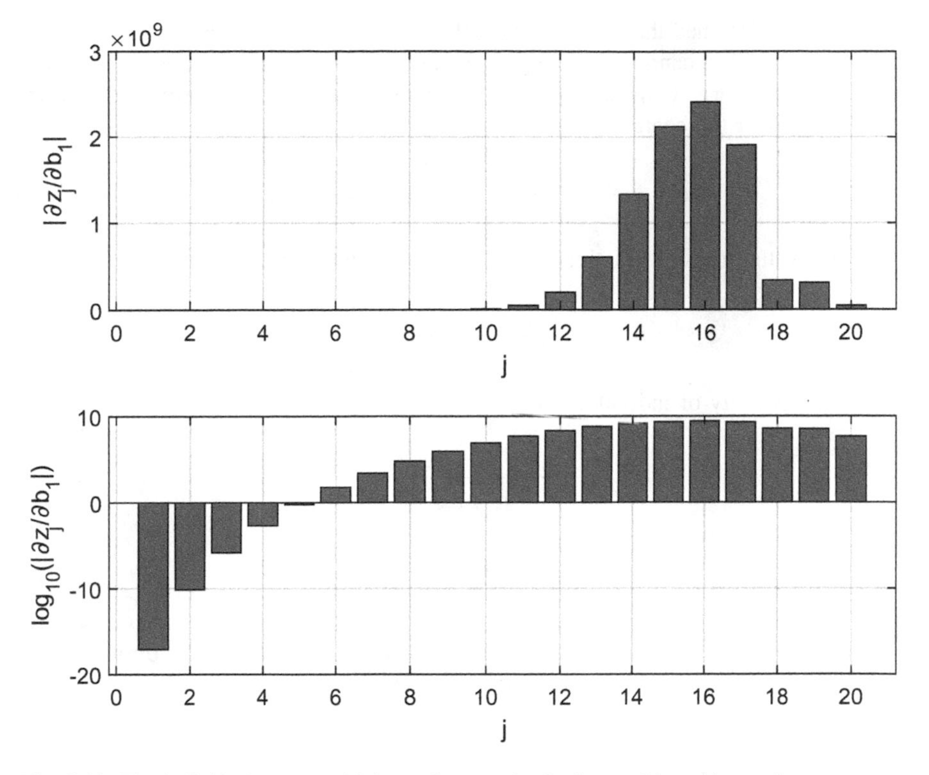

Fig. 7.12 The individual zero sensitivity to b_1 error in absolute and logarithm scale

The high sensitivity can be observed from the 10 zero location to the 20 which is corresponding to the previous zero location plot due to the b_1 coefficient error. The general root sensitivity to the coefficient is below.

$$\frac{\partial z_j}{\partial b_k} = -\frac{z_j^{N-k}}{\prod_{i=1,i\neq j}^{N}(z_j - z_i)} \tag{7.64}$$

The above equation can be applied for the numerator (zeros) or denominator (poles) to derive the sensitivity with respect to the individual coefficients. To reduce the sensitivity, following rules can be developed from the above equation.

- Reduced order N
- Small root magnitude z_j
- Not clustered roots.

Once we designed the optimal filter, the orders (N and M) and roots (poles and zeros) of the filter cannot be changed. In the filter implementation stage, the following direct form transfer function is realized by modified form to minimize sensitivity over the coefficient quantization errors.

$$H(z) = \frac{b_0 + \cdots + b_{M-1}z^{-(M-1)} + b_M z^{-M}}{1 + a_1 z^{-1} + \cdots + a_N z^{-N}} \text{ where } M \le N \tag{7.65}$$

Rewrite the transfer function in terms of poles and zeros as below.

$$H(z) = \frac{b_0\left(1-z_1z^{-1}\right)\left(1-z_2z^{-1}\right)\ldots\left(1-z_Mz^{-1}\right)}{\left(1-p_1z^{-1}\right)\left(1-p_2z^{-1}\right)\ldots\left(1-p_Nz^{-1}\right)} \tag{7.66}$$

The sensitivity of individual numerator and denominator is below.

$$\frac{\partial p_j}{\partial a_k} = -\frac{p_j^{N-k}}{\prod_{i=1,i\neq j}^{N}(p_j-p_i)} \tag{7.67}$$

$$\frac{\partial z_j}{\partial b_k} = -\frac{z_j^{M-k}}{b_0\prod_{i=1,i\neq j}^{M}(z_j-z_i)} \tag{7.68}$$

Without modifying the pole and zero positions, the $H(z)$ can be decomposed into the reduced order functions in cascade connection as below.

$$H(z) = H_{s1}(z)H_{s2}(z)\ldots H_{sN_s}(z) \tag{7.69}$$

The second order is the minimum order for individual sections to maintain the real coefficients. In general, the poles and zeros could be complex number but the filter coefficients are real number. The fundamental second order function is known as second-order section (SOS). The cascade connection with SOS is below (Fig. 7.13).

$$H(z) = \prod_{k=1}^{N_s} \frac{b_{0k} + b_{1k}z^{-1} + b_{2k}z^{-2}}{1 + a_{1k}z^{-1} + a_{2k}z^{-2}} \text{ where } N_s = \left\lfloor \frac{N+1}{2} \right\rfloor \tag{7.70}$$

The poles and zeros should be spread out to minimize the sensitivity in the SOS. Provided that the poles and zeros are complex number, the pole and zero distribution is automatic in the SOS. The real poles and zeros are assigned carefully not

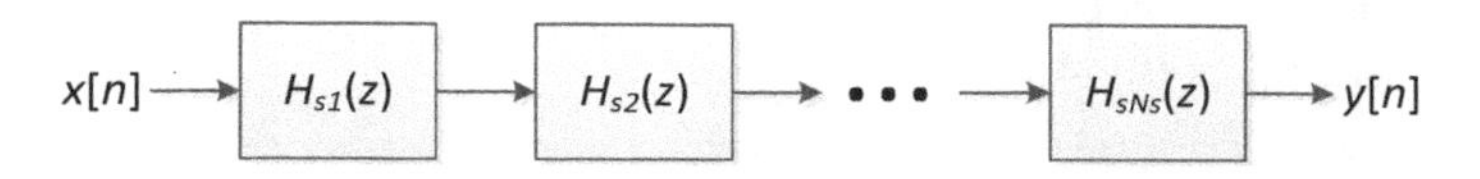

Fig. 7.13 Cascade connection

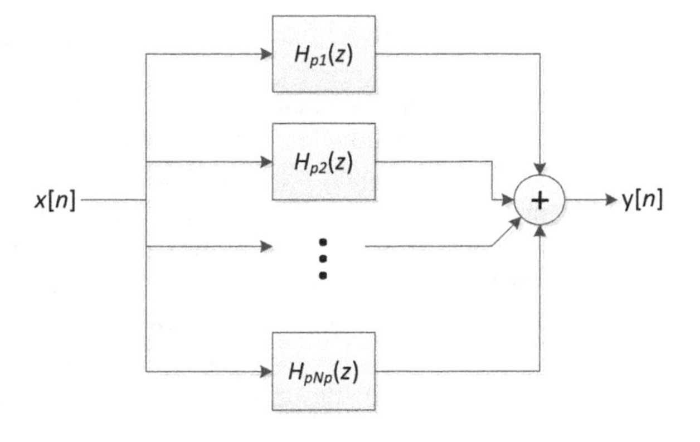

Fig. 7.14 Parallel connection

to close each other. The rule of thumb for pairing pole and zero are given by Leland B. Jackson [6] as below.

- Poles closest to the unit circle ↔ zeros nearest to the poles
- Poles next closest to the unit circle ↔ residual zeros nearest to the poles
- And so on until entire the poles and zeros have been paired for SOS
- Order the SOS from least-peaked to most-peaked in gain (arrange the sections with poles closest to the unit circle last in the cascade).

Also, the direct form $H(z)$ can be decomposed into the partial fractions in parallel connection as below. The addition of each sections produces the final output of the filter.

$$H(z) = H_{p1}(z) + H_{p2}(z) + \cdots + H_{pN_p}(z) \tag{7.71}$$

The parallel connection with the sections is below. The denominator roots of the section present the overall poles in the filter; however, the numerator roots of the section are not related to the overall zeros in Fig. 7.14.

$$H(z) = \sum_{k=1}^{N_p} \frac{c_{0k} + c_{1k}z^{-1}}{1 + a_{1k}z^{-1} + a_{2k}z^{-2}} \text{ where } N_p = \left\lfloor \frac{N+1}{2} \right\rfloor \tag{7.72}$$

The zeros in the direct form require to compute common denominator and corresponding expanded numerators. Therefore, all coefficients in the sections affects every zero. This situation is not recommended due to the high sensitivity of zero locations with respect to the coefficient quantization error. The parallel realization is not prevalent. The cascade implementation shows the further robustness to the quantization error in fixed-point realization. However, the parallel realization still presents considerably less sensitive to coefficient quantization error than the corresponding direct form since the individual sections are not extremely sensitive. Below is the general sensitivity order for each realization. We assume that the below filter is implemented with the high order.

$$\text{Sensitivity}_{\text{Direct}} \gg \text{Sensitivity}_{\text{Parallel}} > \text{Sensitivity}_{\text{Cascade}} \tag{7.73}$$

The SOS in the cascade realization contains the two poles and two zero in each section. The pole locations are important to specify the pass frequency band; therefore, the poles should be placed at the precise positions. The quadratic equation in the denominator is characterized by the quantized coefficients with binary fraction numbers. The roots of the equation demonstrate the limited numbers and pole distributions due to the constrained numeric system. Let's investigate below transfer function. The two pole locations are $re^{j\theta}$ and $re^{-j\theta}$.

$$H(z) = \frac{1}{(1-re^{j\theta}z^{-1})(1-re^{-j\theta}z^{-1})} = \frac{1}{1-2r\cos\theta z^{-1} + r^2 z^{-2}} \tag{7.74}$$

The coefficients of the denominator are $-2r\cos\theta$ and r^2. The coefficients are represented by the binary fraction number with 3 bit below binary point as below.

$$b_0.b_{-1}b_{-2}b_{-3} \text{ where } B = 3 \tag{7.75}$$

The r^2 distribution is below in binary form.

$$r^2 = \{0.000_2\ 0.001_2\ 0.010_2\ 0.011_2\ 0.100_2\ 0.101_2\ 0.110_2\ 0.111_2\}$$

The corresponding decimal numbers are below.

$$r^2 = \{0\ 0.125\ 0.25\ 0.375\ 0.5\ 0.625\ 0.75\ 0.875\}$$

The $r\cos\theta$ distribution is below in binary form.

$$r\cos\theta = \{1.000_2\ 1.001_2\ 1.010_2\ 1.011_2\ 1.100_2\ 1.101_2\ 1.110_2\ 1.111_2 \ldots \\ 0.000_2\ 0.001_2\ 0.010_2\ 0.011_2\ 0.100_2\ 0.101_2\ 0.110_2\ 0.111_2\}$$

The corresponding decimal numbers are below.

$$r\cos\theta = \{-1\ -0.875\ -0.75\ -0.625\ -0.5\ -0.375\ -0.25\ -0.125 \ldots \\ 0\ 0.125\ 0.25\ 0.375\ 0.5\ 0.625\ 0.75\ 0.875\}$$

Note that the $2r\cos\theta$ magnitude is beyond the binary fraction range with given $r\cos\theta$ distribution. As long as the overall product (shown below) is less than one magnitude, the filter is working properly. The multiplication by two is performed by the shift operation.

$$2r\cos\theta z^{-1} \rightarrow |2r\cos\theta y[n-1]| < 1 \tag{7.76}$$

The poles of the given function can be found by the intersections between the real $r\cos\theta$ values and radius r circles in complex domain. Observe that the poles are designed to be $re^{j\theta}$ and $re^{-j\theta}$ in the transfer function. The $r\cos\theta$ is the real value of the $re^{\pm j\theta}$. Below figure shows the real $r\cos\theta$ values, radius r circles, and corresponding intersections.

Figure 7.16 demonstrates the pure pole locations with unit circle. With given binary fraction numbers, the pole distribution displays the sparse population around the ± 1 in complex number domain.

Prog. 7.5 MATLAB program for Figs. 7.15 and 7.16.

```
B = 3;
N = 1000;
theta = linspace(0,2*pi,N);
ggrid = linspace(-1,1,N);
r2 = linspace(0,1-2^(-B),2^B);
r3 = sqrt(r2);
rcos = linspace(-1,1-2^(-B),2^(B+1));
rrad1 = r3'*exp(1j*theta);
rrad2 = 1*exp(1j*theta);
yy1 = rcos'*ones(1,N)+1j*ones(2^(B+1),1)*ggrid;
final = [];
for kk1 = 1:length(r2)
    for kk2 = 1:length(rcos)
        if ((rcos(kk2)).^2 > r2(kk1))
            continue
        end
        y1 = sqrt(r2(kk1)-(rcos(kk2)).^2);
        y2 = -sqrt(r2(kk1)-(rcos(kk2)).^2);

        final = [final [rcos(kk2); y1] [rcos(kk2); y2]];
    end
end

figure,
scatter(final(1,:),final(2,:)), grid, hold on;
plot(rrad2','Color','c'), hold off;
axis equal
figure,
plot(rrad1','Color',[0 0.4470 0.7410]), grid, hold on;
axis equal
plot(rrad2','Color','c')
plot(yy1','Color',[0 0.4470 0.7410]);
scatter(final(1,:),final(2,:)), grid, hold off;
```

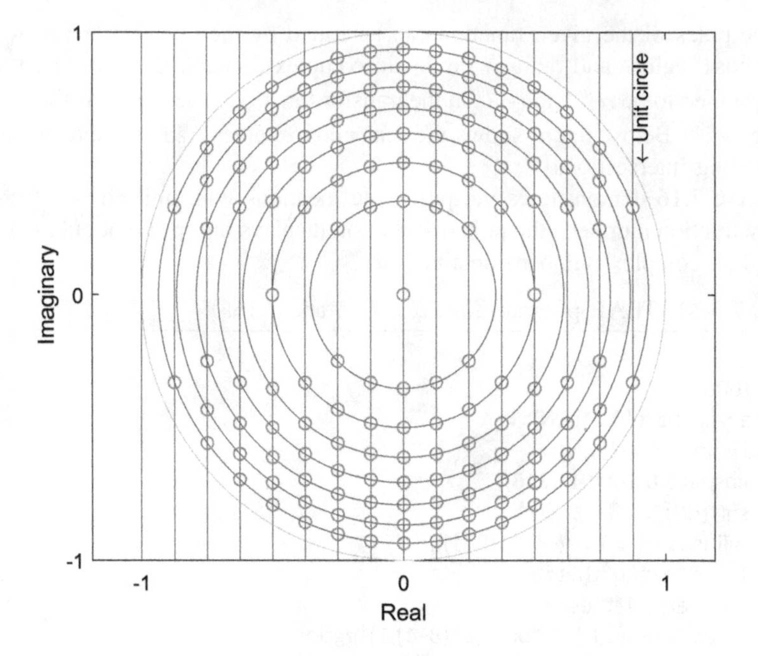

Fig. 7.15 Possible pole locations for Eq. (7.74) with $b_0.b_{-1}b_{-2}b_{-3}$ coefficients

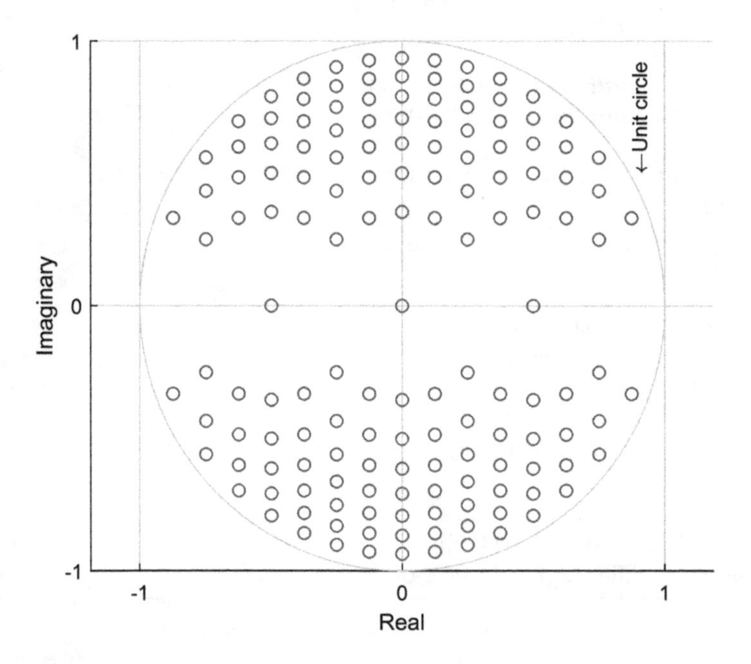

Fig. 7.16 Possible pole locations for Eq. (7.74) based on $b_0.b_{-1}b_{-2}b_{-3}$ coefficients without grid

With 5 bits below binary point, the increasement of bit assignment significantly improves the pole distribution widely. However, the ±1 vicinity still indicates the depletion.

Prog. 7.6 MATLAB program for Fig. 7.17.

```
B = 5;
N = 1000;
theta = linspace(0,2*pi,N);
rrad2 = 1*exp(1j*theta);
a1 = linspace(-1,1-2^(-B),2^(B+1));
a2 = linspace(-1,1-2^(-B),2^(B+1));
[A1,A2] = meshgrid(a1,a2);
LLE = length(a1);
final = [];
for kk1 = 1:LLE
      for kk2 = 1:LLE
            temp1 = roots([1 A1(kk1,kk2)*2 A2(kk1,kk2)]);
            rr = sqrt(temp1.*conj(temp1));
            for kk3 = 1:2
                  if (rr(kk3)>=1)
                        continue
                  end
                  final = [final temp1(kk3)];
            end
      end
end

figure,
scatter(real(final),imag(final)), grid, hold on
plot(rrad2','Color','c'), hold off
axis equal
axis([-1.1860      1.1860     -1.0       1.0]);
yticks([-1 0 1]);
yticklabels({'-1','0','1'});
xticks([-1 0 1]);
xticklabels({'-1','0','1'});
xlabel('Real');
ylabel('Imaginary');
```

The above figures present the rare pole distribution on the real axis. The negative r^2 places the poles on the real axis and the situation is not shown here. For verification purpose, below shows the real pole derivation. The denominator is rewritten below.

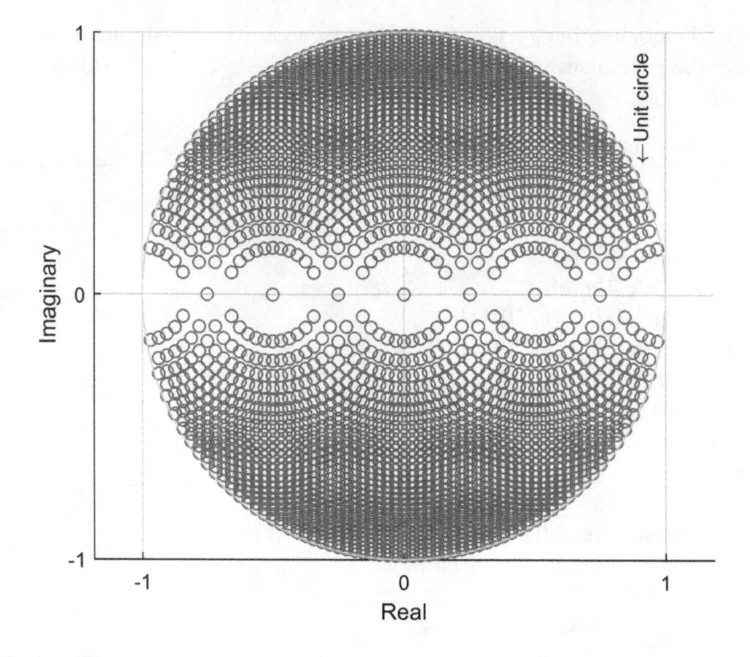

Fig. 7.17 Possible pole locations for Eq. (7.74) based on $b_0.b_{-1}b_{-2}b_{-3}b_{-4}b_{-5}$ coefficients without grid

$$1 - 2r\cos\theta z^{-1} + r^2 z^{-2} = z^{-2}(z^2 - 2r\cos\theta z + r^2) \tag{7.77}$$

Discriminant of the quadratic equation is below.

$$\Delta = 4r^2\cos^2\theta - 4r^2 = 4r^2(\cos^2\theta - 1) = -4r^2\sin^2\theta \tag{7.78}$$

Therefore, the negative r^2 provides the real poles.

$$\begin{cases} \Delta \leq 0 & \text{for } r^2 \geq 0 \\ \Delta \geq 0 & \text{for } r^2 < 0 \end{cases} \tag{7.79}$$

For even pole distribution, we prefer to sample the real and imaginary axis. The pole location $re^{j\theta}$ can be considered as the $r\cos\theta$ and $r\sin\theta$ in real and imaginary axis respectively. The uniform pole distribution in Cartesian coordinate can be realized by an alternative SOS structure known coupled form as below.

$$
\begin{aligned}
H_{coupled}(z) &= \frac{r\sin\theta z^{-1}}{(1-(r\cos\theta+jr\sin\theta)z^{-1})(1-(r\cos\theta-jr\sin\theta)z^{-1})} \\
&= \frac{r\sin\theta z^{-1}}{1-2r\cos\theta z^{-1}+r^2(\cos^2\theta+\sin^2\theta)z^{-2}}
\end{aligned}
\tag{7.80}
$$

The pole $re^{\pm j\theta}$ is denoted by the $(r\cos\theta \pm jr\sin\theta)$ in the coupled form. The individual real and imaginary component are sample by binary fraction numbers as below.

$$
\begin{aligned}
r\cos\theta = \{&1.000_2\ 1.001_2\ 1.010_2\ 1.011_2\ 1.100_2\ 1.101_2\ 1.110_2\ 1.111_2 \ldots \\
&0.000_2\ 0.001_2\ 0.010_2\ 0.011_2\ 0.100_2\ 0.101_2\ 0.110_2\ 0.111_2\}
\end{aligned}
$$

$$
\begin{aligned}
r\cos\theta = \{&-1\ \ -0.875\ \ -0.75\ \ -0.625\ \ -0.5\ \ -0.375\ \ -0.25\ \ -0.125 \ldots \\
&0\ 0.125\ 0.25\ 0.375\ 0.5\ 0.625\ 0.75\ 0.875\}
\end{aligned}
$$

$$
\begin{aligned}
r\sin\theta = \{&1.000_2\ 1.001_2\ 1.010_2\ 1.011_2\ 1.100_2\ 1.101_2\ 1.110_2\ 1.111_2 \ldots \\
&0.000_2\ 0.001_2\ 0.010_2\ 0.011_2\ 0.100_2\ 0.101_2\ 0.110_2\ 0.111_2\}
\end{aligned}
$$

$$
\begin{aligned}
r\sin\theta = \{&-1\ \ -0.875\ \ -0.75\ \ -0.625\ \ -0.5\ \ -0.375\ \ -0.25\ \ -0.125 \ldots \\
&0\ 0.125\ 0.25\ 0.375\ 0.5\ 0.625\ 0.75\ 0.875\}
\end{aligned}
$$

Certain combination of above assignment ruins the filter stability by placing the poles outside of unit circle. We only preserve the poles which is the less than one radius from the coordinate origin as below.

$$
r^2\left(\cos^2\theta+\sin^2\theta\right)<1 \tag{7.81}
$$

The poles of the coupled form can be derived by the intersections between the real $r\cos\theta$ values and the imaginary $r\sin\theta$ values in complex domain. Below figure shows the real $r\cos\theta$ values, the imaginary $r\sin\theta$ values, and corresponding intersections.

Subsequent figure demonstrates the pure pole locations with unit circle. With given binary fraction numbers, the pole distribution demonstrates the uniform population for entire unit circle area in complex domain.

With 5 bits below binary point, the increasement of bit assignment significantly improves the pole distribution widely and evenly. Unlike the SOS case, no prominent spots of pole population are observed around the ± 1 vicinity as shown below.

Prog. 7.7 MATLAB program for Figs. 7.18, 7.19 and 7.20.

```
B = 3;     % For Fig.7.18&19 B = 3 and For Fig.7.20 B=5
N = 1000;
theta = linspace(0,2*pi,N);
xgrid = linspace(-1.1860,1.1860,N);
ygrid = linspace(-1,1,N);
rsin = linspace(-1,1-2^(-B),2^(B+1));
rcos = linspace(-1,1-2^(-B),2^(B+1));
rrad2 = 1*exp(1j*theta);
yy1 = rcos'*ones(1,N)+1j*ones(2^(B+1),1)*ygrid;
xx1 = 1j*rsin'*ones(1,N)+ones(2^(B+1),1)*xgrid;
final = [];
for kk1 = 1:length(rcos)
    for kk2 = 1:length(rsin)
        rr = sqrt(rcos(kk1)^2+rsin(kk2)^2);
        if (rr>=1)
            continue
        end
        final = [final [rcos(kk1); rsin(kk2)]];
    end
end

figure,
scatter(final(1,:),final(2,:)), grid, hold on
plot(rrad2','Color','c'), hold off
axis equal
figure,
plot(xx1','Color',[0 0.4470 0.7410]), grid, hold on
axis equal
plot(rrad2','Color','c')
plot(yy1','Color',[0 0.4470 0.7410]);
scatter(final(1,:),final(2,:)), grid, hold off
```

The disadvantage of the coupled form is the slight increment of the computational requirement.

– Overflow and scaling

The filter output computation requires the multiplications and additions. With fixed-point number, the careful scaling should be involved to avoid the overflow in

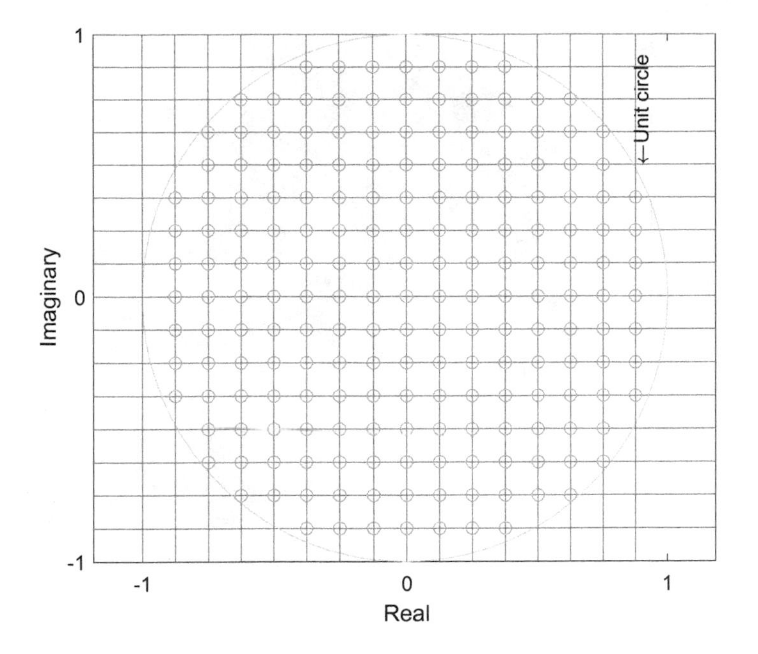

Fig. 7.18 Possible pole locations for Eq. (7.80) with $b_0.b_{-1}b_{-2}b_{-3}$ coefficients

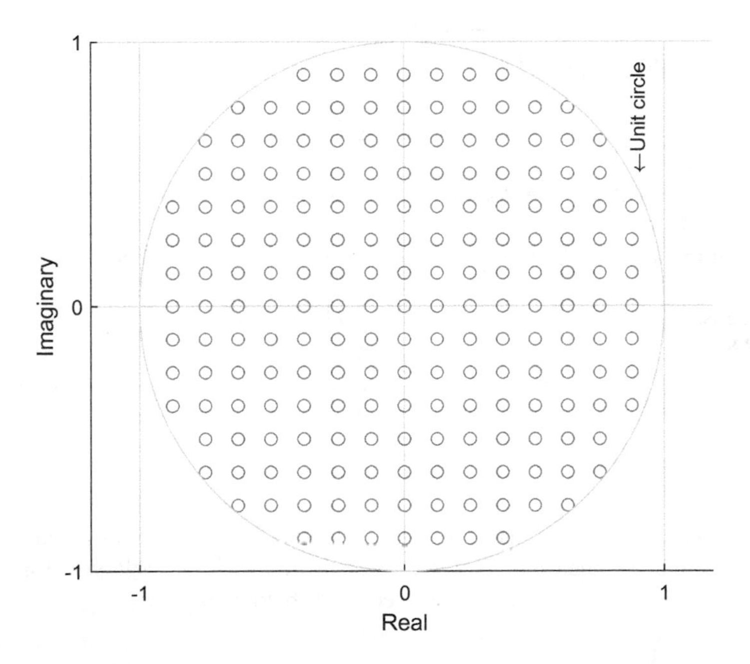

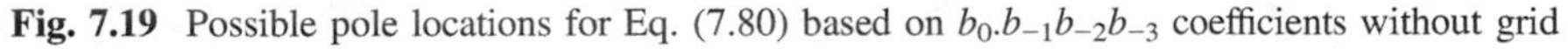

Fig. 7.19 Possible pole locations for Eq. (7.80) based on $b_0.b_{-1}b_{-2}b_{-3}$ coefficients without grid

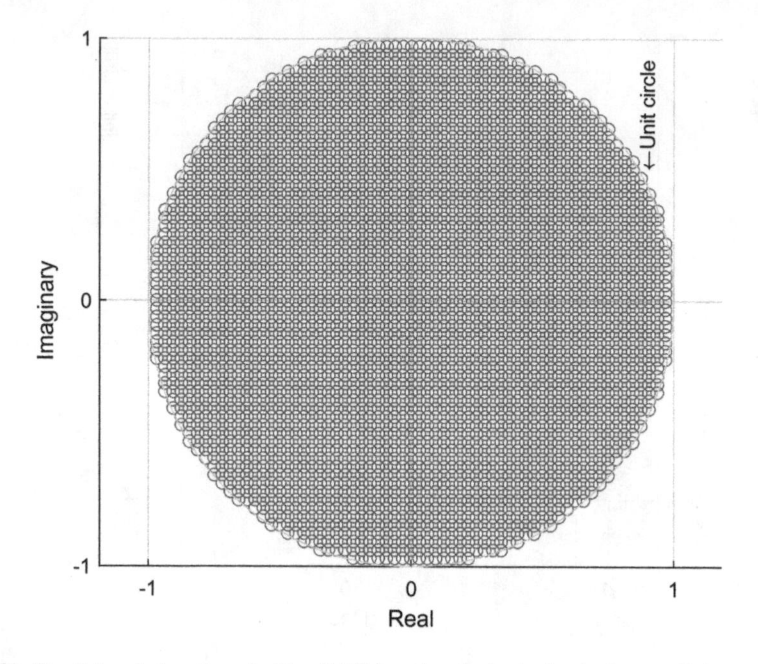

Fig. 7.20 Possible pole locations for Eq. (7.80) based on $b_0.b_{-1}b_{-2}b_{-3}b_{-4}b_{-5}$ coefficients without grid

any arithmetic computation. Note that we use the binary fraction numbers as fixed-point representation. For scaling problem, see below SOS.

$$H(z) = \frac{b_0 + b_1 z^{-1} + b_2 z^{-2}}{1 + a_1 z^{-1} + a_2 z^{-2}} \tag{7.82}$$

The corresponding time domain realization is below.

$$y[n] = b_0 x[n] + b_1 x[n-1] + b_2 x[n-2] - a_1 y[n-1] - a_2 y[n-2] \tag{7.83}$$

Each signal $x[n]$s, $y[n]$s, coefficients and products in the right-hand side of Eq. (7.83) shows less than unity magnitude due to the binary fraction number system. Therefore, the corresponding summation for left-hand side produces the five times of the binary fraction dynamic range as below.

$$-5 \leq y[n] < 5 \tag{7.84}$$

The above overflowed range is the outcome of the given instantaneous time n based on the less than unity assumptions. If the overflow is allowed for outputs, the maximum output y_{max} can be derived from below equations.

$$h[n] \overset{z-trans}{\leftrightarrow} H(z)$$

$$\begin{aligned} |y[n]| &= \left|\sum_{k=0}^{\infty} h[k]x[n-k]\right| \leq \sum_{k=0}^{\infty} |h[k]||x[n-k]| \\ &\leq \sum_{k=0}^{\infty} |h[k]|x_{max} = x_{max}\sum_{k=0}^{\infty} |h[k]| = y_{max} \leq 1 \end{aligned} \tag{7.85}$$

$$\text{where } x_{max} = \max|x[n]|\forall n \;\; y_{max} = \max|y[n]|\forall n$$

The maximum output is the multiplication between maximum input and absolute sum of impulse response. The quantized input is described by the binary fraction number; therefore, the input magnitude is less than one automatically. The absolute sum of the impulse response $h[n]$ is known as L_1-norm of the impulse response. If the L_1-norm is less than one, the overall product for the y_{max} is less than one which can be stored as the binary fraction number.

$$\sum_{k=0}^{\infty} |h[k]| \leq 1 \rightarrow y_{max} \leq 1 \tag{7.86}$$

$$\sum_{k=0}^{\infty} |h[k]| = \frac{1}{s} > 1 \rightarrow \begin{cases} \breve{h}[n] = sh[n] \rightarrow \breve{y}_{max} = 1 \\ \breve{x}[n] = sx[n] \rightarrow \breve{y}_{max} = 1 \end{cases} \tag{7.87}$$

If the L_1-norm is greater than one, we can scale down the impulse response or input signal by factor of the s (=1/L_1-norm) in order to generate the proper dynamic range output.

$$\breve{y}[n] = \sum_{k=0}^{\infty} \breve{h}[k]x[n-k] = \sum_{k=0}^{\infty} h[k]\breve{x}[n-k] = \sum_{k=0}^{\infty} sh[k]x[n-k] \tag{7.88}$$

In the SOS realization, the scale factor s is computed from the impulse response $h[n]$ and the factor is applied over the numerator of the transfer function as below.

$$\breve{H}(z) = sH(z) = s\frac{b_0 + b_1z^{-1} + b_2z^{-2}}{1 + a_1z^{-1} + a_2z^{-2}} \tag{7.89}$$

The time domain realization with scaling is below.

$$\begin{aligned} \breve{y}[n] =& sb_0x[n] + sb_1x[n-1] + sb_2x[n-2] - a_1y[n-1] - a_2y[n-2] \\ =& \breve{b}_0x[n] + \breve{b}_1x[n-1] + \breve{b}_2x[n-2] \\ & - a_1y[n-1] - a_2y[n-2] \leq \breve{y}_{max} = 1 \end{aligned} \tag{7.90}$$

The scaled output $\breve{y}[n]$ is guaranteed to stay within the acceptable dynamic range of binary fraction number. In the given SOS, four additions are performed to deliver

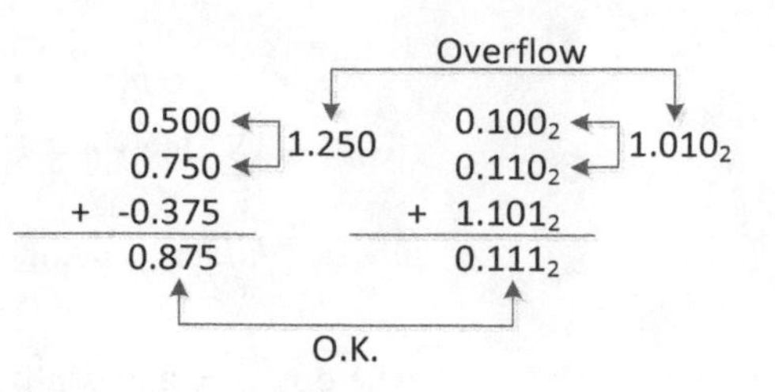

Fig. 7.21 Example of intermediate overflow with binary representation on right

the output. Certain intermediate summation could produce the overflow output; however, the final output $\breve{y}[n]$ is not contaminated by the overflow due to the modular property of binary number. Below example illustrates the three-number addition which induces the intermediate overflow in Fig. 7.21.

The first addition generates the overflow by changing the output sign. The second summation reduces magnitude to override the overflow. The multiple additions with several overflows can yield the proper output as long as the final value is placed within the binary fraction range. Be careful not to use the saturation addition which changes the arithmetic distance. With saturation addition, single overflow pollutes the final output. See below example in Fig. 7.22.

The L_1-norm prevents the overflow from filter output computation; however, the scaling factor L_1-norm significantly reduces the dynamic range or filter gain. The output SNR can be improved by allowing low probability overflow. The L_2-norm shown below provides the decreased scale factor which may lead the overflow with low probability.

$$\frac{1}{s} = l_2 = \sqrt{\sum_k (h[k])^2} \tag{7.91}$$

With the narrowband signal $e^{j\omega_0 n}$, the output is the modified magnitude and shifted phase of the signal based on the DTFT of impulse response as below.

$$\begin{aligned} y[n] &= \sum_{k=0}^{\infty} h[k]x[n-k] = \sum_{k=0}^{\infty} h[k]e^{j\omega_0(n-k)} = e^{j\omega_0 n}\sum_{k=0}^{\infty} h[k]e^{-j\omega_0 k} \\ &= e^{j\omega_0 n}H\left(e^{j\omega_0}\right) = \left|H\left(e^{j\omega_0}\right)\right|e^{j\left(\omega_0 n + \angle H\left(e^{j\omega_0}\right)\right)} \end{aligned} \tag{7.92}$$

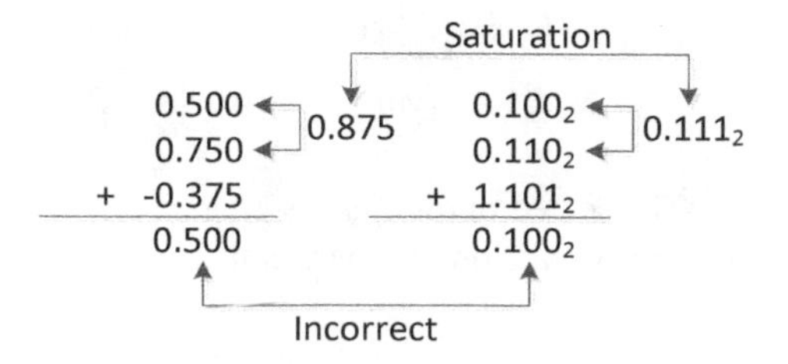

Fig. 7.22 Example of intermediate saturation with binary representation on right

The scaling factor L_∞-norm is the maximum magnitude of the frequency response as below. The scaling by the L_∞-norm presents the less than unity magnitude for any narrowband signal; however, the broadband signal could initiate the overflow.

$$L_\infty = \max_\omega |H(e^{j\omega})| \tag{7.93}$$

Below equation demonstrates the magnitude comparison between norms. The L_∞-norm is the most commonly used scaling factor in practice.

$$L_1 - \text{norm} \geq L_\infty - \text{norm} \geq L_2 - \text{norm} \tag{7.94}$$

In summary, we can choose the one of the norms for the scale factor as below.

$$\frac{1}{s} = \begin{cases} \sum_{n=0}^{\infty} |h[n]| = L_1 \\ \max_\omega |H(e^{j\omega})| = L_\infty \text{ where } H(e^{j\omega}) = \begin{cases} \sum_{n=0}^{\infty} h[n] e^{-j\omega n} \\ H(z)|_{z \to e^{j\omega n}} \end{cases} \\ \sqrt{\sum_{n=0}^{\infty} |h[n]|^2} = L_2 \end{cases} \tag{7.95}$$

For the FIR filter scaling, below is the transfer function.

$$H(z) = b_0 + b_1 z^{-1} + \cdots + b_{N-1} z^{-(N-1)} + b_N z^{-N} \tag{7.96}$$

The time domain realization is below.

$$\begin{aligned} y[n] = b_0 x[n] + b_1 x[n-1] + \cdots \\ + b_{N-1} x[n-(N-1)] + b_N x[n-N] \end{aligned} \tag{7.97}$$

The scaled output can be described by the below equation.

$$\begin{aligned} \breve{y}[n] &= s b_0 x[n] + s b_1 x[n-1] + \cdots + s b_{N-1} x[n-(N-1)] + s b_N x[n-N] \\ &= \breve{b}_0 x[n] + \breve{b}_1 x[n-1] + \cdots + \breve{b}_{N-1} x[n-(N-1)] + \breve{b}_N x[n-N] \end{aligned} \tag{7.98}$$

The scaling factor can be chosen from any norm given below.

$$\frac{1}{s} = \begin{cases} \sum_{k=0}^{N} |b_k| = L_1 \\ \max_\omega |H(e^{j\omega})| = L_\infty \text{ where } H(e^{j\omega}) = \sum_{k=0}^{N} b_k e^{-j\omega k} \\ \sqrt{\sum_{k=0}^{N} |b_k|^2} = L_2 \end{cases} \tag{7.99}$$

The high order filter can be decomposed into the reduced order systems such as SOS with cascade connection. Each SOS should produce the appropriate dynamic range output to avoid the overflow. The individual SOS scaling factor controls the discrete output range independently. Below is the L SOS cascade system.

$$H(z) = GH_1(z)H_2(z)\ldots H_L(z) \tag{7.100}$$

The G is the global gain of the cascade SOS system to improve the computational efficiency. Uniform distribution of the global gain G assists to prevent the peaky SOS gain which amplifies unwanted errors from the previous SOS stage. Rewritten transfer function is below.

$$H(z) = G^{1/L}H_1(z)G^{1/L}H_2(z)\ldots G^{1/L}H_L(z) \tag{7.101}$$

The norm of the stage is derived in the cumulative manner from the initial input $x[n]$ to the given stage output as below. The output range is manipulated by the scaling factor which is the inverse of the norm. Note that the norm function can be L_1, L_∞, or L_2 which regulates output SNR and overflow policy. In this chapter, we compute the norm value numerically with heuristic length. We could use the Cauchy residue theorem [7] or partial fraction expansion [8] for better representation of impulse response by analytic inverse z transform derivation. The L_∞ norm requires the frequency response instead of the impulse response. The * operation indicates the convolution sum.

$$s_1 = \frac{1}{\text{norm}(G^{1/L}h_1[n])}$$

$$s_2 = \frac{1}{\text{norm}(G^{1/L}h_1[n] * G^{1/L}h_2[n])}$$

$$s_L = \frac{1}{\text{norm}(G^{1/L}h_1[n] * G^{1/L}h_2[n] * \ldots * G^{1/L}h_L[n])} \tag{7.102}$$

Figure 7.23 shows the SOS connection and corresponding scale factor scope. The scale factor is applied in current stage and voided in previous stages to leave the latest scale factor only. The subsequent figure illustrates the scale factored SOS connection as well in Figs. 7.24 and 7.25.

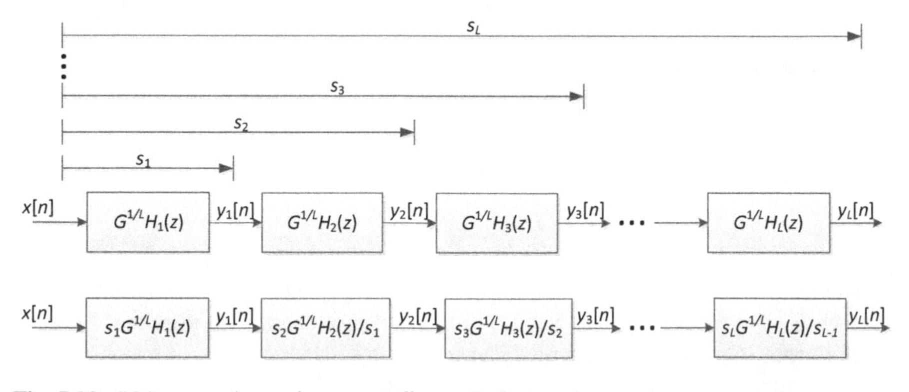

Fig. 7.23 SOS connection and corresponding scale factor scope

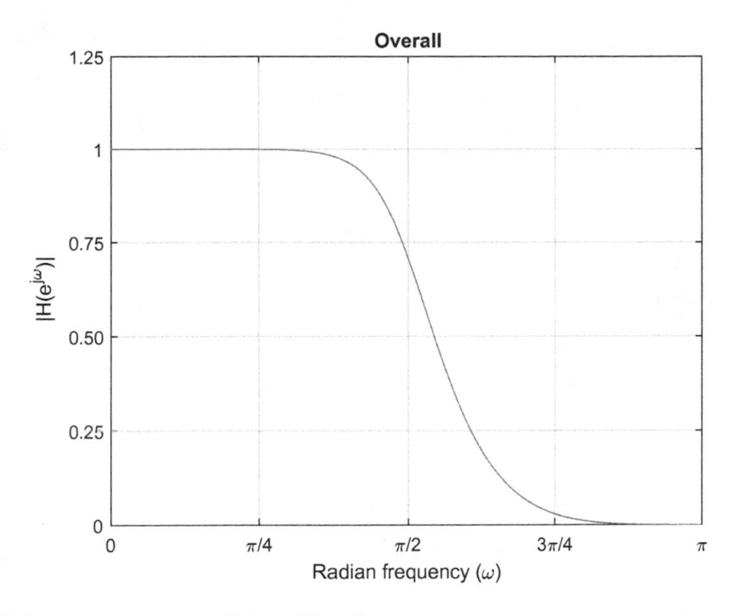

Fig. 7.24 Frequency response of given filter above

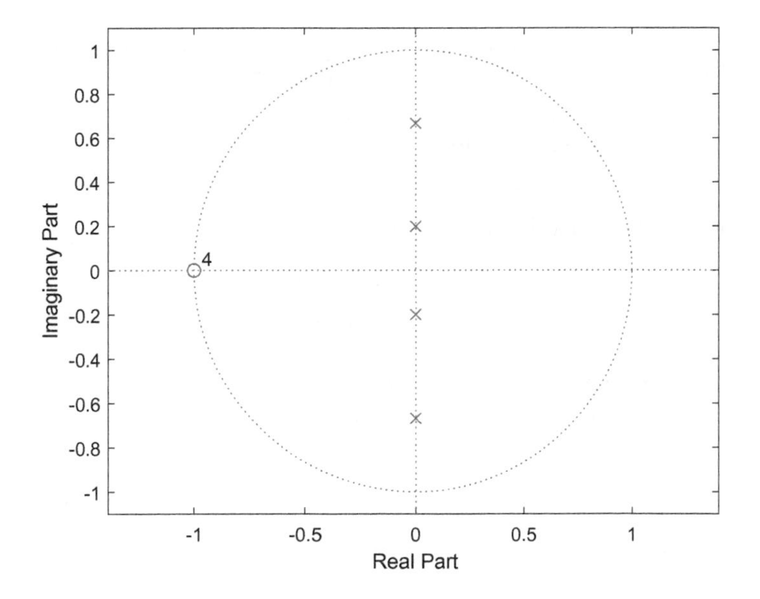

Fig. 7.25 Pole zero plot of given filter above

The mathematical representation of the scaled SOS system is below.

$$H(z) = \left(s_1 G^{1/L} H_1(z)\right)\left(\frac{s_2 G^{1/L} H_2(z)}{s_1}\right)\left(\frac{s_3 G^{1/L} H_3(z)}{s_2}\right)\ldots \left(\frac{s_L G^{1/L} H_L(z)}{s_{L-1}}\right) \tag{7.103}$$

$$= \breve{H}_1(z)\breve{H}_2(z)\breve{H}_3(z)\ldots\breve{H}_L(z)$$

The scale factor is applied on the numerator of the transfer function as below.

$$\breve{H}_k(z) = \frac{s_k}{s_{k-1}}\frac{b_{0k} + b_{1k}z^{-1} + b_{2k}z^{-2}}{1 + a_{1k}z^{-1} + a_{2k}z^{-2}} = \frac{\breve{b}_{0k} + \breve{b}_{1k}z^{-1} + \breve{b}_{2k}z^{-2}}{1 + a_{1k}z^{-1} + a_{2k}z^{-2}} \tag{7.104}$$

The corresponding time domain realization is below.

$$\begin{aligned} y_k[n] = {} & \breve{b}_{0k}x_k[n] + \breve{b}_{1k}x_k[n-1] + \breve{b}_{2k}x_k[n-2] \\ & - a_{1k}y_k[n-1] - a_{2k}y_k[n-2] \end{aligned} \tag{7.105}$$

Example 7.2
Shows the overall scaling procedure for cascade system. The designed filter indicates the LPF with 0.5π cutoff frequency in 4th order system. The design method is Butterworth.

$$H(z) = \frac{0.0940 + 0.3759z^{-1} + 0.5639z^{-2} + 0.3759z^{-3} + 0.0940z^{-4}}{1 + 0.4860z^{-2} + 0.0177z^{-4}}$$

The frequency response is below.

Solution
The SOS implementation is below with global gain.

$$H(z) = 0.0940\frac{1 + 2z^{-2} + z^{-2}}{1 + 0.0396z^{-2}}\frac{1 + 2z^{-2} + z^{-2}}{1 + 0.4465z^{-2}}$$

The corresponding pole zero plot is below.
Distribute the global gain into individual stage as below.

$$H(z) = 0.3066\frac{1 + 2z^{-2} + z^{-2}}{1 + 0.0396z^{-2}}0.3066\frac{1 + 2z^{-2} + z^{-2}}{1 + 0.4465z^{-2}} = H_1(z)H_2(z)$$

The scale factor for cumulative stage is below. Choose one of the norms for factor.

$$\frac{1}{s_1} = \begin{cases} \sum_{k=0}^{\infty} |h_1[k]| = L_1 \approx 1.2515 \\ \max_{\omega} |H_1(e^{j\omega})| = L_\infty = 1.1796 \\ \sqrt{\sum_{k=0}^{\infty} (h_1[k])^2} = L_2 \approx 0.7465 \end{cases}$$

$$\frac{1}{s_2} = \begin{cases} \sum_{k=0}^{\infty} |h[k]| = L_1 \approx 1.6469 \\ \max_{\omega} |H(e^{j\omega})| = L_\infty = 1.0000 \\ \sqrt{\sum_{k=0}^{\infty} (h[k])^2} = L_2 \approx 0.7071 \end{cases}$$

Observe that the $h[n]$ and $H(e^{j\omega})$ for the s_2 is the overall impulse and frequency response, respectively. The scaled SOS is derived as below.

$$\breve{H}_1(z) = 0.3066 s_1 \frac{1 + 2z^{-1} + z^{-2}}{1 + 0.0396 z^{-2}} = \frac{\breve{b}_{01} + \breve{b}_{11} z^{-1} + \breve{b}_{21} z^{-2}}{1 + a_{11} z^{-1} + a_{21} z^{-2}}$$

$$\breve{H}_2(z) = 0.3066 \frac{s_2}{s_1} \frac{1 + 2z^{-1} + z^{-2}}{1 + 0.4465 z^{-2}} = \frac{\breve{b}_{02} + \breve{b}_{12} z^{-1} + \breve{b}_{22} z^{-2}}{1 + a_{12} z^{-1} + a_{22} z^{-2}}$$

The cascade connection for LPF is shown below in Fig. 7.26. The $H_1(z)$ contains the pole pair close to the origin; therefore, the frequency response expects the less peaky response.

For individual scaling factor, the frequency responses to $y_1[n]$ and $y_2[n]$ from input $x[n]$ are shown in Fig. 7.27. The $|H_1(e^{j\omega})||H_2(e^{j\omega})|$ is equivalent to the absolute magnitude of the overall response. The scaling factor only modifies the height of the individual response without altering the shape. The L_1 demonstrates the most conservative scaling which prevents the overflow with confidence. However, the L_2 shows the most liberal scaling which improve the output SNR with sacrificing the overflow.

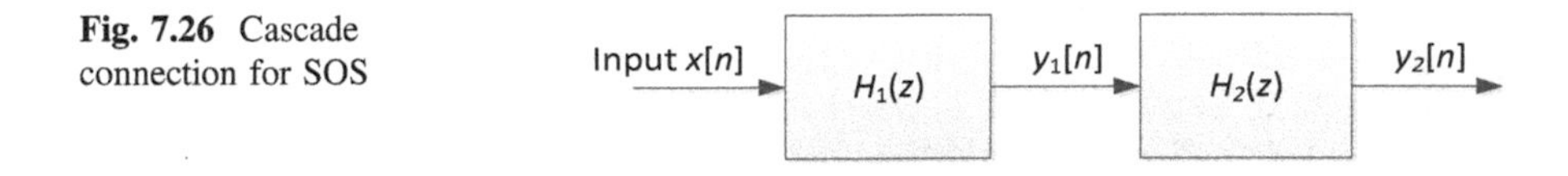

Fig. 7.26 Cascade connection for SOS

Prog. 7.8 MATLAB program for Figs. 7.24, 7.25, and 7.27.

```
[b,a] = butter(4,0.5);
[z,p,k] = tf2zp(b,a);
[h0,w0] = freqz(b,a,2^10);
[sos0,g0] = tf2sos(b,a);
bb = sqrt(g0)*sos0(1,1:3);
aa = sos0(1,4:6);
bbb = conv(bb*sqrt(g0),sos0(2,1:3));
aaa = conv(aa,sos0(2,4:6));
[h1,w1] = freqz(bb,aa,2^10);
[h2,w2] = freqz(bbb,aaa,2^10);
[hn1,n1] = impz(bb,aa,2^10);
[hn2,n2] = impz(bbb,aaa,2^10);
s1_1 = 1/norm(hn1,1); bb1_1 = s1_1*bb;
s1_2 = 1/norm(hn2,1); bbb1_2 = s1_2*bbb;
sinf_1 = 1/max(abs(h1)); bbinf_1 = sinf_1*bb;
sinf_2 = 1/max(abs(h2)); bbbinf_2 = sinf_2*bbb;
s2_1 = 1/norm(hn1,2); bb2_1 = s2_1*bb;
s2_2 = 1/norm(hn2,2); bbb2_2 = s2_2*bbb;
[h3,w3] = freqz(bb1_1,aa,2^10);
[h4,w4] = freqz(bbb1_2,aaa,2^10);
[h5,w5] = freqz(bbinf_1,aa,2^10);
[h6,w6] = freqz(bbbinf_2,aaa,2^10);
[h7,w7] = freqz(bb2_1,aa,2^10);
[h8,w8] = freqz(bbb2_2,aaa,2^10);

figure,
zplane(z,p);
figure,
plot(w1/pi,abs(h0)), grid
xticks([0 1/4 1/2 3/4 1]), ylim([0 1.25])
figure,
subplot(221), plot(w2/pi,abs(h1),w3/pi,abs(h2)), grid
xticks([0 1/4 1/2 3/4 1]), ylim([0 1.75])
subplot(222), plot(w2/pi,abs(h3),w3/pi,abs(h4)), grid
xticks([0 1/4 1/2 3/4 1]), ylim([0 1.75])
legend('|H_1(e^{j\omega})|','|H_1(e^{j\omega})||H_2(e^{j\omega})|')
subplot(223), plot(w2/pi,abs(h5),w3/pi,abs(h6)), grid
xticks([0 1/4 1/2 3/4 1]), ylim([0 1.75])
subplot(224),plot(w2/pi,abs(h7),w3/pi,abs(h8)), grid
xticks([0 1/4 1/2 3/4 1]), ylim([0 1.75])
```

The L_∞ presents the moderate scaling that provides the balance between the output SNR and overflow possibility. ∎

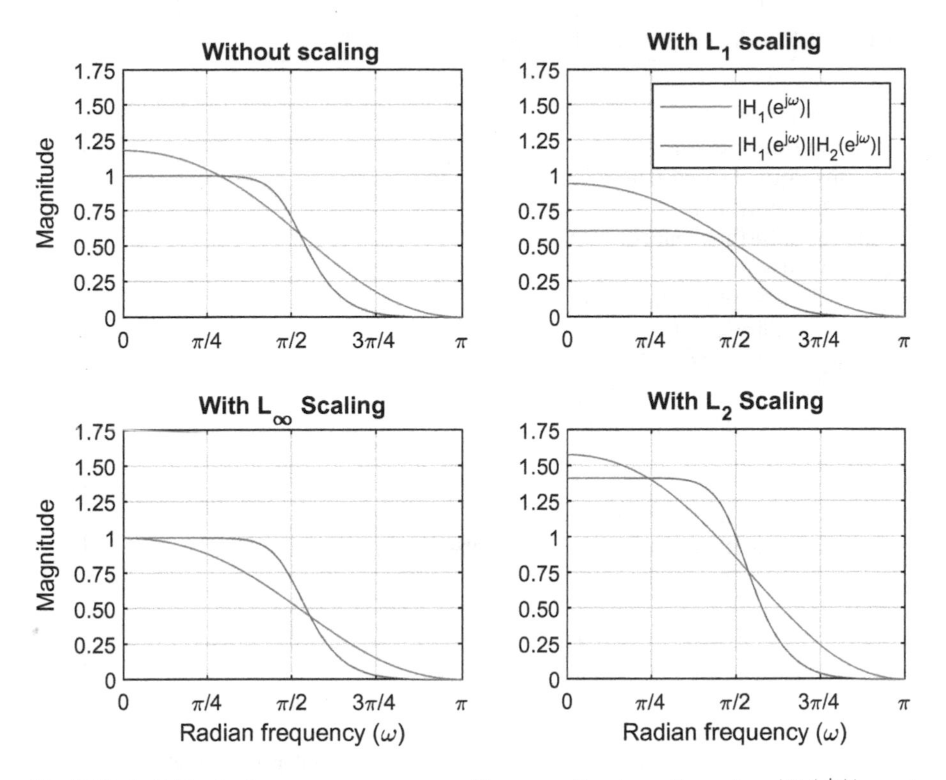

Fig. 7.27 Individual frequency response with or without scaling for $|H_1(e^{j\omega})|$ and $|H_1(e^{j\omega})||H_2(e^{j\omega})|$

Example 7.3
The overflow likelihood for each scaling method can be investigated by simple MATLAB experiment as below. The random signal with zero mean and uniform distribution is provide to the system input of Example 7.2. The input is ranged from −1 to 1 to avoid the overflow and shown below as histogram.

Solution
The subsequent histograms represent the outputs of the LPF SOS for each scaling method. The histogram bin size for overflow range is exaggerated for better representation. The blue and brown bin signifies the $y_1[n]$ and $y_2[n]$ output, respectively.

Prog. 7.9 MATLAB program for Figs. 7.28 and 7.29.

```
rng('default')
xn = rand(2^16,1)*2-1;
nxn = xn./max(abs(xn));
% ... Insert the Prog. 7.8 here for below variables
out0_1 = filter(bb, aa, nxn);
out0_2 = filter(bbb, aaa, out0_1);
out1_1 = filter(bb1_1, aa, nxn);
out1_2 = filter(bbb1_2, aaa, out1_1);
outinf_1 = filter(bbinf_1, aa, nxn);
outinf_2 = filter(bbbinf_2, aaa, outinf_1);
out2_1 = filter(bb2_1, aa, nxn);
out2_2 = filter(bbb2_2, aaa, out2_1);

figure,
edges = [-1.5 -1:1/16:1 1.5];
histogram(nxn,edges,'Normalization','probability'), grid;
set(gca,'YScale','log'), ylim([10^-4 1])
figure,
subplot(221), h1 = histogram(out0_1,edges,'Normalization','probability'),
grid; hold on
h1 = histogram(out0_2,edges,'Normalization','probability');
set(gca,'YScale','log'), ylim([10^-4 1])
subplot(222),h1 = histogram(out1_1,edges,'Normalization','probability'),
grid; hold on
h1 = histogram(out1_2,edges,'Normalization','probability');
set(gca,'YScale','log'), ylim([10^-4 1])
subplot(223),h1 = histogram(outinf_1,edges,'Normalization','probability'),
grid; hold on
h1 = histogram(outinf_2,edges,'Normalization','probability');
legend('y_1[n]','y_2[n]')
set(gca,'YScale','log'), ylim([10^-4 1])
subplot(224),h1 = histogram(out2_1,edges,'Normalization','probability'),
grid; hold on
h1 = histogram(out2_2,edges,'Normalization','probability');
set(gca,'YScale','log'), ylim([10^-4 1])
```

As expected, L_1 shows the least and L_2 illustrates the most overflow probability. The L_∞ is middle and less than the unscaled in terms of overflow likelihood. We can select the scaling method based on the requirement of the overflow and dynamic range. Fundamentally, the scaling does not change the system stability; however, fixed point number scaling could induce the pole relocations due to the sensitivity. After design and scaling, the procedure to check the response and stability is recommended. ∎

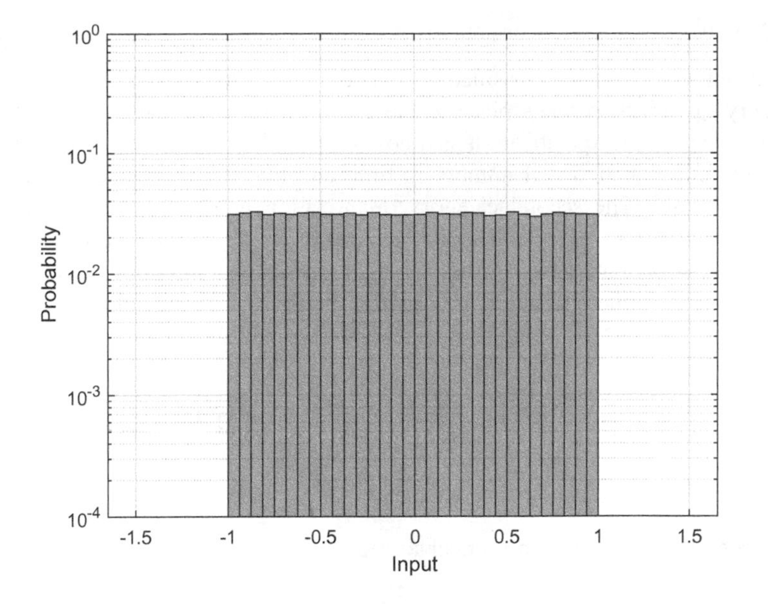

Fig. 7.28 Input signal probability histogram to investigate the overflow

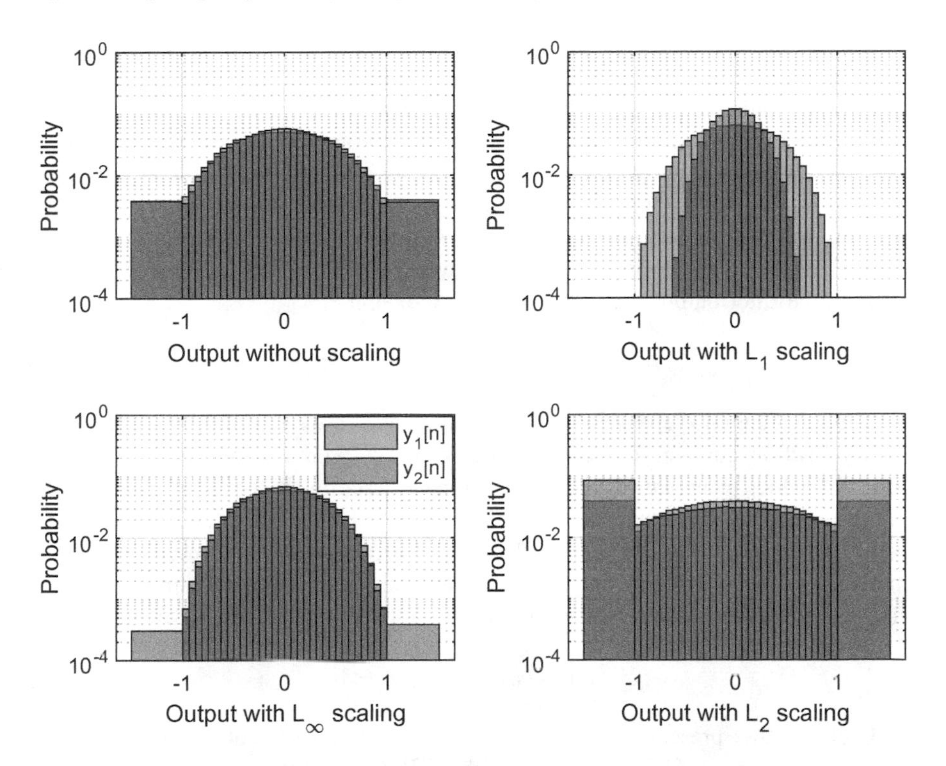

Fig. 7.29 Probability histograms of SOS output $y_1[n]$ and $y_2[n]$ with random input shown in Fig. 7.28

– Overall quantization error sources in filters

The quantization can be classified as two different categories as below. The first category quantization takes place at the signal acquisition stage along with sampling process. The quantization is known as analog to digital conversion. The analog signal which is real number in time and magnitude is discretized into the integer domain. The second category quantization happens at the multiplication stage which requires the double length space to store the product with full precision. If the hardware cannot support the sufficient storage, the multiplication outcome is truncated for further processing. Usually the LSBs are abandoned. The noise variance of two categories are shown below. Note that the data is kept at $B + 1$ bit storage.

1. Input signal quantization from real number to binary fraction number

$$\sigma_{noise}^2 = \frac{X_m^2 2^{-2B}}{12} \tag{7.106}$$

2. Large bit binary fraction number to reduced size binary fraction number

$$\sigma_{noise}^2 = \frac{2^{-2B}}{12} \tag{7.107}$$

The digital filter accepts the quantized signal $x[n]$ and produces the corresponding output $y[n]$. The only quantization related to the digital filter is the second category quantization from multiplication. The first category quantization is concerned at the front-end of the digital filter system. Note that we previously analyzed the SNR for the input signal quantization due to the conversion between the real number to the integer number (binary fraction number). After the analog to digital conversion, the information is processed as quantized and sampled form in the digital filter. We need to investigate the quantization from the multiplication as the SNR in the filter output. Let's see the below example.

$$y[n] = b_0 x[n] + b_1 x[n-1] + b_2 x[n-2] \tag{7.108}$$

The input signal $x[n]$ and individual coefficient b_k are quantized by the $B + 1$ bit binary fraction number. The current output $y[n]$ is produced by the sum of three products which generate the $2B + 2$ bit binary fraction. If the given FIR filter is properly scaled, the output $y[n]$ can be represented by the $2B + 2$ bit number with perfect precision. For any size FIR filter, the output can be described by the manageable length arithmetic operations. However, the IIR filter requires the further size to realize the system without the quantization problem. See below example.

$$y[n] = b_0 x[n] + a_1 y[n-1] \tag{7.109}$$

Each product presents the $2B + 2$ bit outcome and sum of them delivers the current output $y[n]$. Unlike FIR filter, the IIR filter contains the recursive loop which uses the current output as the input to the next output. Therefore, the output size should be increased by $B + 1$ bits for every iterations; otherwise, the output loses information precision. The IIR filter cannot be implemented with the perfect precision due to the recursive property.

The proper quantization is required for FIR and IIR filters because of hardware limitation. Below equation shows the SOS with quantizer for each product computation. The $2B + 2$ bit product outcome is converted to the $B + 1$ bit result by quantizer. Note that the summation for output $y[n]$ is carefully scaled by the scaling factor to reduce the overflow probability.

$$\begin{aligned} y[n] = Q_B(b_0x[n]) + Q_B(b_1x[n-1]) + Q_B(b_2x[n-2]) \\ -Q_B(a_1y[n-1]) - Q_B(a_2y[n-2]) \end{aligned} \tag{7.110}$$

The quantization for the product can be considered as the additive error which is the difference between the real product and quantized product. Subsequent equation demonstrates the quantization with induced error.

$$\begin{aligned} y[n] = (b_0x[n] + e_0[n]) + (b_1x[n-1] + e_1[n]) + (b_2x[n-2] + e_2[n]) \\ -(a_1y[n-1] - e_3[n]) - (a_2y[n-2] - e_4[n]) \end{aligned} \tag{7.111}$$

According to the previous derivation, the variance for the individual quantized error is shown below.

$$\sigma_{e_k}^2 = \frac{2^{-2B}}{12} k = 0, 1, \ldots, 4 \tag{7.112}$$

The quantization error can be represented by single noise source as long as all the noise sources $e_k[n]$ are independent of the input and each other.

$$\begin{aligned} y[n] = b_0x[n] + b_1x[n-1] + b_2x[n-2] - a_1y[n-1] - a_2y[n-2] + e[n] \\ \text{where } e[n] = e_0[n] + e_1[n] + e_2[n] + e_3[n] + e_4[n] \end{aligned} \tag{7.113}$$

The overall quantization noise variance is the addition of individual variances based on the independence condition. The total variance is shown below.

$$\sigma_e^2 = 5\frac{2^{-2B}}{12} \tag{7.114}$$

The $y[n]$ is the weighted combination of input $x[n]$s, output $y[n]$s, and noise $e[n]$; therefore, the generated noise can be regarded as another input to the filter. The filter output is named as $r[n]$ for pure noise input $e[n]$. Since there is no input, the difference equation can be organized as below.

$$r[n] = -a_1r[n-1] - a_2r[n-2] + e[n] \tag{7.115}$$

The output noise variance σ_r^2 is proportional to the absolute squared sum of the impulse response which is derived from noise source only. The higher impulse response energy increases the output noise variance. Also, higher filter order amplifies the output noise variance as below.

$$\sigma_r^2 = \sigma_e^2 \sum_{n=0}^{\infty} |h_{e2r}[n]|^2 = 5\frac{2^{-2B}}{12} \sum_{n=0}^{\infty} |h_{e2r}[n]|^2 \tag{7.116}$$

For general filter, the transfer function is described as below. The filter type decides the numerator and denominator order.

$$H(z) = \frac{b_0 + \cdots + b_{M-1}z^{-(M-1)} + b_M z^{-M}}{1 + a_1 z^{-1} + \cdots + a_N z^{-N}} = \frac{B(z)}{A(z)} \tag{7.117}$$

The output noise variance requires the impulse response from the quantization noise without input signal. The denominator of the transfer function is utilized to compute the impulse response as below. Observe that the numerator contributes to the input signal only.

$$h_{e2r}[n] \overset{z-trans}{\leftrightarrow} \frac{1}{A(z)} \tag{7.118}$$

The given transfer function provides the below output noise variance due to the quantization. The assigned bit size B is only positive factor for the output noise variance; in other words, the increased bit size reduces the noise variance.

$$\sigma_r^2 = (M + 1 + N)\frac{2^{-2B}}{12} \sum_{n=0}^{\infty} |h_{e2r}[n]|^2 \tag{7.119}$$

For FIR filter, there is no recursive output loop; hence, the output noise variance is presented as below.

$$\sigma_r^2 = (M + 1)\frac{2^{-2B}}{12} \tag{7.120}$$

One method to improve the output noise variance is employing the double length accumulator (multiplier and adder) and applying the quantization at the final step. The $2B + 2$ bit products are processed without length modification to cumulate. The accumulated output is quantized at the last step as shown in below. Hence, there is single quantization procedure in any type and order filter.

$$y[n] = Q_B(b_0 x[n] + \cdots + b_M x[n - M] - a_1 y[n - 1] - \ldots a_N y[n - N]) \tag{7.121}$$

The single quantization step is equivalent to the single noise source as below.

$$\begin{aligned} y[n] = b_0 x[n] + \cdots + b_M x[n - M] - a_1 y[n - 1] - \\ \ldots - a_N y[n - N] + e[n] \end{aligned} \tag{7.122}$$

The noise variance is given below for the rounding approximation with uniform distribution.

$$\sigma_e^2 = \frac{2^{-2B}}{12} \tag{7.123}$$

Based on the single quantization noise, the output noise variance is derived as below.

$$\sigma_r^2 = \frac{2^{-2B}}{12} \sum_{n=0}^{\infty} |h_{e2r}[n]|^2 \text{ where } h_{e2r}[n] \overset{z-trans}{\leftrightarrow} \frac{1}{A(z)} \tag{7.124}$$

Compare to the multi-quantization noise performance, the single quantization with double length accumulator significantly improves the output noise variance in the order of filter length.

Example 7.4
Analyze the quantization noise performance. The designed filter indicates the LPF with 0.5π cutoff frequency in 4th order system. The design method is Butterworth.

$$\begin{aligned} H(z) &= \frac{0.0940 + 0.3759z^{-1} + 0.5639z^{-2} + 0.3759z^{-3} + 0.0940z^{-4}}{1 + 0.4860z^{-2} + 0.0177z^{-4}} \\ &= \frac{B(z)}{A(z)} \end{aligned}$$

Solution
The corresponding SOS for the given LPF is below. Note that the G is the global gain.

$$H(z) = 0.0940 \frac{1 + 2z^{-2} + z^{-2}}{1 + 0.0396z^{-2}} \frac{1 + 2z^{-2} + z^{-2}}{1 + 0.4465z^{-2}} = G \frac{B_1(z)}{A_1(z)} \frac{B_2(z)}{A_2(z)}$$

In order to compute the output noise variance, the impulse response for transfer function denominator is specified as below. Any scaling factors do not affect the impulse response since the scaling is applied over the numerator only.

$$h_{e2r}[n] \overset{z-trans}{\leftrightarrow} \frac{1}{A(z)}$$

$$h_{e2r1}[n] \overset{z-trans}{\leftrightarrow} \frac{1}{A_1(z)}$$

$$h_{e2r2}[n] \overset{z-trans}{\leftrightarrow} \frac{1}{A_2(z)}$$

The $h_{e2r}[n]$ is the denominator impulse response for overall LPF. The $h_{e2r1}[n]$ and $h_{e2r2}[n]$ is the first and second section of denominator SOS impulse response, respectively. The output noise variance for multiple quantization is calculated as below. The integer scaling factor in the variance is equivalent to the number of multiplications on the filter difference equation. Observe that the quadruple scaling in the σ_{r1}^2 and σ_{r2}^2 can be further reduced by use of particular coefficient such as 1 or 2. The selected bit length is 15 for B.

$$\sigma_r^2 = 7\frac{2^{-2B}}{12}\sum_{n=0}^{\infty}|h_{e2r}[n]|^2 = 7\frac{2^{-30}}{12}1.2959 = 7.0403\times10^{-10}$$

$$\sigma_{r1}^2 = 4\tfrac{2^{-2B}}{12}\sum_{n=0}^{\infty}|h_{e2r1}[n]|^2 = 4\tfrac{2^{-30}}{12}1.0016 = 3.1093\times10^{-10}$$

$$\sigma_{r2}^2 = 4\frac{2^{-2B}}{12}\sum_{n=0}^{\infty}|h_{e2r2}[n]|^2 = 4\frac{2^{-30}}{12}1.2490 = 3.8774\times10^{-10}$$

The output noise variance for single quantization system is derived as below. The integer scaling is not required; therefore, the variance is significantly lower than the multi quantization counterpart.

$$\sigma_r^2 = \frac{2^{-2B}}{12}\sum_{n=0}^{\infty}|h_{e2r}[n]|^2 = \frac{2^{-30}}{12}1.2959 = 1.0058\times10^{-10}$$

$$\sigma_{r1}^2 = \tfrac{2^{-2B}}{12}\sum_{n=0}^{\infty}|h_{e2r1}[n]|^2 = \tfrac{2^{-30}}{12}1.0016 = 7.7732\times10^{-11}$$

$$\sigma_{r2}^2 = \frac{2^{-2B}}{12}\sum_{n=0}^{\infty}|h_{e2r2}[n]|^2 = \frac{2^{-30}}{12}1.2490 = 9.6935\times10^{-11}$$

The output SNR from the quantization can be calculated by the previously derived output noise variance as below. The σ_{signal}^2 is the output signal variance which can be computed by various methods. We assume that the σ_{signal}^2 is given.

$$SNR = 10\log_{10}\left(\tfrac{\sigma_{signal}^2}{\sigma_{noise}^2}\right) = 10\log_{10}\left(\sigma_{signal}^2\right) - 10\log_{10}\left(\sigma_{noise}^2\right)$$

The calculated SNRs are given as below in Table 7.5.

Table 7.5 SNR of given $H(z)$ for both quantization scenarios

	Multi quantizations	Single quantization
SNR_r	$10\log_{10}\left(\sigma_{signal}^2\right)+91.5241$ dB	$10\log_{10}\left(\sigma_{signal}^2\right)+99.9751$ dB
SNR_{r1}	$10\log_{10}\left(\sigma_{signal}^2\right)+95.0734$ dB	$10\log_{10}\left(\sigma_{signal}^2\right)+101.094$ dB
SNR_{r2}	$10\log_{10}\left(\sigma_{signal}^2\right)+94.1146$ dB	$10\log_{10}\left(\sigma_{signal}^2\right)+100.135$ dB

The single quantization system demonstrates the higher SNR values than the multi quantization system's SNR as we expected. The cascade SOS system presents the increased SNR than the single overall filter system in the individual section. Therefore, we recommend the single quantization SOS cascade filter system with double length accumulator.

■

7.3 FIR Filter Implementations

– FIR realization structures

The FIR filter produces the output $y[n]$ by summing the weighted recent $N + 1$ inputs. The mathematical representation is below. Note that the b_k and $h[k]$ are equivalent mutually for the FIR filter.

$$y[n] = b_0x[n] + b_1x[n-1] + b_2x[n-2] + \cdots + b_Nx[n-N] \tag{7.125}$$

The corresponding block diagram is illustrated in Fig. 7.30 and known as direct form or transversal structure. The signal follows the arrow line for multiplication, addition, and delay. The signal is separated by the delay block in time; in other words, the signal is propagated though the delay block in every synchronized discrete time. For example, the signal before the delay block is the $x[n]$ and after the block is the $x[n - 1]$. The signals over the any wires, additions, and multiplications indicate the certain times to yield the current output $y[n]$.

The MATLAB provides the convolution sum by the *conv* function. The FIR filter realization by using the computer language requires the following steps.

1. Initialize the program by defining below constant and empty variables.
   - $N + 1$ length memory initialized by b_k
   - $N + 1$ length memory initialized by zero for delay components.
2. Shift right the delay component
   - Shift right one memory element at a time from right-end to the left-bound
   - The oldest (rightest) data is removed by the shift right operation

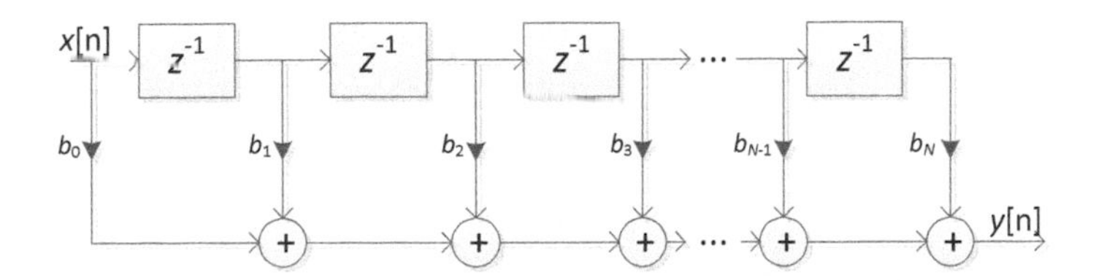

Fig. 7.30 Block diagram for FIR filter in direct form or transversal structure

– The established blank space by the shift operation is stuffed by the new data $x[n]$.

3. Accumulate the products for output $y[n]$

– $y[n] = \sum_{k=0}^{N} b_k x[n-k]$

4. Repeat the step 2 and 3 for all input data.

Below block diagram indicates the execution sequence for step 2 and 3. Further algorithm exploration can reduce the execution complexity; however, overall computational procedure is maintained as shown in Fig. 7.31.

The reversal operations in the block diagram provides the equivalent filter. The procedure is known as transposition [6] and described as following.

– Reverse direction of each signal flow
– Reverse direction of each multiplication and interconnection
– Change junctions to additions and vice versa
– Interchange the input and output signals
– Flipping the diagram horizontally (optional).

The direct form FIR filter is converted to the transposed direct (or transversal) FIR filter as below. Without the horizontal flipping, the input is located on the right as below in Fig. 7.32.

To follow the conventional structure, below figure shows the flipped transposed FIR filter. The computational complexity is identical due to the equal number of additions, multiplications, and delays between the direct and transposed FIR filter as shown in Fig. 7.33.

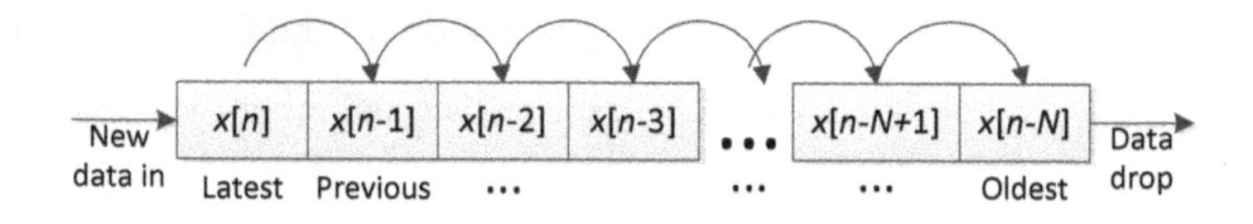

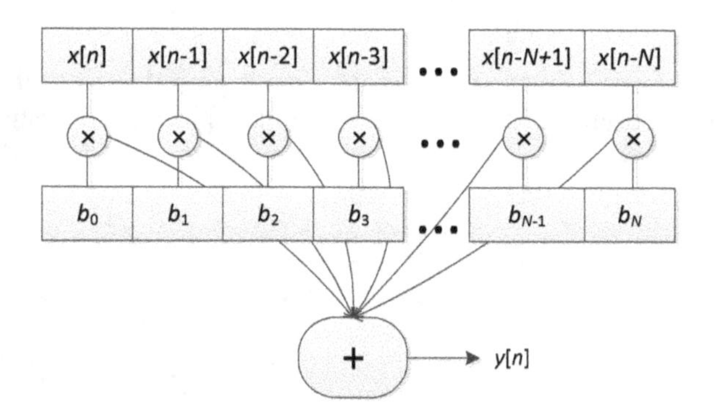

Fig. 7.31 Block diagram for FIR filter realization by computer language

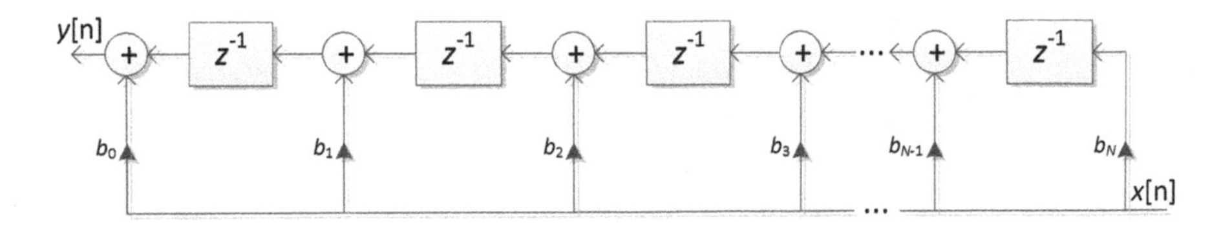

Fig. 7.32 Block diagram for FIR filter realization in transposed direct form

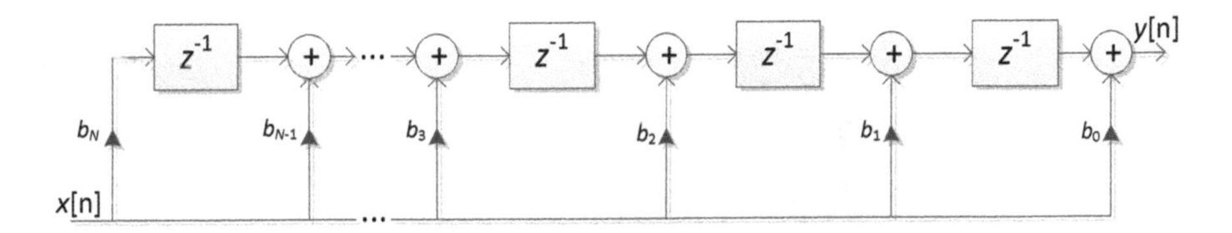

Fig. 7.33 Block diagram for FIR filter realization in flipped transposed direct form

The computational procedure of the transposed FIR filter is illustrated below as block diagram. The direct form performs the computation by shifting and accumulating in the divided stage. The transposed FIR filter executes the shifting and accumulating simultaneously as below in Fig. 7.34.

The circle in the above figure denotes the kth fundamental computational unit which assigns to the subsequent memory $u[n-k]$ with sum of previous memory $u[n-k+1]$ value and b_{N-k} weighted recent value $x[n]$. The computational unit is realized from right-end to the left-bound. The transposed FIR filter implementation by using the computer language requires the following steps.

1. Initialize the program by defining below constant and empty variables.
   - $N+1$ length memory initialized by b_k
   - N length memory initialized by zero for delay components
2. Execute following sequence

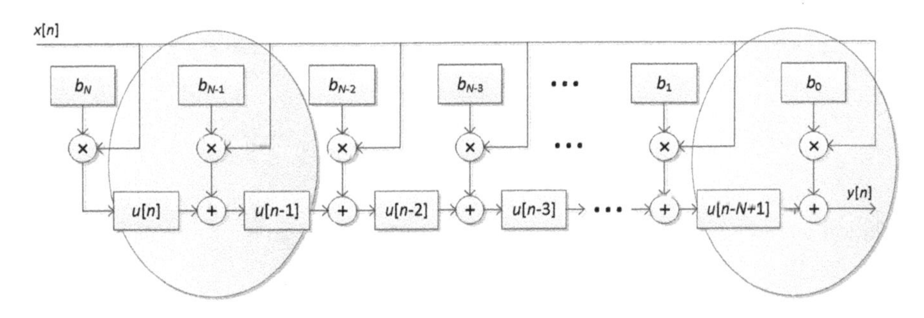

Fig. 7.34 Computational procedure of the transposed FIR filter

- $y[n] = u[n-N+1] + b_0x[n]$
- $u[n-k] = u[n-k+1] + b_{N-k}x[n]$
  for $k = N-1, N-2, \ldots 2, 1$
- $u[n] = b_Nx[n]$

3. Repeat the step 2 for all input data.

According to the previous analysis, we should stay away from the high order filter design because of numerical ill-condition generated by high sensitivity. Let's revisit the sensitivity analysis. The given FIR filter with N order is below.

$$H(z) = b_0 + \cdots + b_{N-1}z^{-(N-1)} + b_Nz^{-N} \tag{7.126}$$

Rewrite the transfer function in terms of zeros as below.

$$H(z) = b_0(1 - z_1z^{-1})(1 - z_2z^{-1})\ldots(1 - z_Nz^{-1}) \tag{7.127}$$

The derived sensitivity of the FIR filter is below.

$$\frac{\partial z_j}{\partial b_k} = -\frac{z_j^{N-k}}{b_0\prod_{i=1,i\neq j}^{N}(z_j - z_i)} \tag{7.128}$$

The filter order and zero locations determine the sensitivity of FIR filter. In fact, the direct form is generally utilized for FIR filter realization because the zeros are likely to be spread evenly for linear phase FIR filter. In addition, the zero-location variation does not produce the filter instability in FIR filter. As long as the frequency response follows the desired specification with tolerable range, the direct form realization is feasible choice. The frequency response of the FIR filter can present the impact of the coefficient quantization error for worst situation. Below is the definition of coefficient quantization error. Note that the b_k and $h[k]$ are equal for FIR filter.

$$Q_B(h[k]) = h[k] + h_e[k] \text{ where } k = 0, 1, \ldots, N \tag{7.129}$$

The frequency response of the FIR filter is given below.

$$\begin{aligned} \breve{H}(e^{j\omega}) &= \sum_{n=0}^{N} Q_B(h[n])e^{-j\omega n} = \sum_{n=0}^{N}(h[n] + h_e[n])e^{-j\omega n} \\ &= H(e^{j\omega}) + H_e(e^{j\omega}) \text{ where } H_e(e^{j\omega}) = \sum_{n=0}^{N} h_e[n]e^{-j\omega n} \end{aligned} \tag{7.130}$$

The range of the frequency response error is derived by using the inequality relation as below.

$$\left|H_e(e^{j\omega})\right| \leq \sum_{n=0}^{N} |h_e[n]|\left|e^{-j\omega n}\right| \leq (N+1)2^{-(B+1)} \tag{7.131}$$

Example 7.5

For below configurations, find maximum frequency response error in magnitude from coefficient quantization. Also calculate the corresponding pass band ripple A_p and stop band attenuation A_{st}

- FIR filter with 31 length (N = 30) and 16-bit quantization
- FIR filter with 31 length (N = 30) and 8-bit quantization

Solution

The type I FIR filter with 31 length (N = 30) and 16-bit quantization (B = 15) presents the 0.00047302 maximum frequency response error in magnitude. The corresponding pass band ripple A_p and stop band attenuation A_{st} are below. Note that the R_p and R_{st} are established to the maximum frequency response error from the quantization.

$$\begin{aligned} A_P &= 40\log_{10}(1+R_p) = 40\log_{10}(1+0.00047302) = 0.0082\text{ dB} \\ A_{st} &= -20\log_{10}R_{st} = -20\log_{10}(0.00047302) = 66.5024\text{ dB} \end{aligned}$$

The identical filter with 31 length (N = 30) and 8-bit quantization (B = 7) presents the 0.1211 maximum frequency response error in magnitude. The corresponding pass band ripple A_p and stop band attenuation A_{st} is below.

$$\begin{aligned} A_P &= 40\log_{10}(1+R_p) = 40\log_{10}(1+0.1211) = 1.9858\text{ dB} \\ A_{st} &= -20\log_{10}R_{st} = -20\log_{10}(0.1211) = 18.3376\text{ dB} \end{aligned}$$

■

The derived maximum frequency response error from the quantization is very conservative limit because of the rare possibility to hit the maximum error for all coefficient quantizations. By comparing the two situations above, sufficiently long bit assignment for quantization exponentially improves the filter performance in term of ripple. The filter length is the linear factor in the maximum error function as Eq. (7.131); hence, the direct form can be used for the filter realization. The increased filter length could be counterbalanced by the higher precision quantization for acceptable magnitude ripples. The FIR filter with very rough quantization or high-order system based on the closely spaced zeros can be considered to realize smaller sets of zeros with cascade structure as shown below.

$$\begin{aligned} H(z) &= H_1(z)H_2(z) \\ &= (b_{01}+b_{11}z^{-1}+b_{21}z^{-2})(b_{02}+b_{12}z^{-1}+b_{22}z^{-2}) \end{aligned} \tag{7.132}$$

The above example consists of two second order FIR filters in cascaded structure to deliver fourth order overall. The FIR filter with linear phase or unit circle zeros provide the further exploration on the realization structure. Numerous variations which are not shown here can be found under the topic of FIR filter realizations.

7.4 IIR Filter Implementations

– IIR realization structures

The general transfer function for the IIR filter is shown again in below. The IIR filter includes the feedforward as well as feedback component simultaneously. The feedforward uses the input signal $x[n]$ with delays and the feedback utilizes the output signal $y[n]$ with delays.

$$H(z) = \frac{b_0 + \cdots + b_{N-1}z^{-(N-1)} + b_N z^{-N}}{1 + a_1 z^{-1} + \cdots + a_M z^{-M}} = \frac{B(z)}{A(z)} \tag{7.133}$$

The given transfer function can be considered as the cascade connection between the $B(z)$ and $1/A(z)$. Therefore, the feedforward realization is followed by the feedback implementation as below. This realization is called direct form I which is the fundamental structure as shown in Fig. 7.35.

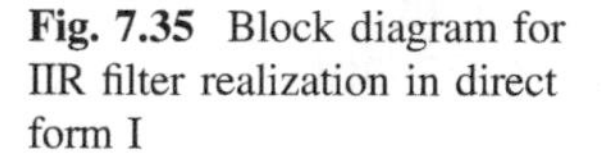

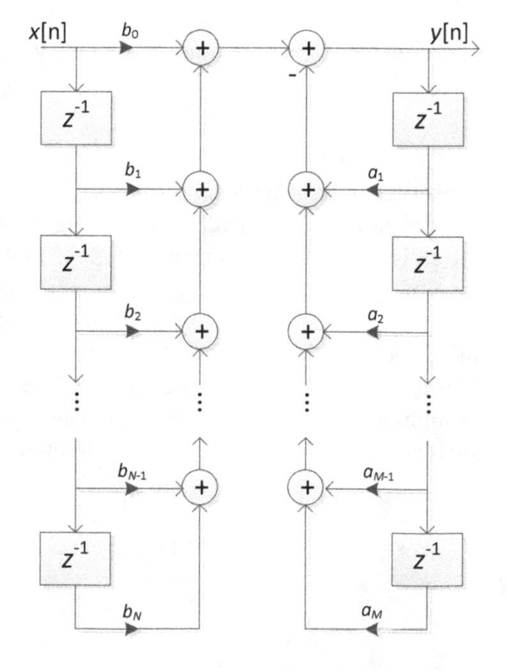

Fig. 7.35 Block diagram for IIR filter realization in direct form I

The IIR filter in direct form I realization by using the computer language requires the following steps.

1. Initialize the program by defining below constant and empty variables.
   - $N + 1$ length memory initialized by b_k
   - M length memory initialized by a_k
   - $N + 1$ length memory initialized by zero for feedforward delays
   - M length memory initialized by zero for feedback delays
2. Shift right the feedforward delays
   - Shift right one memory element at a time from right-end to the left-bound
   - The oldest (rightest) data is removed by the shift right operation
   - The established blank space by the shift operation is stuffed by the new data $x[n]$
3. Accumulate the products for output $y[n]$
   - $y[n] = \sum_{k=0}^{N} b_k x[n-k] - \sum_{k=1}^{M} a_k y[n-k]$
4. Shift right the feedback delays
   - Shift right one memory element at a time from right-end to the left-bound
   - The oldest (rightest) data is removed by the shift right operation
   - The established blank space by the shift operation is stuffed by the new data $y[n]$
5. Repeat the step 2, 3, and 4 for all input data

Below block diagram indicates the execution sequence for step 2, 3, and 4. Further algorithm exploration can reduce the execution complexity; however, overall computational procedure is maintained as shown in Fig. 7.36.

The given transfer function also can be considered as the cascade connection between the $1/A(z)$ and $B(z)$ in reverse order of the direct form I. Therefore, the feedback realization is followed by the feedforward implementation as below Fig. 7.37a. The realization includes the double delay lines which represent the identical signal flow. Merging the delay lines presents simplified structure as the direct form II. The direct form II is shown in Fig. 7.37b and also known as the canonic form which uses the minimum number of delay elements in the realization. Using the transposition property, the direct form II filter is converted to the transposed direct form II (or canonic form) IIR filter as Fig. 7.37c. Transposition reduces the adder number; however, the hardware benefit is limited if the system does not support the three-port adder. Note that the order of numerator and denominator is equal for simple illustration.

The IIR filter in direct form II realization by using the computer language requires the following steps. Assume that M is equal to N here.

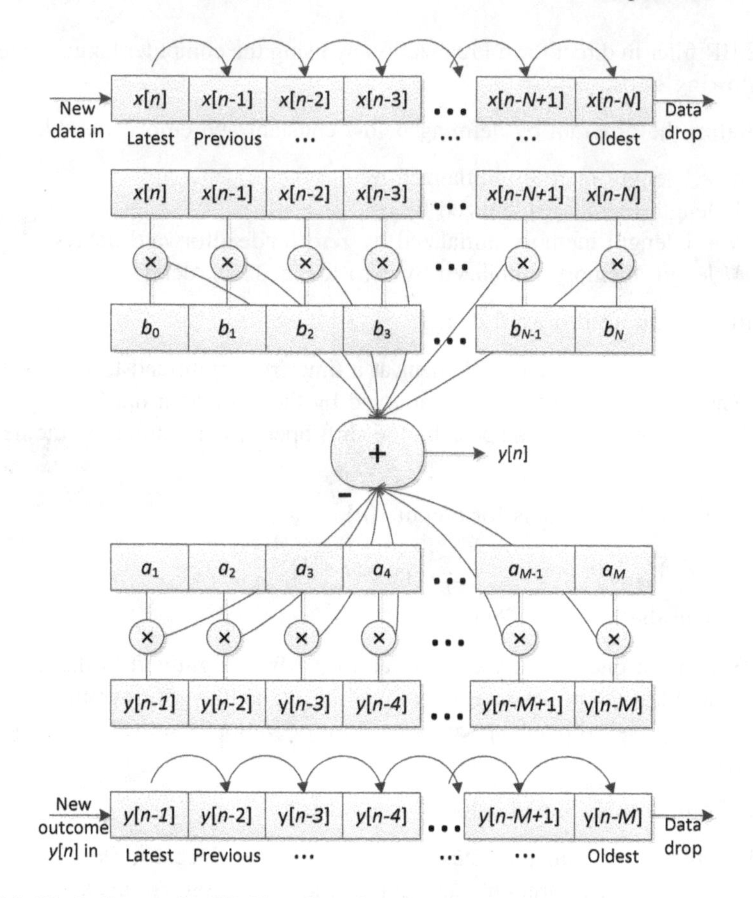

Fig. 7.36 Block diagram for IIR filter direct form I realization by computer language

1. Initialize the program by defining below constant and empty variables.
   - $N + 1$ length memory initialized by b_k
   - M length memory initialized by a_k
   - $N + 1$ (or $M + 1$) length memory initialized by zero for delays
2. Accumulate the products for $u[n]$ and $y[n]$
   - $u[n] = x[n] - \sum_{k=1}^{M} a_k u[n-k]$
   - $y[n] = \sum_{k=0}^{N} b_k u[n-k]$
3. Shift right the delay components
   - Shift right one memory element at a time from right-end to the left-bound
   - The oldest (rightest) data is removed by the shift right operation
   - The established blank space $u[n]$ by the shift operation leave empty. The space will be written by the accumulated products at step 2.

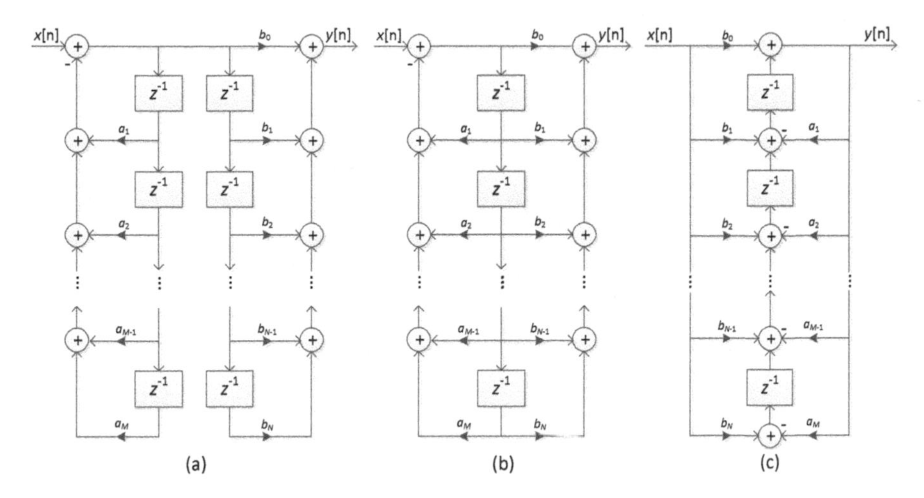

Fig. 7.37 Block diagram for IIR filter realization in direct form I with reverse order (**a**), direct form II (**b**), and transposed direct form II (**c**)

4. Repeat the step 2 and 3 for all input data

Below block diagram below indicates the execution sequence for step 2 and 3. Further algorithm exploration can reduce the execution complexity; however, overall computational procedure is maintained as shown in Fig. 7.38.

The IIR filter in transposed direct form II realization by using the computer language requires the following steps. Assume that M is equal to N here. The circle in the below figure denotes the kth fundamental computational unit for memory $u[n-k]$. The computational unit is executed from right-end to the left-bound.

1. Initialize the program by defining below constant and empty variables.
   - $N+1$ length memory initialized by b_k
   - M length memory initialized by a_k
   - M (or N) length memory initialized by zero for delays
2. Execute following sequence
   - $y[n] = u[n-M+1] + b_0 x[n]$
   - $u[n-k] = u[n-k+1] + b_{N-k}x[n] - a_{M-k}y[n]$
     for $k = M-1, M-2, \ldots 2, 1$ or $k = N-1, N-2, \ldots 2, 1$
   - $u[n] = b_N x[n] - a_M y[n]$
3. Repeat the step 2 for all input data

Previously we investigated the IIR filter sensitivity which presents the significant performance degradation in direct form realization. The cascade SOS structure is recommended to minimize the side effect from the quantization. The rule of thumb for pairing and sequencing was suggested earlier. The SOS structure can use the direct form II with or without the transposition as below Figs. 7.39 and 7.40.

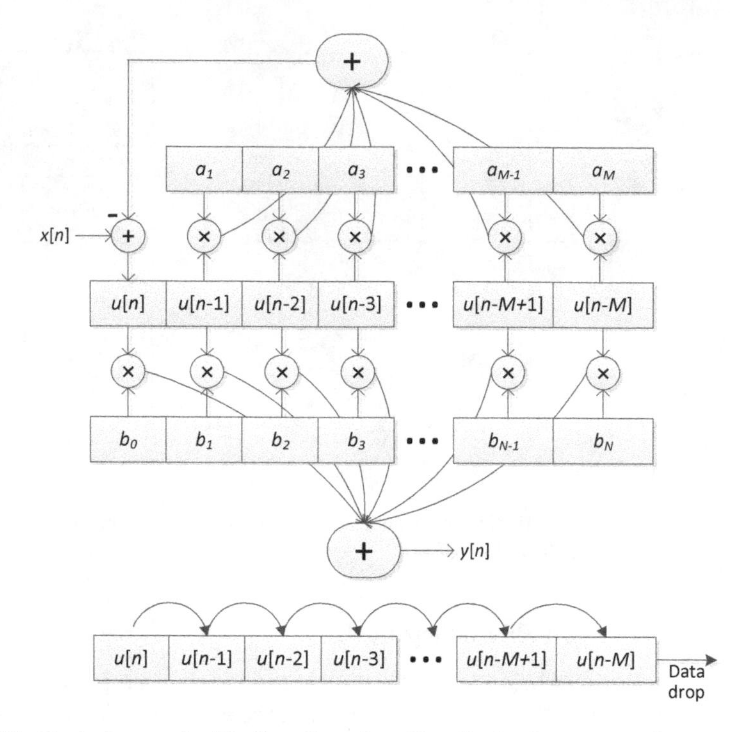

Fig. 7.38 Block diagram for IIR filter direct form II realization by computer language. Assume that M is equal to N

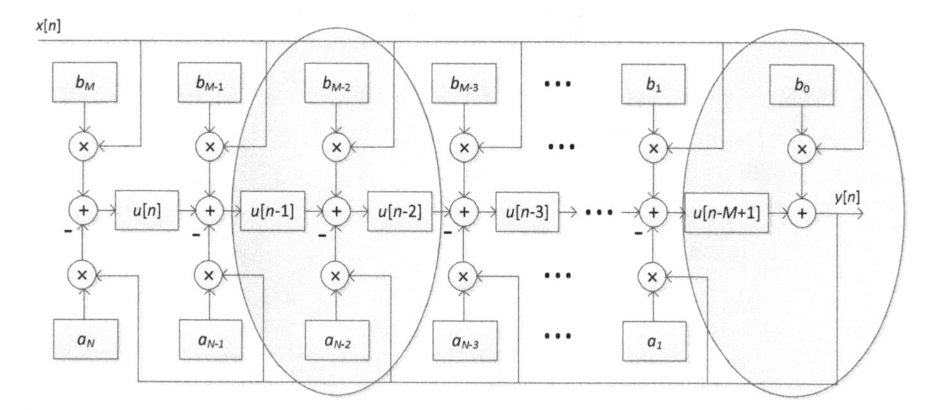

Fig. 7.39 Computational procedure of the transposed direct form II IIR filter

The SOSs are ordered from least-peaked to most-peaked in gain; therefore, the SOS with poles closest to the unit circle is located last in the cascade. In fact, the most-peaked to least-peaked ordering also shows the comparative performance as well. Both consistent orderings are used in the practical realization. The SOSs with

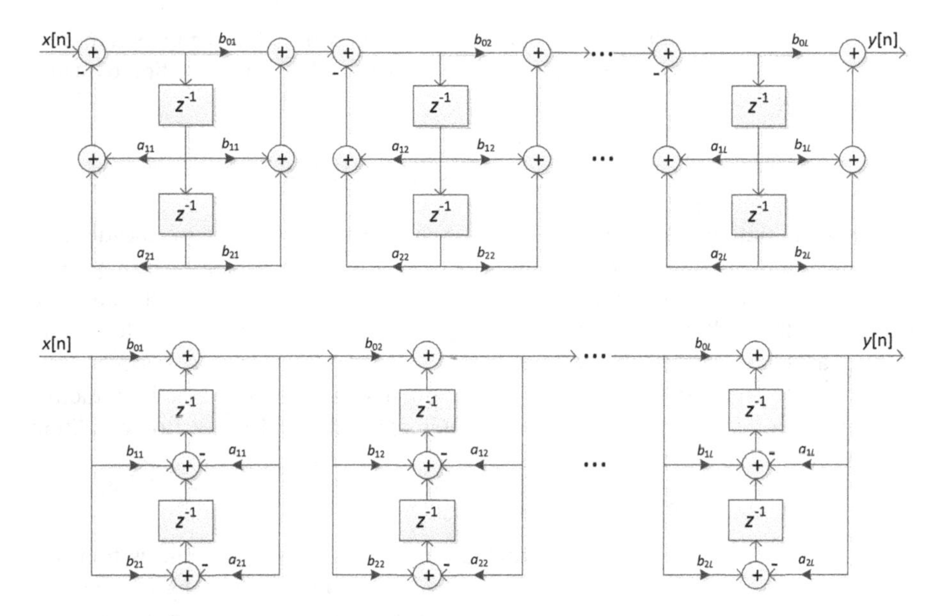

Fig. 7.40 Block diagram for transposed direct form II IIR filter realization in cascade SOS structure with (below) and without (above) transposition

parallel connection are another reasonable choice for the realization in terms of various performance aspects; however, the cascade implementation is more likely used because of zero controllability in coefficient quantization process.

The recursive property of the quantized IIR filter demonstrates the nonlinear characteristic known as limit cycle. The output could persist to fluctuate forever whilst the input stays zero. The limit cycle process is complicated and tricky to understand; hence, we will not try to discuss the limit cycle in detail here. Once we have the unexpected output from the zero input, the possible solutions are below.

- Increase the quantization length
- Use the quantization after the double length accumulation
- Use the saturation overflow

The further information can be found on the papers [9, 10]. The FIR filter cannot provide the limit cycle behavior because of the non-recursive routines in the computation. If the system cannot tolerate the limit cycle, then use the FIR filter.

7.5 Frequency Domain Filter Realization

The time domain filter is realized by the difference equation or convolution sum. The difference equation can apply on the FIR and IIR filter; however, the convolution sum is only valid for the FIR filter. Observe that the difference equation is equivalent to the convolution sum for the FIR filter with delayed inputs. Another implementation choice we can think of is the frequency domain realization by using the following relation.

$$y[n] = x[n] * h[n] \overset{\text{DFT}}{\leftrightarrow} Y[k] = X[k]H[k] \quad (7.134)$$

The convolution sum is identical to the simple multiplication on the discrete frequency domain. The benefits of the frequency domain filtering are complexity and simplicity. We can specify the filter requirement directly on $H[k]$ and realize the filtering by the multiplication. However, the cost of the benefit is the transformation known as DFT. The filter involves the DFT twice from and to between two domains. On the certain condition, the frequency domain filtering is attractive method in terms of complexity. The fast DFT algorithm and realization condition will be introduced in the following sections.

- Fast DFT algorithm

The DFT requires the significant computation for analysis and synthesis transformation. Below revisits the DFT equations. The complex number by exponential part is presented by the cosine and sine function; therefore, the complex number arithmetic increases the complexity.

$$\begin{aligned} X[k] &= \sum_{n=0}^{N-1} x[n]e^{-j\frac{2\pi}{N}kn} \\ x[n] &= \frac{1}{N}\sum_{k=0}^{N-1} X[k]e^{j\frac{2\pi}{N}kn} \end{aligned} \quad (7.135)$$

Note that the complex exponential demonstrates the periodicity property. The kn product in the exponential can be adjusted to the smaller value to improve the computation efficiency as below.

$$e^{-j\frac{2\pi}{N}kn} = e^{-j\frac{2\pi}{N}k(n+N)} = e^{-j\frac{2\pi}{N}k(n+2N)} = e^{-j\frac{2\pi}{N}(k+N)n} = \ldots \quad (7.136)$$

The equivalent matrix representation is shown below for the DFT.

$$\begin{bmatrix} X[0] \\ X[1] \\ \vdots \\ X[N-2] \\ X[N-1] \end{bmatrix} = \begin{bmatrix} 1 & 1 & \cdots & 1 & 1 \\ 1 & e^{-j\frac{2\pi}{N}1} & \cdots & e^{-j\frac{2\pi}{N}(N-2)} & e^{-j\frac{2\pi}{N}(N-1)} \\ \vdots & \vdots & \ddots & \vdots & \vdots \\ 1 & e^{-j\frac{2\pi}{N}(N-2)1} & \cdots & e^{-j\frac{2\pi}{N}(N-2)(N-2)} & e^{-j\frac{2\pi}{N}(N-2)(N-1)} \\ 1 & e^{-j\frac{2\pi}{N}(N-1)1} & \cdots & e^{-j\frac{2\pi}{N}(N-1)(N-2)} & e^{-j\frac{2\pi}{N}(N-1)(N-1)} \end{bmatrix} \begin{bmatrix} x[0] \\ x[1] \\ \vdots \\ x[N-2] \\ x[N-1] \end{bmatrix}$$

(7.137)

$$\begin{bmatrix} x[0] \\ x[1] \\ \vdots \\ x[N-2] \\ x[N-1] \end{bmatrix} = \frac{1}{N}\begin{bmatrix} 1 & 1 & \cdots & 1 & 1 \\ 1 & e^{j\frac{2\pi}{N}1} & \cdots & e^{j\frac{2\pi}{N}(N-2)} & e^{j\frac{2\pi}{N}(N-1)} \\ \vdots & \vdots & \ddots & \vdots & \vdots \\ 1 & e^{j\frac{2\pi}{N}(N-2)1} & \cdots & e^{j\frac{2\pi}{N}(N-2)(N-2)} & e^{j\frac{2\pi}{N}(N-2)(N-1)} \\ 1 & e^{j\frac{2\pi}{N}(N-1)1} & \cdots & e^{j\frac{2\pi}{N}(N-1)(N-2)} & e^{j\frac{2\pi}{N}(N-1)(N-1)} \end{bmatrix}\begin{bmatrix} X[0] \\ X[1] \\ \vdots \\ X[N-2] \\ X[N-1] \end{bmatrix} \tag{7.138}$$

For given problem size N, the DFT involves the N^2 complex multiplications and $N(N-1)$ complex additions. The algorithm complexity is evaluated by the number of multiplications; hence, the DFT complexity is order of N^2. The decomposition in time or frequency considerably reduces the DFT complexity. The following is the decimation-in-time fast Fourier transform (FFT) algorithm [9] which is the fast version of the DFT. The FFT needs the power of two data length to perform the algorithm. The length N in the below DFT is 2^L.

$$\begin{aligned} X[k] &= x[0] + x[1]e^{-j\frac{2\pi}{N}k1} + x[2]e^{-j\frac{2\pi}{N}k2} + \cdots \\ &+ x[N-2]e^{-j\frac{2\pi}{N}k(N-2)} + x[N-1]e^{-j\frac{2\pi}{N}k(N-1)} \end{aligned} \tag{7.139}$$

Divide into the even and odd data part as below.

$$\begin{aligned} &= x[0] + x[2]e^{-j\frac{2\pi}{N}k2} + x[4]e^{-j\frac{2\pi}{N}k4} + x[6]e^{-j\frac{2\pi}{N}k6} + \cdots + x[N-2]e^{j\frac{2\pi}{N}k(N-2)} \\ &\quad + x[1]e^{-j\frac{2\pi}{N}k1} + x[3]e^{-j\frac{2\pi}{N}k3} + x[5]e^{-j\frac{2\pi}{N}k5} + x[7]e^{-j\frac{2\pi}{N}k7} + \cdots + x[N-1]e^{j\frac{2\pi}{N}k(N-1)} \\ &= x[0] + x[2]e^{-j\frac{2\pi}{N/2}k1} + x[4]e^{-j\frac{2\pi}{N/2}k2} + x[6]e^{-j\frac{2\pi}{N/2}k3} \cdots + x[N-2]e^{j\frac{2\pi}{N/2}k\frac{(N-2)}{2}} \\ &\quad + e^{-j\frac{2\pi}{N}k}x[1] + e^{-j\frac{2\pi}{N}k}x[3]e^{-j\frac{2\pi}{N/2}k1} + e^{-j\frac{2\pi}{N}k}x[5]e^{-j\frac{2\pi}{N/2}k2} + e^{-j\frac{2\pi}{N}k}x[7]e^{-j\frac{2\pi}{N/2}k3} + \cdots \\ &\quad + e^{-j\frac{2\pi}{N}k}x[N-1]e^{j\frac{2\pi}{N/2}k\frac{(N-2)}{2}} \end{aligned} \tag{7.140}$$

Organize into the below equation that represents the two distinct $N/2$-point DFTs with even and odd data. The exponential weighting is required over the odd data sequence.

$$\begin{aligned} X[k] &= \sum_{q=0}^{(N/2-1)} x[2q]e^{-j\frac{2\pi}{N/2}kq} + e^{-j\frac{2\pi}{N}k}\sum_{q=0}^{(N/2-1)} x[2q+1]e^{-j\frac{2\pi}{N/2}kq} \\ &= F[k] + e^{-j\frac{2\pi}{N}k}G[k] \end{aligned} \tag{7.141}$$

The $F[k]$ represents the DFT with even data sequence and $G[k]$ denotes the DFT with weighted odd data sequence. In the same manner as above, the $F[k]$ and $G[k]$ can be further decomposed into the even DFT and odd DFT as below. Observe that the $f[q]$ and $g[q]$ is the compressed version of the even and odd of $x[n]$, respectively.

$$
\begin{aligned}
F[k] &= \sum_{q=0}^{(N/2-1)} f[q]e^{-j\frac{2\pi}{N/2}kq} \\
&= \sum_{r=0}^{(N/4-1)} f[2r]e^{-j\frac{2\pi}{N/4}kr} + e^{-j\frac{2\pi}{N/2}k} \sum_{r=0}^{(N/4-1)} f[2r+1]e^{-j\frac{2\pi}{N/4}kr}
\end{aligned} \tag{7.142}
$$

$$
\begin{aligned}
G[k] &= \sum_{q=0}^{(N/2-1)} g[q]e^{-j\frac{2\pi}{N/2}kq} \\
&= \sum_{r=0}^{(N/4-1)} g[2r]e^{-j\frac{2\pi}{N/4}kr} + e^{-j\frac{2\pi}{N/2}k} \sum_{r=0}^{(N/4-1)} g[2r+1]e^{-j\frac{2\pi}{N/4}kr}
\end{aligned} \tag{7.143}
$$

Further decomposition is performed until the data length reaches the minimum as 2. The DFT with length 2 is below.

$$
C[k] = c[0]e^{-j\frac{2\pi}{2}k0} + c[1]e^{-j\frac{2\pi}{2}k1} = c[0] + c[1](-1)^k \tag{7.144}
$$

After the 2-point DFT, the proper weighting and summation provide the next size DFT output. The cascade even odd DFTs result in the N-point DFT with reduced computation. The actual DFT is executed at the first DFT stage (2-point) and later stages perform the simple complex weighting and adding for next computational stage. The linear algebra analysis for the FFT is shown below. Note that the W_N represents the complex exponential as below

$$
W_N = e^{-j2\pi/N}
$$

The length for analysis is 8 which is 2^3.

$$
\mathrm{DFT}_8 \begin{bmatrix} x[0] \\ x[1] \\ x[2] \\ x[3] \\ x[4] \\ x[5] \\ x[6] \\ x[7] \end{bmatrix} = \begin{bmatrix} X[0] \\ X[1] \\ X[2] \\ X[3] \\ X[4] \\ X[5] \\ X[6] \\ X[7] \end{bmatrix} = \begin{bmatrix}
W_8^0 & W_8^0 & W_8^0 & W_8^0 & W_8^0 & W_8^0 & W_8^0 & W_8^0 \\
W_8^0 & W_8^1 & W_8^2 & W_8^3 & W_8^4 & W_8^5 & W_8^6 & W_8^7 \\
W_8^0 & W_8^2 & W_8^4 & W_8^6 & W_8^0 & W_8^2 & W_8^4 & W_8^6 \\
W_8^0 & W_8^3 & W_8^6 & W_8^1 & W_8^4 & W_8^7 & W_8^2 & W_8^5 \\
W_8^0 & W_8^4 & W_8^0 & W_8^4 & W_8^0 & W_8^4 & W_8^0 & W_8^4 \\
W_8^0 & W_8^5 & W_8^2 & W_8^7 & W_8^4 & W_8^1 & W_8^6 & W_8^3 \\
W_8^0 & W_8^6 & W_8^4 & W_8^2 & W_8^0 & W_8^6 & W_8^4 & W_8^2 \\
W_8^0 & W_8^7 & W_8^6 & W_8^5 & W_8^4 & W_8^3 & W_8^2 & W_8^1
\end{bmatrix} \begin{bmatrix} x[0] \\ x[1] \\ x[2] \\ x[3] \\ x[4] \\ x[5] \\ x[6] \\ x[7] \end{bmatrix} \tag{7.145}
$$

The matrix is divided into the even and odd data as below.

$$\begin{bmatrix} X[0] \\ X[1] \\ X[2] \\ X[3] \\ X[4] \\ X[5] \\ X[6] \\ X[7] \end{bmatrix} = \begin{bmatrix} W_8^0 & W_8^0 & W_8^0 & W_8^0 \\ W_8^0 & W_8^2 & W_8^4 & W_8^6 \\ W_8^0 & W_8^4 & W_8^0 & W_8^4 \\ W_8^0 & W_8^6 & W_8^4 & W_8^2 \\ W_8^0 & W_8^0 & W_8^0 & W_8^0 \\ W_8^0 & W_8^2 & W_8^4 & W_8^6 \\ W_8^0 & W_8^4 & W_8^0 & W_8^4 \\ W_8^0 & W_8^6 & W_8^4 & W_8^2 \end{bmatrix} \begin{bmatrix} x[0] \\ x[2] \\ x[4] \\ x[6] \end{bmatrix} + \begin{bmatrix} W_8^0 & W_8^0 & W_8^0 & W_8^0 \\ W_8^1 & W_8^3 & W_8^5 & W_8^7 \\ W_8^2 & W_8^6 & W_8^2 & W_8^6 \\ W_8^3 & W_8^1 & W_8^7 & W_8^5 \\ W_8^4 & W_8^4 & W_8^4 & W_8^4 \\ W_8^5 & W_8^7 & W_8^1 & W_8^3 \\ W_8^6 & W_8^2 & W_8^6 & W_8^2 \\ W_8^7 & W_8^5 & W_8^3 & W_8^1 \end{bmatrix} \begin{bmatrix} x[1] \\ x[3] \\ x[5] \\ x[7] \end{bmatrix} \tag{7.146}$$

Further division is performed for data output as below.

$$\begin{bmatrix} X[0] \\ X[1] \\ X[2] \\ X[3] \end{bmatrix} = \begin{bmatrix} W_8^0 & W_8^0 & W_8^0 & W_8^0 \\ W_8^0 & W_8^2 & W_8^4 & W_8^6 \\ W_8^0 & W_8^4 & W_8^0 & W_8^4 \\ W_8^0 & W_8^6 & W_8^4 & W_8^2 \end{bmatrix} \begin{bmatrix} x[0] \\ x[2] \\ x[4] \\ x[6] \end{bmatrix} + \begin{bmatrix} W_8^0 & W_8^0 & W_8^0 & W_8^0 \\ W_8^1 & W_8^3 & W_8^5 & W_8^7 \\ W_8^2 & W_8^6 & W_8^2 & W_8^6 \\ W_8^3 & W_8^1 & W_8^7 & W_8^5 \end{bmatrix} \begin{bmatrix} x[1] \\ x[3] \\ x[5] \\ x[7] \end{bmatrix} \tag{7.147}$$

$$\begin{bmatrix} X[4] \\ X[5] \\ X[6] \\ X[7] \end{bmatrix} = \begin{bmatrix} W_8^0 & W_8^0 & W_8^0 & W_8^0 \\ W_8^0 & W_8^2 & W_8^4 & W_8^6 \\ W_8^0 & W_8^4 & W_8^0 & W_8^4 \\ W_8^0 & W_8^6 & W_8^4 & W_8^2 \end{bmatrix} \begin{bmatrix} x[0] \\ x[2] \\ x[4] \\ x[6] \end{bmatrix} + \begin{bmatrix} W_8^4 & W_8^4 & W_8^4 & W_8^4 \\ W_8^5 & W_8^7 & W_8^1 & W_8^3 \\ W_8^6 & W_8^2 & W_8^6 & W_8^2 \\ W_8^7 & W_8^5 & W_8^3 & W_8^1 \end{bmatrix} \begin{bmatrix} x[1] \\ x[3] \\ x[5] \\ x[7] \end{bmatrix} \tag{7.148}$$

Extract the common factor as below by using the complex exponential periodicity.

$$\begin{bmatrix} X[4] \\ X[5] \\ X[6] \\ X[7] \end{bmatrix} = \begin{bmatrix} W_8^0 & W_8^0 & W_8^0 & W_8^0 \\ W_8^0 & W_8^2 & W_8^4 & W_8^6 \\ W_8^0 & W_8^4 & W_8^0 & W_8^4 \\ W_8^0 & W_8^6 & W_8^4 & W_8^2 \end{bmatrix} \begin{bmatrix} x[0] \\ x[2] \\ x[4] \\ x[6] \end{bmatrix} + W_8^4 \begin{bmatrix} W_8^0 & W_8^0 & W_8^0 & W_8^0 \\ W_8^1 & W_8^3 & W_8^5 & W_8^7 \\ W_8^2 & W_8^6 & W_8^2 & W_8^6 \\ W_8^3 & W_8^1 & W_8^7 & W_8^5 \end{bmatrix} \begin{bmatrix} x[1] \\ x[3] \\ x[5] \\ x[7] \end{bmatrix} \tag{7.149}$$

Derive the common factors for corresponding row as below.

$$\begin{aligned} \begin{bmatrix} X[0] \\ X[1] \\ X[2] \\ X[3] \end{bmatrix} &= \begin{bmatrix} W_8^0 & W_8^0 & W_8^0 & W_8^0 \\ W_8^0 & W_8^2 & W_8^4 & W_8^6 \\ W_8^0 & W_8^4 & W_8^0 & W_8^4 \\ W_8^0 & W_8^6 & W_8^4 & W_8^2 \end{bmatrix} \begin{bmatrix} x[0] \\ x[2] \\ x[4] \\ x[6] \end{bmatrix} \\ &+ \begin{bmatrix} W_8^0 & 0 & 0 & 0 \\ 0 & W_8^1 & 0 & 0 \\ 0 & 0 & W_8^2 & 0 \\ 0 & 0 & 0 & W_8^3 \end{bmatrix} \begin{bmatrix} W_8^0 & W_8^0 & W_8^0 & W_8^0 \\ W_8^0 & W_8^2 & W_8^4 & W_8^6 \\ W_8^0 & W_8^4 & W_8^0 & W_8^4 \\ W_8^0 & W_8^6 & W_8^4 & W_8^2 \end{bmatrix} \begin{bmatrix} x[1] \\ x[3] \\ x[5] \\ x[7] \end{bmatrix} \end{aligned} \tag{7.150}$$

$$\begin{bmatrix} X[4] \\ X[5] \\ X[6] \\ X[7] \end{bmatrix} = \begin{bmatrix} W_8^0 & W_8^0 & W_8^0 & W_8^0 \\ W_8^0 & W_8^2 & W_8^4 & W_8^6 \\ W_8^0 & W_8^4 & W_8^0 & W_8^4 \\ W_8^0 & W_8^6 & W_8^4 & W_8^2 \end{bmatrix} \begin{bmatrix} x[0] \\ x[2] \\ x[4] \\ x[6] \end{bmatrix} + W_8^4 \begin{bmatrix} W_8^0 & 0 & 0 & 0 \\ 0 & W_8^1 & 0 & 0 \\ 0 & 0 & W_8^2 & 0 \\ 0 & 0 & 0 & W_8^3 \end{bmatrix} \begin{bmatrix} W_8^0 & W_8^0 & W_8^0 & W_8^0 \\ W_8^0 & W_8^2 & W_8^4 & W_8^6 \\ W_8^0 & W_8^4 & W_8^0 & W_8^4 \\ W_8^0 & W_8^6 & W_8^4 & W_8^2 \end{bmatrix} \begin{bmatrix} x[1] \\ x[3] \\ x[5] \\ x[7] \end{bmatrix} \tag{7.151}$$

Combining and simplifying the factors for consistency as below.

$$\begin{bmatrix} X[0] \\ X[1] \\ X[2] \\ X[3] \end{bmatrix} = \begin{bmatrix} W_4^0 & W_4^0 & W_4^0 & W_4^0 \\ W_4^0 & W_4^1 & W_4^2 & W_4^3 \\ W_4^0 & W_4^2 & W_4^0 & W_4^2 \\ W_4^0 & W_4^3 & W_4^2 & W_4^1 \end{bmatrix} \begin{bmatrix} x[0] \\ x[2] \\ x[4] \\ x[6] \end{bmatrix} + \begin{bmatrix} W_8^0 & 0 & 0 & 0 \\ 0 & W_8^1 & 0 & 0 \\ 0 & 0 & W_8^2 & 0 \\ 0 & 0 & 0 & W_8^3 \end{bmatrix} \begin{bmatrix} W_4^0 & W_4^0 & W_4^0 & W_4^0 \\ W_4^0 & W_4^1 & W_4^2 & W_4^3 \\ W_4^0 & W_4^2 & W_4^0 & W_4^2 \\ W_4^0 & W_4^3 & W_4^2 & W_4^1 \end{bmatrix} \begin{bmatrix} x[1] \\ x[3] \\ x[5] \\ x[7] \end{bmatrix} \tag{7.152}$$

$$\begin{bmatrix} X[4] \\ X[5] \\ X[6] \\ X[7] \end{bmatrix} = \begin{bmatrix} W_4^0 & W_4^0 & W_4^0 & W_4^0 \\ W_4^0 & W_4^1 & W_4^2 & W_4^3 \\ W_4^0 & W_4^2 & W_4^0 & W_4^2 \\ W_4^0 & W_4^3 & W_4^2 & W_4^1 \end{bmatrix} \begin{bmatrix} x[0] \\ x[2] \\ x[4] \\ x[6] \end{bmatrix} + \begin{bmatrix} W_8^4 & 0 & 0 & 0 \\ 0 & W_8^5 & 0 & 0 \\ 0 & 0 & W_8^6 & 0 \\ 0 & 0 & 0 & W_8^7 \end{bmatrix} \begin{bmatrix} W_4^0 & W_4^0 & W_4^0 & W_4^0 \\ W_4^0 & W_4^1 & W_4^2 & W_4^3 \\ W_4^0 & W_4^2 & W_4^0 & W_4^2 \\ W_4^0 & W_4^3 & W_4^2 & W_4^1 \end{bmatrix} \begin{bmatrix} x[1] \\ x[3] \\ x[5] \\ x[7] \end{bmatrix} \tag{7.153}$$

The square matrix with full of complex exponential indicates the DFT with 4-point as below. Therefore, the 8-point DFT is performed by 4-point DFT as below.

$$\begin{bmatrix} X[0] \\ X[1] \\ X[2] \\ X[3] \end{bmatrix} = \mathrm{DFT}_4 \begin{bmatrix} x[0] \\ x[2] \\ x[4] \\ x[6] \end{bmatrix} + \begin{bmatrix} W_8^0 & 0 & 0 & 0 \\ 0 & W_8^1 & 0 & 0 \\ 0 & 0 & W_8^2 & 0 \\ 0 & 0 & 0 & W_8^3 \end{bmatrix} \mathrm{DFT}_4 \begin{bmatrix} x[1] \\ x[3] \\ x[5] \\ x[7] \end{bmatrix} \tag{7.154}$$

$$\begin{bmatrix} X[4] \\ X[5] \\ X[6] \\ X[7] \end{bmatrix} = \mathrm{DFT}_4 \begin{bmatrix} x[0] \\ x[2] \\ x[4] \\ x[6] \end{bmatrix} + \begin{bmatrix} W_8^4 & 0 & 0 & 0 \\ 0 & W_8^5 & 0 & 0 \\ 0 & 0 & W_8^6 & 0 \\ 0 & 0 & 0 & W_8^7 \end{bmatrix} \mathrm{DFT}_4 \begin{bmatrix} x[1] \\ x[3] \\ x[5] \\ x[7] \end{bmatrix} \tag{7.155}$$

Exercise the same above procedure to the 4-point DFT.

$$\mathrm{DFT}_4\begin{bmatrix} x[0] \\ x[2] \\ x[4] \\ x[6] \end{bmatrix} = \begin{bmatrix} F[0] \\ F[1] \\ F[2] \\ F[3] \end{bmatrix} = \begin{bmatrix} W_4^0 & W_4^0 & W_4^0 & W_4^0 \\ W_4^0 & W_4^1 & W_4^2 & W_4^3 \\ W_4^0 & W_4^2 & W_4^0 & W_4^2 \\ W_4^0 & W_4^3 & W_4^2 & W_4^1 \end{bmatrix} \begin{bmatrix} x[0] \\ x[2] \\ x[4] \\ x[6] \end{bmatrix} \quad (7.156)$$

The matrix is divided into the even and odd data asbelow.

$$\begin{bmatrix} F[0] \\ F[1] \\ F[2] \\ F[3] \end{bmatrix} = \begin{bmatrix} W_4^0 & W_4^0 \\ W_4^0 & W_4^2 \\ W_4^0 & W_4^0 \\ W_4^0 & W_4^2 \end{bmatrix} \begin{bmatrix} x[0] \\ x[4] \end{bmatrix} + \begin{bmatrix} W_4^0 & W_4^0 \\ W_4^1 & W_4^3 \\ W_4^2 & W_4^2 \\ W_4^3 & W_4^1 \end{bmatrix} \begin{bmatrix} x[2] \\ x[6] \end{bmatrix} \quad (7.157)$$

Further division is performed for data output as below.

$$\begin{bmatrix} F[0] \\ F[1] \end{bmatrix} = \begin{bmatrix} W_4^0 & W_4^0 \\ W_4^0 & W_4^2 \end{bmatrix} \begin{bmatrix} x[0] \\ x[4] \end{bmatrix} + \begin{bmatrix} W_4^0 & W_4^0 \\ W_4^1 & W_4^3 \end{bmatrix} \begin{bmatrix} x[2] \\ x[6] \end{bmatrix} \quad (7.158)$$

$$\begin{bmatrix} F[2] \\ F[3] \end{bmatrix} = \begin{bmatrix} W_4^0 & W_4^0 \\ W_4^0 & W_4^2 \end{bmatrix} \begin{bmatrix} x[0] \\ x[4] \end{bmatrix} + \begin{bmatrix} W_4^2 & W_4^2 \\ W_4^3 & W_4^1 \end{bmatrix} \begin{bmatrix} x[2] \\ x[6] \end{bmatrix} \quad (7.159)$$

Extract the common factor as below by using the complex exponential periodicity.

$$\begin{bmatrix} F[2] \\ F[3] \end{bmatrix} = \begin{bmatrix} W_4^0 & W_4^0 \\ W_4^0 & W_4^2 \end{bmatrix} \begin{bmatrix} x[0] \\ x[4] \end{bmatrix} + W_4^2 \begin{bmatrix} W_4^0 & W_4^0 \\ W_4^1 & W_4^3 \end{bmatrix} \begin{bmatrix} x[2] \\ x[6] \end{bmatrix} \quad (7.160)$$

Derive the common factors for corresponding row as below.

$$\begin{bmatrix} F[0] \\ F[1] \end{bmatrix} = \begin{bmatrix} W_4^0 & W_4^0 \\ W_4^0 & W_4^2 \end{bmatrix} \begin{bmatrix} x[0] \\ x[4] \end{bmatrix} + \begin{bmatrix} W_4^0 & 0 \\ 0 & W_4^1 \end{bmatrix} \begin{bmatrix} W_4^0 & W_4^0 \\ W_4^0 & W_4^2 \end{bmatrix} \begin{bmatrix} x[2] \\ x[6] \end{bmatrix} \quad (7.161)$$

$$\begin{bmatrix} F[2] \\ F[3] \end{bmatrix} = \begin{bmatrix} W_4^0 & W_4^0 \\ W_4^0 & W_4^2 \end{bmatrix} \begin{bmatrix} x[0] \\ x[4] \end{bmatrix} + W_4^2 \begin{bmatrix} W_4^0 & 0 \\ 0 & W_4^1 \end{bmatrix} \begin{bmatrix} W_4^0 & W_4^0 \\ W_4^0 & W_4^2 \end{bmatrix} \begin{bmatrix} x[2] \\ x[6] \end{bmatrix} \quad (7.162)$$

Combining and simplifying the factors for consistency as below.

$$\begin{bmatrix} F[0] \\ F[1] \end{bmatrix} = \begin{bmatrix} W_2^0 & W_2^0 \\ W_2^0 & W_2^1 \end{bmatrix} \begin{bmatrix} x[0] \\ x[4] \end{bmatrix} + \begin{bmatrix} W_4^0 & 0 \\ 0 & W_4^1 \end{bmatrix} \begin{bmatrix} W_2^0 & W_2^0 \\ W_2^0 & W_2^1 \end{bmatrix} \begin{bmatrix} x[2] \\ x[6] \end{bmatrix} \quad (7.163)$$

$$\begin{bmatrix} F[2] \\ F[3] \end{bmatrix} = \begin{bmatrix} W_2^0 & W_2^0 \\ W_2^0 & W_2^1 \end{bmatrix} \begin{bmatrix} x[0] \\ x[4] \end{bmatrix} + \begin{bmatrix} W_4^2 & 0 \\ 0 & W_4^3 \end{bmatrix} \begin{bmatrix} W_2^0 & W_2^0 \\ W_2^0 & W_2^1 \end{bmatrix} \begin{bmatrix} x[2] \\ x[6] \end{bmatrix} \quad (7.164)$$

The square matrix with full of complex exponential indicates the DFT with 2-point as below. Therefore, the 4-point DFT is performed by 2-point DFT as below.

$$\begin{bmatrix} F[0] \\ F[1] \end{bmatrix} = \mathrm{DFT}_2 \begin{bmatrix} x[0] \\ x[4] \end{bmatrix} + \begin{bmatrix} W_4^0 & 0 \\ 0 & W_4^1 \end{bmatrix} \mathrm{DFT}_2 \begin{bmatrix} x[2] \\ x[6] \end{bmatrix} \tag{7.165}$$

$$\begin{bmatrix} F[2] \\ F[3] \end{bmatrix} = \mathrm{DFT}_2 \begin{bmatrix} x[0] \\ x[4] \end{bmatrix} + \begin{bmatrix} W_4^2 & 0 \\ 0 & W_4^3 \end{bmatrix} \mathrm{DFT}_2 \begin{bmatrix} x[2] \\ x[6] \end{bmatrix} \tag{7.166}$$

For odd data, same procedure is performed as below.

$$\mathrm{DFT}_4 \begin{bmatrix} x[1] \\ x[3] \\ x[5] \\ x[7] \end{bmatrix} = \begin{bmatrix} G[0] \\ G[1] \\ G[2] \\ G[3] \end{bmatrix} = \begin{bmatrix} W_4^0 & W_4^0 & W_4^0 & W_4^0 \\ W_4^0 & W_4^1 & W_4^2 & W_4^3 \\ W_4^0 & W_4^2 & W_4^0 & W_4^2 \\ W_4^0 & W_4^3 & W_4^2 & W_4^1 \end{bmatrix} \begin{bmatrix} x[1] \\ x[3] \\ x[5] \\ x[7] \end{bmatrix} \tag{7.167}$$

Also, the 4-point DFT is performed by 2-point DFT as below.

$$\begin{bmatrix} G[0] \\ G[1] \end{bmatrix} = \mathrm{DFT}_2 \begin{bmatrix} x[1] \\ x[5] \end{bmatrix} + \begin{bmatrix} W_4^0 & 0 \\ 0 & W_4^1 \end{bmatrix} \mathrm{DFT}_2 \begin{bmatrix} x[3] \\ x[7] \end{bmatrix} \tag{7.168}$$

$$\begin{bmatrix} G[2] \\ G[3] \end{bmatrix} = \mathrm{DFT}_2 \begin{bmatrix} x[1] \\ x[5] \end{bmatrix} + \begin{bmatrix} W_4^2 & 0 \\ 0 & W_4^3 \end{bmatrix} \mathrm{DFT}_2 \begin{bmatrix} x[3] \\ x[7] \end{bmatrix} \tag{7.169}$$

The 2-point DFT is simple computation as below and no further decomposition is required.

$$\mathrm{DFT}_2 \begin{bmatrix} x[1] \\ x[5] \end{bmatrix} = \begin{bmatrix} W_2^0 & W_2^0 \\ W_2^0 & W_2^1 \end{bmatrix} \begin{bmatrix} x[1] \\ x[5] \end{bmatrix} = \begin{bmatrix} 1 & 1 \\ 1 & -1 \end{bmatrix} \begin{bmatrix} x[1] \\ x[5] \end{bmatrix} \tag{7.170}$$

$$\mathrm{DFT}_2 \begin{bmatrix} x[3] \\ x[7] \end{bmatrix} = \begin{bmatrix} 1 & 1 \\ 1 & -1 \end{bmatrix} \begin{bmatrix} x[3] \\ x[7] \end{bmatrix} \tag{7.171}$$

$$\mathrm{DFT}_2 \begin{bmatrix} x[0] \\ x[4] \end{bmatrix} = \begin{bmatrix} 1 & 1 \\ 1 & -1 \end{bmatrix} \begin{bmatrix} x[0] \\ x[4] \end{bmatrix} \tag{7.172}$$

$$\mathrm{DFT}_2 \begin{bmatrix} x[2] \\ x[6] \end{bmatrix} = \begin{bmatrix} 1 & 1 \\ 1 & -1 \end{bmatrix} \begin{bmatrix} x[2] \\ x[6] \end{bmatrix} \tag{7.173}$$

The 8-point DFT is calculated by 2- and 4-point DFT with weighting and adding. The signal flow graph convention presents the constant multiplication by arrow and the signal addition by junction. The 8-point FFT is shown below in flow graph on Fig. 7.41.

The first stage in the left indicates the 2-point DFT. The second and third stage specify the 4- and 8-point DFT. Each stage requires the N multiplications and N additions. The number of stages is $\log_2 N$; therefore, the total number of multiplications is $N\log_2 N$ that represents the FFT complexity order. The primitive pattern

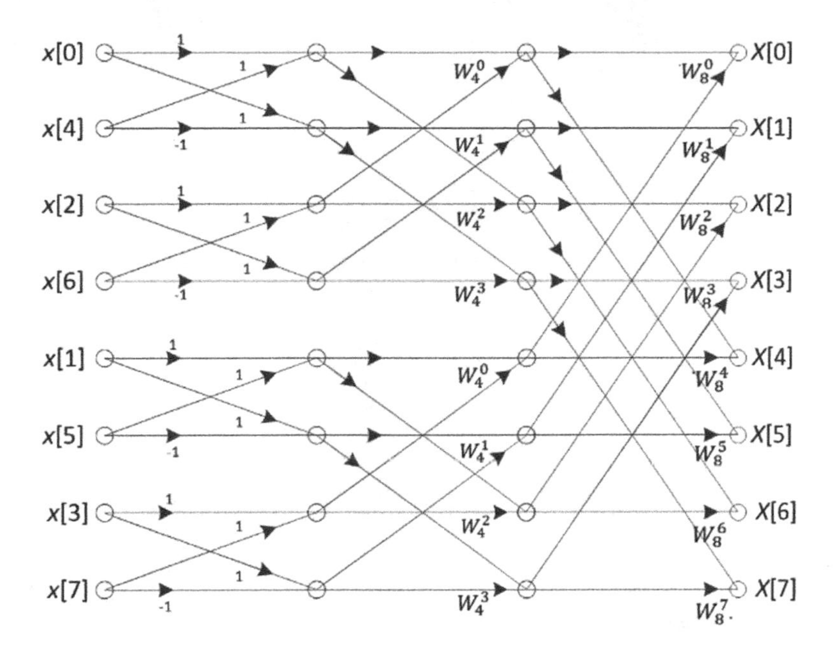

Fig. 7.41 Signal flow graph for 8-point FFT

in the FFT flow graph is the butterfly computation. By exploring the butterfly, we can reduce the computation further. The two complex weights in the butterfly always show the π radian difference as $e^{j\pi}$ which is the -1. The extended butterfly demonstrated below decreased the complex multiplications by using the consistent phase difference as shown in Fig. 7.42.

The FFT flow graph with extended butterfly structure is illustrated below. The number of multiplication in the graph approximates to the $(N/2)\log_2 N$ which is half of the basic FFT structure. The overall complexity of the below FFT structure is still $N\log_2 N$ because the constant factor does not modify the algorithm complexity as shown in Fig. 7.43.

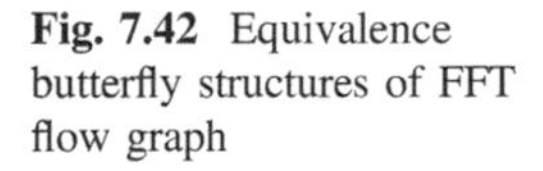

Fig. 7.42 Equivalence butterfly structures of FFT flow graph

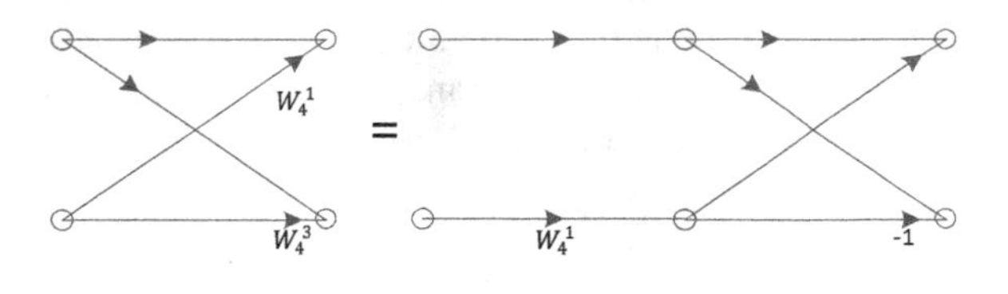

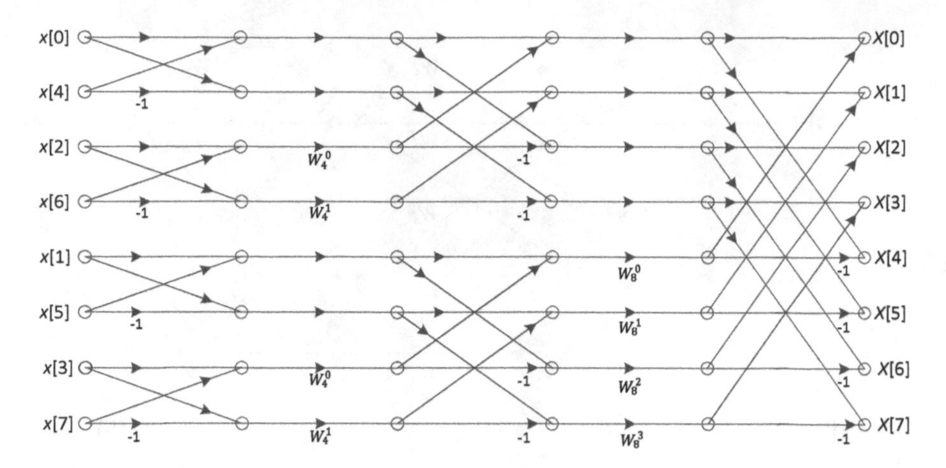

Fig. 7.43 8-point FFT flow graph with extended butterfly structure

Table 7.6 Bit-reversed order for 8-point FFT

Index	Binary	Bit-rev	FFT in	Index	Binary	Bit-rev	FFT in
0	000_2	000_2	0	4	100_2	001_2	1
1	001_2	100_2	4	5	101_2	101_2	5
2	010_2	010_2	2	6	110_2	011_2	3
3	011_2	110_2	6	7	111_2	111_2	7

The FFT algorithm requires the shuffling the input sequence to obtain the output sequence in order. The bit-reversed order turns left and right of the binary index to provide the proper sequencing for FFT input. Table 7.6 shows the bit-reversed order for 8-point FFT.

The above decimation-in-time FFT is derived by dividing the time data into the even and odd sequence. Alternative approach for FFT can be suggested by decomposing the frequency output into the even and odd sequence. The frequency decomposed method is known as decimation-in-frequency FFT [9]. Below matrix analysis shows the derivation procedure for decimation-in-frequency FFT with 8-point data. Start from the 8-point DFT as below.

$$
\begin{aligned}
\mathrm{DFT}_8\begin{bmatrix} x[0] \\ x[1] \\ x[2] \\ x[3] \\ x[4] \\ x[5] \\ x[6] \\ x[7] \end{bmatrix} &= \begin{bmatrix} X[0] \\ X[1] \\ X[2] \\ X[3] \\ X[4] \\ X[5] \\ X[6] \\ X[7] \end{bmatrix} \\
&= \begin{bmatrix}
W_8^0 & W_8^0 & W_8^0 & W_8^0 & W_8^0 & W_8^0 & W_8^0 & W_8^0 \\
W_8^0 & W_8^1 & W_8^2 & W_8^3 & W_8^4 & W_8^5 & W_8^6 & W_8^7 \\
W_8^0 & W_8^2 & W_8^4 & W_8^6 & W_8^0 & W_8^2 & W_8^4 & W_8^6 \\
W_8^0 & W_8^3 & W_8^6 & W_8^1 & W_8^4 & W_8^7 & W_8^2 & W_8^5 \\
W_8^0 & W_8^4 & W_8^0 & W_8^4 & W_8^0 & W_8^4 & W_8^0 & W_8^4 \\
W_8^0 & W_8^5 & W_8^2 & W_8^7 & W_8^4 & W_8^1 & W_8^6 & W_8^3 \\
W_8^0 & W_8^6 & W_8^4 & W_8^2 & W_8^0 & W_8^6 & W_8^4 & W_8^2 \\
W_8^0 & W_8^7 & W_8^6 & W_8^5 & W_8^4 & W_8^3 & W_8^2 & W_8^1
\end{bmatrix}
\begin{bmatrix} x[0] \\ x[1] \\ x[2] \\ x[3] \\ x[4] \\ x[5] \\ x[6] \\ x[7] \end{bmatrix}
\end{aligned}
\tag{7.174}
$$

The matrix is divided into the even and odd frequency output as below.

$$
\begin{bmatrix} X[0] \\ X[2] \\ X[4] \\ X[6] \end{bmatrix} = \begin{bmatrix}
W_8^0 & W_8^0 & W_8^0 & W_8^0 & W_8^0 & W_8^0 & W_8^0 & W_8^0 \\
W_8^0 & W_8^2 & W_8^4 & W_8^6 & W_8^0 & W_8^2 & W_8^4 & W_8^6 \\
W_8^0 & W_8^4 & W_8^0 & W_8^4 & W_8^0 & W_8^4 & W_8^0 & W_8^4 \\
W_8^0 & W_8^6 & W_8^4 & W_8^2 & W_8^0 & W_8^6 & W_8^4 & W_8^2
\end{bmatrix}
\begin{bmatrix} x[0] \\ x[1] \\ x[2] \\ x[3] \\ x[4] \\ x[5] \\ x[6] \\ x[7] \end{bmatrix}
\tag{7.175}
$$

$$
\begin{bmatrix} X[1] \\ X[3] \\ X[5] \\ X[7] \end{bmatrix} = \begin{bmatrix}
W_8^0 & W_8^1 & W_8^2 & W_8^3 & W_8^4 & W_8^5 & W_8^6 & W_8^7 \\
W_8^0 & W_8^3 & W_8^6 & W_8^1 & W_8^4 & W_8^7 & W_8^2 & W_8^5 \\
W_8^0 & W_8^5 & W_8^2 & W_8^7 & W_8^4 & W_8^1 & W_8^6 & W_8^3 \\
W_8^0 & W_8^7 & W_8^6 & W_8^5 & W_8^4 & W_8^3 & W_8^2 & W_8^1
\end{bmatrix}
\begin{bmatrix} x[0] \\ x[1] \\ x[2] \\ x[3] \\ x[4] \\ x[5] \\ x[6] \\ x[7] \end{bmatrix}
\tag{7.176}
$$

Further division is performed as below.

$$\begin{bmatrix} X[0] \\ X[2] \\ X[4] \\ X[6] \end{bmatrix} = \begin{bmatrix} W_8^0 & W_8^0 & W_8^0 & W_8^0 \\ W_8^0 & W_8^2 & W_8^4 & W_8^6 \\ W_8^0 & W_8^4 & W_8^0 & W_8^4 \\ W_8^0 & W_8^6 & W_8^4 & W_8^2 \end{bmatrix} \begin{bmatrix} x[0] \\ x[1] \\ x[2] \\ x[3] \end{bmatrix} + \begin{bmatrix} W_8^0 & W_8^0 & W_8^0 & W_8^0 \\ W_8^0 & W_8^2 & W_8^4 & W_8^6 \\ W_8^0 & W_8^4 & W_8^0 & W_8^4 \\ W_8^0 & W_8^6 & W_8^4 & W_8^2 \end{bmatrix} \begin{bmatrix} x[4] \\ x[5] \\ x[6] \\ x[7] \end{bmatrix} \tag{7.177}$$

$$\begin{bmatrix} X[1] \\ X[3] \\ X[5] \\ X[7] \end{bmatrix} = \begin{bmatrix} W_8^0 & W_8^1 & W_8^2 & W_8^3 \\ W_8^0 & W_8^3 & W_8^6 & W_8^1 \\ W_8^0 & W_8^5 & W_8^2 & W_8^7 \\ W_8^0 & W_8^7 & W_8^6 & W_8^5 \end{bmatrix} \begin{bmatrix} x[0] \\ x[1] \\ x[2] \\ x[3] \end{bmatrix} + \begin{bmatrix} W_8^4 & W_8^5 & W_8^6 & W_8^7 \\ W_8^4 & W_8^7 & W_8^2 & W_8^5 \\ W_8^4 & W_8^1 & W_8^6 & W_8^3 \\ W_8^4 & W_8^3 & W_8^2 & W_8^1 \end{bmatrix} \begin{bmatrix} x[4] \\ x[5] \\ x[6] \\ x[7] \end{bmatrix} \tag{7.178}$$

Extract common factors from matrix and columns by using the complex exponential periodicity as below.

$$\begin{aligned} \begin{bmatrix} X[1] \\ X[3] \\ X[5] \\ X[7] \end{bmatrix} &= \begin{bmatrix} W_8^0 & W_8^0 & W_8^0 & W_8^0 \\ W_8^0 & W_8^2 & W_8^4 & W_8^6 \\ W_8^0 & W_8^4 & W_8^0 & W_8^4 \\ W_8^0 & W_8^6 & W_8^4 & W_8^2 \end{bmatrix} \begin{bmatrix} W_8^0 & 0 & 0 & 0 \\ 0 & W_8^1 & 0 & 0 \\ 0 & 0 & W_8^2 & 0 \\ 0 & 0 & 0 & W_8^3 \end{bmatrix} \begin{bmatrix} x[0] \\ x[1] \\ x[2] \\ x[3] \end{bmatrix} \\ &+ W_8^4 \begin{bmatrix} W_8^0 & W_8^0 & W_8^0 & W_8^0 \\ W_8^0 & W_8^2 & W_8^4 & W_8^6 \\ W_8^0 & W_8^4 & W_8^0 & W_8^4 \\ W_8^0 & W_8^6 & W_8^4 & W_8^2 \end{bmatrix} \begin{bmatrix} W_8^0 & 0 & 0 & 0 \\ 0 & W_8^1 & 0 & 0 \\ 0 & 0 & W_8^2 & 0 \\ 0 & 0 & 0 & W_8^3 \end{bmatrix} \begin{bmatrix} x[4] \\ x[5] \\ x[6] \\ x[7] \end{bmatrix} \end{aligned} \tag{7.179}$$

Simplify the matrix for consistency.

$$\begin{bmatrix} X[0] \\ X[2] \\ X[4] \\ X[6] \end{bmatrix} = \begin{bmatrix} W_4^0 & W_4^0 & W_4^0 & W_4^0 \\ W_4^0 & W_4^1 & W_4^2 & W_4^3 \\ W_4^0 & W_4^2 & W_4^0 & W_4^2 \\ W_4^0 & W_4^3 & W_4^2 & W_4^1 \end{bmatrix} \begin{bmatrix} x[0] \\ x[1] \\ x[2] \\ x[3] \end{bmatrix} + \begin{bmatrix} W_4^0 & W_4^0 & W_4^0 & W_4^0 \\ W_4^0 & W_4^1 & W_4^2 & W_4^3 \\ W_4^0 & W_4^2 & W_4^0 & W_4^2 \\ W_4^0 & W_4^3 & W_4^2 & W_4^1 \end{bmatrix} \begin{bmatrix} x[4] \\ x[5] \\ x[6] \\ x[7] \end{bmatrix} \tag{7.180}$$

$$\begin{aligned} \begin{bmatrix} X[1] \\ X[3] \\ X[5] \\ X[7] \end{bmatrix} &= \begin{bmatrix} W_4^0 & W_4^0 & W_4^0 & W_4^0 \\ W_4^0 & W_4^1 & W_4^2 & W_4^3 \\ W_4^0 & W_4^2 & W_4^0 & W_4^2 \\ W_4^0 & W_4^3 & W_4^2 & W_4^1 \end{bmatrix} \begin{bmatrix} W_8^0 & 0 & 0 & 0 \\ 0 & W_8^1 & 0 & 0 \\ 0 & 0 & W_8^2 & 0 \\ 0 & 0 & 0 & W_8^3 \end{bmatrix} \begin{bmatrix} x[0] \\ x[1] \\ x[2] \\ x[3] \end{bmatrix} \\ &+ \begin{bmatrix} W_4^0 & W_4^0 & W_4^0 & W_4^0 \\ W_4^0 & W_4^1 & W_4^2 & W_4^3 \\ W_4^0 & W_4^2 & W_4^0 & W_4^2 \\ W_4^0 & W_4^3 & W_4^2 & W_4^1 \end{bmatrix} \begin{bmatrix} W_8^4 & 0 & 0 & 0 \\ 0 & W_8^5 & 0 & 0 \\ 0 & 0 & W_8^6 & 0 \\ 0 & 0 & 0 & W_8^7 \end{bmatrix} \begin{bmatrix} x[4] \\ x[5] \\ x[6] \\ x[7] \end{bmatrix} \end{aligned} \tag{7.181}$$

Combine the diagonal square matrix with column vector.

$$\begin{bmatrix} X[1] \\ X[3] \\ X[5] \\ X[7] \end{bmatrix} = \begin{bmatrix} W_4^0 & W_4^0 & W_4^0 & W_4^0 \\ W_4^0 & W_4^1 & W_4^2 & W_4^3 \\ W_4^0 & W_4^2 & W_4^0 & W_4^2 \\ W_4^0 & W_4^3 & W_4^2 & W_4^1 \end{bmatrix} \begin{bmatrix} W_8^0 x[0] \\ W_8^1 x[1] \\ W_8^2 x[2] \\ W_8^3 x[3] \end{bmatrix} + \begin{bmatrix} W_4^0 & W_4^0 & W_4^0 & W_4^0 \\ W_4^0 & W_4^1 & W_4^2 & W_4^3 \\ W_4^0 & W_4^2 & W_4^0 & W_4^2 \\ W_4^0 & W_4^3 & W_4^2 & W_4^1 \end{bmatrix} \begin{bmatrix} W_8^4 x[4] \\ W_8^5 x[5] \\ W_8^6 x[6] \\ W_8^7 x[7] \end{bmatrix} \tag{7.182}$$

Using the DFT linearity, merge the data sequence.

$$\begin{bmatrix} X[0] \\ X[2] \\ X[4] \\ X[6] \end{bmatrix} = \mathrm{DFT}_4 \begin{bmatrix} x[0] \\ x[1] \\ x[2] \\ x[3] \end{bmatrix} + \mathrm{DFT}_4 \begin{bmatrix} x[4] \\ x[5] \\ x[6] \\ x[7] \end{bmatrix} = \mathrm{DFT}_4 \begin{bmatrix} x[0]+x[4] \\ x[1]+x[5] \\ x[2]+x[6] \\ x[3]+x[7] \end{bmatrix} \tag{7.183}$$

$$\begin{aligned} \begin{bmatrix} X[1] \\ X[3] \\ X[5] \\ X[7] \end{bmatrix} &= \mathrm{DFT}_4 \begin{bmatrix} W_8^0 x[0] \\ W_8^1 x[1] \\ W_8^2 x[2] \\ W_8^3 x[3] \end{bmatrix} + \mathrm{DFT}_4 \begin{bmatrix} W_8^4 x[4] \\ W_8^5 x[5] \\ W_8^6 x[6] \\ W_8^7 x[7] \end{bmatrix} \\ &= \mathrm{DFT}_4 \begin{bmatrix} W_8^0 x[0] + W_8^4 x[4] \\ W_8^1 x[1] + W_8^5 x[5] \\ W_8^2 x[2] + W_8^6 x[6] \\ W_8^3 x[3] + W_8^7 x[7] \end{bmatrix} = \mathrm{DFT}_4 \begin{bmatrix} W_8^0 (x[0] - x[4]) \\ W_8^1 (x[1] - x[5]) \\ W_8^2 (x[2] - x[6]) \\ W_8^3 (x[3] - x[7]) \end{bmatrix} \end{aligned} \tag{7.184}$$

Now, we have 4-point DFT to obtain the 8-point DFT. Apply the identical analysis to the reduced size DFT as below.

$$\mathrm{DFT}_4 \begin{bmatrix} x[0]+x[4] \\ x[1]+x[5] \\ x[2]+x[6] \\ x[3]+x[7] \end{bmatrix} = \begin{bmatrix} X[0] \\ X[2] \\ X[4] \\ X[6] \end{bmatrix} = \begin{bmatrix} W_4^0 & W_4^0 & W_4^0 & W_4^0 \\ W_4^0 & W_4^1 & W_4^2 & W_4^3 \\ W_4^0 & W_4^2 & W_4^0 & W_4^2 \\ W_4^0 & W_4^3 & W_4^2 & W_4^1 \end{bmatrix} \begin{bmatrix} x[0]+x[4] \\ x[1]+x[5] \\ x[2]+x[6] \\ x[3]+x[7] \end{bmatrix} \tag{7.185}$$

The even frequency sequence can be calculated by the 2-point DFT as below.

$$\begin{aligned} \begin{bmatrix} X[0] \\ X[4] \end{bmatrix} &= \mathrm{DFT}_2 \begin{bmatrix} x[0]+x[4] \\ x[1]+x[5] \end{bmatrix} + \mathrm{DFT}_2 \begin{bmatrix} x[2]+x[6] \\ x[3]+x[7] \end{bmatrix} \\ &= \mathrm{DFT}_2 \begin{bmatrix} x[0]+x[4]+x[2]+x[6] \\ x[1]+x[5]+x[3]+x[7] \end{bmatrix} \end{aligned} \tag{7.186}$$

$$\begin{aligned}
\begin{bmatrix} X[2] \\ X[6] \end{bmatrix} &= \mathrm{DFT}_2 \begin{bmatrix} W_4^0(x[0]+x[4]) \\ W_4^1(x[1]+x[5]) \end{bmatrix} + \mathrm{DFT}_2 \begin{bmatrix} W_4^2(x[2]+x[6]) \\ W_4^3(x[3]+x[7]) \end{bmatrix} \\
&= \mathrm{DFT}_2 \begin{bmatrix} W_4^0(x[0]+x[4]) + W_4^2(x[2]+x[6]) \\ W_4^1(x[1]+x[5]) + W_4^3(x[3]+x[7]) \end{bmatrix} \\
&= \mathrm{DFT}_2 \begin{bmatrix} W_4^0((x[0]+x[4]) - (x[2]+x[6])) \\ W_4^1((x[1]+x[5]) - (x[3]+x[7])) \end{bmatrix}
\end{aligned} \tag{7.187}$$

The odd frequency sequence can be calculated by the 2-point DFT as below.

$$\mathrm{DFT}_4 \begin{bmatrix} W_8^0(x[0]-x[4]) \\ W_8^1(x[1]-x[5]) \\ W_8^2(x[2]-x[6]) \\ W_8^3(x[3]-x[7]) \end{bmatrix} = \begin{bmatrix} X[1] \\ X[3] \\ X[5] \\ X[7] \end{bmatrix} = \begin{bmatrix} W_4^0 & W_4^0 & W_4^0 & W_4^0 \\ W_4^0 & W_4^1 & W_4^2 & W_4^3 \\ W_4^0 & W_4^2 & W_4^0 & W_4^2 \\ W_4^0 & W_4^3 & W_4^2 & W_4^1 \end{bmatrix} \begin{bmatrix} W_8^0(x[0]-x[4]) \\ W_8^1(x[1]-x[5]) \\ W_8^2(x[2]-x[6]) \\ W_8^3(x[3]-x[7]) \end{bmatrix} \tag{7.188}$$

$$\begin{aligned}
\begin{bmatrix} X[1] \\ X[5] \end{bmatrix} &= \mathrm{DFT}_2 \begin{bmatrix} W_8^0(x[0]-x[4]) \\ W_8^1(x[1]-x[5]) \end{bmatrix} + \mathrm{DFT}_2 \begin{bmatrix} W_8^2(x[2]-x[6]) \\ W_8^3(x[3]-x[7]) \end{bmatrix} \\
&= \mathrm{DFT}_2 \begin{bmatrix} W_8^0(x[0]-x[4]) + W_8^2(x[2]-x[6]) \\ W_8^1(x[1]-x[5]) + W_8^3(x[3]-x[7]) \end{bmatrix}
\end{aligned} \tag{7.189}$$

$$\begin{aligned}
\begin{bmatrix} X[3] \\ X[7] \end{bmatrix} &= \mathrm{DFT}_2 \begin{bmatrix} W_4^0 W_8^0(x[0]-x[4]) \\ W_4^1 W_8^1(x[1]-x[5]) \end{bmatrix} + \mathrm{DFT}_2 \begin{bmatrix} W_4^2 W_8^2(x[2]-x[6]) \\ W_4^3 W_8^3(x[3]-x[7]) \end{bmatrix} \\
&= \mathrm{DFT}_2 \begin{bmatrix} W_4^0 W_8^0(x[0]-x[4]) + W_4^2 W_8^2(x[2]-x[6]) \\ W_4^1 W_8^1(x[1]-x[5]) + W_4^3 W_8^3(x[3]-x[7]) \end{bmatrix} \\
&= \mathrm{DFT}_2 \begin{bmatrix} W_4^0\left(W_8^0(x[0]-x[4]) - W_8^2(x[2]-x[6])\right) \\ W_4^1\left(W_8^1(x[1]-x[5]) - W_8^3(x[3]-x[7])\right) \end{bmatrix}
\end{aligned} \tag{7.190}$$

The individual 2-point DFT is the last computation stage which is calculated as below.

$$\begin{aligned}
\begin{bmatrix} X[0] \\ X[4] \end{bmatrix} &= \mathrm{DFT}_2 \begin{bmatrix} x[0]+x[4]+x[2]+x[6] \\ x[1]+x[5]+x[3]+x[7] \end{bmatrix} \\
&= \begin{bmatrix} 1 & 1 \\ 1 & -1 \end{bmatrix} \begin{bmatrix} x[0]+x[4]+x[2]+x[6] \\ x[1]+x[5]+x[3]+x[7] \end{bmatrix}
\end{aligned} \tag{7.191}$$

$$\begin{aligned}
\begin{bmatrix} X[2] \\ X[6] \end{bmatrix} &= \mathrm{DFT}_2 \begin{bmatrix} W_4^0((x[0]+x[4]) - (x[2]+x[6])) \\ W_4^1((x[1]+x[5]) - (x[3]+x[7])) \end{bmatrix} \\
&= \begin{bmatrix} 1 & 1 \\ 1 & -1 \end{bmatrix} \begin{bmatrix} W_4^0((x[0]+x[4]) - (x[2]+x[6])) \\ W_4^1((x[1]+x[5]) - (x[3]+x[7])) \end{bmatrix}
\end{aligned} \tag{7.192}$$

$$\begin{bmatrix} X[1] \\ X[5] \end{bmatrix} = \mathrm{DFT}_2 \begin{bmatrix} W_8^0(x[0]-x[4]) + W_8^2(x[2]-x[6]) \\ W_8^1(x[1]-x[5]) + W_8^3(x[3]-x[7]) \end{bmatrix}$$
$$= \begin{bmatrix} 1 & 1 \\ 1 & -1 \end{bmatrix} \begin{bmatrix} W_8^0(x[0]-x[4]) + W_8^2(x[2]-x[6]) \\ W_8^1(x[1]-x[5]) + W_8^3(x[3]-x[7]) \end{bmatrix} \tag{7.192}$$

$$\begin{bmatrix} X[3] \\ X[7] \end{bmatrix} = \mathrm{DFT}_2 \begin{bmatrix} W_4^0\left(W_8^0(x[0]-x[4]) - W_8^2(x[2]-x[6])\right) \\ W_4^1\left(W_8^1(x[1]-x[5]) - W_8^3(x[3]-x[7])\right) \end{bmatrix}$$
$$= \begin{bmatrix} 1 & 1 \\ 1 & -1 \end{bmatrix} \begin{bmatrix} W_4^0\left(W_8^0(x[0]-x[4]) - W_8^2(x[2]-x[6])\right) \\ W_4^1\left(W_8^1(x[1]-x[5]) - W_8^3(x[3]-x[7])\right) \end{bmatrix} \tag{7.194}$$

The decimation-in-frequency FFT flow graph is illustrated below in Fig. 7.44. The number of multiplication in the graph approximates to the $(N/2)\log_2 N$ which indicates $N\log_2 N$ overall complexity .

The decimation-in-frequency FFT accepts the sequence in order and generates output in bit-reversed order. Proper relocation is required to obtain the sequential frequency information.

– FFT based filter (block convolution)

The filter realization in frequency domain exploits the benefit of FFT algorithm due to the reduced complexity. The complexity for the convolution sum is proportional to the N^2 in which the filter length and data length are both N. Below equation revisits the convolution sum.

$$y[n] = \sum_{k=0}^{N-1} h[k]x[n-k] = \sum_{k=0}^{N-1} x[k]h[n-k] \text{ for } 0 \leq n \leq 2N-2 \tag{7.195}$$

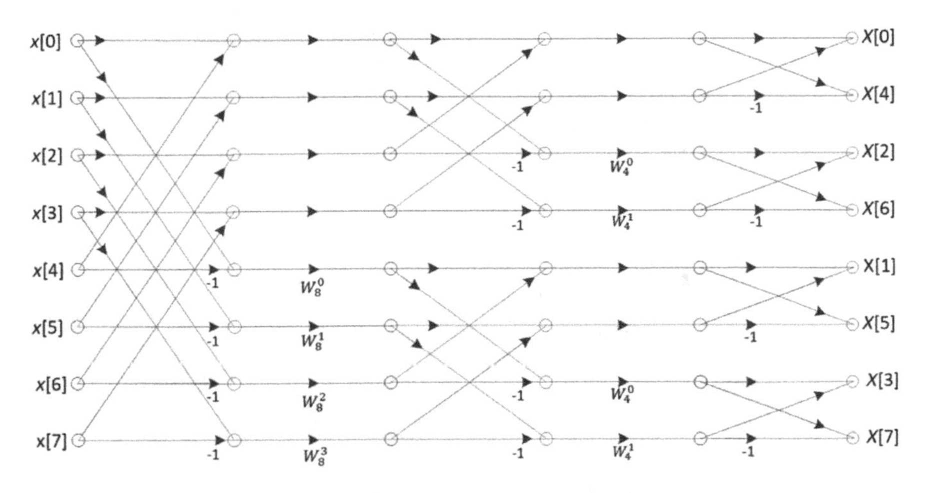

Fig. 7.44 8-point FFT flow graph for decimation-in-frequency

The filter with convolution sum is only applied for the FIR filter because the IIR filter presents the infinite length for impulse response $h[n]$. With infinite length $h[n]$, the convolution sum cannot be calculated unless the $h[n]$ is truncated. Observe that the truncated IIR filter is the FIR filter fundamentally. The IIR filter is realized by the difference equation with recursive inputs. For complexity analysis, the matrix representation for convolution sum is shown below.

$$\begin{bmatrix} y[0] \\ y[1] \\ y[2] \\ \vdots \\ y[2N-3] \\ y[2N-2] \end{bmatrix} = \begin{bmatrix} h[0] & 0 & 0 & \cdots & 0 & 0 \\ h[1] & h[0] & 0 & \cdots & 0 & 0 \\ h[2] & h[1] & h[0] & \cdots & 0 & 0 \\ \vdots & \vdots & \vdots & \ddots & \vdots & \vdots \\ 0 & 0 & 0 & \cdots & h[N-1] & h[N-2] \\ 0 & 0 & 0 & \cdots & 0 & h[N-1] \end{bmatrix} \begin{bmatrix} x[0] \\ x[1] \\ x[2] \\ \vdots \\ x[N-2] \\ x[N-1] \end{bmatrix} \tag{7.196}$$

As the problem size linearly increases in convolution sum, the computational requirement rises in quadratic sense. The FFT complexity denotes the $N\log_2 N$ complexity; therefore, the filer realization based on FFT presents substantial performance advantage for increased problem size. Note that the 1024 N indicate 1048576 for N^2 and 10240 for $N\log_2 N$.

In Chap. 4, we discussed the frequency domain filter by using the DFT. In the same manner, the FFT length should be limited to the power of two and greater than or equal to $N + L - 1$ as below. With this condition, the multiplication of the FFT processed signal is equivalent to the convolution sum output according to the DFT properties. Note that the FFT length should be consistent for entire processing.

$$\begin{array}{l} x[n] : N \text{ length} \xrightarrow{\text{FFT with } N+L-1 \text{length}} X[k] \\ h[n] : L \text{ length} \xrightarrow[\text{FFT with } N+L-1 \text{length}]{} H[k] \end{array} \Big\} \xrightarrow{\text{Multiplication}} X[k]H[k]$$

$$\xrightarrow{\text{Inverse FFT with } N+L-1 \text{ length}} x[n] * h[n] \tag{7.197}$$

The inverse FFT can be realized by the forward FFT algorithm by using the following relation. Simply conjugating the input and output of the FFT with constant weighting provides the inverse FFT operation.

$$\begin{aligned} X[k] &= \sum_{n=0}^{N-1} x[n] e^{-j\frac{2\pi}{N}kn} = \text{DFT}_N(x[n]) = \text{FFT}_N(x[n]) \quad \text{for } N = 2^L \\ x[n] &= \frac{1}{N} \sum_{k=0}^{N-1} X[k] e^{j\frac{2\pi}{N}kn} = \frac{1}{N} \left(\sum_{n=0}^{N-1} X^*[k] e^{-j\frac{2\pi}{N}kn} \right)^* \\ &= \frac{1}{N} \left(\text{DFT}_N(X^*[k]) \right)^* \end{aligned} \tag{7.198}$$

In general, the data for the filter shows the infinite length with real-time processing requirement. The continuous sampling of analog signal creates the current data $x[n]$ for unlimited time span. The data obtaining policy determines the latency performance of the filter output. Possible policies are below.

- Sample-based processing: Immediate processing with current data
- Frame-based processing: Every N-length data is processed
- Entirety-based processing: Complete data is processed once.

Along with the data obtaining policies, the filter can be realized in time or frequency domain as below.

- Time-domain filter: Use the difference equation (FIR and IIR)
- Frequency-domain filter: Use the FFT and multiplication (FIR only).

Table 7.7 organizes the data obtaining policies and filter realization methods with pros and cons.

We cannot wait for the data end to obtain the complete data set toward single convolution sum because of the substantial latency. The offline processing allows to use the single convolution sum for complete filter output. In general, the data with fixed size is processed with $h[n]$ for partial convolution sum. The sample-based and frame-based processing execute the convolution sum with one and N (>1) data length, respectively. The corresponding output also presents the identical size with given input data length. Note that the augmented output length from convolution is generated by the transition.

The realization method can select the time domain or frequency domain to implement the filter. The time domain filter employs the difference equation as below. The constant coefficient as and bs are derived by the filter design methods.

$$y[n] = b_0 x[n] + \cdots + b_M x[n-M] - a_1 y[n-1] - \cdots a_N y[n-N] \qquad (7.199)$$

The FIR filter only contains the coefficient bs with previous and current input data. The FIR difference equation is equivalent to the convolution sum in which the $h[n]$ indicates the constant coefficient bs. The IIR filter includes the coefficient as and bs with previous and current input output data. The sample-based and frame-based filter in time domain well connect the fragmented outputs into continuous structure. The final states from previous filter computation are maintained toward the initial states for current computation in order to remove the transition. Therefore, the post-processing to consolidate the segmented outputs is not necessary.

The frequency domain filter only selects the frame-based realization. Unlike time domain filter, there is no way to deliver the final states to initial states between the frame output computations. Hence, in every filter computation with given N-length data, we assume that the initial states are zeros. The frame outputs are required to be interwoven for continuous streamed output due to the zero initial states. The block convolution unites the framed outputs into an appropriate way as single convolution output. The block convolution is not dedicated for the frequency domain filter. The

Table 7.7 Data obtaining policies and filter realization methods

Policy	Time domain realization	Frequency domain realization
Sample based	→ Obtain current $x[n]$ → Perform difference equation → Get current $y[n]$ → Shift input buffer for previous inputs → Shift output buffer for previous outputs (IIR only) ↑ Repeat	Not possible
	Good: Very low latency, No post-processing	
	Bad: High complexity	
Frame based	→ Obtain current data frame N-length → Perform difference equation for N times → Get current output frame → Maintain last input states ($x[n-1]$, $x[n-2]$, etc.) for next frame processing → Maintain last output states ($y[n-1]$, $y[n-2]$, etc.) for next frame processing (IIR only) ↑ Repeat	→ Computing $H[k]$ with N length (Execute once) → Obtain current data frame N length → Perform FFT on data frame with N length → Calculate $X[k]H[k]$ → Compute IFFT on $X[k]H[k]$ to get filter output → Assemble with previous output frame as post-processing ↑ Repeat from obtain data
	Good: Minor latency, No post-processing	Good: Minor latency, Low complexity
	Bad: High complexity, Input and output frame	Bad: Input and output frame, Post-processing, 2^L length
Entirety based	→ Obtain complete data → Perform difference equation for all data → Get complete output	Procedure is identical to the frame-based frequency domain realization except the longer data length and post-processing
	Good: High latency, No post-processing	Good: High latency, Low complexity, No post-processing
	Bad: High complexity, Input and output frame with long length	Bad: Input and output frame with long length, 2^L length

block convolution can apply for the any filter realizations in which the input is framed without the state transfer.

The overlap-add method [9] adds the overlap samples between consecutive frame outputs. For example, N-length input $x_m[n]$ and L-length $h[n]$ provide the convolution output with $N + L - 1$ length. Overlap and add the $L - 1$ samples between the adjacent frames over the end to end. The figure below demonstrates the overall procedure for overlap-add method where the N is 12 and L is 5. The convolution sum equation is shown below for mth frame.

$$y_m[n] = \sum_{k=0}^{\infty} x_m[k]h[n-k] = \sum_{k=n-L+1}^{n} x_m[k]h[n-k] \tag{7.200}$$

Based on the given parameters, the $h[n-k]$ escapes from and enters into the first and second frame, correspondingly at $n = 12$. The discrete time 12, 13, 14, and 15 provide the partial convolution outputs and the sum of two convolutions is identical to the unframed convolution as shown in the figure below. Therefore, the last $L-1$ elements of preceding and the first L-1 elements of subsequent output frame are overlapped and added to complete the transition convolution output as shown in Fig. 7.45.

The FFT reduces the complexity of the convolution sum at overlap-add method. The FFT length is chosen carefully to avoid the output corruption. Fundamentally, the convolution sum realized by the FFT shows the circular property; therefore, the output edges are connected and rounded for inappropriate FFT length. The FFT length is greater than or equal to the convolution length as well as power of two. The above example requires the 16 FFT length because of $N + L - 1$ convolution length.

The above figure presents the given random data $x[n]$ and framed data $x_1[n]$ & $x_2[n]$. The corresponding outputs are displayed below by locating the $L - 1$

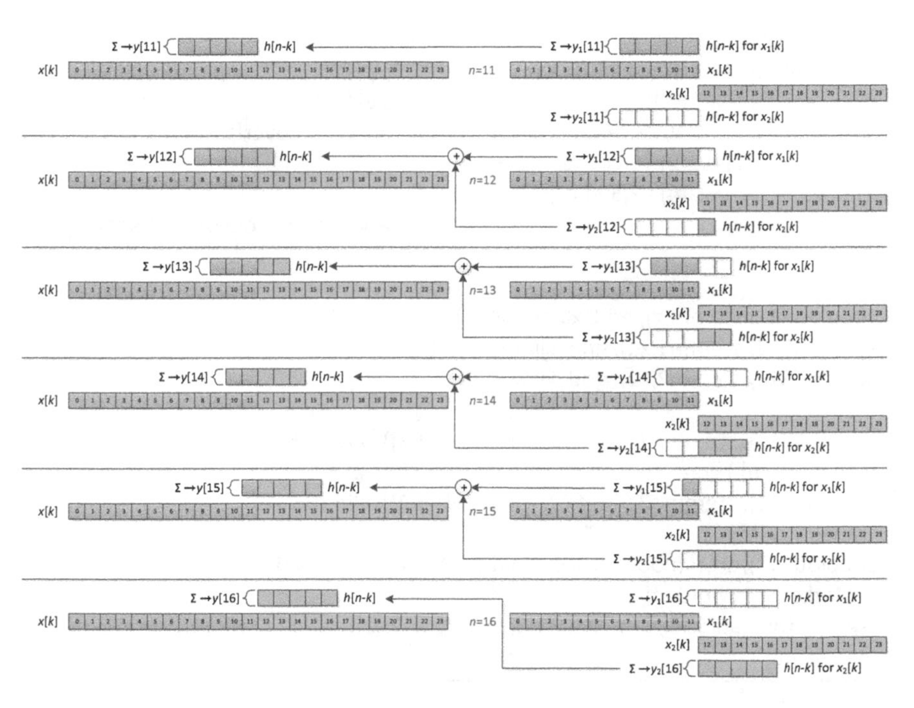

Fig. 7.45 Overlap-add method where the N is 12 and L is 5. Left is continuous convolution and right is block convolution (overlap-add method)

samples overlapped. Note that the outputs are computed by using the FFT method. The overlapped outputs are added for final output as $y[n]$ in the figure. For comparison purpose, the time domain convolution outputs are overlaid with overlap-add output. The indistinguishable difference is observed between the outputs.

Prog. 7.10 MATLAB program for Figs. 7.46 and 7.47.

```
rng('default');
N = 24;                    % Whole data length
H = 5;                     % h[n] length
FN = N/2 + H - 1;      % FFT length (assume that N is divided into two frames)
n =    0:N-1;              % Time sequence for whole data
n1 = 0:N/2-1;             % Time sequence for x1[n]
n2 = N/2:N-1;             % Time sequence for x2[n]
no1 = 0:FN-1;             % Time sequence for y1[n]
no2 = FN-(H-1):N+H-1-1;        % Time sequence for y2[n]
noo = 0:N+H-2;                    % Time sequence for y[n]
hn = ones(1,H);                  % h[n]
HK = fft(hn,FN);                 % H[k]
x = randi([-5,5],1,N);         % x[n]
x1 = x(1:N/2);                    % x1[n]
x2 = x(N/2+1:N);                  % x2[n]
X1K = fft(x1,FN);                % X1[k]
X2K = fft(x2,FN);                % X2[k]
y1 = real(ifft(X1K.*HK));                        % y1[n]
y2 = real(ifft(X2K.*HK));                        % y2[n]
yo = [y1(1:N/2) y1(N/2+1)+y2(1) ...             % y[n] from overlap-add
      y1(N/2+2)+y2(2) y1(N/2+3)+y2(3) ...
      y1(N/2+4)+y2(4) y2(5:end)];
yc = conv(hn,x);                                   % y[n] from convolution sum

figure,
subplot(311),stem(n,x), grid, xlim([0 N-1]), ylabel('x[n]')
subplot(312),stem(n1,x1), grid, xlim([0 N-1]), ylabel('x_1[n]')
subplot(313),stem(n2,x2), grid, xlim([0 N-1]), ylabel('x_2[n]')
figure,
subplot(311), stem(no1,y1), grid, xlim([0 N+H-2]), ylim([-5 20])
ylabel('y_1[n]')
subplot(312), stem(no2,y2), grid, xlim([0 N+H-2]), ylim([-5 20])
ylabel('y_2[n]')
subplot(313), stem(noo,yo), grid, hold on, h = stem(noo,yc);
h.Marker = '*'; hold off
legend('Overlap-add','Convolution'), xlim([0 N+H-2]), ylim([-5 20])
ylabel('y[n]')
```

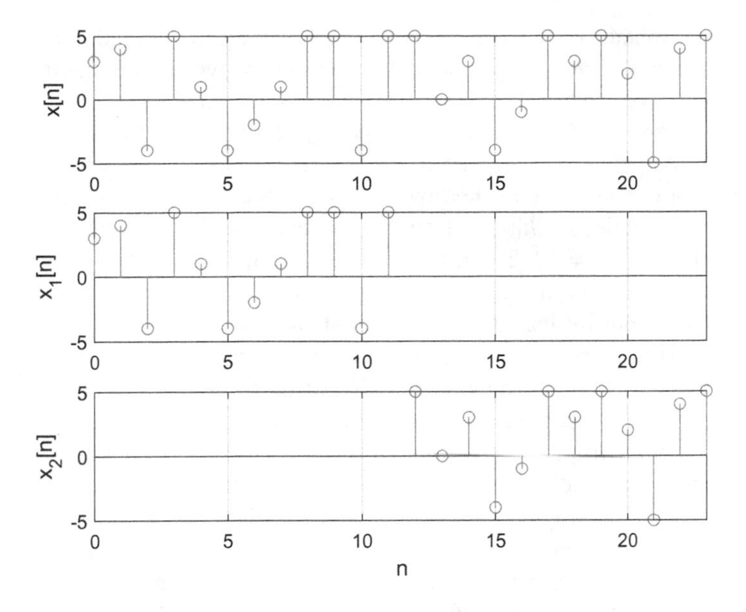

Fig. 7.46 Given random data $x[n]$ and framed data $x_1[n]$ and $x_2[n]$ for overlap-add method

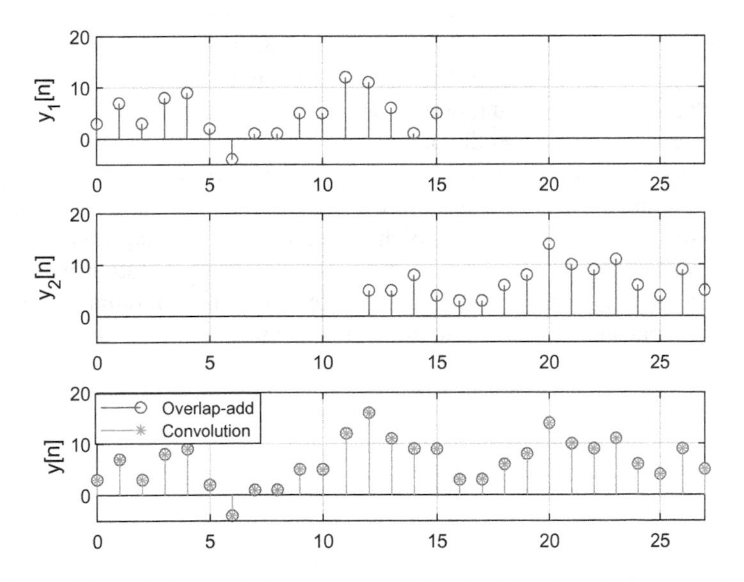

Fig. 7.47 Individual block convolution output of $x_1[n]$ and $x_2[n]$ and continuous convolution and overlap-added block convolution output

One disadvantage of the overlap-add method is the extra operation to merge the framed outputs. We can remove the add operations by overlapping the framed data. The overlap-save method [9] overlaps the $L - 1$ length data for efficient output transition where the N-length input frame $x_m[n]$ and L-length $h[n]$ provide the convolution output with $N + L-1$ length. The figure below demonstrates the overall procedure for overlap-save method with 12 N and 5 L. The $h[n - k]$ enters into the second frame while the $h[n - k]$ still holds the first frame entirely due to the overlapped data at $n = 12$. The discrete time 12, 13, 14, and 15 provide the complete outputs from the first frame convolution and the second frame convolution prepares the output by increasing the overlapping area. As soon as the complete overlapping is initiated, the second frame convolution delivers the output whereas the first frame convolution outputs are discarded. Therefore, the first $L - 1$ elements of subsequent data frame are attached over the preceding data frame end to generate the uninterrupted outputs. In the framed output, the first and last $L - 1$ elements are abandoned, and the middle N elements are taken for final output. Note that the convolution length of the overlap-save method is $N + (L - 1) + (L - 1)$. No inter-frame operations are necessary for final output as shown in Fig. 7.48.

The conventional convolution length for the above example is 20 because of overlapped data. With given configuration, the FFT length should be 32. The first and last $L - 1$ outputs are not in use; therefore, the last $L - 1$ outputs are allowed to intrude into the first $L - 1$ positions by circular convolution property. The above example requires the 16 FFT length because of 16 data length as N.

The above figure presents the given random data $x[n]$ and framed data $x_1[n]$ & $x_2[n]$. The last or first 4 $(L - 1)$ data samples are duplicated and overlapped in the framed data. The corresponding outputs are displayed below by locating the $L - 1$ samples overlapped. Note that the outputs are computed by using the 16-point FFT method. The last 4 output elements of first frame are adopted as the final output and the first 4 output elements of second frame are discarded. For comparison purpose, the time domain convolution outputs are overlaid with overlap-save output. The very first 4 outputs are not equal to the time domain convolution as a result of circular convolution with N period. From the second frame, all outputs are identical to the time domain convolution outputs. Shifting with zero padding saves the lost 4 outputs.

Prog. 7.11 MATLAB program for Figs. 7.49 and 7.50.

```
rng('default');
N = 28;      % Whole data length (frame length 12 + overlap 4)
H = 5;       % h[n] length
FN = 16;     % FFT length (frame length 12 + h[n] length 5 - 1 -> power of two)
n =   0:N-1;                 % Time sequence for whole data
n1 = 0:15;                   % Time sequence for x1[n] 0 ~ 12+4-1
n2 = 12:27;                  % Time sequence for x2[n] 12 ~ N-1
no1 = 0:FN-1;                % Time sequence for y1[n] 0 ~ FFT length - 1
no2 = FN-4:N+H-6;            % Time sequence for y2[n] (same length as y1[n]
with 4 overlap)
noo = 0:27;                  % Time sequence for y[n]
hn = ones(1,H);             % h[n]
HK = fft(hn,FN);           % H[k]
x = randi([-5,5],1,N);     % x[n]
x1 = x(1:16);               % x1[n] with 4 overlap
x2 = x(13:end);             % x2[n] with 4 overlap
X1K = fft(x1,FN);          % X1[k]
X2K = fft(x2,FN);          % X2[k]
y1 = real(ifft(X1K.*HK));          % y1[n]
y2 = real(ifft(X2K.*HK));          % y2[n]
yo = [y1 y2(5:end)];                 % y[n] from overlap-save
yc = conv(hn,x);                       % y[n] from convolution sum

figure,
subplot(311),stem(n,x), grid, xlim([0 N-1]), ylabel('x[n]')
subplot(312),stem(n1,x1), grid, xlim([0 N-1]), ylabel('x_1[n]')
subplot(313),stem(n2,x2), grid, xlim([0 N-1]), ylabel('x_2[n]')
figure,
subplot(311), stem(no1,y1), grid, xlim([0 27]), ylim([-5 20])
ylabel('y_1[n]')
subplot(312), stem(no2,y2), grid, xlim([0 27]), ylim([-5 20])
ylabel('y_2[n]')
subplot(313), stem(noo,yo), grid, hold on
h = stem(noo,yc(1:28));
h.Marker = '*'; hold off
legend('Overlap-save','Convolution')
xlim([0 27]); ylim([-5 20]); ylabel('y[n]')
```

The block convolution permits to use the FFT for convolution sum. The clever arrangement for data or output overcomes the initial state disruption between the framed convolution sum. Since the filter with FFT reduces the complexity, the increased segment size provides the better performance in terms of computation. However, the high N generates the extra latency in filter processing.

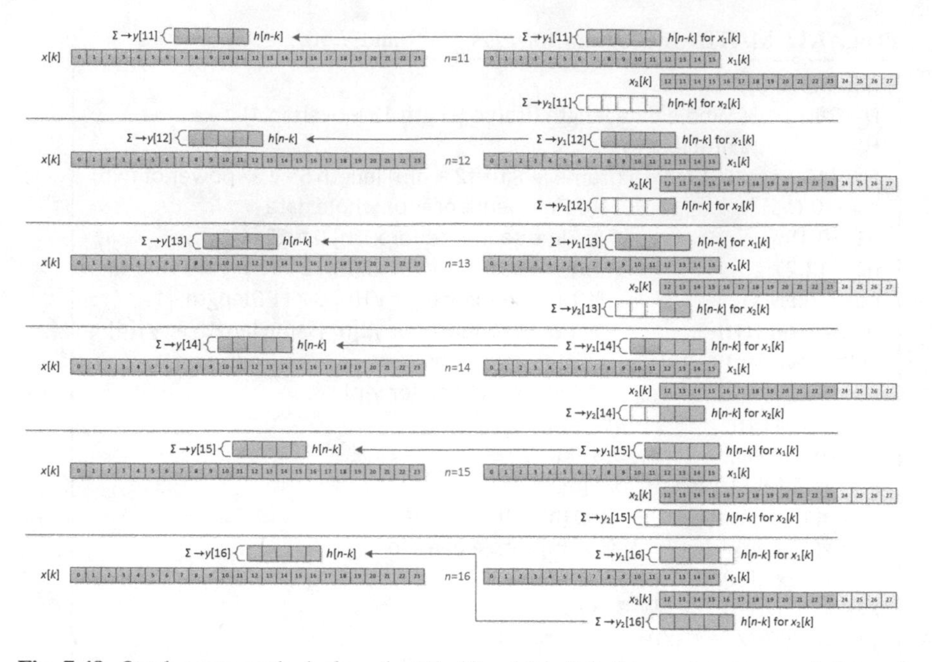

Fig. 7.48 Overlap-save method where the N is 12 and L is 5. Left is continuous convolution and right is block convolution (overlap-save method)

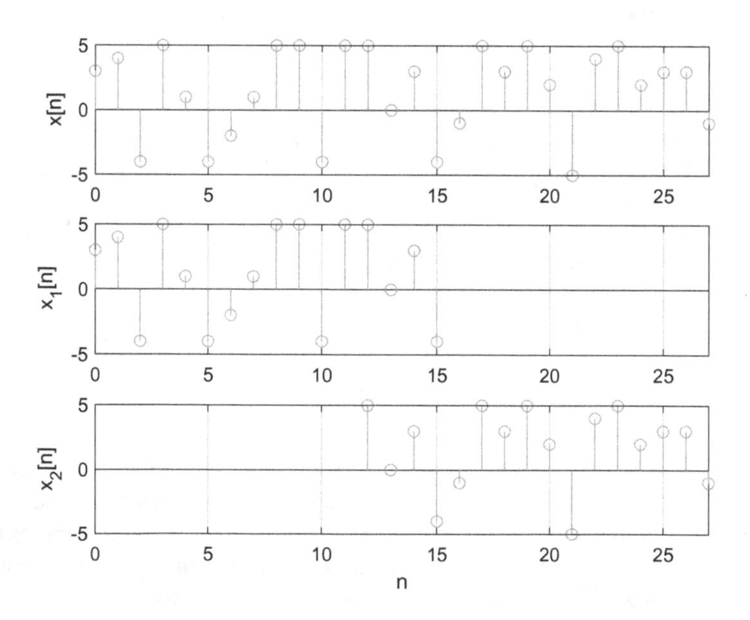

Fig. 7.49 Given random data $x[n]$ and framed data $x_1[n]$ and $x_2[n]$ for overlap-save method

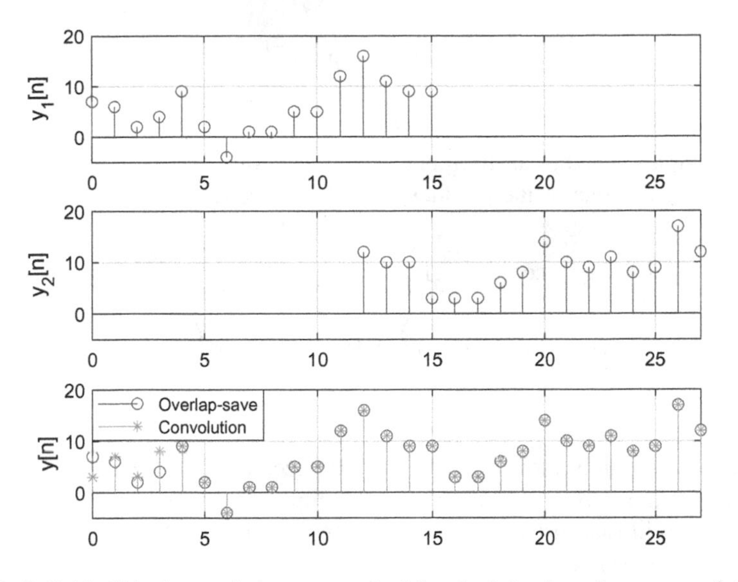

Fig. 7.50 Individual block convolution output of $x_1[n]$ and $x_2[n]$ and continuous convolution and overlap-saved block convolution output

– Quantization effect

In previous sections, we discussed the implementation matters on time domain filter such as sampling, quantization, and scaling. With floating-point number, we assume that the problems are ignored and restricted to the fixed-point number. The DFT and FFT also experiences the similar troubles for realization except the sampling. The complex weight and sum with fixed-point number require the quantization and scaling for proper transform output. The filter in frequency domain extensively uses the FFT algorithm; hence, the FFT analysis over the quantization and scaling is essential to understand the filter performance. Below shows the DFT in matrix form again for quick reference.

$$\begin{bmatrix} X[0] \\ X[1] \\ \vdots \\ X[N-2] \\ X[N-1] \end{bmatrix} = \begin{bmatrix} 1 & 1 & \cdots & 1 & 1 \\ 1 & e^{-j\frac{2\pi}{N}1} & \cdots & e^{-j\frac{2\pi}{N}(N-2)} & e^{-j\frac{2\pi}{N}(N-1)} \\ \vdots & \vdots & \ddots & \vdots & \vdots \\ 1 & e^{-j\frac{2\pi}{N}(N-2)1} & \cdots & e^{-j\frac{2\pi}{N}(N-2)(N-2)} & e^{-j\frac{2\pi}{N}(N-2)(N-1)} \\ 1 & e^{-j\frac{2\pi}{N}(N-1)1} & \cdots & e^{-j\frac{2\pi}{N}(N-1)(N-2)} & e^{-j\frac{2\pi}{N}(N-1)(N-1)} \end{bmatrix} \begin{bmatrix} x[0] \\ x[1] \\ \vdots \\ x[N-2] \\ x[N-1] \end{bmatrix} \tag{7.201}$$

For single output $X\,[1]$, the weighted sum is presented as below according to the DFT matrix.

$$X[1] = x[0] + x[1]e^{-j\frac{2\pi}{N}1} + x[2]e^{-j\frac{2\pi}{N}2} + \cdots + x[N-2]e^{-j\frac{2\pi}{N}(N-2)} + x[N-1]e^{-j\frac{2\pi}{N}(N-1)} \quad (7.202)$$

The complex weight can be represented as below with complex input x. The below example is one of the product in X [1] at $n = 1$.

$$x[1]e^{-j\frac{2\pi}{N}1} = \left(\mathrm{Re}(x[1])\cos\left(-\frac{2\pi}{N}1\right) - \mathrm{Imag}(x[1])\sin\left(-\frac{2\pi}{N}1\right)\right) + j\left(\mathrm{Re}(x[1])\sin\left(-\frac{2\pi}{N}1\right) + \mathrm{Imag}(x[1])\cos\left(-\frac{2\pi}{N}1\right)\right) \text{ for } x[1] \in \mathbb{C} \quad (7.203)$$

Since the multiplication doubles the bit size, the quantization truncates the product output for $B + 1$ bit size. The complex number product executes the multiplication in four instances; therefore, the four quantizations are performed for single complex weighting as below.

$$Q_B\left(x[1]e^{-j\frac{2\pi}{N}1}\right) = \left(Q_B\left(\mathrm{Re}(x[1])\cos\left(-\frac{2\pi}{N}1\right)\right) - Q_B\left(\mathrm{Imag}(x[1])\sin\left(-\frac{2\pi}{N}1\right)\right)\right) + j\left(Q_B\left(\mathrm{Re}(x[1])\sin\left(-\frac{2\pi}{N}1\right)\right) + Q_B\left(\mathrm{Imag}(x[1])\cos\left(-\frac{2\pi}{N}1\right)\right)\right) \quad (7.204)$$

The quantization can be described by the random noise addition as below. We assume that the random noise follows the uniform distribution in probability density function.

$$Q_B\left(x[1]e^{-j\frac{2\pi}{N}1}\right) = \left(\left(\mathrm{Re}(x[1])\cos\left(-\frac{2\pi}{N}1\right) + e_1[1,1]\right) - \left(\mathrm{Imag}(x[1])\sin\left(-\frac{2\pi}{N}1\right) + e_2[1,1]\right)\right) + j\left(\left(\mathrm{Re}(x[1])\sin\left(-\frac{2\pi}{N}1\right) + e_3[1,1]\right) + \left(\mathrm{Imag}(x[1])\cos\left(-\frac{2\pi}{N}1\right) + e_4[1,1]\right)\right) \quad (7.205)$$

The random noise function $e_{1,2,3,4}[n, k]$ uses the parameter n and k for time and frequency sample respectively. The individual noise variance is given below for the approximation with uniform distribution according to the previous section result. We assume the zero-mean condition.

$$\mathrm{Var}\left(e_{1,2,3,4}[1,1]\right) = \mathbb{E}\left(\left(e_{1,2,3,4}[1,1]\right)^2\right) = \frac{2^{-2B}}{12} \tag{7.206}$$

The squared magnitude of the complex error e [1, 1] is below.

$$|e[1,1]|^2 = (e_1[1,1] - e_2[1,1])^2 + (e_3[1,1] + e_4[1,1])^2 \tag{7.207}$$

Using the uncorrelated and independent condition, the expected value of the e [1, 1] squared magnitude can be represented as below.

$$\mathbb{E}\left(|e[1,1]|^2\right) = 4\frac{2^{-2B}}{12} = \frac{2^{-2B}}{3} \tag{7.208}$$

The expected value of output error squared magnitude for the given $k = 1$ is below. Note that the error is uncorrelated and independent for sample to sample in time.

$$\mathbb{E}\left(|R[1]|^2\right) = \sum_{n=0}^{N-1}\mathbb{E}\left(|e[n,1]|^2\right) = N\frac{2^{-2B}}{3} \tag{7.209}$$

For any k value, we can obtain the general output as below. The N times in the output error squared magnitude average can be decreased by the number of unit product $e^{-j\frac{2\pi}{N}0}$; however, the overall the average is proportional to N.

$$\begin{aligned}
Q_B(X[k]) &= \sum_{n=0}^{N-1} Q_B\left(x[n]e^{-j\frac{2\pi}{N}kn}\right)\\
&= \left(\sum_{n=0}^{N-1}\left(Q_B\left(\mathrm{Re}(x[n])\cos\left(-\frac{2\pi}{N}kn\right)\right) - Q_B\left(\mathrm{Imag}(x[n])\sin\left(-\frac{2\pi}{N}kn\right)\right)\right)\right)\\
&\quad +j\left(\sum_{n=0}^{N-1}\left(Q_B\left(\mathrm{Re}(x[n])\sin\left(-\frac{2\pi}{N}kn\right)\right) + Q_B\left(\mathrm{Imag}(x[n])\cos\left(-\frac{2\pi}{N}kn\right)\right)\right)\right)\\
&Q_B(\mathrm{Re}(X[k])) = \mathrm{Re}(X[k]) + \left(\sum_{n=0}^{N-1}(e_1[n,k] - e_2[n,k])\right)\\
&Q_B(\mathrm{Imag}(X[k])) = \mathrm{Imag}(X[k]) + \left(\sum_{n=0}^{N-1}(e_3[n,k] + e_4[n,k])\right)\\
&\mathbb{E}\left(|R[k]|^2\right) = \sum_{n=0}^{N-1}\mathbb{E}\left(|e[n,k]|^2\right) = N\tfrac{2^{-2B}}{3}
\end{aligned} \tag{7.210}$$

where

$$|e[n,k]|^2 = (e_1[n,k] - e_2[n,k])^2 + (e_3[n,k] + e_4[n,k])^2$$

The double-length accumulator improves the quantization noise performance. The individual noise sources from the complex weight are converted into the single source by using the double-length accumulator. The four products in the complex weight are accumulated along with time sequence then quantized as below.

$$\begin{aligned} Q_B(X[k]) &= Q_B\left(\sum_{n=0}^{N-1} x[k]e^{-j\frac{2\pi}{N}kn}\right) \\ = Q_B\left(\sum_{n=0}^{N-1}\mathrm{Re}(x[n])\cos\left(-\tfrac{2\pi}{N}kn\right)\right) &- Q_B\left(\sum_{n=0}^{N-1}\mathrm{Imag}(x[n])\sin\left(-\tfrac{2\pi}{N}kn\right)\right) \\ +j\left(Q_B\left(\sum_{n=0}^{N-1}\mathrm{Re}(x[n])\sin\left(-\tfrac{2\pi}{N}kn\right)\right)\right. &+ \left.Q_B\left(\sum_{n=0}^{N-1}\mathrm{Imag}(x[n])\cos\left(-\tfrac{2\pi}{N}kn\right)\right)\right) \\ Q_B(\mathrm{Re}(X[k])) &= \mathrm{Re}(X[k]) + (e_1[k] - e_2[k]) \\ Q_B(\mathrm{Imag}(X[k])) &= \mathrm{Imag}(X[k]) + (e_3[k] + e_4[k]) \end{aligned} \tag{7.211}$$

The uncorrelated and independent condition provides the following average magnitude squared of the output error.

$$\begin{aligned} \mathbb{E}\left(|R[k]|^2\right) &= \mathbb{E}\left((e_1[k] - e_2[k])^2 + (e_3[k] + e_4[k])^2\right) \\ &= 4\frac{2^{-2B}}{12} = \frac{2^{-2B}}{3} \end{aligned} \tag{7.212}$$

The FFT employs the complex exponential periodicity for reduced complexity. With length limitation, the extended butterfly decreases the computational stage as well as burden in FFT algorithms. Below is the one extended butterfly which performs one complex weighting and negating with two input and two output ports as shown in Fig. 7.51.

From the FFT output, the one X[k] value is connected to the two butterflies in the previous stage. The two butterflies are joined to the four butterflies in the preceding stage as shown in below FFT signal flow as shown in Fig. 7.52.

The power of two pattern involves the butterfly numbers in each stage from the output such as one, two, four, eight, etc. The total number of butterflies can be calculated by using the binary number representation. The 8-point FFT requires the three stages as $\log_2 8$ and MSB stands for the output stage.

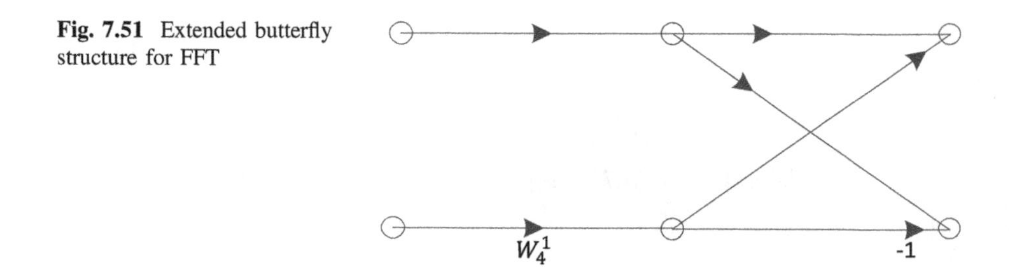

Fig. 7.51 Extended butterfly structure for FFT

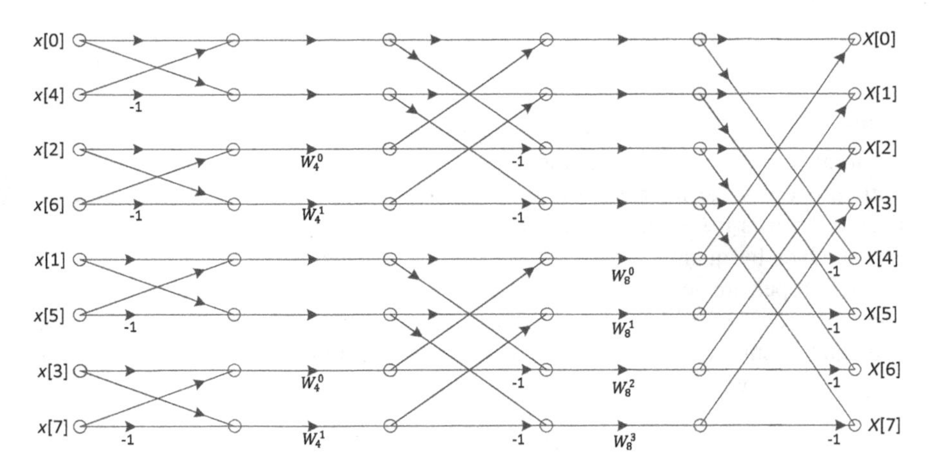

Fig. 7.52 8-point FFT flow graph with extended butterfly structure

$$0111_2 = 1000_2 - 1_2 = \left(2^3 - 1\right)_{10} = 7$$

The seven butterflies are engaged for the given 8-point FFT output at kth frequency. For general N-point FFT, the number of butterflies for each output is below.

$$\underbrace{0}_{2^L}\underbrace{1}_{2^{L-1}}\cdots\underbrace{1}_{2^2}\underbrace{1}_{2^1}\underbrace{1}_{2^0} = \underbrace{1}_{2^L}\underbrace{0}_{2^{L-1}}\cdots\underbrace{0}_{2^2}\underbrace{0}_{2^1}\underbrace{0}_{2^0} - 1 = N - 1 \text{ where } L = \log_2 N \tag{7.213}$$

The complex weighting in each butterfly provides four multiplications to deliver the following variance with zero mean condition.

$$\sigma_B^2 = 4\frac{2^{-2B}}{12} = \frac{2^{-2B}}{3} \tag{7.214}$$

The overall output noise at individual output port is equivalent to the propagated noise sum. With the uncorrelated noise source condition, the mean-squared magnitude of the output noise is calculated by the product of individual variance over the weight counts on the signal propagation path as below.

$$\mathbb{E}\left(|R[k]|^2\right) = (N-1)\sigma_B^2 = (N-1)\frac{2^{-2B}}{3} \tag{7.215}$$

The mean-squared magnitude of the output noise could be further reduced by considering the unit weight on the FFT input stage as below.

$$\mathbb{E}\left(|R[k]|^2\right) = \left(N - \frac{N}{2} - 1\right)\sigma_B^2 = \left(\frac{N}{2} - 1\right)\frac{2^{-2B}}{3} \tag{7.216}$$

The cascade connection between FFT stages does not promote the double-length accumulator benefit on output noise figures. To obtain the perfect precision for FFT output, every stage increases the storage size by $B + 1$ bits; otherwise, the quantization error is inevitable. In general, the butterfly outputs are stored in the $B + 1$ bit variables by quantization for next stage processing.

The proper pre-scaling over DFT input prevents the output overflow. The DFT and FFT provide the identical computational result; therefore, the pre-scaling policy is equal to both transforms. The fixed-point number by $B + 1$ bit binary fraction are ranged as below.

$$-1 \leq x[n] \leq 1 - 2^{-B} \tag{7.217}$$

The corresponding DFT range is indicated as below.

$$|X[k]| \leq \sum_{n=0}^{N-1} |x[n]| \left|e^{-j\frac{2\pi}{N}kn}\right| < N \tag{7.218}$$

The input is divided by N to maintain the output range below the binary fraction limits as below.

$$|x[n]| < \frac{1}{N} \rightarrow |X[k]| < 1 \tag{7.219}$$

The increased DFT length N produces the following results for signal and noise power.

$$N \uparrow \Rightarrow \begin{cases} x[n]\text{range} \downarrow \Rightarrow \text{Decrease signal power} \\ \mathbb{E}\left(|R[k]|^2\right) = N\frac{2^{-2B}}{3} \uparrow \Rightarrow \text{Increase noise power} \end{cases} \tag{7.220}$$

The SNR from fixed-point DFT is deteriorate by the extended DFT length. Oppenheim and Schafer [9] derived the SNR for fixed-point DFT as below.

$$\frac{\mathbb{E}\left(|X[k]|^2\right)}{\mathbb{E}\left(|R[k]|^2\right)} = \frac{1}{N^2 2^{-2B}} \tag{7.221}$$

The signal power is derived by the signal probability distribution function and DFT equation. The derivation assumed that the DFT is performed on the $B + 1$ bit fixed-point number. The SNR equation represents that the SNR decreases as N^2 sense. To maintain the constant SNR, the doubled N should be counterbalanced by the added one bit on quantization. The DFT with double-length accumulator presents the identical SNR as above since both N factors in signal and noise power are removed by double-length accumulator. Alternative scaling procedure for FFT shows the better performance in SNR by adapting 1/2 attenuator at each FFT stage rather than a large scaling at input stage. We can find the detail procedures and results from the Oppenheim and Schafer [9]. Also, Oppenheim and Schafer analyzed the coefficient quantization in the FFT. The SNR from the coefficient quantization by $B + 1$ bits is below.

$$\frac{\mathbb{E}\left(|X[k]|^2\right)}{\mathbb{E}\left(|R[k]|^2\right)} \propto \frac{6}{\log_2 N 2^{-2B}} \tag{7.222}$$

The major consequence is that the SNR decreases very gently with N which is inversely proportional to $\log_2 N$; therefore, doubling N produces only a slight decrease in the SNR. Further analysis can be found on the reference [6, 9].

7.6 Problems

1. Calculate the SNR in round quantization. Assume that the probability density function is uniform as shown below.

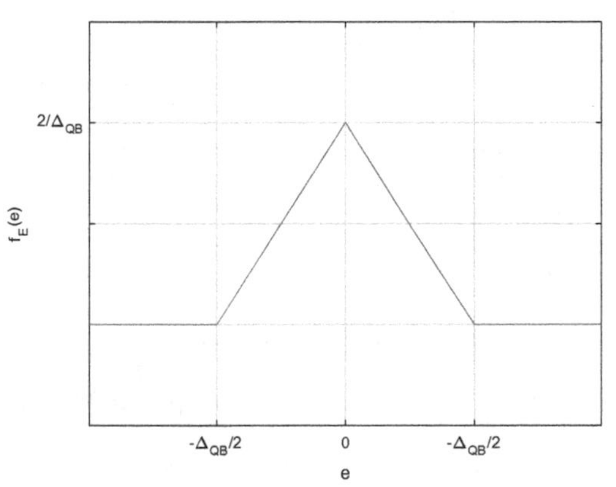

2. Shows the overall scaling procedure for cascade system. The designed filter indicates the LPF with $\pi/3$ cutoff frequency in 4th order system. The design method is Butterworth.

$$H(z) = \frac{0.0261 + 0.1043z^{-1} + 0.1565z^{-2} + 0.1043z^{-3} + 0.0261z^{-4}}{1 - 1.3066z^{-1} + 1.0305z^{-2} - 0.3624z^{-3} + 0.0558z^{-4}}$$

3. The random signal with zero mean and uniform distribution is provide to the system input of Problem 2. The input is ranged from -1 to 1 to avoid the overflow. Investigate the overflow likelihood for each scaling method by simple MATLAB experiment.

4. Analyze the quantization noise performance. The designed filter indicates the LPF with $\pi/3$ cutoff frequency in 4th order system. The design method is Butterworth.

$$H(z) = \frac{0.0261 + 0.1043z^{-1} + 0.1565z^{-2} + 0.1043z^{-3} + 0.0261z^{-4}}{1 - 1.3066z^{-1} + 1.0305z^{-2} - 0.3624z^{-3} + 0.0558z^{-4}}$$

5. For below configurations, find maximum frequency response error in magnitude from coefficient quantization. Also calculate the corresponding pass band ripple A_p and stop band attenuation A_{st}.
   - FIR filter with 30 length ($N = 29$) and 8-bit quantization
   - FIR filter with 60 length ($N = 59$) and 9-bit quantization.

References

1. IEEE Standard for Floating-Point Arithmetic. IEEE Std 754-2019 (Revision of IEEE 754-2008), 1−84 (2019). https://doi.org/10.1109/ieeestd.2019.8766229
2. Wikipedia: Moment (mathematics) (2020). https://en.wikipedia.org/w/index.php?title=Moment_(mathematics)&oldid=959967484
3. Wilkinson, J.H.: The evaluation of the zeros of ill-conditioned polynomials. Part I. Numerische Mathematik **1**(1), 150–166 (1959). https://doi.org/10.1007/BF01386381
4. Wilkinson, J.H.: Rounding Errors in Algebraic Processes By J.H. Wilkinson. Prentice-Hall, (1963)
5. Wikipedia: Wilkinson's polynomial (2020). https://en.wikipedia.org/w/index.php?title=Wilkinson%27s_polynomial&oldid=954862605
6. Jackson, L.B.: Digital Filters and Signal Processing, 3rd edn. Kluwer Academic Publishers, Boston (1996)
7. Wikipedia: Residue theorem (2020). https://en.wikipedia.org/w/index.php?title=Residue_theorem&oldid=959312933
8. Wikipedia: Partial fraction decomposition (2020). https://en.wikipedia.org/w/index.php?title=Partial_fraction_decomposition&oldid=954972624
9. Oppenheim, A.V., Schafer, R.W.: Discrete-Time Signal Processing. Prentice Hall (1989)
10. Parks, T.W., Burrus, C.S.: Digital Filter Design. Wiley (1987)

Chapter 8
Filters with MATLAB

The previous chapters have provided the knowledges on the digital filter design, analysis, and realization based on the digital signal understanding. The MATLAB assists the all procedures by computing, deriving, and visualizing the mathematics and algorithms. The robust and strong functions and toolboxes in the MATLAB are essential instruments for novice students as well as professional engineers. The MATLAB usage shown in proceeding chapters are focused on the theory understanding and verification. This chapter extends the MATLAB programming for practical problems such as optimization, fixed-point number, etc. This chapter also introduces the system objects which are designed for dynamic iterative computations that process large streams of data in segments. The MATLAB coder for non-MATLAB environment will be explored for rapid deployment system. The MATLAB is very extensive technical language for various applications. This chapter delivers quick introductions only and further information can be found on MATLAB support documents at MathWorks webpages. Note that this chapter provides in-text examples without numbering. The problem section is not generated explicitly in the chapter end.

8.1 Fundamental Filter Design Methods

The MATLAB provides two programming approaches as textual and graphical environment. We have used the textual environment as script-based programming by using the command or editor window. The MATLAB also gives the graphical environment known as Simulink. The intuitive blocks and wiring present the convenient programming style in Simulink. The flexible and extensive supports from the Simulink expand the graphical blocks to the script programming and beyond. Further information on Simulink can be found at MathWorks webpage [1].

The MATLAB contains the numerous toolboxes which collect the discrete functions for individual application area. By adding the specific toolboxes, we can

K. Kim, *Conceptual Digital Signal Processing with MATLAB*,
Signals and Communication Technology,
https://doi.org/10.1007/978-981-15-2584-1_8

perform the technical computing for the application with reduced efforts on programming. The discrete function in MATLAB is the realization of mathematical algorithm. The fundamental mathematical algorithms from the conventional study areas are continuously appended and updated for toolboxes biannually. The MATLAB also provides apps for the window-based interactive application to design and analysis the technical computing task. The below diagram shows the overall MATLAB configuration for programming environments as Fig. 8.1.

The digital filter design, analysis, and realization can be executed in any MATLAB programming level. In this book, we exclude the Simulink environment which should be investigated by dedicated chapters due to the comprehensiveness. Below present the filter related programming levels.

- Primitive MATLAB programming for mathematical algorithm
- Toolbox function
- MATLAB apps.

The mathematical algorithm programming utilizes the MATLAB primitive functions to solve the equations in numerical or symbolical way. The established equations are realized by the MATLAB in step by step. For example, the least squared error frequency domain FIR filter design (integral squared error approximation criterion) requires the following procedure. The theoretical background can be found at Chap. 6. See Fig. 8.2.

The MATLAB realization is demonstrated in below. The inverse DTFT is performed by the symbolic math and converted to the numerical computation by using the *subs* and *eval* function. The frequency domain analysis is prepared by the DFT (FFT) with symmetric impulse response in order to obtain the real DFT outcome. The advantage of the primitive MATLAB programming is the flexible and profound understanding on the mathematical algorithm. The disadvantage of the method is requirement of extra verification steps to prove the written MATLAB program.

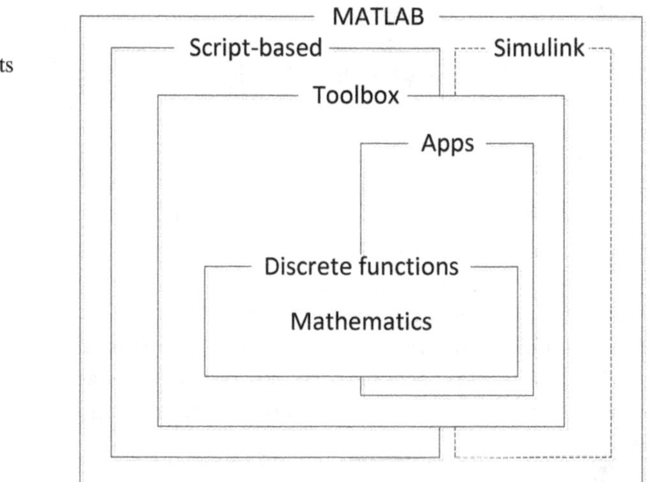

Fig. 8.1 MATLAB configuration for programming environments

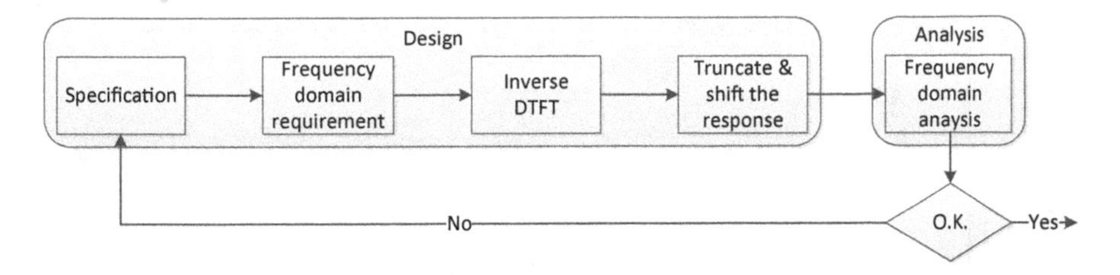

Fig. 8.2 Example of MATLAB algorithm (least squared error frequency domain FIR filter design)

Prog. 8.1 Least squared error frequency domain FIR filter design.

```
syms w n                  % Declare symbols
wc = 3*pi/8;         % The center of the pass band (pi/4) and stop band (pi/2) edge
N = 16;                   % FIR filter length = N+1
NN = 2^13;               % DFT(FFT) length
N1 = NN/2+1;             % DFT(FFT) half position
N2 = N/2;
nnn = (-N2:1:N2);        % Impulse response positions

hd0 = int(exp(1j*w*n),-wc,wc)/(2*pi);  % Inverse DTFT of frequency response
hd1 = simplify(hd0);                    % Simplify the symbolic math
b1 = subs(hd1,n,nnn+eps);               % Place values onto the symbolic math
b2 = eval(b1);                          % Compute the symbolic math
b3 = [b2 zeros(1,NN-length(b2))];       % Extend the impulse response to the
                                        DFT(FFT) length
b4 = circshift(b3,-8);                  % Circular shift for real DFT(FFT) value
B2 = real(fft(b4,NN));                  % Frequency analysis

findex = linspace(0,2*pi,NN);
ff = findex(1:N1);

figure(1)
subplot(211), stem(nnn,b2), grid
subplot(212),plot(ff,B2(1:N1)), grid
...
```

The program output is illustrated below in Fig. 8.3. Note that the decoration code is not shown above but the grid and labels are displayed at below figure. Once the response is acceptable, we can use the $h[n]$ as the filter coefficients; otherwise, the specification is modified to meet the requirement.

The IIR filter is also designed and analyzed by primitive MATLAB programming as below. The symbolic math is employed in the Butterworth model and bilinear transform stage. When the filter coefficient values are determined, the

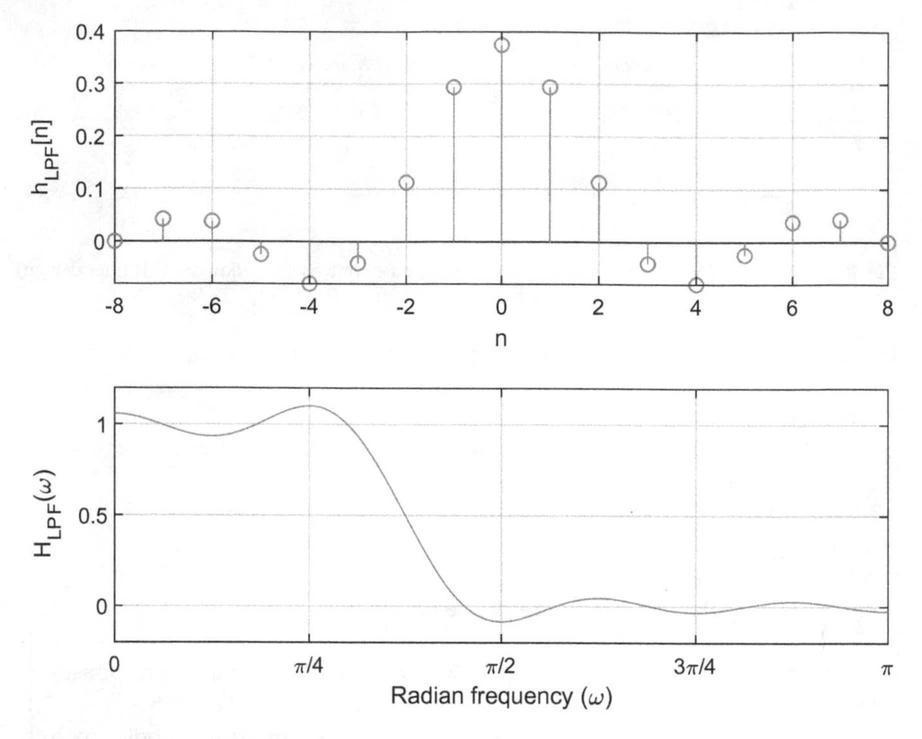

Fig. 8.3 Output of **Prog. 8.1**

analysis stage is executed by the numerical computation. The code realization of the below diagram is omitted due to the replication from Chap. 6. See Fig. 8.4.

The toolbox functions serve the extensive tasks on filter design, analysis, and realization stage. Tables 8.1, 8.2, 8.3, and 8.4 organize the dedicated filter related functions in part. The detail and further information on the function can be obtained from the MATLAB support documents [2].

Below example shows the FIR filter design and analysis based on the least-square approximation with hamming window. Observe that the *fir1* function uses the normalized frequency. The individual function performs the filter design

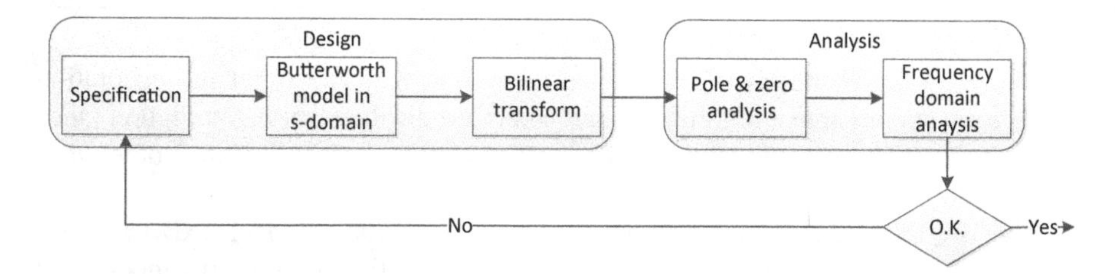

Fig. 8.4 Example of MATLAB algorithm (Butterworth IIR filter design)

Table 8.1 MATLAB functions for IIR filter design

Function name	Function task
butter	Butterworth filter design
cheby1	Chebyshev Type I filter design
cheby2	Chebyshev Type II filter design
ellip	Elliptic filter design
yulewalk	Recursive digital filter design
maxflat	Generalized digital Butterworth filter design
lpc	Linear prediction filter coefficients
prony	Prony method for filter design
stmcb	Compute linear model using Steiglitz-McBride iteration
invfreqz	Identify discrete-time filter parameters from frequency response data

Table 8.2 MATLAB functions for FIR filter design

Function name	Function task
fir1	Window-based FIR filter design with least-squares approximation
fir2	Frequency sampling-based FIR filter design
firls	Least-squares linear-phase FIR filter design
firpm	Parks-McClellan optimal FIR filter design
fircls	Constrained-least-squares FIR multiband filter design
fircls1	Constrained-least-squares linear-phase FIR lowpass and highpass filter design
cfirpm	Complex and nonlinear-phase equiripple FIR filter design

Table 8.3 MATLAB functions for filter analysis in frequency domain

Function name	Function task
freqz	Frequency response of digital filter
fvtool	Open filter visualization tool
grpdelay	Average filter delay (group delay)
phasedelay	Phase delay of digital filter
phasez	Phase response of digital filter
unwrap	Correct phase angles to produce smoother phase plots
zerophase	Zero-phase response of digital filter
zplane	Zero-pole plot for discrete-time systems

and analysis automatically. Instead of *freqz*, the *fvtool* function provides the interactive menu to select the various analysis tasks.

Prog. 8.2 FIR filter design and analysis based on the least-square approximation with hamming window.

```
wc = 3*pi/8;      % The center of the pass band (pi/4) and stop band (pi/2) edge
N = 16;           % FIR filter length = N+1
b = fir1(N,wc/pi);  % Filter design
freqz(b,1);         % Filter analysis
```

The *freqz* function presents the magnitude and phase response in frequency domain for the given filter as below. The plot decoration is given by default. After the filter design, the filter execution with input data stream is performed by realization functions listed at Table 8.5. The coefficients *a*s and *b*s with input data are placed into the function arguments and the returned outcome presents the filter output. The filter type (FIR or IIR) selects the usable functions for filter execution in Table 8.5.

Apps in the MATLAB are interactive applications written to perform specific computing tasks. The app is a self-contained program with a user interface that automates a task. All the operations required to complete the task are performed within the app. The Apps tab of the MATLAB Toolstrip shows the currently installed apps. Below table consolidates the filter design and analysis related apps in Table 8.6.

The filter designer app is launched by clicking the app icon in the tab or typing the *filterDesigner* in the command window. We specify the response type, design method, and design specifications for desired filter. The derived filter coefficients can be analyzed and verified by the filter designer app in numerous performance factors. The finalized filter design is converted to the specific structure as well as exported to the MATLAB workspace. Below figure shows the filter designer app window in Fig. 8.6.

The filter builder app is launched by clicking the app icon in the tab or typing the *filterBuilder* in the command window. The filter builder app provides a graphical interface to the object-oriented filter design method to find the best algorithm for the desired response. The designed filter can be operated as the system object which delivers the flexibility and expandability for dynamic systems. Similar to the filter designer, we specify the various parameters such as response type, design method, and design specifications for desired filter. The derived filter coefficients can be analyzed and verified by clicking the view filter response button in the app. The finalized filter design is converted to the specific structure as well as exported to the MATLAB workspace. Below figure shows the filter builder app window in Fig. 8.7.

Table 8.4 MATLAB functions for filter analysis in time domain

Function name	Function task
impz	Impulse response of digital filter
impzlength	Impulse response length
stepz	Step response of digital filter

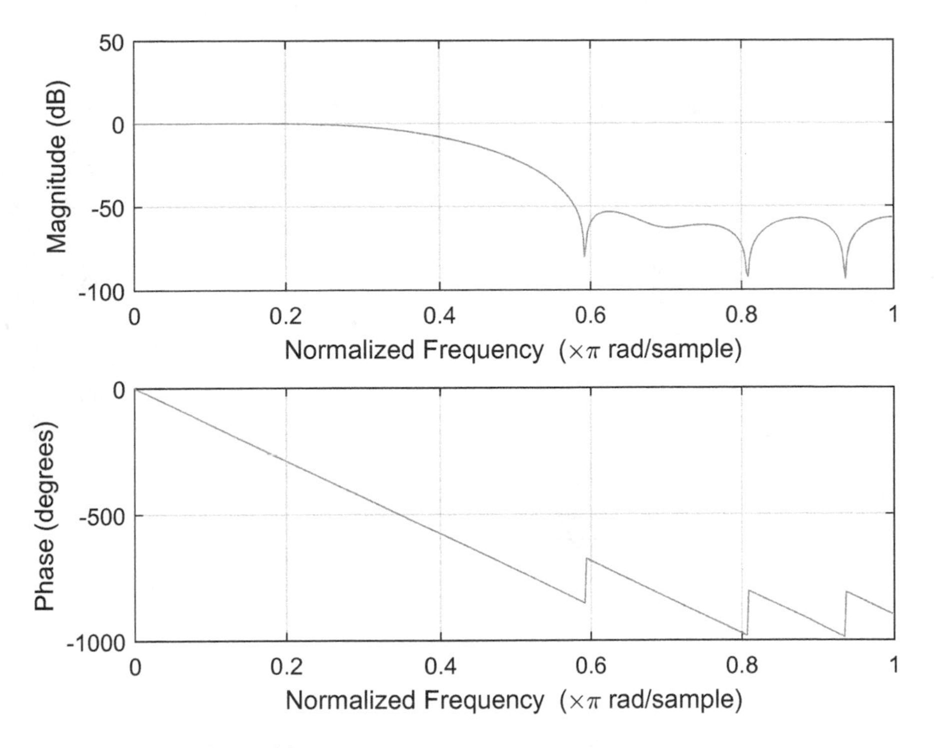

Fig. 8.5 Output of **Prog. 8.2**

Table 8.5 MATLAB functions for filter realization

Function name	Function task
conv	Convolution and polynomial multiplication. FIR only
filter	1-D digital filter. FIR and IIR
filtfilt	Zero-phase digital filtering. FIR and IIR
fftfilt	FFT-based FIR filtering using overlap-add method. FIR only

Table 8.6 MATLAB apps for filter design and analysis

Apps name	Function task
filter designer	Design filters starting with algorithm selection
filter builder	Design filters starting with frequency and magnitude specifications
window designer	Design and analyze spectral windows

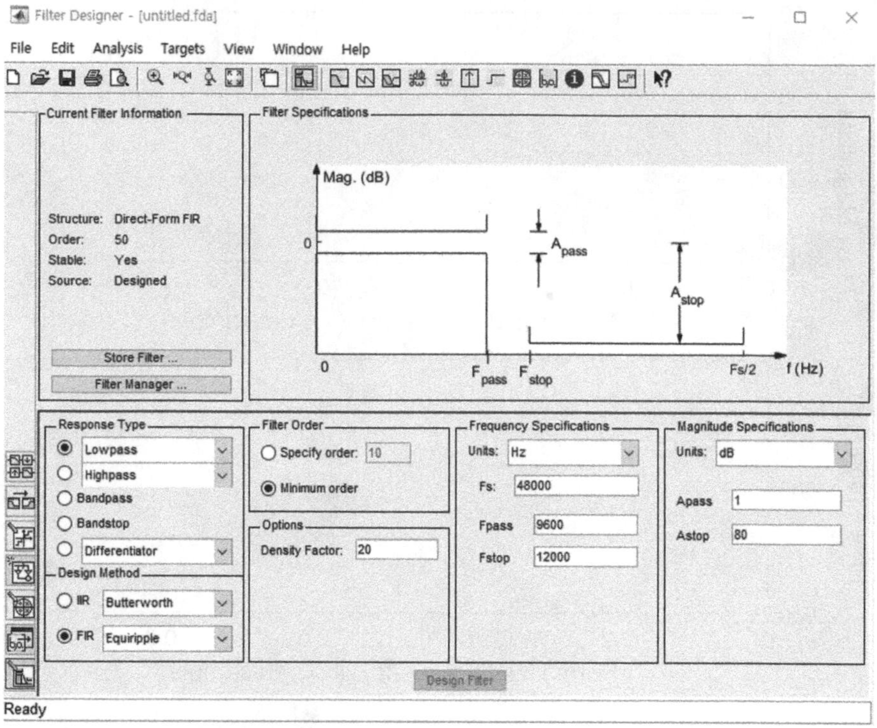

Fig. 8.6 Filter designer app

The window designer app is launched by clicking the app icon in the tab or typing the *windowDesigner* in the command window. The window designer app enables us to design and analyze spectral windows by investigating how the behavior of a window changes as a function of its length and other parameters. The length and shape of window determine the spectral estimation and FIR filter performance. We specify the various parameters such as window type, length, and sampling for desired spectral performance. The corresponding spectral window analyzes and verifies the performance with parametric outputs for instance leakage factor, relative sidelobe attenuation, and mainlobe width. The finalized window design is exported to the MATLAB workspace. Below figure shows the window designer app screenshot in Fig. 8.8.

Obviously, the MATLAB lends the customizing capability on the functions and apps. We can create any functions and apps for our purpose and share with other MATLAB and non-MATLAB users. The stand-alone application can be written for most computing environments. By mounting the designated hardware, the designed filter can perform the real-time processing with streaming input and output data. The conventional computer can support the digital filter system over audio data streaming with relaxed latency requirement. The previous chapters have used the MATLAB to derive and verify the theoretical and mathematical equations for

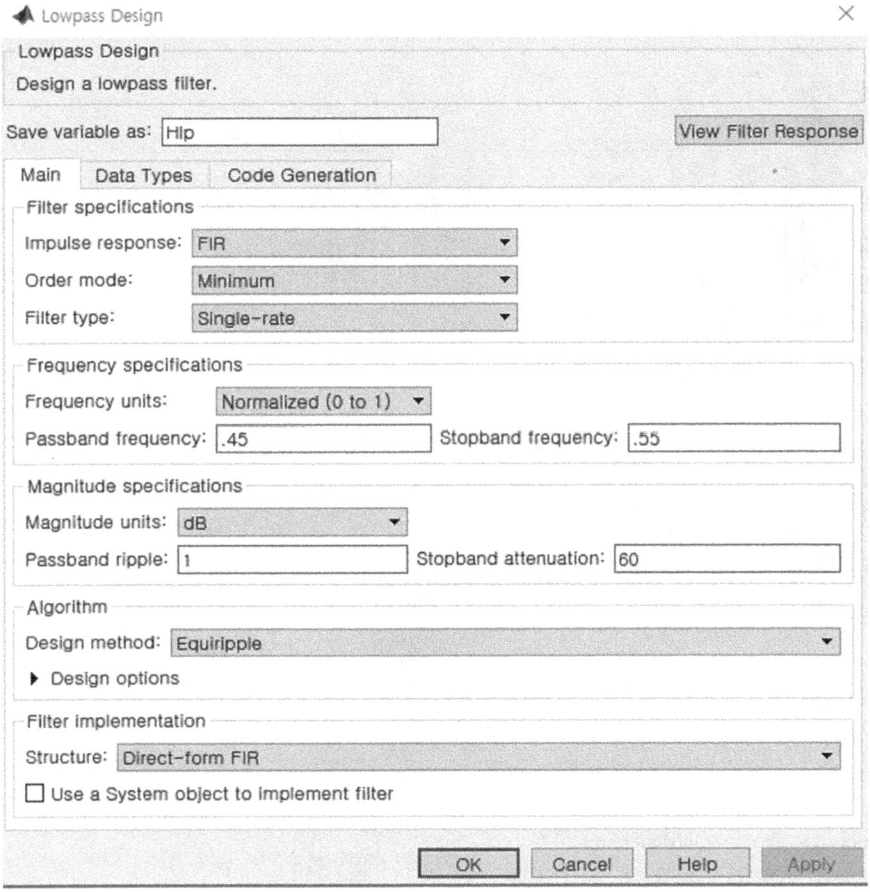

Fig. 8.7 Filter builder app

digital signal processing. The MATLAB expandability and adaptability reach beyond the typical computer system by using the extended concept and toolbox usage. Next sections will introduce some further usability of MATLAB.

8.2 Advanced Filter Design Methods

The MATLAB functions perform the digital filter design, analysis, and realization. The digital filter system prefers to use the frame-based iterative processing because of latency, memory, and complexity. The function-based programming is difficult to handle the iterative processing since the programming requires the explicit trace of data positions and retention of internal states. The frequent parameter modifications within the filter design and analysis stage increase the programming

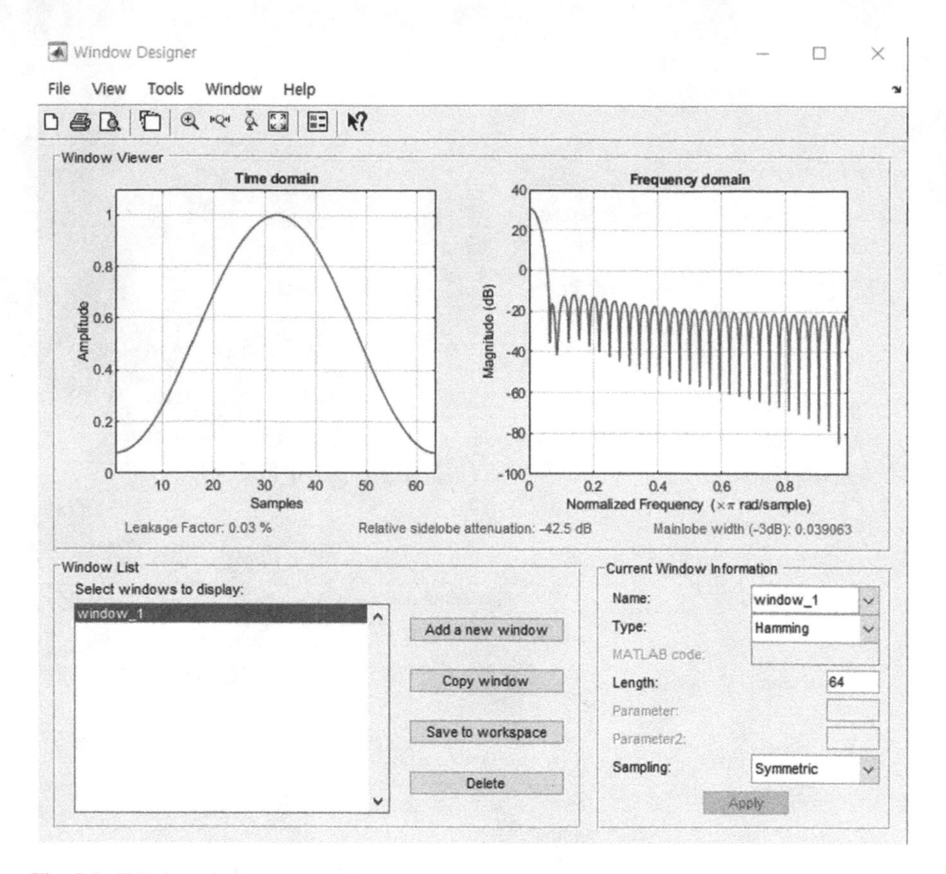

Fig. 8.8 Window designer app

complexity exponentially. The explicit and fragmented programming prevent the smooth conversion to other language for filter realization. The object-oriented programming paradigm is required for encapsulation to hide data and bind together with functions. The encapsulated object can provide the standardized code generation for optimized and verified performance.

The MATLAB introduced the system object as a dedicated MATLAB object. System objects are designed for implementing and simulating dynamic systems with inputs that change over time. The dynamic system generates the output signals which depends on the current input values as well as the previous output values. The system object stores the previous output values as internal states for next computational iteration. The state delivery between the consecutive iterations allow us to use the framed processing in segmented data. The framed processing with system object gives simplified programming method that use loops efficiently.

The example MATLAB code below demonstrates FIR LPF based on the functions. The frame size is 1024 samples with 100 frames for whole random data stream. The 160 order FIR filter performs low pass filtering with 0.2π cutoff frequency. The *while* loop executes the iterative processing with explicit trace of data positions. The internal state handover is performed by the *state* variable within the input and output arguments of *filter* function. Without the internal state delivery, the filter output ends are corrupted by the transition. The periodogram [3] function displays the spectral power distribution with window size as frame size. The framed output sequence is cumulated on the self-resizing array. The function-based programming shows the transparent environment where you must specify the detail information obviously.

Prog. 8.3 FIR LPF realization from MATLAB functions for frame-based processing.

```
frameSize = 1024;               % Frame size
N = frameSize*100;              % Data size
numFrame = N/frameSize;         % Number of frames

L = 160;                        % FIR order
wc = 0.2;                       % Cutoff frequency (normalized)
index = 1;

input = randn(1,N);             % Generate full input data
b = fir1(L,wc);                 % FIR filter coefficients
states = zeros(1,length(b)-1);  % Internal states
output = [];                    % Output buffer

while index <= (N - frameSize + 1)
      x = input(index:index + frameSize - 1);      % Framed data
      [y,states] = filter(b,1,x,states);           % Filtered data
      output = [output y];                         % Output buffer
      index = index + frameSize;                   % Data position index
end

figure,
subplot(211), periodogram(input,[],frameSize)   % Plot input spectral power
subplot(212), periodogram(output,[],frameSize) % Plot output spectral power
```

The MATLAB program produces the below plot for input and output spectral distribution. The random signal demonstrates the wide and flat frequency distribution. The filter output shown in second subplot depicts the concentrated spectral distribution toward the low frequencies as we expected in Fig. 8.9.

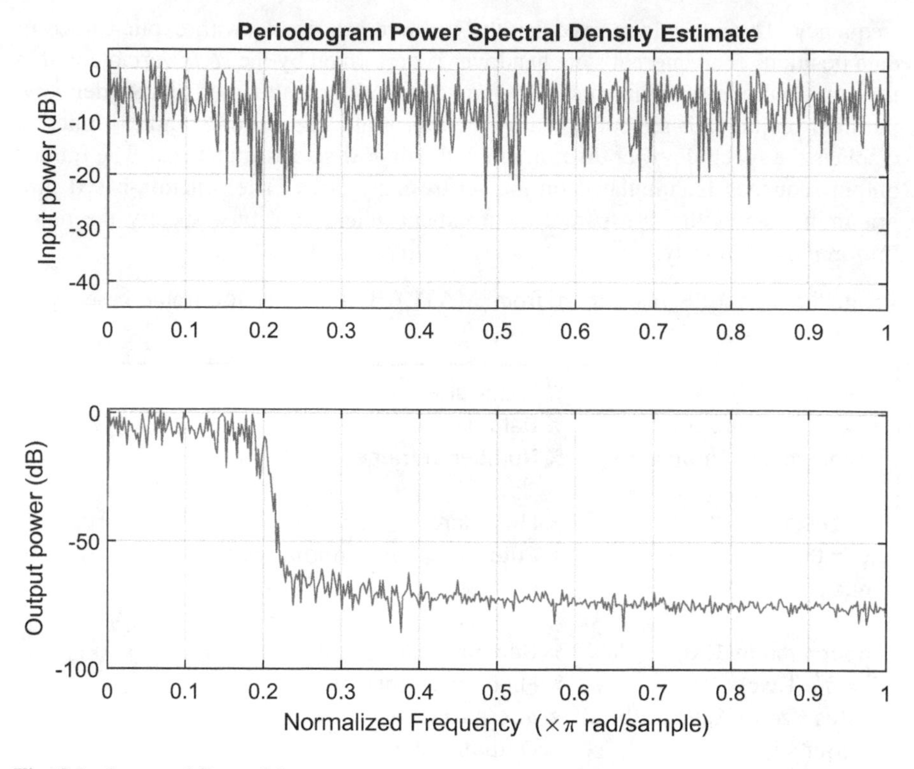

Fig. 8.9 Output of **Prog. 8.3**

The system object-based programming is shown below to realize the identical task as the previous code job. For the time being, we skip the detail explanation about the system object. In overall, the program is neat by suppressing arguments and states. Various reserved system objects provide the extensive filter design, analysis and realization.

Prog. 8.4 FIR LPF realization from MATLAB system object for frame-based processing.

```
frameSize = 1024;                 % Frame size
N = frameSize*100;                % Data size
L = 160;                          % FIR order
wc = 0.2;                         % Cutoff frequency (normalized)

sr = dsp.SignalSource(randn(N,1),frameSize);     % Input generator
sink1 = dsp.SignalSink;                          % Output buffer
sink2 = dsp.SignalSink;                          % Input buffer

fir = dsp.FIRFilter(fir1(L,wc));                 % Filter system object

 while ~isDone(sr)
      input = sr();                              % Generate framed input
      filteredOutput = fir(input);               % Filtered data
      sink1(filteredOutput);                     % Output buffer
      sink2(input);                              % Input buffer
 end

figure,
subplot(211), periodogram(sink2.Buffer,[],frameSize)
subplot(212), periodogram(sink1.Buffer,[],frameSize)
```

Similar to the previous program, the system object-based code produces Fig. 8.10 for input and output spectral distribution. The figures are not perfectly matched mutually because of the random number generation; however, the overall distributions are closely comparable to the previous figure.

The system object-based filter design requires the specification, design, analysis, and realization stage. The specification stage selects the filter type, parameters, algorithm, and customization to generate the specification object. The design stage creates the filter system object based on the specification object. The analysis stage presents the time and frequency domain information about the system object filter as well as the realization cost. The realization stage implements the designed filter in the iterative loop for frame-based processing. If the measured performance does not meet the requirements, the procedure returns to the specification stage to adjust the various parameters and methods. The overall diagram is illustrated as Fig. 8.11.

The specification stage defines the filter response by placing the proper suffix after *fdesign* prefix. The possible responses for the *fdesign* object are organized in

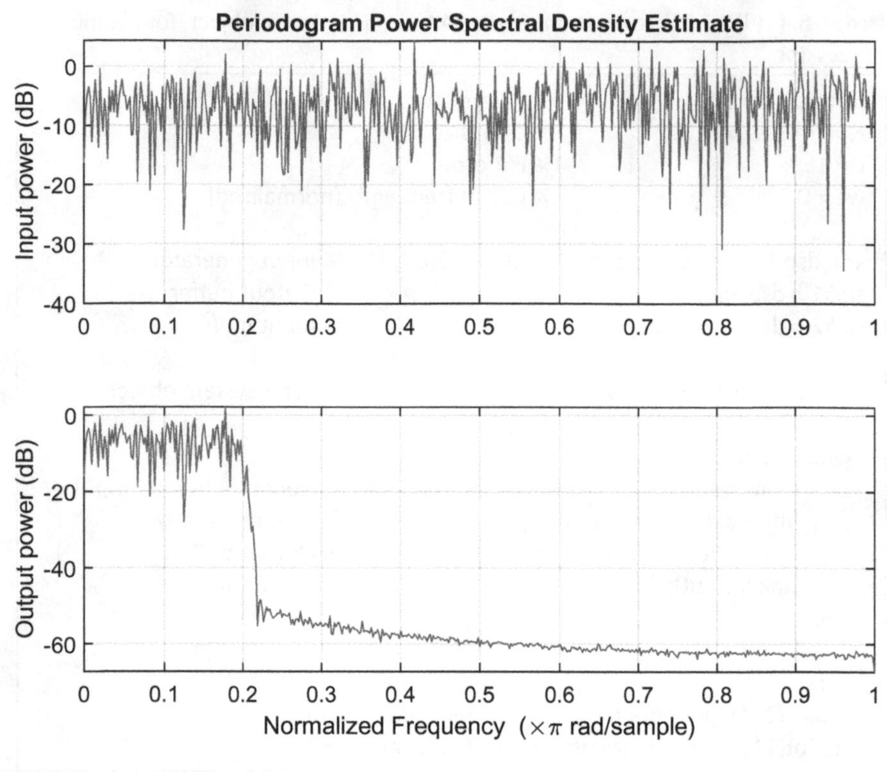

Fig. 8.10 Output of **Prog. 8.4**

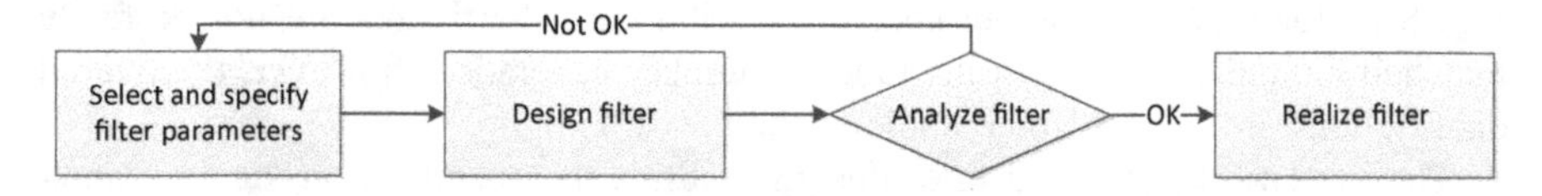

Fig. 8.11 Overall procedures for MATLAB filter design from system object method

Table 8.7. Suitable toolboxes are necessary for the responses. The specification stage is the first stage for the system object filter design because the response defines the following stage parameters and variables.

The following stages are investigated by applying the design example based on the given parameters from the subsequent table in Table 8.8.

The lowpass response creates the specification object 'dfilter' by *fdesign.lowpass* command as below. The *set* function returns the possible specifications for lowpass filter response. Choose one of the specifications for filter design. The specification

Table 8.7 List of responses (suffix) for the *fdesign* object

Filter designer response name	Filter designer response function	Signal processing toolbox required?	DSP system toolbox required?
arbmag	Arbitrary magnitude filters	Yes	
bandpass	Bandpass filters	Yes	
bandstop	Bandstop filters	Yes	
differentiator	Differentiators	Yes	
highpass	Highpass filters	Yes	
hilbert	Hilbert transformers	Yes	
lowpass	Lowpass filters	Yes	
arbgrpdelay	Arbitrary group delay filters	Yes	Yes
arbmagnphase	Arbitrary magnitude and phase filters	Yes	Yes
audioweighting	Audio weighting filters	Yes	Yes
ciccomp	CIC compensators	Yes	Yes
comb	Comb filters	Yes	Yes
decimator	Decimators	Yes	Yes
fracdelay	Fractional delay filters	Yes	Yes
halfband	Halfband filters	Yes	Yes
interpolator	Interpolators	Yes	Yes
isinclp	Inverse-sinc lowpass filters	Yes	Yes
notch	Notching filters	Yes	Yes
nyquist	Nyquist filters	Yes	Yes
parameq	Parametric equalizer filters	Yes	Yes
peak	Peaking filters	Yes	Yes
polysrc	Polynomial sample-rate converters	Yes	Yes
octave	Octave-band and fractional-octave-band filters	Yes	Yes
rsrc	Rational factor sample-rate converters	Yes	Yes

definition can be found by *help* or *doc* command with *fdesign.lowpass*. Without selecting the specification, the first entry will be chosen by default. Note that the command executions are performed in the command window.

Table 8.8 Given filter specification

Filter specifications	Value
Filter type	Low pass filter
Filter realization	IIR with Butterworth
Filter length	Optimization
Passband frequency (f_p)	1000 Hz;$\omega_p = \pi/4$
Pass magnitude	0 dB
Pass band ripple (peak to peak)Δ_p	1 dB
Stopband frequency (f_{st})	2000 Hz;$\omega_{st} = \pi/2$
Stop band attenuation in decibels Δ_{st}	40 dB
Sampling frequency (f_s)	8000 Hz

Prog. 8.5 Specification object creation for LPF.

```
>> dfilter = fdesign.lowpass;                % Response is lowpass
>> set(dfilter,'Specification')

ans =

   18×1 cell array

      {'Fp,Fst,Ap,Ast'}
      {'N,F3dB'         }
      {'Nb,Na,F3dB'     }
      {'N,F3dB,Ap'      }
      {'N,F3dB,Ap,Ast'}
      {'N,F3dB,Ast'     }
      {'N,F3dB,Fst'     }
      {'N,Fc'           }
      {'N,Fc,Ap,Ast'    }
      {'N,Fp,Ap'        }
      {'N,Fp,Ap,Ast'    }
      {'N,Fp,F3dB'      }
      {'N,Fp,Fst'       }
      {'N,Fp,Fst,Ap'    }
      {'N,Fp,Fst,Ast' }
      {'N,Fst,Ap,Ast' }
      {'N,Fst,Ast'      }
      {'Nb,Na,Fp,Fst' }

   >>
```

The MATLAB provides the auto-completion and auto-listing by tab keystroke. For example, the *set* function argument is completed by 'S<tab> and listed by 'N<tab> as shown below. Navigate the specifications by arrow keys as shown in Fig. 8.12.

The *set* function decides the specification by additional argument as below. The *designmentods* function presents the possible design methods for the given response and specification. The below example shows the window method as the candidate.

Prog. 8.6 Filter design method selection example.

```
>> set(dfilter,'Specification','N,Fc')              % Set specification
>> designmethods(dfilter,'Systemobject',true)    % Find available algorithms


Design Methods that support System objects for class fdesign.lowpass (N,Fc):


window

  >>
```

The specification change provides the distinct subset of the design methods. After the specification selection, we can assign the actual values for the specification object by using the *setspecs* function. The specification frequencies and ripples are derived from the design example table.

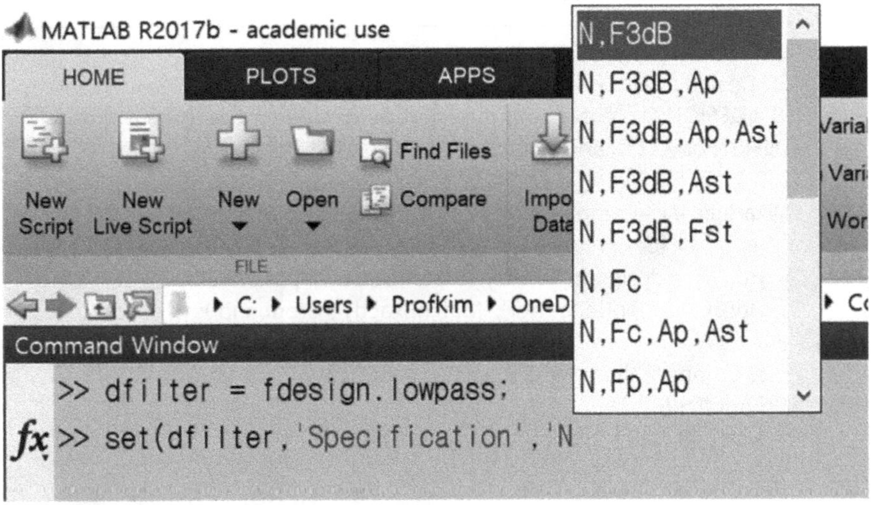

Fig. 8.12 Auto-completion and auto-listing example in command window

Prog. 8.7 Assign specification values.

```
>> set(dfilter,'Specification','Fp,Fst,Ap,Ast')        % Set specification
>> setspecs(dfilter,0.25,0.5,1,40)                     % Specification
>> designmethods(dfilter,'Systemobject',true)          % Find available algorithms

Design Methods that support System objects for class fdesign.lowpass
(Fp,Fst,Ap,Ast):


butter
cheby1
cheby2
ellip
equiripple
ifir
kaiserwin
multistage

   >>
```

The further information about the specification object with the filter design method can be obtained by *help* function as below.

Prog. 8.8 Specification object help.

```
>> help(dfilter,'butter')
 DESIGN Design a Butterworth IIR filter.
    HD = DESIGN(D, 'butter') designs a Butterworth filter specified by the
    FDESIGN object D, and returns the DFILT/MFILT object HD.
 ...
    % Example #1 - Compare passband and stopband MatchExactly.
       h      = fdesign.lowpass('Fp,Fst,Ap,Ast', .1, .3, 1, 60);
       Hd     = design(h, 'butter', 'MatchExactly', 'passband');
       Hd(2) = design(h, 'butter', 'MatchExactly', 'stopband');

       % Compare the passband edges in FVTool.
       fvtool(Hd);
       axis([.09 .11 -2 0]);

   >>
```

The *design* function with specification object and additional arguments defines the filter system object as below. The filter is built by the Butterworth method with direct form II SOS structure. The *design* function displays the detail filter properties as return. By calling the filter system object, we can perform the filter operation for the desired specification.

Prog. 8.9 The *design* function example for specification object.

```
>> filt = design(dfilter,'butter','filterstructure','df2sos','Systemobject',true)

filt =

  dsp.BiquadFilter with properties:

                   Structure: 'Direct form II'
             SOSMatrixSource: 'Property'
                   SOSMatrix: [13×6 double]
                 ScaleValues: [14×1 double]
           InitialConditions: 0
    OptimizeUnityScaleValues: true

  Show all properties

  >>
```

The analysis stage evaluates the filter system object over various aspects. The *measure* function provides the frequency, magnitude, and ripple performance. The *cost* function presents the computation and memory requirement. The *info* function exhibits the filter structure, stability, and phase linearity.

Prog. 8.10 Analysis of filter system object.

```
>> measure(filt)

ans =
Sample Rate          : N/A (normalized frequency)
Passband Edge        : 0.25
3-dB Point           : 0.27666
6-dB Point           : 0.29956
Stopband Edge        : 0.5
Passband Ripple      : 0.98666 dB
Stopband Atten.      : 40 dB
Transition Width : 0.25

>> cost(filt)

ans =
   struct with fields:

                  NumCoefficients: 12
                        NumStates: 6
    MultiplicationsPerInputSample: 12
          AdditionsPerInputSample: 12

>> info(filt)

ans =
   6×59 char array

    'Discrete-Time IIR Filter (real)                            '
    '-------------------------------                            '
    'Filter Structure      : Direct-Form II, Second-Order Sections '
    'Number of Sections    : 3                                  '
    'Stable                : Yes                                '
    'Linear Phase          : No                                 '

  >> fvtool(filt)
```

The *fvtool* with filter system object shows the extensive analysis on the filter. Figure 8.13 illustrates the magnitude response with given specifications as automated and dotted lines. The ribbon menu on the figure can choose the phase, group delay, phase delay, impulse response, step response, pole/zero, coefficients, filter information, and etc. plots.

After the analysis satisfies the specification, the realization stage implements the system object filter in the loop for frame-based processing. Below code realizes the LPF system with random input data. The filter system object is located within the loop for 1024 samples per frame processing. The filter system object is created by the *fdesign* and following functions. The code is written in the MATLB editor.

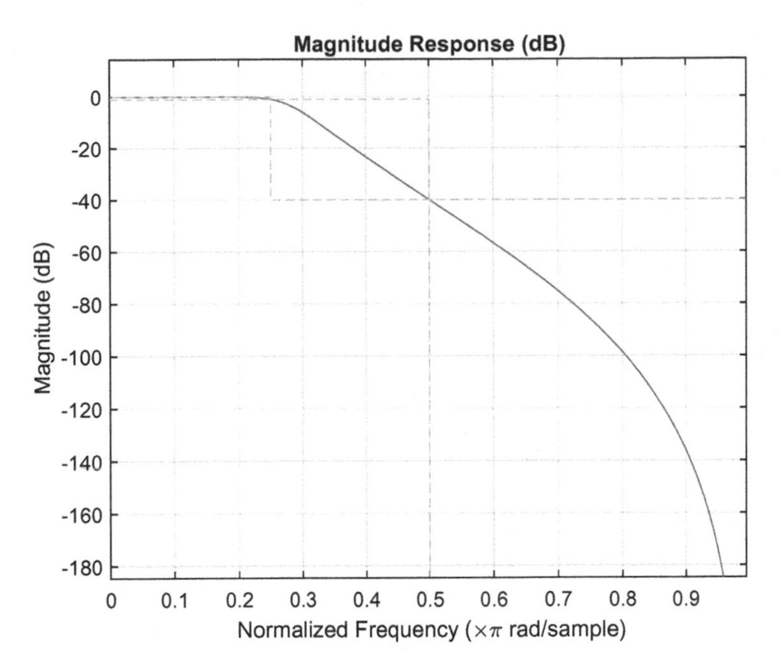

Fig. 8.13 Magnitude response from *fvtool* function with filter system object

Prog. 8.11 Filter realization with system object for **Table 8.6** specification.

```
frameSize = 1024;            % Frame size
N = frameSize*100;           % Data size

sr = dsp.SignalSource(randn(N,1),frameSize);     % Input generator
sink1 = dsp.SignalSink;                          % Output buffer
sink2 = dsp.SignalSink;                          % Input buffer

dfilter = fdesign.lowpass;                       % Response is lowpass
set(dfilter,'Specification','Fp,Fst,Ap,Ast')     % Set specification
setspecs(dfilter,0.25,0.5,1,40)                  % Specification
filt = design(dfilter,'butter','filterstructure','df2sos','Systemobject',true)

while ~isDone(sr)
      input = sr();                              % Generate framed input
      filteredOutput = filt(input);              % Filtered data
      sink1(filteredOutput);                     % Output buffer
      sink2(input);                              % Input buffer
end

figure,
subplot(211), periodogram(sink2.Buffer,[],frameSize)
subplot(212), periodogram(sink1.Buffer,[],frameSize)
```

The spectral power distributions are illustrated below in Fig. 8.14 for the input and output of the filter. The random white noise is represented by the wide and flat frequency distribution. The filter output shown in second subplot depicts the concentrated spectral distribution toward the low frequencies as we expected. The spectral power starts to decrease from $\pi/4$ and the decline slope is mild to meet the given specifications.

The MATLAB provides many system objects in toolboxes. Tables 8.9, 8.10, 8.11, 8.12, and 8.13 organize and categorize the partial system objects for filter related subjects. Further list and information can be found on MathWorks webpage [2].

Below program represents the filtering task from mp3 signal to the audio device. The input data is obtained from the mp3 file reading by the system object *dsp.AudioFileReader*. The frame size is defined at the file reading system object and automatically applied to the data filtering and audio playing. The output data is played over the default audio device by the system object *audioDeviceWriter*. The sampling rate is delivered from the file reading system object. The LPF is published by the system object *dsp.LowpassFilter* which provides simple procedure with less

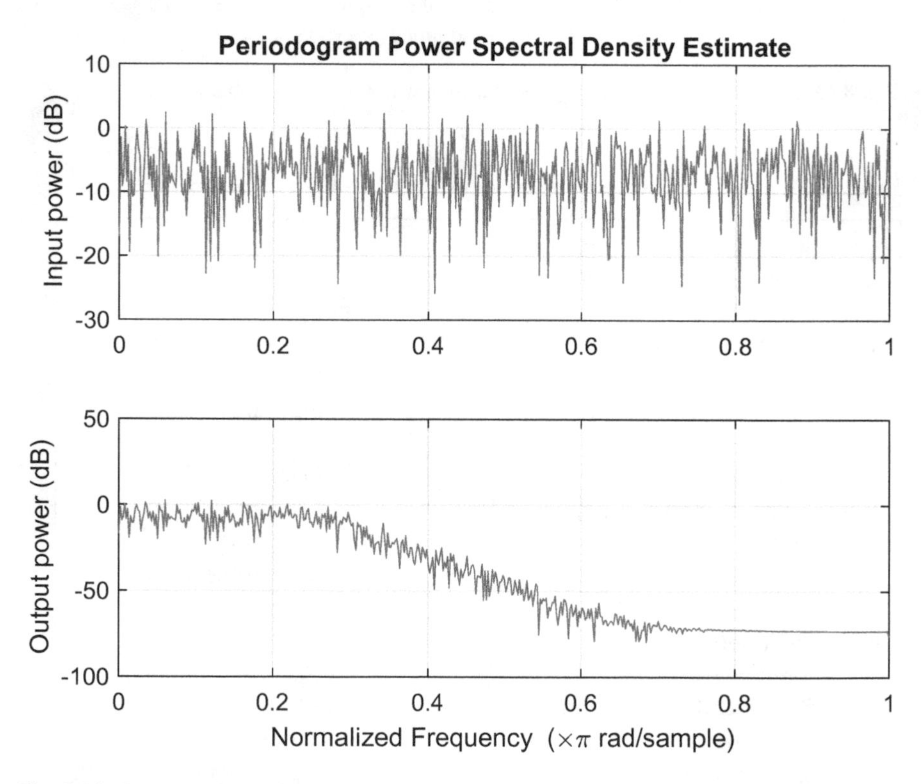

Fig. 8.14 Output of **Prog. 8.11**

Table 8.9 Signal generation system objects

dsp.SignalSource	Import variable from workspace
dsp.SineWave	Generate discrete sine wave

Table 8.10 Signal input and output system objects

audioDeviceReader	Record from sound card
audioDeviceWriter	Play to sound card
dsp.AudioFileReader	Stream from audio file
dsp.AudioFileWriter	Stream to audio file

Table 8.11 Scope and data logging system objects

dsp.TimeScope	Time domain signal display and measurement
dsp.SpectrumAnalyzer	Display frequency spectrum of time-domain signals
dsp.SignalSink	Log simulation data in buffer

Table 8.12 Filter system objects

dsp.Differentiator	Direct form FIR fullband differentiator filter
dsp.FIRFilter	Static or time-varying FIR filter
dsp.HighpassFilter	FIR or IIR highpass filter
dsp.LowpassFilter	FIR or IIR lowpass filter
dsp.AllpassFilter	Single section or cascaded allpass filter
dsp.BiquadFilter	IIR filter using biquadratic structures
dsp.IIRFilter	Infinite impulse response (IIR) filter

Table 8.13 Transform system objects

dsp.FFT	Discrete Fourier transform
dsp.IFFT	Inverse discrete Fourier transform (IDFT)

freedom on design method. The complete parameter list for the given system objects can be found by the MATLAB documents and help [1, 2]. Note that the auto-completion and auto-list are available for system object arguments. The iterative loop by *while* statement executes the frame-based data flow and filtering. By selecting the input variable for *adw* object, you can hear the raw or filtered audio.

Prog. 8.12 Filtering task from mp3 file to the audio device.

```
frame = 1024;                              % Samples per frame

afr = dsp.AudioFileReader('speech_dft.mp3','SamplesPerFrame',frame);
% Read data from file
adw = audioDeviceWriter('SampleRate', afr.SampleRate);
% Write data to audio device
L1 = dsp.LowpassFilter('SampleRate',afr.SampleRate,...
     'FilterType','FIR','PassbandFrequency',afr.SampleRate/8,...
     'StopbandFrequency',afr.SampleRate/4,'PassbandRipple',1,...
     'StopbandAttenuation',60);
% LPF with 1/4 wpa and 1/2 wst

while ~isDone(afr)
     audio = afr();                        % Get data
     out = L1(audio);                      % Filtering
     adw(out);                             % Filtered data to audio device
     %adw(audio);                          % Raw data to audio device
end

release(afr);
release(adw);
```

Fixed-Point Designer

The MATLAB stores and processes numeric variables as double-precision floating-point values by default. Additional data types are text, integer and single-precision floating-point values. The powerful PC processors support the double-precision algorithm for low latency and high throughput processing by using the complex hardware architecture. Therefore, the MATLAB algorithms are executed rapidly on the PC environment. The MATLAB algorithms extends the application areas beyond the desktop computers in order to be the highly deployable language of technical computing.

The hardware complexity is proportional to the data length and type complication. The reduced data length or fixed-point data type requires the less complicated hardware than the counterpart for double-precision floating-point data processing. The conventional embedded hardware is optimized for the fixed-point and single-precision algorithms to meet the cost and power budget. The MATLAB provides the fixed-point designer [4] for extensive tool to analyze and simulate the data type conversion in the MATLAB algorithms. The fixed-point designer allows the automatic conversion from double-precision algorithms to single precision or fixed point. The fixed-point designer recommends the optimal data types, length, and scaling. Also, we can customize the detail data attributes for instance rounding, overflow, and etc. The simulation presents the algorithm output variation from the data conversions to understand the algorithm sensitivity. The fixed-point designer provides apps that lead the data conversion process and enable to compare fixed-point results with floating-point baselines.

The fixed-point designer is the large toolbox and this section only shows the brief introduction. A fixed-point data type is defined by the word length, the binary point position with signed or unsigned property. The binary point position is interpreted by the scaled fixed-point values. A binary representation of a generalized fixed-point number is shown in Fig. 8.15.

In above, the word length is *wl* and fraction length is 4. The fixed-point designer performs the automatic conversion process based on the function or app. Below example shows the representation conversion by using the fixed-point designer function. The *fi* function creates the fixed-point numeric object from the double-precision floating-point number with word length and (un)signed property. The undefined fraction length provides the best-precision fraction length by MATLAB. In the first case, the equal word and fraction length indicate that the binary point is located at the left of MSB. By placing the *bin* suffix, the fixed-point numeric object shows the full binary representation. The second case for integer number demonstrates the 16 word length and 9 fraction length. The sum of two vectors creates the new fixed-point numeric object with expanded word and fraction length to embed the output. The MATLAB automatically adjust and traces the lengths and properties of the objects to preserve the proper numeric information.

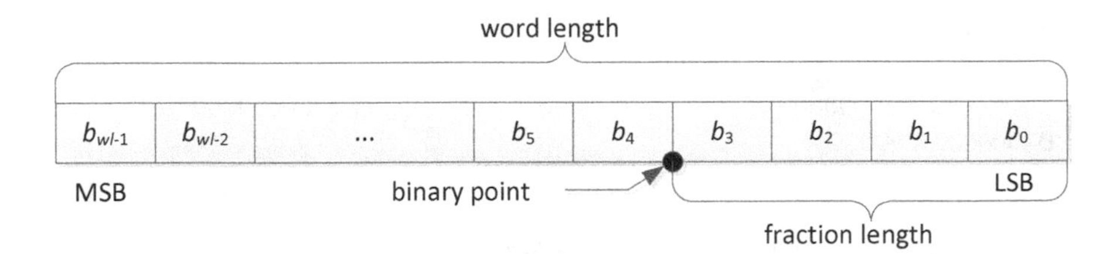

Fig. 8.15 Binary representation of a generalized fixed-point number

Prog. 8.13 Representation conversion by using the fixed-point designer function.

```
>> a = [1/4 1/8 1/16 1/32]     % Double precision floating-point number
a =
    0.2500    0.1250    0.0625    0.0313
>> b = fi(a,1,16)     % Convert to signed 16 bits number (word length is 16
bits) and fraction length is determined by MATLAB automatically
b =
    0.2500    0.1250    0.0625    0.0313
          DataTypeMode: Fixed-point: binary point scaling
            Signedness: Signed
            WordLength: 16
        FractionLength: 16
>> b.bin              % Show us the binary representation of the number
ans =
    '0100000000000000   0010000000000000   0001000000000000
0000100000000000'
>> c = [4 8 16 32]                % Double precision floating-point number
c =
     4     8    16    32
>> d = fi(c,1,16)     % Convert to signed 16 bits number (word length is 16
bits) and fraction length is determined by MATLAB automatically
d =
     4     8    16    32
          DataTypeMode: Fixed-point: binary point scaling
            Signedness: Signed
            WordLength: 16
        FractionLength: 9
>> d.bin              % Show us the binary representation of the number
ans =
    '0000100000000000   0001000000000000   0010000000000000
0100000000000000'
>> e = b+d
e =
    4.2500    8.1250    16.0625    32.0313
          DataTypeMode: Fixed-point: binary point scaling
            Signedness: Signed
            WordLength: 24
        FractionLength: 16
>> e.bin
ans =
    '000001000100000000000000   000010000010000000000000
000100000001000000000000   001000000000100000000000'
>>
```

If we want to declare the fixed-point numeric object, we can use the *numerictype* function. For the automatic conversion process, we prepare the target function as below with *myfilter.m* file name. The function includes the persistent variable to perform the filter design process once. The second execution of the function skips the *dsp.FIRFilter* system object by the *if* statement.

Prog. 8.14 Target function for automatic conversion process in fixed-point designer.

```
function [y] = myfilter(x)
%#codegen
% For code generation, you should put the codegen hashtag in above
persistent d1         % Static variable for function
if isempty(d1)        % If it is not defined, the function compute filter
coefficient
    d1 = dsp.FIRFilter('CoefficientsDataType','Custom','Numerator',...
        fir1(51,0.1),'FullPrecisionOverride',false,...
        'ProductDataType','Full precision','AccumulatorDataType',...
        'Custom','OutputDataType','Custom');    % Actucal filter
end
y = d1(x);          % Filter execution
end
```

The test code for *myfilter* function is below. We name the file as *BAfilter.m*. The *tic* and *toc* function measure the execution time between the functions. The *soundsc* function plays the vector as sound.

Prog. 8.15 Test code for automatic conversion process of **Prog. 8.14**.

```
clear all
clc

data = load('mtlb.mat');                % Load sound
x = data.mtlb;
Fs = data.Fs;
Q = length(x);

tic
y = myfilter(x);                        % Filter algorithm
toc

soundsc(double(y),Fs);
```

When you execute the *BAfilter.m* program, you can hear the low frequency MATLAB pronunciation with execution time in seconds. For automatic fixed-point conversion, execute the *fixed-point converter* in APPS ribbon tab as Fig. 8.16. Be careful that the target function and file should be located at the English named directory in all levels; otherwise, the conversion process will be failed.

Select the *myfilter.m* file for conversion and click the next button located at the lower right corner. In the next stage as below, you select test code and click the autodefine input type button. Click the next again in Fig. 8.17.

In the analyze tab, activate the *log data for histogram* and click the analyze button Fig. 8.18.

The MATLAB suggests the optimal fixed-point configuration for each variables and computations. The *numerictype* function contains the three arguments as signed property, word length, and fraction length in Fig. 8.19.

Once you click the proposed type, you can observe the histogram for coverage. If you accept the suggestion, click convert tab in Fig. 8.20.

Activate the *log inputs and outputs for comparison plots* in the test tab and click the test for final test in Fig. 8.21.

You can see the difference between the floating-point and fixed-point configurations for input and output as Figs. 8.22 and 8.23.

Click the next for finalizing the process as Fig. 8.24.

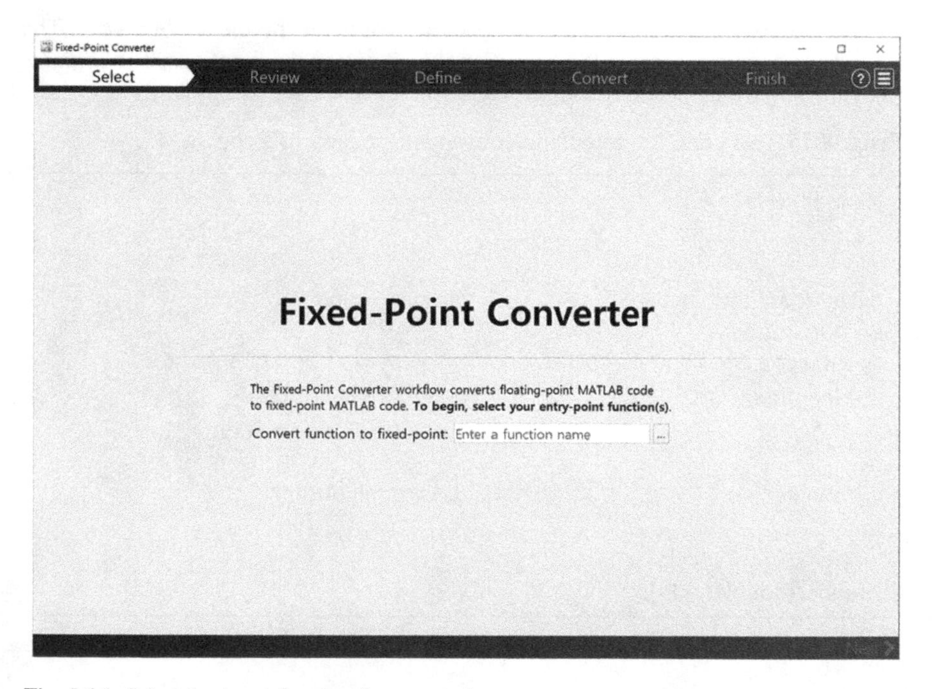

Fig. 8.16 Select the target function for conversion

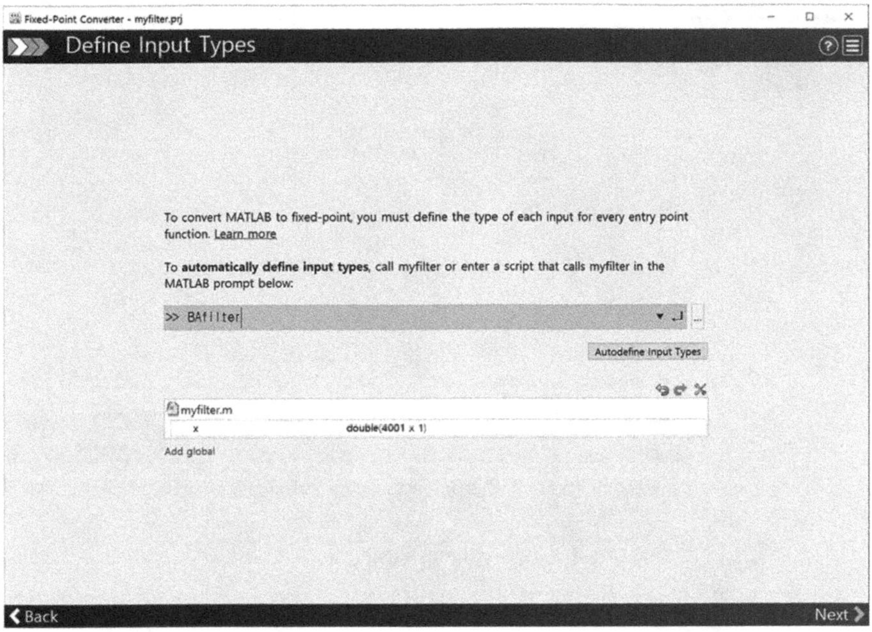

Fixed-Point Converter - myfilter.prj
Define Input Types
To convert MATLAB to fixed-point, you must define the type of each input for every entry point function. Learn more
To automatically define input types, call myfilter or enter a script that calls myfilter in the MATLAB prompt below:
>> BAfilter
Autodefine Input Types
myfilter.m
x
double(4001 x 1)
Add global
Back
Next

Fig. 8.17 Select the test code for conversion

Fixed-Point Converter - myfilter.prj
Convert to Fixed Point
SETTINGS
ANALYZE
Source Code
myfilter
Analyze ranges using simulation
Specify design ranges
Test bench
BAfilter.m
Log data for histogram
Show code coverage
Analyze ranges using derived range analysis
Timeout (minutes)
Quick derived range analysis
Analyze Ranges
function [y] = myfilter(x)
%#codegen
% For code generation, you should put the codegen hashtag in above
persistent d1 % Static variable for function
if isempty(d1) % If it is not defined, the function compute filter coefficient
d1 = dsp.FIRFilter('CoefficientsDataType','Custom','Numerator',fir1(51,0.1),'FullPrecisionOv
'ProductDataType','Full precision','AccumulatorDataType','Custom',...
'OutputDataType','Custom'); % Actual filter
end
y = d1(x); % Filter execution
Variables
Function Replacements
Output
Variable
Type
Sim Min
Sim Max
Whole ...
Proposed Type
Log...
Max Diff
Input
x
4001 x 1 dou...
No
Output
y
4001 x 1 dou...
No
Persistent
d1
dspcodegen....
CustomProductData...
double
No
Back

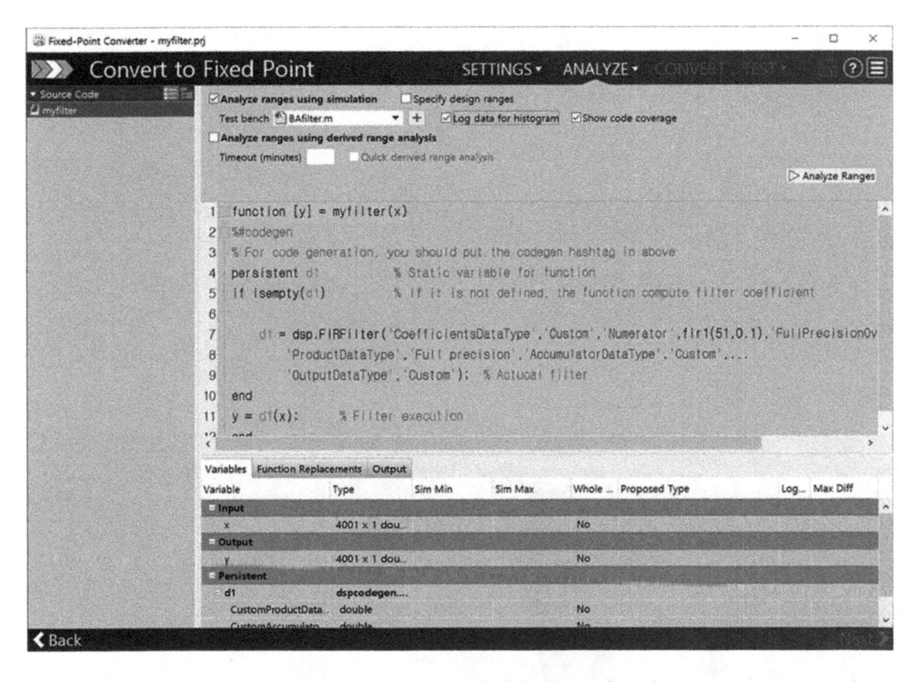

Fig. 8.18 Activate the *log data for histogram* and click the analyze button

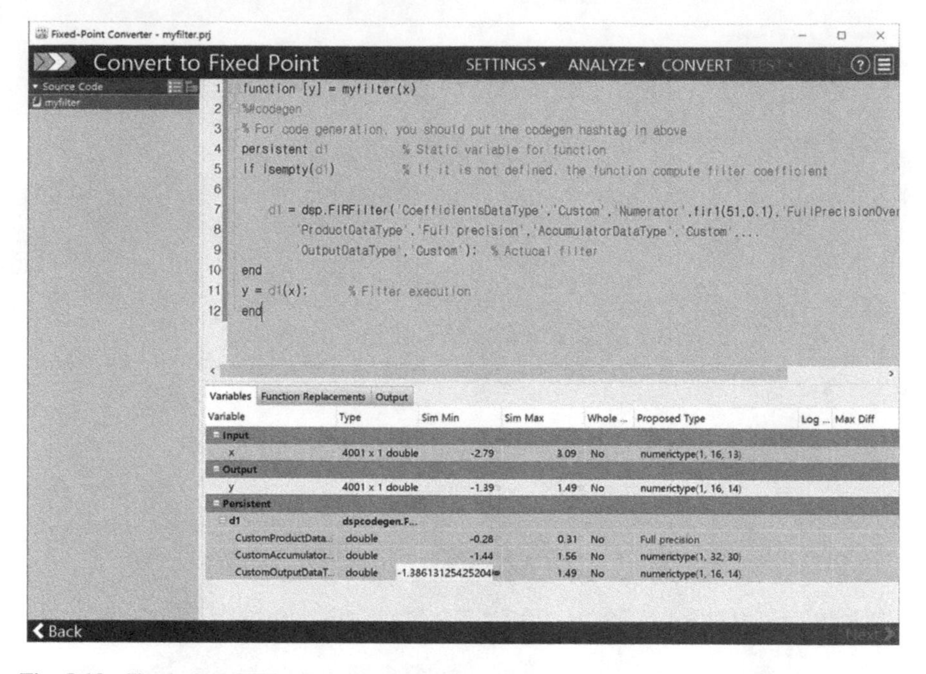

Fig. 8.19 Check the MATLAB suggestions for optimal fixed-point configurations

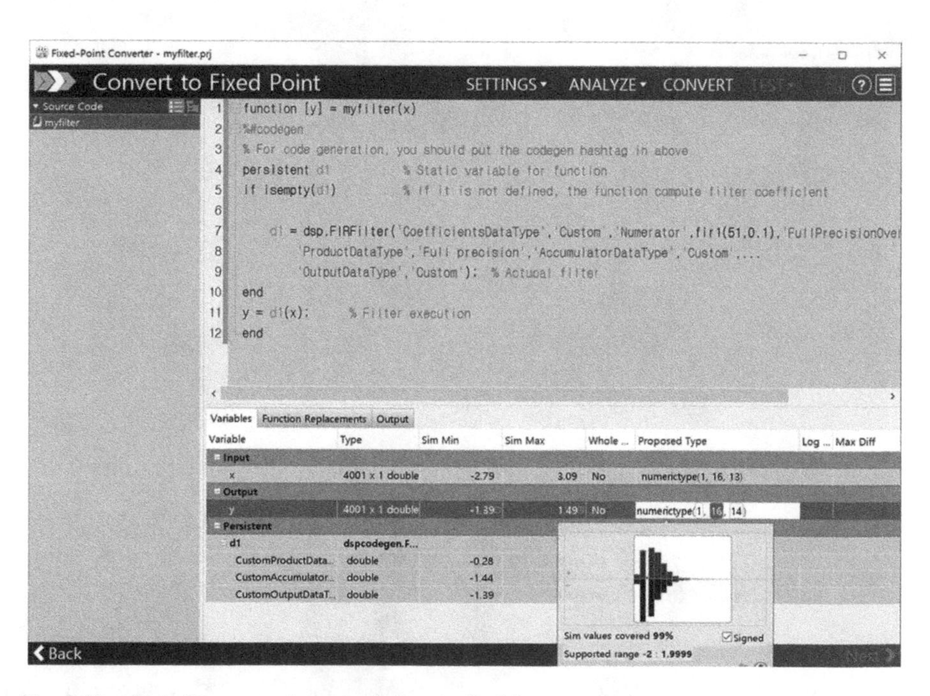

Fig. 8.20 Click the proposed type and observe the histogram for coverage

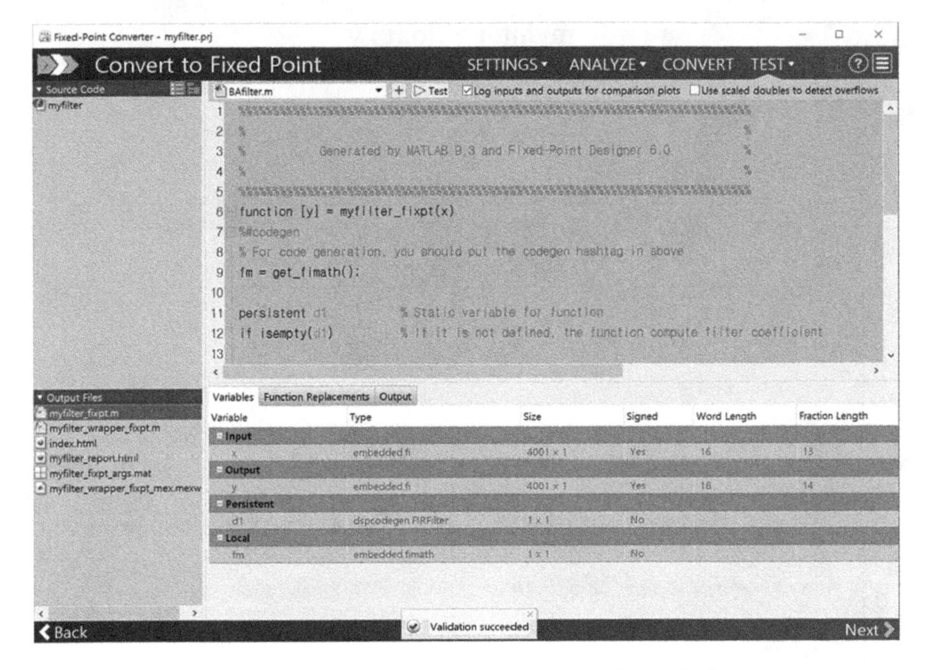

Fig. 8.21 Activate the *log inputs and outputs for comparison plots* and click the test

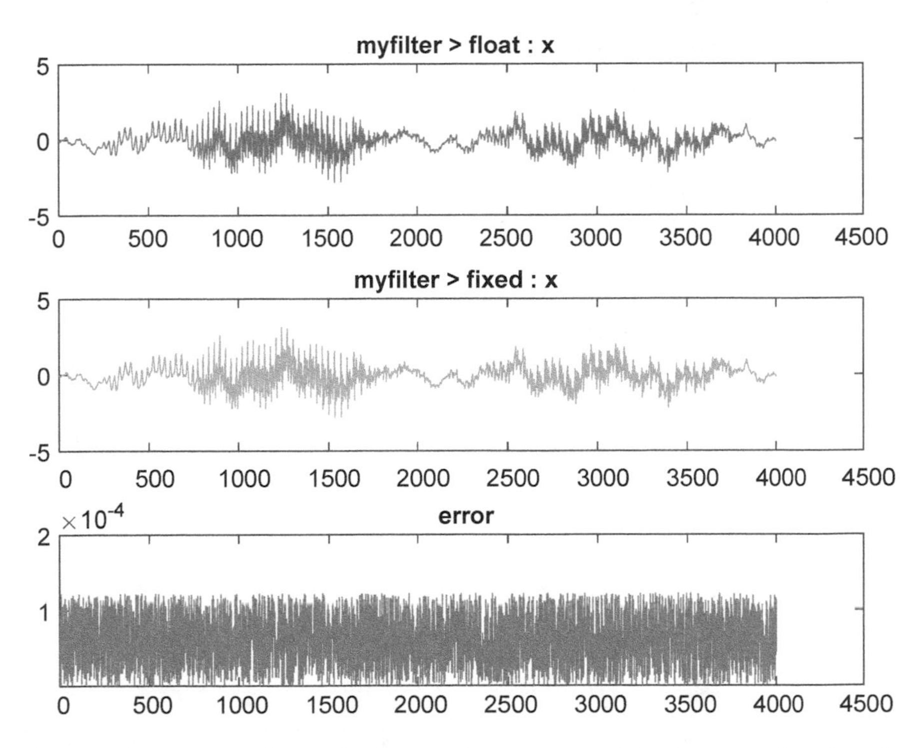

Fig. 8.22 Observe the difference between the floating-point and fixed-point configurations for input

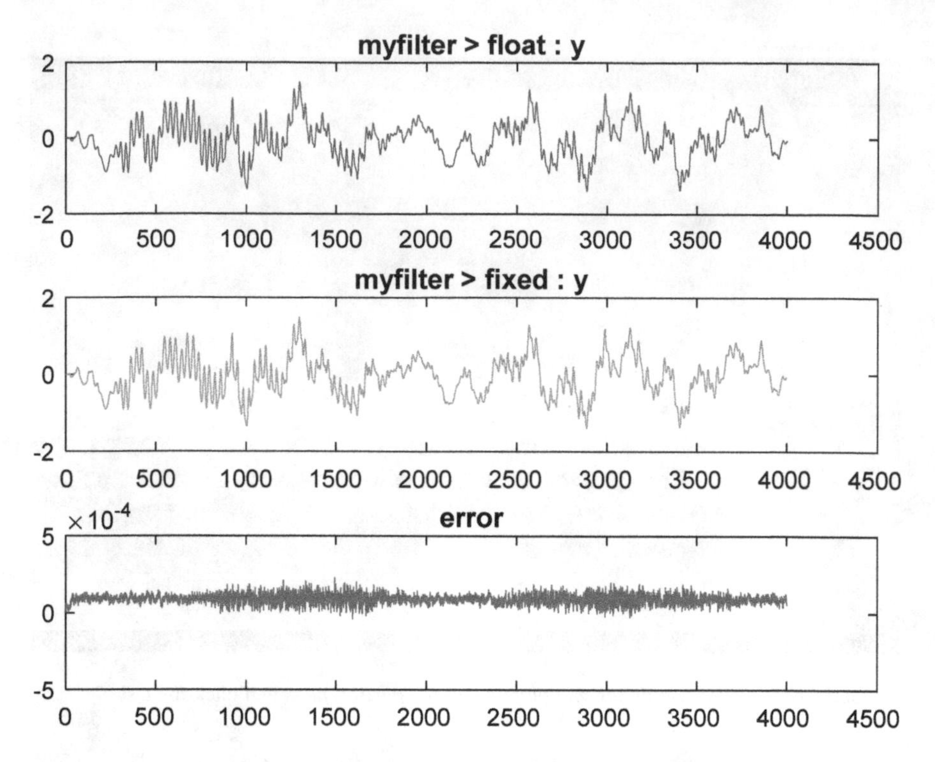

Fig. 8.23 Observe the difference between the floating-point and fixed-point configurations for ouput

We can find the overall configuration and profile by opening the conversion report located at *your_directory/codegen/myfilter/fixpt* as *myfilter_report.html* or by clicking the *Reports* at finish workflow window in Fig. 8.25.

FIR Filter Using Fixed-Point Numbers

Frequently, the system is designed with analog-to-digital converter (ADC) and digital-to-analog converter (DAC) for real-time processing. Along with fixed-point processor, the complete arithmetic can be implemented by using the fixed-point operations. In this situation, we carefully manipulate the word and fraction length. Below is the system with 12-bit ADC and DAC for processing the audio signal in Fig. 8.26.

To simulate the 12-bit ADC, let's load audio signal as below.

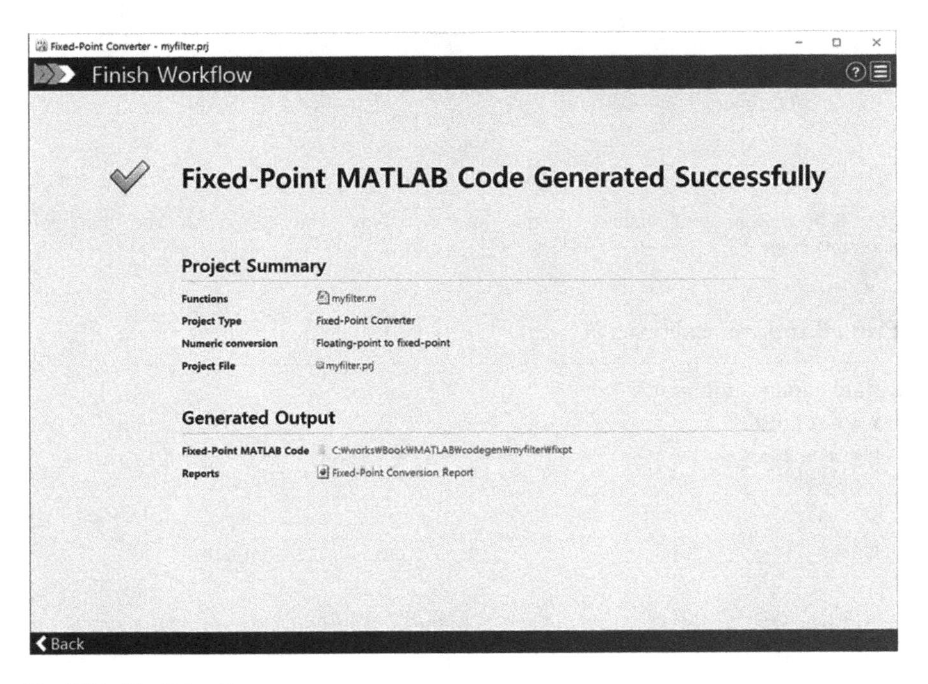

Fig. 8.24 Finalizing the process

Web Browser - Fixed-Point Report for myfilter

Fixed-Point Report for myfilter

Location: file:///C:/works/Book/MATLAB/codegen/myfilter/fixpt/myfilter_report.html

Generated on 2019-01-14 00:27:09

The following table shows fixed point instrumentation results

Fixed-Point Report *myfilter*

Simulation Coverage	Code
100%	function [y] = myfilter(x) %#codegen % For code generation, you should put the codegen hashtag in above persistent d1 % Static variable for function
Once	if isempty(d1) % If it is not defined, the function compute filter coefficient d1 = dsp.FIRFilter('CoefficientsDataType','Custom','Numerator',fir1(51,0.1),'FullPrecisionOverride',false,... 'ProductDataType','Full precision','AccumulatorDataType','Custom',... 'OutputDataType','Custom'); % Actucal filter end
100%	y = d1(x); % Filter execution end

Variable Name	Type	Sim Min	Sim Max	Static Min	Static Max	Whole Number	ProposedType (Best For WL = 16)
d1.CustomProductDataType	double	-0.2771001800981165	0.30702402721824185			No	Full precision
d1.CustomAccumulatorDataType	double	-1.4421415649078628	1.5555617131390753			No	numerictype(1, 32, 30)
d1.CustomOutputDataType	double	-1.38613125425204	1.4859812251439668			No	numerictype(1, 16, 14)
x	double 4001 x 1	-2.791591901558386	3.0930538827611227			No	numerictype(1, 16, 13)
y	double 4001 x 1	-1.38613125425204	1.4959812251439668			No	numerictype(1, 16, 14)

Fig. 8.25 Overall configuration and profile for fixed-point conversion

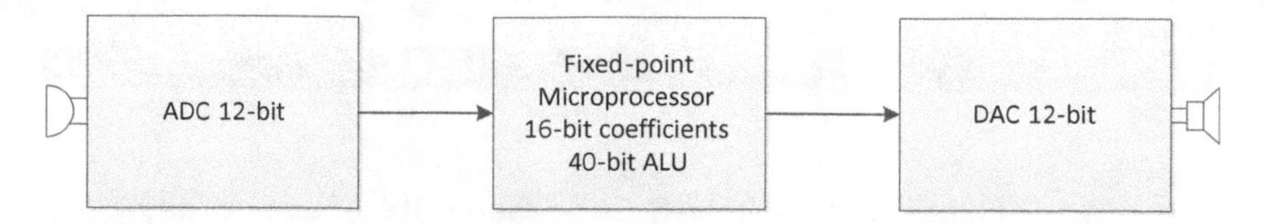

Fig. 8.26 System configuration for real-time FIR filter with ADC/DAC and fixed-point microprocessor

Prog. 8.16 Load audio signal.

```
data = load('mtlb.mat');                % Load sound
x = data.mtlb;
L = length(x);
whos x
>>
Name         Size               Bytes   Class     Attributes

  x          4001x1              32008   double
```

The loaded x variable is a floating-point number with double precision. Change the variable to 12-bit integer as below.

Prog. 8.17 Change the variable to 12-bit integer.

```
tx = fi(x,1,12);
tx.get
    % Now you can see that the variable has the 12-bit WL and 9-bit FL as
below
ans =
  struct with fields:
...
                        WordLength: 12
                      FractionLength: 9
...
ix = fi(tx.int,1,12,0);         % Change the 12-bit numbers to integer
ix.WordLength                   % Check word length
ans =
     12
ix.FractionLength               % Check fraction length
ans =
      0
```

The MATLAB function computes the FIR Type I LPF impulse response as below. The cutoff frequency is $\pi/4$.

Prog. 8.18 Design and analysis the FIR filter.

```
N = 60;                          % FIR order
NN = 2^12;                       % Frequency samples
nn = 0:1:N;                      % Impulse response index
LPFhn = fir1(N,0.25);            % Type I FIR filter
[H,W] = freqz(LPFhn,1,NN);       % Frequency analysis
figure,
subplot(211), stem(nn,LPFhn), grid
subplot(212), plot(W,20*log10(abs(H))), grid
```

The impulse response and frequency response are given as Fig. 8.27.

The FIR filter coefficients can be converted to the 16-bit integer number as below.

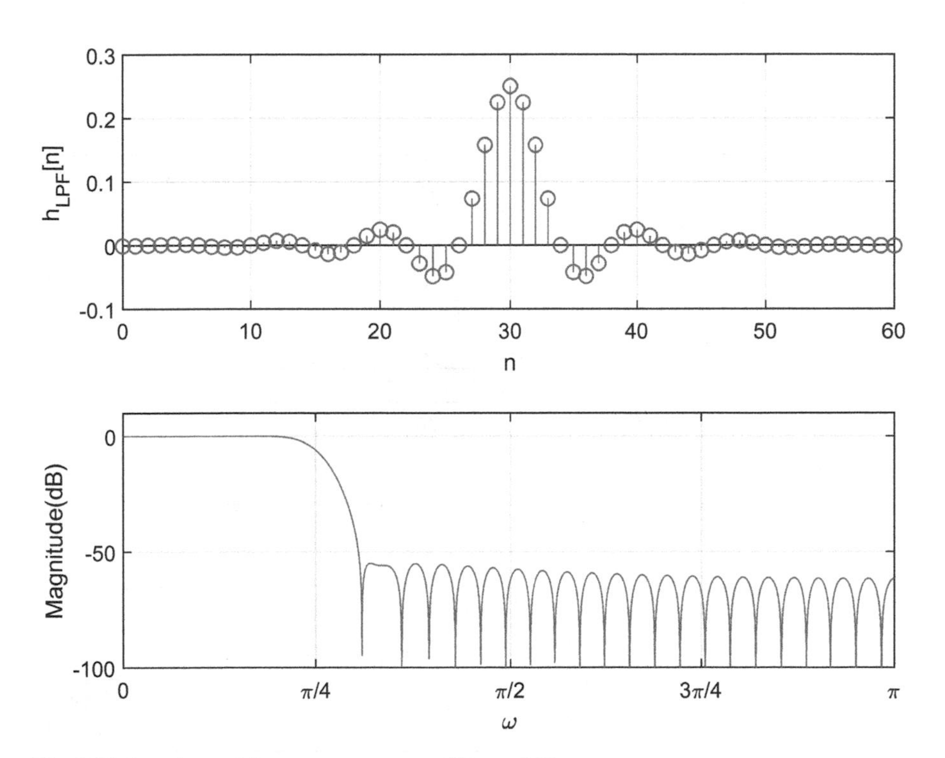

Fig. 8.27 Impulse and frequency response of **Prog. 8.18**

Prog. 8.19 FIR filter coefficient conversion to 16-bit integer number.

```
tLPFhn = fi(LPFhn,1,16);                % Convert to fixed-point 16-bit #
iLPFhn = fi(tLPFhn.int,1,16,0);         % Convert to 16-bit integer
iLPFhn.WordLength                       % Check word length
ans =
     16
iLPFhn.FractionLength                   % Check fraction length
ans =
    0
[H1,W1] = freqz(double(iLPFhn),1,NN); % Frequency analysis
figure,
subplot(211), stem(nn,iLPFhn), grid
subplot(212), plot(W1,20*log10(abs(H1))), grid
```

Designed LPF has all integer numbers which is greater than the original coefficient values. The 16-bit integer impulse response and corresponding frequency response are shown in Fig. 8.28.

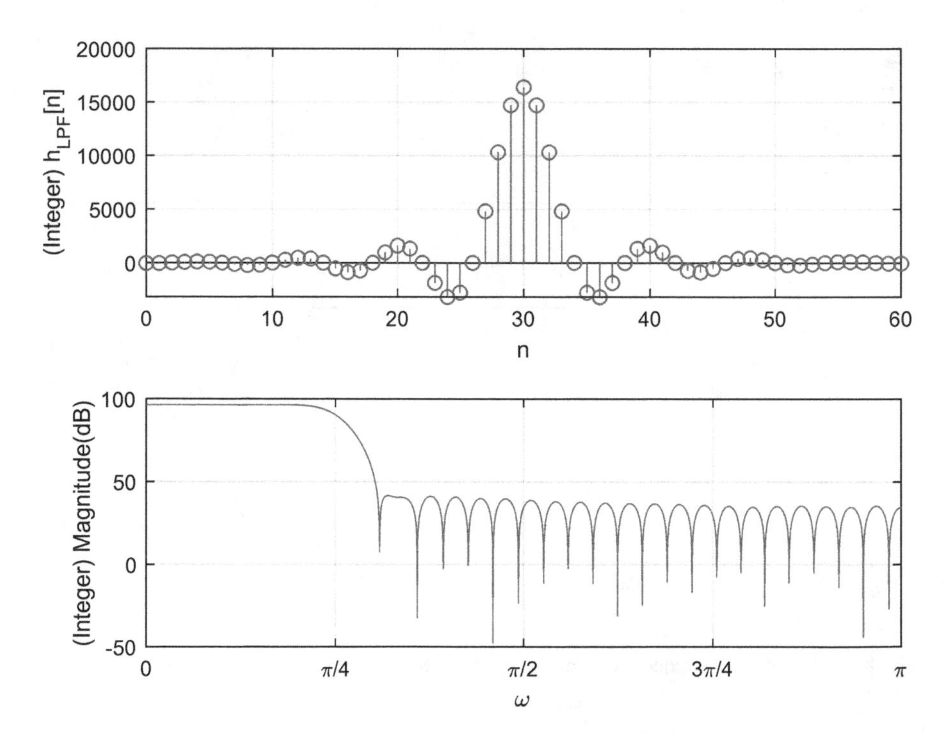

Fig. 8.28 Output of **Prog. 8.19**

The gain of the integer LPF is significantly higher than the original LPF; therefore, the filter output is required to be scale down afterward. Perform the filter operation as below.

Prog. 8.20 Perform the integer filter operation.

```
iy = filter(iLPFhn,1,ix);                          % Perform filter operation
iy.get
ans =
   struct with fields:
...
                                   Signedness: 'Signed'
                                  WordLength: 34
                              FractionLength: 0

...
nn2 = 0:1:(L-1);
figure,
stem(nn2(end-100:end),[ix(end-100:end) iy(end-100:end)]),grid
% Plot last 100
```

The input and output of the filter are illustrated in Fig. 8.29. The output magnitude is significantly higher than the input amplitude because of the integer FIR coefficients. We need to compute the integer filter gain from the coefficients for scale down.

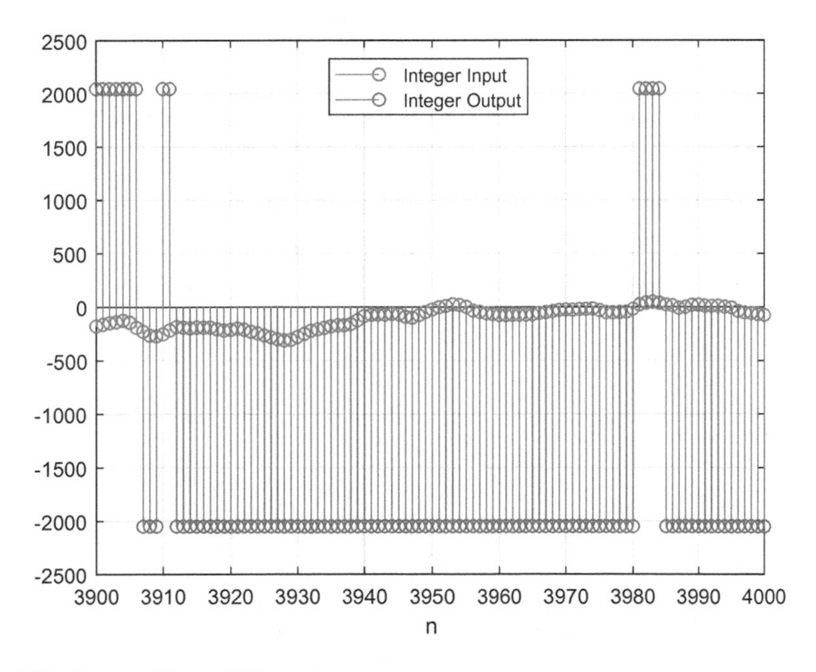

Fig. 8.29 Output of **Prog. 8.20**

The FIR filter gain is below.

$$Gain = \sum_{k=0}^{L-1} |h[k]| \tag{8.1}$$

Instead of direct scaling, the bit shift is performed based on the gain as below. The number of bit shift is represented by the power of two factor for binary number system. Note that the $\lfloor \ \rfloor$ is floor function.

$$Bit\,Shift = \lfloor \log_2(Gain) \rfloor \tag{8.2}$$

Below code creates the new fixed-point variable *yout* with *WL* word length and *FL* fraction length. The binary values are assigned from the integer filter output to new fixed-point variable. Both variable sizes are identical, but the binary point locations are distinct. Therefore, the binary representations are identical, but the decimal representations are different as factor of bit shift. Note that we cannot change the word and fraction length of the existing fixed-point variables.

Prog. 8.21 Apply filter gain to adjust output magnitude.

```
GAIN = sum(abs(iLPFhn));                          % Gain computation
WL = iy.WordLength;                               % New word length
FL = iy.FractionLength + floor(log2(double(GAIN)));   % New fraction length
% log2(gain) represents the gain by factor of 2
yout = fi(size(iy),1,WL,FL);                      % New fixed-point variable
yout.bin = iy.bin;                                % Binary copy
figure,
stem(nn2(end-100:end),[ix(end-100:end) yout(end-100:end)]), grid
```

Display the input and scaled output as Fig. 8.30.

Observe that the scale of the input and output is comparable now. However, the output word length is still 34 bits. Change to 12-bit for DAC.

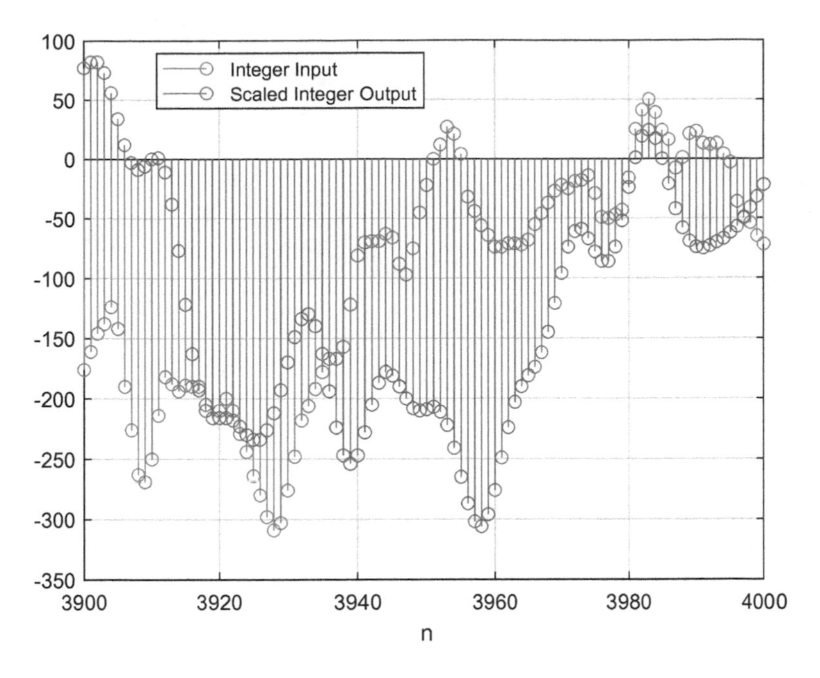

Fig. 8.30 Output of **Prog. 8.21**

Prog. 8.22 Change the output size to 12-bit integer.

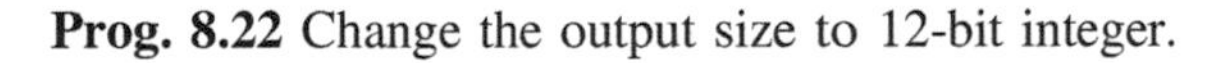

```
yfinal = fi(yout,1,12);                    % Convert to 12-bit integer number
yfinal.WordLength
ans =
    12
yfinal.FractionLength
ans =
    0
figure,
stem(nn2(end-100:end),[ix(end-100:end) yfinal(end-100:end)]),grid
```

The output is modified to the 12-bit integer. Verify output by graph in Fig. 8.31.

Let's hear the filtered sound.

Prog. 8.23 Listen the filtered sound.

```
soundsc(double(yfinal),data.Fs);          % Play sound
```

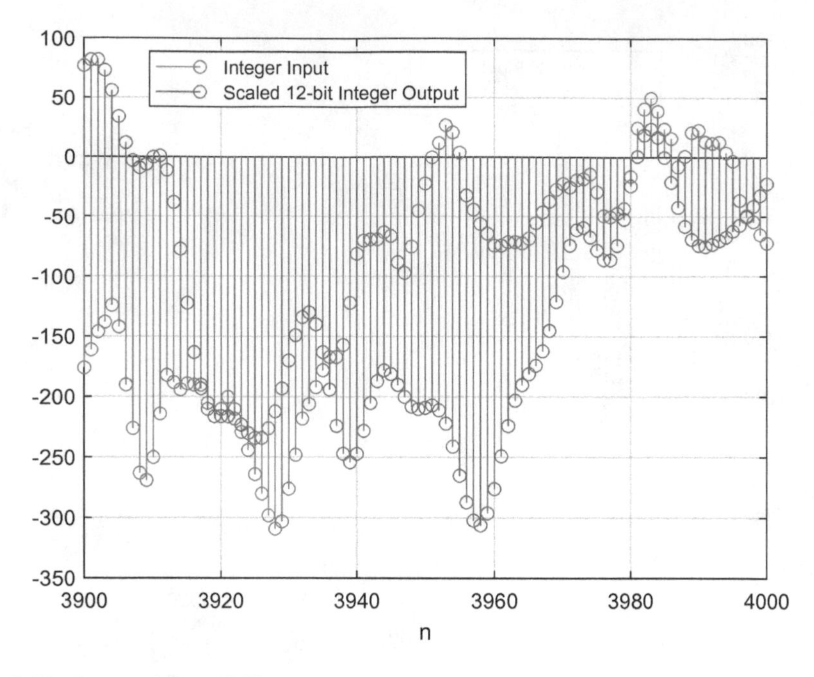

Fig. 8.31 Output of **Prog. 8.22**

The above integer FIR filter with integer data can apply for processing the filter system based on the binary fraction numbers. The procedure is identical as below.

1. ADC input → 12-bit binary fraction number
2. FIR filter coefficient → compute L_∞ – norm → Scaling → 16-bit binary fraction number
3. FIR filtering (full precision)
4. Filter output → 12-bit binary fraction number.

We can select the different scaling policy such as L_1-, L_∞-, and L_2-norm to avoid the output overflow. In this example, the L_∞-norm is utilized. The subsequent MATLAB program realizes the equivalent filtering task with binary fraction representation. Note that the fraction length in the fixed-point number should be one bit less than the word length because of the sign bit in MSB location. The FIR filter computation is the arithmetic of consecutive multiplications and additions. The full precision filter operation indicates that the MATLAB automatically adjusts the data width for processing and storing without losing any information. We can manipulate the precision by modifying the properties of the fixed-point objects. The *ProductMode* and *SumMode* property can be specified as *FullPrecision*, *KeepLSB*, *KeepMSB*, or *SpecifyPrecision*. The further information is beyond the book scope and can be found at the MathWorks documentations [5]. Remember that the auto-completion and auto-listing by tab keystroke are convenient tool to input the object property.

Prog. 8.24 Filter operation based on the binary fraction numbers. (ADC/DAC → 12 bits and $h[n]$ → 16 bits).

```
data = load('mtlb.mat');                  % Load sound
x = data.mtlb;                            % Extract data
L = length(x);                            % Data length
N = 60;                                   % Filter order
NN = 2^12;                                % Frequency sampling
nn = 0:1:N;                               % Filter time index
n = 0:1:L-1;                              % Data time index
WW = linspace(0,2*pi,NN);                 % FFT frequency index

tx = fi(x,1,12);                          % 12-bit ADC
bx = fi(size(tx),1,12,11);                % 12-bit for binary fraction #
bx.bin = tx.bin;

LPFhn = fir1(N,0.25);                     % LPF FIR h[n]

l1 = sum(abs(LPFhn));                     % L1-norm
l2 = sqrt(sum(LPFhn.^2));                 % L2-norm
[H,W] = freqz(LPFhn,1,NN);
Linf = max(abs(H));                       % Linf-norm (selected)

bLPFhn = fi(LPFhn/Linf,1,16,15);          % h[n] with binary fraction #
[H1,W1] = freqz(double(bLPFhn),1,NN);

by = filter(bLPFhn,1,bx);                 % Filtering
yfinal = fi(by,1,12,11);                  % Convert to 12-bit binary fraction #

figure,
subplot(311), plot(n,bx), grid
subplot(312), plot(W1,20*log10(abs(H1))), grid
subplot(313), plot(n,yfinal(1:L)), grid

xxx = fft(double(bx),NN);
yyy = fft(double(yfinal),NN);

pxxx = 10*log10(xxx.*conj(xxx)/(2*pi*NN));   % Spectral density
pyyy = 10*log10(yyy.*conj(yyy)/(2*pi*NN));   % Spectral density

figure,
subplot(211), plot(WW(1:NN/2+1),pxxx(1:NN/2+1)), grid
subplot(212), plot(WW(1:NN/2+1),pyyy(1:NN/2+1)), grid
```

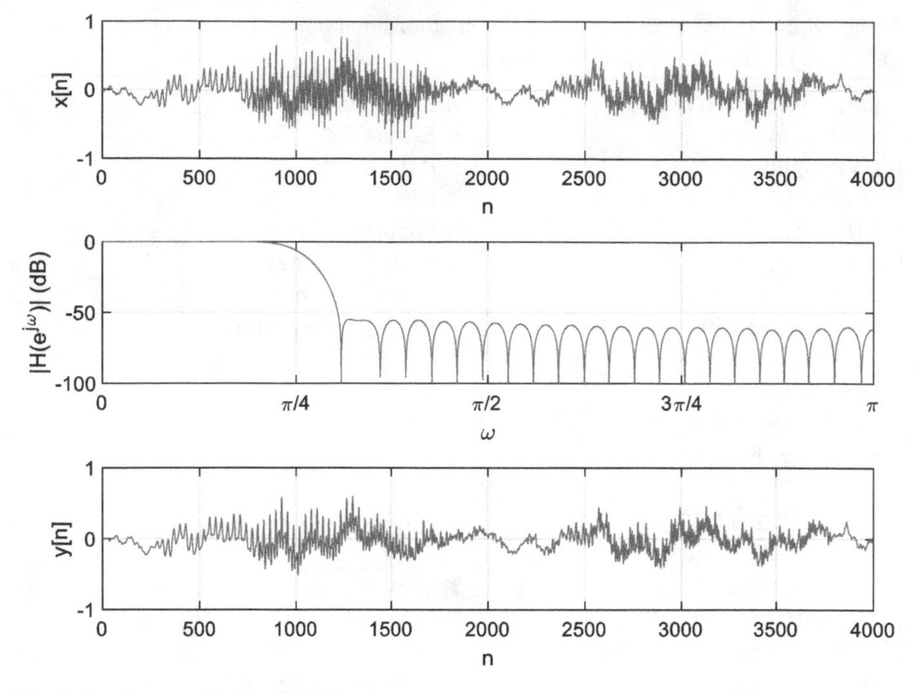

Fig. 8.32 Output of **Prog. 8.24**. Part 1

Figure 8.32 shows the input, frequency response, and output of the FIR filter system with binary fraction numbers. The cutoff frequency is $\pi/4$ and the L_∞-norm is applied for scaling. The range of the input and output is placed within the binary fraction number. No overflow signs are observed.

To verify the filter processing, Fig. 8.33 demonstrates the power spectral density for input and output. The output spectral power distribution presents the substantially reduced magnitude from $\pi/4$; hence, the filter removes the high frequency components efficiently.

The conventional programing with embedded processor requires to represent the all data and coefficient as integer number for binary fraction processing. The intrinsic functions from the processor manufactures support fraction mode arithmetic with various properties such as saturation, output width, and etc. The typical standard arithmetic operators (+, *, etc.) cannot use the given hardware fully. The filter for binary fraction numbers should be built with integer number and intrinsic functions. The dedicated compiler is essential as well.

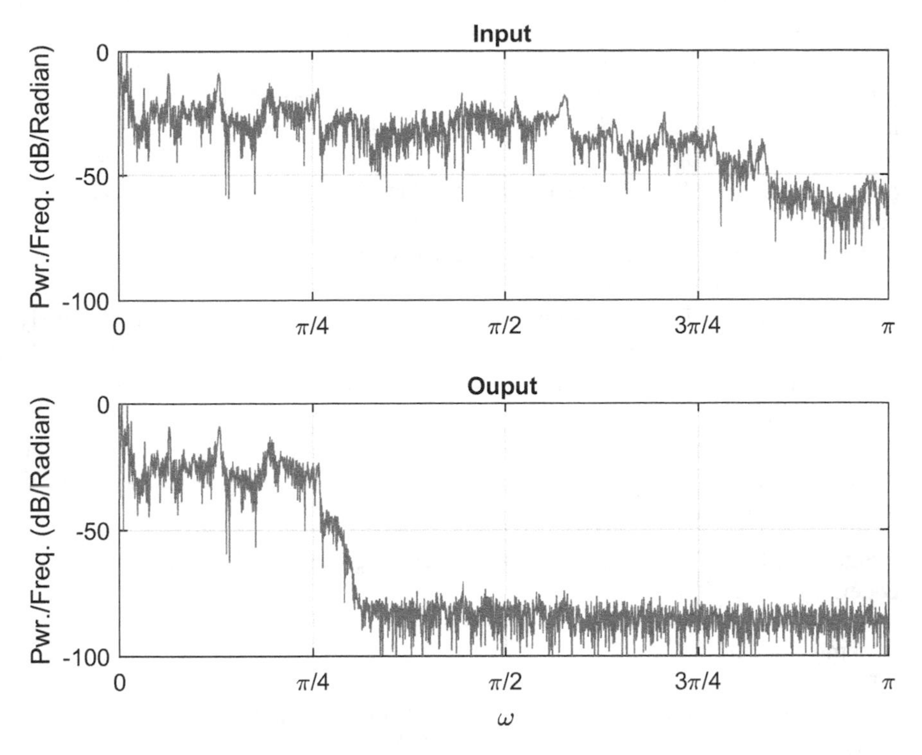

Fig. 8.33 Output of **Prog. 8.24**. Part 2

8.3 MATLAB Code Generation

All of the filter designs and executions have been performed on the MATLAB environment. Even the mathematical derivation procedures are realized in the MATLAB symbolic math. The designed filters are required to be executed in the various testbeds and targets for deployment. Conventional deployment method is the rewriting the MATLAB program manually as the C and C++ code for a variety of hardware platforms. The code conversion process is tedious and time-consuming task with multiple verifications for accuracy. The overall procedure is illustrated as Fig. 8.34.

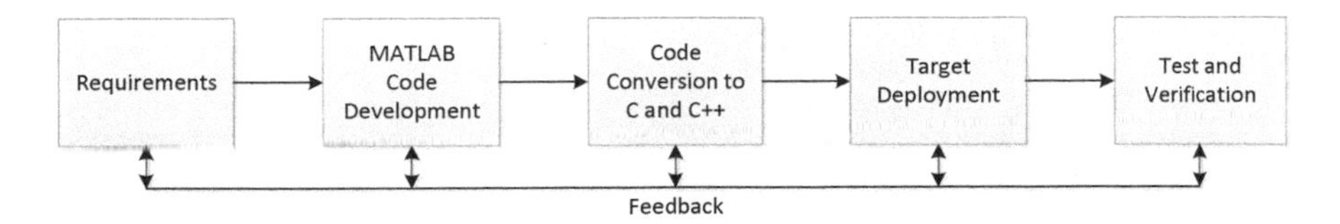

Fig. 8.34 Overall procedure for MATLAB code deployment on target system

The MATLAB provides the MATLAB coder [6] to generate C and C++ code from MATLAB program. The generated code can be integrated with the existing projects as source code or libraries in readable and portable format. The mathematically complicated algorithms are developed in MATLAB environment and the interface programs are written by the C and C++ code. As the part of combining procedure, the MATLAB coder generates the C and C++ code. We can also compile the generated code as a MEX-function to use in the MATLAB environment for verification and acceleration. The MEX stands for 'MATLAB executable.' The MEX subroutines can be called by the command line as if they are MATLAB functions. The MATLAB codes beyond the MATLAB environment are extremely extensive and fast-growing area in MATLAB world. However, this section only provides the code generation for MEX-function to execute the algorithm fast. Further information can be found at MATLAB documentations.

The code generation only supports the MATLAB functions not a general script. We have to separate the algorithms into the individual functions. Before we perform the code generation, check the compiler as below.

Prog. 8.25 Check compiler for MEX-function generation by using MATLAB coder.

```
>> mex -setup
Error using mex
No supported compiler or SDK was found. You can install the freely available
MinGW-w64 C/C++
compiler; see Install MinGW-w64 Compiler. For more options, visit
http://www.mathworks.com/support/compilers.
```

If we have any compatible compiler, the MATLAB returns the compiler name, otherwise we have to install the software. The above is the error message with no available compiler. The open-source compiler for MATLAB is MinGW64. Install the compiler by following procedure.

HOME ribbon tab → Add-Ons → Get Add-Ons → Search 'gcc' → Choose and install 'MATLAB Support for MinGW-w64 C/C++ Compiler'.

Now, you have proper environment. Below is the example filter function.

Prog. 8.26 Example filter for MEX-function.

```
function [y] = myfilter1(x)
%#codegen
% For code generation, you should put the codegen hashtag in above
persistent d1 % Static variable for function
if isempty(d1) % If it is not defined, the function compute filter coefficient
     d1 = dsp.FIRFilter('Numerator',fir1(50,0.1)); % Actual filter
end
y = d1(x); % Filter execution
end
```

In the filter function, the filter coefficient is computed once due to the static variable and conditional statement. Not all MATLAB intrinsic functions are compatible to the code generation. Visit the website [7] for compatibility. Potential our filter selections for code generation are below.

dsp.FIRFilter, dsp.IIRFilter, dsp.HighpassFilter, dsp.LowpassFilter.

See the manual [8] for further information on the above functions. Check the suitability by typing the following instruction in the command window.

Prog. 8.27 Check suitability for code generation.

```
coder.screener('myfilter1')
```

If there are no problems, you can see the below window as Fig. 8.35.

Ready to generate and compile the c code for the function.

Prog. 8.28 Generate and compile the c code for the myfilter1.m function.

```
codegen myfilter1 -args {randn(4001,1)}
```

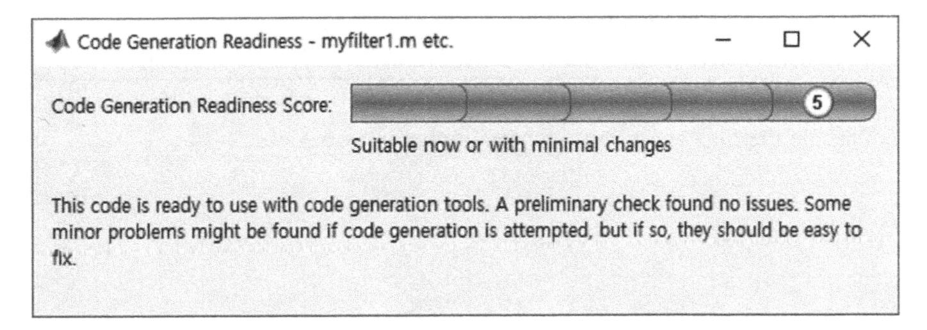

Fig. 8.35 Check readiness for code generation

The last argument specifies the input argument for my filter function. Note that the argument size must correspond with caller argument size. The 4001 is the size of MATLB sound.

Prog. 8.29 Test code for MATLAB function.

```
data = load('mtlb.mat'); % Load sound
x = data.mtlb;
Fs = data.Fs;

tic
y = myfilter1(x); % Filter algorithm (Change this for compiled version)
toc

figure,
subplot(211), spectrogram(x,128,120,128,Fs)
title('Input')
subplot(212), spectrogram(y,128,120,128,Fs)
title('Output')
```

Figure 8.36 shows the spectrogram [3] of the input and output signal. The spectrogram is the frequency plot in terms of time. The input spectrogram illustrates the wide frequency distribution and the filtered output spectrogram presents the low frequency centric configuration according to the LPF property. The sampling frequency is 7418 Hz and the cutoff frequency is the 370.9 Hz (0.1 normalized frequency). Around the cutoff frequency, the magnitude is significantly decreasing on the output spectrogram. The filter is working properly.

The *tic* & *toc* function measures the execution time in second. Execute the above code and check the execution time. The example execution time is below. Note that above code is pure MATLAB code and the execution time is statistical.

Prog. 8.30 Execution time of the MATLAB function from **Prog. 8.29**.

```
Elapsed time is 0.265705 seconds.
>>
```

Change the filter function with MEX-function as below.

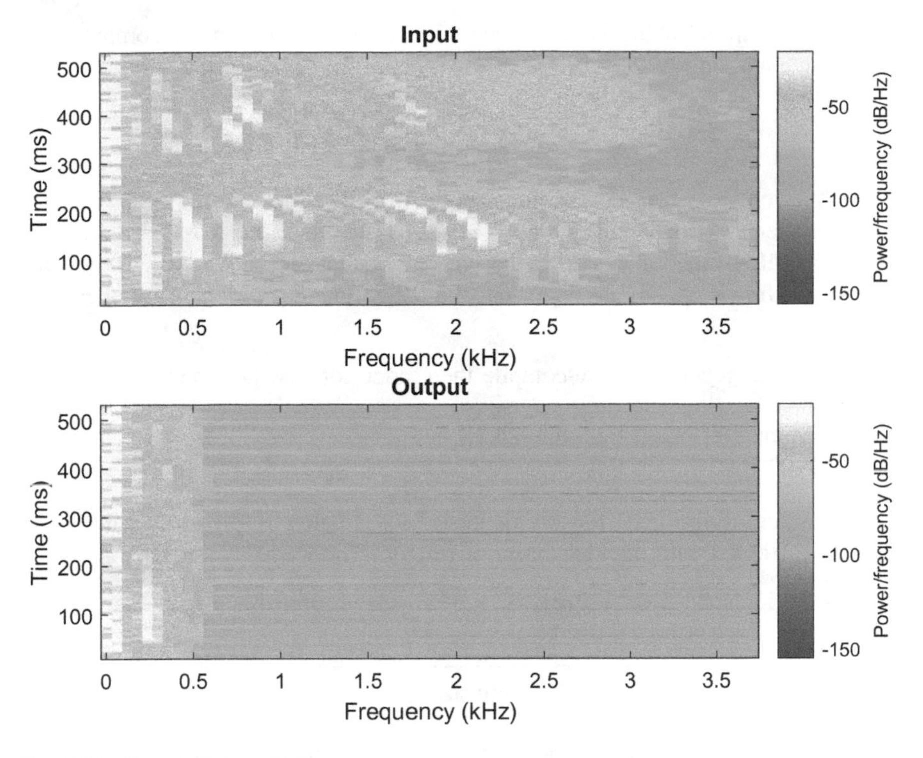

Fig. 8.36 Output of **Prog. 8.30**

Prog. 8.31 Change **Prog. 8.29** for MEX-function execution.

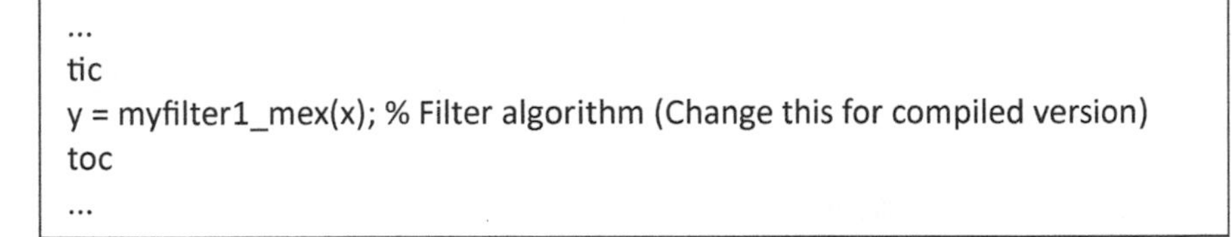

```
...
tic
y = myfilter1_mex(x); % Filter algorithm (Change this for compiled version)
toc
...
```

The execution time with MEX-function is shown below. The execution time varies with trials due to the resource assignments of operating system.

Prog. 8.32 Execution time of the MEX-function from **Prog. 8.31**.

```
Elapsed time is 0.001380 seconds.
>>
```

The speedup from the code generation and compiling can be computed by following equation.

$$\text{Speedup} = \frac{\text{Execution time with MATLAB code}}{\text{Execution time with MEX code}} \tag{8.3}$$

With given execution times, the speedup from the MEX is the 192.54 which indicates that the MEX code is 192.54 times faster than the MATLAB intrinsic code. The MEX-function can be applied for the system object as well. For the 1024 sample size frame, we have to regenerate the MEX function with new argument size as below.

Prog. 8.33 Re-generate and re-compile the c code for new argument size.

```
codegen myfilter1 -args {randn(1024,1)}
```

The system object code for the identical task is below. Observe that the audio data is substituted with the longer length version. The frame size is 1024 samples.

Prog. 8.34 Identical task as **Prog. 8.29** with system object and frame-based processing.

```
FrameSize = 1024;                % Frame size

reader = dsp.AudioFileReader('handel.ogg','SamplesPerFrame',FrameSize);
% Audio file reader object
P = audioDeviceWriter(reader.SampleRate);    % Audio device writer object

ttime = 0;             % Execution time initialization

while ~isDone(reader)
      x = reader();
      tic;
      y = myfilter1(x); % Filter algorithm (Change this for compiled version)
      ttime = ttime + toc;
      P(y);
end

ttime

ttime =
      0.2069
>>
```

The execution time is accumulated for multiple frame times because of the frame-based executions. The audio device uses the default device. We can modify the device by setting the object properties. Remember to use the auto-completion and auto-listing by applying the tab key.

Prog. 8.35 Modify **Prog. 8.34** for MEX-function execution.

```
...
while ~isDone(reader)
    ...
    y = myfilter1_mex(x); % Filter algorithm (Change this for compiled version)
    ttime = ttime + toc;
    ...
end

ttime

ttime =
    0.0203
>>
```

The execution time of the MEX function can be obtained by changing the filter function as above. The speedup of the MEX based on the system object is 10.2. The reduced performance results from the segmented execution which requires the overhead to initialize and finalize the executions. Overall, the MEX-function provides the significant speedup. The explicit code generation and system integration can be found at the MATLAB documentation [6].

References

1. MathWorks: MATLAB Homepage (2020). https://www.mathworks.com/products/matlab.html
2. MathWorks: MATLAB Help Center (2020). https://www.mathworks.com/help/
3. Stoica, P., Moses, R.L.: Introduction to Spectral Analysis. Prentice Hall (1997)
4. MathWorks: Fixed-Point Designer (2020). https://www.mathworks.com/help/fixedpoint/index.html
5. MathWorks: fimath ProductMode and SumMode (2020). https://www.mathworks.com/help/fixedpoint/ug/using-fimath-productmode-and-summode.html
6. MathWorks: MATLAB Coder (2020). https://www.mathworks.com/help/coder/index.html
7. MathWorks: Functions and Objects Supported for C/C++ Code Generation (2020). https://www.mathworks.com/help/simulink/ug/functions-and-objects-supported-for-cc-code-generation.html
8. MathWorks: Fixed-Point Filters (2020). https://www.mathworks.com/help/dsp/fixed-point-filters.html?s_tid=CRUX_lftnav

Appendix A
MATLAB Fundamentals

The MATLAB provides the extensive tutorials for all its products in various format. This appendix briefly introduces the MATLAB fundamentals based on the MATLAB Primer. You can observe the primitive MATLAB capabilities in short. Once you click the MATLAB icon, you will get the following screen as Fig A.1.

We can observe the four window spaces as below.

- Current Folder: Your working directory and files.
- Command Window: Enter commands for immediate execution.
- Workspace: Variables and data.
- Editor-Untitled: Once click the New button, the editor window provides the script space for programming.

Most of the icon buttons and ribbon menus are intuitive and self-explanatory for its functions. We can execute the MATLAB commands by

- Type the command in command window and put the <enter> key
- Write the program in the editor window and put the <ctrl>+<enter> key

The script program also can be executed by clicking the run button. One big advantage of the MATLAB programming is the worry-free variables. If we need the variables, just create and use them. Below example creates the a and b variable with initial values in command window. As soon as we type the 'a=1' and <enter>, the a variable is created with initial value one. When we compute the value without the assignment, the output value is allocated at the temporary variable *ans*. See the 'a+b' command in below. All created variables can be seen in the workspace window.

K. Kim, *Conceptual Digital Signal Processing with MATLAB*,
Signals and Communication Technology,
https://doi.org/10.1007/978-981-15-2584-1

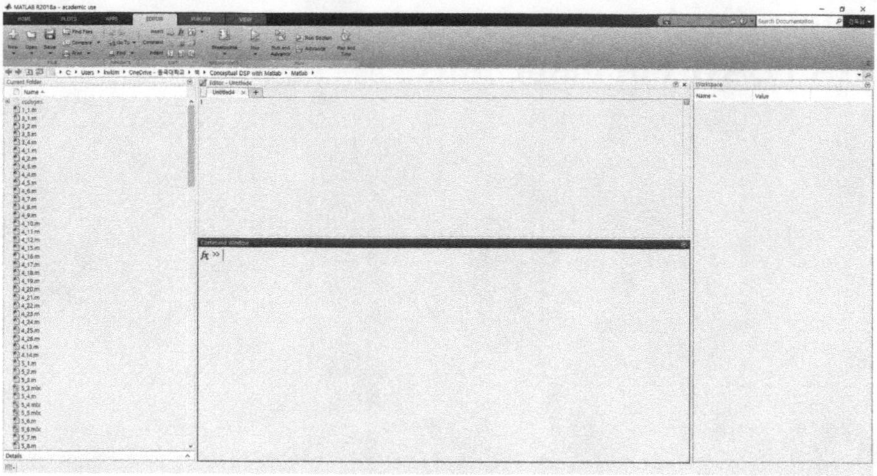

Fig. A.1 MATLAB screen

Prog. A.1 Variable creation and initialization

```
>> a = 1
a =
       1
>> b = 2
b =
       2
>> a+b
ans =
       3
>>
```

To clear the command window, type the *clc*. The second MATLAB advantage is the strong supports for the linear algebra operations based on the matrix and array. Just like variable, we can create the matrix and array as below.

Prog. A.2 Creation, initialization, and arithmetic of matrix variables

```
>> a = [1 2 3]
a =
      1      2      3
>> b = [4, 5, 6]
b =
      4      5      6
>> c = [7; 8; 9]
c =
      7
      8
      9
>> d = [1 2 3; 4 5 6; 7 8 9]
d =
      1      2      3
      4      5      6
      7      8      9
>> e = d*a
Error using   *
Incorrect dimensions for matrix multiplication. Check that the number of
columns in the first matrix matches the number of rows
in the second matrix. To perform elementwise multiplication, use '.*'.
 >> e = a*d
e =
     30     36     42
  >>
```

The most arithmetic operators in MATLAB perform the linear algebra operations which require to match the dimensions between the matrices. Table A.1 shows the arithmetic operator list. The addition and subtraction are working for the constant, matrix, and element-wise operation.

Other operators for logic and relation are listed below as Table A.2.

The matrix can be generated by the MATLAB functions. The part of functions is shown below in Table A.3.

Below example demonstrates the matrix generation from the function. The first command *rng*(*'default'*) places the random number generator setting to their default values so that they produce the same random numbers.

Table A.1 Arithmetic operator list of MATLAB

Symbol	Role
+	Addition
−	Subtraction
*	Matrix multiplication
/	Matrix right division
\	Matrix left division
^	Matrix power
‘	Complex conjugate transpose
.*	Element-wise multiplication
./	Element-wise right division
.\	Element-wise left division
.^	Element-wise power
.’	Transpose

Table A.2 Logical and relational operator list of MATLAB

<table>
<tr><th>Symbol</th><th>Role</th></tr>
<tr><td>==</td><td>Equal to</td></tr>
<tr><td>~=</td><td>Not equal to</td></tr>
<tr><td>></td><td>Greater than</td></tr>
<tr><td>>=</td><td>Greater than or equal to</td></tr>
<tr><td><</td><td>Less than</td></tr>
<tr><td><=</td><td>Less than or equal to</td></tr>
<tr><td>&</td><td>Logical AND</td></tr>
<tr><td>|</td><td>Logical OR</td></tr>
<tr><td>&&</td><td>Logical AND (with short-circuiting)</td></tr>
<tr><td>||</td><td>Logical OR (with short-circuiting)</td></tr>
<tr><td>~</td><td>Logical NOT</td></tr>
</table>

Table A.3 Special matrix generation functions

Function	Role
zeros	Create array of all zeros
ones	Create array of all ones
rand	Uniformly distributed random numbers
eye	Identity matrix

Prog. A.3 Matrix generation example from function

```
>> rng('default')
>> aa = rand(3,3)
aa =
      0.8147     0.9134     0.2785
      0.9058     0.6324     0.5469
      0.1270     0.0975     0.9575
>> bb = ones(3,2)
bb =
       1     1
       1     1
       1     1
>>
```

The matrix is also produced by the concatenation which joins the arrays to make larger ones. See below example.

Prog. A.4 Matrix generation example by concatenation

```
>> rng('default')
>> bb = randi(10,2,3)
bb =
       9     2     7
      10    10     1
>> cc = [bb bb]
cc =
       9     2     7     9     2     7
      10    10     1    10    10     1
>> dd = [bb, bb]
dd =
       9     2     7     9     2     7
      10    10     1    10    10     1
>> ee = [bb; bb]
ee =
       9     2     7
      10    10     1
       9     2     7
      10    10     1
   >>
```

Use the indexing, we can access or modify selected elements of an array. The most general method is to indicate row and column number which starts from one. A single column-wise numbering can specify the matrix position as well. The matrix element can be modified by assigning the specific value to the element.

Prog. A.5 Matrix indexing and modification

```
>> bb = randi(100,4,4)
bb =
      23     54     11     82
      92    100     97     87
      16      8      1      9
      83     45     78     40
>> bb(2,4)
ans =
      87
>> bb(14)
ans =
      87
>> bb(2,4)=0
bb =
      23     54     11     82
      92    100     97      0
      16      8      1      9
      83     45     78     40
>> bb(2,1:3)
ans =
      92    100     97
>> bb(:,2)
ans =
      54
     100
       8
      45
>> bb(end,:)
ans =
      83     45     78     40
>> cc = 0:5:30
cc =
       0      5     10     15     20     25     30
>> dd = 1:10
dd =
       1      2      3      4      5      6      7      8      9     10
>>
```

The colon operator allows to refer the multiple elements in the matrix as the *start:end* form. The colon alone indicates all of the elements in that dimension. The *end* indicator denotes the last row or column number. The general colon operator creates an equally spaced vector using the form *start:step:end*. Missing the step value indicates the default step value of one.

MATLAB provides a large number of functions which perform various tasks. The input and output argument format can be found by using the *help* or *doc* function as below. Note that *doc* function opens the new window for information.

Prog. A.6 MATLAB help

```
>> help max
 max      Largest component.
     ...
     [Y,I] = max(X) returns the indices of the maximum values in vector I.
     If the values along the first non-singleton dimension contain more
     than one maximal element, the index of the first one is returned.
     ...
     Example:
            X = [2 8 4; 7 3 9]
            max(X,[],1)
            max(X,[],2)
            max(X,5)
     ...
 >>
```

One example of function usage is to find the minimum, negative, and zero locations of the given function below.

$$y = x^2 - 1 \quad \text{for} - 10 \le x \le 10$$

Find the x values which satisfy the following conditions.

$$\min_x (x^2 - 1)$$
$$x^2 - 1 < 0$$
$$x^2 - 1 = 0$$

The MATLAB numerically computes the values based on the discretized input values for the given range. The continuous input is divided into the discrete input with 0.1 step by using the colon operator. The output is calculated by the element-wise operation. The *min* and *find* function perform the designated tasks. Note that the fine-grain results can be derived from the reduced step size for input.

In general, the function supports the various argument styles. See help for further information. Below example is written by the script editor.

Prog. A.7 Numerically solve the equation

```
x = -10:0.1:10;              % Input values
y = x.^2-1;                  % Output equation

[k1,l1] = min(y)             % Find the min value at k1 and location l1
[k2] = find(y<0)             % Find the locations for negative output
y(k2)                        % Corresponding negative values
[k3] = find(y==0)            % Find the zero locations
y(k3)                        % Verify the zero outputs

k1 =
    -1
l1 =
   101
k2 =
    92    93    94    95    96    97    98    99   100   101
102   103   104   105   106   107   108   109   110
ans =
  Columns 1 through 13
   -0.1900   -0.3600   -0.5100   -0.6400   -0.7500   -0.8400   -
0.9100   -0.9600   -0.9900   -1.0000   -0.9900   -0.9600   -0.9100
  Columns 14 through 19
   -0.8400   -0.7500   -0.6400   -0.5100   -0.3600   -0.1900
k3 =
    91   111
ans =
     0     0
```

The third advantage of the MATLAB is the powerful visualization tools. The impressive 2-D and 3-D plots are available with various decoration functions to display extensive information. Labeling, legend, range, and etc. are specified by the MATLAB functions. Using the TeX markup, Geek letters and special characters can be added to the plot. Some of the list are arranged below in Table A.4.

Table A.4 Special text markup for special characters

Char. sequence	Symbol	Char. sequence	Symbol
^{ }	Superscript	\delta	δ
_{ }	Subscript	\epsilon	ϵ
\bf	Bold font	\theta	θ
\it	Italic font	\pi	π
\alpha	α	\omega	ω
\beta	β	\Omega	Ω
\gamma	γ	\rightarrow	→

Below example shows the line and stem plots.

Prog. A.8 2-D plot example

```
x = 0:pi/100:2*pi;              % Input
y1 = sin(x);                    % sine function
y2 = cos(x);                    % cosine function

figure,
plot(x,y1,x,y2), grid                    % Line plot for sine and cosine
title('Example 1')                       % Title
legend('sin(\omega)','cos(\omega)');     % Legend with special char
xticks([0 pi/2 pi 3*pi/2 2*pi])          % Tick position
xticklabels({'0','\pi/2','\pi','3\pi/2','2\pi'})      % Tick label
xlim([0 2*pi])                                        % x display range
ylim([-1.5 1.5])                                      % y display range
xlabel('\omega')                                      % x label
ylabel('Magnitude')                                   % y label

figure,
subplot(211), stem(x,y1), grid           % Subplot with stem
title('Example 2')
ylabel('sin(\omega)')
xticks([0 pi/2 pi 3*pi/2 2*pi])
xticklabels({'0','\pi/2','\pi','3\pi/2','2\pi'})
xlim([0 2*pi])
subplot(212), stem(x,y2), grid
ylabel('cos(\omega)')
xticks([0 pi/2 pi 3*pi/2 2*pi])
xticklabels({'0','\pi/2','\pi','3\pi/2','2\pi'})
xlim([0 2*pi])
xlabel('\omega')
```

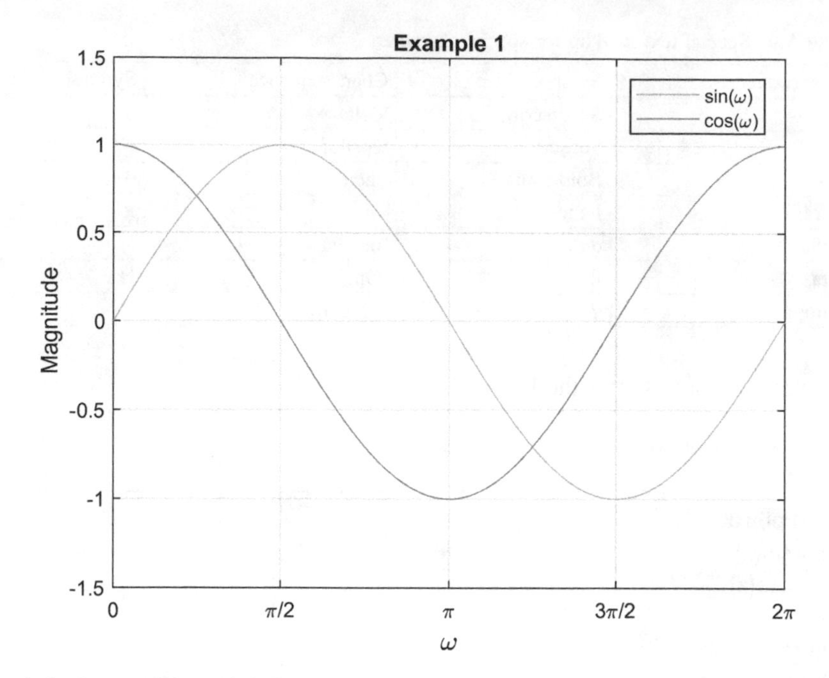

Fig. A.2 Output of **Prog. A.8**. Part 1

The simple sine and cosine plots are displayed for 2π range as below (Fig A.2).

The *subplot* function allows us to place the more than one plot in the single figure as Fig A.3.

The MATLAB also gives the versatile 3-D plot features. Below example presents the 3-D Gaussian plot according to the following equation.

$$z = f(x, y) = e^{-((x-x_0)+(y-y_0))}$$

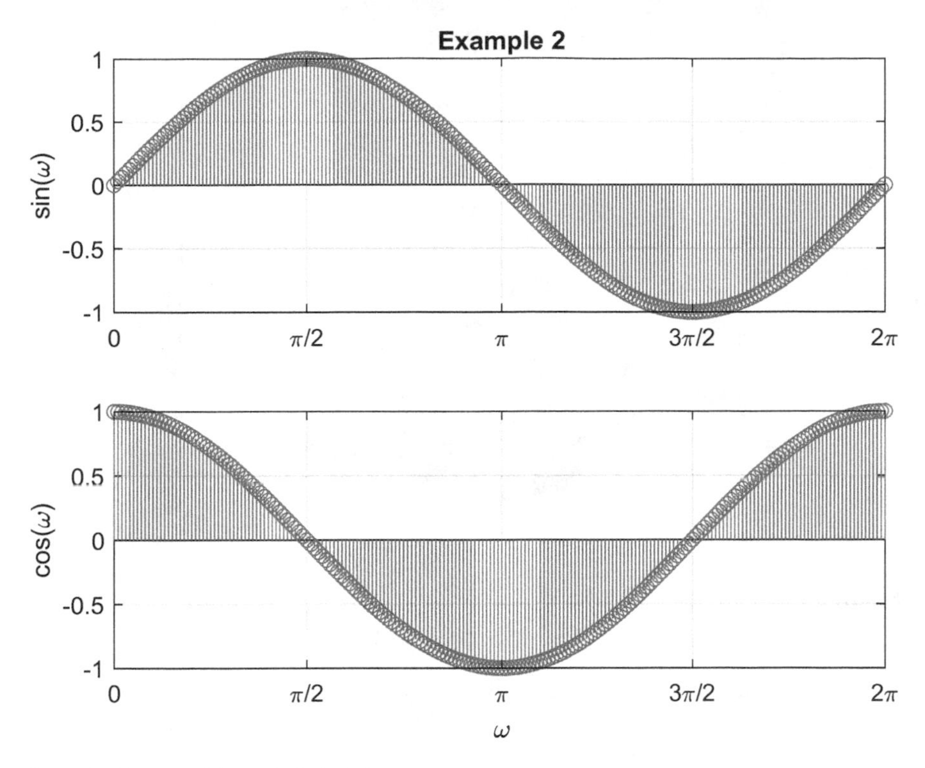

Fig. A.3 Output of **Prog. A.8**. Part 2

Prog. A.9 3-D plot example

```
[X,Y] = meshgrid(-2:0.1:2);                  % Cartesian grid
Z = exp(-(X.^2+Y.^2));                       % 2-D Gaussian function Z = f(X,Y)

figure,
surf(X,Y,Z)                                  % 3D surface plot
title('3-D Normal Distribution')             % Title
xticks([-2 -1 0 1 2])                        % Tick position
xticklabels({'-2','-1','x_0','1','2'})       % Tick label
xlabel('x')                                  % x axis label
yticks([-2 -1 0 1 2])                        % Tick position
yticklabels({'-2','-1','y_0','1','2'})       % Tick label
ylabel('y')                                  % y axis label
zlabel('e^{-((x-x_0)^2+(y-y_0)^2)}')         % z axis label
colormap('gray')                             % Color bar color configuration
colorbar                                     % Color bar
```

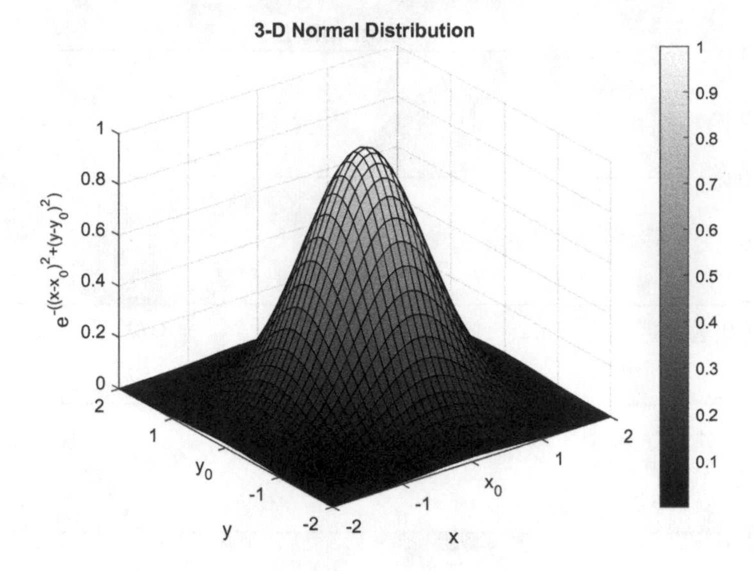

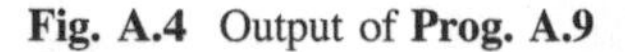

Fig. A.4 Output of **Prog. A.9**

Prior to create the 3-D plot, we must generate the input coordinates for the 3-D equation. The *meshgrid* function returns 2-D grid coordinates based on the coordinates contained in input vector(s). The *X* and *Y* are the square grid coordinates with single input grid size. The element-wise operation completes the equation calculation followed by the 3-D plot function. Below figure shows the 3-D surface plot. We can change the viewpoint by clicking the rotate 3D button and dragging the figure (Fig A.4).

We can create the MATLAB function by generating the new script for function as shown in Fig A.5.

The created editor includes the function format as below. Decide function name, write function code, and update the comment. Finally save the function as the function name with *m* extension.

Prog. A.10 Created new function format

```
function [outputArg1,outputArg2] = untitled2(inputArg1,inputArg2)
%UNTITLED2 Summary of this function goes here
%   Detailed explanation goes here
outputArg1 = inputArg1;
outputArg2 = inputArg2;
end
```

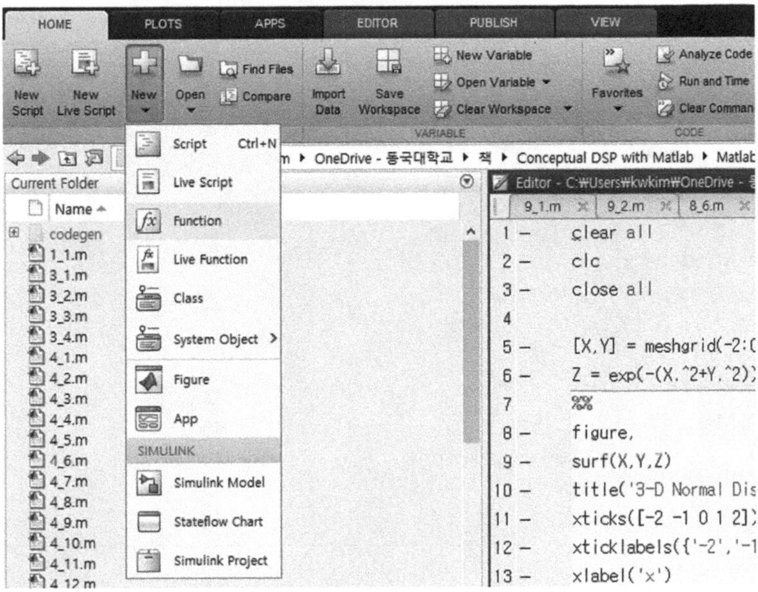

HOME PLOTS APPS EDITOR PUBLISH VIEW

New Script | New Live Script | New | Open | Find Files | Compare | Import Data | Save Workspace | New Variable | Open Variable | Clear Workspace | Favorites | Analyze Code | Run and Time | Clear Commands

VARIABLE CODE

OneDrive - 동국대학교 ▸ 책 ▸ Conceptual DSP with Matlab ▸ Matlab

Current Folder

Name

codegen
1_1.m
3_1.m
3_2.m
3_3.m
3_4.m
4_1.m
4_2.m
4_3.m
4_4.m
4_5.m
4_6.m
4_7.m
4_8.m
4_9.m
4_10.m
4_11.m
4_12.m

Script Ctrl+N
Live Script
Function
Live Function
Class
System Object
Figure
App
SIMULINK
Simulink Model
Stateflow Chart
Simulink Project

Editor - C:\Users\kwkim\OneDrive -

9_1.m | 9_2.m | 8_6.m

```
1 -  clear all
2 -  clc
3 -  close all
4
5 -  [X,Y] = meshgrid(-2:(
6 -  Z = exp(-(X.^2+Y.^2))
7    %%
8 -  figure,
9 -  surf(X,Y,Z)
10 - title('3-D Normal Dis
11 - xticks([-2 -1 0 1 2])
12 - xticklabels({'-2','-1
13 - xlabel('x')
```

Fig. A.5 New script for function

The simple sum function is written for example as below.

Prog. A.11 Function example

```
function [z] = Mysum(x,y)
% Compute the sum of two variables or arrays
% z = x+y
% Example
% x = [1 2 3];
% y = [3 5 7];
% z = Mysum(x,y)
% Created by Keonwook Kim
% Jan. 23rd 2019
z = x+y;
end
```

Save the function as *Mysum.m* in the working directory. The *help* function with *Mysum* replies the comment in the function as shown below. Execute the comment example for verification.

Prog. A.12 Function help and execution

```
>> help Mysum
   Compute the sum of two variables or arrays
   z = x+y
   Example
   x = [1 2 3];
   y = [3 5 7];
   z = Mysum(x,y)
   Created by Keonwook Kim
   Jan. 23rd 2019
>>     x = [1 2 3];
   y = [3 5 7];
   z = Mysum(x,y)
z =
       4     7    10
>>
```

We can execute the part of MATLAB program code as section. Sometime, this kind of execution is required for debugging. The double percent sign (%%) divides the program into the sections as shown in Fig A.6. By moving the cursor, the selected section is highlighted by yellow color as shown below. Typing the <ctrl> +<enter> executes the chosen section. MATLAB provides the advanced debugging tools and the section run is the part of them.

In command window, the command history is recalled by clicking the <up> key and the history is narrowed by typing the character letters as shown in Fig A.7. The quick access to the history makes the multiple executions easy for various purpose.

Followings are the frequently used commands.

clear all: Clear all variables
clc: Clear the command window
clf: Clear the current figure
close all: Close all figure windows
lookfor 'keyword': Search all MATLAB files for keyword

Further information about the MATLAB and toolboxes can be referred from the MathWorks documentations.

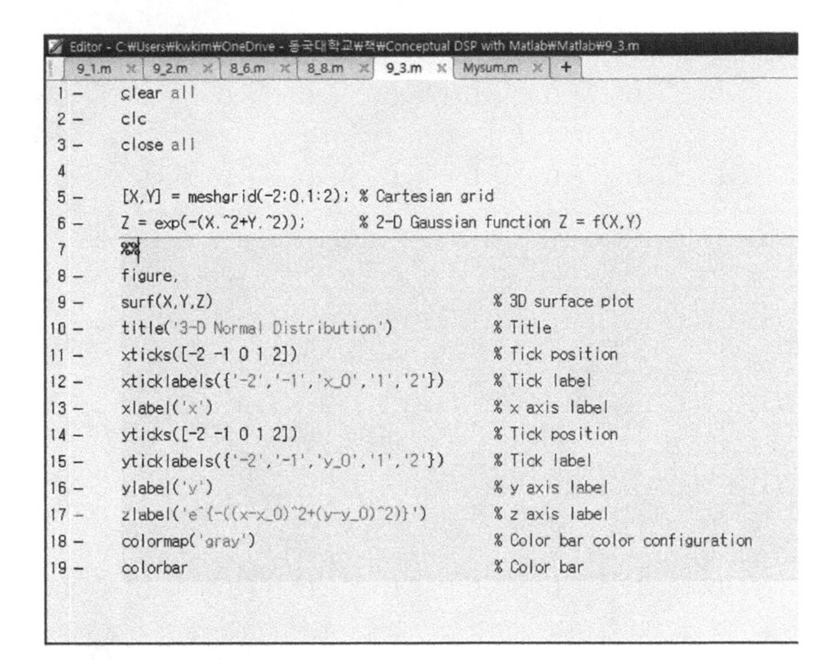

Fig. A.6 Section division

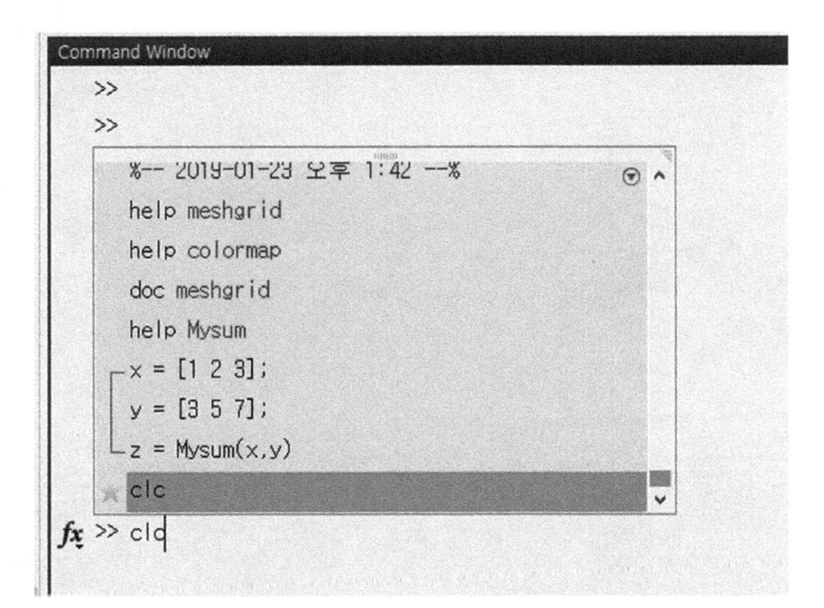

Fig. A.7 Recall command history

Appendix B
MATLAB Symbolic Math Toolbox

Symbolic math performs exact computation with variables which have no given value as symbols. Symbolic math toolbox provides functions for solving and manipulating symbolic math equations. The toolbox delivers functions in subjects such as calculus, linear algebra, differential equations, equation simplification, and equation manipulation. The computations can be realized either analytically or using variable-precision arithmetic. The MATLAB live script editor presents the optimal environment to create and run the symbolic math code. We can share the symbolic work with other users as live scripts or convert them to HTML or PDF for publication. We can generate the new live script by selecting the New → Live Script in the menu as below Fig B.1.

The live script is divided into the text and code area. The text area (white) describes the background, algorithm, explanation, analysis, etc. with plain text. The code area (gray) locates the MATLAB code for algorithm execution. The MATLAB outcome is embedded at the live script immediate right or below according to your choice as shown in Fig B.2.

K. Kim, *Conceptual Digital Signal Processing with MATLAB*,
Signals and Communication Technology,
https://doi.org/10.1007/978-981-15-2584-1

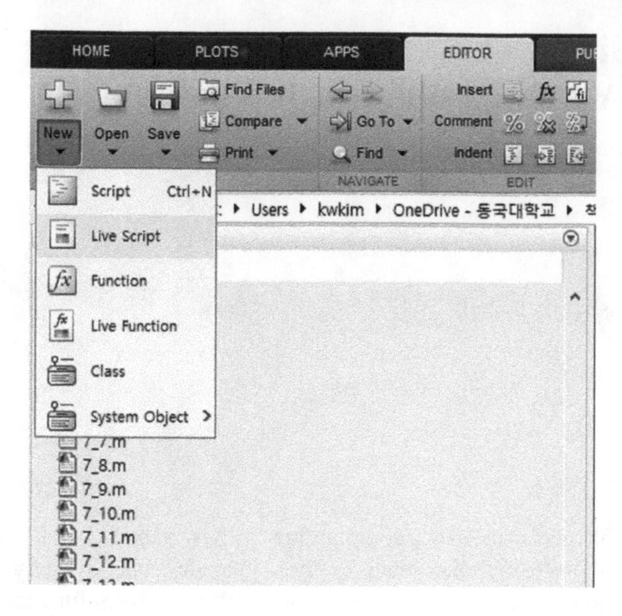

Fig. B.1 Generate live script

The buttons under the LIVE EDITOR ribbon menu specify the file, text, code, and execution related information. The buttons are intuitive and self-explanatory. Example live script code is below. Note that the below is not the symbolic math code. Execute the code by clicking <ctrl>+<enter>.

Prog. B.1 Example of live script code for MATLAB programming

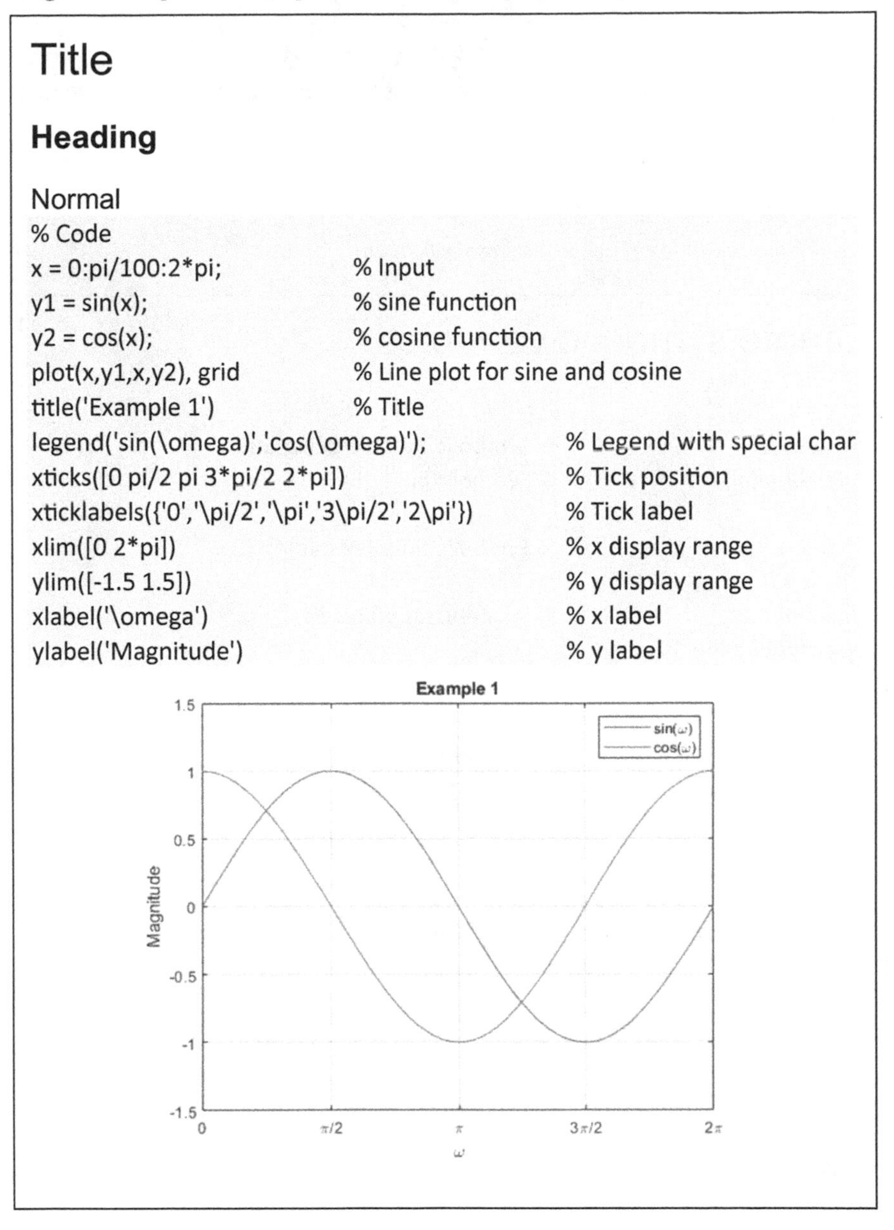

Title

Heading

Normal

```
% Code
x = 0:pi/100:2*pi;                 % Input
y1 = sin(x);                       % sine function
y2 = cos(x);                       % cosine function
plot(x,y1,x,y2), grid              % Line plot for sine and cosine
title('Example 1')                 % Title
legend('sin(\omega)','cos(\omega)');                 % Legend with special char
xticks([0 pi/2 pi 3*pi/2 2*pi])                      % Tick position
xticklabels({'0','\pi/2','\pi','3\pi/2','2\pi'})     % Tick label
xlim([0 2*pi])                                       % x display range
ylim([-1.5 1.5])                                     % y display range
xlabel('\omega')                                     % x label
ylabel('Magnitude')                                  % y label
```


Most type of MATLAB code can be executed in the live script. The graphic environment of live script gives the powerful presentation for the symbolic math code which use the mathematical symbols and characters frequently. Now, we will explore the symbolic math toolbox with live script editor. Below example shows the symbolic math example.

Fig. B.2 Live editor ribbon menu

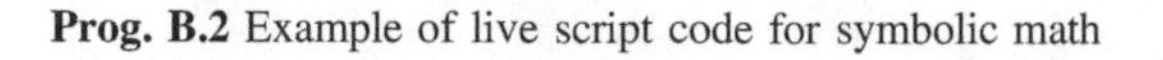

Prog. B.2 Example of live script code for symbolic math

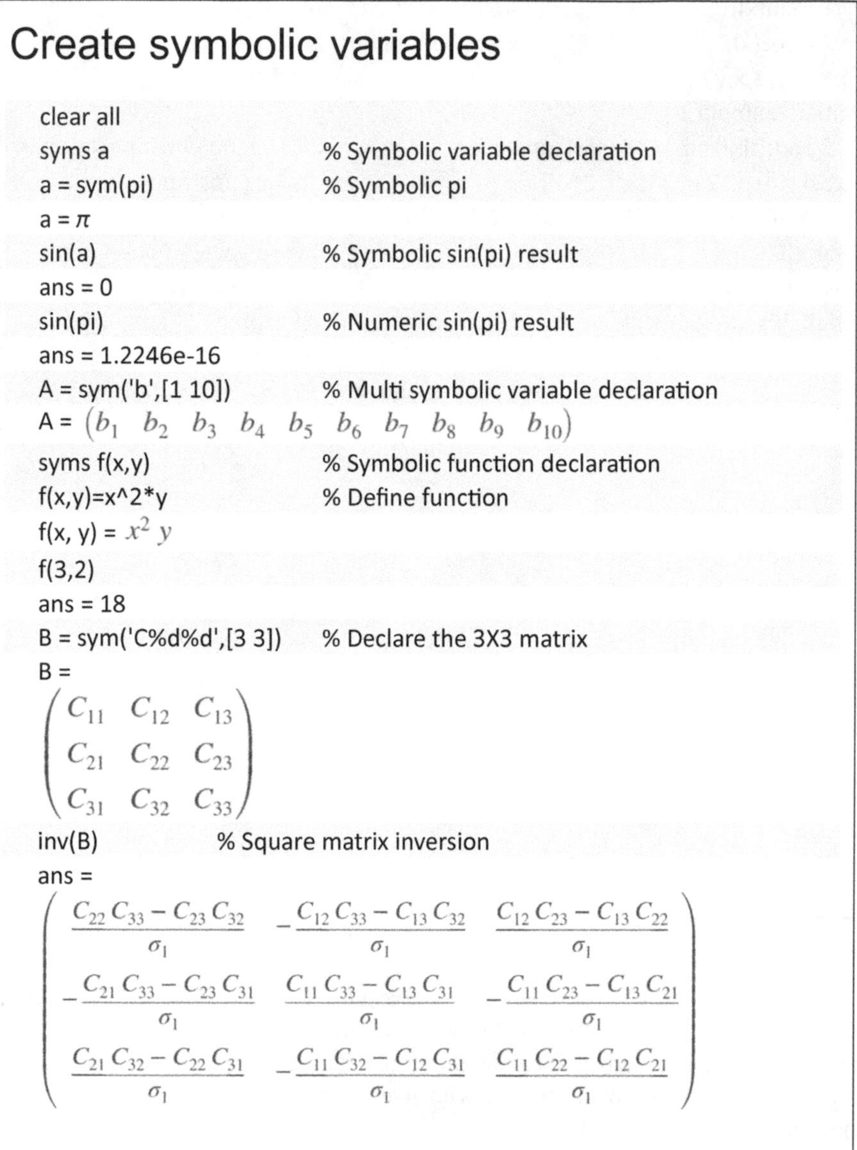

Create symbolic variables

```
clear all
syms a                      % Symbolic variable declaration
a = sym(pi)                 % Symbolic pi
```

a = π

```
sin(a)                      % Symbolic sin(pi) result
```

ans = 0

```
sin(pi)                     % Numeric sin(pi) result
```

ans = 1.2246e-16

```
A = sym('b',[1 10])         % Multi symbolic variable declaration
```

A = $\begin{pmatrix} b_1 & b_2 & b_3 & b_4 & b_5 & b_6 & b_7 & b_8 & b_9 & b_{10} \end{pmatrix}$

```
syms f(x,y)                 % Symbolic function declaration
f(x,y)=x^2*y                % Define function
```

f(x, y) = $x^2\, y$

```
f(3,2)
```

ans = 18

```
B = sym('C%d%d',[3 3])      % Declare the 3X3 matrix
```

B =

$$\begin{pmatrix} C_{11} & C_{12} & C_{13} \\ C_{21} & C_{22} & C_{23} \\ C_{31} & C_{32} & C_{33} \end{pmatrix}$$

```
inv(B)             % Square matrix inversion
```

ans =

$$\begin{pmatrix} \frac{C_{22}\,C_{33} - C_{23}\,C_{32}}{\sigma_1} & -\frac{C_{12}\,C_{33} - C_{13}\,C_{32}}{\sigma_1} & \frac{C_{12}\,C_{23} - C_{13}\,C_{22}}{\sigma_1} \\ -\frac{C_{21}\,C_{33} - C_{23}\,C_{31}}{\sigma_1} & \frac{C_{11}\,C_{33} - C_{13}\,C_{31}}{\sigma_1} & -\frac{C_{11}\,C_{23} - C_{13}\,C_{21}}{\sigma_1} \\ \frac{C_{21}\,C_{32} - C_{22}\,C_{31}}{\sigma_1} & -\frac{C_{11}\,C_{32} - C_{12}\,C_{31}}{\sigma_1} & \frac{C_{11}\,C_{22} - C_{12}\,C_{21}}{\sigma_1} \end{pmatrix}$$

where

$$\sigma_1 = C_{11}\,C_{22}\,C_{33} - C_{11}\,C_{23}\,C_{32} - C_{12}\,C_{21}\,C_{33} + C_{12}\,C_{23}\,C_{31} + C_{13}\,C_{21}\,C_{32} - C_{13}$$

The symbolic math on MATLAB requires to declare the symbolic variable before use. The *sym* creates numbered symbolic variables or symbolic variables. The *syms* command is shorthand for the *sym* syntax. The above code demonstrates the various situations for symbolic variable declarations. The $\sin(\pi)$ computation with numeric and symbolic π results in the different outcome due to the precision limitation. We can define the multi-variable function by using the symbolic representation.

Prog. B.3 Symbolic computation example

Perform symbolic calculus

```
clear all
syms x y
f1 = sin(x)^2+cos(y)^2
```

f1 = $\cos(y)^2 + \sin(x)^2$

```
diff(f1)                % Differentiate
```

ans = $2\cos(x)\sin(x)$

```
diff(f1,y)              % Differentiate over y
```

ans = $-2\cos(y)\sin(y)$

```
int(f1)                 % Integral (indefinite)
```

ans =

$$-\frac{\sin(2x)}{4} - x\left(\sin(y)^2 - \frac{3}{2}\right)$$

```
int(f1,y)               % Integral (indefinite) over y
```

ans =

$$\frac{\sin(2y)}{4} + y\left(\sin(x)^2 + \frac{1}{2}\right)$$

```
int(f1,-2,2)            % Integral (definite) over x from -2 to 2
```

ans =

$$4\cos(y)^2 - \frac{\sin(4)}{2} + 2$$

The above code performs the symbolic computations for simple calculus. The symbolic math toolbox supports extensive calculus functions such as differentiation, integral, vector operations, series, limits, transformation, and etc.

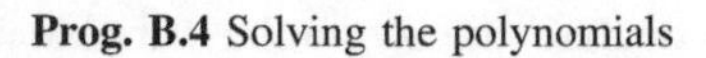

Prog. B.4 Solving the polynomials

Solve equations

```
clear all
syms x y z
z1 = x^3-6*x^2
```

z1 = $x^3 - 6\,x^2$

```
z2 = 6-11*x
```

z2 = $6 - 11\,x$

```
solve(z1==z2)
```

ans =

$$\begin{pmatrix} 1 \\ 2 \\ 3 \end{pmatrix}$$

```
solve(z1-z2==0)
```

ans =

$$\begin{pmatrix} 1 \\ 2 \\ 3 \end{pmatrix}$$

```
z3 = 6*x^2 - 6*x^2*y + x*y^2 - x*y + y^3 - y^2
```

z3 = $-6\,x^2\,y + 6\,x^2 + x\,y^2 - x\,y + y^3 - y^2$

```
solve(z3==0,y)
```

ans =

$$\begin{pmatrix} 1 \\ 2\,x \\ -3\,x \end{pmatrix}$$

```
[x,y,z] = solve(z == 4*x, x == y, z == x^2 + y^2)
```

x =

$$\begin{pmatrix} 0 \\ 2 \end{pmatrix}$$

y =

$$\begin{pmatrix} 0 \\ 2 \end{pmatrix}$$

z =

$$\begin{pmatrix} 0 \\ 8 \end{pmatrix}$$

We can solve the multi or single variable polynomials as above. The symbolic math toolbox provides the equation solvers for polynomials and differential equations.

Prog. B.5 Manipulation of polynomials

Manipulation of polynomials

```
syms x y
f1 = (x^2- 1)*(x^4 + x^3 + x^2 + x + 1)*(x^4 - x^3 + x^2 - x + 1)
```

f1 = $(x^2-1)\ (x^4-x^3+x^2-x+1)\ (x^4+x^3+x^2+x+1)$

```
expand(f1)                    % Expand polynomial
```

ans = $x^{10}-1$

```
f2 = x^3 + 6*x^2 + 11*x + 6
```

f2 = $x^3+6\,x^2+11\,x+6$

```
factor(f2)                    % Factor polynomial
```

ans = $(x+3\quad x+2\quad x+1)$

```
f3 = x^5 + x^4 + x^3 + x^2 + x
```

f3 = $x^5+x^4+x^3+x^2+x$

```
horner(f3)                    % Nested representation for polynomial
```

ans = $x\ (x\ (x\ (x\ (x+1)+1)+1)+1)$

```
f4 = 2*x^2 - 3*x + 1
```

f4 = $2\,x^2-3\,x+1$

```
aa = subs(f4,1/3)             % Substitute the x with 1/3
```

aa =

$$\frac{2}{9}$$

```
eval(aa)                      % Compute the value
```

ans = 0.2222

```
f5 = x^2*y + 5*x*sqrt(y)
```

f5 = $x^2\,y+5\,x\,\sqrt{y}$

```
subs(f5,x,3)                  % Substitute the x with 3
```

ans = $9\,y+15\,\sqrt{y}$

Manipulations for polynomial can be performed as shown above. The toolbox presents the extensive simplification methods for instance expand, decomposition, factorization, and etc. If we do not need a particular form of expressions, use *simplify* function to shorten mathematical expressions. The *simplify* works universally; otherwise, we apply the specific manipulations on the equation to derive the specific form.

Prog. B.6 FIR filter design by using symbolic math

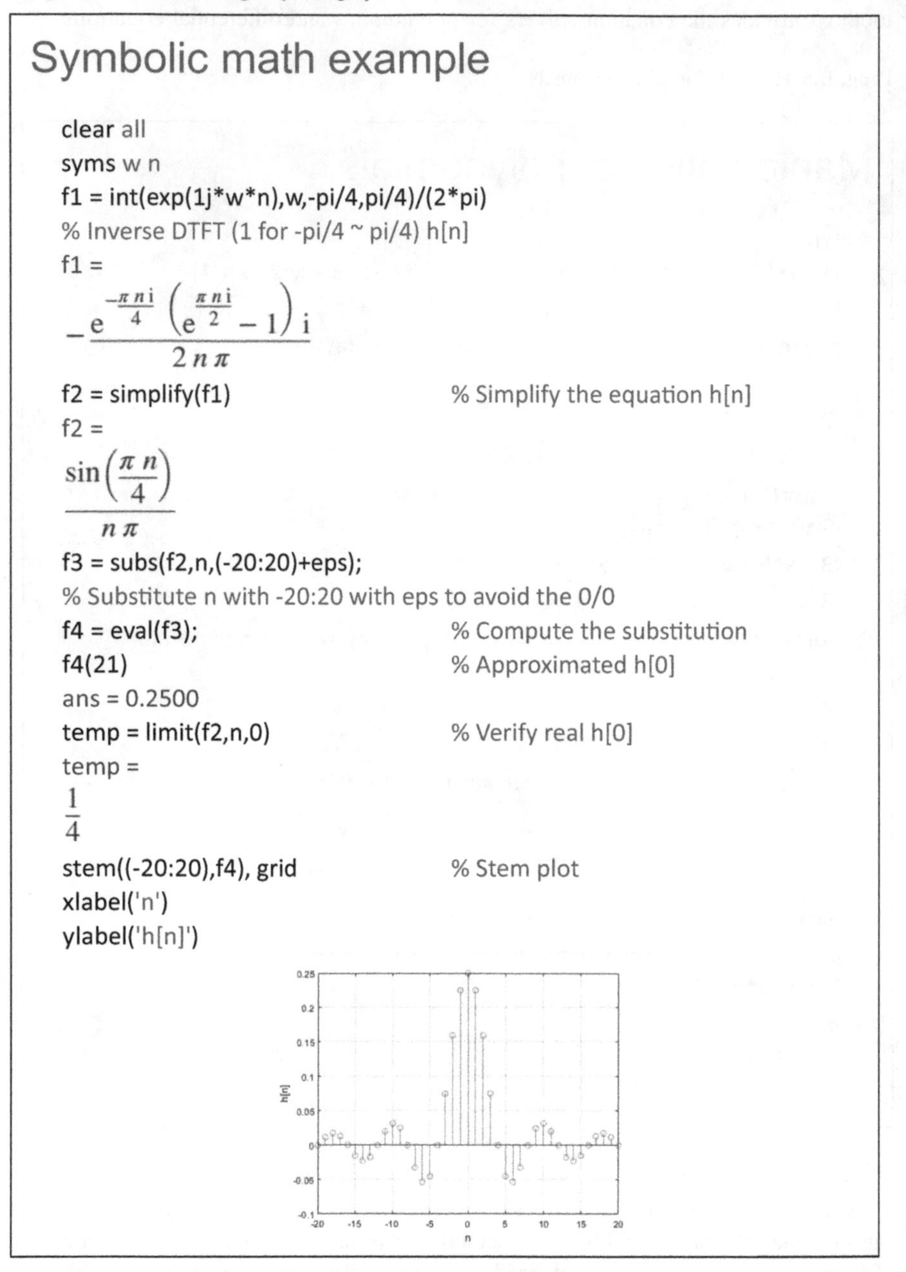

Symbolic math example

```
clear all
syms w n
f1 = int(exp(1j*w*n),w,-pi/4,pi/4)/(2*pi)
% Inverse DTFT (1 for -pi/4 ~ pi/4) h[n]
f1 =
```

$$-\frac{e^{\frac{-\pi n i}{4}}\left(e^{\frac{\pi n i}{2}}-1\right)i}{2n\pi}$$

```
f2 = simplify(f1)                    % Simplify the equation h[n]
f2 =
```

$$\frac{\sin\left(\frac{\pi n}{4}\right)}{n\pi}$$

```
f3 = subs(f2,n,(-20:20)+eps);
% Substitute n with -20:20 with eps to avoid the 0/0
f4 = eval(f3);                       % Compute the substitution
f4(21)                               % Approximated h[0]
ans = 0.2500
temp = limit(f2,n,0)                 % Verify real h[0]
temp =
```

$$\frac{1}{4}$$

```
stem((-20:20),f4), grid              % Stem plot
xlabel('n')
ylabel('h[n]')
```


The above code performs the inverse DTFT to design the LPF with cutoff frequency $\pi/4$. The corresponding impulse response is displayed as stem plot. Note that the *simplify* function makes the equation in simple form. The substitution is executed to replace the n symbolic variable with range of integer values. In zero replacement condition, the equation represents the NaN (Not-a-Number) outcome. To avoid the NaN output, the integer values are added with very small value *eps* (2.2204×10^{-16}) which is the distance from 1.0 to the next larger double-precision number. The zero input result is verified with the *limit* function which computes the output for infinitesimal approach to given input value.

Prog. B.7 Plot in symbolic math

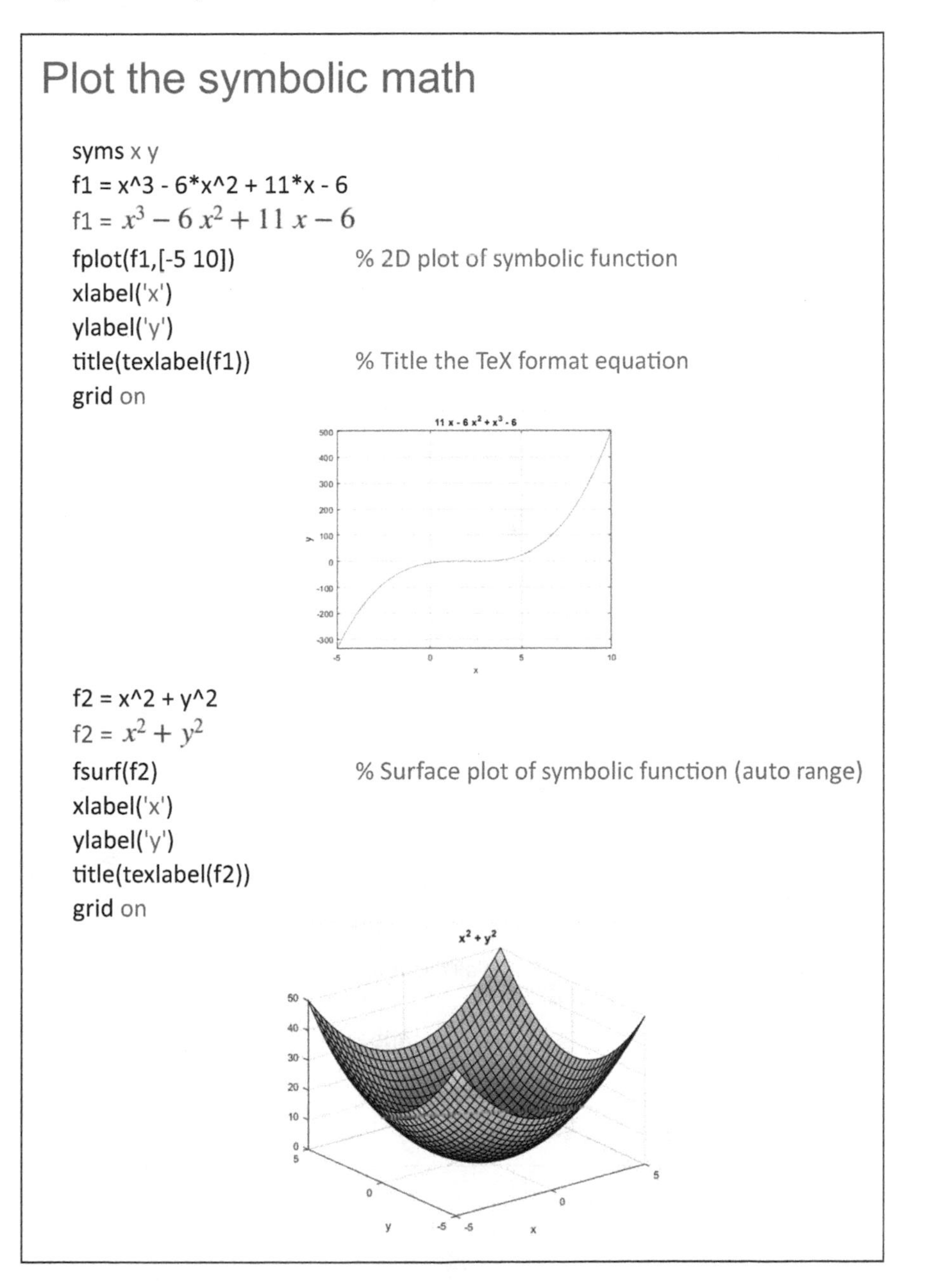

Plot the symbolic math

```
syms x y
f1 = x^3 - 6*x^2 + 11*x - 6
```

$$f1 = x^3 - 6\,x^2 + 11\,x - 6$$

```
fplot(f1,[-5 10])            % 2D plot of symbolic function
xlabel('x')
ylabel('y')
title(texlabel(f1))          % Title the TeX format equation
grid on
```



```
f2 = x^2 + y^2
```

$$f2 = x^2 + y^2$$

```
fsurf(f2)                    % Surface plot of symbolic function (auto range)
xlabel('x')
ylabel('y')
title(texlabel(f2))
grid on
```

Again, there is no given value to the symbolic variables. This condition is also true for drawing the plot in symbolic math. We only assign the symbolic function to the designated plot functions to illustrate the plot as shown above.

Prog. B.8 Assumptions on variables

Assumptions

```
clear all
syms n x
f1 = sin(2*n*pi)                 % Given function1
f1 = sin(2 π n)
simplify(f1)                     % Simplify function1
ans = sin(2 π n)
assume(n,'integer')              % Assume that n is integer
simplify(f1)                     % Re-simplify function1
ans = 0
f2 = x^4==1                      % Given function2
f2 = x^4 = 1
solve(f2,x)                      % Solve function2
ans =
```

$$\begin{pmatrix} -1 \\ 1 \\ -\mathrm{i} \\ \mathrm{i} \end{pmatrix}$$

```
assume(x,'real')                 % Assume that x is real
solve(f2,x)                      % Re-solve function2
ans =
```

$$\begin{pmatrix} -1 \\ 1 \end{pmatrix}$$

```
assumeAlso(x>0)                  % Assume again x is real and positive
solve(f2,x)                      % Re-solve function2
ans = 1
assumptions                      % Show all assumptions
ans = (x ∈ ℝ   0 < x   n ∈ ℤ)
assume([x n],'clear')            % Clear assumptions on x and n
assumptions                      % Show all assumptions again
  ans =
Empty sym: 1-by-0
```

When the symbolic variables are declared, there is no assigned conditions. Therefore, the variable could be any value and number set. We can apply certain conditions on the variables to specify the output range. See above code for example.

Index

K. Kim, *Conceptual Digital Signal Processing with MATLAB*,
Signals and Communication Technology,
https://doi.org/10.1007/978-981-15-2584-1

CPSIA information can be obtained
at www.ICGtesting.com
Printed in the USA
LVHW082043051120
670842LV00002B/2